Geotechnical, Geological and Earthquake Engineering

GEOTECHNICAL, GEOLOGICAL AND EARTHQUAKE ENGINEERING

Volume 35

For further volumes:
http://www.springer.com/series/6011

Fabio Taucer • Roberta Apostolska
Editors

Experimental Research in Earthquake Engineering

EU-SERIES Concluding Workshop

Springer

Editors
Fabio Taucer
Joint Research Centre, Institute for the Protection and Security of the Citizen (IPSC), European Laboratory for Structural Assessment Unit, TP 480
European Commission
Ispra
Varese
Italy

Roberta Apostolska
Joint Research Centre, Institute for the Protection and Security of the Citizen (IPSC), European Laboratory for Structural Assessment Unit, TP 480
European Commission
Ispra
Varese
Italy

ISSN 1573-6059 ISSN 1872-4671 (electronic)
Geotechnical, Geological and Earthquake Engineering
ISBN 978-3-319-10135-4 ISBN 978-3-319-10136-1 (eBook)
DOI 10.1007/978-3-319-10136-1

Library of Congress Control Number: 2014956381

Springer Cham Heidelberg New York Dordrecht London

Printed on acid-free paper

Springer International Publishing Switzerland is part of Springer Science+Business Media (www.springer.com)

Preface

The construction of earthquake resistant buildings was based for centuries on the observation of earthquake damage to the built environment and the adoption of corrective measures during reconstruction. As a result, the areas affected by earthquake disasters have served as large fields for testing construction practice, albeit in an uncontrolled way. The advent of laboratory-controlled experiments, in combination with field observations, allowed during the last decades a fast advancement towards what we know today as modern earthquake engineering, which required experimental tests to be representative of real conditions in the field. It meant the use of full-scale specimens, accurate control of boundary conditions and reliable readings of response, which implied continuous evolution of the infrastructure supporting the experimental tests, still today one of the main challenges in earthquake engineering.

Shaking tables and reaction frames/walls coupled with hydraulic actuators have been in use for more than half a century for carrying out experimental tests in earthquake engineering. This has permitted the development of new design procedures and innovative technologies for earthquake protection, as well as the calibration of numerical models. In Europe, experimental testing has been at the fore-front of pre-Normative research in support of the European standards for construction, in particular of EN 1998 (Eurocode 8: Design of structures for earthquake resistance), through the participation of smaller scale labs, mostly for testing of sub-assemblages and components, up to large laboratories for full-scale validation and demonstration tests. Geotechnical centrifuges have also been used in earthquake engineering for studying wave propagation and soil-structure interaction phenomena, for both surface and embedded structures.

In an effort towards capitalizing on the existing infrastructures in Europe for experimental testing in earthquake engineering, the European Commission financed the SERIES Project (Seismic Engineering Research Infrastructures for European Synergies, www.series.upatras.gr) under grant agreement n° 227887 of the Research Infrastructures Programme in the Seventh Framework Programme. The project was coordinated by Prof Michael N. Fardis of the University of Patras. It aimed at fostering a sustainable culture of co-operation among all research infrastructures, by taking advantage of their complementarities while at the same time bringing the less advanced infrastructures to the levels of the most advanced ones. A major part of the project was devoted to transnational access of users to a world-class portfolio combining Europe's largest facility for pseudo-dynamic testing, four diverse shake tables and two centrifuges. The project also envisioned joint research activities towards new fundamental technologies and innovative techniques, promoting efficient and joint use of the research infrastructures.

In this volume, experts from Europe, USA and China present their work in the three areas addressed by the SERIES Project: networking, transnational access and joint research activities. The networking activities include the development of a public distributed database of past, present and future test results, distributed testing capabilities, hybrid simulation, telepresence, and protocols for qualification of Research Infrastructures. The results of transnational access activities are presented for a number of projects among the 27 carried out at the seven large-scale infrastructures of SERIES, ranging from the retrofit of existing to the design of new reinforced concrete, steel, masonry and wood structures, as well as soil-structure interaction and wave propagation. The joint research activities explored novel techniques for better control of fast tests or special applications, new sensing and instrumentation systems, data assimilation in equipment-specimen models for better test control and optimisation of testing campaigns, as well as experimental studies of soil-structure interaction.

The SERIES concluding workshop on "Earthquake Engineering Research Infrastructures" was held at the Joint Research Centre of the European Commission at Ispra, Italy on May 28–30, 2013, jointly with the George E. Brown, Jr. Network for Earthquake Engineering Simulation (NEES, USA). The workshop attracted a large audience to listen to renowned experts from around the world presenting close to 55 invited contributions. The event was dedicated to the memory of Prof Roy Severn, who established and led the EQUALS facility at the University of Bristol and co-ordinated seismic infrastructure projects which preceded SERIES in past Framework Programmes, notably ECOEST I and II and ECOLEADER.

We gratefully acknowledge the support of Ms Claudia Carniel from the Joint Research Centre, and Ms Vassia Vayenas and Dr Dionysis Biskinis from the University of Patras for their contribution to the organisation of the SERIES Workshop at Ispra.

In Memory of Prof Roy Severn

Professor Roy Severn CBE FREng FICE initiated and led the formation of the strong network of European earthquake engineering laboratories that has underpinned European research in this field for nearly 25 years, leading to EU funded projects including ECOEST1, ECOEST2, ECOLEADER, CASCADE, FUDIDCOEEF and SERIES. All these projects promoted the development and application of large scale shaking table, reaction wall and centrifuge experiments, enabling a large number of EU researchers to access a network of world class facilities. The resulting research has led to many advances and has underpinned the development of Eurocode 8.

Roy joined the University of Bristol as a lecturer in 1956, becoming a Pro Vice Chancellor, twice Dean of Engineering, Head of the Department of Civil Engineering and Director of the Earthquake Engineering Research Centre, before retiring in 1995. He was elected President of the Institution of Civil Engineers in 1990–1991.

Roy Severn was born in Hucknall Nottinghamshire on 6th September 1929. He eventually moved with his family to Great Yarmouth, spending his final 6th form year at Great Yarmouth Grammar School.

At the end of the war, in 1947, he read mathematics at the Royal College of Science in London. He played cricket, rugby and soccer and as a direct consequence met Professor Sammy Sparkes of Imperial College when they both played for the

Wasps Rugby Club. Sparkes persuaded Roy that a career in Civil Engineering would be sensible. At that time Roy also met Deryck N. de G. Allen who was later to become his PhD supervisor at Imperial College. Allen was a student of Sir Richard Southwell who was developing numerical relaxation techniques for the solution of large sets of equations. The Consulting Engineers, Binnie, Deacon and Gourley had a commission to design the Dokan Arch Dam near Baghdad. They asked Allen for help so he and Roy took on the task and spent many hours with hand calculators grinding out numerical stress analysis solutions.

In 1956, after National Service with the Royal Engineers, during which he served as a 2nd Lieutenant in Egypt, Cyprus and Aden, Roy was offered a lectureship in civil engineering at Bristol. That same year, the Institution of Civil Engineers set up an Arch Dams Committee, chaired by Sir Angus Paton, to investigate the potential of the emerging finite element technique. Roy was the most junior member of the committee and was allocated the task of developing numerical earthquake response analysis techniques. This was the starting point of earthquake engineering research at the University of Bristol.

Although a mathematician by background, Roy recognised the importance of large scale experimentation and prototype observations in seeding and validating theory and analysis. In 1981, he won the Telford Gold Medal of the Institution of Civil Engineers with Alan Jeary and Brian Ellis of the Building Research Establishment for their pioneering work on forced vibration testing of embankment dams using a novel eccentric mass exciter system. Then, in 1984, Roy became a member of the Civil Engineering Committee of the UK Science and Engineering Research Council. He won the bid from that committee to set up a 15 t capacity, six degree of freedom shaking table capable of testing large structural components and models with simulations of real earthquakes. At the same time, Roy promoted the SERC's initial funding of the UK Earthquake Engineering Field Investigation Team (EEFIT).

In 1990, the European Commission invited Roy to co-ordinate the large earthquake engineering facilities in LNEC (Lisbon), ISMES (Italy), NTU Athens and Bristol in a joint programme to calibrate and improve the performance of shaking tables. This project put Europe's shaking table facilities amongst the best in the world and led to the sequence of EU funded projects listed above. The ELSA reaction wall facility at JRC Ispra, the shaking tables at CEA Saclay and EUCENTRE Pavia, and centrifuges at Cambridge and IFSTTAR Nantes subsequently joined the laboratory network, making their facilities available to researchers from across the EU. Roy continued to co-ordinate this European laboratory-based earthquake engineering research until 2005—some 10 years after his official retirement—leaving behind a thriving, world class, research community that continues to underpin improvements in global earthquake safety.

In 1981, Roy was elected a fellow of Royal Academy of Engineering. In 1992, he was awarded a CBE for services to Civil Engineering. In 1997 he delivered the 6th Mallet-Milne Lecture entitled "Structural Response Prediction Using Experimental Data".

Roy was still working right to the end. His last big project was a book of the history of the Faculty of Engineering at the University. He, very generously, funded its production and donated the royalties to the University to set up a scholarship fund for students.

Roy Severn died on 25 November 2012, aged 83. He is survived by his wife Hilary, daughters Fiona and Elizabeth and his five grandchildren.

The European Research Infrastructures on earthquake engineering will always remember Roy Severn as the initiator of a long and successful period of European research and collaboration. We all acknowledge Roy for his continuous leadership and European spirit.

Bristol, UK — Colin Taylor
Patras, Greece — Michael N. Fardis
Ispra, Italy — Artur V. Pinto
August 2014

Acknowledgment

This volume has been prepared and published with the support of the Joint Research Centre, the European Commission's in-house science service.

As the Commission's in-house science service, the Joint Research Centre's mission is to provide EU policies with independent, evidence-based scientific and technical support throughout the whole policy cycle. Working in close cooperation with policy Directorates-General, the JRC addresses key societal challenges while stimulating innovation through developing new methods, tools and standards, and sharing its know-how with the Member States, the scientific community and international partners.

Contents

Contributors

Giuseppe Abbiati Department of Civil, Environmental and Mechanical Engineering, University of Trento, Trento, Italy

Ioannis Anastasopoulos Laboratory of Soil Mechanics, National Technical University of Athens, Athens, Greece

Gabriela M. Atanasiu Faculty of Civil Engineering and Building Services, Technical University Gheorghe Asachi of Iasi, Iasi, Romania

Stefano Aversa Dipartimento di Ingegneria, Università di Napoli Parthenope, Napoli, Italy

Juan J. Aznárez University Institute SIANI, Universidad de Las Palmas de Gran Canaria, Las Palmas, Spain

Pierre-Yves Bard ISTerre/IFSTTAR, Université de Grenoble, Grenoble, France

Darko Beg Faculty of Civil and Geodetic Engineering, University of Ljubljana, Ljubljana, Slovenia

Dionisio Bernal Civil and Environmental Engineering, Center for Digital Signal Processing, Northeastern University, Boston, MA, USA

Katrin Beyer Earthquake Engineering and Structural Dynamics Laboratory (EESD), School of Architecture, Civil and Environmental Engineering (ENAC), École Polytechnique Fédérale de Lausanne (EPFL), Switzerland Earthquake Engineering, Lausanne, Switzerland

Anthony Blakeborough Department of Engineering Science, University of Oxford, Oxford, UK

Julijana Bojadjieva Institute of Earthquake Engineering and Engineering Seismology, IZIIS, Ss Cyril and Methodius, Skopje, Republic of Macedonia

Alessio Bonelli Department of Civil, Environmental and Mechanical Engineering, University of Trento, Trento, Italy

Luis Fabian Bonilla Department of Geotechnics, Environment, Natural hazards and Earth sciences, French Institute of Science and Technology for Transport, Development and Networks (IFSTTAR), Marne la Vallée, France

Anna Bosi Vienna Consulting Engineers ZT GmbH, Vienna, Austria

George D. Bouckovalas National Technical University of Athens (NTUA), Athens, Greece

Dionysios Bournas Department of Civil Engineering, University of Nottingham, Nottingham, UK

Stathis Bousias University of Patras, Patras, Greece

Claude Boutin ENTPE/LGCB – LTDS - CNRS UMR 5513, Université de Lyon, Vaulx-en-Velin, France

Francesca Bozzano Department of Earth Sciences and Research Centre for Geological Risks (CERI), "Sapienza" University of Rome, Roma, Italy

Alberto Bretschneider Department of Geotechnics, Environment, Natural hazards and Earth sciences, French Institute of Science and Technology for Transport, Development and Networks (IFSTTAR), Marne la Vallée, France

Brian M. Broderick Department of Civil, Structural and Environmental Engineering, Trinity College, Dublin, Ireland

Oreste Salvatore Bursi Department of Civil, Environmental and Mechanical Engineering, University of Trento, Trento, Italy

Paulo Candeias Department of Structures, National Laboratory for Civil Engineering (LNEC), Lisbon, Portugal

Enrico Cazzador Department of Civil, Environmental and Mechanical Engineering, University of Trento, Trento, Italy

Rosario Ceravolo Department of Structural, Geotechnical and Building Engineering, Politecnico of Turin, Turin, Italy

Laura Di Chiacchio Department DICAM, University of Bologna, Bologna, Italy

Ambra Chioccariello University of Genova, Genova, Italy

Ulas Cilingir University of Cambridge (UCam), Cambridge, UK

Ema Coelho Department of Structures, National Laboratory for Civil Engineering (LNEC), Lisbon, Portugal

Riccardo Conti Dipartimento di Ingegneria, Università degli Studi Niccolò Cusano, Roma, Italy

António A. Correia Department of Structures, National Laboratory for Civil Engineering (LNEC), Lisbon, Portugal

Daniele Corritore Department of Engineering, University Roma Tre, Rome, Italy

Alfredo Campos Costa Department of Structures, National Laboratory for Civil Engineering (LNEC), Lisbon, Portugal

Raffaele De Risi Department of Structures for Engineering & Architecture, University of Naples Federico II, Naples, Italy

Luca de Sanctis Dipartimento di Ingegneria, Università di Napoli Parthenope, Napoli, Italy

Hervé Degée University of Liège, Liège, Belgium

Vasilis K. Dertimanis Laboratory for Earthquake Engineering, School of Civil Engineering, National Technical University of Athens, Athens, Greece

Luigi Di Sarno Department of Engineering, University of Sannio, Benevento, Italy

Matthew Dietz Department of Civil Engineering, University of Bristol, Bristol, UK

Seref Diler SCHÖCK Company, Baden-Baden, Germany

Uwe E. Dorka Steel and Composite Section, Department of Civil and Environmental Engineering, University of Kassel, Kassel, Germany

Vasileios Drosos Laboratory of Soil Mechanics, National Technical University of Athens, Athens, Greece

Kemal Edip Institute of Earthquake Engineering and Engineering Seismology, IZIIS, Ss Cyril and Methodius, Skopje, Republic of Macedonia

Ahmed Y. Elghazouli Department of Civil and Environmental Engineering, Imperial College London, London, UK

Mustafa Erdik Department of Earthquake Engineering, KOERI, Bogazici University, Istanbul, Turkey

Sandra Escoffier Department of Geotechnics, Environment, Natural hazards and Earth sciences, French Institute of Science and Technology for Transport, Development and Networks (IFSTTAR), Marne la Vallée, France

Feroz Farazi Department of Information Engineering and Computer Science, University of Trento, Trento, Italy

Christos Fidas University of Patras, Patras, Greece

Dejan Filipovski Institute of Earthquake Engineering and Engineering Seismology, IZIIS, Ss Cyril and Methodius, Skopje, Republic of Macedonia

Dora Foti Department of Civil and Environmental Engineering (DICA), Technical University of Bari, Bari, Italy

Mihail Garevski Institute of Earthquake Engineering and Engineering Seismology, IZIIS, Ss Cyril and Methodius, Skopje, Republic of Macedonia

Giada Gasparini Department of Civil, Chemical, Environmental and Materials Engineering (DICAM), University of Bologna, Bologna, Italy

George Gazetas Laboratory of Soil Mechanics, National Technical University of Athens, Athens, Greece

Jamie Goggins College of Engineering and Informatics, National University of Ireland, Galway, Ireland

Pablo Guindos Fraunhofer Wilhelm Klauditz Institut (WKI), Braunschweig, Germany

Selim Günay Department of Civil and Environmental Engineering, University of California, Berkeley, CA, USA

Stuart K. Haigh University of Cambridge (UCam), Cambridge, UK

Rashedul Hasan Department of Civil, Environmental and Mechanical Engineering, University of Trento, Trento, Italy

Andreas Heiduschke HESS TIMBER GmbH & Co. KG, Kleinheubach, Germany

Charles Heron Department of Engineering, Schofield Centre, University of Cambridge, Cambridge, UK

Alan Hunt Department of Civil, Structural and Environmental Engineering, Trinity College, Dublin, Ireland

Erdin Ibraim Department of Civil Engineering, University of Bristol, Bristol, U.K.

Ioannid Ioannidis University of Patras, Patras, Greece

Salvador Ivorra Department of Civil Engineering, Universidad de Alicante, Alicante, Spain

Andreas Jäger Wienerberger AG, Vienna, Austria

Dimitris K. Karamitros National Technical University of Athens (NTUA), Athens, Greece

Bohumil Kasal Fraunhofer Wilhelm Klauditz Institut (WKI), Braunschweig, Germany

Ilias Kotinas University of Patras, Patras, Greece

Igor Lanese Eucentre—European Centre for Training and Research in Earthquake Engineering, Pavia, PV, Italy

João Leite Department of Civil Engineering, University of Minho and ISISE, Guimarães, Portugal

Alain Le Maoult CEA Saclay, DEN/DANS/DM2S/SEMT/EMSI, Gif sur Yvette Cedex, France

Luca Lenti Department of Geotechnics, Environment, Natural hazards and Earth sciences, French Institute of Science and Technology for Transport, Development and Networks (IFSTTAR), Marne la Vallée, France

Zhenbao Li Beijing Key Lab of Earthquake Engineering and Structural Retrofit, Beijing University of Technology, Beijing, China

Jean Louis Chazelas Department of Geotechnics, Environment, Natural hazards and Earth sciences, French Institute of Science and Technology for Transport, Development and Networks (IFSTTAR), Marne la Vallée, France

Paulo B. Lourenço Department of Civil Engineering, ISISE, University of Minho, Guimarães, Portugal

Suikai Lu Consulting Engineer, Vienna, Austria

Xilin Lu State Key Laboratory of Disaster Reduction in Civil Engineering, Tongji University, Shanghai, China

Gopal S. P. Madabhushi SDepartment of Engineering, Schofield Centre, University of Cambridge, Cambridge, UK

Orlando Maeso University Institute SIANI, Universidad de Las Palmas de Gran Canaria, Las Palmas, Spain

Rosa Maria Stefania Maiorano Dipartimento di Ingegneria, Università di Napoli Parthenope, Napoli, Italy

Marialaura Malena Department of Engineering, University Roma Tre, Rome, Italy

Igor Markovski Institute of Earthquake Engineering and Engineering Seismology, IZIIS, Ss Cyril and Methodius, Skopje, Republic of Macedonia

Ignacio Lamata Martínez Department of Engineering Science, University of Oxford, Oxford, UK

Salvatore Martino Department of Earth Sciences and Research Centre for Geological Risks (CERI), "Sapienza" University of Rome, Roma, Italy

Zhu Mei Department of Civil, Environmental and Mechanical Engineering, University of Trento, Trento, Italy

Luis Mendes Structures Department, Laboratório Nacional de Engenharia Civil, Lisbon, Portugal

Slobodan Micajkov Institute of Earthquake Engineering and Engineering Seismology, IZIIS, Ss Cyril and Methodius, Skopje, Republic of Macedonia

Arkam Mohammad Zeki Mohamad Department of Engineering, University Roma Tre, Rome, Italy

Francisco Javier Molina Joint Research Centre (JRC), Institute for the Protection and Security of the Citizen, European Laboratory for Structural Assessment Unit, European Commission, Ispra, Italy

Philippe Mongabure CEA Saclay, DEN/DANS/DM2S/SEMT/EMSI, Gif sur Yvette Cedex, France

Christophe Mordant University of Liège, Liège, Belgium

Khalid M. Mosalam Department of Civil and Environmental Engineering, University of California, Berkeley, CA, USA

Harris P. Mouzakis Laboratory for Earthquake Engineering, School of Civil Engineering, National Technical University of Athens, Athens, Greece

Primoz Moze Faculty of Civil and Geodetic Engineering, University of Ljubljana, Ljubljana, Slovenia

Paolo Negro Joint Research Centre (JRC), Institute for the Protection and Security of the Citizen, European Laboratory for Structural Assessment Unit, European Commission, Ispra, Italy

Gerard O'Reilly College of Engineering and Informatics, National University of Ireland, Galway, Ireland

Ferran Obón Santacana Steel and Composite Section, Department of Civil and Environmental Engineering, University of Kassel, Kassel, Germany

M. Yener Ozkan Department of Civil Engineering, Middle East Technical University, Ankara, Turkey

Luis Padron University Institute SIANI, Universidad de Las Palmas de Gran Canaria, Las Palmas, Spain

Vasiliki Palieraki Department of Structural Engineering, National Technical University of Athens (NTUA), Athens, Greece

Fabrizio Paolacci Department of Engineering, University Roma Tre, Rome, Italy

Achilleas G. Papadimitriou University of Thessaly (UTh), Volos, Greece

Pierre Pegon Joint Research Centre (JRC), Institute for the Protection and Security of the Citizen, European Laboratory for Structural Assessment Unit, European Commission, Ispra, Italy

Simone Peloso Eucentre—European Centre for Training and Research in Earthquake Engineering, Pavia, PV, Italy

Sarah Petry Earthquake Engineering and Structural Dynamics Laboratory (EESD), School of Architecture, Civil and Environmental Engineering (ENAC), École Polytechnique Fédérale de Lausanne (EPFL), Switzerland Earthquake Engineering, Lausanne, Switzerland

Kyriazis Pitilakis Department of Civil Engineering, Research Unit of Soil Dynamics and Geotechnical Earthquake Engineering, Aristotle University of Thessaloniki, Thessaloniki, Greece

Martin Poljansek Joint Research Centre (JRC), Institute for the Protection and Security of the Citizen, European Laboratory for Structural Assessment Unit, European Commission, Ispra, Italy

Tiberiu Polocoser Technical University of Braunschweig, Braunschweig, Germany

Stanislav Pospisil Institute of Theoretical and Applied Mechanics, Academy of Science of the Czech Republic, Prague, Czech Republic

Alberto Prestininzi Department of Earth Sciences and Research Centre for Geological Risks (CERI), "Sapienza" University of Rome, Roma, Italy

Andrea Prota University of Naples, Naples, Italy

Ioannis N. Psycharis Laboratory for Earthquake Engineering, School of Civil Engineering, National Technical University of Athens, Athens, Greece

Frederic Ragueneau LMT-Cachan, ENS-Cachan/CNRS, Cachan, France

Zoran T. Rakicevic Institute of Earthquake Engineering and Engineering Seismology, IZIIS, Ss Cyril and Methodius, Skopje, Republic of Macedonia

Julio A. Ramirez NEEScomm Center, Purdue University, West Lafayette, USA

Md Shahin Reza Department of Civil, Environmental and Mechanical Engineering, University of Trento, Trento, Italy

Ilaria Ricci Department of Civil, Chemical, Environmental and Materials Engineering (DICAM), University of Bologna, Bologna, Italy

Benjamin Richard CEA, DEN, DANS, DM2S, SEMT, Laboratoire d'Etudes de Mécanique Sismique, Gif-sur-Yvette, France

Emmanouil Rovithis Institute of Engineering Seismology & Earthquake Engineering (EPPO-ITSAK) Dassyliou Str, Eleones, Pylaia, Thessaloniki, Greece

Norbert Rüther Fraunhofer Wilhelm Klauditz Institut (WKI), Braunschweig, Germany

Selman Saglam Department of Civil Engineering, Adnan Menderes University, Aydin, Turkey

Suhaib Salawdeh College of Engineering and Informatics, National University of Ireland, Galway, Ireland

Gabriele Scarascia-Mugnozza Department of Earth Sciences and Research Centre for Geological Risks (CERI), "Sapienza" University of Rome, Roma, Italy

Stefen Scheer SCHOECK France SARL, Entzheim, France

Logan Schwan ENTPE/LGCB – LTDS - CNRS UMR 5513, Université de Lyon, Vaulx-en-Velin, France

Jean-François Semblat Department of Geotechnics, Environment, Natural hazards and Earth sciences, French Institute of Science and Technology for Transport, Development and Networks (IFSTTAR), Marne la Vallée, France

Veronika Sendova Institute of Earthquake Engineering and Engineering Seismology, IZIIS, Ss Cyril and Methodius, Skopje, Republic of Macedonia

Vlatko Sesov Institute of Earthquake Engineering and Engineering Seismology, IZIIS, Ss Cyril and Methodius, Skopje, Republic of Macedonia

Stefano Silvestri Department of Civil, Chemical, Environmental and Materials Engineering (DICAM), University of Bologna, Bologna, Italy

Franc Sinur Faculty of Civil and Geodetic Engineering, University of Ljubljana, Ljubljana, Slovenia

Zhenyun Tang Beijing Key Lab of Earthquake Engineering and Structural Retrofit, Beijing University of Technology, Beijing, China

Fabio Taucer Joint Research Centre (JRC), Institute for the Protection and Security of the Citizen, European Laboratory for Structural Assessment Unit, European Commission, Ispra, Italy

Colin Taylor Department of Civil Engineering, University of Bristol, Bristol, UK

Marco Tondelli Earthquake Engineering and Structural Dynamics Laboratory (EESD), School of Architecture, Civil and Environmental Engineering (ENAC), École Polytechnique Fédérale de Lausanne (EPFL), Switzerland Earthquake Engineering, Lausanne, Switzerland

Michele Tricarico Dipartimento di Ingegneria, Università di Napoli Parthenope, Napoli, Italy

Tomaso Trombetti Department of Civil, Chemical, Environmental and Materials Engineering (DICAM), University of Bologna, Bologna, Italy

Grigorios Tsinidis Department of Civil Engineering, Research Unit of Soil Dynamics and Geotechnical Earthquake Engineering, Aristotle University of Thessaloniki Thessaloniki, Greece

Deniz Ulgen Department of Civil Engineering, Mugla Sitki Kocman University, Mugla, Turkey

Shota Urushadze Institute of Theoretical and Applied Mechanics, Academy of Science of the Czech Republic, Prague, Czech Republic

Giulia Viggiani Dipartimento di Ingegneria Civile, Università di Roma Tor Vergata, Rome, Italy

Elizabeth Vintzileou Department of Structural Engineering, National Technical University of Athens (NTUA), Athens, Greece

Dun Wang Research institute of Structural Engineering and Disaster Reduction, Tongji University, Shanghai, China

Martin S. Williams Department of Engineering Science, University of Oxford, Oxford, UK

Ahmet Yakut Department of Civil Engineering, Middle East Technical University, Ankara, Turkey

Orestis Zarzouras Laboratory of Soil Mechanics, National Technical University of Athens, Athens, Greece

Zbigniew Zembaty University of Technology, Opole, Poland

Bin Zhao State Key Laboratory of Disaster Reduction in Civil Engineering, Tongji University, Shanghai, China

Maurizio A. Zola Consultants of P&P LMC, Seriate, BG, Italy

Chapter 1
The George E. Brown, Jr., Network for Earthquake Engineering Simulation (NEES): Reducing the Impact of EQs and Tsunamis

Julio A. Ramirez

1.1 Introduction

In November 1998, the National Science Board approved the George E. Brown Jr., Network for Earthquake Engineering Simulation (NEES) for construction with funds totalling $ 82 million from the National Science Foundation (NSF) Major Research Equipment and Facilities Construction (MREFC) appropriation. Construction occurred during the period 2000–2004. As part of its contribution to the National Earthquake Hazards Reduction Program, the National Science Foundation (NSF) funds NEES operations (Award # CMMI-0927178) as well as many of the research projects that are conducted in NEES facilities. NEEScomm houses the headquarters of operations of a nationwide network of 14 laboratories. Each of these university-based laboratories enables researchers to explore a different aspect of the complex way that soils and structures behave in response to earthquakes and tsunamis. The laboratories are available not just to researchers at the universities where they are located, but to investigators throughout the USA who are awarded grants through NSF's annual NEES Research (NEESR) Program and other NSF programs.

NEES laboratories are also used for research conducted or funded by other federal, state, and local agencies, by private industry, and by international researchers under the partnerships that NEES has cultivated with research facilities and agencies in Japan, Taiwan, Canada, and China. To date, more than 400 multi-year, multi-investigator projects have been completed or are in progress at NEES sites. These projects are yielding a wealth of valuable experimental data and continue to produce transformational research and outcomes that impact engineering practice from analytical models to design guidelines and codes. The family of NEES researchers, educators, and students encompasses an ever increasing group of universities, industry partners, and research institutions in the US and abroad. Project

J. A. Ramirez (✉)
NEEScomm Center, Purdue University, West Lafayette, USA
e-mail: ramirez@purdue.edu

F. Taucer, R. Apostolska (eds.), *Experimental Research in Earthquake Engineering*,
Geotechnical, Geological and Earthquake Engineering 35,
DOI 10.1007/978-3-319-10136-1_1

teams and the NEEScomm team have developed a rich set of resources for research and education.

This paper provides a sample of the breadth of the activities of researchers, students, educators, and practitioners collaborating in NEES (http://nees.org/). More information about the activities presented here and others can be found in Buckle and Ramirez (2010), in the 2009–2010 NEES Facility Project Highlights (NEEScomm, 2010), 2010–2011 NEES Activity Highlights (NEEScomm, 2011) at: https://nees.org/about/neescommunications/neesprojecthighlights, and in the retrospective study on NEES requested by IDA Science and Technology Policy Institute of Washington DC on behalf of the National Science Foundation, "STPI Report—NEES Self-Study Report for Science and Technology Policy Institute (STPI)/IDA", at: https://nees.org/resources/3234/supportingdocs.

1.2 Research Accomplishments

Today, NEES is a vibrant collaboratory consisting of world-class laboratories and cyberinfrastructure with its collaboration platform, NEEShub at: http://nees.org/, representing hundreds of millions of dollars of investment. The NEES collaboratory serves tens of thousands of users from over 180 different countries. The NEEShub now has 5700 registered users with thousands of data downloads per quarter and more than 59,000 general users from over 182 nations (Fig. 1.1).

Research at NEES facilities has resulted in a wealth of information produced by large-scale testing facilities such as tsunami wave basins, centrifuges, and

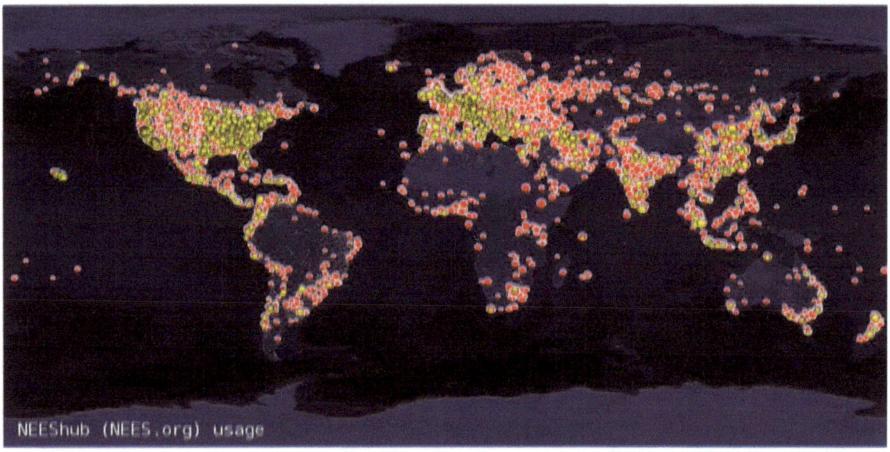

Fig. 1.1 NEEShub global usage. *Red dots* represent researchers and students browsing NEEShub, watching videos, and taking courses while performing 840,656 web and 38,854 tool sessions between August 2010 and April 2013. *Yellow dots* represent users who are running simulations. *Dot size* corresponds to the number of users at a location

Fig. 1.2 NEESR-GC: simulation of the seismic performance of nonstructural systems. NSF Award #: 0721399PI: Emmanuel "Manos" Maragakis. Co-PIs: Robert Reitherman, Steven French, Andre Filiatrault, Tara Hutchinson. Research Sites: University of Nevada, Reno; Hyogo Earthquake Engineering Research Center, aka "E-Defense," Miki, Japan; NEES, University at Buffalo. (Reproduced by permission of PI: Emmanuel "Manos" Maragakis, Dean of Engineering, University of Nevada, Reno)

shake tables. This research community has advanced the understanding of seismic phenomena and laid the groundwork for improvements in design and construction practices to enhance resilience.

A research team led by Manos Maragakis, dean and professor of engineering at the University of Nevada, Reno, has recently completed a major examination of the seismic responses of nonstructural systems. They focused on ceiling-piping-partition systems: identifying their major failure mechanisms, quantifying how their failure affects critical facilities (Fig. 1.2). This grand challenge project provides experimental data and experimentally based simulation tools that will allow better understanding of the seismic behaviour of nonstructural systems.

The Sacramento-San Joaquin River Delta in USA supplies fresh water to 25 million people in southern and central California and irrigates the breadbasket of the San Joaquin Valley. Some 1100 miles of levees protect the region from inundation and serve as a protective lifeline for California agriculture and nearly two-thirds of the state's population. The aging Delta levees are a known flood risk, and many levees are composed of loose granular soils that are susceptible to liquefaction during earthquakes. However, no one knows exactly how the peat underlying many of the levees will behave in an earthquake. A team of researchers led by Scott Brandenberg, a professor of civil and environmental engineering at the University of California, Los Angeles (UCLA) have partnered with the California Department of Water Resources (DWR) to determine how the Delta's peat soil, which serves as the base of the levees, will affect these earthen structures during an earthquake. The research team conducted full-scale shake tests on model levees. To conduct the shake tests, the research team built a model-scale levee six feet tall and 40 ft wide on an island located within the Delta (Fig. 1.3). Unlike existing levees, which were constructed in a haphazard, non-engineered manner, the model levee was constructed

Fig. 1.3 NEESR-II: evaluation of seismic levee deformation potential by destructive cyclic field testing NSF award #: 0830081, PI name: Scott J. Brandenberg, Co-PI: Jonathan P. Stewart, Robb E.S. Moss, Research Sites: University of California, Los Angeles and University of California. (Reproduced by permission of PI: Prof. Scott J. Brandenberg, Department of Civil and Environmental Engineering, University of California, Los Angeles)

from clay that was carefully compacted and reinforced with geogrids. Furthermore, the clay was unsaturated and therefore not susceptible to liquefaction like many Delta levees are.

Other NEES research has developed or validated new seismic protection systems, design methods, or simulation tools that enable engineers to improve the seismic performance of structures. After an earthquake, infrastructure damage can be difficult to detect. In particular, cracks and breaks in underground water pipelines often can't be identified without excavating the pipes. Yet securing lifelines is a high priority for post-quake repairs; undetected or inaccurate information about damage to water-delivery networks can lead to epidemic outbreaks in communities hit by natural disasters.

Earthquake engineers are seeking to design "smart" self-sensing concrete pipes that are able to send wireless messages to signal where and how they were damaged. In a four-year study, Radoslaw Michalowski, professor of civil engineering at the University of Michigan, conducted large-scale tests on buried segmental concrete pipelines to understand exactly how water pipes fail during earthquakes.

These damage-assessment tests mark the first steps toward the development of strategies for wirelessly detecting damage and monitoring the structural health of pipelines (Fig. 1.4). Using data collected from the experiments, Michalowski and his team are now developing protocols for rapid, wireless damage assessment, obviating the need for pipe excavation.

Many of the projects conducted in the NEES laboratories have prompted, or laid the groundwork for, improvements in model building codes and in design and construction practices, enhancing societal resilience to earthquakes and tsunamis.

Fig. 1.4 NEESR-SG: damage detection and health monitoring of buried pipelines after earthquake-induced ground movement. NSF Award #: 0724022. PI name: Radoslaw Michalowski. Co-PIs: W. Jason Weiss, Russell Green, Jerome Lynch, Aaron Bradshaw Research Site: Cornell. (Reproduced by permission of PI: Radoslaw Michalowski, Department of Civil and Environmental Engineering, University of Michigan)

Facilitating these outcomes has been the dissemination of NEES findings through publications, NEEShub at nees.org, and NEES Education, Outreach and Training (EOT) activities. NEES research has been cited in more than 1700 publications according to data collected to June 2012, including rising numbers of refereed journal articles. The distribution of these publications as of July 2011 is shown in Fig. 1.5.

In 2012 the National Science Foundation commissioned a number of studies to help elucidate the view of the broader research community regarding grand challenges ahead for earthquake engineering and the type of earthquake research infrastructure needed to meet those challenges. The findings of those studies are in the NRC Report *Grand Challenges in Earthquake Engineering Research: A Community Workshop Report* (available at: http://www.nap.edu/catalog.php?record_id=13167) and the Report *2020 Vision for Earthquake Engineering Research: Report on an OpenSpace Technology Workshop on the Future of Earthquake Engineering* (available at: https://nees.org/resources/1636).

1.3 NEES Cyberinfratructure and the NEEShub

Linking the NEES experimental facilities to each other, to NEEScomm, and to off-site users is the NEES cyberinfrastructure. This unique system of information technology resources enables researchers participating on-site or remotely to collect, view, process, and store data from NEES experiments, to conduct numerical

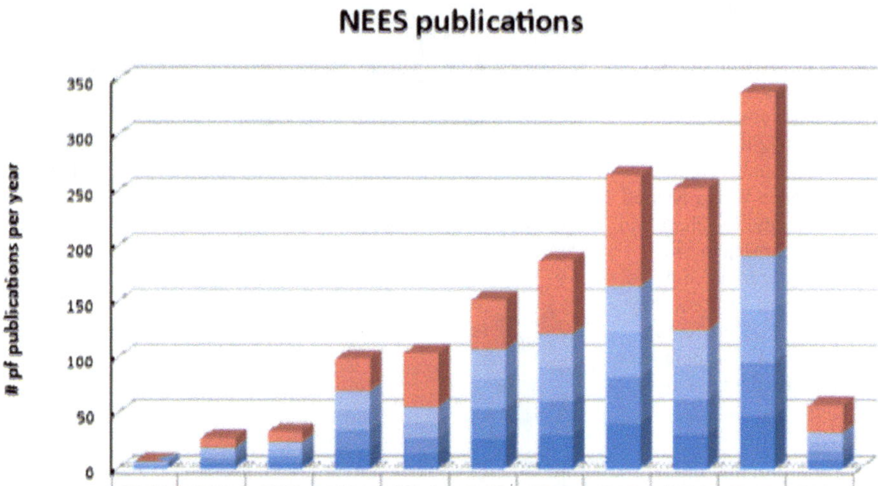

	2001	2002	2003	2004	2005	2006	2007	2008	2009	2010	2011
Journals	0	8	9	29	48	44	65	99	128	147	24
Other	6	19	24	70	56	108	122	165	125	192	33

Fig. 1.5 Publications resulting from NEES work as of July 2011

simulation studies, and to perform hybrid (combined experimental and numerical) testing involving one or more NEES equipment sites. At the heart of this system is NEEShub, a platform designed to facilitate information exchange and collaboration among earthquake engineering researchers, educators, students, practitioners, and stakeholders. Accessed via the NEES website, https://nees.org/, NEEShub is powered by HUBzero software developed at Purdue University.

NEEShub features the NEES Project Warehouse (Fig. 1.6), a curated, centralized data repository used to store and share research results. When launched in 2009, NEEScomm prioritized a strong partnership with the NEES sites and targeted what had been a seriously deficient central data repository and cyberinfrastructure for collaboration. Since the first release of the NEEShub "cloud" platform less than three years ago, the community has actively responded to our user-focused cyberinfrastructure improvements with a pace of file and directory creation that has increased sevenfold. Today, the NEES-curated central repository of research data—Project Warehouse—features a vastly populated repository of NEES research data and showcases over 1.9 M data files and folders that engineers can search, sort, download, and manipulate. NEEShub also stores and shares a variety of other earthquake engineering resources, including publications, databases (Browning et al. 2013), computational models, simulation software, educational materials, and data management and visualization tools. Some of these resources and tools have been developed by NEES staff, while others have been contributed by the earthquake engineering community. NEES solicits and welcomes such contributions from the United States and abroad.

In addition to enabling sharing and collaboration that can accelerate advances in earthquake risk reduction, NEEShub is also helping to disseminate these advances

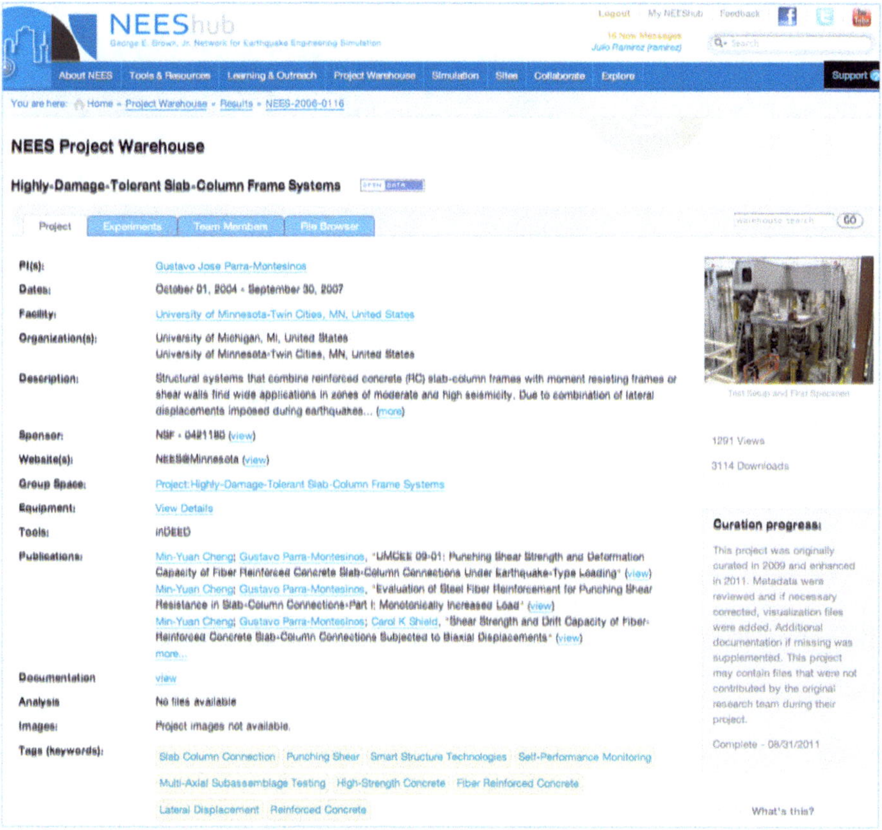

Fig. 1.6 NEES data repository: project warehouse

and to build the workforces needed to discover and implement research findings. NEES is enabling students to learn earthquake engineering through involvement in research projects, undergraduates through NEES' annual Research Experiences for Undergraduates program, and graduate students by directly assisting NEES investigators. In a recent survey, NEEScomm found that at least 559 graduate students, including 191 PhD candidates, have been trained through participation in NEES research. Many of those receiving PhDs now hold faculty positions at major research universities worldwide.

1.4 International Collaborations

NEES has cultivated partnerships with research facilities and agencies in Japan, Taiwan, Canada, and China (Ramirez 2010). The development of a Memorandum of Understanding with The National Research Institute for Earth Science and Disaster

Fig. 1.7 Testing the NEES-Wood Capstone building on the E-defense shake table. (Reproduced by permission of PI: John W. van de Lindt, Colorado State University)

Prevention (NIED) on earthquake engineering research using E-Defense and NEES Facilities represents an important accomplishment with significant realizations in the collaborative research arena. Japan's E-Defense shake table, operated by NIED, is the world's largest multi-degree shake table. In September 2005, the NSF and the Japanese Ministry of Education, Culture, Sports, Science, and Technology (MEXT) signed a memorandum concerning cooperation in the area of disaster prevention research. NSF-supported NEESR projects addressing the seismic performance of midrise wood frame buildings, steel frames, and base-isolated structures utilized both NEES facilities and E-Defense during the 2009–2010 timeframe. An example of the successes is the testing on July 14, 2009, of a six-story condominium building on the shake table at the E-Defense facility, located in the city of Miki, north of Kobe (Fig. 1.7). This was the culminating experiment of the National Science Foundation (NSF) multi-year NEESWood project under the direction of Prof. John van de Lindt. The enabling agreement originally was intended to last five years. NSF and NIED supported the extension of this program for another 5-year term.

Thus, on 7 June 2010, the Memorandum of Understanding with The National Research Institute for Earth Science and Disaster Prevention (NIED) on earthquake engineering research using E-Defense and NEES Facilities was renewed for up to five more years.

Acknowledgement NEES Operations is managed through a cooperative agreement between the National Science Foundation and Purdue University for the period of FY 2010–2014 [NSF Award (0927178) from the Civil, Mechanical and Manufacturing Innovation (CMMI) Division] under the supervision of Dr. Joy M. Pauschke, Program Director for the George E. Brown, Jr. Network for Earthquake Engineering Simulation (NEES) Operations and Research Programs in the Directorate for Engineering. The findings, statements and opinions presented in this paper are those of the author and do not necessarily represent those of the National Science Foundation.

References

Buckle IA, Ramirez JA (2010) NEES Research Highlights. Proc. 9th US National & 10th Canadian Conference, Toronto, Canada

Browning J, Pujol S, Eigenmann R, Ramirez JA (2013) NEEShub databases. Concr Int 35(4): 55–60

Ramirez JA (2010) International collaboration—networking the earthquake engineering re-search community into a global framework. In Proc. 9th US National & 10th Canadian Conference, Toronto, Canada

Chapter 2
A Faceted Lightweight Ontology for Earthquake Engineering Research Projects and Experiments

Rashedul Hasan, Feroz Farazi, Oreste Salvatore Bursi and Md Shahin Reza

2.1 Introduction

The inventor of the Web, Tim Berners-Lee, envisioned a more organized, well connected and well integrated form of its data that are suitable for humans to read and for machines to understand (Berners-Lee 1999; Berners-Lee et al. 2001). This new form of the Web is called the Semantic Web. With the invention of the Semantic Web, computing paradigm is experiencing a shift from databases to Knowledge Bases (KB), where ontologies play a major role in enabling inferencing that can make hidden facts unconcealed to produce better results for users.

Moreover, KB-based systems provide a mechanism to manage information and semantics thereof that can make systems semantically interoperable and as such can exchange and share data between them. To overcome the interoperability issues and to exploit the benefits offered by the state of the art technologies, we moved to the KB-based system.

In fact, we have developed an ontology named as Earthquake Engineering Research Projects and Experiments (EERPE) using a faceted approach that gives emphasis on research project management and experiments. Following the validation of the ontology by a domain expert, it was published in the knowledge representation language RDF and integrated to the generic ontology WordNet (http://wordnet.

R. Hasan (✉) · O. S. Bursi · M. S. Reza
Department of Civil, Environmental and Mechanical Engineering,
University of Trento, Trento, Italy
e-mail: md.hasan@unitn.it

O. S. Bursi
e-mail: oreste.bursi@unitn.it

M. S. Reza
e-mail: Md.Shahin.Reza@unitn.it

F. Farazi
Department of Information Engineering and Computer Science,
University of Trento, Trento, Italy
e-mail: farazi@disi.unitn.it

F. Taucer, R. Apostolska (eds.), *Experimental Research in Earthquake Engineering*,
Geotechnical, Geological and Earthquake Engineering 35,
DOI 10.1007/978-3-319-10136-1_2

princeton.edu/). The experimental data coming from, inter alia, cyclic and pseudo-dynamic tests were also published in RDF. We used Jena (http://jena.apache.org/), OWLIM (http://www.onto-text.com/owlim) and Sesame (http://www.openrdf.org/) tools for publishing, storage and management, respectively. Finally, integrating the tools, ontologies and data, we developed a system to evaluate the effectiveness of the approach.

The rest of the paper is organized as follows. Section 2.2 depicts an ontology based information management approach. Section 2.3 describes the ontology development methodology that has been followed to build the ontology. Section 2.4 provides a brief description of the ontology representation languages RDF and OWL. In Sect. 2.5, we present existing ontology/thesaurus relevant for this work and as such worth discussing them. Section 2.6 provides the ontology integration. Section 2.7 presents the experimental set-up whilst Sect. 2.8 illustrates evaluation results that show the effectiveness of EERPE ontology. In Sect. 2.9 we briefly describe related work and, finally, in Sect. 2.10 we present conclusions.

2.2 Approach

Figure 2.1 describes an ontology based information management system development approach that involves standard three-tier architecture. KB works as a back-end of the system hosting ontologies represented in RDF while query processing, reasoning and inference mechanisms are incorporated in the business logic layer. User queries and corresponding results are shown in the User Interface (presentation) layer. However, for ontology development we follow the DERA methodology (Giunchiglia and Dutta 2011), for ontology representation in RDF we use Jena and for ontology integration we implemented a facet based algorithm (see Sect. 2.6).

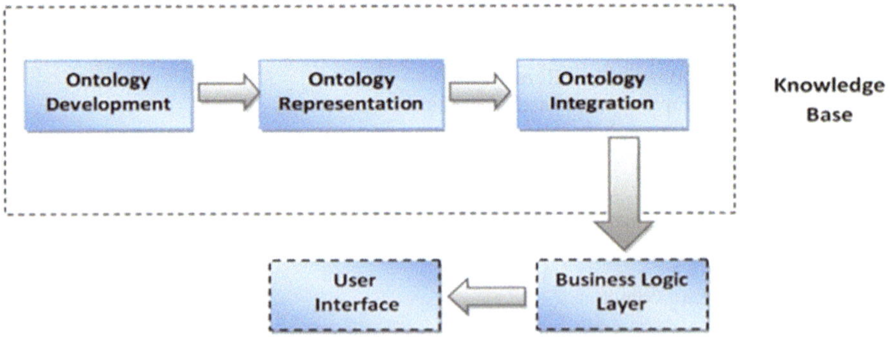

Fig. 2.1 Ontology based systems development approach

2.3 Ontology Development

We use the DERA methodology for ontology development. In fact it is known that it is extendable and scalable (Giunchiglia and Dutta 2011) and ontologies such as GeoWordNet were developed following this approach (Giunchiglia et al. 2010a).

DERA methodology allows for building domain specific ontologies. Domain is an area of knowledge in which users are interested in. For example, earthquake engineering, oceanography, mathematics and computer science can be considered as domains. In DERA, a domain is represented as a 3-tuple ***D***=<***E, R,*** *A*>, where ***E*** is a set of entity-classes that consists of concept sand entities; ***R*** is a set of relations that can be held between concepts and entities and ***A*** is a set of attributes of the entities.

In this three basic components concepts, relations and attributes are organized into facets; hence, the ontology is based on faceted methodology. Facet is a hierarchy of homogeneous concepts describing an aspect of a domain. Ranganathan SR, who was an Indian mathematician-librarian, was the first to introduce faceted approach capable of categorizing books in the libraries (Ranganathan 1967).

Note, however, that a domain can alternatively be called as domain ontology. Henceforth in this paper it will be referred to as domain ontology. Among the macro-steps to develop each component of a domain ontology, we used the following ones.

In the first step (identification) towards building an ontology, we identified the atomic concepts of terms collected from research papers, books, existing ontological resources and experts belonging to the Earthquake Engineering domain giving emphasis on research projects and experiments aspects. We found terms such as device, shaker, experiment, dynamic test, etc., and identified the atomic concept for each of them. We use WordNet as Knowledge Base. The term device has 5 different concepts in it. In our case, we selected the one that has the following description: "*device—(an instrumentality invented for a particular purpose)*". We have found 193 atomic concepts. In the second step (analysis) we analyzed the concepts, i.e. we studied their characteristics to understand the similarity and differences between them. Once the analysis was completed, in the third step (synthesis) we organized them into some facets according to their characteristics. For example, shaker is more specific than device, dynamic testing is more specific than experiment and we assigned the following relationship between them: shaker *IS_A* device and dynamic test *IS_A* experiment. In this way, we built 11 facets. Device and experiment facets are shown in Fig. 2.2 In the fourth step (standardization), we marked concepts with a preferred name in the cases of availability of synonymous terms. For example, while experiment and test are referred to the same concept, we assigned the former term as the preferred one. Finally, the ontology was validated by a domain expert.

Device

- Shaker
- Hammer
- Active Structural device
- Passive Structural device
  - Hydraulic damper
  - Electrical damper
  - MR damper
  - Friction damper

Experiment

- Static
  - Cyclic test
  - Monotonic test
- Dynamic
  - PSD (Pseudo-dynamic) test with substructuring
  - Shaking table test
  - Shaker-Based Test
  - Hammer-Based test

Fig. 2.2 Device and experiment facets

2.4 Ontology Representation

2.4.1 RDF

The Resource Description Framework (RDF) is a data model used to represent information about resources in the World Wide Web (WWW) and can be used to describe the relationships between concepts and entities. It is a framework to describe metadata on the web. Three types of things are in RDF: resources (entities or concepts) that exist in the real world, global names for resources (i.e. URIs) that identify entire web sites as well as web pages, and RDF statements (triples, or rows in a table) (Klyne 2004). Each triple includes a subject, an object and a predicate. RDF is designed to represent knowledge in a distributed way particularly concerned with meaning. The following RDF statements describe the resources 'Hammer' and 'Damper'.

```
<rdf:Descriptionrdf:about="http://earthquake.linkeddata.it/resource/Hammer">
        <rdfs:subClassOfrdf:resource="http://earthquake.linkeddata.it/resource/Device"/>
        <ontology:desciptionrdf:datatype="http://www.w3.org/2001/XMLSchema#string">A hand
        tool with a heavy rigid head and a handle; used to deliver an impulsive force by strik-
        ing</ontology:desciption>
        <rdf:typerdf:resource="http://www.w3.org/2002/07/owl#Class"/>
</rdf:Description>

   <rdf:Descriptionrdf:about="http://earthquake.linkeddata.it/resource/Damper">
        <rdfs:subClassOfrdf:resource="http://earthquake.linkeddata.it/resource/Device"/>
            <ontology:desciptionrdf:datatype="http://www.w3.org/2001/XMLSchema#string">A
        device that decreases the amplitude of electronic, mechanical, acoustical or aerodynamic
        oscillations</ontology:desciption>
        <rdf:typerdf:resource="http://www.w3.org/2002/07/owl#Class"/>
   </rdf:Description>
```

The above example represented relationship between 'Hammer' and 'Device' concepts; and the rdfs:subClassOf property is used to relate the former class to its more generic later class.

2.4.2 OWL

Web Ontology Language is designed to represent comparatively complex ontological relationships and to overcome some of the limitations of RDF such as representation of specific cardinality values and disjointness relationship between classes (Giunchiglia et al. 2010b). The language is characterized by formal semantics and RDF/XML based serializations for the web. As an ontology representation language, OWL is essentially concerned with defining terms that can be used in RDF documents, i.e., classes, properties and instances (Antoniou and van Harmelen 2004). It serves two purposes: first, it identifies current documents as an ontology and second it serves as a container metadata regarding the ontology. This language focuses on reasoning techniques, formal foundations and language extensions. OWL uses URI references as names and constructs these URI references in the same manner as that used by RDF. The W3C allows OWL specification includes the definition of three variants of OWL, with different levels of expressiveness. These are OWL Lite, OWL DL and OWL Full, ordered by increasing expressiveness.

2.5 Existing Ontology/Thesaurus

2.5.1 WordNet

WordNet (Miller et al. 1990) is an ontology that consists of more than 100 thousand concepts and 26 different kinds of relations e.g., hyponym, synonym, antonym, hypernyms and meronyms. It was created and is being maintained at the Cognitive Science Laboratory of Princeton University. The most obvious difference between WordNet and a standard dictionary is that its concepts are organized into hierarchies, like professor *IS_A* kind of person and person *IS_A* kind of living thing. It can be used for knowledge-based applications. It is a generic knowledge base and as such does not have good coverage for domain specific applications. It has been widely used for a number of different purposes in information systems including word sense disambiguation, information retrieval and automatic text summarization.

2.5.2 NEES Thesaurus

The Network for Earthquake Engineering Simulation (NEES) is one of the leading organizations for Earthquake Engineering in USA. They developed the earthquake engineering thesaurus; it is based on Narrower and Broader terms. It contains

NEES Earthquake Engineering Ontology by Broader Term		
Broader Term	Term	Narrower Term
AASHTO_2001	AASHTO_LRFD_Bridge_Design_Specifications	
Acceleration	Peak_Base_Acceleration	
Actuator	Dynamic_Actuator	
Actuator	Static_Actuator	
Axial_Load	Cyclic_Axial_Load	
Bearings	Preformed_Fabric_Pads	Cotton_Duck_Bearing_Pads

Fig. 2.3 NEES thesaurus

around 300 concepts and we have integrated in our ontology 75 concepts from NEES. Figure 2.3 depicts a small portion of NEES thesaurus.

2.6 Ontology Integration

Developed facets include concepts that were selected from NEES thesaurus to be incorporated into our ontology. This integration was accomplished in fact when we built the facets. In this Section, we describe how we integrated our developed ontology with Wordnet. Basically, we applied the semi-automatic ontology integration algorithm proposed in Farazi et al. (2011). In particular, we implemented the following macro steps:

- ***Facet concept identification:*** For each facet, the concept of its root node is manually mapped to WordNet, in the case of availability.
- ***Concept Identification:*** For each atomic concept *C* of the faceted ontology, it checks if the concept label is available in WordNet. In the case of availability, it retrieves all the concepts connected to it and maps with the one residing in the sub-tree rooted at the concept that corresponds to the facet root concept.
- ***Parent Identification:*** In the case of unavailability of a concept it tries to identify parent. For each multiword concept label it checks the presence of the header, and if it is found within the given facet, it identifies it as a parent. For instance, in WordNet it does not find hydraulic damper for which damper is the header and that is available there in the hierarchy of device facet. Therefore, it recognizes the damper with the description *"damper, muffler—(a device that decreases the amplitude of electronic, mechanical, acoustical, or aerodynamic oscillations)"* as the parent of the hydraulic damper.

2.7 Experimental Set-Up

Figure 2.4 represents the architecture of our KB-based information management system that uses semantic tools and technologies. We published our developed domain specific ontology into RDF using Jena, a Semantic Web tool for publishing

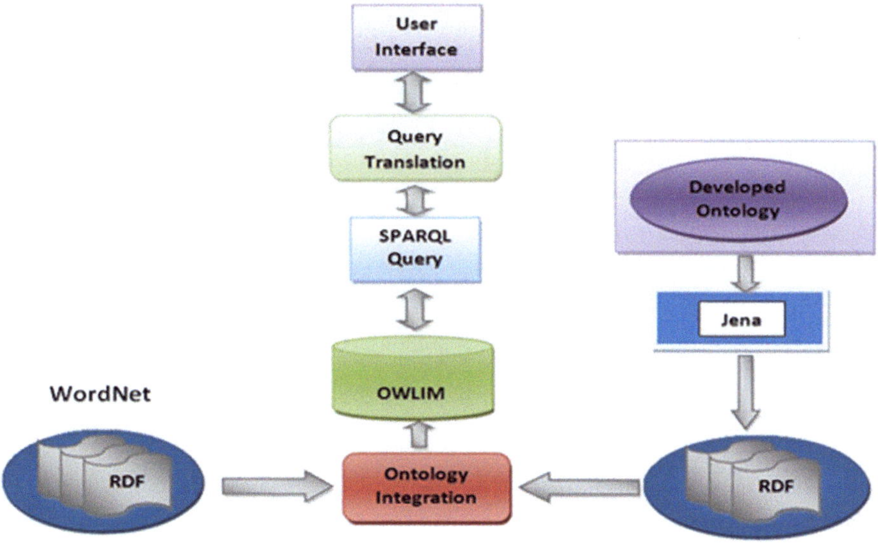

Fig. 2.4 KB-based system architecture

and managing ontologies and integrated it with WordNet RDF using the approach described in Sect. 2.6. To increase the coverage we integrated our developed ontology with WordNet. The outcome of the ontology integration was put in OWLIM triple store that was queried using Sesame API. For translating user natural language queries into SPARQL queries, we developed a few lines of code written in java. SPARQL is a query language specific to query RDF representations. It allows add, update and delete of RDF data.

2.8 Results

In this Section, we describe basically what advantages users can get with KB-based systems over traditional DB systems. In particular, we performed synonym search and more specific concept search.

Synonym Search When a concept is represented with two or more terms, they are essentially synonymous that are represented in RDF with *owl: equivalent Class*. For example, test and experiment represent the same concept and in the ontology they are encoded accordingly with equivalent relation. Therefore, user query for test can also return experiment, (Fig. 2.5) because they are semantically equivalent.

More Specific Concept Search In our ontology concept hierarchies are represented using *rdfs: subClassOf*. For example, hammer and damper are more specific concepts of device, hence, they are represented as follows: hammer *rdfs: subClassOf*device; and damper *rdfs: subClassOf* device. Moreover, hydraulic damper is more specific than damper and it is encoded as hydraulic damper *rdfs:subClassOf*

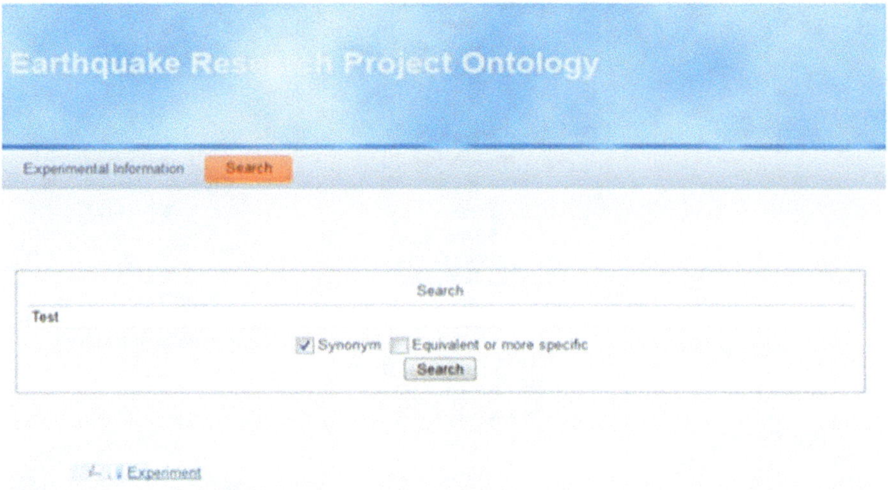

Fig. 2.5 Synonymous relationship

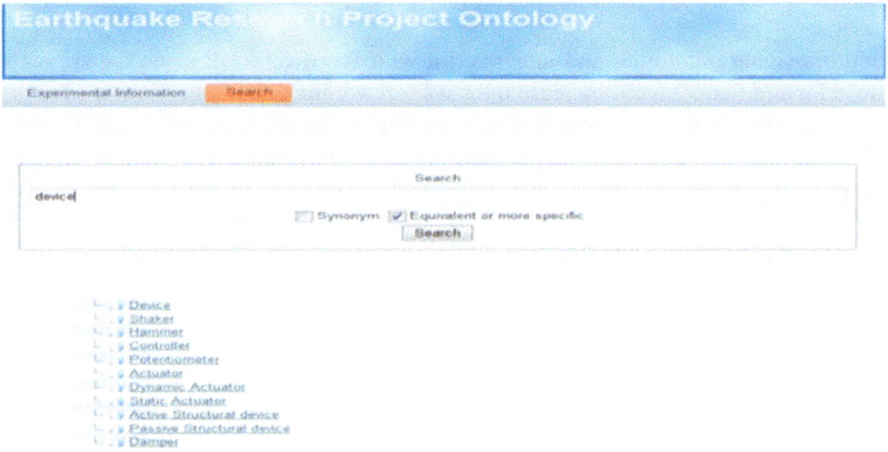

Fig. 2.6 Transitive relationship of device

damper. Note that *rdfs: subClassOf* is a transitive relation. Using OWL inference engine, we can utilize the power of transitivity and for a given concept we can retrieve all more specific concepts that are directly or indirectly connected by *rdfs:subclassof*. Therefore, a search for device retrieved all of its more specific concepts as shown in Fig. 2.6.

2.9 Related Work

NEES ontology has been developed in the domain of earthquake engineering. However, it is mainly a thesaurus encoding broader and narrower relations that cannot capture ontological details. For instance, it cannot be clarified in thesaurus whether a relation between two concepts is *IS_A* or *PART_OF*. As a result ontologies represented as thesaurus might lead to some unexpected results. DBPedia is an example that uses broader/narrower relations and ended up establishing connections between Telecommunication, and Flora and Fauna. In contrast, the ontology developed in this paper does not suffer from this issue; rather it provides better clarification because it exploits ontological relations.

2.10 Conclusion

In this paper, we provided a detailed description of the development of Earthquake Engineering Projects and Experiments ontology. We followed DERA methodology for building this domain specific ontology. We exploited an ontology integration algorithm that was employed to incorporate our ontology into WordNet. It helped to increase the coverage of the Knowledge Base. On top of the integrated ontology, i.e., put in an OWLIM store of an experimental setting, we experimented the semantic and ontological capabilities of the developed system and interesting results were found.

References

Antoniou, van Harmelen (2004) Grigoris Antoniou and Frank van Harmelen: a Semantic Web Primer. MIT Press
Berners-Lee T (1999) Weaving the web. Orion business books, London
Berners-Lee T, Hendler JA, Lassila O (2001) The semantic web. Sci Am J 284(5):34–43
Farazi Feroz et al (2011) A faceted ontology for a semantic geo-catalogue. The semanic web: research and applications. Springer, Berlin, pp 169–182
Giunchiglia F, Biswanath D (2011) DERA: a faceted knowledge organization framework
Giunchiglia F, Farazi F, Tanca L, de Virgilio R (2010a) The semantic web languages. Semantic web information management. Springer Verlag, Berlin
Giunchiglia F, Maltese V, Farazi F, Dutta B (2010b) GeoWordNet: a resource for geo-spatial applications. The semantic web: research and applications. Springer, Berlin, pp 121–136
Giunchiglia F et al (2012) A facet-based methodology for the construction of a large-scale geospatial ontology. J Data Semant 1(1):57–73
Grigoris A, van Harmelen F (2004) Semantic web primer. MIT Press, USA
Klyne G, Jeremy JC (2004) Resource description framework (RDF): concepts and abstract syntax. W3C Recommendation, 2004. World Wide Web Consortium. http://w3c.org/TR/rdf-concepts
Miller GA, Beckwith R, Fellbaum CD, Gross D, Miller K (1990) WordNet: an online lexical database. J Lexicograph 3(4):235–244
Ranganathan SR (1967) Prolegomena to library classification. Asia Publishing House, New York

Related Work

[illegible]

Chapter 3
The SERIES Virtual Database: Architecture and Implementation

Ignacio Lamata Martínez, Ioannid Ioannidis, Christos Fidas, Martin S. Williams and Pierre Pegon

3.1 Introduction

The European earthquake engineering community is highly fragmented, lacking an operational unity to maximise research effectiveness. Rapid increases in computational resources and connectivity have contributed to address such problems efficiently and to fill the technological gap between European laboratories. These laboratories have typically worked independently, storing their experimental data in proprietary or, sometimes, not formal structures. This complicates the dissemination and reuse of information, which has motivated the development of the SERIES virtual database (SVDB): an infrastructure to make laboratory experimental data available in a formalized way whose treatment can be automated.

The European situation has strongly affected the architecture of such infrastructure. One of the main challenges is the fact that the 22 participant laboratories are very different from each other: they have different objectives, engineering infrastructures, capabilities, working languages, hardware and software platforms and they create and utilise data from many different types of experiments.

Existing data repositories and sharing facilities were studied before starting the development of the SVDB. The most relevant earthquake engineering repository is the one developed by the US Network for Earthquake Engineering Simulation (NEES 2012; Hacker et al. 2011; Pejša 2012). NEES is composed of a headquarter and 14 laboratories distributed around the United States, with different shared facilities. The NEES data repository is a centralised data repository to store and share experimental data through a well-defined interface, and supported by a number of data management tools.

P. Pegon (✉)
European Commission, Joint Research Centre (JRC)
e-mail: pierre.pegon@jrc.ec.europa.eu

I. Lamata Martínez · M. S. Williams
University of Oxford, Oxford, UK
e-mail: ignacio.lamata@eng.ox.ac.uk

I. Ioannidis · C. Fidas
University of Patras, Patras, Greece

F. Taucer, R. Apostolska (eds.), *Experimental Research in Earthquake Engineering*,
Geotechnical, Geological and Earthquake Engineering 35,
DOI 10.1007/978-3-319-10136-1_3

Another data repository was developed in UK-NEES, a collaborative grid-based research network established between the Universities of Oxford, Bristol and Cambridge in the United Kingdom (UNKEES 2013). The UK-NEES data repository was a simple centralised database to store experimental results from the three institutions.

The JRC also developed a Web accessible database to store results of tests conducted at ELSA (European Laboratory for Structural Assessment). Its data model served as a foundation for the SVDB exchange data model (Bosi and Pegon 2009).

In the light of the differences in institutional practice within the SERIES consortium, the decision was taken to develop a virtual database instead of a centralised repository. The virtual database provides access to multiple distributed sources of information by using a single, centralised gateway. The experience to the end user should be similar to accessing a single data repository.

The virtual database aims to create the necessary infrastructure for data integration between the 22 participant laboratories. This involves inculcating a common understanding of the data structure and data exchange methods within the European earthquake engineering community as well as providing access to experimental results for the earthquake engineering community worldwide. This work also encourages the automation of data processing by providing systems that treat information in a standard way and store it in a formal, common format.

In this paper, we describe the architecture and implementation of the SERIES virtual database, linking the 22 leading European earthquake engineering research institutions participating in the European FP7 project SERIES. Section 3.2 discusses the architecture of the SERIES virtual database, presenting its main characteristics. Section 3.3 describes the implementation of the elements part of the virtual database. Section 3.4 explains the general security architecture and finally Sect. 3.5 presents some conclusions.

3.2 Architecture of the Virtual Database

The Series Virtual Database (SVDB) is the infrastructure that allows access to the SERIES European laboratories' data through a single user interface. The system developed is described as a "virtual database", meaning that it provides single-point access to different distributed data sources (many data sources are accessed using a common method).

3.2.1 Characteristics of the SERIES Virtual Database

The main characteristics of the SVDB architecture are discussed below:

1. It is **decentralised**. The SVDB has no centralised repository to store data, herein every institution in the virtual database is a different node acting as a source of information. Nodes are responsible and ultimate owners of the data they produce, which makes institutions behind those nodes to work for their own benefit

and avoids the need to motivate a dispersed scientific community. All the information is stored and controlled locally and nodes decide what and what not to share. A decentralised data architecture is more appropriate for institutions that have inter-state political and legal differences within Europe and do not enjoy a complete political and territorial unity. Data decentralisation also encourages the enhancement of the technical computing infrastructure in each node, constructing a technical balance between nodes within the SVDB.

2. It provides a **centralised access**. The access to every node in the SVDB is provided by a single Web interface (also called DAP or Data Access Portal). This single access is desirable since it provides a single route to the data and enables operations (such as searching) over the data as a whole instead of accessing each node's data separately. Another main advantage is that users only need to become familiar with one single interface. For the end users, the decentralised nature of the SVDB is transparent: the system acts as if it was a big centralised repository. A centralised access can become a single point of failure: even if all the distributed repositories are available, they cannot be accessed if the centralised access is off-line. This situation occurs in centralised repositories as well, and can be corrected easier in a distributed environment by providing redundant systems to access the SVDB.
3. The centralised access performance is enhanced by a **central site**, which hosts the DAP and provides facilities to increase performance of the data access. To do so, the central site stores metadata of every node's data. When an end user access the DAP, it is not necessary to connect to every node in the virtual database (which would make the access unacceptably slow) but to use the local metadata information instead. Extended information about the communication between the central site can be found in Sect. 3.2.3.
4. The SVDB is **node-oriented**. The primary component is the node, not the central site. Nodes are the providers of information and they are managed and controlled by their respective institutions. The central site does not provide a method to submit data to the repository. On the contrary, it is the node that provides the necessary mechanisms to store and access its data. This aspect is a fundamental difference between this and other systems governed by a centralised component. This approach allows at the same time promoting sharing of data as well as coping with the strong sense of ownership laboratories usually have for the test results they produce.
5. The SVDB architecture is **service oriented**. This means nodes provide information by following a Service Oriented Architecture (SOA). SOA is an architectural paradigm in distributed systems that focuses on organizing and utilising distributed capabilities of heterogeneous systems under the control of different owners (MacKenzie et al. 2006). This reference model allows interoperability between different systems, which satisfies the European requirements perfectly. Herein, nodes make available some services to provide access to their data. In the SVDB, nodes are servers that provide access to their respective data to the central site—which acts as client of every node. The node services are discussed in Sect. 3.2.2.

3.2.2 Services Provided by the Nodes

The service oriented architecture of the SVDB is implemented by means of SOAP Web Services, which are "a software system designed to support interoperable machine-to-machine interaction over a network" (Web Services Architecture 2013). SOAP (Simple Object Access Protocol) is a fundamentally—but not mandatorily, stateless framework to exchange structured, XML-based messages between two entities (SOAP 2013) and it is standardised as a W3C recommendation (W3C 2012). SOAP makes use of WSDL (Web Services Description Language (WSDL 2013)), that is an XML-based language to describe Web Services. The WSDL is what formally defines the list of services that a node provides and the data structures that are exchanged in each possible request.

In the SVDB, a common list of services has been defined by means of a WSDL structure, which is basically a piece of software to define operations, message exchanged and data types for input and output data. To be part of the SVDB, every node must provide the operations defined in a WSDL file. Hence, the real data integration between nodes is achieved by defining a common list of services that each node must provide. An example of service is "getProject", which returns a project data based on a provided project identifier.

Web Services allow interoperability of heterogeneous systems. Nodes are free to choose their internal technical structure, as long as they provide the necessary list of services for the SVDB defined in the WSDL structure. From the SVDB point of view, it is only required that nodes provide their information in a common format but how this information is obtained is up to each node. This provides great flexibility to nodes, since different types of databases, operating systems, programming languages and hardware can be used.

3.2.3 Communication Between the Nodes and the Central Site

The central site needs to keep an updated version of each node's data to make it available through the DAP. For this reason, the central site contacts each of the nodes in the SVDB overnight, and collects all the necessary information. The central site is in charge of triggering this updating process and managing errors coming from it. With this operational schema, nodes are released from many administrative tasks. How updated the information is in the DAP directly depends on updating policies in the central site, which can adjust the updating procedure differently for each node, updating information more often for more active nodes in the virtual database. Additional information of the updating process can be found in (Lamata Martinez et al. 2013).

Since all nodes in the SVDB provide their information in a common format (specified by the WSDL), the central site can query nodes in the exact same way, simplifying the operations that it has to do to collect and treat information. This process is easily scalable to any potential new node in the network, and the effort to

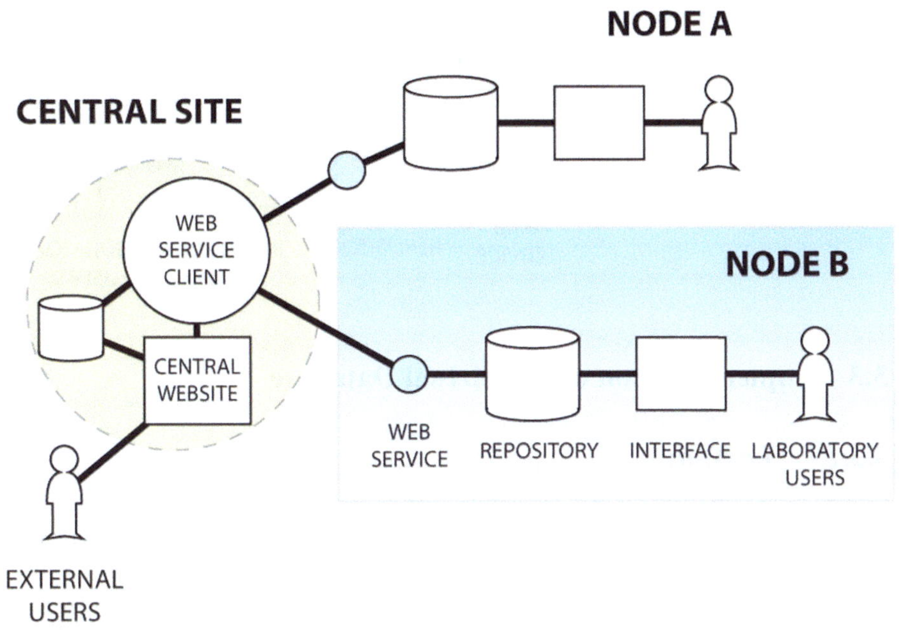

Fig. 3.1 SERIES virtual database global architecture

include a new node in the SVDB is minimum. Figure 3.1 depicts the SVDB architecture, where the nodes' information is accessed by the central site and supplied to external users (the general public, the earthquake engineering community).

3.2.4 Large Files Download

The central site provides access to every node's information within the SERIES virtual database by means of a Web interface called the DAP. To speed up the process of accessing each node's information, the central site stores some metadata. However, the central site is not a centralised repository, so it does not store large files such as documents, videos, images and signals, which are still served by each particular node. When an external user requests a large file, it is transparently redirected to the specific node hosting the file. This process delegates the task of delivering large files to the different nodes, which should accept direct connections from external users. An extended discussion of the operation of large files download can be found in (Lamata Martinez et al. 2013).

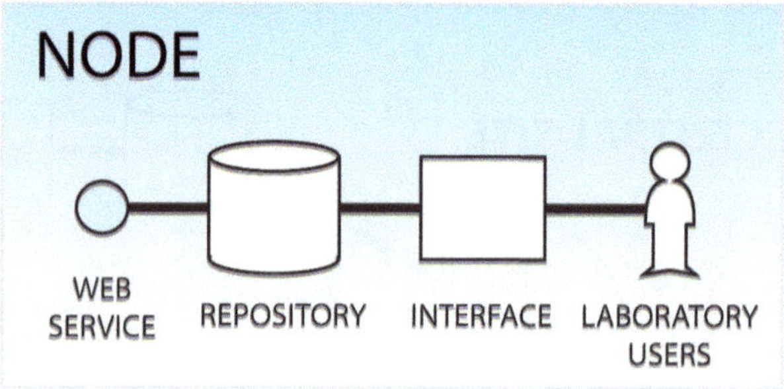

Fig. 3.2 Internal design of the node

3.3 Implementation of the Virtual Database

3.3.1 The Node

The node is the main source of information and, therefore, the main component of the SVDB. Each node can use heterogeneous architectures internally as long as they comply with the list of services defined in the WSDL. At the University of Oxford, an initial node structure has been formalised, consisting of the design depicted in Fig. 3.2. In this figure, the node is divided in three layers: the Web Services, that makes the node's information available for external systems; the repository, that stores the node's experimental data; and the interface, that allows node's local users to interact with the database in a user-friendly way.

How these elements are implemented (or even they are implemented at all) is the decision of the node. However, software implementing the repository, the Web Services and the software component to allow download of large files has been developed at the University of Oxford, and can be easily reused by other nodes. In practice, most of the nodes in the SVDB are reusing these components instead of developing their own ones. The software at Oxford has been developed by using a MySQL database management system, Java and Apache Tomcat.

The database developed at Oxford has been based on the data user requirements defined for SERIES (Lamata Martinez et al. 2010). The model defines four hierarchically organised levels: Project, Specimen, Experiment/Computation and Signal. A fifth independent level can be considered, containing metadata of media resources: documents, videos and images, as depicted in Fig. 3.3. One of the most important aspects is the privacy attribute of some of the elements in the database. Three levels are defined to mark database information: private (for information that must not be shared), partner (for information that can be shared only within SVDB partners) and public (for information that can be shared with the whole earthquake engineering community).

The Oxford Web Services implement the Web Services according to the defined SERIES WSDL and mediate with the local database to obtain the requested information. A Java interface has been created to interact with the database in an abstract way, so any code accessing the database can access it in a uniform way. The Web

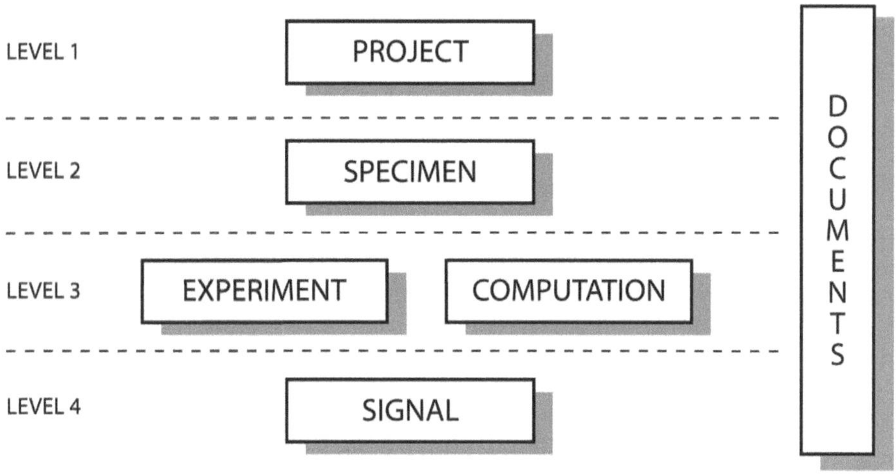

Fig. 3.3 The four-level database schema with a fifth vertical layer

Services are in charge of filtering the information according to its privacy level, so they will only send to the central site information marked as "partner" or "public". Information marked as "private" is invisible for any external system. The Web Services are divided into the following groups:

- Testing Services, to check the correct operation of the node and its interaction with the central site. These services are especially useful to trace possible errors when a node joins in the SVDB.
- Data Services, to deliver the information from the local repository or database.
- Updating Services, related to the updating process of the central site.
- Control Services, to control the system operation or to obtain system information.

Finally, the Oxford component to download large files works as an independent piece of software that interacts with the local database. When an external user requires the download of a file, they access this component which validates the request, identifies the element of the database to be downloaded and sends it to the end user. Documents, videos and images are sent in their established format. As for signals, the download component will perform some data transformation to deliver the data.

3.3.2 The Central Site

The central site is the main access point to all data available from the SERIES partners, including downloading and searching. However, no actual data are stored on it but only some metadata retrieved from the distributed SERIES laboratories.

The central site consists of:

- Data Access Portal (D.A.P.), which is the main Web interface for accessing the available projects of the SERIES partners. The main purpose of D.A.P. is to act as an information space where users are able to explore, download and search for the information they need. Information on the D.A.P. is presented and structured according to the data user requirements defined for SERIES (Lamata Martinez et al. 2010). Two different privacy levels are available at the D.A.P. (public to the whole earthquake engineering community or restricted to SERIES partners). Projects for SERIES partners require a login process. The D.A.P. provides functionalities for downloading the actual data, redirecting external users in a transparent way to the specific laboratory which owns the data. Download is permitted only for users that have accessed the site and agreed to the terms of use. The D.A.P. was designed following a user centered design approach (UCD) and any complex security process involved is transparent to the end user.
- Web Services client: which is the communication layer between all the nodes and the central site. This is the component that communicates overnight to every node in the SVDB as explained in Sect. 3.2.3.
- Metadata database: hosts all the metadata necessary to provide the information through the D.A.P, which is obtained by means of the Web Services client. This database gathers the metadata for all the nodes in the SVDB.

These three elements can be seen in the central site component of Fig. 3.1.

The central site resides at the University of Patras and has been developed and maintained by the Human Computer Interaction group of the Department of Electrical and Computer Engineering.

3.4 Security

The security of the SVDB has been supported by public-key certificates, which are based on public key cryptography (Schneier 1996). This method uses two complementary keys, one "public" and another one "private", which are mathematically bound. Obtaining one key from the other is protected by the mathematical problem of factoring the product of two large prime numbers.

The main security measures implemented in the SVDB architecture are:

- Communication between the central site and the nodes is encrypted by using the HTTPS protocol. This also helps to determine if the exchanged information was altered during its transmission.
- The identity of the central site is verified by means of public-key certificates, before accepting any Web Service request.
- When an external user requests the download of a large file, the central site has to create a special link for the node to accept such request. That special link can be only generated by the central site (which is guaranteed by public cryptography) and has a time expiration.

Security on the node side mainly depends on each node's security policies. Some institutions have very severe security policies whereas some others are more permissive. As a general rule, most of the SVDB nodes provide firewall restrictions (either via software or via hardware) and some make use of a DMZ infrastructure (Demilitarized Zone) to protect SVDB public services from the rest of the internal network. The components most likely to be attacked are the Web Services and the download component, because they have to be publically exposed. Security restrictions have been applied to these components so they are only able to read from the local database, which can be also located on a different machine that applies additional security rules. Some nodes make use of virtualization techniques to isolate software in a "virtual machine" and minimize security implications.

The SERIES Central Site is exposed to a number of virtual threats, from hacking and malicious code to denial of service attacks. Two main external entities interact with the central site: external users (Website visitors, who retrieve information from the central site) and laboratories (which provide information to the central site via Web Services). Access from/to the central site requires traffic supervision on incoming and outgoing connections, especially to know when and over which ports someone is trying to grant unauthorized access and what changes have been performed on the file system. For this purpose, an Intrusion Detection System (IDS) is used to monitor the network traffic and to log suspicious activity such as violation of computer security policies. Also, data replications mechanisms have been applied to the central site in the case of a failure.

3.5 Conclusions

The SERIES Virtual Database creates an infrastructure to enable data integration between the leading earthquake engineering institutions in Europe. Such virtual database consists of a group of independent, geographically distributed data sources that are accessed in a unified manner via a Web interface. To communicate the distributed data sources (also called nodes) with the centralised access (also called the central site), Web Services are used, which are a standardised method for machines to exchange information on the Web. Nodes are owners of the information they produce and can have heterogeneous hardware and software configurations. To speed up the process of presenting information to external users, the central site keeps a local copy of each node's metadata which updates overnight by connecting to each node in the SVDB. Security of the communication between nodes and central site firmly rely on public key certificates.

Acknowledgments The research leading to these results has received funding from the European Community's Seventh Framework Programme [FP7/2007–2013] under grant agreement n 227887.We thank all SERIES project partners who contributed to the database development through discussion, trials and feedback.

References

Bosi A, Pegon P (2009) The ELSA database and what can be done regarding SERIES networking activities. JRC56446. EUR Report 23931 EN, Luxembourg (Luxembourg)
Hacker TJ, Eigenmann R, Bagchi S, Irfanoglu A, Pujol S, Catlin A, Rathje E (2011) The NEEShub cyberinfrastructure for earthquake engineering. Comput Sci Eng 13(4):67–78
Lamata Martinez I Pegon P, Bosi A, Dietz M, Bousias S, Fidas C, Wenzel H (2010) SERIES deliverable D2.1: distributed database—review of beneficiary current data formats http://www.series.upatras.gr/userfiles/file/Public_Deliverables/Deliverable_2_1.pdf. Accessed July 2013
Lamata Martinez I, Ioannidis I, Pegon P, Williams M, Blakeborough A (2013) The process and future of data integration within the European earthquake engineering laboratories. J Comput Civ Eng 10.1061/(ASCE)CP.1943-5487.0000308. Accessed 10 April 2013
MacKenzie CM, Laskey K, McCabe F, Brown PF, Metz R (2006) OASIS reference model for service oriented architecture 1.0. http://docs.oasis-open.org/soa-rm/v1.0/soa-rm.pdf. Accessed July 2013
NEES (2012) NEEShub: A platform for research, collaboration and education.http://nees.org. Accessed June 2012
Pejša S (2012) NEEScomm requirements for curation and archiving of research data. https://nees.org/resources/4759/. Accessed Dec 2012
Schneier B (1996) Applied cryptography: protocols, algorithms, and source code in C, 2nd edn. Wiley ISBN: 978–0471117094
SOAP (2013) Simple Object Access Protocol. http://www.w3.org/TR/soap/. Accessed July 2013
UKNEES (2013) UK Network for Earthquake Engineering Simulation. http://www-civil.eng.ox.ac.uk/research/structdyn/presentations/uknees.html. Accessed June 2013
W3C (2012) World Wide Web Consortium. http://www.w3.org/. Accessed July 2013
Web Services Architecture (2013) http://www.w3.org/TR/ws-arch/. Accessed July 2013
WSDL (2013) Web Services Description Language Version 2.0 Part 1: Core Language. http://www.w3.org/TR/wsdl20/. Accessed July 2013

Chapter 4
The SERIES Virtual Database: Exchange Data Format and Local/Central Databases

Anna Bosi, Ilias Kotinas, Ignacio Lamata Martínez, Stathis Bousias, Jean Louis Chazelas, Matthew Dietz, Rashedul Hasan, Gopal S. P. Madabhushi, Andrea Prota, Anthony Blakeborough and Pierre Pegon

4.1 Introduction

SERIES (NA1 Networking Activity) targeted at creating a European platform (the SERIES Virtual database, SVD) for wide sharing of experimental data and knowledge across the field of earthquake engineering and amongst academia, research and industry, which could be maintained and enhanced in time.

A. Bosi (✉)
Vienna Consulting Engineers ZT GmbH, Vienna, Austria
e-mail: Bosi@vce.at, bosianna78@gmail.com

I. Kotinas · S. Bousias
University of Patras, Patras, Greece

I. Lamata Martínez · A. Blakeborough
University of Oxford, Oxford, UK

J. L. Chazelas
IFSTTAR, Nante, France

M. Dietz
University of Bristol, Bristol, United Kingdom
e-mail: m.dietz@bristol.ac.uk

R. Hasan
University of Trento, Trento, Italy

G. S. B. Madabhushi
University of Cambridge, Cambridge, United Kingdom

A. Prota
University of Naples, Naples, Italy

P. Pegon
European Laboratory for Structural Assessment Unit, European Commission,
Joint Research Centre (JRC), Institute for the Protection and Security
of the Citizen (IPSC), Via Enrico Fermi 2749, Ispra VA 21027, Italy
e-mail: pierre.pegon@jrc.ec.europa.eu

F. Taucer, R. Apostolska (eds.), *Experimental Research in Earthquake Engineering*,
Geotechnical, Geological and Earthquake Engineering 35,
DOI 10.1007/978-3-319-10136-1_4

Before SERIES, despite European earthquake engineering laboratories generated large amounts of data either in the experimental facilities (shake tables, centrifuges, reaction walls), or by in-situ (field) tests, few laboratories had adopted the approach of a database for storing their test results, with the majority saving data in a fragmented and unstructured way and without any strategy for the long term preservation of data or for its effective dissemination. Hence, the dissemination and use of the experimental results outside the laboratory where they were produced was problematic.

To overcome this situation, a common effort led by the Joint Research Centre, the University of Oxford and the University of Patras was undertaken to develop a prototype database with enhanced characteristics, which could then be implemented at all collaborating sites. The approach was not that of building a central database in which local ones would either migrate or merge their data into, but to provide centralised access to individual database nodes distributed over the network, and which would be able to dialog with a central portal in a uniform manner.

The most important elements of the SERIES virtual database developed by this joint effort are (Ioannidis et al. 2011, 2012):

- Data Access Portal. It provides a centralized access to all the projects the SERIES laboratories make public. The Data Access Portal presents the information of the available projects, by following the structure of the Exchange Data Format. It is at the discretion of each individual laboratory to select which projects or project results to make public.
- Exchange Data Format. This is the format in which the data and other information is stored (locally) and presented by the Data Access Portal.
- Local database. It is the local repository where data is stored, following the defined Exchange Data Format.
- Web Services. An ensemble of software tools which allow the exchange of content and configuration between the Data Access Portal and the local databases.

The present paper presents the SERIES virtual database from both the point of view of an external user and that of an internal one. Details on the architecture and implementation of the SERIES virtual database can be found in a companion paper (Lamata Martínez et al. 2013a). Section 4.2 describes the data format that is used in the communication between every SERIES participant and the central site containing the Data Access Portal. Section 4.3 discusses methods used by the different SERIES participants to interact with their respective local databases. Section 4.4 explains the SERIES virtual database from the perspective of external users and how they can take advantage of this SERIES infrastructure. Finally, Sect. 5.3 presents some conclusions.

4.2 The Exchange Data Format

The Exchange Data Format (EDF) is the format in which data are presented via the Data Access Portal as well as the format in which data (with some additional fields) are stored locally at individual sites. Therefore, it has to be both comprehensive in satisfying internal user needs and understandable by external users. The Exchange Data Format has been designed to:

- Host data produced by experiments and/or simulations.
- Be suitable for any experimental data type: data produced by centrifuges, reaction walls, shaking tables, etc.
- Allow storing data along with all other types of information (documents, images, etc.) which are useful to describe, repeat or simulate the experiments under the same conditions (Lamata Martínez et al. 2010).
- Allow for data accessibility restrictions: projects can be public, restricted only to partners or, completely private (accessible only to the laboratory where have been produced).

A critical analysis of data formats pre-dating that of SERIES was performed at the beginning of the project. In view of the lack of many alternatives, the analysis practically was limited to the NEES (US Network for Earthquake Engineering Simulation) and the ELSA (Joint Research Centre) repositories (Bosi and Pegon 2009). To cover the desirable characteristics of the existing repositories and further extend their potential, a four-tier hierarchical structure, consisting of *Project*, *Specimen*, *Experiment/Computation* and *Signal* (Fig. 4.1) has then been selected for the Exchange Data Format.

Project level includes general information on the project: infrastructures and persons involved and scope of the project. For the sake of uniformity, most of the fields have a fixed list of possible entries from which the user can chose. This allows for avoiding typos or using different naming for same objects, while simplifying retrieval of data and information via the search functionality. Examples of fixed fields at the *Project* level are '*project main focus*', '*role*', '*institution name*', '*infrastructure*'.

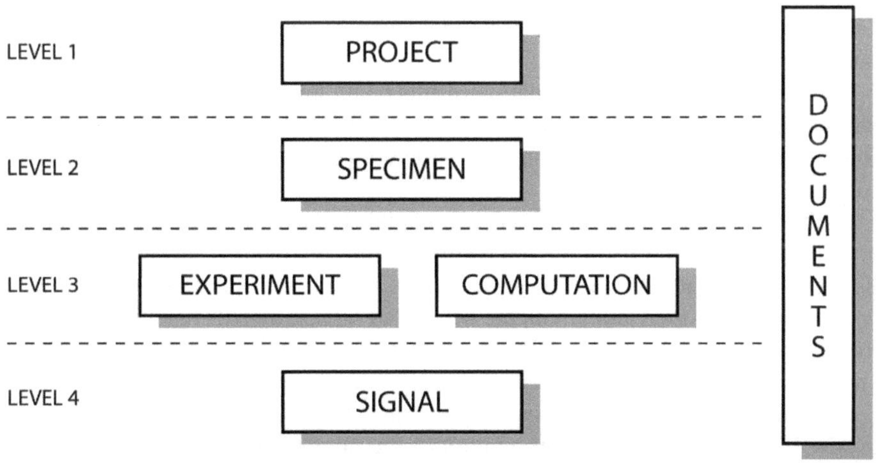

Fig. 4.1 The four-tier hierarchical structure of the Exchange Data Format. (Lamata Martínez et al. 2013b)

A project usually includes testing of more than one physical (or numerical) structure (a short bridge pier and a tall one, several masonry structures made by different kinds of clay…) identified as *Specimens*. It is also possible to test the same structure but in different "states": for example the structure in its original state and then after different types of retrofitting. While it may be argued that, in this case, all tests are performed on the same specimen, the hierarchical structure of the database demands that retrofitted specimens are included as new specimen. At this level, the physical and mechanical characteristics of the specimen are specified. Each structure is subdivided into structural elements (as for example beam, column, jacketing system…). Nominal mechanical properties and, when experimentally measured, also actual ones can be specified.

In the case of a physical experiment, the same specimen is usually subjected to several types of tests that differ by the type of load imposed (quasi-static test, pseudo-dynamic test, shake table test, impact hammer test, etc. with or without substructuring, in-situ or in laboratory), by the location of the loading and/or by the configuration of the sensors. The original load time-histories and the effective inputs used on the different experiments must be explicitly identified. For example, in case of seismic experiments, the same accelerogram can be used several times by changing its intensity, or a different one may be used for each test. The original signals are preserved by providing some information on their nature (natural for accelerogram, natural-normalized for natural accelerogram normalized in the intensity, natural-modified for natural accelerogram modified according to Eurocode, etc.) and the peak excitation. The original load signal can then be scaled in intensity or applied in different directions: this represents the effective input that has also to be provided.

The laboratory database residing at each local site adopts the very same Exchange Data Format, with the addition of some extra fields which allow the description of the characteristics and configuration of devices and sensors employed in testing. As this information is considered meaningful to (and in some cases, understandable by) only the laboratory personnel that performed the experiment, it is not made available to external users.

In the case of numerical simulation results being introduced in the database, the computer system and software used must be specified, along with detailed information on issues regarding modelling the structure (models, assumptions, hysteresis rules, etc.).

At the bottom of the hierarchy is the *Signal* level. Each signal is delivered together with data regarding its units, the nature of the signal (*force*, *acceleration* etc.), the location and the associated time sequence. In the local site database each *measured* signal is reported along with the associated sensor. Signals resulting from data processing or computation (for instance modal frequency, target displacement for a pseudo dynamic algorithm, inter-storey drift, etc.) are stored as *computed* ones.

Numerical values of signal measurements are stored and exchanged using either an ASCII human-readable representation of the numerical series, i.e. a serialized, space-separated series of the alphanumeric representation of one or more numbers, or by using binary IEEE 754-2008 standard format (Lamata Martínez et al. 2010).

The design of the Exchange Data Format allows for additional documentation, photos, and videos to be stored at each level.

4.3 Local Site Management of SERIES Virtual Database

The standardization of the Exchange Data Format has been an iterative process involving all laboratories, especially for the part concerning the definition of a common naming which could accommodate the heterogeneity of the data encountered in the different laboratories. Once the Exchange Data Format had been defined, it was implemented in a MySQL database (MySQL database 2013) and tested with real experimental data. Figure 4.2 depicts the corresponding *Experiment* level of the Exchange Data Format. MySQL Workbench was initially used to input information into the database, although using this generalized user-interface for data manipulation appeared to be tedious and error prone, considering that just one complete experiment may consist of 250 or more interconnected records comprising signals, sensors, configurations, materials and other metadata. Therefore, a formal process definition for the automatic conversion of laboratory data into the common format and specialized tools for its implementation have been developed, consisting of two main logical layers:

- An intermediate portable experiment format enabling the expression and storage of proprietary experimental structures in a common specification.
- Specialized interfaces and tools that allow the local users to automatically import the portable experiment files and easily manage the database.

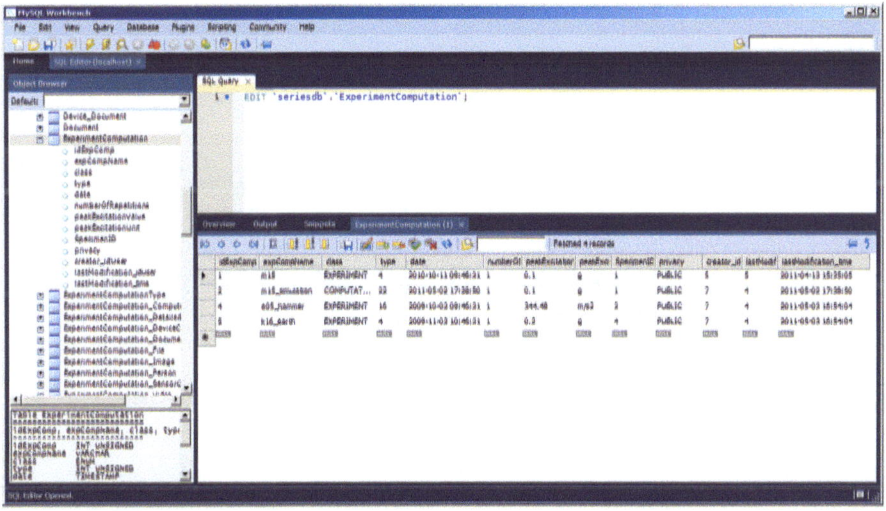

Fig. 4.2 Experiment level in MySQL Workbench: on the *left* side, the other fields of the database

The portable experiment format has been developed in XML and is accompanied by a specialized platform, called "Data Import Platform (DIP)" to allow the automatic import of portable experiment files and management of the local database. An additional tool, called "GUIDE" has been created as a Web interface for database management and users' administration at local sites.

4.3.1 Local Data Processing and Data Import Platform (DIP)

The task of storing proprietary experimental data into the local databases would be tedious and error-prone without a formal process definition and the necessary technology for its implementation. To facilitate the conversion and importing process, an enhanced portable XML specification of the Exchange Data Format was developed, providing an intermediate language between the proprietary data structures used by local laboratories and the SERIES database. The XML schema describing this language was designed as a mapping of the structure of the real database allowing for automatic conversion scripts to express laboratory experiments without significant prior knowledge of the database structure. The schema expresses the necessary data constraints and relations present in the database, along with additional helper variables to represent inner data relations. These include the cardinality of data records and their relations as foreign keys, according to the database terminology. A fragment of the XML schema and a corresponding fragment of a portable XML experiment are presented in Tables 4.1 and 4.2.

Table 4.1 Sample part of the XML Schema: It describes how a proper XML data file should be formatted

```
...
<edf:element name="signal" parentTables="experimentcomputation
sensorconfiguration_sensor signaltimes"
childTables="detailedloadingcharacteristic originalloadingsignal">
<edf:complexType>
<edf:element name="idSignal" type="int" size="10" numeric="1"
unsigned="1" primary="true" nullable="false" autoUpdated="true"
referencedBy="originalloadingsignal.signalID" />
<edf:element name="signalLabel" type="string" size="25"
nullable="false" autoUpdated="false"naturalKey="true" />
<edf:element name="attribute" type="string" size="20"
nullable="false" autoUpdated="false"/>
   ...
</edf:complexType>
<edf:attribute name="serial" type="integer" />
</edf:element>
...
```

Table 4.2 Sample part of an XML data file: the XML data file holds the actual data

```
...
<signal serial="2">
<signalLabel>m15 gefran</signalLabel>
<attribute>Internal Step Counter</attribute>
<physicalQuantity>Number</physicalQuantity>
<type>Computed</type>
<unit>m</unit>
<location>nord 1 floor</location>
<valueVector>-1.201661934 -1.201065324 -1.206541137 1.202002825 -
6.39337E-05 -1.202855051 ...</valueVector>
<expCompID>1</expCompID>
<repetitionNumber>1</repetitionNumber>
</signal>
...
```

The database value constraints are expressed by the *numeric, nullable, size, allowedValues, default* attributes. Foreign key relationships are specified for fields with the *ref* and *referencedBy* attributes. The *naturalKey* attribute defines a field which should uniquely identify a record, as a natural key outside the database. Finally, the *serial* attribute is used to allow the expression of cardinality for similar records.

Since the internal data structures used by each local laboratory are informally specified, unpublished and often inconsistent, the responsibility of their conversion to the proposed XML format relies with the individual laboratory. In realizing this task, a possible route is that deployed and distributed by the Structures Laboratory, University of Patras, consisting of a well-defined process and the accompanying scripts to convert an internal proprietary MATLAB structure to XML. The process involves usage of an intermediate comma-separated (CSV) mapping file as a reference for the conversion process and MATLAB script files to transform the initial structures to a portable XML experiment file.

The portable XML files are finally uploaded and the enclosed data records are automatically imported into the database using the DIP platform, presented next.

The scope of the Data Import Platform (DIP) is to facilitate the import process enabled by the portable XML experiment format, as well as to provide a convenient database manipulation interface for local laboratories. Additionally, since each laboratory may use a slightly different version of the SERIES database for its own needs, it is engineered to allow easy local modifications decoupling the business logic from the underlying database model. This was achieved using the same XML schema as the specification of the underlying model and removing specific data dependencies from the code. A block diagram of the DIP platform operation is illustrated in Fig. 4.3.

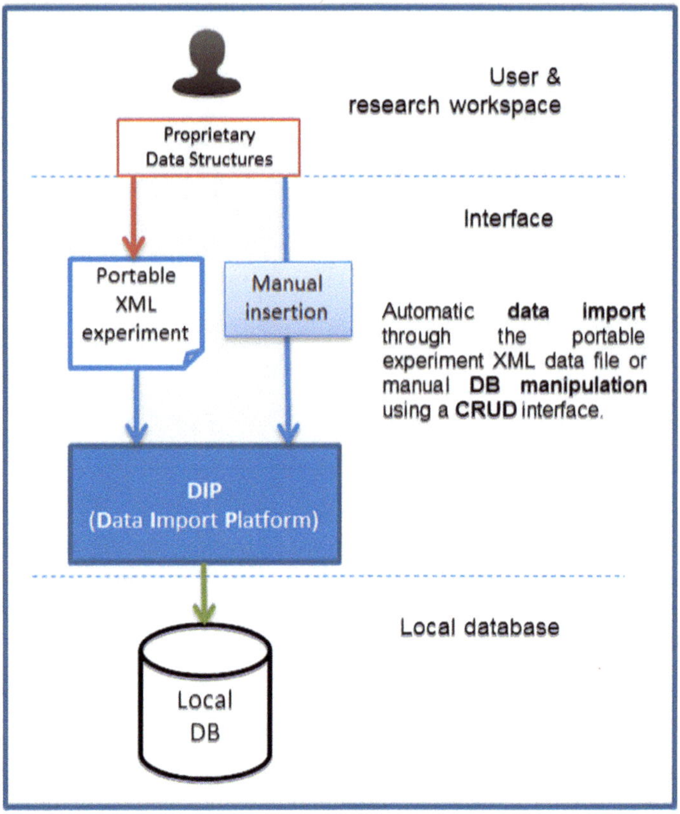

Fig. 4.3 Data Import Platform (DIP) block diagram

The DIP tool offers two main functionalities, the Import space and the Database Manipulation space. The first allows users to upload their portable XML experiment files and import the enclosed data into the database. The enclosed data is checked for conflicts with existing database records and users may select between overwriting/merging with existing data, inserting duplicate or ignoring conflicting records. The experimental data is also checked for validation errors before the import into the database. Figure 4.4 illustrates a part of the conflict-resolution step.

The *Database Manipulation* space provides a database interface to create, replace, update and delete records and individual fields, forming a fully featured CRUD interface (in the databases terminology, from the initials of the above operations). Users are able to manually add projects, specimens, experiments and related metadata or edit and delete existing ones. Convenient functions for rapid data manipulation, like multiple selections, have also been introduced. Apart from the modifications to the database, users may also use the provided interface to navigate the database, using drop-down menus, tree-like navigation, hierarchical filters and

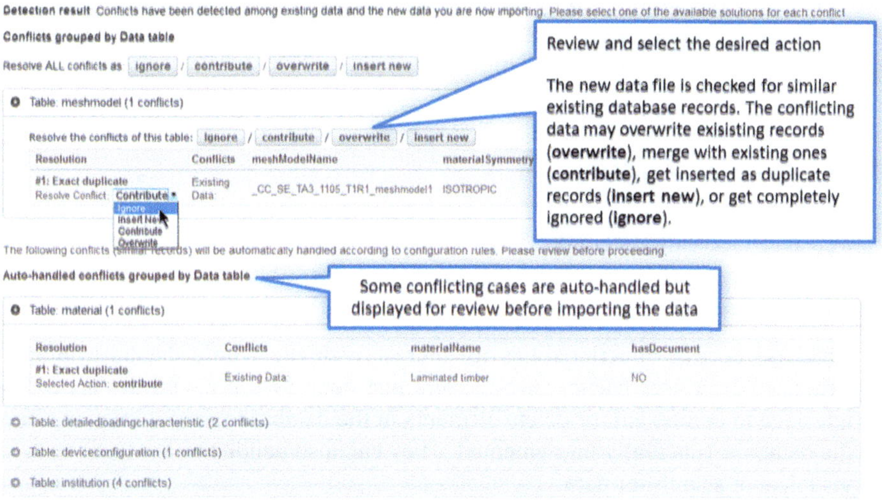

Fig. 4.4 Conflict-resolution step during the XML import process

field searching. The main interface elements of the database manipulation space are illustrated in the Fig. 4.5.

A multi-user authorization system has also been incorporated in order to provide different levels of access to privileged and non-privileged users. More specifically, the user roles Viewer (guest), Contributor, Editor, Publisher and Admin are provided with increasing manipulation privileges from just viewing to adding, editing, publishing data and managing users (in the same order the roles were listed above). Concluding, the web-based DIP platform along with the portable data specification

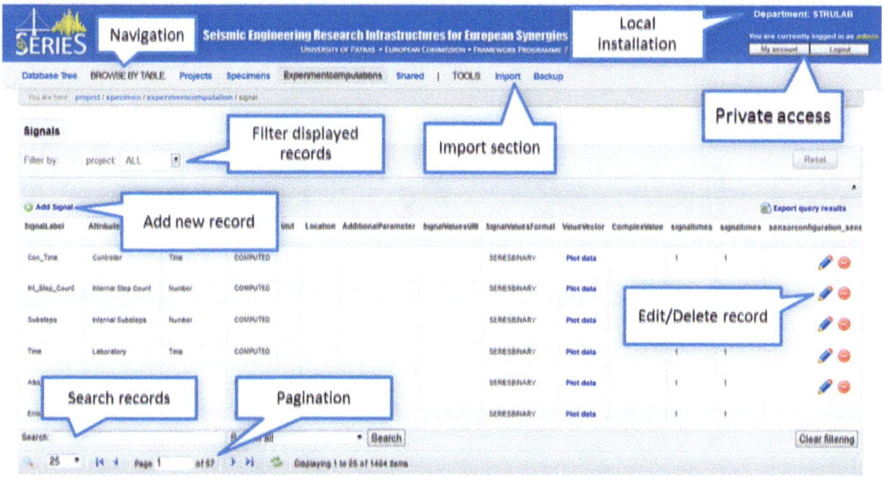

Fig. 4.5 Data Import Platform (DIP) block diagram

provide a flexible solution to populate local databases either standardizing the local experimental data format and automatically importing data into the database or manually using a database CRUD interface.

4.3.2 GUIDE Interface

A Web interface called "Guide" has been developed by the University of Naples and the University of Trento under the supervision of the University of Oxford. This interface is designed to enable:

- Database access: functionalities to interact with the whole database internal structures in a user-friendly way. Users just need to use a visually appealing interface to create, edit or delete elements in the database without knowing how the database is actually implemented. They can also conduct other tasks such as visualise or search for data.
- Management of local users: Guide allows different local users to access the database. Every user has a role assigned (administrator, contributor or guest) that enables them to use different functionalities within the interface. For instance, guest users can only visualise data, but they cannot modify any information.
- Advance tools: to extend the functionality of the system by supporting data migration, automatic input of large sets of information, visualisation of signal data, etc.

Figure 4.6 presents two screens of GUIDE, corresponding to the *Project* level visualization and the login page. For each project, the relevant information (that is

Fig. 4.6 GUIDE log in page and project page

required by the Exchange Data Format) is specified, together with the privacy restriction. The interface enables browsing of existing projects, as well as the creation of new projects (button on the left side).

4.4 SERIES Virtual Database for an External User

Implementing local databases based on the Data Exchange Format allows SERIES laboratories to store their data in a (common) structured way. However, the access to these databases would still be limited for external users. In order to provide the necessary access to shared data, thus increasing their value and promoting collaboration, the concept of SERIES virtual database. was introduced. Users can access individual databases distributed over Europe by using a single centralized Web interface also known as Data Access Portal (Lamata Martínez et al. 2013a).

4.4.1 SERIES Virtual Database Working Principle

The distributed architecture, depicted at Fig. 4.7, is named "the SERIES virtual database" since, although it is presented like a centralized database to external users

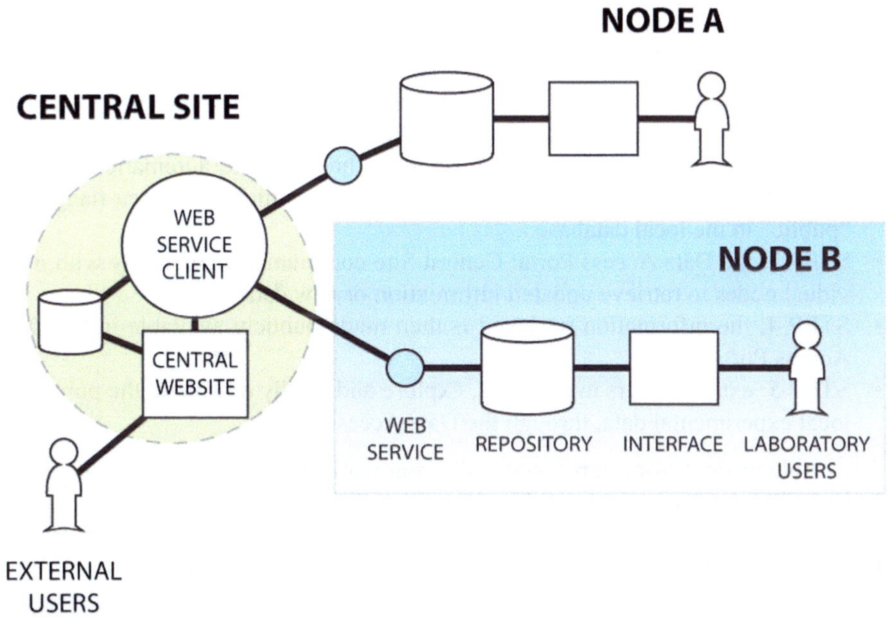

Fig. 4.7 SERIES virtual database: local databases, web services and Data Access Portal. (Lamata Martínez et al. 2013b)

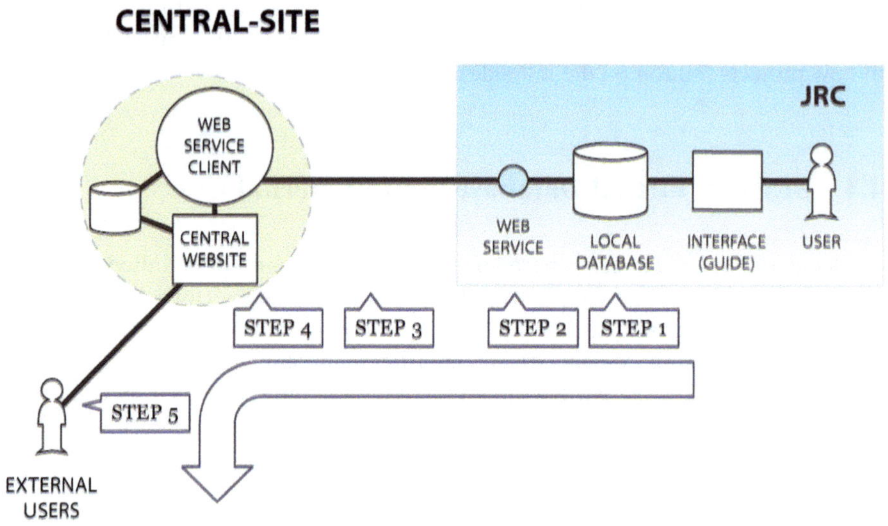

Fig. 4.8 Data workflow: from the local database to the Data Access Portal

through the Data Access Portal, it is actually a time-evolving aggregated collection of experimental data, which are regularly retrieved and updated from local distributed repositories. The aggregation of publicly shared data is performed by the Web Services installed at each local node and their communication with the Central Site.

The data flow from SERIES laboratories to the external user is depicted in Fig. 4.8 and can be broken down into the following steps:

- STEP 1: a laboratory produces experimental data and stores them in its local database. At this stage only the local users can access the data.
- STEP 2: the Web Services implemented at the local site automatically make available for the central site the experimental data which have been flagged as "public" in the local database.
- STEP 3: the Data Access Portal Central Site communicates regularly with individual nodes to retrieve updated information or new data.
- STEP 4: the information retrieved is then made publicly available in the Data Access Portal.
- STEP 5: external users may access, explore and finally download the published local experimental data, through the Data Access Portal.

The whole process from step 2 to step 4 is automatically performed without human intervention. Internal users provide data according to the Exchange Data Format (described in the paragraph 2), while external users are presented with a user-friendly representation of the same format through the Data Access Portal.

4.4.2 *The Data Access Portal*

The SERIES Virtual Database can be accessed by an external user at the Data Access Portal (DAP) at www.dap.series.upatras.gr. The interface simulates that of the SERIES portal with the difference of a left column, which actually presents a breakdown list of available test results from the laboratories participating in SERIES (the presentation order is selected by the user after the Laboratory name, the Project name or the date). Information about the SERIES database and its format (EDF) and well as a user's manual is available at this level (Fig. 4.9). Navigation around all available data produced by the SERIES laboratories and flagged by them as "public", is open without restriction at any level. The information offered may be characterized as general (information about the project and contributors—see Fig. 4.10), or detailed (when referring to *Specimen*, *Experiment*, *Computation* or *Signals* level—Figs. 4.11, 4.12, 4.13 and 4.14).

Fig. 4.9 SERIES Data Access Portal: the tree-based representation of published data

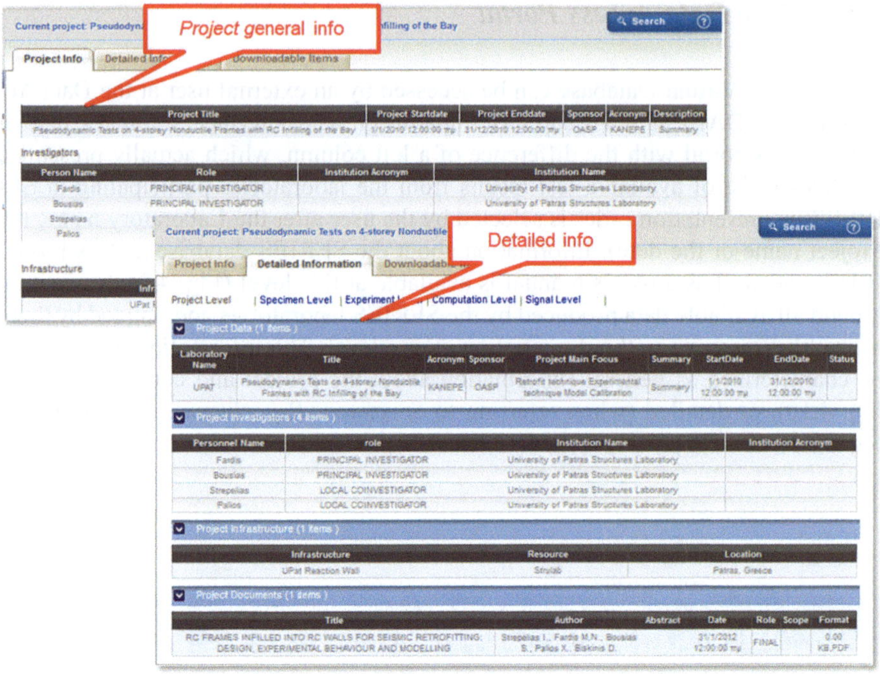

Fig. 4.10 Navigation at the Central Site: project level. General and detailed information as specified in the EDF

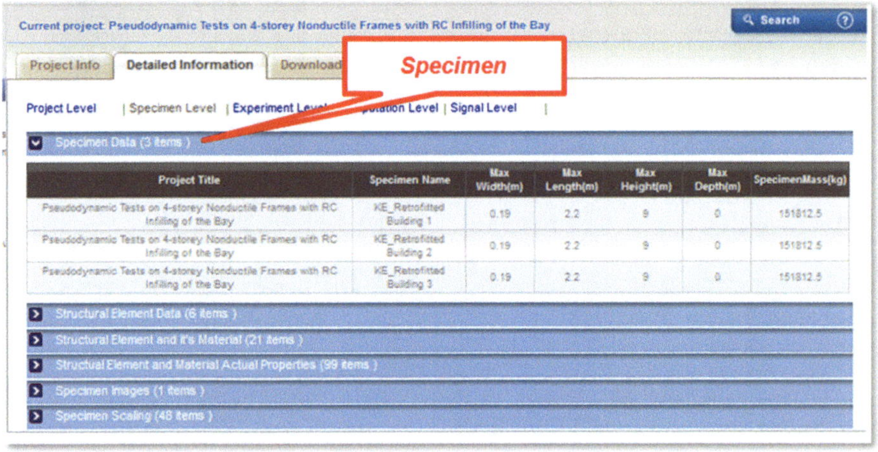

Fig. 4.11 *Specimen* level: three specimens have been tested. The characteristics of these specimens can be seen collapsing the folders above (*Structural Element Data, Structural Element and it's Material, etc.*)

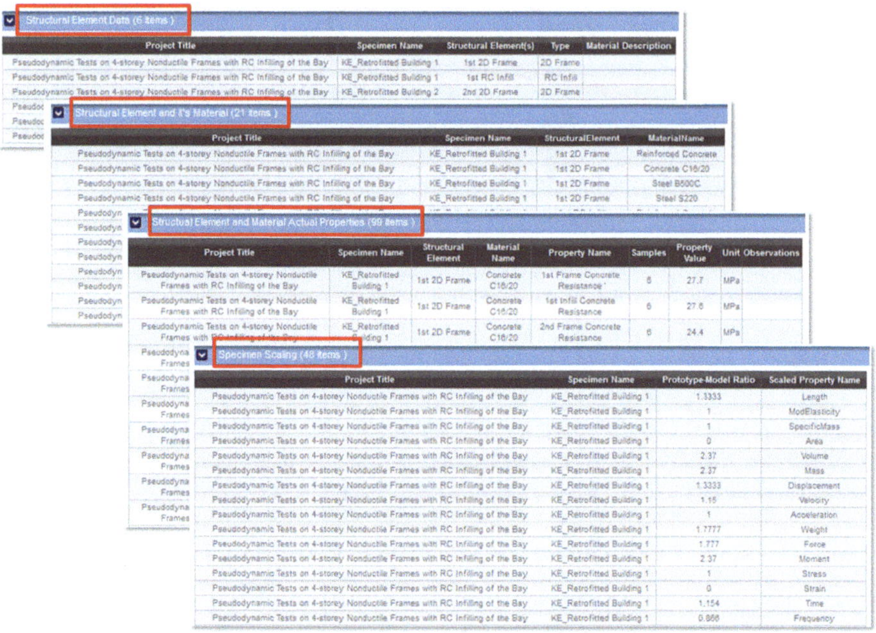

Fig. 4.12 *Specimen* level: expanded view of specimen characteristics

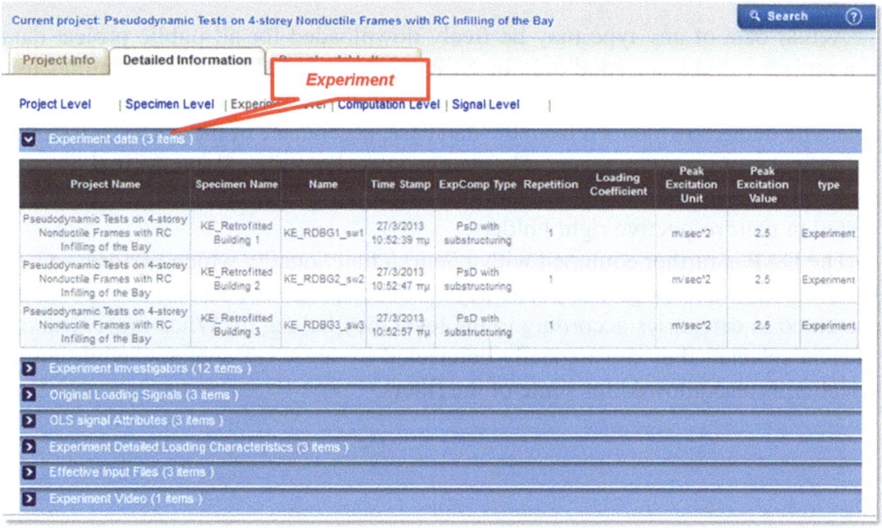

Fig. 4.13 *Experiment* level: three pseudo-dynamic tests have been performed

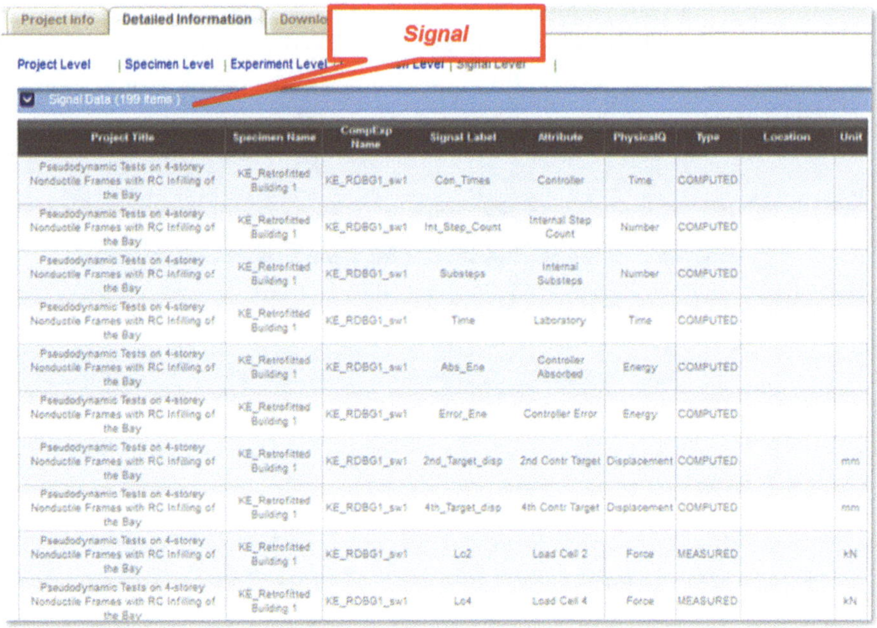

Project Title	Specimen Name	CompExp Name	Signal Label	Attribute	PhysicalQ	Type	Location	Unit
Pseudodynamic Tests on 4-storey Nonductile Frames with RC Infilling of the Bay	KE_Retrofitted Building 1	KE_RDBG1_sw1	Con_Times	Controller	Time	COMPUTED		
Pseudodynamic Tests on 4-storey Nonductile Frames with RC Infilling of the Bay	KE_Retrofitted Building 1	KE_RDBG1_sw1	Int_Step_Count	Internal Step Count	Number	COMPUTED		
Pseudodynamic Tests on 4-storey Nonductile Frames with RC Infilling of the Bay	KE_Retrofitted Building 1	KE_RDBG1_sw1	Substeps	Internal Substeps	Number	COMPUTED		
Pseudodynamic Tests on 4-storey Nonductile Frames with RC Infilling of the Bay	KE_Retrofitted Building 1	KE_RDBG1_sw1	Time	Laboratory	Time	COMPUTED		
Pseudodynamic Tests on 4-storey Nonductile Frames with RC Infilling of the Bay	KE_Retrofitted Building 1	KE_RDBG1_sw1	Abs_Ene	Controller Absorbed	Energy	COMPUTED		
Pseudodynamic Tests on 4-storey Nonductile Frames with RC Infilling of the Bay	KE_Retrofitted Building 1	KE_RDBG1_sw1	Error_Ene	Controller Error	Energy	COMPUTED		
Pseudodynamic Tests on 4-storey Nonductile Frames with RC Infilling of the Bay	KE_Retrofitted Building 1	KE_RDBG1_sw1	2nd_Target_disp	2nd Contr Target	Displacement	COMPUTED		mm
Pseudodynamic Tests on 4-storey Nonductile Frames with RC Infilling of the Bay	KE_Retrofitted Building 1	KE_RDBG1_sw1	4th_Target_disp	4th Contr Target	Displacement	COMPUTED		mm
Pseudodynamic Tests on 4-storey Nonductile Frames with RC Infilling of the Bay	KE_Retrofitted Building 1	KE_RDBG1_sw1	Lc2	Load Cell 2	Force	MEASURED		kN
Pseudodynamic Tests on 4-storey Nonductile Frames with RC Infilling of the Bay	KE_Retrofitted Building 1	KE_RDBG1_sw1	Lc4	Load Cell 4	Force	MEASURED		kN

Fig. 4.14 *Signal* level: each signal is delivered together with data regarding its units, the nature of the signal, the location and the associated time sequence

Actual data of any type may be freely downloaded for all public project data. If a project is to be accessed only by SERIES partners, downloading requires user authentication (managed at the Data Access Portal). Nevertheless, regardless of the data type being downloaded from the database, acceptance of the Terms and Conditions displayed is a prerequisite. The statement declares that all intellectual property rights in the data, including, but not limited to, copyright and database rights are vested in their respective right holders.

The DAP is further equipped with a *Search* functionality which performs a keyword-based search (Fig. 4.15). The keywords forming the basis for the search are presented in categories according to the level they belong to. When more than one filter is selected they are logically connected by an AND operator, while multi-selection is also allowed for some of the filters.

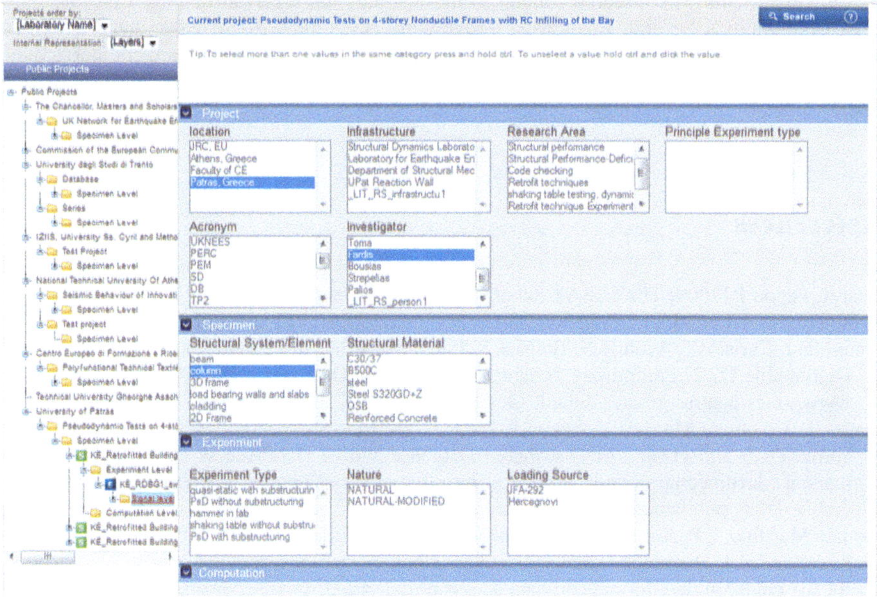

Fig. 4.15 Search functionality implemented at the Data Access Portal

4.5 Conclusions

The paper describes the principle and associated elements which constitute the SERIES virtual database. An Exchange Data Format that could host heterogeneous experimental data and provide all the information needed to reproduce a test, has been developed and agreed. Additionally, a portable specification of the Exchange Data Format in XML and user-friendly tools which allow data navigation or/and automatic data insertion have been developed and are currently under evaluation.

Data stored at local sites is made accessible to external users by means of the Data Access Portal hosted at the University of Patras. In this way a centralised access to database nodes that are distributed over a network and are able to dialog with a central portal in a uniform manner, is provided.

The SERIES virtual database enables a wider sharing of data and knowledge and ultimately, offers an unprecedented service to the earthquake engineering community. SERIES users will be able to have access to a wide database of experimental data and information, without violating the ownership of the data that will remain with the local laboratory where data have been produced.

This platform is to be maintained and enhanced well beyond the end of SERIES project, to serve as a reference point for the earthquake engineering community worldwide.

Acknowledgments The research leading to these results has received funding from the European Community's Seventh Framework Programme (FP7/2007–2013) under grant agreement n° 227887. The coordinator Prof. Fardis, University of Patras and all the partners that actively contributed to the database development with suggestions and fruitful discussions are greatly acknowledged.

References

Bosi A, Pegon P (2009) The ELSA database and what can be done regarding SERIES networking activities, JRC56446. EUR Report 23931 EN, Luxembourg (Luxembourg)

Ioannidis I, Christos F, Avouris N, Bousias S, Lamata Martínez I, Pegon P, Bosi A (2011) SERIES Deliverable D2.2. Preliminary version of Distributed Database e Data Access Portal. http://www.series.upatras.gr/sites/default/files/Deliverable_D2.3.pdf. Accessed May 2010

Ioannidis I, Lamata Martínez I, Pegon P, Fidas C, Bousias S, Avouris N (2012) SERIES Deliverable D2.5 2nd version of Distributed Database and of Data Access Portal including user manual, documentation and guidelines. http://www.series.upatras.gr/sites/default/files/Deliverable_D2.5_public.pdf. Accessed May 2010

Lamata Martínez I, Pegon P, Bosi A, Dietz M, Bousias S, Fidas C, Wenzel H (2010) SERIES Deliverable D2.1. Distributed database: review of beneficiary current data formats -Specifications for the common data format and data access portal. http://www.series.upatras.gr/sites/default/files/pri/drafts/Deliverable2_1_0.pdf. Accessed May 2010

Lamata Martínez I, Ioannidis I, Fidas C, Williams M, Pegon P (2013a) The SERIES virtual database: architecture and implementation, submitted

Lamata Martínez I, Ioannidis I, Pegon P, Williams M, Blakeborough A (2013b) The process and future of data integration within the European earthquake engineering laboratories. J Comput Civ Eng 10.1061/(ASCE)CP.1943-5487.0000308 (Apr. 10, 2013)

MySQL database (2013) http://www.mysql.com/. Accessed July 2013

Chapter 5
Qualification of Seismic Research Testing Facilities in Europe

Maurizio A. Zola and Fabio Taucer

5.1 Introduction

5.1.1 Why the Qualification?

The Qualification process is defined in EN ISO 9000: 2000 as a Process to demonstrate the ability to fulfil specified requirements.

As far as the industrial sector the answer to the question is taken from the International Standard ISO/IEC 17011 (2005) dealing with Conformity assessment:

- In the regulatory sector, government authorities implement laws covering the approval of products (including services) **for reasons of safety, health, environmental protection, fraud prevention or market fairness**. In the **voluntary** sector, many lines of industry have, both within an economy as well as globally, set up systems for conformity assessment and approval, aiming at **achieving a minimum technical level, enabling comparability, and also ensuring competition on equal terms**. A prerequisite for trade on equal terms is that any product (including services), accepted formally in one economy, must also be free to circulate in other economies **without having to undergo extensive re-testing, re-inspection, re-certification, etc**. This should be the case regardless of whether the product (including services) falls wholly or partly under the regulatory sector.
- In today's society it is often required **to state objectively conformity of products (including services) to specified requirements**. Conformity assessment bodies

M. A. Zola (✉)
Consultants of P&P LMC, Seriate, BG, Italy
e-mail: maurizio.zola@gmail.com

F. Taucer
European Laboratory for Structural Assessment Unit, European Commission,
Joint Research Centre (JRC), Institute for the Protection and Security of the Citizen (IPSC),
Via Enrico Fermi 2749, Ispra VA 21027, Italy
e-mail: fabio.taucer@jrc.ec.europa.eu

F. Taucer, R. Apostolska (eds.), *Experimental Research in Earthquake Engineering*,
Geotechnical, Geological and Earthquake Engineering 35,
DOI 10.1007/978-3-319-10136-1_5

(CABs) can objectively state such conformity. These CABs perform conformity assessment activities that include certification, inspection, testing and calibration.
- It is important for the purchaser, regulator and the public **to know that these CABs are competent** to perform their tasks. For that reason there is **an increasing demand for impartial verification** of their competence. Such **verification is done by authoritative accreditation bodies** that are impartial in relation to both the CABs and their clients, and which normally operate in a non-profit distributing manner.
- A system to accredit CAB conformity assessment services should provide confidence to the purchaser and regulator. Such a system should facilitate cross-border trade, as pursued by trade authorities and organizations. **The ultimate goal is to achieve one-stop accreditation and one-stop conformity assessment.**
- A "cross border" trade facilitating system can work well if **accreditation bodies and CABs all operate to globally accepted requirements in an equivalent manner** and take into account the interests of all parties concerned.
- On these bases ISO/IEC 17011 International Standard specifies the general requirements for accreditation bodies. Peer evaluation mechanisms have been created at regional and international levels, through which assurance is provided that accreditation bodies are operating in accordance with ISO/IEC 17011 Standard. Those who have passed such an evaluation can become members of mutual recognition arrangements.
- **These mutual recognition arrangement members facilitate the one-stop process**, through recognition, promotion and acceptance of each other's accredited conformity assessments. This means that a CAB in an economy should not need to be accredited more than once for the same scope by different accreditation bodies.

In the European Community the requirements for the conformity assessment concerning the free circulation of products are stated in REGULATION (EC) No 765/2008 (2008).

A sound system of National Accreditation Bodies is already operating in Europe under the European cooperation for Accreditation (EA) peer evaluation system. On a worldwide basis, ILAC (International Laboratory Accreditation Cooperation) and IAF (International Accreditation Forum) are the reference. International Associations deal with the mutual recognition agreements between the National Accreditation and Certification Bodies.

The following definitions are taken from the Dutch Accreditation Council document RvA/T031 (2010) respectively of research activity and research and development activity.

- The carrying out of observations/tests on a research object, based on a research plan, and the interpretation of the results thereof.
- The research, within the field of competence which in its **design and/or execution is not of a repetitive nature**, and the relevant research techniques in which the body has proven experience and the required expertise. **The field of research is primarily determined by the problem** with which the body has experience of and is faced with. The research field must be based on the research methods/

techniques or on the inspection methods/techniques as specified in the list of already recognized activities, applied standard methods within the body. The development of new research techniques, within the research field, may form a part of the research.

As far as research activities, from a technical point of view the qualification is a tool that guarantees the reliability of testing research laboratories through the combination of two main requirements: technical competence and quality assurance. Reliability is translated in practice into

- Repeatability, the principle that experimental activities repeated on the same specimen in the same laboratory lead to the same results, and
- Reproducibility, the principle that experimental activities repeated on the same specimen in different laboratories lead to the same results.

Repeatability and reproducibility can be more easily achieved if common standard test procedures are used. Moreover there is the need to establish a common language on the basis of international standards, thus helping avoiding mismatches and misunderstandings.

Besides establishing the general reliability of structural testing in Europe, a common platform for qualification will significantly enhance the expertise of testing facilities, as a result of the continuous benchmarking of similar laboratories.

It is expected that the mutual acknowledgement of European research infrastructures in earthquake engineering through the qualification will enhance their standing with respect to their American or Japanese counterparts, promoting a unified EU policy on acceptance criteria for products and techniques.

5.1.2 The SERIES Project

The SERIES project has received funding from the European Community's Seventh Framework Programme [FP7/2007–2013] under grant agreement n° 227887 (2009).

The SERIES project aimed at bridging the following two gaps of Research and Technological Development (RTD) in experimental earthquake engineering and structural dynamics:

a. between Europe and the US and Japan, and
b. between European countries with high seismicity but less advanced RTD infrastructures on one hand and some more technologically advanced but not so seismic Member States on the other.

The goal was pursued by integrating the entire European RTD community in earthquake engineering via the following list of activities:

- A concerted program of Networking Activities, fostering a sustainable culture of co-operation among all research infrastructures and teams active in European earthquake engineering;

- A distributed database of test results, pooling data from the beneficiary research infrastructures and others, accessible and maintained by a virtual research community after the project's end;
- Tele-presence and geographically distributed concurrent testing at the research infrastructures;
- Standards, protocols and criteria for qualification of RTD infrastructures in earthquake engineering;
- Enhancement of human resources by training new users and beneficiary technical/research personnel in courses on good practices in operation and use of research infrastructures.

In the framework of the Project there were Transnational access Activities (TA), Joint Research Activities (JRA) and Networking Activities (NA); the networking activities of the Project aimed at enhancing the services provided by the research infrastructures, transcending their current extreme fragmentation, through the following:

- Common European standards and protocols for similar research infrastructures and
- Qualification criteria for European research infrastructures in earthquake engineering.

5.1.3 NA2 Networking Activity—Qualification of Research Infrastructures

5.1.3.1 General Description of the Activity

NA2 was the Networking Activity of the SERIES Project, which had as a primary goal to create the conditions leading to the qualification (in the form of mutual accreditation) of Structural Testing Research Laboratories specialising in earthquake engineering and equipped for large scale testing.

Among the other Partners participating to the SERIES Project, the following nine European Countries and a European Joint Research Centre were contributing to the development of the NA2 activities:

- France—CEA (Paris)
- Germany—University of Kassel
- Greece—University of Athens, University of Patras
- Italy—P&P LMC srl (Bergamo), EUCENTRE (Pavia), University of Trento, University of Naples
- Portugal—LNEC (Lisboa)
- Republic of Macedonia—IZIIS (Skopje)
- Romania—University of Iasi
- United Kingdom—University of Bristol
- Turkey—University of Ankara, University of Istanbul, and
- EC Joint Research Centre in Ispra (Italy).

The activities in NA2 addressed the assessment criteria for technical competence of large European research infrastructures—and of similar ones in the future—and developed the basis for their mutual accreditation, focusing on seismic experimentation through real-time, pseudo-dynamic and quasi-static testing with shaking table and reaction wall, and on-site testing and monitoring. Moreover, special testing techniques, such as hybrid techniques, and devoted testing apparatuses, such as seismic bearing and damper testing equipment, were considered.

The activity was broken down in four Tasks.

- Task 1: Evaluation and impact of qualification of experimental facilities in Europe
- Task 2: Assessment of testing procedures and standards requirements
- Task 3: Criteria for instrumentation and equipment management
- Task 4: Development and implementation of a common protocol for qualification

As far as the two main requirements for the qualification of the laboratories, quality management system and technical competence, a sound experience is available in the application of international standards (EN ISO 9001: 2008 and ISO/IEC 17025: 2005). Therefore it was decided to conduct the activities with reference to these standards to get a contribution from European Accreditation Bodies and Standardisation Organisations to reach an agreement to issue a final Common Protocol for the qualification of research infrastructures in earthquake engineering.

5.1.3.2 Task NA2.1: Evaluation and Impact of Qualification of Experimental Facilities in Europe

The task included:

- a critical analysis of the problems that limit the free circulation of the products of the European Industry, the solution of which will be promoted by the qualification of structural laboratories, and
- a study of the issues—of technical, quality or commercial relevance—which in fact constitute obstacles to mutual accreditation.

The accreditation of research infrastructures faces at present the following obstacles:

- while the qualification of CABs performing structural tests on products for the civil structures industry is covered by Regulation 765, there are no standards covering the qualification of RTD infrastructures in earthquake engineering;
- The existing International or European Standards dealing with certification or accreditation do not specifically cover RTD infrastructures;
- There are no International or European standards or regulations requesting the qualification of RTD infrastructures in the field of structural and seismic engineering testing.

Final Users (Industry) and Accreditation or Standardisation Organisations were directly involved, through questionnaires and meetings. The task ended with a technical report including a detailed roadmap towards the common protocol for mutual accreditation among the laboratories (Zola and Taylor 2010).

5.1.3.3 Task NA2.2: Assessment of Testing Procedures and Standards Requirements

The focus of this task was on Shaking Table and Reaction Wall facilities, which, owing to their special character and relatively small number, their differences in technical solutions, the wide variety of testing procedures among their operators and the limited knowledge of the technology used by the facilities among users in the industrial sector, do not lend themselves to fully harmonised approaches.

The task carried out a survey that aimed at checking the compliance of the different approaches used with the requirements of applicable standards or reference recommendations (concerning, for instance, the selection of the seismic input and how it conforms to reference spectra, as well as the measurement, assessment and documentation of the technical aspects affecting the reliability of the tests). The work was based on a comparative analysis and critical assessment of present procedures and equipment in the different laboratories for the execution of tests. Quality Management approaches were investigated as well.

The conclusions of the assessment of testing procedures and standards requirements have led to a critical analysis of the requirements imposed by official standardization and accreditation organisations, National and European, based on ISO/IEC 17011 (2005).

5.1.3.4 Task NA2.3: Criteria for Instrumentation and Equipment Management

This task treated the management criteria of instrumentation (testing equipment, measuring instrumentation, acquisition systems and processing tools), including the issues related to the calibration of instruments (periodicity and technical conditions), their maintenance, the estimation of measurement uncertainty and the instrumentation implementation on specimens.

As in Task NA2.2, a critical assessment of the methods currently being adopted was made, after having exchanged information between the different laboratories regarding the procedures used, and cross visits of the laboratory operators to the cooperating facilities on the occasion of important or benchmark tests.

As in Task NA2.2 the assessment of testing and instrumentation management procedures led to a critical analysis of the requirements imposed by official standardization and accreditation organisations, National and European.

5.1.3.5 Task NA2.4: Development and Implementation of a Common Protocol for Qualification

The starting point of this task were the International Workshops on the "Qualification of research infrastructures" and on the "Role of research infrastructures in performance-based earthquake engineering", held in Ohrid (Republic of Macedonia) in

September 2010, in conjunction with the 14th European Conference on Earthquake Engineering (Zola and Taylor 2010; Kurc et al. 2010).

A factor which is related to the construction and use of large-scale facilities is the application of the structural Eurocodes. In detail, with the Eurocodes earthquake engineering moved from prescriptive codes to performance-based codes (Performance Based Engineering—PBE). In prescriptive codes, the designer has little or no opportunity to take a rational engineering approach to design provisions. With performance-based codes, the designer can use any safety strategy he wishes provided that adequate safety can be demonstrated. In general terms, a prescriptive code states how a structure is to be constructed whereas a performance-based code states how a structure is to perform under a wide range of conditions. The performance-based analysis and design of structures needs large amount of experimental data relevant to several limit states. As a result, structural research testing represents a key point for this new approach (Taucer and Pinto 2010; Ozturk 2010).

Concerning the last point, as stated above, there is a lack of specific Standards covering RTD seismic testing and special seismic testing with multi-axial large shaking tables, quasi-static and pseudo-dynamic techniques or hybrid experimental & mathematical modelling techniques; moreover the taking in of the testing approach, as a consequence of too rigid accreditation requirements and procedures, is often unpractical, due to the complexity of the tests and to the hazardous character of the specimen response: in many situations, testing procedures may require real-time adjustments suggested by engineering judgments based on the acquired results of the experiments.

The accreditation, as called for by REGULATION (EC) No 765/2008 (2008), is based on the so called fixed scopes of accreditation, published by the national accreditation bodies (NABs). This system for defining the scope permits on one hand, an accurate description of the accredited activities of the CAB and on the other ensures that an appropriate evaluation of its competence is carried out each time an additional activity is added to the scope.

However, this method of describing a scope of accreditation is considered too restrictive by most research infrastructures:

- There is a difference between listing specific activities that an organization is competent to carry out, and defining its competence to apply its knowledge, skills, and experience to different activities.
- A fixed scope does not readily enable additional or modified activities to be added to the scope of a research infrastructure without further assessment, even where competence in this general area has already been demonstrated.
- In practice, end-users are not always experienced in seismic testing standards. They are often faced with significant difficulties to understand the content of the very long, detailed and technical activities of seismic testing. A more generic approach might help end-users to promote the use of the scope of accreditation.

5.2 Acknowledgement of the European Situation

5.2.1 Introduction

The questionnaires were prepared and distributed starting in January 2010 and the collection of the filled questionnaires ended in March 2010.

Thirty five European Large Testing Facilities replied to the questionnaire:

- As far as the state of the certification of the quality system the situation seems not uniform: 40 % of the laboratories have a quality system, while 66 % of laboratories do not have any certification of the quality system nor any accreditation for the performance of experimental tests.
- 46 % of the laboratories declare to have a demand for qualification, but only 29 % indicate a reference standard, mainly ISO 9001.
- The main part of the laboratories (more than 80 %) seems to cover the general and technical requirements coming from EN ISO/IEC 17025 standard.

Six National Accreditation Bodies replied to the questionnaires:

- One of the main issues as far as the "General requirements for the accreditation of Laboratories" has been the necessity to make reference to REGULATION (EC) No 765/2008 (2008).
- Regulation 765 states that accreditation is part of an overall system, including conformity assessment and market surveillance, designed to assess and ensure conformity with the applicable requirements; moreover, the particular value of accreditation lies in the fact that it provides an authoritative statement of the technical competence of bodies whose task is to ensure conformity with the applicable requirements.
- Regulation 765 was issued in order to develop a comprehensive framework for accreditation and to lay down at European level the principles for its operation and organization.
- A system of accreditation which functions by reference to binding rules helps to strengthen mutual confidence between Member States as regards the competence of conformity assessment bodies and consequently the certificates and test reports issued by them.
- In order to avoid multiple accreditations, to enhance acceptance and recognition of accreditation certificates and to carry out effective monitoring of accredited conformity assessment bodies, conformity assessment bodies should request accreditation by the national accreditation body of the Member State in which they are established.
- Regulation 765 provides for the recognition of a single organization at European level in respect of certain functions in the field of accreditation. The European cooperation for Accreditation (the EA) manages a peer evaluation system among national accreditation bodies from Member States and other European countries. Therefore EA is the first body recognized under the Regulation and by Member

States to ensure that their national accreditation bodies seek and maintain membership of the EA for as long as it is so recognized.
- ISO 17025 Standard covers also the accreditation of a Facility that conducts non-standard testing and research at large scale on a base of flexible scope (EA-2/15 2008).

Four Standardization Organizations replied to the questionnaires:

- No specific standards were identified covering qualification and accreditation of seismic tests facilities; the outcome was that reference should be made to the general approach to certification of quality system by ISO 9001 and to the accreditation of laboratories by EN ISO/IEC 17025.
- The following Standards covering seismic testing were identified with specific reference to:
  - Nuclear Power Plants
  - Electro-technical Equipment
  - Telecommunication Equipment
- No specific standards were identified for the seismic testing of Civil Structures and Buildings. Some standards were found for the measurement of vibrations on buildings and evaluation of their effects and some standards are covering the seismic testing of components for civil structural applications.
- As far as standards covering the management of experimental facilities and measuring equipment, some ISO standards were identified. No standards were found specifically devoted to the management of data acquisition systems.
- ISO standards are covering data acquisition and processing, including frequency analysis, statistics, modal analysis. Moreover specific ISO standards are covering the accuracy (trueness and precision) of measurement methods and results.

Twelve Industrial Companies replied to the questionnaires:

- As far as the free circulation of the products the industry is requiring Facilities with quality management system, if possible with the certification, and with the accreditation for seismic tests performance. Moreover, reference should be made to the certification of the quality system by ISO 9001 and to the accreditation of laboratories by EN ISO/IEC 17025.
- A special request is the outcome of the telecommunication industry that is relaying on the QuEST Forum that has developed the TL 9000 quality management system (QMS). Built on ISO 9001 and eight quality principles, TL 9000 is designed specifically for the communications industry.
- Legally independent Facilities are preferred and Facility's policies and procedures to ensure the protection of its customer's confidential information and proprietary rights are strongly envisaged.
- Standardized test methods are preferred and the uncertainty of test measurements should be addressed.
- Opinions and interpretations of a test should be in the test report only on request.

5.3 The Common Protocol

5.3.1 The Road Map

Using as baseline the outcomes of Tasks NA2.1 to NA2.3 and the conclusions of the discussions held at the International Workshop in Ohrid, the following conclusions were drawn:

- *ISSUES*
  - Qualification of a Research and Technological Development (RTD) infrastructure in earthquake engineering implies a process different from the qualification of a Conformity Assessment Body;
  - International or European Standards dealing with certification or accreditation are not specifically devoted to the RTD laboratories;
  - Lack of International or European standards or regulation requesting the qualification of RTD infrastructures;
  - Lack of international recognition of the large capacities and associated human resources of the European RTD infrastructures;
  - Lack of co-operation between
    a. more technologically advanced but less seismic countries and
    b. less technologically advanced but more seismic countries.
- *OBSTACLES*
  - Lack of specific Standards for the qualification of RTD infrastructures;
  - Lack of specific Standards covering RTD seismic testing;
  - Lack of specific Standards covering special seismic testing with single and multi-axial large shaking tables, quasi-static and pseudo-dynamic techniques or hybrid experimental & mathematical modelling techniques and on-site testing;
  - Lack of a qualification-oriented mentality of the high level management of the RTD infrastructures;
  - Underestimation by the laboratory staff of the benefits of an official qualification of the RTD infrastructures;
  - Reduced investment capabilities of the RTD infrastructures.

A road map towards a Common Protocol for Qualification was conceived:

1. Evaluation of the suitability of the General **Management** Requirements of EN ISO/IEC 17025 for RTD infrastructures;
2. Evaluation of the suitability of the General **Technical** Requirements of EN ISO/IEC 17025 for RTD infrastructures;
3. Identification of **Specific Technical** Requirements (STR) for RTD seismic testing;
4. Identification of Specific Technical Requirements relevant to documentation and data sharing to guarantee repeatability and reproducibility of test results aiming to the development of a **Common European Database**;
5. Issue of a **draft Common Protocol** for Qualification with respect to the General Management and Technical Requirements;

6. Drafting of RTD testing procedures;
7. Issue of a technical specification for the development of a Common European database to guarantee repeatability and reproducibility of seismic testing;
8. **Implementation on a voluntary basis of the draft Common Protocol** in some laboratories of the SERIES Partners;
9. Development of the **Final Common Protocol** for Qualification.

5.3.2 The Draft Common Protocol Implementation

After the completion of the first two points of the road map it was recognized that: the general management and technical requirements of EN ISO/IEC 17025 are suitable also for the RTD infrastructures, but not for the requirements of clause 5.4.2—Selection of methods.

RTD infrastructures therefore need to include additional activities in their scope of accreditation on the basis that their competence has been evaluated not only to carry out activities in accordance with previously evaluated procedures, but also for the development and validation of their procedures in accordance with a pre-established system. This kind of scope is defined "flexible scope" by EA document 2/15 (2008).

To lead to the qualification with "flexible scope" of the RTD infrastructures, in accordance with EA 2/15, Specific Technical Requirements for the RTD seismic testing were identified with reference to the requirements of clauses 5.4.3, 5.4.4 and 5.4.5 of EN ISO/IEC 17025.

The adoption of technical annexes to the Common Protocol dealing with Specific Technical Requirements is accounting for the need to cover the special requirements for the performance of research tests in the seismic engineering field.

After the exercise for the implementation of the Draft Common Protocol made by seven large European research testing facilities, as stated in point 8 of the road map, three activities of the RTD Facilities were identified:

- **Research Engineering Activities**: general specifications for the research, design and planning of the research, test specifications, supervision of experimental research activities, mathematical modelling, experimental data processing, design of the experimental rig, research reporting. The research engineering activities may be summarized as follows:

1. general study of the problem and identification of the research activities needed;
2. supervision of the experimental activities;
3. experimental data processing, results interpretation and final reporting.

- **Measurement Activities**: these activities are performed in accordance to standard or internal methods; the internal methods should be validated;
- **Research Testing Activities**: the experimental testing activities are performed by following a test specification issued as a result of the research design and planning performed by the research engineer. The research testing activities are

concluded by issuing the test report; the test report should include the raw data and pre-processed data.

As far as the qualification of the research testing facilities the following approach may be adopted:

- **Research Engineering Activities** may be covered by the certification of the Management System after ISO 9001;
- **Measurement Activities** may be covered by the accreditation of the Laboratory after ISO/IEC 17025;
- **Research Testing Activities** may be covered by the accreditation of the Laboratory after ISO/IEC 17025 with flexible scope.

Thus the qualification of a research testing facility performing research testing activities should be covered by the application of the Common Protocol for Qualification.

The output for a research testing laboratory, just as it is for a laboratory which is accredited with fixed or flexible scope, is the crude result of the research tests. A characteristic difference is that in the case of a research testing laboratory the development of the relevant testing method also falls within the scope of the accreditation. Accordingly a research testing laboratory is operating on the base of a test specification or a research plan issued by the Customer who is responsible of the research engineering activities.

5.3.3 The Final Version of the Common Protocol

The publication of the following documents concluded the NA2 activity of the SERIES Project (Zola et al. 2013). Please refer http://www.series.upatras.gr/public_documents in Work Package 3:

- Common Protocol for the Qualification of Research Infrastructures in Earthquake Engineering and Technical Annexes
- Annex 01—Check List for the Performance of the Audit to the Research Testing Facilities
- Annex 02—Specific Technical Requirements for Seismic Research Tests by Shaking Table
- Annex 03—Specific Technical Requirements for Seismic Research Tests by On-site Testing
- Annex 04—Specific Technical Requirements for Seismic Research Tests by Reaction Wall Testing
- Annex 05—Specific Technical Requirements for Data Acquisition and Processing

These documents have been drafted making reference to docs. ISO 9001 and ISO/IEC 17025; the justification of this choice is as follows:

- Concerning the two main requirements for the qualification of laboratories, quality management system and technical competence, a sound experience is

available in the application of ISO 9001 and ISO/IEC 17025; this choice will allow to fulfil the requirements of the last task of NA2, by proposing to European Accreditation and Standardisation Organisations the Common Protocol on which base it is envisaged to reach an agreement to issue a guiding document for the qualification of research infrastructures;
- Four technical Annexes were developed to cover the Specific Technical Requirements for seismic research testing, which are proposed to European Standardization Organizations as a base on which to reach an agreement to issue technical documents for the performance of seismic research tests;
- In the realm of globalization of economic activities the International Standards serve as comparison between various enterprises around the world and they are applicable to the manufacturing and services industries and are a necessary tool for mutual recognition.

It should be remembered that the aforementioned International Standards are suitable for the Laboratories to demonstrate that:

- They have a Management System to guarantee the Quality of the performed activities;
- They have the technical competence to perform the tests;
- They are able to produce technically correct and reliable results.

Moreover these International Standards cover the accreditation of tests:

- Performed after Standard methods
- Performed after non standard methods and
- Performed after internal methods.

Annex 01 is composed of the following six sections addressing:

1. Issues and remarks relevant to the qualification of Research Infrastructures.
2. Audit performance.
3. Current state of certification, accreditation, and management procedures within the Facility. This could be used for a preliminary self-evaluation.
4. Current state of the Facility's management system.
5. Current state of the Facility's general technical procedures.
6. Specific test procedures that are applied.

Sections of Annex 01 apply for the qualification of the following activities:

- Sections 1 through 4 apply to research engineering activities
- Sections 1 through 5 apply to measurement activities
- Sections 1 through 6 apply to research testing activities.

The four Technical Annexes to the Common Protocol, from 2 to 5, are intended to cover the Specific Technical Requirements for the performance of research experimental tests in the seismic field.

5.4 Conclusions

5.4.1 General Requirements

As far as the general requirements for the qualification of experimental facilities the following conclusions can be drawn:

- Most of the testing facilities do not have any official certification, nevertheless they declared to have a Quality Management System;
- The laboratories declared that they are often requested to be certified.
- The laboratories declared to fulfil most of the requirements for accreditation after EN ISO/IEC 17025;
- The accreditation of the Conformity Assessment Bodies (CAB) for the free circulation of products in the European Community is ruled by REGULATION (EC) No 765/2008 setting out the requirements for accreditation and market surveillance;
- Regulation (EC) No 765/2008 makes reference to the European cooperation for Accreditation (the EA) for peer tests of the national accreditation bodies of Member States;
- Regulation (EC) No 765/2008 makes reference to ISO 9001 and EN ISO/IEC 17025 for certification and accreditation, respectively;
- No specific standards cover the qualification of facilities for the performance of seismic tests;
- There are Standards that cover the seismic testing of equipment for Nuclear Power Plants, electro-technical and telecommunication applications; no specific standards cover seismic testing of Civil Structures and Buildings, either on site or on laboratory models. Some standards cover the seismic testing of structural components;
- Some ISO standards cover the management of both experimental facilities and measuring equipment and data acquisition and processing;
- Industry requires qualified Facilities;
- Industry also requires Standardized test methods and an estimate of the uncertainty of test measurements.

Irrespective of the activity, research or services to industry, the testing facilities are requested to be qualified.

5.4.2 Specific Technical Requirements

As far as the specific technical requirements, the two main experimental procedures that should be dealt with are:

- Shaking table testing and
- Point load testing.

Shaking table testing is mainly referred to the base excitation of a specimen, whereas point load testing is referred to static, quasi-static and dynamic excitation by the application of loads through actuators directly connected to the tested specimen.

For the shaking table testing many international standards are available to perform shock and vibration tests on electrical and mechanical equipment, but there are no specific standards for the testing of civil structures and components. There are many design standards dealing with structures made of concrete, steel and wood, but no standards exist covering experimental testing.

For point load testing there are no standards for equipment or structures.

5.4.3 European Standard Development

The Common Protocol for the qualification of large research infrastructures was presented to the Laboratory Committee of EA in Oslo and to the 1st Conference of the European Programme for Critical Infrastructures Protection (ERNCIP) in Ispra (Taucer and Zola 2012).

The Common Protocol for the qualification and Annex 1 (Check List for the Performance of the Audit to the Research Testing Facilities) could be the starting point for the development of a Guide for the qualification of Research Testing Facilities in Europe.

The four Technical Annexes to the Common Protocol could be a technical proposal for the development of technical standards for the performance of research experimental tests in the seismic field.

Acknowledgments The research leading to these results has received funding from the European Community's Seventh Framework Programme (FP7/2007–2013) under grant agreement n° 227887.

References

EA-2/15 (2008) EA Requirements for the accreditation of flexible scopes

EN ISO 9000 (2000) Quality management systems—Fundamentals and vocabulary

EN ISO 9001 (2008) Quality management systems—requirements

Grant agreement No. 227887 (2009) Integrating activity—combination of collaborative project and coordination and support action—series—seismic engineering research infrastructures for European synergies—Annex I: "Description of Work"—29 January 2009

ISO/IEC 17011 (2005) Conformity assessment—General requirements for accreditation bodies accrediting conformity assessment bodies

ISO/IEC 17025 (2005) General requirements for the competence of testing and calibration laboratories

Kurç Ö, Sucuoğlu H, Molinari M, Zanon G (2010) Qualification of testing procedures, instrumentation and equipment management in large testing facilities in earthquake engineering. 14th ECEE -International workshop on qualification of research infrastructures—Ohrid (Republic of Macedonia)—2 September 2010

Ozturk B (2010) Current state of art of the role of research infrastructures on performance based earthquake engineering and future perspectives for Europe. 14th ECEE-International workshop on role of research infrastructures in performance-based earthquake engineering—Ohrid (Republic of Macedonia)—2 September 2010

Regulation (EC) No 765/2008 of the European parliament and of the council of 9 July 2008 setting out the requirements for accreditation and market surveillance relating to the marketing of products and repealing Regulation (EEC) No 339/93

RvA/T031 (2010)—Dutch Accreditation Council (Raad voor Accreditatie)—Research and Development

Taucer F, Pinto A (2010) How can experimental testing contribute to performance based earthquake engineering. 14th ECEE -International Workshop on Role of Research Infrastructures in performance-based earthquake engineering—Ohrid (Republic of Macedonia)—2 September 2010

Taucer F, Zola M (2012) Qualification of large seismic research infrastructures in Europe. Presented at 1st conference on European programme for critical infrastructures protection—EU-wide certification and accreditation for security solutions, Ispra (Italy)—12th—13th December 2012

Zola M, Taylor C (2010) Impact of the qualification of seismic testing facilities.– 14th ECEE -International workshop on qualification of research infrastructures—Ohrid (Republic of Macedonia)—2 September 2010

Zola M, Taucer F, Bonzi A (2013) Deliverable 3.4—Common protocol for the qualification of research infrastructures in earthquake engineering and technical annexes, seismic engineering research infrastructures for European Synergies, January 2013

Chapter 6
Towards Faster Computations and Accurate Execution of Real-Time Hybrid Simulation

Khalid M. Mosalam and Selim Günay

This chapter reports three recent developments aimed towards faster computations and more accurate execution of real-time hybrid simulations (RTHS). The first of these developments is a standalone RTHS system which can accommodate integration time steps as small as 1 ms. The fast execution feature eliminates the approximations that would be introduced by the application of a predictor-corrector smoothing technique and increases the applicability range of explicit integration methods. The second development is the use of an efficient equation solver in RTHS which reduces computation time. This efficient solver, which decreases the computation time by factorizing the Jacobian of the system of linear algebraic equations only once at the beginning of the simulation, is especially beneficial in RTHS which involves analytical substructures with large number of degrees of freedom (DOF). The third development is a novel use of a three-variable control (TVC) for RTHS on a shaking table configuration. Although the TVC, which employs velocity and acceleration control in addition to the typical displacement control, is commonly used in conventional shaking table tests, this development is the first application of TVC in RTHS.

6.1 Introduction

Hybrid simulation is a testing method for examining the seismic response of structures using a hybrid model comprised of both physical and numerical substructures. Because of the unique feature of the method to combine physical testing with numerical simulations, it provides an opportunity to investigate the seismic response

---

K. M. Mosalam (✉) · S. Günay
Department of Civil and Environmental Engineering,
University of California, Berkeley, CA 94720, USA
e-mail: mosalam@berkeley.edu

F. Taucer, R. Apostolska (eds.), *Experimental Research in Earthquake Engineering,*
Geotechnical, Geological and Earthquake Engineering 35,
DOI 10.1007/978-3-319-10136-1_6

of structures in an efficient and economically feasible manner. It is this feature of the method which made it gain widespread use in recent years.

Hybrid simulation can be divided into three categories as follows: (1) Slow hybrid simulation (HS) in a discrete actuator configuration, (2) Real-time hybrid simulation (RTHS) in a discrete actuator configuration, and (3) RTHS in a shaking table configuration. In the first category, slow HS, the experimental substructure(s) is directly connected to actuator(s), physical mass generally does not exist and the test rate is slower than the computed velocity. Starting from the initiation of the method in (Takanashi et al. 1975), considerable amount of research has been conducted on conventional slow HS in the last forty years including the works of Mahin and Shing (1985), Nakashima et al. (1988) and Elkhoraibi and Mosalam (2007) among others. Second category, RTHS, is similar to the first one except that the loading is applied to the experimental substructure(s) at a rate equal to the computed velocity. Conventional HS with slow rates of loading is sufficient for substructure testing in most of the cases where rate effects are not important. However, for rate-dependent materials and devices, such as viscous dampers or friction pendulum bearings, RTHS becomes essential. Dynamic actuators and a digital servo-mechanism have been used by Nakashima et al. (1992) as the first progress of RTHS. After the development of actuator-delay compensation methods by Horiuchi et al. (1999), research on RTHS gained momentum. Rapid development of computing technologies and control methods increased the number of RTHS research in recent years (Mercan and Ricles 2007; Bonnet et al. 2008; Bursi et al. 2008; among others). Unlike the first two categories, the experimental substructure(s) is not connected to actuator(s), but located on shaking table(s) in the third HS category, RTHS on a shaking table. Although conventional shaking table testing is well-established in many laboratories and there is considerable amount of accumulated experience on RTHS in the recent years, research and developments on the combination of these two, RTHS on a shaking table, is limited. The use of shaking tables in RTHS, similar to the approach presented in this chapter, where the analytical substructure represents a lower portion of the hybrid structure, has been reported in (Igarashi et al. 2000; Neild et al. 2005; Lee et al. 2007; Ji et al. 2009). Shao et al. (2011) used a shaking table together with dynamic actuators, where it was possible to model analytical substructures both on top and bottom of the experimental substructure. Recently, Nakata and Stehman (2012) proposed a method that allows the modeling of analytical substructures above the experimental substructure where the latter is tested on the shaking table. It should be mentioned that an increase in smart shaking table RTHS applications is expected in the coming years, especially with the development of shaking table grids in various laboratories around the world, such as Tongji University or the NEES facilities at the University of Nevada, Reno. Günay and Mosalam (2013) demonstrated that RTHS on a shaking table configuration can be used as an effective and economical testing method and as an alternative to conventional shaking table tests for electric substation equipment.

This chapter presents three recent developments on the last two HS categories mentioned above, namely RTHS in actuator and shaking table configurations. Details of these developments are discussed in the subsequent sections.

6.2 Development I: Standalone RTHS System

The first development is an RTHS system that can accommodate integration time steps as small as 1 ms. This system is developed at the Structures Laboratory of the Civil Engineering Department of the University of California, Berkeley. The components of the developed RTHS system, namely the shaking table, controller, data acquisition (DAQ) system, and the test specimen (in this case, an insulator post of a high voltage vertical switch in an electric substation), are shown in Fig. 6.1. As indicated in the figure, the DAQ system also represents the computational platform via the digital signal processor (DSP) module. An outline of the communication process between the components of the developed RTHS system is as follows:

1. The computations are conducted on the DSP card (computational platform).
2. The displacement computed by the DSP card is physically transferred to the controller with a standard BNC to BNC cable.
3. The controller commands the computed displacement to the actuator connected to the uniaxial shaking table.
4. At the completion of the displacement command within the allocated time, that is equal to the integration time step for real-time compatibility, the algorithm proceeds with step 1 above for the next time step, using the force feedback acquired by the DAQ system.

The fast execution time of an integration time step, i.e. completion of all four steps listed above and shown in Fig. 6.1, as small as 1 ms is realized by a combination

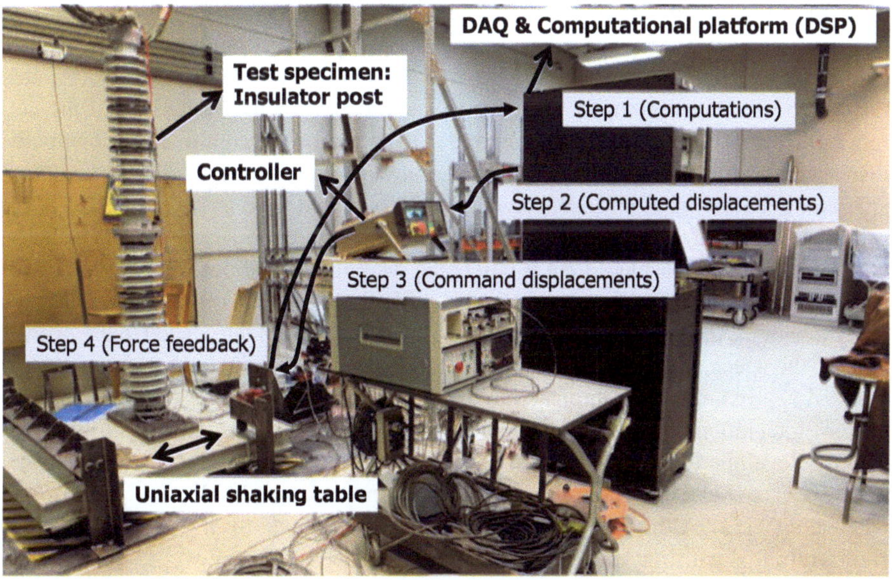

Fig. 6.1 Components of developed RTHS systems and communication between the components

of the computation power introduced by the DSP card (communication step 1), the physical cable transfer between the computational platform and the controller (communication step 2), and the real-time compatible PID control technology of the controller together with the servo-hydraulic system (communication step 3).

Explicit integration methods are more suitable than the implicit ones for RTHS. Implicit methods require at least two iterations, where the number of iterations is in general dictated by the convergence checks, while the explicit ones require no iterations. It is important to note that the number of iterations, if any, in an integration scheme used in RTHS should be constant and small in order to be able to allocate a fixed and practically applicable time for the completion of each iteration. Moreover, the displacement increments in all iterations of an integration time step should be as close as possible to each other to avoid velocity and acceleration oscillations within the integration time step. Although implicit integration methods with constant number of iterations have been developed for HS, e.g. Schellenberg et al. (2009), the small number of iterations has the potential of introducing larger numerical errors since the convergence criteria are not controlling the advancement of the integration in this case. A limiting factor which restricts the applicability of explicit integration is the conditional stability criterion, $\Delta t \leq T_n/\pi$, where Δt is the integration time step and T_n is the period of the highest mode of vibration of the structure. In that regard, the ability to use integration time steps as small as 1 ms within the developed RTHS system allows the use of real-time compatible explicit integration for a broader range of analytical and experimental substructure configurations.

Another advantage introduced by the 1 ms integration time step is the assurance of continuous movement of the actuator of the smart shaking table, since a command is sent to the controller at every millisecond. It is noted that the controller operates at a rate such that it expects a command every millisecond. Since the developed RTHS system is capable of executing integration time steps equal to a millisecond, the computed displacement is directly sent to the controller as the command. If it were not possible to complete the integration time step in a millisecond, it would not be possible to send the computed displacement as the command. This is attributed to the fact that at the end of the current millisecond, where the next command is expected by the controller, the computed displacement would still be in the determination stage. In this case, a predictor-corrector smoothing algorithm, e.g. Mosqueda et al. (2005), would need to be invoked to determine the command to the controller and to assure continuous movement of the actuator, while the computed displacement is still being determined. Although the predictions in these predictor-corrector algorithms can be sufficiently accurate due to the use of high order functions, they can still be considered as approximations introduced to the loading path of the specimen. An example of such an approximation is illustrated in Fig. 6.2. Such an approximation is eliminated in the developed RTHS system with the ability to execute integration time steps corresponding to the operating rate of the controller.

The developed RTHS system was utilized extensively for RTHS of electrical disconnect switches as reported in Mosalam and Günay (2013b) and Günay and Mosalam (2013). Although the RTHS system is developed for a shaking table configuration, it is worth noting that there are no restrictions that prevent the use of the developed system for RTHS in an actuator configuration.

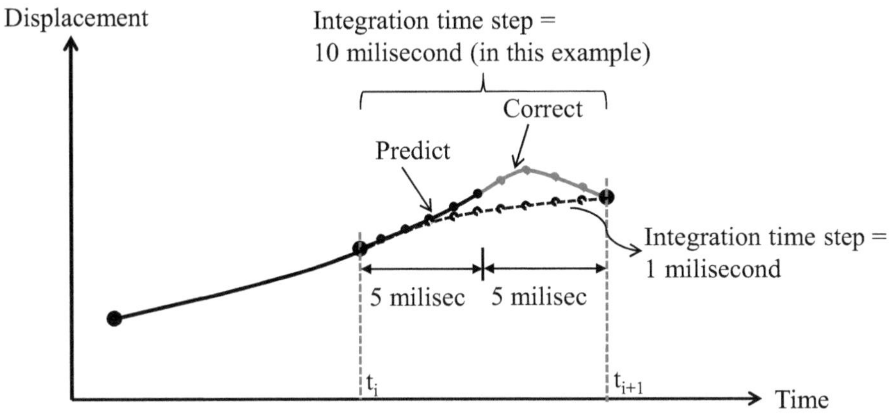

Fig. 6.2 Approximation in the loading path due to a predictor-corrector smoothing algorithm

6.3 Development II: An Efficient Equation Solver

A proper computational platform is one of the essential requirements of a HS system. In the first development, the DSP card served as the computational platform, which was an important factor for the RTHS capability to accommodate integration time steps as small as 1 ms. Similarly, real-time functioning Simulink models, such as those operating on a workstation that runs Mathworks xPC real-time operating system provide the same fast computational platform option. However, as opposed to the advantage offered by the fast computation capability, these computational platforms require case-specific programming and therefore do not offer a versatile usage. Furthermore, they require a considerable training period to be completed by a novice HS user. The Open System for earthquake engineering simulation (OpenSees, McKenna et al. 2010) offers relatively slower computations compared to the mentioned fast computational platforms. However, OpenSees possesses the basic features such as being a well-established, versatile, research-oriented, open-source and HS-compatible platform, which makes it one of the most suitable candidates to be utilized as a computational platform in HS. The second development presents an efficient equation solver to increase the computation speed of OpenSees, especially for the RTHS of structures possessing analytical substructures with a large number of DOF.

In order to provide a background to the explanation of the second development, the HS scheme of one of the most straightforward and HS-compatible numerical integration methods, namely the Explicit Newmark method (Newmark 1959), is presented first for the two DOF hybrid structure in Fig. 6.3.

The equations of motion of the hybrid structure shown in Fig. 6.3 subjected to a ground motion excitation $\ddot{u}_g$ are represented as follows:

$$\mathbf{m\ddot{u}} + \mathbf{c\dot{u}} + \mathbf{f} = \mathbf{p} \tag{6.1}$$

Where

$$\mathbf{m} = \begin{bmatrix} m_1 & 0 \\ 0 & m_2 \end{bmatrix}, \mathbf{\ddot{u}} = \begin{bmatrix} \ddot{u}_1 \\ \ddot{u}_2 \end{bmatrix}, \mathbf{c} = \begin{bmatrix} c_{11} & c_{12} \\ c_{21} & c_{22} \end{bmatrix}, \mathbf{\dot{u}} = \begin{bmatrix} \dot{u}_1 \\ \dot{u}_2 \end{bmatrix}, \mathbf{f} = \begin{bmatrix} -f_a + f_e \\ f_a \end{bmatrix}, \mathbf{p} = -\begin{bmatrix} m_1 \\ m_2 \end{bmatrix} \ddot{u}_g$$

In the above definitions, m_1 and m_2 are the masses corresponding to DOF 1 and 2, respectively, c_{11}, c_{12}, c_{21} and c_{22} are the viscous damping coefficients, $\ddot{u}_1$, $\ddot{u}_2$, $\dot{u}_1$ and $\dot{u}_2$ are the accelerations and velocities at DOF 1 and 2, respectively, and f_a and f_e are the restoring forces of the analytical and experimental springs, respectively.

The algorithm for the application of Explicit Newmark integration for the response determination of the hybrid structure in Fig. 6.3 is presented in Fig. 6.4. The most time consuming part in this algorithm is step "3e", where a linear system of equations is solved. This is especially true for a hybrid system with a large number of DOF in the analytical substructure. Therefore, in order to reduce the computation time, the utilized efficient solver seeks to reduce the computation time by reducing the duration of step 3e. Solution of the linear set of equations in this step requires the coefficient matrix, i.e. the Jacobian matrix, $\mathbf{m}_{\text{eff}}$, to be factorized. Although it is observed in Fig. 6.4 that $\mathbf{m}_{\text{eff}}$ is constant and does not change during the full course of integration, standard equation solvers execute this factorization for each integration time step, which leads to an unnecessary increase of the computation duration. However, there is a special, efficient solver in OpenSees, defined by the Tcl (Ousterhout 1994) syntax, "algorithm Linear-factorOnce", which does this factorization only once at the beginning of the simulation.

The Jacobian matrix is not constant during the entire time history for all the integration methods, e.g. Implicit Newmark, which includes the tangent stiffness matrix. Hence, the efficient solver cannot be used together with such integration methods. It is noted that standard solvers in conventional structural analysis software do not employ the efficient solver to avoid such potential of misuse.

The efficient solver is used for the RTHS of the hybrid structure in Fig. 6.5, which possesses a generic analytical framed substructure with varying number of bays and stories. Figure 6.6 compares the computation time per integration time step with the efficient and standard solvers as a function of the number of DOF.

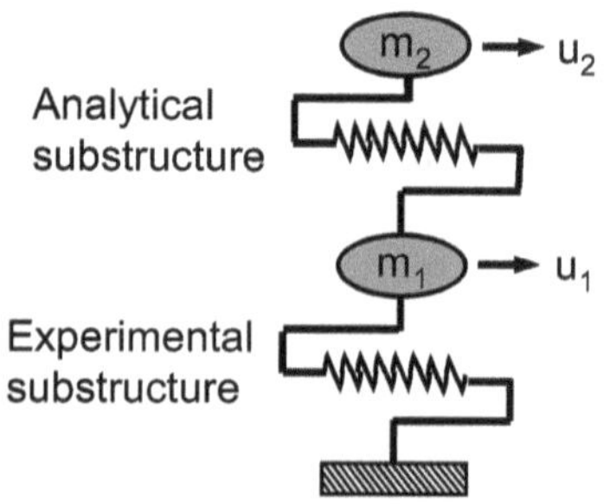

Fig. 6.3 A two DOF system consisting of experimental and analytical substructures

1. Determine the initial values of response variables: $\mathbf{u}_0, \dot{\mathbf{u}}_0, \ddot{\mathbf{u}}_0$
2. Calculate the effective mass: $\mathbf{m}_{\text{eff}} = \mathbf{m} + \Delta t \gamma \mathbf{c}$

   Δt: integration time step, γ: integration parameter
3. For each time step i; $1 \leq i \leq N$, N: total number of steps
   a. Compute the displacement: $\mathbf{u}_i = \mathbf{u}_{i-1} + \Delta t \times \dot{\mathbf{u}}_{i-1} + \left(\Delta t^2/2\right) \times \ddot{\mathbf{u}}_{i-1}$

   b1. Compute the force $f_{a,i}$ corresponding to the displacement $u_{2,i} - u_{1,i}$ from the constitutive relationship of the analytical spring

   b2. Apply the displacement $u_{1,i}$ to the experimental spring and measure the corresponding force $f_{e,i}$

   b3. Determine $\mathbf{f}_i$ from $f_{e,i}$ and $f_{a,i}$ by using the above definition for $\mathbf{f}$

   c. Compute the predicted velocity: $\tilde{\dot{\mathbf{u}}}_i = \dot{\mathbf{u}}_{i-1} + \Delta t(1-\gamma)\ddot{\mathbf{u}}_{i-1}$

   d. Compute the effective force: $\mathbf{p}_{\text{eff}} = \mathbf{p}_i - \mathbf{f}_i - \mathbf{c}\tilde{\dot{\mathbf{u}}}_i$

   e. Compute the acceleration by solving the linear system of equations: $\mathbf{m}_{\text{eff}} \ddot{\mathbf{u}}_i = \mathbf{p}_{\text{eff}}$

   f. Compute the velocity: $\dot{\mathbf{u}}_i = \tilde{\dot{\mathbf{u}}}_i + \Delta t \gamma \ddot{\mathbf{u}}_i$

   g. Increment i and go to step a

Fig. 6.4 Algorithm for Explicit Newmark integration for HS of the two DOF hybrid structure

It is to be noted that the computation times in this figure are obtained from pure analytical simulations where the experimental substructure is replaced with another analytical column. The number of DOF is varied by changing the number of stories for a fixed number of four bays. Figure 6.6 clearly demonstrates the reduction in computation time due to the efficient solver, where it is observed that the efficient solver results in about 2.5 times reduced computation time compared to the standard solver. It was possible to conduct RTHS of the hybrid structure in Fig. 6.5 with the efficient solver accurately up to 720 DOF while this number was limited to only 240 DOF with the standard solver.

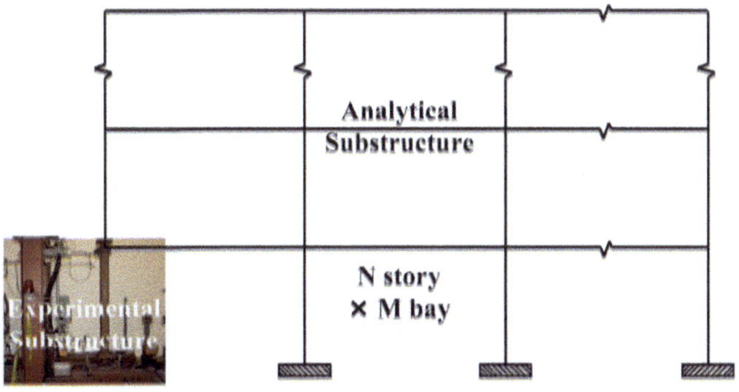

Fig. 6.5 Hybrid structure with varying DOF analyzed and tested with different equation solvers

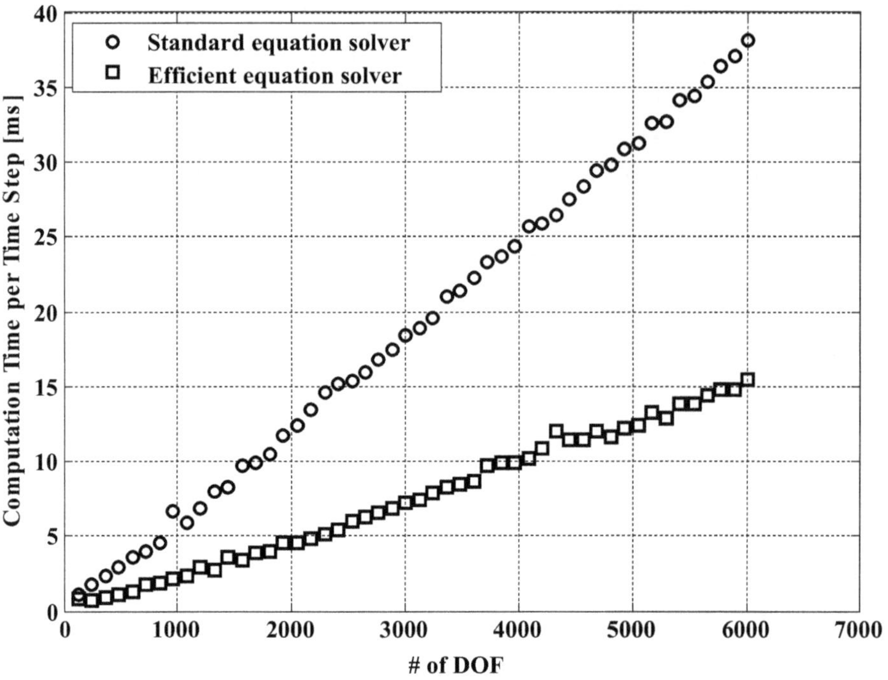

Fig. 6.6 Computation time reduction due to the efficient equation solver

6.4 Development III: TVC Implementation

In the third development, an advanced control method, namely the Three Variable Control (TVC), developed by the MTS Corporation, is implemented in the HS system of the nees@berkeley laboratory for RTHS on a shaking table configuration. Proper displacement tracking may not be adequate for the accuracy of RTHS on a shaking table configuration, because a small displacement error may translate into larger acceleration errors for high frequencies according to the relationship between the pseudo spectral acceleration ($Sa = \omega^2 Sd$) and the spectral displacement (Sd). If the natural frequency (ω) of a specimen is high, then the force feedback used in the computations will be erroneous since the force feedback of a specimen on a shaking table strongly depends on the acceleration of the shaking table. An example of such phenomenon is shown in Fig. 6.7, where in Fig. 6.7a it is observed that the feedback displacements satisfactorily follow the command displacements during the full time history with a slight overshoot at the displacement peaks as shown in Fig. 6.7b. This slight overshooting of displacements translates into large acceleration differences for periods smaller than 0.4 s. as shown in the 1 % damped response spectra in Fig. 6.7c.

There is an inherent delay in the actuator response, which can be eliminated by using the feed-forward gain offered by a servo-hydraulic controller, which does not

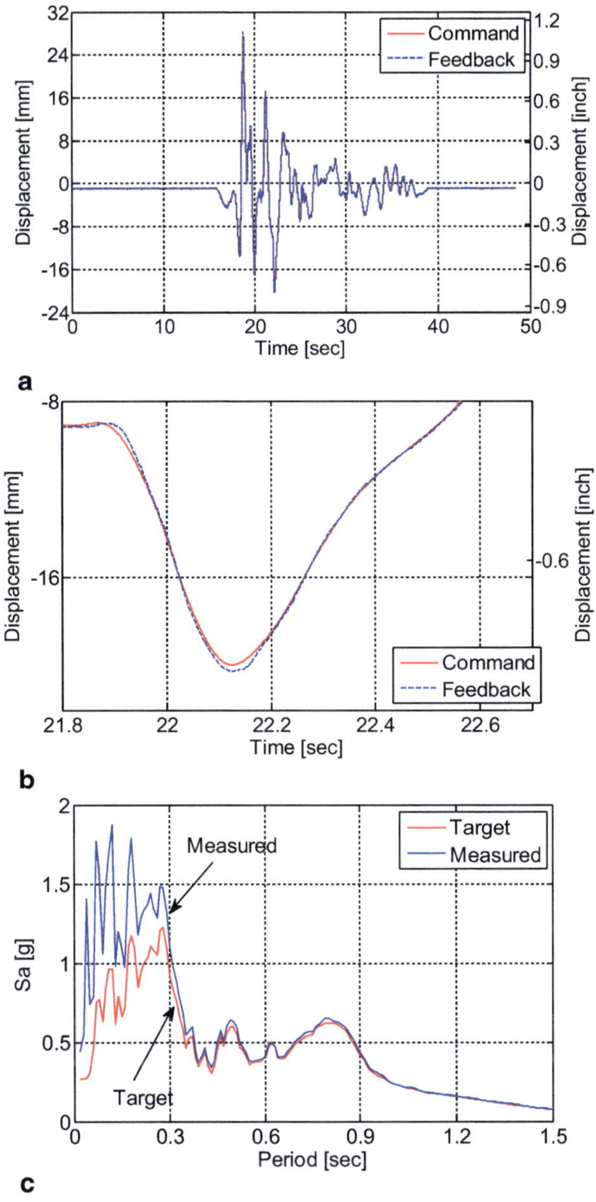

Fig. 6.7 Acceleration mismatch at small periods despite satisfactory displacement tracking. **a** Satisfactory displacement tracking. **b** Slight overshooting of a displacement peak. **c** Acceleration response spectra of target and measured accelerations

necessarily offer additional features for a shaking table control. The feed-forward gain is multiplied by the rate of the command, i.e. command velocity, and added to the servo-valve command. This additional valve command minimizes or completely eliminates the time delay (Fig. 6.8). The effectiveness of a similar feed-forward error compensation is demonstrated in Mosalam and Günay (2013b). However, as shown in Fig. 6.8b, the feed-forward gain introduces overshooting of peak

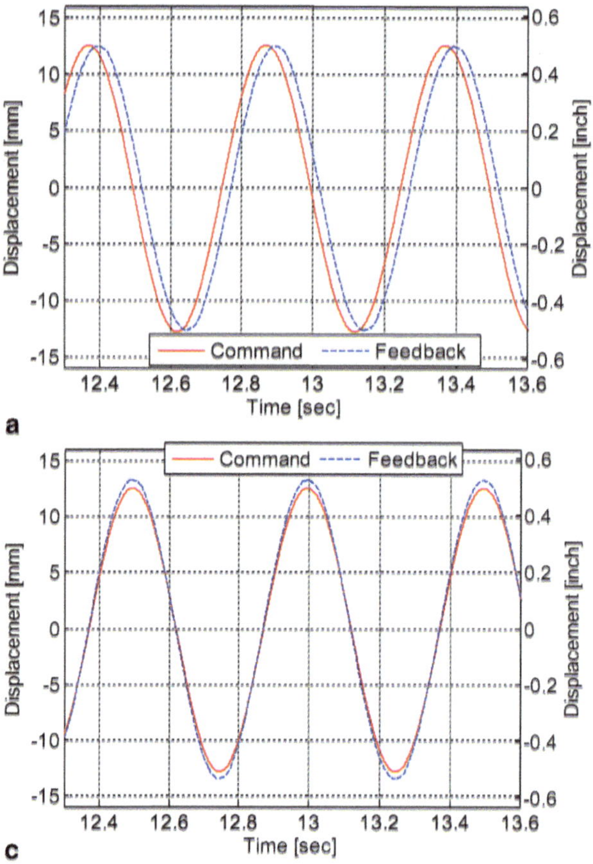

Fig. 6.8 Eliminating time delay by feed-forward gain and resulting peak displacement overshoot. **a** Without feed-forward gain. **b** With feed-forward gain

displacements, which can be explained by the fact that the command rate is close to zero around the peak displacements, leading to almost no feed-forward benefit around these regions. It is this same reason that is responsible for the overshooting of displacement peaks in Fig. 6.7b and the corresponding acceleration errors in Fig. 6.7c. It is noted that the time delay does not introduce a problem on a conventional shaking table configuration as long as the target displacements and accelerations are realized as intended, such as the case shown in Fig. 6.8a. However, phase lag induces not only errors but also negative damping which may lead to instability during RTHS (Nguyen and Dorka 2009), the reasons of which are explained in (Mosalam and Günay 2013a; Mosqueda et al. 2005). Accordingly, there is a need for an advanced control to improve the acceleration response at high frequencies without sacrificing the elimination of the time-delay. TVC is an adequate control method that serves well for this purpose. Since the HS controller of the nees@berkeley laboratory is not a shaking table controller, TVC is implemented in this controller for the improvement of RTHS on a shaking table configuration. This implementation is achieved by adopting and modifying the TVC Simulink model provided by MTS to the HS system of nees@berkeley and by using a feature of the controller to bypass the servo-valve control to an outside resource, while maintaining all other controller features. Implementation is not only conducted for HS, but also for a conventional shaking table configuration. Details of the implementation process are not presented herein due to space limitations.

TVC is a modified version of state variable control enhanced with additional features. In addition to the conventional displacement feedback, two additional kinematic variables, namely velocity and acceleration, are used in the control process. TVC is currently being utilized for the control of many conventional shaking tables including those in the NEES facilities of University of California at San Diego (Luco et al. 2010), University of Nevada, Reno, The University at Buffalo and the PEER shaking table at the University of California, Berkeley. The internal structure of TVC is presented in Thoen (2010). A modified version of this scheme, as adopted for the implementation described herein, is shown in Fig. 6.9. As shown in this figure, the adopted scheme consists of four main parts: (1) reference generator, (2) feedback generator, (3) Delta-P stabilization term and (4) determination of servo-valve command.

The reference generator is used to provide the command, where the reference input in terms of a specific kinematic variable (displacement, velocity, or acceleration) is converted to the command input in terms of the other kinematic variables. Although the original TVC scheme can use all of the mentioned kinematic variables as the reference input, the modified scheme herein considers only displacement, because displacement is the quantity conventionally computed and imposed to the experimental substructure in HS. Furthermore, considering displacements as the reference input better suits the existing HS system of nees@berkeley. After a low-pass filter, the reference displacement input is differentiated once and twice to obtain the velocity and acceleration commands, respectively. It is noted that the low-pass filter is not utilized in the implementation for RTHS to avoid altering the computed displacement and only considered for the implementation for a conventional shaking table configuration.

The feedback generator is utilized to determine the feedback in terms of the considered kinematic variables. Displacement and acceleration measurements obtained from displacement and acceleration sensors are used by the reference generator to determine the displacement, velocity and acceleration feedbacks to be used in the control process. Displacement feedback is obtained as a combination of the low-pass filtered displacement measurement and the double integrated acceleration measurement after high-pass filtering. Acceleration feedback is obtained as a combination of the high-pass filtered acceleration measurement and the double differentiated displacement measurement after low-pass filtering. Velocity feedback is determined as a combination of the differentiated low-pass filtered displacement measurement and integrated high-pass filtered acceleration measurement. In this manner, the displacement and acceleration measurements are primarily used for low-to-medium and high frequencies, respectively, which correspond to the most accurate measurement range of the corresponding sensors.

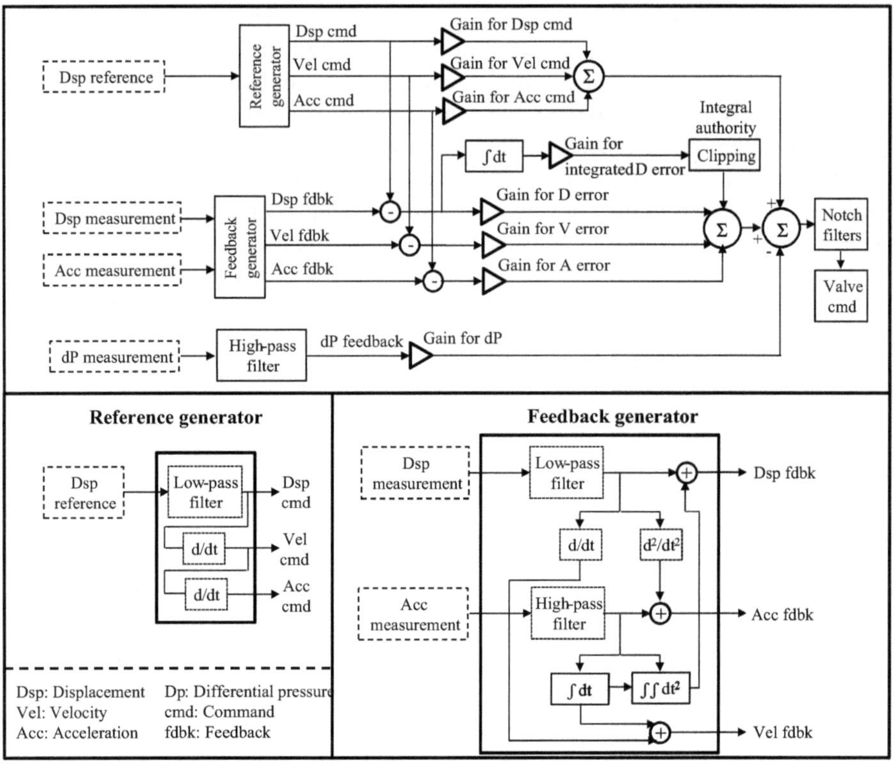

Fig. 6.9 Schematic of TVC as implemented in nees@berkeley

Three types of gains are applied to the commands and feedbacks obtained from the reference and feedback generators. First of these types, the proportional gain, is the gain applied on the displacement, velocity and acceleration errors, defined as the difference in the respective command and feedback terms. Gains applied to the commands constitute the second type, which can be regarded as the feed-forward gain terms. The third type of gain is the integral gain, the one applied on the integrated displacement error. All the resultant values after the application of the gains are normalized by the full scale values and then summed.

The third part in the adopted TVC scheme, the Delta-P stabilization, is employed to suppress the effect of the oil-column. It is known that the hydraulic fluid in the actuator chambers is compressible and therefore acts as a linear spring. This spring and all the mass that slides together with the actuator, e.g. the table mass, specimen mass, and actuator piston mass, form a single degree of freedom (SDOF) system (Conte and Trombetti 2000) with a natural frequency designated as the oil-column frequency. The Delta-P gain acts as damping that reduces and preferably eliminates the response of this SDOF system, which otherwise alters the hybrid simulation by introducing artificial forces. The Delta-P measurement in the TVC scheme is the differential pressure between the actuator chambers, measured by a differential pressure sensor after high-pass filtering. The Delta-P gain applied to the Delta-P measurement normalized by its full scale value is added to the resultant of all other terms mentioned in the previous paragraph. The last step before the determination of the servo-valve command is the application of the notch filters, which are filters that attenuate or amplify the response in a specific frequency range. In addition to the Delta-P gain, the notch filters are sought to be used herein to eliminate the oil-column effect. However, since a significant improvement is not observed due to the application of the notch filters, they are not employed in the current HS implementation.

Finally, it is beneficial to note that one other kinematic variable, which is the third derivative of displacement and designated as the jerk, is utilized in the original TVC scheme. This term is neglected in the implementation herein because the jerk related gain is mainly beneficial to boost the frequency above the oil column frequency (Thoen 2010), which is about 25 Hz in the present system. Therefore, the jerk is not beneficial for the objectives of this development, which is the reduction of the overshooting of the accelerations in the high frequency range, approximately 3 Hz and higher in the present study.

The improvement of the acceleration responses as a result of the TVC scheme is shown in Fig. 6.10. The response as a result of to the tuning of the TVC parameters for RTHS, where delay is eliminated between command and feedback, is shown in Fig. 6.10a. A further improved acceleration response is shown on Fig. 6.10b, which presents the tuning for a conventional shaking table configuration, where delay is allowed. With further tuning efforts, it may be possible to improve the acceleration response, especially for the case when delay is eliminated.

After the sufficiently acceptable tuning shown in Fig. 6.10, TVC implementation is used in RTHS of 230-kV electrical vertical disconnect switches, Fig. 6.11a, where the experimental substructure, Fig. 6.11b, is a jaw side post insulator of the disconnect switch and the analytical substructure consists of a 3D model of the steel support structure and the remaining insulators, Fig. 6.11c. In addition to the TVC utilization, the conducted RTHS features other novel HS characteristics, first of which is the 3D analytical substructure, where the bandwidth of the Jacobian matrix is increased due to the presence of 12 DOF per element. This situation introduces a computational time challenge in RTHS. In order to be able to overcome this challenge, the efficient solver mentioned in the second development is utilized together with the RTHS compatible α-Operator-Splitting integration (Combescure and Pegon 1997). It is to be noted that it was not possible to use Explicit Newmark integration since the infinite frequency modes introduced by the massless rotational DOF violated the stability criterion.

The ratios of the Sa of accelerations measured on the platform to the Sa of computed accelerations at the top of the support structure, at the corner node where the experimental substructure is attached to the analytical substructure, are plotted in Fig. 6.12 for the cases of displacement control and TVC. Similar to the case of the conventional shaking table configuration, TVC is successful in bringing the measured accelerations closer to the computed values, which increases the accuracy of RTHS by reducing the force feedback errors. This is also valid at the natural period

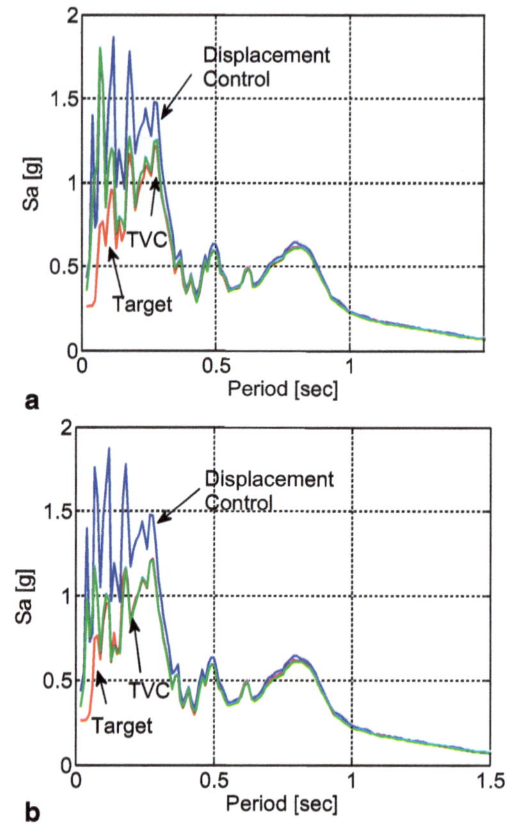

Fig. 6.10 Improved acceleration response due to TVC. **a** Improvement in acceleration response: Time-delay eliminated. **b** Further improvement in acceleration response: Time-delay allowed

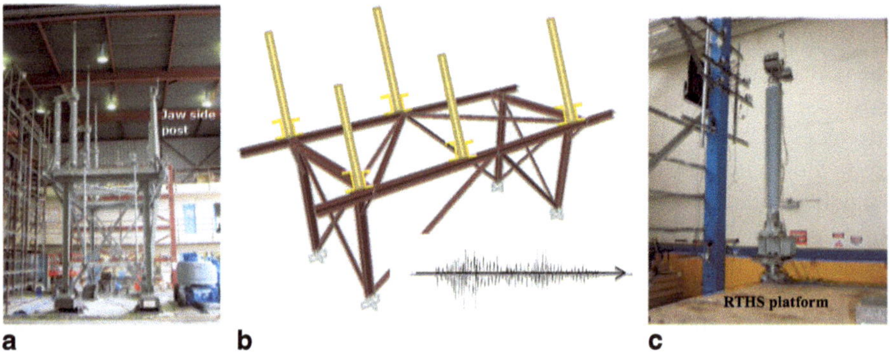

Fig. 6.11 Hybrid model of the electrical disconnect switch tested with RTHS using TVC. **a** 230-kV vertical switch (porcelain posts) tested on a shaking table. **b** Analytical substructure of the switch (3D braced steel frame as the support structure + other posts). **c** Experimental substructure (polymer composite post)

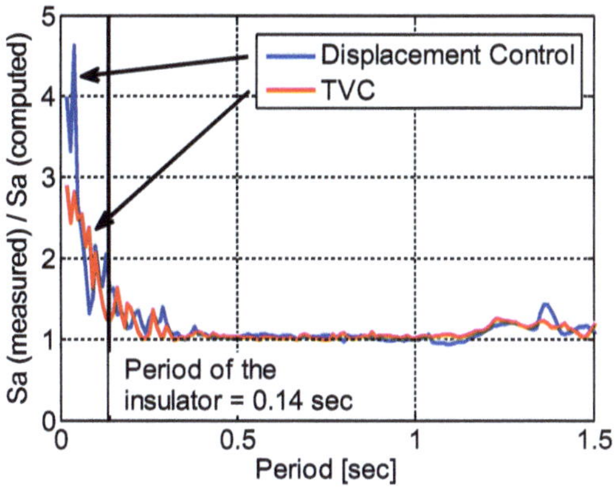

Fig. 6.12 Improvement in acceleration response in RTHS of the disconnect switch due to TVC

of the tested insulator post, which is represented with the vertical line in Fig. 6.12. As mentioned previously, although it may be possible to further reduce the measured accelerations to bring the ratios closer to 1.0 in the high frequency range, i.e. for periods smaller than about 0.3 s, the current results are indeed indicative of the benefits of the TVC, validating the objective of the implementation of TVC in HS system of the nees@berkeley laboratory.

6.5 Closure

This chapter presented three recent developments aimed at faster and more accurate execution of RTHS. As the first development, a standalone RTHS system is developed which can execute integration time steps of 1 millisecond and can accordingly benefit from the advantages of such fast execution. Towards the objective of the improvement of the more general and versatile computational platform OpenSees, an efficient linear equation solver is utilized in the second development to decrease computation time. The additional challenge introduced by the control of accelerations in RTHS on a shaking table configuration is addressed in the third development by the adaptation of the three-variable control.

Acknowledgements Special thanks are due to Mr. Brad Thoen of MTS Corporation for providing the original Simulink model for TVC and to Dr. Frank McKenna for coding the efficient solver in OpenSees.

References

Bonnet PA, Williams MS, Blakeborough A (2008) Evaluation of numerical time-integration schemes for real time hybrid testing. Earthq Eng Struct Dyn 37(13):1467–1490

Bursi OS, Gonzalez-Buelga A, Vulcan L, Neild SA, Wagg DJ (2008) Novel coupling Rosenbrock-based algorithms for real-time dynamic substructure testing. Earthq Eng Struct Dyn 37(3):339–360

Combescure D, Pegon P (1997) α-Operator splitting time integration technique for pseudodynamic testing error propagation analysis. Soil Dyn Earthq Eng 16:427–443

Conte JP, Trombetti TL (2000) Linear dynamic modeling of a uni-axial shaking table system. Earthq Eng Struct Dyn 29(9):1375–1404

Elkhoraibi T, Mosalam KM (2007) Towards error-free hybrid simulation using mixed variables. Earthq Eng Struct Dyn 36(11):1497–1522

Günay S, Mosalam KM (2013) Seismic performance evaluation of high voltage disconnect switches using real-time hybrid simulation: II. Parametric Study. Earthq Eng Struct Dyn 43(8):1223–1237

Horiuchi T, Inoue M, Konno T, Namita Y (1999) Real-time hybrid experimental system with actuator delay compensation and its application to a piping system with energy absorber. Earthq Eng Struct Dyn 28(10):1121–1141

Igarashi A, Iemura H, Suwa T (2000) Development of substructured shaking table test method. 12th world conference on earthquake engineering. Auckland, New Zealand

Ji X, Kajiwara K, Nagae T, Enokida R, Nakashima M (2009) A substructure shaking table test for reproduction of earthquake responses of high-rise buildings. Earthq Eng Struct Dyn 38(12):1381–1399

Lee S, Parka E, Mina K, Park J (2007) Real-time substructuring technique for the shaking table test of upper substructures. Eng Struct 29(9):2219–2232

Luco J, Ozcelik O, Conte J (2010) Acceleration tracking performance of the UCSD-NEES shake table. J Struct Eng 136(5):481–490

Mahin SA, Shing PSB (1985) Pseudodynamic method for seismic testing. J Struct Eng ASCE 111(7):1482–1503

McKenna F, Fenves GL, Filippou FC (2010) OpenSees. http://opensees.berkeley.edu

Mercan O, Ricles JM (2007) Stability and accuracy analysis of outer loop dynamics in real-time pseudodynamic testing of SDOF systems. Earthq Eng Struct Dyn 36(11):1523–1543

Mosalam KM, Günay MS (2013a) Hybrid simulations: theory, applications, and future directions. Adv Mater Res 639–640:67–95

Mosalam KM, Günay S (2013b) Seismic performance evaluation of high voltage disconnect switches using real-time hybrid simulation: I. System development and validation. Earthq Eng Struct Dyn 43(8):1205–1222

Mosqueda G, Stojadinovic B, Mahin S (2005) Implementation and accuracy of continuous hybrid simulation with geographically distributed substructures, Report No: UBC/EERC 2005–02, Earthquake Engineering Research Center, University of California, Berkeley

Nakashima M, Ishii K, Kamagata S, Tsutsumi H, Ando K (1988) Feasibility of pseudo-dynamic test using substructuring techniques. In: Proceedings of the Ninth World Conference on Earthquake Engineering. Tokyo, Japan

Nakashima M, Kato M, Takaoka E (1992) Development of real-time pseudo-dynamic testing. Earthq Eng Struct Dyn 21(1):79–92

Nakata N, Stehman M (2012) Substructure shake table test method using a controlled mass: formulation and numerical simulation. Earthq Eng Struct Dyn 41(14):1977–1988

Neild S, Stoten D, Drury D, Wagg D (2005) Control issues relating to real-time substructuring experiments using a shaking table. Earthq Eng Struct Dyn 34(9):1171–1192

Newmark NM (1959) A method of computation for structural dynamics. ASCE J Eng Mech Div 85(EM3):67–94

Nguyen NV, Dorka U (2009) Adaptive error compensation based on online system identification for real-time substructure testing. In: Proceedings of the 3rd international conference on advances in experimental structural engineering. San Francisco, CA

Ousterhout J (1994) Tcl and the Tk Toolkit. Addison-Wesley, Reading, Massachusetts, USA

Schellenberg AH, Mahin SA, Fenves GL (2009) Advanced implementation of hybrid simulation. Report No: PEER 2009/104. Pacific earthquake engineering research center, University of California, Berkeley

Shao X, Reinhorn AM, Sivaselvan MV (2011) Real-Time hybrid simulation using shake tables and dynamic actuators. ASCE J Struct Eng 137(7):748–760

Takanashi K, Udagawa K, Seki M, Okada T, Tanaka H (1975) Non-linear earthquake response analysis of structures by a computer-actuator on-line system (details of the system). Trans. of the Architectural Institute of Japan, 229:77–83

Thoen BK (2010) 469D Seismic digital control software, MTS Systems Corporation

Chapter 7
Pseudo-Dynamic Testing Based on Non-linear Dynamic Substructuring of a Reinforced Concrete Bridge

Giuseppe Abbiati, Oreste Salvatore Bursi, Enrico Cazzador, Rosario Ceravolo, Zhu Mei, Fabrizio Paolacci and Pierre Pegon

7.1 Introduction

In order to estimate the seismic performance of an old concrete viaduct for both the *as built* and the *retrofitted* conditions at Serviceability Limit State (SLS) and Ultimate Limit States (ULS) a comprehensive set of continuous-time hybrid simulations was conceived within the RETRO TA (Taucer 2011) of the SERIES European research project (Fardis 2009). The structure is characterized by twelve couples of portal piers, which support two independent roadways. The views of the viaduct

G. Abbiati (✉)
University of Trento, Trento, Italy
e-mail: giuseppe.abbiati@unitn.it

O. S. Bursi · E. Cazzador · Z. Mei
Department of Civil, Environmental and Mechanical Engineering,
University of Trento, Trento, Italy
e-mail: oreste.bursi@unitn.it

E. Cazzador
e-mail: enrico.cazzador@unitn.it

Z. Mei
e-mail: zhu.mei@unitn.it

R. Ceravolo
Department of Structural, Geotechnical and Building Engineering,
Politecnico of Turin, Turin, Italy
e-mail: rosario.ceravolo@polito.it

F. Paolacci
Department of Engineering, University Roma Tre, Rome, Italy
e-mail: fabrizio.paolacci@uniroma3.it

P. Pegon
European Laboratory for Structural Assessment Unit, European Commission, Joint Research Centre (JRC), Institute for the Protection and Security of the Citizen (IPSC), Via Enrico Fermi 2749, 21027 Ispra VA, Italy
e-mail: pierre.pegon@jrc.ec.europa.eu

F. Taucer, R. Apostolska (eds.), *Experimental Research in Earthquake Engineering*,
Geotechnical, Geological and Earthquake Engineering 35,
DOI 10.1007/978-3-319-10136-1_7

a b

Fig. 7.1 **a** Views of the Rio Torto viaduct portal piers. **b** Views of the Rio Torto viaduct concrete decks

portal piers and of the concrete deck are illustrated in Fig. 7.1. The installation of a couple of friction pendulum bearing isolators -one per column- interposed between the cap beam for each pier frame and the deck was proposed to achieve Eurocode 8 seismic-performance requirements (Eurocode 1998 Part 2 2006).

Hence, the hybrid simulation set-up is described first: two of the twelve piers, i.e., the Physical Substructures (PSs), are loaded through dynamic actuators, whilst the remaining piers and the deck, i.e., the NS, are simulated numerically. Then, the OpenSEES (OpenSEES 2009) Reference Model (RM) of the bridge is introduced to support the design of the experimental campaign. Since time history analyses exhibited appreciable nonlinearities in the dynamic response at SLS, in order to obtain a NS suitable for testing purposes, a rigorous reduction of the aforementioned OpenSEES RM is proposed. In this particular case, a component mode wise synthesis approach was considered. With reference to the piers, 3-DoFs superelements of each pier portal frame obtained via the Guyan method (Guyan 1965) were extended to the nonlinear regime by means of a modified Bouc-Wen spring. In order to provide stiffness and mass matrices for the preliminary linear substructuring, a linear ANSYS FE model was set. With regard to isolator elements, a state space model based on the formulation proposed by Mostaghel (1999) was implemented. Since isolators contributed to most of the energy dissipation, linear reduced piers were adopted in the isolated case. A comparison in terms of transversal displacements, velocities and accelerations is presented as validation of the reduced model of the bridge. Lastly, a testing procedure aimed at simulating a consistent degradation of both PS and NS is presented. It was based on recursive offline identification and model updating sessions occurring at the end of each Pseudo-Dynamic (PsD) test run.

7.2 Main Characteristics of Hybrid Simulations

The performance of the bridge with and without seismic isolation system was estimated by means of the continuous-PsD method. Two of the twelve piers with relevant isolators -PSs- were loaded through dynamic actuators at the ELSA labora-

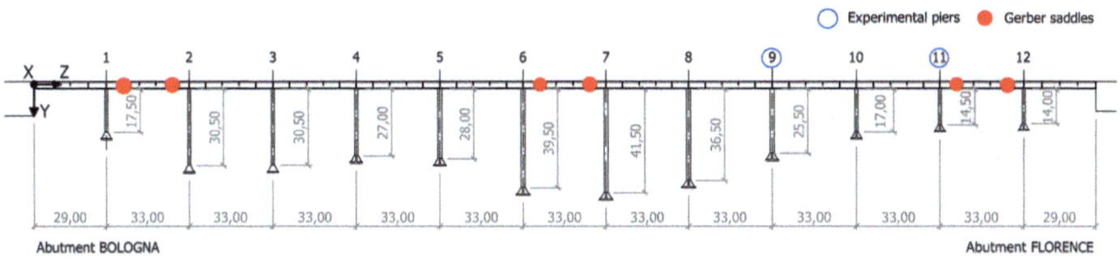

Fig. 7.2 Structural scheme and main dimensions (in m) of the Rio Torto viaduct

tory of the Joint Research Centre of the European Commission at Ispra, Italy, whilst the remaining ten piers and the deck as well -NSs- were numerically modelled and solved. Figure 7.2 depicts the whole emulated system and highlights Piers #9 and #11, that is, the PSs.

In order to comply with the capacity of actuators and more general restrictions imposed by the experimental facility, 1:2.5 scale specimens of Piers #9 and #11 were considered for the PsD tests. Because deck masses play an important role during testing, the scale factor for forces and time was S^2 and S, respectively (Kumar et al. 1997). As shown in Fig. 7.3, which depicts the experimental set-up in the non-isolated case, both physical and numerical piers are coupled to the linear deck at the Center of Gravity (CG) of its own cross section.

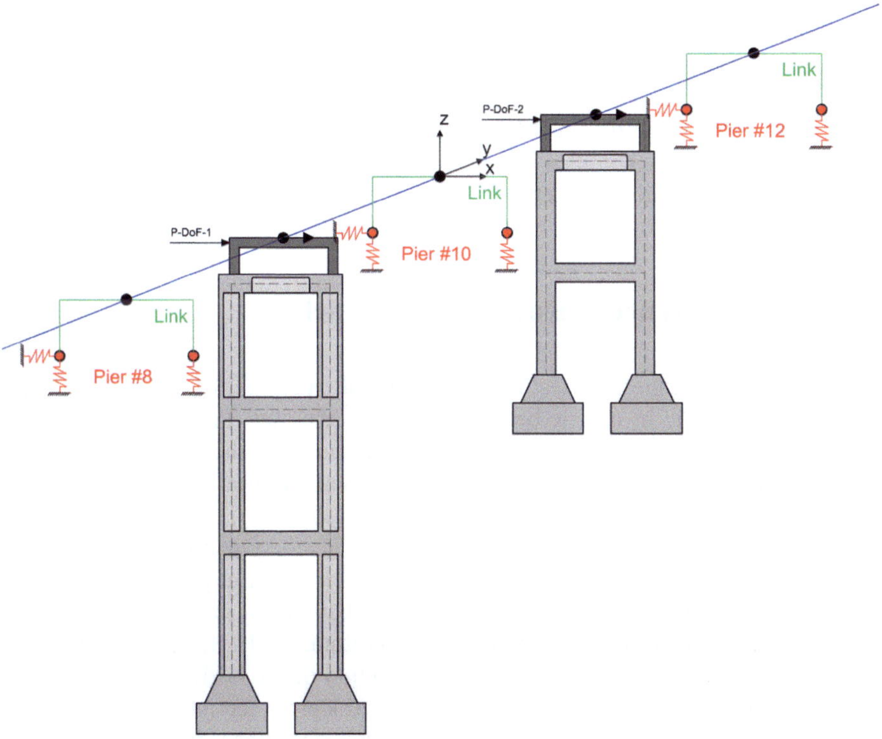

Fig. 7.3 Physical-DoFs localisation

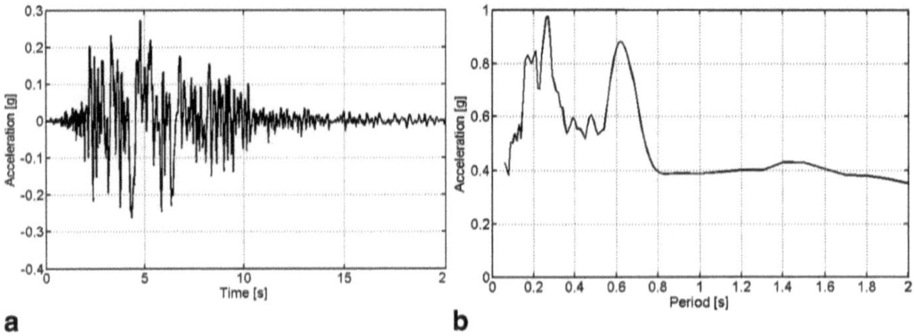

Fig. 7.4 N-S component of the Emilia earthquake –May 2012– selected for the ULS. **a** accelerogram. **b** acceleration response spectra

Vertical loads were applied to the physical piers by means of force controlled hydraulic actuators according to nominal values considered in the OpenSEES RM. Two records form the Mirandola station corresponding to the Emilia (Italy) earthquake of May 29th, 2012 were selected owing to their seismological characteristics, i.e. PGA and duration. The East-West (EW) component of the earthquake with 2.56 m/s^2 PGA was considered for the SLS, whilst the North-South component with 2.67 m/s^2 PGA was assumed for the ULS. For both the E-W and N-S accelerograms, a significant amplification was observed for low periods, i.e. between 0.50 and 1.00 s. The N-S component exhibited spectral accelerations of about 0.40 g in the period range from 1.00 to 1.50 s. The ULS record and its relevant acceleration response spectrum are depicted in Fig. 7.4..

In this context, the PM interfield-parallel time integration algorithm (Pegon and Magonette 2002; Bonelli et al. 2008) was implemented: its subcycling capabilities allow for handling the complex NS with a coarse time step, whilst the PS advances with the controller fine time step.

7.3 The Reference OpenSEES FE Model

In order to support the PsD test design, a refined OpenSEES fiber-based FE RM able to simulate the hysteretic behaviour of piers was set (Paolacci and Giannini 2012). The Kent-Scott-Park model was used to emulate the concrete behaviour (Kent and Park 1971), while the Menegotto-Pinto model was adopted for the steel reinforcement (Menegotto and Pinto 1973). As a result, the *Concrete01* OpenSEES material was considered for concrete, whilst the *Steel02* OpenSEES material was adopted to model steel reinforcement. The *hysteretic* material provided by OpenSEES was used to simulate the nonlinear shear behaviour of transverse beams. The piers were considered fully constrained at the bottom, whilst the abutments at both sides were released along the longitudinal direction of the bridge. To take into account the offset distance between the CG of the deck cross section and the cap beam axis, each pier was connected to the deck through rigid links. In detail, each rigid link was considered fixed to the deck and hinged to the relevant pier, as shown in

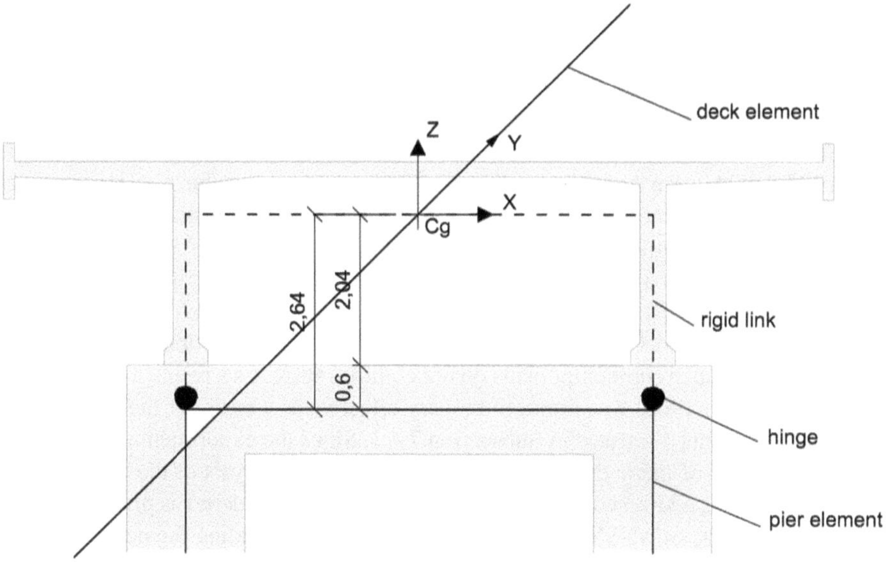

Fig. 7.5 Details of the FE modelling of the pier-deck connection (dimensions in m)

Fig. 7.5. Only flexural behaviour of the deck was taken into account. Since this refined FE model was capable of reproducing the complex behaviour of the full-emulated system, a comprehensive set of time history analyses was carried out to simulate the dynamic response of the bridge at different limit states.

Figures 7.6a and b report the hysteretic loops relevant to Piers #9 and #11 (PSs) for the ULS; each transversal displacement was measured at the cap beam level, whilst forces correspond to base reactions along the same X direction.

In principle, hysteretic loops of tall piers, such as Pier #9, are more jagged than those of short piers, such as Pier #11.

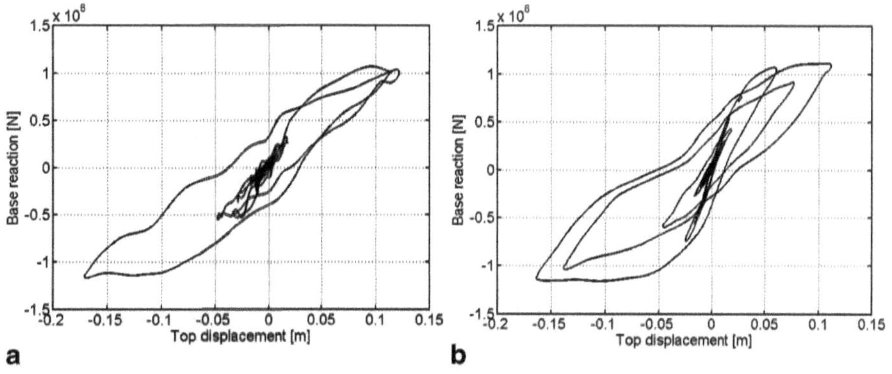

Fig. 7.6 Hysteretic loop relevant to transversal behaviour of **a** Pier #9 at ULS. **b** to Pier #11 at ULS

7.4 State Space Reduction of the RM

In order to provide stiffness and mass matrices for model reduction, an 832-DoFs ANSYS Linear Model (LM) of the bridge was implemented. The same constraint conditions, material properties and equivalent viscous damping of 5% of the OpenSEES RM were considered. Figures 7.7a and b show the two eigenmodes of the ANSYS LM carrying most of the modal mass in the X transversal direction along which the seismic loading acted.

The two transversal bending modes of the deck, see Fig. 7.7, govern the global dynamics response of the system and entail a substantial in-plane deformation of piers. More precisely, the deck pulls piers along the transversal direction. Consequently, higher modes of such sub-components do not come on stage. As a result, a static Guyan reduction (Guyan 1965) was found to be very effective as a basis for the synthesis of piers superelements discussed in Subsection 7.4.1. Since the experimental set-up does not allow for out-of-plane displacements of piers, a deep analysis of the internal constraint setting was addressed to prove that piers could be considered as plane superelements. As a result, an ANSYS Simplified Model (SM) embedding the new constraint setting was set; in particular, relative rotations between deck and piers were released, whilst out-of-plane displacements of piers were fixed. The first five eigenfrequencies of the ANSYS RM, the ANSYS SM and the OpenSEES RM are reported in Table 7.1:

ANSYS linear models are in agreement with the OpenSEES RM. In particular, the ANSYS SM lends itself for an effective Guyan reduction of piers as plane superelements. This implementation aspect facilitated the tuning of each single substructured pier as a stand-alone Multiple-Input-Multiple-Output (MIMO) system with a reduced number of DoFs.

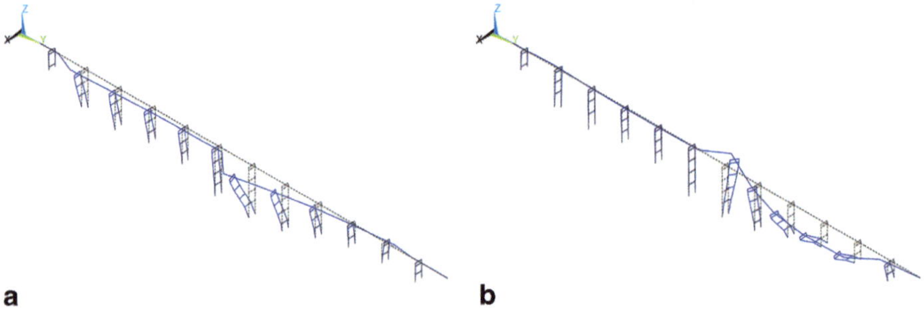

Fig. 7.7 ANSYS FE model of the Rio Torto viaduct **a** 2nd eigenmode at 0.64 Hz. **b** 4th eigenmode at 1.1 Hz

Table 7.1 Eigenfrequency comparison between bridge numerical models

Mode	OpenSEES RM	ANSYS RM	ANSYS SM
	[Hz]	[Hz]	[Hz]
1	0.6035	0.6254	0.6227
2	0.6590	0.6452	0.6433
3	0.6687	0.7017	0.7004

7.4.1 Nonlinear Dynamic Substructuring of Piers

Since the seismic load excites mainly the lowest global eigenmodes of the viaduct, piers can be considered as being pulled along the X transversal direction indicated in Fig. 7.7. Therefore, an effective pier superelement was obtained via the Guyan method applied to the ANSYS SM; in-plane translational interface DoFs between the piers and the relevant rigid link elements were retained, whilst the remainder was condensed. Moreover, the proposed tailoring allowed for the implementation of isolator elements without any change to the pier superelements. According to the relevant formulation, the static condensation reads:

$$[u]=\begin{bmatrix} u_R \\ u_L \end{bmatrix}=\begin{bmatrix} I \\ \phi_R \end{bmatrix}\cdot[u_R]=[T]\cdot[u_R] \tag{7.1}$$

where:

- u_R: master DoFs
- u_L: slave DoFs
- ϕ_R: boundary node functions

Figure 7.8a represents the 3-DoFs pier superelement, which allows for easy inclusion of the isolator elements, as shown in Fig. 7.8b.

Since no local eigenmodes are excited, the proposed reduction is almost exact in the linear range. In the ease of local modes, a further refinement of the reduction could be based on the Craig-Bampton method (Craig and Bampton 1968). As highlighted by Paolacci and Giannini (2012) during the seismic event most of the damage was concentrated within the piers. The proposed pier superelement lent itself to a lightweight extension to the nonlinear range through modified Bouc-Wen springs. In particular, three by three matrices resulting from linear substructuring of each single pier were assumed as basis for the formulation of nonlinear state-space models capable of reproducing the hysteretic energy dissipation typical of such structural elements. In detail, the k_{11} element of the reduced stiffness matrix of Eq. (7.2) and corresponding to the transversal displacement was removed and the relevant restoring force was taken into account by means of a modified Bouc-Wen hysteretic spring characterized by a softening elastic component. Hence, the nonlinear extension of the state space model reads,

$$\begin{bmatrix} k_{11} & k_{12} & k_{13} \\ k_{21} & k_{22} & k_{23} \\ k_{31} & k_{32} & k_{33} \end{bmatrix}\begin{bmatrix} u_{R,1} \\ u_{R,2} \\ u_{R,3} \end{bmatrix}\Rightarrow\begin{bmatrix} 0 & k_{12} & k_{13} \\ k_{21} & k_{22} & k_{23} \\ k_{31} & k_{32} & k_{33} \end{bmatrix}\begin{bmatrix} u_{R,1} \\ u_{R,2} \\ u_{R,3} \end{bmatrix}+\begin{bmatrix} f_1 \\ 0 \\ 0 \end{bmatrix} \tag{7.2}$$

$$\dot{f}_1=\left(\rho\cdot A/\left(1+\alpha\cdot x_1^2\right)-(\beta\cdot sgn(\dot{x}_1\cdot f_1)+\gamma)\,|\,f_1\,|^n\right)\cdot\dot{x}_1 \tag{7.3}$$

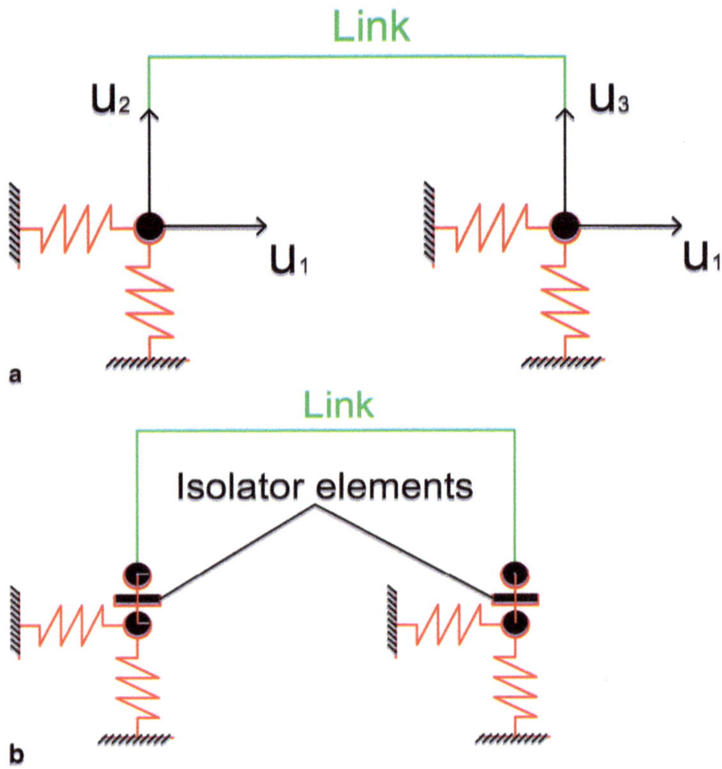

Fig. 7.8 **a** Superelement representing the generic substructured pier without isolators. **b** Inclusion of isolator elements within the pier superelement endowed with isolators

where A, β, γ and n are parameters of the Bouc-Wen model. A was assumed equal to the k_{11} element of the linear initial tangent stiffness matrix, whilst ρ was introduced to represent its average degradation. Since the reduction of the global bridge entailed the tuning of all twelve piers, in order to decrease the computational burden of the resulting set of optimization problems γ was set equal to zero and n to one. The softening factor depending on the α parameter was introduced according to the material properties of the OpenSEES RM. It is evident from Eq. (7.2) that possible nonlinear interactions between horizontal and vertical springs of Fig. 7.8 were neglected. Nonetheless, this assumption was proven to be satisfactory. With regard to the nonlinear identification of parameters, each substructured pier was considered as a stand-alone MIMO system. Interface forces calculated by OpenSEES RM were considered as input applied to the coupling DoFs of the substructured pier, whilst cap beam level displacements, velocities, accelerations and dissipated energies were considered as output. With respect to the reference OpenSEES solution, a penalty function was defined as a weighted sum of the Normalized Root Mean Square Errors (NRMSEs) on the transversal displacements and on the absorbed energy of the substructured pier. In greater detail, NRMSE reads,

Table 7.2 Identified parameters of reduced piers for both SLS and ULS

Pier	SLS			ULS		
	ρ	α [1/m²]	β [1/m]	ρ	α [1/m²]	α [1/m]
1	1.00	198.71	0.00	0.55	199.59	0.66
2	0.82	54.70	1.54	0.50	0.03	0.94
3	0.88	89.58	1.36	0.50	0.05	0.84
4	0.64	61.54	1.46	0.52	37.06	1.23
5	0.64	78.23	1.63	0.50	28.69	1.24
6	0.86	84.01	1.14	0.61	0.06	0.82
7	0.67	10.06	0.92	0.59	6.30	0.53
8	0.75	48.30	0.68	0.70	34.13	1.24
9	0.78	191.34	1.19	0.50	23.00	1.00
10	0.99	199.06	0.00	0.53	165.58	1.58
11	0.66	175.68	1.05	0.50	193.81	2.50
12	0.91	198.86	0.02	0.56	199.60	0.00

$$\mathrm{NRMSE} = \frac{\sqrt{\frac{1}{n}\sum_{i=1}^{n}(y_{red} - y_{ref})^2}}{y_{ref,\max} - y_{ref,\min}} \tag{7.4}$$

Where the subindeces *red* and *ref* stand for reduced and reference quantities, respectively. The MatLAB pattern search algorithm was used to minimize such penalty function. Table 7.2 summarize the identified parameters for both the SLS and ULS.

It is interesting to note that an appreciable stiffness degradation occurs at ULS. With reference to Pier #9, Fig. 7.9 compares both the transversal displacement and the dissipated energy of the reduced model with the OpenSEES RM; the results are satisfactory.

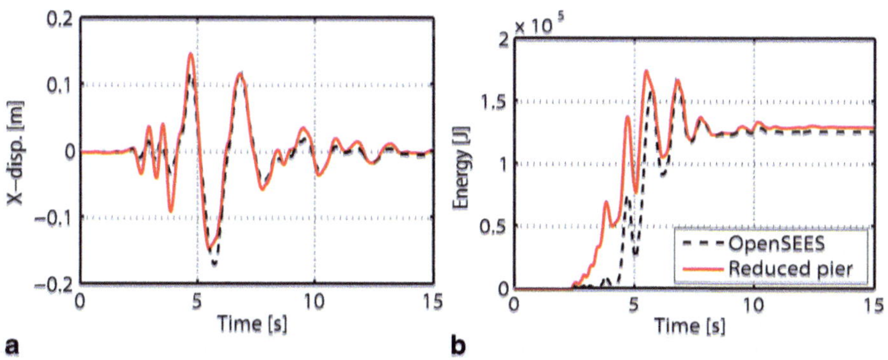

Fig. 7.9 Validation of the reduced model of Pier #9: **a** displacement and **b** dissipated energy histories at ULS

7.4.2 Nonlinear Dynamic Substructuring of Isolator Elements

The hysteretic model of nonlinear friction pendulum isolators used in the isolated case, shown in Fig. 7.10a, was based on the bilinear model proposed by Mostaghel (1999) and shown in Fig. 7.10b. The aforementioned bilinear model was used to reproduce the nonlinear behaviour of the two-node *singleFPBearing* element implemented in the OpenSEES RM (2009).

The slip behaviour of the element was considered thanks to the state space variable *u*. The system of differential equations governing the behaviour of the bilinear system depicted in Fig. 7.10c reads:

$$\begin{cases} m\cdot\ddot{x}+c\cdot\dot{x}+\alpha k x+(1-\alpha)ku=\bar{P}_0\cdot p(t) \\ \dot{u}=\dot{x}\left(\bar{N}(\dot{x})\bar{M}(u-\delta)+M(\dot{x})N(u+\delta)\right) \end{cases} \tag{7.5}$$

where $N, M, \bar{N}$ and $\bar{M}$ are defined as linear combinations of the standard signum function. The integration of Eq. (7.5) defines the response of any nondegenerating hysteretic bilinear system under a given load. In the present case, no mass and damping contributions of isolators were considered. Since OpenSEES recorders provide relative displacements and restoring forces of isolator elements, the identification of the nonlinear parameters of the single isolator was associated to the following optimisation problem:

$$\left\{\hat{\alpha},\hat{k},\,\hat{\delta}\right\}=\min_{\alpha,k,\delta}\left\|\mathbf{r}_{ref}-\mathbf{r}_{red}(\alpha,k,\delta)\right\| \tag{7.6}$$

$$\begin{cases} r_{red,i}=\alpha\cdot k\cdot x_{ref,i}+(1-\alpha)\cdot k\cdot u_i \\ u_i=\sum_{j=1}^{i}\dot{x}_{ref,j}\begin{bmatrix}\bar{N}\left(\dot{x}_{ref,j}\right)\bar{M}\left(u_j-\delta\right)+\ldots \\ M\left(\dot{x}_{ref,j}\right)N\left(u_j+\delta\right)\end{bmatrix}dt \end{cases} \tag{7.7}$$

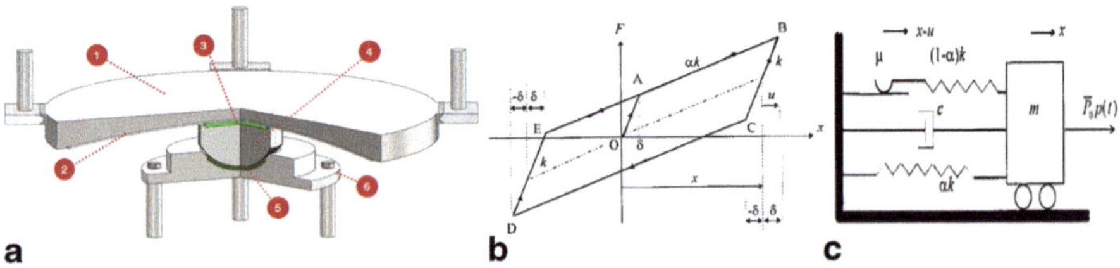

Fig. 7.10 Representation of a SDoF bilinear hysteretic system. **a** single sliding surface friction pendulum system (courtesy of ALGA Spa). **b** Hysteretic loop for the bilinear system. **c** bilinear S-DoF system after Mostaghel (1999)

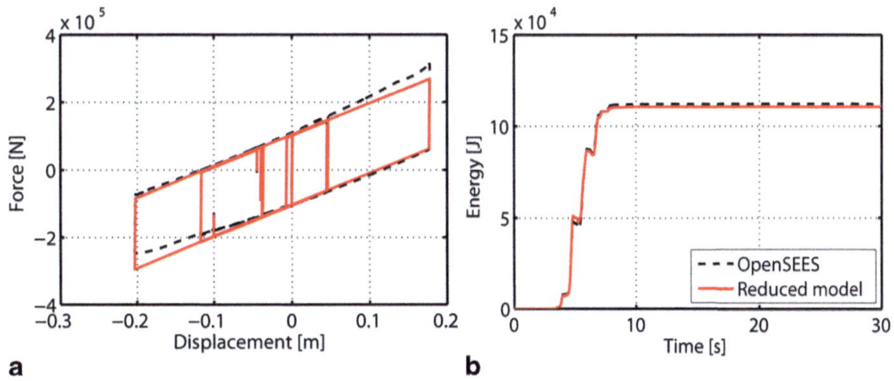

Fig. 7.11 Validation of the reduced isolator element in Pier #9 at ULS: **a** Hysteretic loop and **b** Time history of dissipated energy

where r_{red} is the restoring force vector from the simplified bilinear model, r_{ref} is the restoring force vector from the OpenSEES simulation, x_{ref} is the relative displacement vector from OpenSEES, $\dot{x}_{ref}$ is the relative velocity vector from OpenSEES. Identified parameters read,

$$\hat{k} = 2.03e8\ N/m,\ \hat{\alpha} = 0.0046 \text{ and } \hat{\delta} = 0.00050\ m$$

Figure 7.11 compares the hysteretic loop and the dissipated energy history of the reduced and the OpenSEES isolator element located on Pier #9 at ULS.

The same quality of matching was achieved for all piers. Although the effect of variable vertical loading was neglected, simplified bilinear models reproduced satisfactorily the complex behaviour of OpenSEES isolator elements.

7.5 Validation of the Reduced Model of the Bridge

The validation of the nonlinear reduction of the OpenSEES RM is discussed herein. With reference to the OpenSEES RM solution, NRMSE was introduced as a dimensionless error measure on kinematic quantities, such as displacements, velocities and accelerations. Tables 7.3 and 7.4 summarized NRMSEs on kinematic quantities in the transverse direction obtained from time history analyses conducted at both limit states.

With regard to the isolated case, piers were supposed to remain in the linear regime. Consequently, the nonlinear extension based on the modified Bouc-Wen spring was not needed and all nonlinearities were concentrated on isolators.

With reference to Pier #9, Figs. 7.12 and 7.13 compare the cap beam level transversal displacements obtained from the global reduced model of the bridge with the OpenSEES reference solution:

Table 7.3 NRMSE values for the non-isolated case

Pier	SLS			ULS		
	Disp. [m]	Vel. [m/s]	Acc. [m/s^2]	Disp. [m]	Vel. [m/s]	Acc. [m/s^2]
1	0.06	0.06	0.06	0.11	0.06	0.04
2	0.05	0.05	0.05	0.07	0.06	0.03
3	0.05	0.05	0.05	0.06	0.05	0.03
4	0.05	0.05	0.05	0.06	0.05	0.03
5	0.05	0.05	0.05	0.06	0.05	0.03
6	0.05	0.05	0.05	0.06	0.05	0.03
7	0.03	0.03	0.03	0.05	0.03	0.02
8	0.03	0.03	0.03	0.05	0.03	0.02
9	0.03	0.03	0.03	0.08	0.03	0.02
10	0.05	0.05	0.05	0.13	0.04	0.02
11	0.04	0.04	0.04	0.15	0.04	0.03
12	0.04	0.04	0.04	0.08	0.03	0.02

Table 7.4 NRMSE values for the isolated case

Pier	SLS			ULS		
	Disp. [m]	Vel. [m/s]	Acc. [m/s^2]	Disp. [m]	Vel. [m/s]	Acc. [m/s^2]
1	0.09	0.07	0.08	0.06	0.08	0.07
2	0.05	0.05	0.05	0.04	0.05	0.06
3	0.06	0.04	0.05	0.09	0.05	0.08
4	0.08	0.07	0.08	0.06	0.07	0.08
5	0.10	0.09	0.07	0.07	0.08	0.08
6	0.06	0.07	0.08	0.07	0.08	0.09
7	0.06	0.07	0.07	0.08	0.08	0.08
8	0.06	0.06	0.07	0.07	0.08	0.08
9	0.09	0.09	0.06	0.07	0.08	0.08
10	0.07	0.08	0.08	0.06	0.06	0.06
11	0.10	0.07	0.07	0.07	0.07	0.08
12	0.07	0.06	0.06	0.06	0.07	0.08

Figures 7.12 and 7.13 qualitatively confirm the NRMEs collected in Tables 7.3 and 7.4, which emphasize a slight degradation in the quality of the solution for the isolate case. The different mass matrix formulations of the OpenSEES and the reduced models can explain this phenomenon. In fact, to avoid matrix ill-conditioned cases, a consistent formulation was adopted for the ANSYS based reduced model, whilst OpenSEES allows for lumped mass matrices only. Since hysteretic damping does not occurs in the linear range, the dynamic response of higher eigenmodes, sensitive to small mass changes, deteriorate kinematic matching. According to

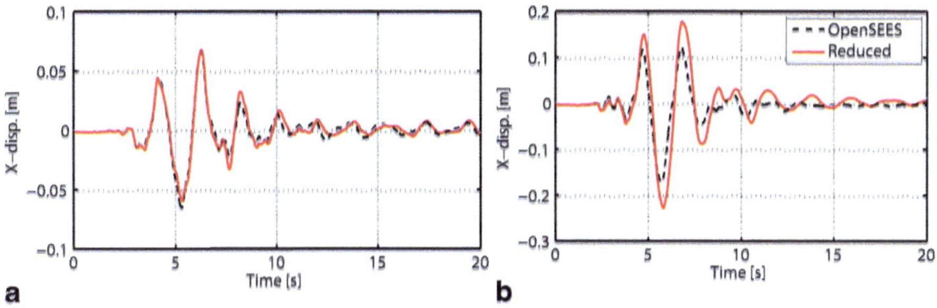

Fig. 7.12 Comparison between OpenSEES and the reduced model: Cap beam level transversal displacement response of Pier #9 in the non-isolated case at **a** SLS and **b** ULS

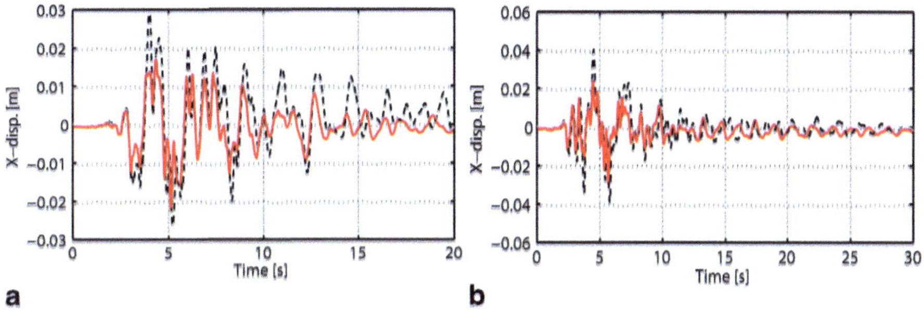

Fig. 7.13 Comparison between OpenSEES and the reduced model: Cap beam level transversal displacement response of Pier #9 in the isolated case at **a** SLS and **b** ULS

Section 7.2, the proposed reduced model is suitable for the foreseen experimental implementation where only scaled specimens of Piers #9 and #11 replace relevant superelements in the NS as depicted in Fig. 7.3.

7.6 Simulation of a Consistent Degradation of the Bridge

To date, only a few attempts aimed at simulating a consistent degradation of PS and NS during a PsD test campaign have been carried out. In particular, the work of Kwon et al. 2013 and Yang et al. 2012, were limited to numerical simulations. An offline testing procedure is proposed herein to take into account degradation. In order to propagate the damage experienced by specimens to the overall emulated system, hybrid simulations are performed at increasing levels of PGA with offline model updating of the NSs according to PS data from one run to the next. Two-dimensional OpenSEES fiber based FE models of experimental piers are taken as reference for damage identification. In particular, the displacements measured at the cap beam level of the physical pier at a generic PsD test run are imposed on

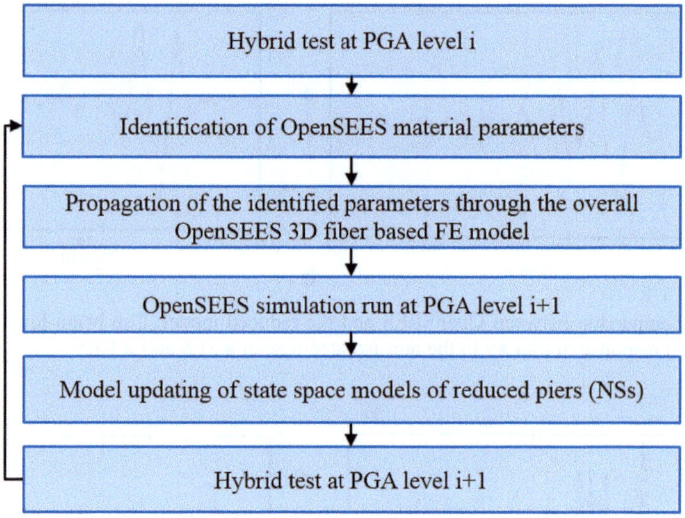

Fig. 7.14 Recursive identification and model updating tasks

the OpenSEES model of the specimen through a nonlinear static analysis; then a penalty function is defined as the NRMSE of the calculated restoring force with respect to the measured one. The minimization of such penalty function leads to the estimation of the concrete yielding strength parameter of the *concrete01* OpenSEES material. In fact, after an accurate sensitivity analysis conducted with respect to all the material parameters, fpc was considered as an effective measure of the degradation level. This implies that the mechanical properties of reinforcing bars are known and fixed in the analysis. fpc can be easily applied to the overall OpenSEES RM in order to simulate the dynamic response of the bridge from one PsD test to the next. The flowchart depicted in Fig. 7.14 summarizes the main steps of the proposed updating procedure:

7.7 Conclusions

The assessment of the seismic performance of an old 400 m span reinforced concrete bridgef by means of continuous-time hybrid simulations was conceived within the RETRO project within the framework of Transnational Access of the SERIES FP7 project. A complex fiber-based FE OpenSEES model of the viaduct was set to support the PsD test design. Nonetheless, implementation issues relevant to the typical solving time of the Numerical Substructure (NS) dictated by the controller time step rendered complex FE fiber models not suitable for testing purposes. Therefore, a rational design of a suitable NS was conceived; it was based on a model reduction of the whole emulated bridge. In order to provide stiffness and mass matrices an additional ANSYS Linear Model (LM) was produced. Moreover, a deep study aimed at discarding out-of-plane displacements of piers was addressed to set internal constraints. As a result, nonlinear subcomponents, i.e. piers and isolators, were reduced to standalone subsystems. Resulting state space models were deemed as suitable for time integration with deterministic solving time. In particular, 3-DoFs pier plane superelements were extended to the nonlinear range by means of a modified Bouc-Wen spring, whilst isolators were synthetized with the bilinear hysteretic model suggested by Mostaghel (1999). According to previous analyses of Paolacci and Giannini (2012) a linear deck was considered. Time history analyses both at serviceability and ultimate limit states proved the suitability of the reduction with respect to the OpenSEES Reference Model (RM). Moreover, the component wise reduction approach made the accommodation of physical specimens straightforward according to the experimental set-up. Finally, a testing procedure aimed at simulating a consistent degradation of physical and numerical piers, which was based on recursive identification and tuning sessions, was proposed.

Acknowledgement The research leading to these results has received funding from the European Community's Seventh Framework Programme [FP7/2007–2013] for access to the EUROPEAN LABORATORY FOR STRUCTURAL ASSESSMENT of the EUROPEAN COMMISSION—JOINT RESEARCH CENTRE under grant agreement n° 227887. Opinions are those of the authors and do not represent those of the European Commission.

References

Bonelli A, Bursi OS, He L, Magonette G, Pegon P (2008) Convergence analysis of a parallel interfield method for heterogeneous simulations with substructuring. IJNME 75(7):800–825

Craig R, Bampton M (1968) Coupling of substructures in dynamic analysis, *American Institute of Aeronautics and Astronautics Journal*, Vol. 6 (7)

Eurocode 8-Part 2, (2006) Design of Structures for Earthquake Resistance: Bridges

Fardis MN (2009) Seismic Engineering Research Infrastructures for European synergies, SERIES project, Grant number: 227887, 2009–2013

Guyan RJ (1965) Reduction of stiffness and mass matrices, *American Institute of Aeronautics and Astronautics Journal*, 3(2), 380

Kent DC, Park R (1971) Flexural members with confined concrete. J Struct Div ASCE 97:1969–1990

Kwon OS, Kammula V (2013) Model updating method for substructure pseudo-dynamic hybrid simulation. Earthquake Engineering & Structural Dynamics, 42(13), 1971–1984. doi:10.1002/eqe.2307

Kumar S, Itoh Y, Saizuka K, Usami T (1997) Pseudodynamic testing of scaled models. J Struct Eng 123(4):524–526. doi:10.1061/(ASCE)0733-9445(1997)123:4(524)

Kwon OS, Kammula V (2013) Model updating method for substructure pseudo-dynamic hybrid simulation. Earthquake Engineering & Structural Dynamics, 42(13), 1971–1984. doi:10.1002/eqe.2307

Menegotto M, Pinto PE (1973) Method of analysis for cyclically loaded R.C. plane frames including change in geometry and non-linear behavior of elements under combined normal force and bending. Proc. of IABSE symposium on resistance and ultimate deformability of structures acted on by well defined repeated loads, vol 13, 1973; 15–22

Mostaghel N (1999) Analytical description of pinching, degrading hysteretic systems. J Eng Mech 125(2):216–224. doi:10.1061/(ASCE) 0733-9399 (1999) 125:2(216)

OpenSees (2009) Open system for earthquake engineering simulation. http://opensees.berkeley.edu. Accessed 10 Jan 2011

Paolacci F, Giannini R (2012) An experimental and numerical investigation on the cyclic response of a portal frame pier belonging to an old reinforced concrete viaduct. Earthquake Engineering & Structural Dynamics, 41(6), 1109–1127. doi:10.1002/eqe.1175

Pegon P, Magonette G (2002) Continuous PSD testing with nonlinear substructuring: presentation of a stable parallel inter-field procedure. Technical report 1.02.167, E.C., JRC, ELSA, Ispra, Italy

Taucer F (2011) Seismic engineering research infrastructures for European synergies: activities of the Joint Research Centre in large scale experimental testing within the framework of the eurocodes. In: 7th International conference on structural defects and rehabilitation, Fortaleza, Brazil, 2–4 June 2011

Yang YS, Tsai KC, Elnashai AS, Hsieh TJ (2012) An online optimization method for bridge dynamic hybrid simulations. Simul Model Pract Theory 28:42–54. doi:10.1016/j.simpat.2012.06.002

Chapter 8
Geographically Distributed Continuous Hybrid Simulation Tests Using Shaking Tables

Ferran Obón Santacana and Uwe E. Dorka

8.1 Introduction

Geographically distributed testing has been one important goal of the FP7 SERIES project. The envisioned platform was designed to be able to deal with different algorithms and communication protocols so that its users or the facilities participating in it would not be constrained by these aspects. The platform also goes beyond the mere protocol and algorithm freedom, and should prove the possibility to perform continuous time-scaled distributed tests with non-linear substructures, including large numerical models. With this in mind, several activities were performed at the University of Kassel with major institutions in the Earthquake Engineering field, not only within Europe but also worldwide: University of Oxford in the United Kingdom, the University of California at Berkeley (USA), the Hybrid Simulation Testing Center (HYSTEC) in South Korea and the National Research Center for Earthquake Engineering (NCREE) in Taiwan. These facilities were also selected due to their participation in the creation of the communication softwares that are being used within the community: OpenFresco and NSEP.

The *continuous* time-scaled distributed tests that were performed included a non-linear experimental substructure located at the University of Kassel and a simple numerical model at the different "remote" or numerical facilities. An efficient substructure algorithm was used in order to perform the time integration, while the communication and the exchange of data was assigned to OpenFresco or to the NSEP protocols.

F. Obón Santacana (✉) · U. E. Dorka
Steel and Composite Section, Department of Civil and Environmental Engineering,
University of Kassel, Kassel, Germany
e-mail: ferran.obon@uni-kassel.de

U. E. Dorka
e-mail: uwe.dorka@uni-kassel.de

F. Taucer, R. Apostolska (eds.), *Experimental Research in Earthquake Engineering*,
Geotechnical, Geological and Earthquake Engineering 35,
DOI 10.1007/978-3-319-10136-1_8

8.2 Dorka's Substructure Algorithm

One of the basic elements of the success on having performed continuous distributed tests was the use of Dorka's substructure algorithm which has been successfully applied in both civil and aerospace engineering fields (Bayer et al. 2005; Dorka and Heiland 1991; Dorka and Füllekrug 1998; Dorka 2002; Dorka et al. 2006, 2007; Roik and Dorka 1989). The underlying layer of the substructure algorithm is a discrete formulation of the dynamic equilibrium equation, which is given by:

$$M\frac{d^2u}{dt}+C\frac{du}{dt}+Ku=p(t)+f_r+f_s \tag{8.1}$$

where *M*, *C* and *K* are respectively the mass, damping and stiffness matrices of the numerical model and *u* and *p(t)* are the displacement and loading vectors. The terms f_r (numerical non-linear restoring force) and f_s (force measured at the interface between numerical model and substructure specimen) take care of the restoring forces coming from substructures.

A general solution of the previous equation can be obtained if a finite element discretisation in the time domain and shape functions with three supporting points[1] (3-step scheme) as in (Zienkiewicz 1977) are used. It is worth to point out that nearly all major time stepping algorithms can be obtained if different weighting functions are used when performing the time integration. If the displacement is discretised in the time domain, Eq. (8.1) results into:

$$u^{n+1} = K_e^{-1}\cdot\left\{f_{eff}^n + {f_*}^n + {f_e}^n\right\} \tag{8.2}$$

$$K_e^{-1} = K + a_0M + a_1C \tag{8.3}$$

$$f_{eff}^n = M(a_0u^n + a_2\dot{u}^n + a_3\ddot{u}^n) + C(a_0u^n + a_4\dot{u}^n + a_5\ddot{u}^n) \tag{8.4}$$

where $f_*=f_r+f_s+p$, *u* is the discretised displacement vector in the time domain, K_e is the effective stiffness matrix, f_{eff} is the effective load vector and f_e is the error force (see Sect. 8.2.2). The super-index *n* is used to differentiate the time steps. The abbreviations introduced are:

$$a_0=\frac{1}{\beta\Delta t^2};\ a_1=\frac{\gamma}{\beta\Delta t};\ a_2=\frac{1}{\beta\Delta t};\ a_3=\frac{1}{2\beta}-1\triangleright;\ a_4=\frac{\gamma}{\beta}-1;\ a_5=\frac{\Delta t}{2}\left(\frac{\gamma}{\beta}-2\right)$$

The different parameters such as time intervals, stability, accuracy and numerical damping are analysed in a deeper way in (Dorka 2002). In conclusion, the only unconditionally stable implicit 3-step algorithm without artificial damping and with the least numerical softening is the so-called *Newmark-β* (γ=0.5; *β*=0.25).

[1] The formulation can also be extended to four supporting points.

8.2.1 Sub-Stepping

In Eq. (8.2) there is also the term f_s^{n+1} (inside f_*^n) which takes care of the reaction of the subsystem. Since it is only available through measurement, the use of iteration methods leads to high frequency oscillations during a test. To deal with this problem, a sub-stepping approach has been used. In (Dorka 2002) in Eq. (8.5) the implicit algorithm is expressed as a form of a general linear control equation within each time step.

$$u^{n+1} = u_0{}^{n+1} + K_e{}^{-1}(f_s^{n+1} + f_r^{n+1}) \tag{8.5}$$

$$u_0^{n+1} = K_e^{-1} \cdot \{f_{eff}^n + p^{n+1} + f_e{}^n\} \tag{8.6}$$

where u_0 is the initial vector (updated at the beginning of the time step) and f_r, f_s the calculated and measured force vectors respectively. By applying a sub-stepping technique, the displacements present in Eq. (8.5) are introduced to the specimen in small increments as specified in the ramp function, forming the inner loop in Fig. 8.1b:

$$x^t = x^{t-\Delta t}\left(1 - \frac{i}{k_{sub}}\right) + x^t\left(\frac{i}{k_{sub}}\right) \tag{8.7}$$

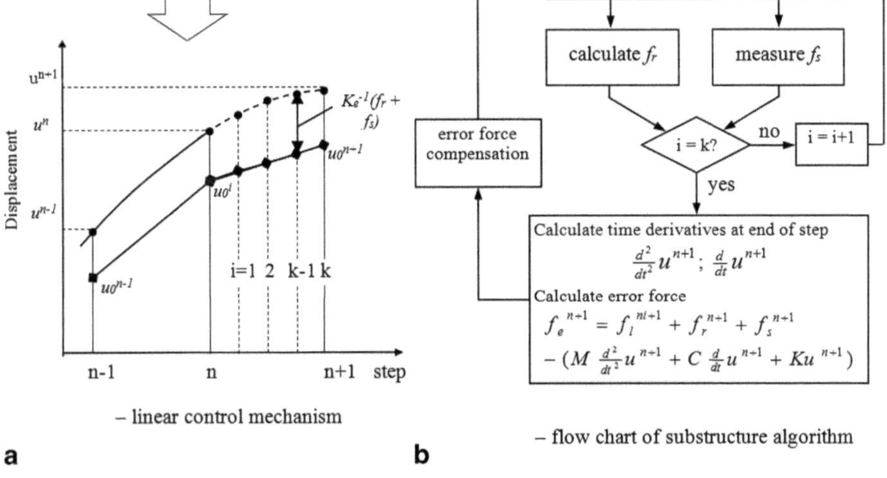

Fig. 8.1 Flowchart of Dorka's substructure algorithm

Where k_{sub} is the number of sub-steps and i denotes the current sub-step. In this way, for each sub-step i, the displacement of the subsystem is computed, applied to the test and then the measured; the calculated non-linear restoring forces are given as an input to the algorithm.

8.2.2 Error Force Compensation

The number of sub-steps k_{sub} (Fig. 8.1a) is an important factor for stability and accuracy (when $k_{sub} \rightarrow \infty$ Eq. (8.5) and (8.6) give the exact value), but there are also other sources like the positioning error of actuators, that can make the test unstable. In order to minimise their effect and their tendency to destabilise the test, a PID compensator is used to minimise the error:

$$f_e^n = -P\left[e^n + I\Delta t\sum_i^n e^n + \frac{D}{\Delta t}(e^n - e^{n-1})\right] \tag{8.8}$$

where e is the equilibrium error defined as the sum of all dynamic forces, whether they are internal or external, at the end of the time step. P, I and D are adequately chosen proportional, integral and derivative constants of the error compensator, and f_e^n is the compensation force, which is added at the beginning of the next step as a load to the system. Figure 8.1b summarises how the algorithm runs in displacement control.[2]

8.2.3 Adaptation of the Algorithm for Multiple Testing Sites

To perform distributed testing the algorithm must be adapted in order to accommodate the necessary data exchange between facilities. Since the restoring forces are treated separately this is achieved by splitting the algorithm into two natural parts: the computation of the explicit displacement and time derivatives (in the numerical facility) and the sub-stepping part (the experimental facility), as shown schematically in Fig. 8.2.

The exchange of data requires however some consideration due to the nature of the matrices and vectors involved in this process. Since the restoring forces are vectors with all the elements equal to zero except in those degrees of freedom where these forces are applied, there is no need to send the whole vector, only the relevant parts. This effectively decreases the amount of data to be exchanged and reduces the order of the operations to be performed in the sub-stepping process. This also applies to the gain matrix, K_e^{-1}, sent to the experimental facility only at the beginning of the test. The sub-stepping process can be rewritten as:

[2] It is possible to formulate it also in velocity and acceleration control.

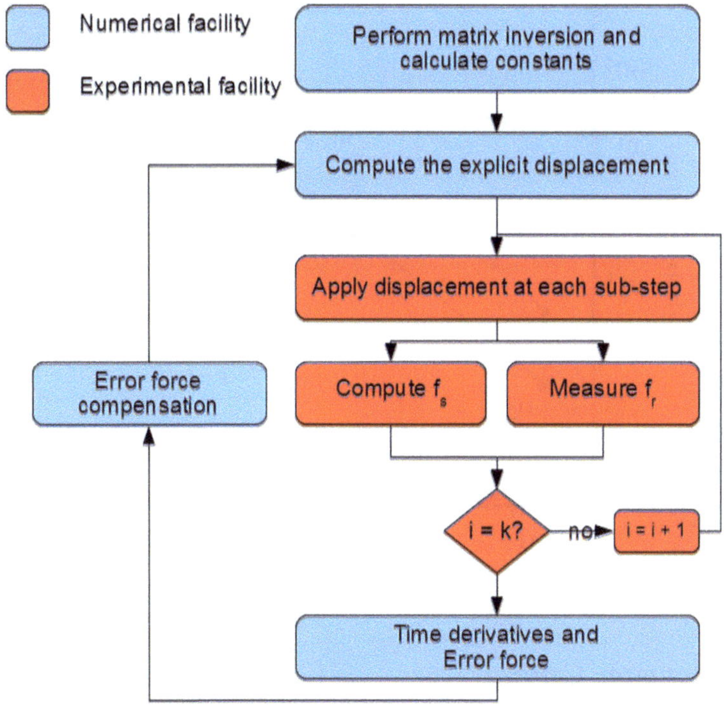

Fig. 8.2 Adapting Dorka's substructure algorithm for multiple testing sites

$$\tilde{u}^{n+1} = \tilde{u}_0^n \left(1 - \frac{j}{k_{sub}}\right) + \tilde{u}_0^{n+1}\left(\frac{j}{k}\right) + \tilde{K}_e^{-1}\left(\tilde{f}_r + \tilde{f}_s\right) \tag{8.9}$$

Where the superscript "~" denotes the sub-matrix or sub-vector that contains the coupling nodes. From here there are several possibilities for the data communication depending on the number of experimental facilities:

- There is one experimental facility with one or more subsystems: The communication can be performed at the step level, that is, the data exchange takes place after computing the explicit displacement and after finishing the sub-stepping process.
- There are multiple experimental facilities: the data has to be exchanged between the different experimental facilities at the sub-step level due to the presence of off-diagonal coefficients in the gain matrix which describe the interaction between the subsystems.
- Note that if the influence between the substructures is negligible (the off-diagonal coefficients are several orders of magnitude smaller than the ones in the diagonal) the communication can still be performed at the step level as described above.

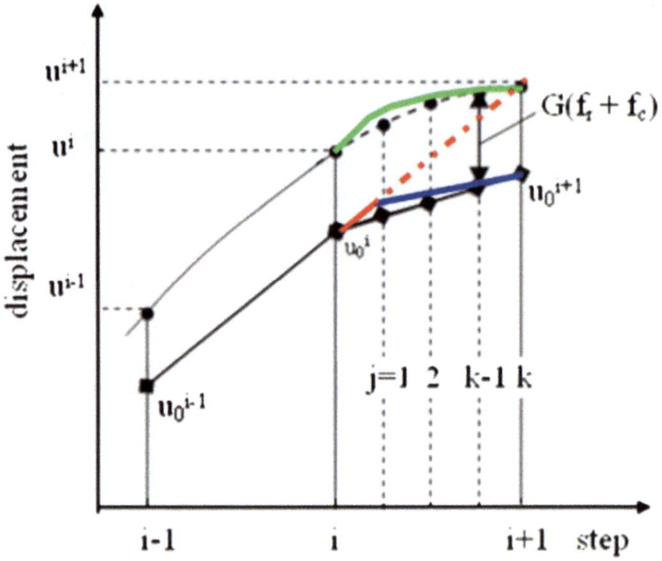

Fig. 8.3 Special actuator control for continuous distributed testing

8.2.4 Actuator Control

Since the data is exchanged between the two facilities through the internet and the speed is not constant (jitter), the hydraulic actuators require special attention to ensure a smooth and continuous motion throughout the test. In the case of not receiving the required displacement update at the beginning of the step, the actuator will continue with the ramp function that was present at the end of the previous sub-stepping process (red line in Fig. 8.3) and when an update is available it will move to the desired point (blue line). Although this may introduce some error it is negligible and it will be taken care of by the error compensation. This is because only the last displacement and coupling force of the sub-stepping loop are sent to the numerical facility for time derivatives and error force compensation calculations.

8.3 Continuous Time-Scaled Geographically Distributed Tests with Non-Linear Experimental Substructures

8.3.1 Description of the Test Set-Up

The tests were performed using the shaking table present in the laboratory of the University of Kassel (UNIKA). The test set-up consisted of a friction device, the UHYDE-*fbr*, and a Tuned Mass Damper (TMD). However, since the time scaled

Fig. 8.4 Test set-up at the University of Kassel. (See Table 8.1)

tests cannot accommodate dynamic masses, the TMD was fixed using four bracings to restrain movements in the x and y direction. This leaves the friction device as the only active device in the test set-up. The table was driven by a single actuator with a capacity of 200 kN (*x-direction*). Figures 8.4 and 8.5 show the test set-up while Table 8.1 summarises the components and sensors used.

The actuator, the sub-stepping process and the data acquisition are delegated to the ADwin Pro II system (ADwin 2013). This electronic equipment features a real-time operating system and multitasking capabilities as well as digital and analog inputs/outputs. It exchanges data, using the UDP protocol, with the computer used for communication purposes (the one running OpenFresco or PNSE) through an Ethernet cable. This was purely a security decision, since complete isolation from the out-side world was desired for the controller. This introduces another layer of communication, with a negligible impact on performance, since the communication time is between 0.2 and 0.3 *ms*.

8.3.2 The Friction Device UHYDE-fbr

The UHYDE-*fbr*, Uwe's Hysteretic Device, *f* for friction and *br* for bridges, is a friction device that dissipates energy as a result of solid friction. It was developed

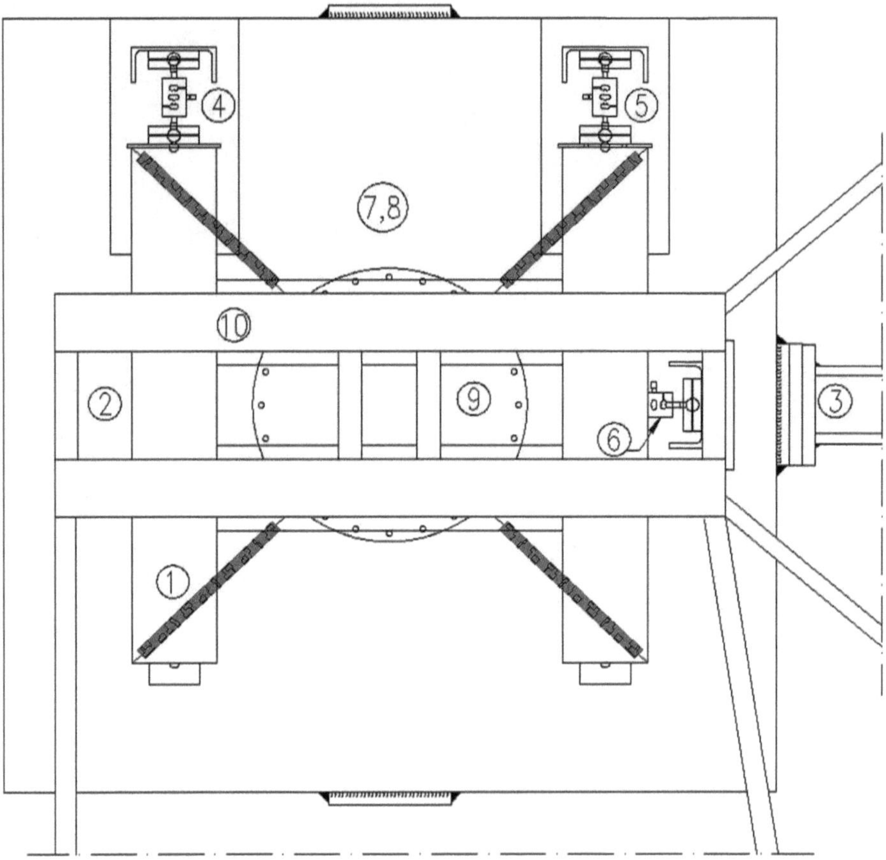

Fig. 8.5 Position of the sensors and components of the test set-up

by Uwe E Dorka for the protection of structural systems against dynamic actions. As can be seen in Fig. 8.6, there are two steel plates and a group of bronze inserts in the patented sliding mechanism (Dorka 1995).

This device shows an ideal friction behaviour, with the friction force easily adjusted varying the gas pressure in the chamber. If this device is made semi-active (a pressure control is implemented) a range of different force-displacement characteristics can be achieved.

Table 8.1 List of sensors and elements

Label	Name	Location	Model	Range	Sensitivity
Displacement gauges					
1	DTMDY	TMD (y-direction)	HBM W10TK	± 10 mm	80 mV/ (V * 10 mm)
2	DTMDX	TMD (x-direction)	HBM W100K	± 100 mm	80 mV/ (V * 100 mm)
3	DCYLX	Cylinder (x-direction)			
Force gauges					
4	LCY1	y-direction	HBM S9	± 10 kN	2 mV/ (V * 10 kN)
5	LCY2				
6	LCX	x-direction			
Pressure sensors					
7	PMEAS	Valve		2 bar	
8	PCTRL	Valve		2 bar	
Other components					
3	Hydraulic cylinder in the x-direction (Sandner 2013)				
9	Friction device UHYDE-fbr				
10	Tuned mass damper				
11	Steel braces				

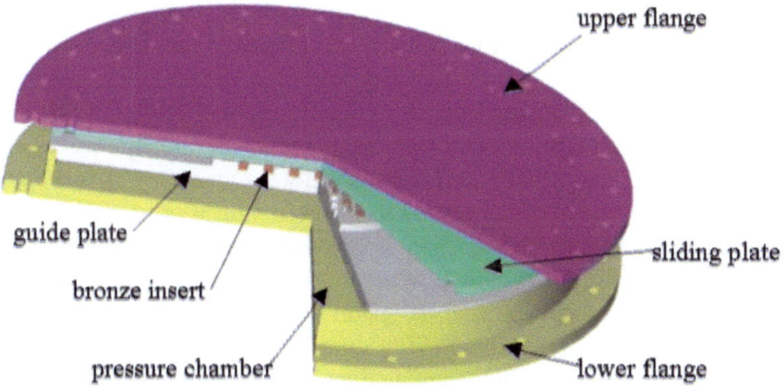

Fig. 8.6 The UHYDE-*fbr*

8.3.3 Numerical Models

Two experiments were initially envisioned. The first would help in validating the actuator control and the software set-up, while the second would be targeted at performing continuous geographically distributed hybrid tests. The numerical model of

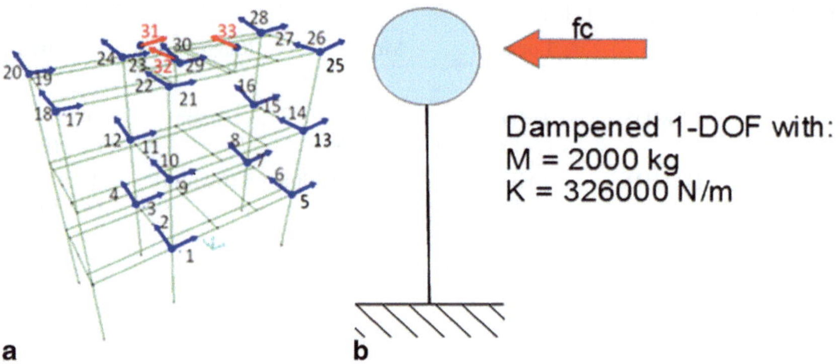

Fig. 8.7 Numerical models: steel frame (Khanlou et al. 2011) and SDOF

a steel frame with 33 degrees of freedom (Fig. 8.7, left) was used for the first experiment, while a damped single degree of freedom model with the friction device acting on it was used for substructure testing in the second experiment (Fig. 8.7, right).

8.3.4 Continuous Geographically Distributed Tests Using OpenFresco and NSEP Protocol

For the tests, two popular communication softwares were tested: the Open-source Framework for Experimental Setup and Control (OpenFresco) and the Networked Structural Experiment Protocol (NSEP).

OpenFresco is a NEES software actively developed at the University of Berkeley in California. It provides the necessary layers in order to perform both local and distributed hybrid simulation using an object oriented programming approach (C++). OpenFresco features a multi-tier client-server architecture as described in (Schellenberg et al. 2008, 2009, 2011) and integrates OpenSSL for secure communications. For the purpose of the tests, OpenFresco v2.6 was used in both sites to avoid compatibility problems (see Chap. 4 in (Obón Santacana and Dorka 2012b)). To communicate with the ADwin system, the GenericTCP experimental control facility from OpenFresco was used as specified in (Obón Santacana and Dorka 2012b). It is worth noting that this facility does not depend on a specific hardware since it connects to the controller using a client/server approach.

On the other side, the NSEP protocol was developed during the ISEE (Internet-based Simulation on Earthquake Engineering) project in the National Research Center for Research on Earthquake Engineering in Taiwan. It was designed to perform geographically distributed tests and it is written in C++ and features a star topology network. After a redesign in 2011, the integration of other facilities and hardware became more flexible. NSEP uses the Transmission Control Protocol/Internet Protocol (TCP/IP), designed to guarantee reliable data transfer between different hosts, to communicate with different components or modules assuming single precision

Table 8.2 Average round trip time with University of Kassel

Facility	Average round trip time (ms)
University of Oxford	38 ms
University of California	197 ms
National Research Center for Earthquake Engineering	343 ms
Hybrid Simulation Testing Center	382 ms

data types for floating point operations. For a more in-depth documentation on how-to implement the NSEP protocol the reader should refer to (Wang et al. 2007; Wang 2011). Table 8.2 summarises, independently of the protocol used, the average round trip time between the University of Kassel and the different institutions that participated in the test.

8.3.4.1 Validation of the Actuator Control with HYSTEC

With HYSTEC, the numerical model of the steel frame (Fig. 8.7) was used to evaluate the performance and the control of actuators using internet in order to exchange data. Due to the slow motion, the gain value of the PID control in the actuator had to be decreased and therefore positioning errors were encountered as shown in Fig. 8.8. Through the use of adaptive phase lag compensation (Nguyen and Dorka 2008, 2009; Nguyen et al. 2011), and the number of control steps (a control step is a sub-division in time between the displacement updates applied to the cylinder), this position error was reduced during the tests as shown in Fig. 8.9.

8.3.4.2 Continuous Time-Scaled Continuous Tests

The SDOF system was used for sub-structure testing, using the same sine-sweep excitation with each facility as summarised in Table 8.3. During the first test, the pressure in the air chamber at the UHYDE-*fbr* was kept constant at 0.25 bar, while on the second test the pressure was suddenly dropped to 0.1 bar to simulate a *breakage*. This translated into a sudden force drop and thinner hysteresis loops (Fig. 8.9).

Note that although the UHYDE-*fbr* device dissipates energy through solid friction, four springs are installed in the TMD. This explains the "post-yield" stiffness effect shown in Fig. 8.9. The reader should refer to (Obón Santacana and Dorka 2013) for the complete set of results.

It is noted that with the NSEP protocol, the advantage of having a smaller round trip time could not be exploited and the test had to always run at the speed of 500 ms per sub-step. Both protocols yield, as expected, the same results, as shown in Fig. 8.10.

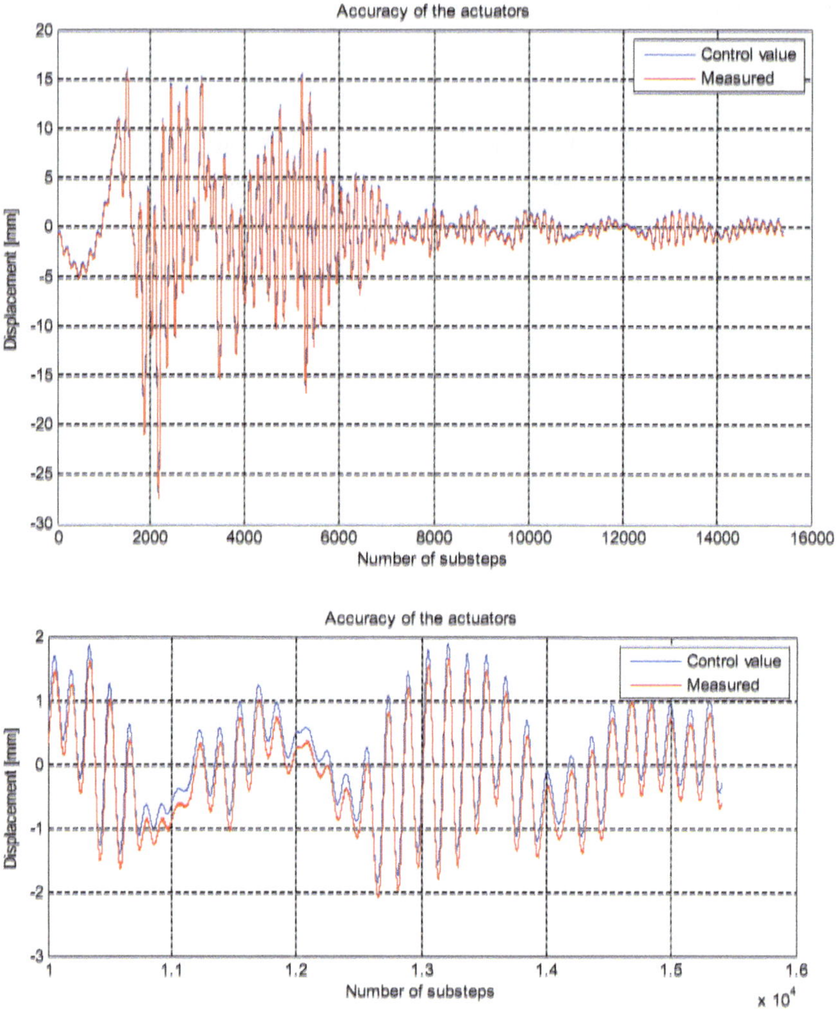

Fig. 8.8 Error positioning of the actuators at UNIKA

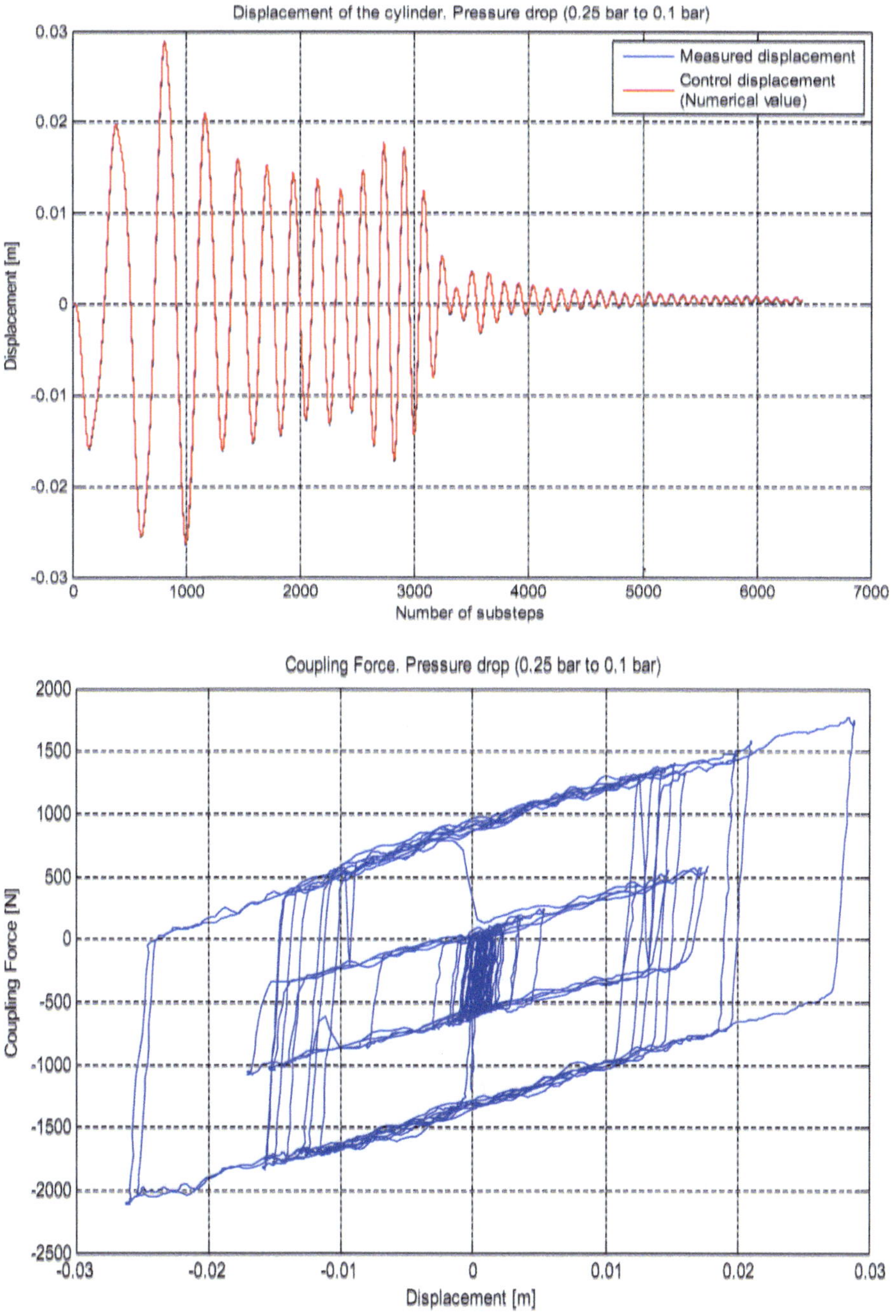

Fig. 8.9 Tests with the University of California and sudden pressure drop in the friction device at 2500 sub-steps

Table 8.3 Protocols tested

Institution	Communication software tested
University of Oxford	OpenFresco, NSEP
University of California	OpenFresco
National Research Center for Earthquake Engineering	NSEP

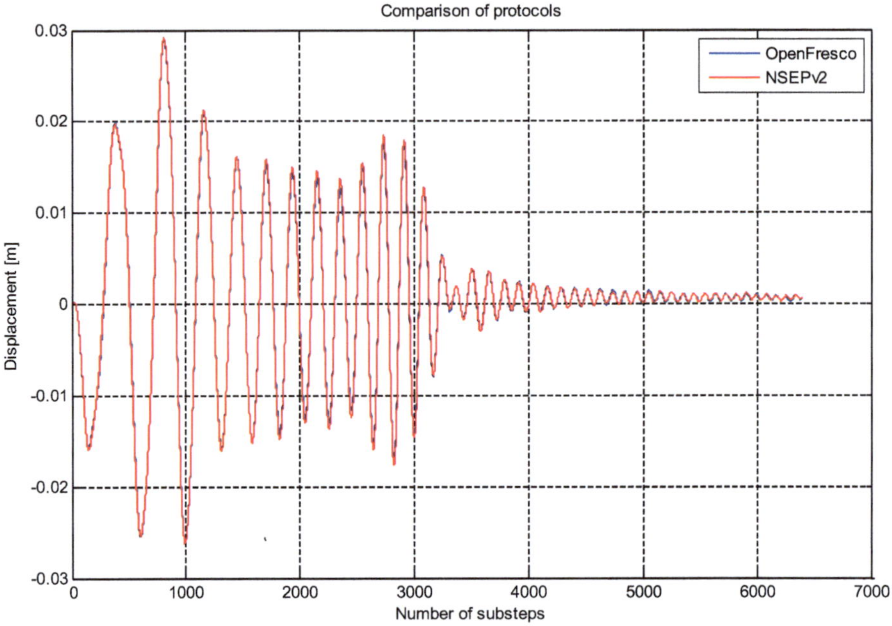

Fig. 8.10 Comparison between NSEP and OpenFresco. Pressure drop in the UHYDE-*fbr*

8.4 Large Numerical Models in Continuous Hybrid Simulation

As a prove of concept and in order to assess the extensibility of the platform to large and complex numerical models, purely numerical simulations were conducted at the University of Kassel with the collaboration of the Middle East Technical University (METU). These preliminary tests included all the electronic equipment present on the previous tests, leaving out only the hydraulic cylinders.

The preparation of the tests was conducted in two parts. The first phase was a parallel adaptation of Dorka's substructure algorithm in order to take advantage of multiple cores and high performance computers. As shown in (Obón Santacana and Dorka 2011a, b, 2012a) this allowed using large and complex models with several thousands of degrees of freedom. The second phase was a collaboration

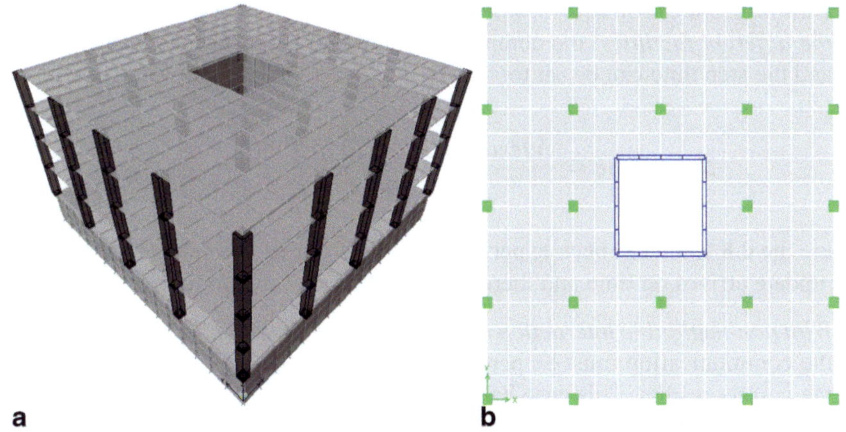

Fig. 8.11 A five storey building with 10,000 equations (provided by METU)

Table 8.4 Required step-time for different numerical model sizes

Size of the numerical models	Step-time	
	Regular desktop	Linux cluster
1000 DOF	0.7ms	–
10,000 DOF	250ms	87ms
50,000 DOF	Not possible	452ms

with METU, who provided numerical models with different complexity and with multiple substructures (UHYDE-*fbr* devices), as shown in Fig. 8.11.

The numerical tests were performed using the numerical models presented in (Obón Santacana and Dorka 2013) using only eight nodes with 32 cores each of the Linux Cluster at the University of Kassel and the Intel MKL libraries and a regular desktop with an Intel core i5 using the OpenBLAS libraries. A custom TCP/IP protocol was implemented in order to exchange data. Table 8.4 summarises the required step-time:

8.5 Conclusions

Through the use of a single degree of freedom system and the UHYDE-*fbr* non-linear device, it has been proven that it is possible to perform time-scaled *continuous* geographically distributed tests also on the transcontinental scale. The key element, the substructure algorithm developed by Dorka, can be adapted, without much effort, to work with the different available protocols and it offers great flexibility and performance. Depending on the number of experimental facilities and/or the interaction between these subsystems, two situations are possible:

- *With only one substructure or negligible interaction between the subsystems of the different facilities*: the communication could be performed at the step level and the step time can be set to:

$$Step\ Time = \frac{Average\ Round\ Trip\ Time}{0.8} * n_{Substeps}$$

where the 0.8 value has been experimentally determined and selected to ensure that the update arrives on time and therefore actuators move continuously.

- *With non-negligible interaction between the subsystems of the different facilities*: the communication must be performed at the sub-step level in order to achieve the correct results. This case has not been addressed in the current study and it should be analysed in future work.

Two communication protocols were also evaluated and despite of using different floating point precision (NSEP uses single precision while OpenFresco assumes double precision) the results showed no difference in this respect (Fig. 8.10). It cannot be said the same, however, regarding the maximum speed that they offer since the NSEP protocol always had to work with a scale factor of 200 regardless of the distance between the facilities. Using Dorka's substructure algorithm and only one experimental facility, the time scale factors achieved considering 10 ms to be real time, are shown in Table 8.5.

The substructure algorithm also showed a great versatility when being ported to large numerical facilities. Using a custom and in-house built TCP/IP communication protocol and high performance linear algebra libraries, large numerical facilities and complex models were used for the first time in the field of hybrid simulation. In this field, further improvements are still being investigated in order to reduce the amount of RAM memory required to run large models in regular desktops and increase performance altogether through the use of sparse routines and further parallel optimisations.

Table 8.5 Achieved time scale factors

With	Time scale factor
OpenFresco	
University of Oxford (UK)	20
University of California, Berkeley (USA)	100
Hybrid Simulation Testing Center, HYSTEC (S. Korea)	200
NSEP	
University of Oxford (UK)	200
National Center for Research on Earthquake Engineering, NCREE (Taiwan)	

Acknowledgement The research leading to these results has received funding from the European Community's Seventh Framework Programme [FP7/2007-2013] for access to the laboratory of the University of Kassel (UNIKA) under grant agreement n° 227887.

References

ADwin (2013) Jäger Computergesteuerte Messtechnik GmbH, Rheinstraße 2–4, D-64653 Lorsch

Bayer V, Dorka UE et al (2005) On real-time pseudo dynamic sub-structure testing: algorithm, numerical and experiment results. Aerosp Sci Technol 9:223–232

Dorka UE (1995) Friction device for protection of structural systems against dynamic actions. Patent Number 5456047. United States Patent

Dorka UE (2002) Hybrid experimental—numerical simulation of vibrating structures. Proceedings of the International Workshop WAVE 2002: 183–191. Okayama, Japan

Dorka UE, Füllekrug U (1998) Algorithmen für real-time pseudo-dynamische Substrukturtests. Report of the DFG project SubPSD-Algorithmen, project number Do 360/7, University of Kaiserlautern, Germany

Dorka UE, Heiland D (1991) Fast online earthquake simulation using a novel pc supported measurement and control concept. Proceedings 4th Int. Conf. Structural Dynamics, Southhampton: 636–645

Dorka UE, Queval JC et al (2006) Real-time sub-structure testing on distributed shaking tables in CEA Saclay. Proceedings of 4th world conference on structural control and monitoring, San Diego, USA

Dorka UE, Queval JC et al (2007) Substructure testing on distributed shaking tables. Proceedings of the 2nd international conference on Advances in Experimental Structural Engineering (2AESE), Shanghai, China

Khanlou NM, Dorka UE, Ristic D (2011) Sub-structure tests on non-linear TMD using a hydraulic shaking table. The 2011 International Conference on Earthquakes and Structures (ICEAS'11), Seoul, Korea, September 18–23

Nguyen VT, Dorka UE (2008) Phase lag compensation in substructure testing based on online system identification. Proceedings of the the 14th world conference on earthquake engineering, October 12–17, 2008, Beijing, China

Nguyen VT, Dorka UE (2009) Adaptive compensations based on online system identification for real-time substructure testing. Proceedings of the 3AESE conference, San Francisco, USA

Nguyen VT, Dorka UE, Phan TV (2011) Adaptive phase lag compensation in real-time substructure testing of non-linear tuned mass damper using hydraulic shaking table. The 2011 International Conference on Earthquakes and Structures (ICEAS'11), Seoul, Korea, September 18–23

Obón Santacana F, Dorka UE (2011a) Use of large numerical models and high performance computers in geographicall distributed seismic tests. In: Fardis M. Rakicevic, Z (eds) Role of research infrastructures in performance-based earthquake engineering (SERIES Workshop). Springer Netherlands

Obón Santacana F, Dorka UE (2011b) Large numerical models in continuous hybrid simulation. The 2011 International Conference on Earthquakes and Structures (ICEAS'11), Seoul, Korea, September 18–23

Obón Santacana F, Dorka UE (2012a) Effects of large numerical models in continuous hybrid simulation. Proceedings of the the 15th world conference on earthquake engineering, September 12–17, 2012, Lisbon, Portugal

Obón Santacana F, Dorka UE (eds) (2012b) Specifications for distributed testing (connectivity tests, implementation). SERIES Deliverable D2.4

Obón Santacana F, Dorka UE (eds) (2013) Report on distributed testing. SERIES Deliverable D2.6

Roik K, Dorka UE (1989) Fast online earthquake simulation of friction damped systems. SFB151 Report No. 15, Ruhr-University Bochum, Germany

Sandner (2013) Sandner Messtechnik GmbH, Am Winkelgraben 4, 64584 Biebesheim, Germany
Schellenberg A, Kim H et al (2008) Environment independent implementation of a software framework for fast local and geographically distributed hybrid simulations. The 14th world conference on earthquake engineering, Beijing (S16–02-005)
Schellenberg A, Kim H et al (2009) OpenFRESCO command language manual. Version: 2.6
Schellenberg A, Takahashi Y, McKenna F (2011) Architecture & Tcl-Commands. http://nees.org/resources/2698/download/OPFW_Architecture__Tcl_Schellenberg.pdf. Accessed Nov 2014
Wang KJ (2011) PNSE platform for networked structural experiment. http://ws2.ncree.org/~kjwang/NSEP/NSEP.html. Accessed January 2013
Wang KJ, Keh-Chyuan T et al (2007) ISEE: internet-based simulation for earthquake engineering—Part II: the application protocol approach. Earthq Eng Struct Dyn (Wiley InterScience), 36(15):2307–2323
Zienkiewicz OC (1977) The finite element method. McGraw-Hill, Maidenhead

Chapter 9
Hybrid Simulations of a Piping System Based on Model Reduction Techniques

Md Shanin Reza, Giuseppe Abbiati, Alessio Bonelli and Oreste Salvatore Bursi

9.1 Introduction

The highly important role of piping systems, e.g., transferring dangerous fluids like oil and gas, in many industries such as petrochemical plants calls for necessary measures to ensure their safe operations, since a single failure can trigger serious accidental chains. Nevertheless, significant damages of piping systems and their components causing severe consequences both to human lives and to the environment during recent earthquakes have been reported (Sezen and Whittaker 2006; Krausmann et al. 2010; Zare and Wilkson 2010; Paolacci et al. 2013a) which inspired researchers to conduct seismic performance evaluation of these systems/components; see, for example, Touboul et al. (2006), Bursi et al. (2012) and Reza (2013). However, until now few experimental investigations -mainly through shaking table tests- have been performed on such structures at full-scale under realistic seismic loading; see DeGrassi et al. (2008) and Otani et al. (2011). All these tests displayed a good performance of piping systems which remained below yield limits under design level earthquakes, while authors confirmed an over-conservativeness of relevant design standards; see, among others, Touboul et al. (2006), Otani et al. (2011) and Paolacci et al. (2011, 2013b). To this end and in order to evaluate its seismic performance, the University of Trento, Italy, undertook an experimental test campaign on a typical full-scale industrial piping system through pseudo-dynamic and real time tests.

M. S. Reza (✉) · G. Abbiati · A. Bonelli · O. S. Bursi
University of Trento, Trento, Italy
e-mail: Md.Shahin.Reza@unitn.it

G. Abbiati
e-mail: giuseppe.abbiati@unitn.it

A. Bonelli
e-mail: alessio.bonelli@unitn.it

O. S. Bursi
e-mail: oreste.bursi@unitn.it

F. Taucer, R. Apostolska (eds.), *Experimental Research in Earthquake Engineering*,
Geotechnical, Geological and Earthquake Engineering 35,
DOI 10.1007/978-3-319-10136-1_9

Pseudo-Dynamic (PDT) and Real Time (RT) Testing with Dynamic Substructuring (DS) (Mahin and Shing 1985; Shing et al. 1996; Bursi et al. 2008, 2011) are hybrid testing techniques (HSDS) in which the overall response of a structure is evaluated by combining the experimental PS) -which is generally the most critical part of the structure- with the numerical response of a Numerical Substructure(NS).Thus, these methods allow for testing of structures, essentially without a size limit, by the use of actuators and controllers. Such experiments have successfully been performed to test various systems (Melo et al. 2001) including civil structures (Pinto et al. 2004; Braconi et al. 2008a, b). Unlike RTs which are conducted at real time, PDTs are generally carried out at an extended time scale, typically 50–200 times slower than the actual earthquake time, requiring inertia and damping forces to be numerically modelled. However, because rate dependent effects can be neglected for steel elements (Tanaka 2012), thanks to experimental measurements of restoring forces combined with numerical evaluations of other contributions, the PDT method can potentially reproduce the actual response of a structure under dynamic loading.

Nevertheless, due to lumped mass idealization, so far PDTs have been considered inadequate for systems characterized by distributed masses. In fact, to apply the PDT technique to a distributed mass system, the needs for spatial discretization entails a large number of actuators leading to an ineffective test method. However, implementations of hybrid tests on piping systems are challenging mainly because these systems with distributed masses are subjected to distributed earthquake forces. Hence, the work presented in this study brings a novelty in the PDT technique, by enabling its applicability to structures endowed with distributed masses. Relevant hybrid experiments on the piping system were implemented by means of model reduction techniques, which were employed for the reduction of both the PS and earthquake forces to coupling nodes.

A description of the aforementioned test campaign is offered herein. Implementations of PDTs and RTs with DS are presented in detail along with the application and effectiveness of adopted component mode synthesis methods. Finally, the performance of a piping system characterized by an internal pressure of 3.2 MPa under several levels of earthquake loading is analyzed and discussed.

9.2 Main Characteristics and FE Analysis of the Piping System Under Investigation

9.2.1 General Dimensions

A typical industrial piping system placed on a steel support structure illustrated in Fig. 9.1 was considered for this study; its general dimensions and other geometrical properties were taken from DeGrassi et al. (2008). The piping network contained 8 and 6 inch straight pipes, several elbows, a Tee joint and an EN 1092-1 Standard PN 40 weld-neck bolted flange joint. The pipes were of API 5L Gr. X52 material (yield and ultimate strengths: 418 and 554 MPa, respectively) and were filled with water at an internal pressure of 3.2 MPa, corresponding to 80 % of the maximum al-

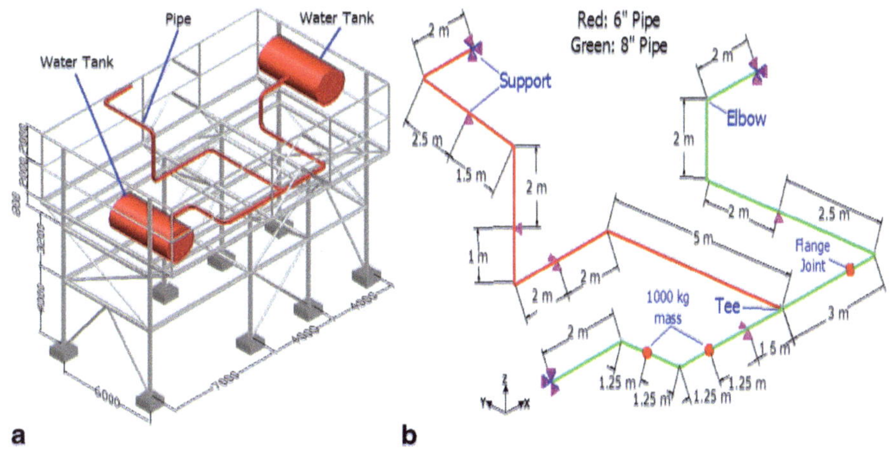

Fig. 9.1 **a** A 3D model of the piping system and the support structure (dimensions are in mm); **b** specifications and dimensions of the piping system after DeGrassi et al. (2008)

lowable pressure of the piping network. Specifications and dimensions of the piping system are presented in Fig. 9.1b.

9.2.2 FE Modelling and Modal Analysis

A 3D FE model of the piping system was developed in ANSYS (2007). Straight beam elements with pipe sections were used to model all pipes including elbows. Two 1000 kg masses, employed to take into account valves etc., were applied in two relevant joints through MASS21 elements. Pipe material density was increased to take into account water mass. Although elbows were modelled using straight elements, flexibilities (see EN 13480-3 (2002)) of these elements were adjusted according to an ABAQUS-based (Hibbit et al. 2003) FE analysis. Each elbow had a radius R equal to 1.5 times the outer diameter d_{out} of connecting pipes; moreover, we

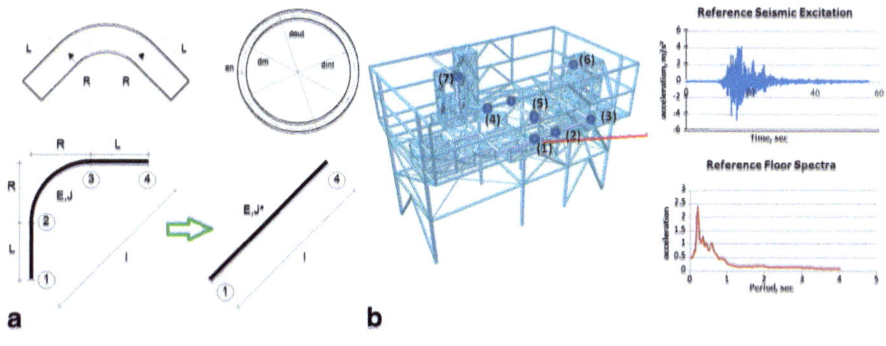

Fig. 9.2 **a** Elbow geometry and equivalent straight elbow element; **b** support structure and reference floor accelerogram with relevant spectrum generated at point (*1*)

Table 9.1 Elbow properties considered in the piping system model

Property	8″ Elbow		6″ Elbow	
	Original	Modified	Original	Modified
Thickness, e_n (mm)	8.18	6.61	7.11	4.35
Flexibility factor, k_B	6.84	1.35	5.97	2.46
Moment of inertia, J^* (mm^4)	3.02×10^7	2.49×10^7	1.17×10^7	7.53×10^6

considered that the effect of flexibility of an elbow spreads across a distance l equal to two times the mean diameter of the pipe as depicted in Fig. 9.2a. The adjusted geometry and properties of elbow elements considered in the FE model are reported in Table 9.1.

A modal analysis of the piping system model showed that the first 20 modes were able to excite about 87 % of the total mass of the piping system in the x direction. In particular, see Fig. 9.1b, response in the x direction was mainly of interest, therefore, during the experiments, earthquake loading was applied to that direction. Due to reduction requirements described later, the 1st and 2nd modes were considered as the two main modes; they excited about 36 and 12 % masses, respectively, in the x direction.

9.2.3 *Selection of Input Earthquake Loading*

The steel support structure of the piping system shown in Fig. 9.1a typically acts as a dynamic filter which causes amplifications of input earthquakes at different structure locations. Therefore, in order to select a realistic earthquake input both for analyses and experimental tests, earthquake accelerations were generated on elevated floors of the support structure through time history analyses carried out on an FE model of the structure subjected to a base input, i.e., the north-south El Centro earthquake with 0.29 g Peak Ground Acceleration (PGA). A reference floor accelerogram was thus chosen as shown in Fig 9.2b, relevant PGA was about 4.13 m/s^2 and the period T at maximum amplification was around 0.2 s which was close to the natural frequency of the piping system. To comply with performance-based earthquake engineering Italian Standards (Norme Techniche 2008), PGA of the reference floor accelerogram was magnified corresponding to both serviceability (SLO, SLD) and ultimate limit states (SLV, SLC) as listed in Table 9.2.

Table 9.2 PGAs corresponding to serviceability and ultimate limit states

Limit states			PGA (m/s^2)	PGA (g)
Serviceability limit states	SLO	Operational limit state	0.77	0.08
	SLD	Damage limit state	1.1	0.11
Ultimate limit states	SLV	Safe life limit state	4.13	0.42
	SLC	Collapse limit state	5.88	0.60

9.3 Substructuring and System of Equations of Motion

9.3.1 Substructuring

As already mentioned, during a hybrid test, a structure is divided into two parts: (i) a PS which is physically built in the lab and loaded through actuators; (ii) a NS that is solved via software; the two substructures exchange information through the coupling degrees of freedom (DoFs) they mutually share. Hence, for substructuring purposes, it was required to find proper coupling points and ensure the compatibility and equilibrium conditions at these nodes. In order to be compatible with the available actuator control system, two coupling DoFs at two nodes were chosen where two MOOG actuators were mounted. The nodes were selected in the xy plane -most of the pipe runs in this plane- at the positions of bending moments close to zero found from time history analyses relevant to the horizontal x direction. Both the piping system with the two substructures and coupling nodes are illustrated in Fig. 9.3a, whereas specifications and dimensions of the PS are depicted in Fig. 9.3b.

During experiments, displacement commands were applied to the two coupling nodes of the PS in the x direction. To better reflect this testing situation, an FE model was considered where the two coupling nodes were constrained to move together in the x direction, thus satisfying the compatibility condition. The other movements were kept free except movements along y and z, which were constrained. This model was considered as the Reference Model (RM). It was found that, by satisfying the compatibility condition of the two coupling nodes in the x direction only, we were potentially able to reproduce seismic response of the piping system by means of experimental tests. In this respect, the root mean square error e_{RMS} between time history responses of the RM and of the Continuous Model (CM) of the piping system was calculated as,

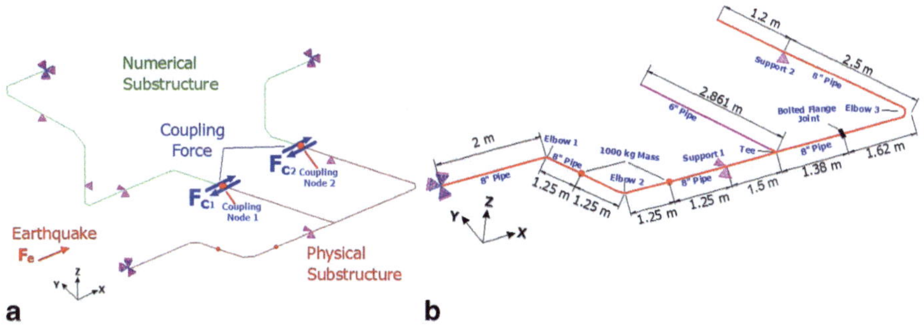

Fig. 9.3 **a** PS, NS and relevant coupling nodes; **b** Specifications and dimensions of the PS

$$e_{RMS} = \sum_{i=1}^{N} \frac{1}{N} \cdot \frac{\sqrt{\sum_{j=1}^{M} x_{CM,ij}^2} - \sqrt{\sum_{j=1}^{M} x_{RM,ij}^2}}{\sqrt{\sum_{j=1}^{M} x_{CM,ij}^2}} \tag{9.1}$$

where x_{CM} and x_{RM} define time history responses of the CM and RM, respectively; in detail, N is the number of total DoF considered and M is the length of time histories in terms of number of samples. The average value of e_{RMS} of all nodes was found to be about 0.22.

9.3.2 System of Equations of Motion

The system of equations of motion of the emulated system subjected to earthquake forces -neglecting damping for simplicity- can be written as,

$$\boldsymbol{M}^N \ddot{\boldsymbol{u}}^N + \boldsymbol{K}^N \boldsymbol{u}^N + \boldsymbol{M}^N \boldsymbol{I}\ddot{u}_g + \boldsymbol{M}^P \ddot{\boldsymbol{u}}^P + \boldsymbol{M}^P \boldsymbol{I}\ddot{u}_g = \boldsymbol{F}_C \tag{9.2}$$

$$\boldsymbol{M}^N \ddot{\boldsymbol{u}}^N + \boldsymbol{K}^N \boldsymbol{u}^N + \boldsymbol{M}^N \boldsymbol{I}\ddot{u}_g + \boldsymbol{M}^P \boldsymbol{I}\ddot{u}_g = \boldsymbol{F}_C \tag{9.3}$$

for a PDT and RT, respectively, with DS, where $\boldsymbol{M}$ and $\boldsymbol{K}$ are mass and stiffness matrices; $\ddot{u}$ and $\boldsymbol{u}$ are acceleration and displacement vectors; $\boldsymbol{I}$ is a matrix composed of one and zero that projects earthquake forces to desired DoFs; $\ddot{u}_g$ is the ground acceleration from the earthquake; N and P corresponds to the NS and PS, respectively. $\boldsymbol{F}_C$ defines the coupling force vector. In a greater detail, $\boldsymbol{F}_C$ reads,

$$\boldsymbol{F}_C = -\boldsymbol{K}^P \boldsymbol{u}^P$$

$$\boldsymbol{F}_C = -\boldsymbol{M}^P \ddot{\boldsymbol{u}}^P - \boldsymbol{K}^P \boldsymbol{u}^P$$

in Eqs. (9.2) and (9.3), respectively. Note that the expression of $\boldsymbol{F}_C$ indicates that inertia and damping forces of the PS are numerically modelled in case of a PDT.

The reader may note that in order to solve Eqs. (9.2) and (9.3), earthquake forces needed to be applied to all DoFs of the PS, which required an unfeasible task of applying one actuator per DoF. Therefore, given the actuator availability, an earthquake force vector having two elements corresponding to the two coupling DoFs was considered in order to solve the system of equations of motion. These two forces that synthesized the overall earthquake forces on the PS were then summed to the measured coupling forces. In case of a PDT, two by two mass and damping matrices were also required to be numerically modelled. To overcome this problem, we applied model reduction techniques, which allowed for reducing the number of DoFs of a system preserving properties of the original model; a discussion on these techniques is provided in the following Section.

9.4 Model Reduction of Physical Substructure and Earthquake Forces

Reduction of earthquake forces and of the PS were deemed necessary to carry out hybrid tests. In this respect, mode synthesis or model reduction techniques, in order to retain a subset ***uR*** of the total DoFs, truncate the remainder ***uL*** while retaining important properties, e.g., mode shapes of interest of the original model. The relationship between the two models can be expressed as,

$$\boldsymbol{u} = \begin{bmatrix} \boldsymbol{u}_R & \boldsymbol{u}_L \end{bmatrix}^{\mathrm{T}} = \boldsymbol{T}\ \boldsymbol{u}_R \tag{9.4}$$

where $\boldsymbol{T}$ is a linear transformation matrix that allows for the transformation between the original and the reduced model.

A number of model reduction techniques are available in literature, e.g., Guyan method (Guyan 1965), dynamic IRS method (O'Callahan 1989), SEREP method (O'Callahan et al. (1989), Craig Bampton (CB) method (Craig and Bampton 1968) and its variants. Several reduction methods were investigated for hybrid tests and, on the basis of their effectiveness, the SEREP and CB techniques were implemented, both of which are briefly discussed herein.

9.4.1 SEREP Reduction

The System Equivalent Reduction Expansion Process (SEREP) developed by O'Callahan et al. (1989) is an effective dynamic reduction that allows the reduced system to retain arbitrary modes of interest of the original model with exact eigenvalues. It utilizes computation of eigenvectors to produce the transformation between master and slave coordinates. Let us split the mass normalized eigenvectors $\boldsymbol{\Phi}$ of our system in retained F***R*** and truncated F***L*** eigenmodes -column wise- and relevant retained and truncated DoFs -row wise-:

$$\boldsymbol{\Phi} = [\boldsymbol{\Phi}_R \boldsymbol{\Phi}_L] = \begin{bmatrix} \boldsymbol{\Phi}_{RR} & \boldsymbol{\Phi}_{RL} \\ \boldsymbol{\Phi}_{LR} & \boldsymbol{\Phi}_{LL} \end{bmatrix} \tag{9.5}$$

The SEREP transformation matrix $\boldsymbol{T}_{SE}$ can then be defined as,

$$T_{SE} = \boldsymbol{\Phi}_R \boldsymbol{\Phi}_{RR}^{-1} \tag{9.6}$$

One may note that transformation between original and reduced models requires only manipulations of retained eigenvectors entailing reduced computational costs. The SEREP reduction was useful for RT where all contributions of coupling forces, i.e., inertia, damping and restoring force, were experimentally measured and only earthquake forces needed to be reduced. Since an RT aimed at the reproduction of dynamics of the emulated Global System (GS), modes of the global piping system model were considered for the reduction, thus potentially better approximating the actual response of the piping system; Mode #1 and Mode #2, which excited maximum masses in the x direction, were retained to perform this reduction.

9.4.2 Craig–Bampton Reduction

The CB reduction technique was originally developed by Craig and Bampton (1968); it produces a reduced model of the substructure whose mode shape information consists of all free modes expressed in physical coordinates and a deleted set of elastic modes expressed in modal coordinates. The reduction can be defined as,

$$\boldsymbol{u} = \begin{Bmatrix} \boldsymbol{u}_R \\ \boldsymbol{u}_L \end{Bmatrix} = \begin{bmatrix} \boldsymbol{I} & \boldsymbol{0} \\ \boldsymbol{B} & \boldsymbol{\Phi} \end{bmatrix} \begin{Bmatrix} \boldsymbol{u}_R \\ \boldsymbol{q} \end{Bmatrix} = \boldsymbol{T}_{CB} \begin{Bmatrix} \boldsymbol{u}_R \\ \boldsymbol{q} \end{Bmatrix} \tag{9.7}$$

in which $\boldsymbol{u}_R$ and $\boldsymbol{u}_L$ are coupling and internal DoFs respectively; $\boldsymbol{B}$ and $\boldsymbol{\Phi}$ are rigid body vector and fixed-based mode shapes, respectively; $\boldsymbol{q}$ contains the modal coordinates; $\boldsymbol{T}_{CB}$ is the CB transformation matrix. The CB reduction seemed to be very appropriate for PDT testing. With this reduction, the overall response of the PS was obtained by combining two contributions: (i) the response of internal nodes owing to static movements of the coupling nodes; (ii) the response of internal nodes owing to local dynamic modes of interest of the PS. Note that the measured restoring force during an experiment was the contribution of (i), whereas the contribution of (ii) was numerically considered. As a result, overall interaction between PS and NS could be taken into account and we were potentially able to reproduce the behaviour of the emulated system through a PDT. Mode #1 and Mode #3 of the PS that contained maximum participation masses in x were preserved for this reduction.

The effectiveness of the aforementioned reduction techniques were assessed through several time history simulations carried out on the Reduced Model of the piping system (NS + reduced PS) and results were compared with those of the Reference Model. The CB and SEREP techniques were found to be the most effective for hybrid tests; in detail, they were characterized by e_{RMS} values calculated on the PS DoFs of 0.01 and 0.02, respectively.

9.5 Integration Schemes and Hardware-Software Architecture

9.5.1 Integration Methods

The successful execution and quality of a PDT or RT depends, among other factors, on the integration scheme that solves the system of equations of motions. In order to carry out the RTs, both the LSRT2 method developed by Bursi et al. (2008) and the Modified Generalized-α(MG-α) method developed by Abbiati et al. (2012) were employed, whereas PDTs were carried out by the LSRT2 method. These monolithic algorithms are both RT compatible and offer second order accuracy and linear stability with proper selections of relevant parameters.

9.5.1.1 The LSRT2 Method

The L-Stable Real-Time compatible algorithm with two stages (LSRT2) results to be more competitive than popular Runge-Kutta methods in terms of stability, accuracy and ease of implementation (Bursi et al. 2008). This method is unconditionally stable for uncoupled problems and entails a moderate computational cost for real-time performance. Let us consider the following state space representation of our system:

$$\dot{y}_n = g(y_n, t_n) = \bar{M}^{-1}(f_n - \bar{K}\, y_n) \tag{9.8}$$

where,

$$\bar{M} = \begin{bmatrix} I & 0 \\ 0 & M_N \end{bmatrix}; \bar{K} = \begin{bmatrix} 0 & -I \\ K_N & C_N \end{bmatrix}; y_n = \begin{Bmatrix} u_n \\ \dot{u}_n \end{Bmatrix}; f_n = \begin{Bmatrix} 0 \\ f_n^{e} \quad -f_n^{c} \end{Bmatrix}$$

in which K_N, C_N and M_N are stiffness, damping and mass matrices of the NS, whereas: f_n^e and f_n^c are external load vector and the restoring force vector measured on the PS, respectively; u_n and $\dot{u}_n$ are displacement and velocity vectors. Hence, the LSRT2 algorithm can be summarized in algorithmic form as follows:

Stage 1:

$$y_{n+\alpha_2} = y_n + \alpha_{21} k_1;\; k_1 = [I - \gamma \Delta t J]^{-1} g(y_n, t_n)\Delta t \tag{9.9}$$

Stage 2:

$$y_{n+1} = y_n + b_1 k_1 + b_2 k_2; k_2 = [I - \gamma \Delta t J]^{-1} g(y_{n+\alpha_2}, t_{n+\alpha_2})\Delta t \tag{9.10}$$

where $y_{n+\alpha_2}$ and y_{n+1} are displacement commands sent to actuators at $n+\alpha_2$ and $n+1$ time steps, respectively; relevant measured forces are fed back to $g(y_n, t_n)$ and $g(y_{n+\alpha 2}, t_{n+\alpha 2})$ at $n+\alpha_2$ and $n+1$ time steps, respectively.

In order to preserve A-Stability, the Jacobian J was evaluated on the GS. In order to achieve L-stability, second order accuracy and to reduce algorithmic damping in the low frequency range, the following parameters are recommended for the LSRT2 method:

$$\gamma = 1-\sqrt{2}/2, \alpha_2 = \alpha_{21} = 1/2, \gamma_{21} = -\gamma, b_1 = 0, b_2 = 1 \tag{9.11}$$

9.5.2 Modified Generalized-α Method

The balance equation of the MG-α method proposed by Abbiati et al. (2012) and based on the Generalized-α (G-α) method of Jansen et al. (2000), is forced at the end of the time step by means of additional state variables or velocity-like quantities. These quantities satisfy the following recurrence relationship:

$$\alpha_m v_{n+1} + (1-\alpha_m) v_n = \alpha_f \dot{y}_{n+1} + (1-\alpha_f)\dot{y}_n \tag{9.12}$$

in which parameters α_m, α_f and γ can be expressed in terms of ρ_∞, i.e. the spectral radius for an infinite time-step length:

$$\alpha_m = 1/2(3-\rho_\infty)/(1+\rho_\infty),\ \alpha_f = 1/(1+\rho_\infty),\ \gamma = 1/2 + \alpha_m - \alpha_f \tag{9.13}$$

If ρ_∞ is chosen to be zero, the method annihilates the highest frequency components of the response. Conversely, if ρ_∞ is chosen to be unitary, all frequency components of the response are preserved in the linear problem. The MG-α is L-Stable for $\rho_\infty < 1$ and RT compatible.

9.5.3 Modification of the NS and Delay Compensation for RTs

With regard to RT, a critical limitation was posed by hydraulic actuators. Due to several factors, such as delay and insufficient hydraulic power, high frequency operation set limits of about ± 10 mm to maximum strokes of actuators for 6 Hz. Hence, it was not possible to run a RT on the piping system with high earthquakes PGA values. In order to obtain a real time compatible structure, the NS of the piping system was modified by adding masses to several nodes thus reducing the overall eigenfrequencies of the piping system to about 1.10 Hz. Additionally, to compensate actuator delays, the delay over prediction method developed by Wu et al. (2013) was implemented and incorporated with Simulink (Simulink 2012) models of relevant algorithms. This allowed us to carry out RTs on the piping system, partly different from the original structure, with low PGA earthquakes.

9.5.4 Hardware-Software Architecture

Integration schemes were developed in MATLAB/Simulink environment in a Host PC where they were compiled and then sent to an xPC target (real time software) via a LAN connection. During experimental tests, integration algorithms solved the system of equations of motion in the xPC target and estimated displacement commands for the PS. These displacement commands were written locally to the xPC target, which instantaneously copied these signals to an MTS controller (MTS 2008) through a SCRAMNET (a shared memory between the Host PC and the controller). The controller then commanded two MOOG actuators (capacity: 1000 kN force, ±250 mm stroke) to move the coupling DoFs to desired positions and read back reaction forces from actuators. Measured reaction forces were instantaneously copied to the xPC target via SCRAMNET. The hardware-software scheme for the hybrid tests is presented in Fig. 9.4.

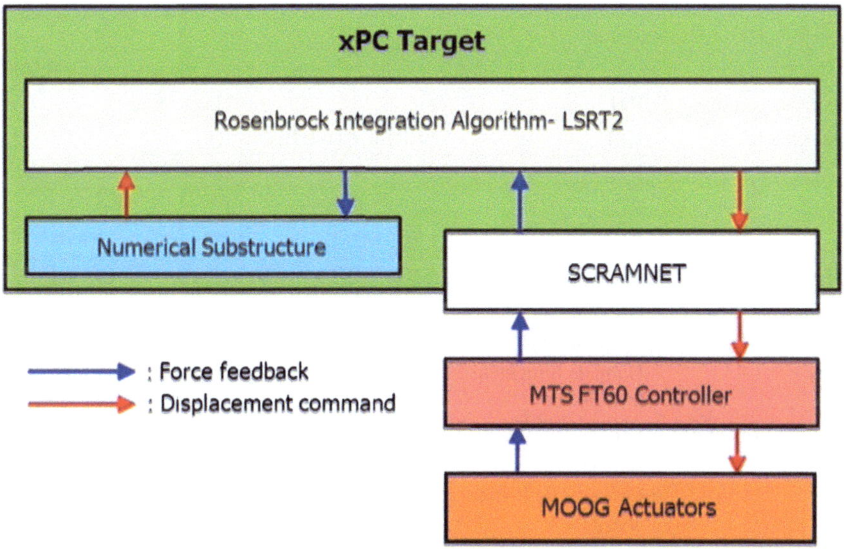

Fig. 9.4 Hardware-software architecture for hybrid tests with the LSRT2 integration scheme

9.6 Test Program and Experimental Set-Up

As reported in Table 9.3, a number of PDTs and RTs were carried out. RTs were conducted with low PGA values and handled a similar structure owing to the limitations underlined in Subsect. 9.5.3. PDTs were performed with the CB reduction, while both the SEREP and CB reductions were adopted to perform RTs. All PDDS were carried out at a time scale extended 50 times with respect to the actual earthquake time.

In addition to hybrid tests, in order to characterize dynamic properties of the PS and to have an estimation of actual damping to be used in the NS, four Identification Tests (IDTs) were performed on the PS. In all tests, an internal pressure of 3.2 MPa was used and earthquake loading was applied in the horizontal x direction depicted in Fig. 9.3b.

The test specimen corresponds to the PS described in Subsect. 9.3.1. The experimental set-up was placed on the reaction floor of the Materials and Structural Testing laboratory (LPMS) of the University of Trento. A schematic arrangement of the test set-up is depicted in Fig. 9.5a, while the actual set-up is presented in Fig. 9.5b. In order to measure strains, displacements and rotations in different positions, the test specimen was instrumented with several strain gauges and displacement transducers. Data were acquired by 4 Spider8 acquisition systems and by an MTS FT60 controller. IDTs were carried out using ten accelerometers and a National Instruments data acquisition system.

Table 9.3 Hybrid test program

Test case				PGA (g)
Identification tests	IDT	–	Hammer test	–
Real time tests	RT1	SEREP + LSRT2	RT	0.02
	RT2	CB + LSRT2	RT	0.02
	RT3	SEREP + MG-α	RT	0.02
	RT4	CB + MG-α	RT	0.02
Elastic tests	Elastic test, ET	CB + LSRT2	PDT	0.04
Serviceability limit state tests	Operational limit state test, SLOT	CB + LSRT2	PDT	0.08
	Damage limit state test, SLDT	CB + LSRT2	PDT	0.11
Ultimate limit state tests	Safe life limit state test, SLVT	CB + LSRT2	PDT	0.42
	Collapse limit state test, SLCT	CB + LSRT2	PDT	0.60

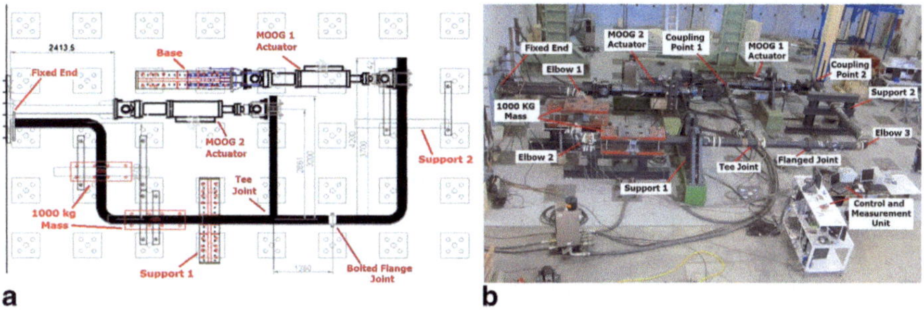

Fig. 9.5 Test set-up **a** schematic; **b** actual

9.7 Main Experimental Results

All PDTs and RTs listed in Table 9.3 were successfully executed. IDTs carried out on the PS confirmed a damping of 0.5 %, which was used in the NS during experiments.

The piping system and its components exhibited a favorable performance under all limit state earthquakes. In fact, it was observed that, even under the Collapse Limit State (SLCT) earthquake, the piping system remained below its yield limits without any leakage, and only limited strains and rotations were found in different components. In all tests, maximum strain was found in Elbow #2 as can be noted in Fig. 9.6a; in greater detail, one may observe that the maximum elbow strain at SLCT was about 950 μm/m, which was well below its yield strain, i.e., 2019 μm/m.

Significant amplifications of input earthquake PGAs were found in the piping system as depicted in Fig. 9.6b, where accelerations of Coupling Node #1 at SLCT are presented. One may note that the maximum acceleration at SLCT was about twice of the corresponding input PGA. Moreover, relevant Fourier spectra illustrate that the dynamic responses of the piping system were dominated by its lower modes corresponding to first (5.87 Hz) and second (6.47 Hz) eigenfrequencies.

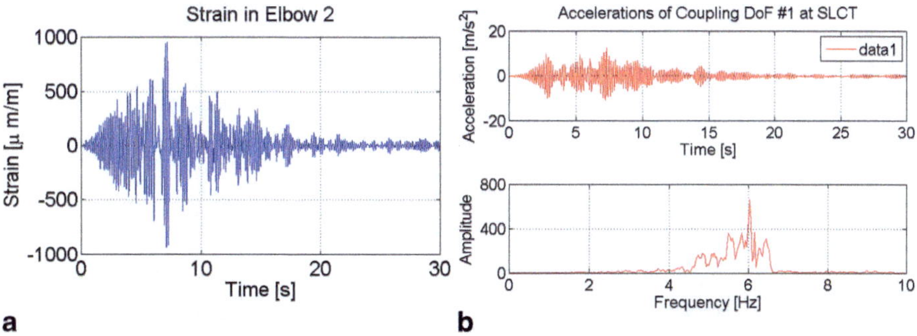

Fig. 9.6 **a** Strain histories in Elbow #2; **b** acceleration history and relevant spectra of Coupling Node #1 at SLCT

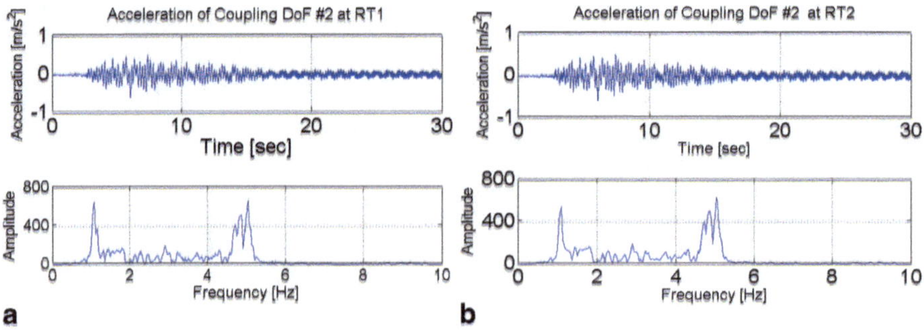

Fig. 9.7 Accelerations of Coupling DoF #2 at: **a** RT1; **b** RT2

All RTs exhibited similar responses of the piping system. In this respect, acceleration histories of Coupling Node #2 from RT1 and RT2 are presented in Fig. 9.7; one can note that the input PGA was amplified about three times in both tests.

According to the selection of eigenmodes characterizing the SEREP reduction in the case of the modified NS, one may observe from relevant Fourier spectra that the system response was dominated by Modes #2 and 3#, both at about 1.10 Hz.

Additionally, both LSRT2 and MG-α integration schemes were proved to be effective for RTs; they entailed similar experimental responses, which showed a favorable agreement with relevant numerical simulation results, as depicted in Fig. 9.8.

A comparison between numerical and experimental results exhibited a favourable agreement; e_{RMS} calculated according to Eq. (9.1) between experimental and numerical responses of Coupling DoF #2 at SCLT was found to be about 11 %, while its value was about 30 % for all RTs. Nevertheless, the above-mentioned agreement between numerical and experimental results justified effectiveness of the CB and the SEREP reduction techniques, as was predicted analytically.

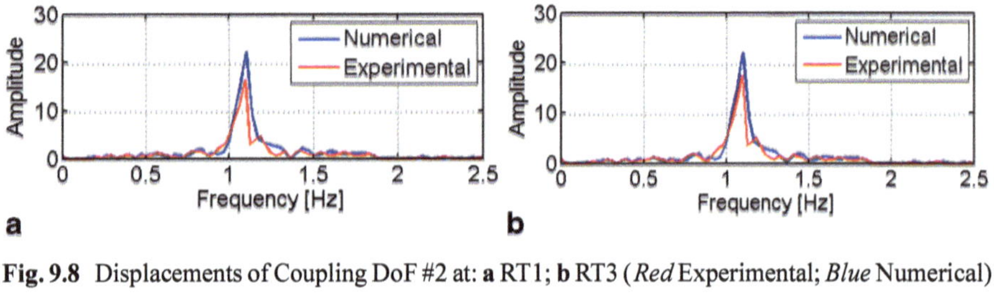

Fig. 9.8 Displacements of Coupling DoF #2 at: **a** RT1; **b** RT3 (*Red* Experimental; *Blue* Numerical)

9.8 Conclusions

This paper presented a seismic performance evaluation of a typical industrial piping system at full-scale through pseudo-dynamic and real-time tests under Serviceability and Ultimate limit state earthquake levels suggested by performance based earthquake engineering Standards. All tests were conducted with the presence of an internal pressure of 3.2 MPa and considering a 0.5% damping found through identification tests on the test specimen.

The long-held considered inadequacy of PDT technique for distributed mass systems was overcome through successful execution of relevant tests on the piping system. The so called mode synthesis methods, in particular, SEREP and Craig-Bampton, were employed for implementations of hybrid tests. Moreover, two real-time compatible monolithic time integration algorithms namely, the LSRT2 and the MG-α method, were employed to carry out PDTs and RTs. Effectiveness of reduction techniques and time integration schemes were justified through the successful execution of relevant hybrid tests.

A favourable seismic performance of the piping system and its critical components was observed during experiments; they remained below their yield limits without any leakage even for the Collapse Limit State. Furthermore, a favourable agreement between experimental and numerical results was observed for all tests.

Acknowledgements The research leading to these results has received funding from the European Community's Seventh Framework Programme [FP7/2007–2013] for access to the laboratory of the University of Trento under grant agreement n° 227887. Finally, we thank Dr. Nie and Dr. Hofmayer from Brookhaven National Laboratory for the provision of some earthquake records.

References

Abbiati G, Bursi OS, Xu G, Wu B (2012) Novel partitioned time integration schemes for DAE systems based on generalized-a methods. In: Proceedings of the 5th European Conf. on Struct. Control, Genoa, 18–20 June

ANSYS (2007) Release 11.0 documentation for ANSYS. Online Manual, ANSYS, Inc

Braconi A, Bursi OS, Fabbrocino G, Salvatore W, Tremblay R (2008a) Seismic performance of a 3D full-scale high-ductility steel-concrete composite moment-resisting structure—part I: design and testing procedure. Earthq Eng Struct Dyn 37(14):1609–1634

Braconi A, Bursi OS, Fabbrocino G, Salvatore W, Taucer F, Tremblay R (2008b) Seismic performance of a 3D full-scale high-ductile steel-concrete composite moment-resisting frame—Part II: test results and analytical validation. Earthq Eng Struct Dyn 37(14):1635–1655

Bursi OS, Buelga AG, Vulcan L, Neild SA, Wagg DJ (2008) Novel coupling Rosenbrock-based algorithms for real-time dynamic substructure testing. Earthq Eng Struct Dyn 37:339360

Bursi OS, Jia C, Vulcan L, Neild SA, Wagg DJ (2011) Rosenbrock-based algorithms and subcycling strategies for real-time nonlinear substructure testing. Earthq Eng Struct Dyn 40(1):1–19

Bursi OS, Reza MS, Paolacci F, Kumar A (2012) Seismic performance of bolted flange joints in piping systems for oil and gas industries. 15th World Conference on Earthquake Engineering (15WCEE), Lisbon, Portugal, 24–28 September 2012

Craig R, Bampton M (1968) Coupling of substructures for dynamic analysis. AIAA J 6(7):1313–1319

DeGrassi G, Nie J, Hofmayer C (2008) Seismic analysis of large-scale piping systems for the JNES-NUPEC ultimate strength piping test program. U.S. NRC. NUREG/CR-6983, BNL-NUREG-81548-2008

EN 13480-3 (2002) Metallic industrial piping—part 3: design and calculation

Guyan RJ (1965) Reduction of stiffness and mass matrices. Am Inst Aeronaut Astronaut J 3(2):380

Hibbit HD, Karlsson BI, Sorensen (2003) Theory manual, ABAQUS, version 6.3. Providence, RI. USA

Jansen KE, Whiting CH, Hulbert GM (2000) A generalized-α method for integrating the filtered Navier–Stokes equations with a stabilized finite element method. Comput Method Appl Mech Eng 190(3–4):305–319. doi:10.1016/S0045-7825(00)00203-6

Krausmann E, Cruz AM, Affeltranger B (2010) The impact of the 12 May 2008 Wenchuan earthquake on industrial facilities. J Loss Prev Process Ind 23(2):242–248 (ISSN 0950-4230)

Mahin SA, Shing PB (1985) Pseudodynamic method for seismic testing. J Struct Eng 111(7): 1482–1503 (304)

Melo FJMQ de, Carneiro JAO, Lopes HR, Rodrigues JFD, Gomes JFS (2001) The dynamic analysis of piping systems using pseudo-dynamic techniques. J Strain Anal 36(5), 441–451

MTS (2008) Models FlexTest 40/60/100/200 controller hardware. MTS Systems Corporation

Norme Techniche (2008) Norme Techniche per le costruzioni. DM Infrastrutture

O'Callahan JC (1989) A new procedure for and improved reduced system IRS. In IMAC VII, Las Vegas

O'Callahan JC, Avitabile P, Riemer R (1989) System equivalent reduction expansion process (SEREP). In: Proceedings of the 6th International Modal Analysis Conference, Las Vegas, 29–37 January 1989

Otani A, Nakamura I, Takada H, Shiratori M (2011) Consideration on seismic design margin of elbow in piping. In: Proceedings of the ASME 2011 Pressure Vessels 237 and Piping Division Conference PVP2011 July 17–21, 2011, Baltimore, Maryland, USA

Paolacci F, Reza MS, Bursi OS (2011) Seismic design criteria of refinery piping systems. COMPDYN 2011 -III ECCOMAS Thematic Conference on computational methods in structural dynamics and earthquake engineering, Corfu, Greece, 26–28 May 2011

Paolacci F, Giannini R, Angelis MD (2013a) Seismic response mitigation of chemical plant components by passive control systems. J Loss Preve Process Ind. doi:10.1016/j.jlp.2013.03.003

Paolacci F, Reza MS, Bursi OS, Gresnigt N, Kumar A (2013b) Main issues on the seismic design of industrial piping systems and components. In: Proceedings of the ASME 2013 pressure vessels & piping division conference, PVP 2013, 14–18 July, 2013, Paris, France (in print)

Pinto AV, Pegon P, Magonette G, Tsionis G (2004) Pseudo-dynamic testing of bridges using nonlinear substructuring. Earthq Eng Struct Dyn 33:1125–1146

Reza MS (2013) Seismic safety evaluation of industrial piping systems and components under serviceability and ultimate limit state conditions. PhD thesis, University of Trento, Trento, Italy

Sezen H, Whittaker AS (2006) Seismic performance of industrial facilities affected by the 1999 Turkey earthquake. J Perform Constr Facil 20(1), 28–36

Shing PB, Nakashima M, Bursi OS (1996) Application of pseudodynamic test method to structural research. Earthq Spectra EERI 12(1):29–54

Simulink v 8.0 (2012) The MathWorks, Inc

Tanaka Y (2012) Study on analogy between dynamic load and displacement-controlled load. Int J Press Vessels Pip 90–91:37–45

Touboul F, Blay N, Sollogoub P, Chapuliot S (2006) Enhanced seismic criteria for piping. Nucl Eng Des 236:1–9

Wu B, Wang Z, Bursi OS (2013) Actuator dynamics compensation based on upper bound delay for real-time hybrid simulation. Earthq Eng Struct Dyn. doi:10.1002/eqe.2296

Zare MR, Wilkson S (2010) Resilience of wastewater pipelines in earthquakes. In: Proceedings of the 9th U.S. National and 10th Canadian Conference, Toronto, Ontario, Canada, 25–29 July 2010

Chapter 10
A Support Platform for Distributed Hybrid Testing

Ignacio Lamata Martínez, Ferran Obón Santacana, Martin S. Williams, Anthony Blakeborough and Uwe E. Dorka

10.1 Introduction

Large-scale testing continues to play an important role in earthquake engineering, generating research results that lead to improved safety and security of European society. However, realistic tests of large or complex structural systems are constrained by the capacities of existing laboratories.

Distributed hybrid testing enables the use of resources at different locations to perform more complex, larger-scale tests than are possible in most individual laboratories. In a distributed experiment, the system under test is split into several sub-structures which can be tested or simulated in different locations. Data are passed between the sub-structures at each time-step to ensure that the distributed experiment realistically simulates the full system under test.

There are multiple benefits in conducting an experiment in a distributed manner but it also involves some difficulties: organizing and planning distributed experiments entails much more complexity than what is involved in a single-laboratory hybrid test. In a distributed test, the experimental procedure and the systems to support the experiment execution must be closely linked, and multi-site experiment coordination can entail many technical and social challenges (De la Flor et al. 2010).

Moreover, the difficulty of tracing errors caused by the distributed environment is greater than in a local one. Errors can occur in any of the participants of the experiment or in the communication between them, and tracing where and why something failed can be extremely complicated. When a system raises an error during an

I. Lamata Martínez (✉) · M. S. Williams · A. Blakeborough
Department of Engineering Science, University of Oxford, Oxford, UK
e-mail: ignacio.lamata@eng.ox.ac.uk

F. Obón Santacana · U. E. Dorka
Institute of Structural Engineering, University of Kassel, Kassel, Hesse, Germany
e-mail: ferran.obon@uni-kassel.de

F. Taucer, R. Apostolska (eds.), *Experimental Research in Earthquake Engineering*,
Geotechnical, Geological and Earthquake Engineering 35,
DOI 10.1007/978-3-319-10136-1_10

experiment, it can be detected and fixed. Unfortunately, not every error is raised and recognised at its source, and many of them have to be traced through several distributed systems or devices. Even when seemingly no major or obvious error happened, bugs in programs or unexpected situations might give the impression of a test being conducted successfully yet produce incorrect results. Tracing where the test failed involves following all testing steps carefully, and sometimes the erroneous situation is impossible, or not easily reproducible. This all points to the importance of a platform to support testing activities, dividing them into two groups: high-level activities (experiment planning, organization, participant location…) and low-level activities (the actual data exchange).

Several programs exist to conduct distributed tests, such as UK-NEES DHT (developed at the University of Oxford (Ojaghi et al. 2010)), OpenFresco (developed at the University of California, Berkeley (Schellenberg et al. 2010)) or the Platform for Networked Structural Experiments (developed at the National Center for Research on Earthquake Engineering, Taiwan (Yang et al. 2004; Wang et al. 2004)). All of these programs expect to communicate with their homologous program at the other end, so it is not possible to use UK-NEES DHT at one end and OpenFresco at the other, for instance. This means that service integration in distributed testing software is null, which is one of the motivations for the creation of a common framework called Celestina[1].

The Celestina framework is explained in Sect. 10.2 of the paper. In Sect. 10.3 we describe a first implementation of the framework and its evaluation through a series of distributed experiments. In a typical substructured test, nodes at Oxford and Kassel Universities were used to simulate the response of a 33-degree-of-freedom steel frame fitted with a tuned mass damper, with both nodes conducting testing (in simulation) according to instructions from a Celestina-based program running in Oxford. Finally, conclusions and future developments are presented in Sect. 10.4.

10.2 The Celestina Framework

The Celestina framework is a specification to support high-level activities such as identification and location of participants, compatibility verification of participants, experimental planning and results collection.

When conducting an experiment on the Celestina framework, a computer works as a manager sending orders to the rest of the experiment participants to instruct them how to proceed and organise the experimental workflow. The experiment plan is defined only by the manager, so researchers just need to specify the experiment steps at one time at one place. The manager will contact every participant to instruct them specifically what they have to do at every workflow step.

[1] As a side note, Celestina gets its name from a 15th century Spanish novel (Comedia de Calisto y Melibea by Fernando de Rojas), where a woman called Celestina manages to get a lady to fall completely in love with a man who she initially rejected.

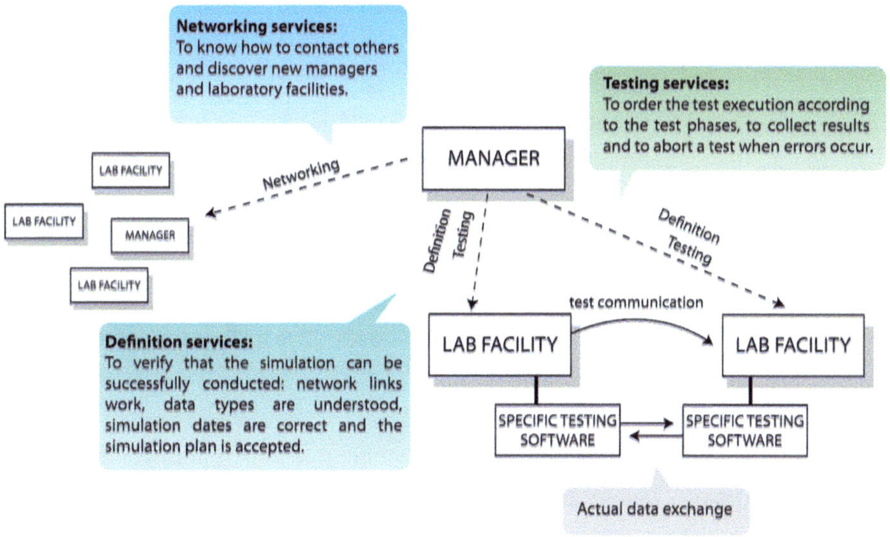

Fig. 10.1 Celestina services overview

The Celestina framework specifies what to do but not how to do it. The experiment manager sends orders about what to do but it is actually the decision of each experiment participant how to conduct the instructions given by the manager. Likewise, the Celestina framework does not dictate how the actual data exchange of low-level testing activities is conducted between participants, it only defines the data types that have to be exchanged at each experiment step.

To enable understanding between the manager and the experiment participants sharing laboratory facilities, the specification defines a set of services. These services must be implemented by any testing software integrated into the Celestina framework, but their specific implementation is actually decided by the software. The Celestina services are divided into three groups, as depicted in Fig. 10.1:

- Networking services: deal with operations to locate and discover new Celestina nodes in the network (the experiment participants), identification of nodes and the resources they share.
- Definition services: deal with compatibility, preparation of an experiment and the verification that the experiment is feasible and can be achieved. It involves verification of connectivity, checking that the data types used in potentially heterogeneous systems are compatible and validating an agreed experiment plan.
- Testing services: deal with the actual experiment execution. This involves establishing the communication, running the data exchange over the network and collecting the results.

These services will be discussed in the following sections.

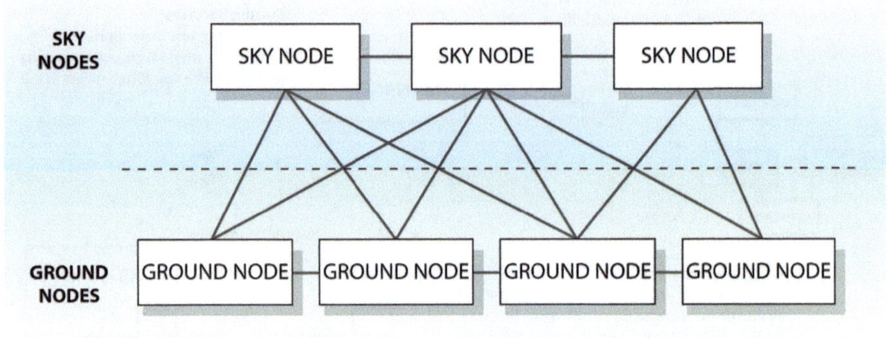

Fig. 10.2 P2P Celestina network divided into sky and ground nodes

10.2.1 Networking Services

The Celestina framework is configured in a hybrid peer to peer network (P2P) operated above the Internet infrastructure. A peer to peer network is a network where all the machines within it share some of their hardware resources and where any machine is accessible by any other machine in the network without intermediaries (Schollmeier 2010). The Celestina hybrid P2P network is divided into sky nodes, which are pure managers in charge of controlling the experiment plan, and ground nodes, which are in charge of the simulation execution. Sky and ground nodes are connected as depicted in Fig. 10.2. In terms of the network, both sky and ground nodes are treated equally. The only practical exception is that sky nodes are expected to be online all the time and do not have the necessity of sharing laboratory facilities.

Every Celestina node has a unique ID that identifies the node within the network. Since the network is totally unstructured and decentralised, every node in the network keeps a list of known nodes (also called the *friendbook*). Friendbooks are exchanged between nodes in one-to-one basis when they get in contact. When a node joins in the Celestina network, it should know the contact details of at least another node (preferably a sky node) already in the network, so its contact details are populated within the network and it starts creating a larger friendbook. To make friendbook information persistent, a file, database or other structure must be used.

10.2.2 Definition Services

The definition services verify the feasibility of an experiment, in terms of the understanding of the participants on the nature and configuration of the experiment, the possibility of establishing network connections and an agreement on the terms the experiment to be conducted. Each service in the definition services is commanded by a sky node and executed by the respective ground nodes, which return the results once completed. The main services are discussed below.

One of the services verifies the link connection between two nodes. This is normally done by means of a ping command, which is considered a low-level network command, but in practice there is no restriction of how nodes test the link. Typically two nodes will be ordered to execute a *ping* command to the other node, and after executing it, the results will be returned to the sky node. Note that this verification is just a first verification and is only illustrative: ping might work in one way but not in the other, and it does not really reflect accurately the possibility of a test being conducted or the time-step that can be achieved.

Another service verifies data compatibility to check if two nodes understand each other in terms of exchanged data types. To do so, the sky node sends two values to a node (normally similar to the ones that will be used during the experiment) together with an operation. The node will then send this information to the second node (so the network link is verified at a higher level), which operates the two values with the operation and returns the result. The first node then verifies that the received result is exactly the expected result and sends this to the sky node. An example is a sky node sending the values "7.803" and "124.986" with the operation "multiplication". The first ground node multiplies the two values and gets the expected result. Then it sends the information to the second ground node which performs the multiplication and sends the result back. This checks the possibility that different machines understand values in different ways, so the result sent by the second ground node is different from what is expected.

A third service assures that the experiment plan is agreed between all participants. Ground nodes must be able to analyse the experiment plan sent by the sky node and determine if they are able to conduct it. For example, if a ground node is only capable to conduct a two-site test, it will reject any multi-site test. Similarly, if the experiments use unknown data types or the date to conduct the experiment is not suitable, the experiment plan can be rejected.

10.2.3 Testing Services

The testing services are the most complex of the three groups of services. They define an experiment workflow and, unlike the network or definition services, they have to be called in a strict order. The experiment workflow defines the following phases:

- Test locking: This prevents other tests from running at the same time.
- Establishment of connections: Every node that needs to communicate with others will establish network connections (if necessary) and leave them open ready to start the test.
- Preparation: During this phase, nodes will process all instructions in the experiment to have them ready for the final data exchange. Researchers might need to conduct some additional tasks in the laboratory to get the physical devices ready.
- Experiment execution: This is when the experiment takes place and the actual data exchange occurs. At this point, no task in the computers should interfere while the experiment is running. The experiment will finish either because an error arises or an "end-of-experiment" signal is transmitted.
- Result collection: After the sky node is informed of the end of the test, it will attempt to collect results (mainly experimental data) from each of the participants. Every participant will normally generate a signal file for each command they managed during the test.
- Test unlocking and closing resources: Once the experiment is finished, every node receives the order of cleaning resources and reverting to a stable initial state, ready to conduct a new test.

10.2.4 Services Implementation in the Nodes

There are two types of implementations in the Celestina framework. Every ground node in Celestina has to implement the three groups of services and provide mechanisms to conduct each of them (nodes are free to decide how to implement them in practice). Sky nodes have to implement methods to call ground-node services, managing orders and data involved in the process. The nodes should normally provide a user-friendly interface for the operator.

While it is very useful to have many different implementations of ground-node software, sky-node software is not as important—it can be implemented once and be reused in many places. The reason for this is that ground-node software might need to implement different testing behaviors, but the main operation of sky nodes does not often vary.

The next section will discuss how the Celestina framework can be currently used.

10.3 A First Celestina Implementation

The Celestina specification has been implemented in a functional software by using Java and Web Services, and tested in a distributed environment between the universities of Oxford and Kassel. This implementation includes both the Celestina high-level activities as well as the data exchange required in low-level activities. It is very flexible in terms of the experiment plans that can be conducted. Therefore,

	NCREE	OpenFresco	Celestina (with DEBUG)	Celestina (FastJava)	Direct Communication
Measured (average, s/step)	-	-	0.18823	0.03637	0.03613
TOTAL (ms/step)	200	36	188	36	36

Fig. 10.3 Celestina performance compared to other testing software (average time required per test step—latency not guaranteed)

all the experiment steps can be configured at the sky node, and the ground nodes will execute these steps automatically without requiring any code modification. For instance, an experiment could be configured to send two "float" values from a node to another, and receive three "float" values as a response. Should it be decided that the experiment uses "double" values instead of "float" and to send one value and receive four, this can be configured in the experiment plan without requiring any code modification. Ground nodes will just adapt themselves to the experiment situation.

Several experiments were conducted to validate the operation of the Celestina framework and its implementation. Nodes at Oxford and Kassel were used to simulate the response of a 33-degree-of-freedom steel frame fitted with a tuned mass damper, with both nodes conducting testing (in simulation) according to the instructions given by a Celestina sky node installed at Oxford. Two different implementations of Celestina were tested ("with Debug" and "FastJava"). Figure 10.3 shows the performance of the Celestina implementations compared to other testing software (NCREE and OpenFresco) under similar testing circumstances. The "direct communication" shows the performance of the numerical testing program by itself—with no additional software. As the figure shows, there is no significant overhead added by the operation of the Celestina FastJava implementation.

In order to use Celestina and take advantage of the framework, testing software has to implement the Celestina specification and be aware of its existence. Existing software can be operated under Celestina, but if it does not implement any of the services, the real benefit of the framework is rather limited.

Current options to use the framework are the following:

- Using the Celestina implementation created by the University of Oxford and validated with the University of Kassel. This is the implementation that has been used to validate the whole framework and to demonstrate the capabilities of service integration as well as the execution of low-level testing activities without significant overhead.
- Integrate existing software or create a new one following the Celestina specification. Once the software is aware of the specification, it can take full advantage of the service integration.

The success of Celestina is bound to the commitment of testing software to implement the specification.

10.4 Conclusions

In this paper we have discussed the Celestina framework, which is a specification to support high-level activities (such as experiment planning and verification of an experiment feasibility) as well as to promote the integration of heterogeneous testing software. Celestina defines three groups of services that must be implemented by every node conducting experiments in the Celestina network. Two types of nodes, sky and ground, are considered, and they are integrated in a P2P network. Sky nodes are managers of the experiment whereas ground nodes share their laboratory facilities and conduct the actual data exchange.

Several compatibility tests are conducted under Celestina, such as testing the network link or verifying the data types understanding. A first implementation has been developed at the University of Oxford and tested with the University of Kassel, to validate the framework and demonstrate its capability of conducting distributed testing without significant overhead.

The next efforts should focus on promoting a framework to be used in existing or new testing software and to improve and create new implementations.

Acknowledgments The research leading to these results has received funding from the European Community's Seventh Framework Programme [FP7/2007-2013] for access to the laboratory of the University of Kassel (UNIKA) under grant agreement n° 227887.

References

De la Flor G Ojaghi M Lamata Martínez I Jirotka M Williams MS Blakeborough A (2010) Reconfiguring practice: the interdependence of experimental procedure and computing infrastructure in distributed earthquake engineering. Philos Trans Royal Soc A 368:4073–4088

Ojaghi M, Lamata Martínez I, Dietz M, Williams MS, Blakeborough A, Crewe A, Taylor C, Madabhushi G, Haigh S, Ali A (2010) UK-NEES—Distributed hybrid testing between Bristol, Cambridge and Oxford Universities: connecting structural dynamics labs to a geotechnical centrifuge. 9th U.S. National and 10th Canadian conference on earthquake engineering (in press paper 1024)

Schellenberg A, Kim H, Mahin SA (2010) OpenFresco. http://nees.org/resources/openfresco. Accessed July 2013

Schollmeier R (2001) A definition of peer-to-peer networking for the classification of peer-to-peer architectures and applications, p2p, pp 0101. First international conference on peer-to-peer computing (P2P'01)

Wang SJ, Wang KJ, Yang YS, Cheng WC, Yeh CC, Tsai KC (2004) Networked pseudo-dynamic testing part II: application protocol approach. 13th world conference on earthquake engineering vancouver, B.C., Canada, Paper No. 1548

Yang YS, Wang KJ, Wang SJ, Hsu CW, Tsai KC, Hsieh SH (2004) Networked pseudo-dynamic testing part I: database approach. 13th world conference on earthquake engineering vancouver, B.C., Canada, Paper No. 1910

Chapter 11
Substructuring for Soil Structure Interaction Using a Shaking Table

Matthew Dietz, Zhenyun Tang, Colin Taylor and Zhenbao Li

11.1 Introduction

In a conventional seismic analysis, incipient waves applied at the foundation of a structure induce dynamic loads within the superstructure. Dynamic soil structure interaction (SSI) analyses recognize that the foundation is affected by interplay between the dynamic characteristics of the underlying soil and those of the overlying superstructure. Theoretical results (e.g. Wolf 1985; Mylonakis and Gazetas 2000) indicate that SSI is sometimes beneficial and sometimes detrimental to structural performance. However, experimental evidence relating to SSI systems is scarce. The capacities of conventional testing apparatus makes full scale testing of SSI systems (including both superstructure and soil-foundation systems) under earthquake loading difficult to achieve. As a result, simplified models are tested at reduced scales (e.g. Chen et al. 2006; Pitilakis et al. 2008) producing uncertainty in the appropriate interpretation and application of results.

Real-time dynamic substructuring (RTDS) works to circumnavigate such issues and as such has garnered a great deal of interest since the pioneering study of Nakashima et al. (1992). Conceptually, real-time dynamic substructuring is a method of dynamically testing an entire structure with only a part of that structure represented physically in the laboratory. The structure is split into two coupled parts; the unpredictable component of interest is tested experimentally, while the remainder is tested numerically. Thus, the mass of the structure can be much larger than the capacity of the test apparatus since much of that mass exists only numerically.

M. Dietz (✉) · C. Taylor
Department of Civil Engineering, University of Bristol, Bristol, UK
e-mail: m.dietz@bristol.ac.uk

Z. Tang · Z. Li
Beijing Key Lab of Earthquake Engineering and Structural Retrofit,
Beijing University of Technology, Beijing, China

F. Taucer, R. Apostolska (eds.), *Experimental Research in Earthquake Engineering,*
Geotechnical, Geological and Earthquake Engineering 35,
DOI 10.1007/978-3-319-10136-1_11

SSI systems are an obvious candidate for RTDS testing. Konagai and Ahsan (2002) modeled a soil-foundation system numerically and a superstructure physically while Heath et al. (2008) modeled a superstructure numerically and the soil-foundation system physically. However, the work is in its infancy and further validation of the test method is required. This chapter establishes a framework for SSI substructuring. Prevalent control strategies are examined and found wanting. A new controller is developed that provides enhanced performance. Finally, the output of a conventional, entirely physical SSI shaking table tests is compared to that obtained by different substructure test variants in order to validate the method. The chapter begins with a description of the benchmark SSI test used for validation purposes.

11.2 Benchmark SSI test

A benchmark shaking table test will be used as a source of experimental data relating to an SSI system in order to validate the substructuring test methodology. The benchmark test was conducted as part of the New Methods of Mitigating the Seismic Risk of Existing Foundations (NEMISREF) project funded by the 5th framework of the European Commission.

11.2.1 Experimental Components

The term 'shear stack' refers to the flexible-walled hollow box designed and built to enable geodynamic testing at the University of Bristol. The latest incarnation on the shear stack is a direct descendant of Crewe et al.'s (1995) apparatus. However, it is closer in size to Dar's (1993) small shear stack: 1.2 m long as opposed to 5 m, 0.55 m wide as opposed to 1 m, and 0.8 m deep as opposed to 1.2 m, significantly reducing the costs and timescales associated with testing. Design details are presented in Fig. 11.1.

The apparatus consists of eight aluminium rings, rectangular in plan, which are stacked alternately with rubber sections. The rubber sections span only the end-walls of the stack, not the side-walls, giving a fundamental frequency when empty of 6 Hz, significantly less than Dar's (1993) apparatus. The aluminium rings are constructed from box section to minimise inertia. The stack is secured to the shaking table by its base and shaken horizontally lengthways. Its floor is roughened to aid the transmission of shear waves; the internal end walls are similarly treated to enable complementary shear stresses. The internal side walls are lubricated for plane strain. Not pictured in Fig. 11.1 are the rigid steel restraining frame nor the system of bearings used to prevent unwanted motion in the x and z-directions.

The stack works in conjunction with the shaking table at the University of Bristol's Bristol Laboratory of Advanced Dynamics Engineering (BLADE) facility. The

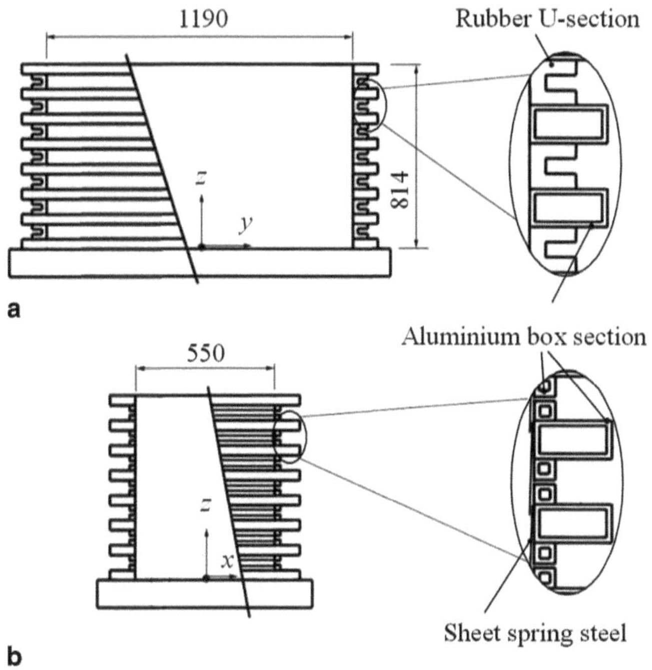

Fig. 11.1 Shear stack soil container **a** Long section, **b** Cross section

BLADE shaking table consists of a 3 m by 3 m platform driven by eight 70 kN servo-hydraulic actuators to give full control of motion in all six degrees of freedom. Despite this capability, the shear stack is designed to be subjected solely to uniaxial dynamic excitation in the *y*-direction.

Dry Hostun S28 sand is used as the test soil. The average mass of sand deposited across the reported tests is 722 kg (±1%) implying a void ratio of 0.76 (±2%). The spread results from difficulties in depositing and leveling the sand close to the model structure, accidental spilling of sand outside of the shear stack, and inaccuracies of the crane scale used to measure mass. The fundamental frequency and damping of the shear stack when filled to a depth of 798 mm and excited with a random (white noise) waveform at a strain level of 0.003% in the *x*-direction was recorded as 26.5 Hz and 8%, respectively. The depth of sand used during the test was 750 mm.

The dimensions of the single degree-of-freedom model of the Euroseistest building (Pitilakis et al. 1999) are detailed in Fig. 11.2. The mass of the model is measured as 130 kg, of which 80 kg are attributable to the 'foundation' and the remaining 50 kg to the 'structure'. The direction of shaking is from left to right in the cross section detailed on the left of the figure. An array of steel columns are fixed to the foundation and pinned to the structure. The model was configured with twelve 5 mm thick columns. The resonant frequency and damping of this model configuration measured by securing the model foundation to the shaking table is 22.5 Hz and

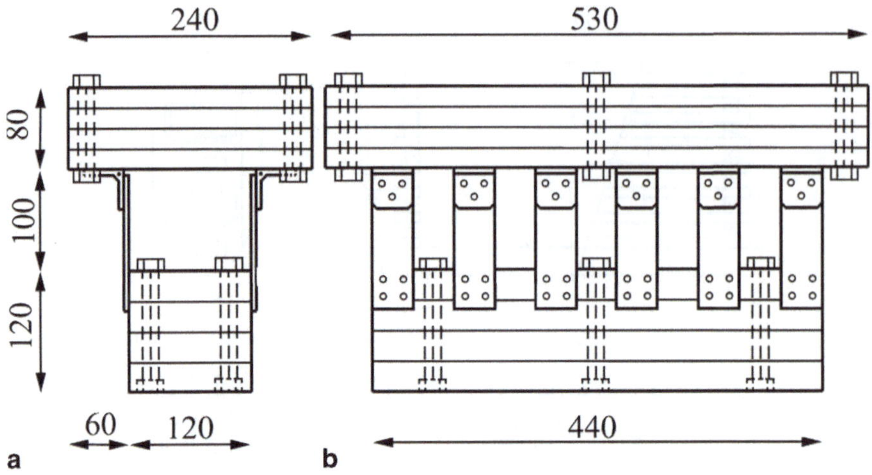

Fig. 11.2 The geometry of the benchmark model structure **a** in direction of shaking, **b** cross section

2.4 % respectively. The foundation was embedded by 100 mm into the soil deposit. An image of the benchmark test taken prior to testing is displayed as Fig. 11.3.

11.2.2 Experiment

Prior to testing, a digital spectrum analyser was used to measure the resonant properties of the system. A random signal from the analyser's inbuilt signal generator was used to excite the shaking table along its *y*-axis with an RMS acceleration of around 0.03 g. The measured FRF is presented in Fig. 11.4 wherein resonant frequency and modal damping values are also inscribed. The 22.5 Hz fixed-base

Fig. 11.3 The benchmark test

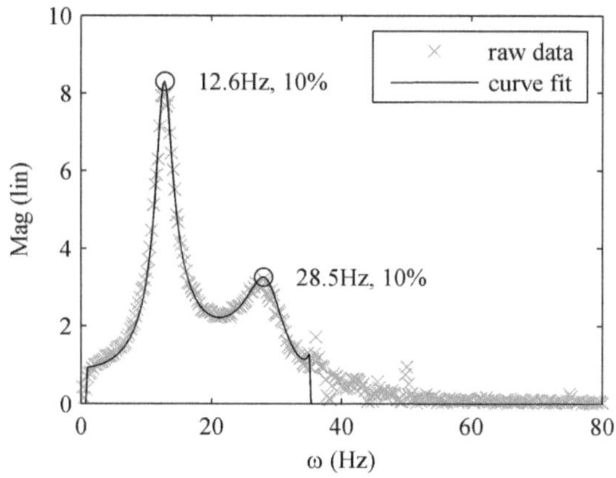

Fig. 11.4 Frequency response function measured across the benchmark SSI system

response of the model structure is not apparent in the FRF. Instead a peak at 12.6 Hz is apparent which represents the resonant frequency of the SSI system.

The shear stack is excited in the *y*-direction by the two earthquake time histories: the Y-Y component of Friuli-San Rocco earthquake (15/9/76) and the N-S component of the Vrancea earthquake (4/4/77). In order to excite both the model (which resonates at 22.5 Hz) and the soil (which resonates at small strain magnitudes at 26.5 Hz) time was scaled by a factor of 5.4 ($1/\sqrt{l}$ where l is the scale factor for length). The 16.75 s Friuli-San Rocco earthquake was replayed over 3.1 s; the 40 s Vrancea earthquake, over 7.4 s.

11.3 RTDS Test Method

A generalized shaking table RTDS system comprising of a numerical model of a soil column and a single degree of freedom physical substructure is shown fully-assembled in Fig. 11.5a and decomposed into its physical and numerical components in Fig. 11.5b. Herein, subscripts N and P are used to denote the numerical model and physical substructure respectively, and m, c, and k, represent mass, damping and stiffness.

The numerical model of the soil column is acted upon by both an excitation load $d(t)$ and an experimentally measured reaction force f generated by the physical substructure inertia and measured at the physical-numerical interface. A numerical interface displacement y_N is calculated under these loads. The interface displacement is used to derive the command signal y_u sent to the transfer system, here a shaking table, which is subsequently imposed using an outer loop controller. The transfer system responds to the command signal and imparts a physical displacement y_P to the physical substructure which, for accuracy, must be commensurate with the interface displacement y_N.

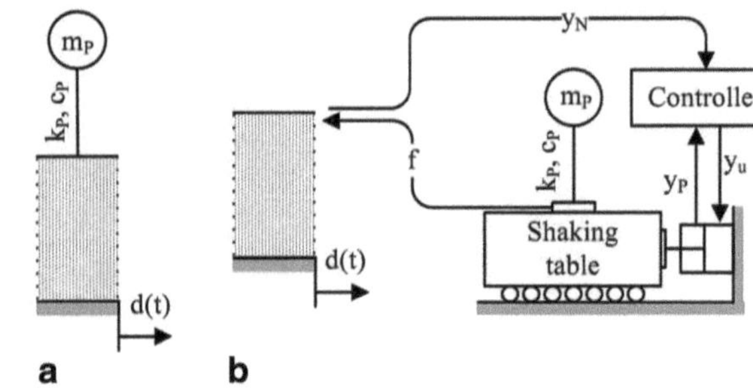

Fig. 11.5 Generalized RTDS for SSI **a** emulated system, **b** RTDS system

Since the focus here is on validation of the SSI substructuring test methodology rather than attaining a high level of realism, linear elastic properties are adopted for the numerical model of the soil. The SSI substructuring control loop is then as depicted in Fig. 11.6, from which linear expressions governing the relationships between the system components can be synthesized (Eqs. (11.1)–(11.4)):

$$y_N = G_{Nd} d - G_{Nf} f \tag{11.1}$$

$$y_u = G_{controller} y_N \tag{11.2}$$

$$y_P = G_{table} y_u \tag{11.3}$$

$$f = G_P y_P \tag{11.4}$$

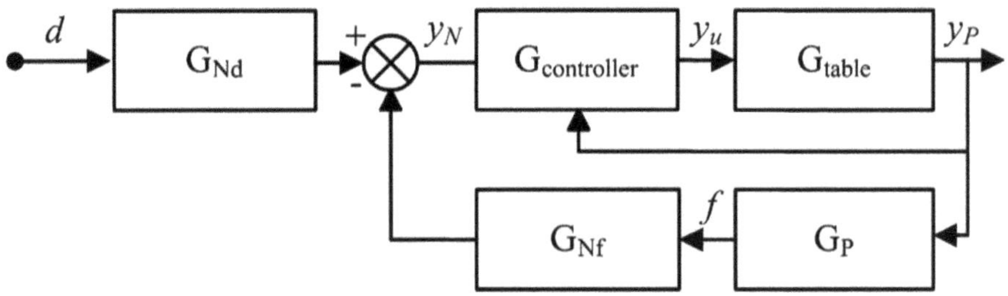

Fig. 11.6 Block diagram for generalized SSI substructuring

11.3.1 Control Strategy

The successful application of the dynamic substructuring test method requires that the displacements of the shaking table track those of the interface outputted from the numerical model. For this, an outer loop controller is required to compensate the dynamics of the shaking table.

11.3.1.1 Proprietary Shaking Table Control

The dynamic motion of the platform of the BLADE shaking table is controlled digitally via a proprietary Instron Labtronic 8800 system. Due to their commercial sensitivity, the precise workings of the digital controller in its entirety are unavailable. However, it is clear that signals from eight linear variable differential transformers (LVDTs), one per actuator, are transformed and fed into six proportional-integral-derivative-lag (PIDL) controllers, one per axis (i.e. x, y, z, roll, pitch yaw). The predominant action of these controllers is via the proportional gain; small integral gain proves to be sufficient to correct any steady-state error and derivative and lag gains are set to zero. The geometric transformation which converts measured (i.e. actuator) to modal (i.e. axis) positions is conducted via an Instron Accelerator controller which also converts modal servovalve drives into physical valve drives. The precise nature of these transformations is unknown.

The analogue input ports of the digital controller provide the ability to use externally generated command signals to drive the shaking table. Thus, an outer-loop controller can be used to drive the shaking table directly whilst retaining the desirable features of the proprietary controller (transformations, drive limits, system health monitoring, etc). Here, the utilized RTDS controllers are complied alongside the numerical model onto a dSpace D1103 real time control board using the Matlab tool Simulink. Command and feedback signals are routed via the I/O connector panel of the dSpace control board.

To evaluate the dynamics of the BLADE shaking table (including the proprietary controller), a sine sweep test was conducted. Figure 11.7 shows the experimentally measured transfer function alongside the frequency domain characteristics of a fourth-order transfer function model of Eq. (11.5). The model provides a reasonable fit to the experimentally measured data up to around 15 Hz.

$$G_{table}(s) = \frac{7.53e7}{(s+60.43)(s^2+41.3s+62.9^2)(s+305.2)} \tag{11.5}$$

11.3.1.2 Delay Compensation

The most prevalent outer-loop control scheme that is intended to enhance performance of RTDS tests is delay compensation. Delay compensation recognizes that a natural delay is inherent within a RTDS control loop due to the inability

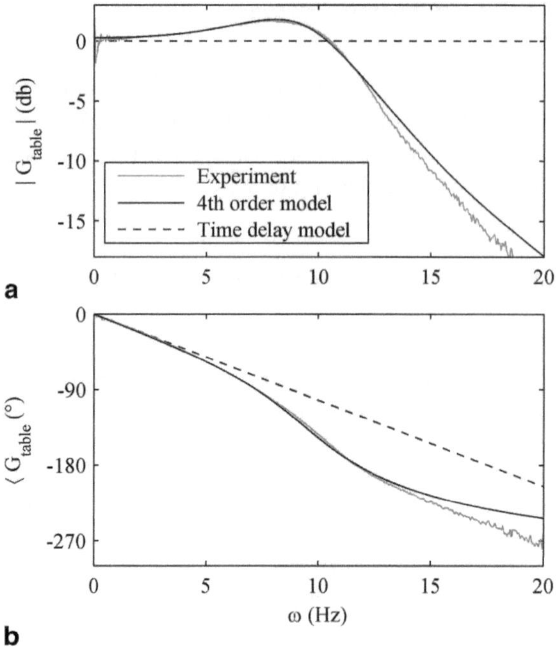

Fig. 11.7 Alternative shaking table models

of the transfer system to react instantaneously to a change in state prescribed by the numerical model. As such, the motions of the transfer system lag behind those intended by τ seconds adding negative damping to a RTDS control loop and resulting in poor performance. These adverse effects can be negated by extrapolating y_N forward in time by an amount equivalent to τ and using the predicted values to drive the transfer system, thereby compensating for the delay. On this basis, numerous alternative formulations of delay compensation have been proposed.

The conception at the core of delay compensation is that the dynamics inherent in the transfer system are encapsulated by the following 'pure delay' model:

$$y_P(t) = y_N(t-\tau) \tag{11.6}$$

The Padé approximation can be used to obtain a rational transfer function from Eq. (11.6) (Chi et al. 2011). Specifying τ as 28.5 ms produces the pure delay transfer function presented in Fig. 11.7. The fit to the experimental data at frequencies above 4 Hz is poor and, as such, the pure-delay model is a poor choice to represent the dynamical characteristics of the shaking table.

As a result, the application of delay compensation to shaking table RTDS has adverse consequences. By way of demonstration, Fig. 11.8 presents the y_P/y_N bode plots derived by simulating the $G_{controller}G_{table}$ outer control loop of Fig. 11.6. Here, a 20 Hz sine sweep is used as the reference excitation (applied as y_N) and the force feedback loop is neglected. Equation (11.5) is taken to represent the shaking table and three alternative formulations for delay compensation are considered: a computationally undemanding single step delay compensation (Horiuchi et al.

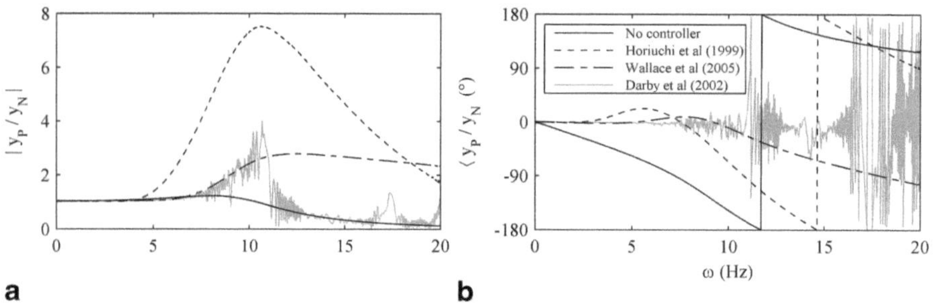

Fig. 11.8 Delay compensated shaking table substructuring

1999), a multi step delay compensation which removes some of the restrictions of the single step technique (Wallace et al. 2005), and a variable delay compensation technique intended to account for time varying delays (Darby et al. 2002). (Details of these alternative strategies are left to be expounded upon within the original publications.)

In Fig. 11.8, an optimal frequency response would be associated with a magnitude of unity and phase of zero. Delay compensation is found to be practicable for shaking table RTDS across narrow, low-frequency bandwidths due to the lack of fit between the underlying pure-delay model and the dynamic characteristics of the shaking table. The frequency-dependent magnitude and phase errors that exist for shaking tables are beyond the scope of delay compensation.

11.3.1.3 Full State Control via Simulation

Shaking table RTDS demands control strategies that are capable of improving on the performance offered by delay compensation. Much attention has been focused on developing mathematical models to represent the dynamics of servo-hydraulic systems (e.g. Williams et al. 2001; Zhao et al. 2003). If such models are inverted, they can be used to cancel the magnitude and phase errors of a transfer system (Carrion et al. 2009). This conception is used herein to develop a new controller for shaking table RTDS. Based on Inverse Dynamics Compensation via Simulation (Tagawa and Fukui 1994) and using Full State Feedback control the new substructuring controller is called Full State Compensation via Simulation (FSCS). Figure 11.9 illustrates the detail of FSCS control scheme, wherein a parallel, real-time simulation loop of the shaking table (the 'virtual' shaking table) is used to generate a noise-free, inverse dynamics control signal. The simulation loop incorporates the fourth-order transfer function model of Eq. (11.5). Since the focus of this chapter is on SSI substructuring, further details regarding FSCS control will be reported elsewhere.

A performance assessment for FSCS is obtained again via simulation of the Fig. 11.6 outer loop (i.e. no force feedback) and applying a sine sweep input as the reference excitation (y_N). Results are contrasted with the 'no controller' case of Fig. 11.8 in Fig. 11.10. FSCS can successfully compensate for the dynamics of the shaking table.

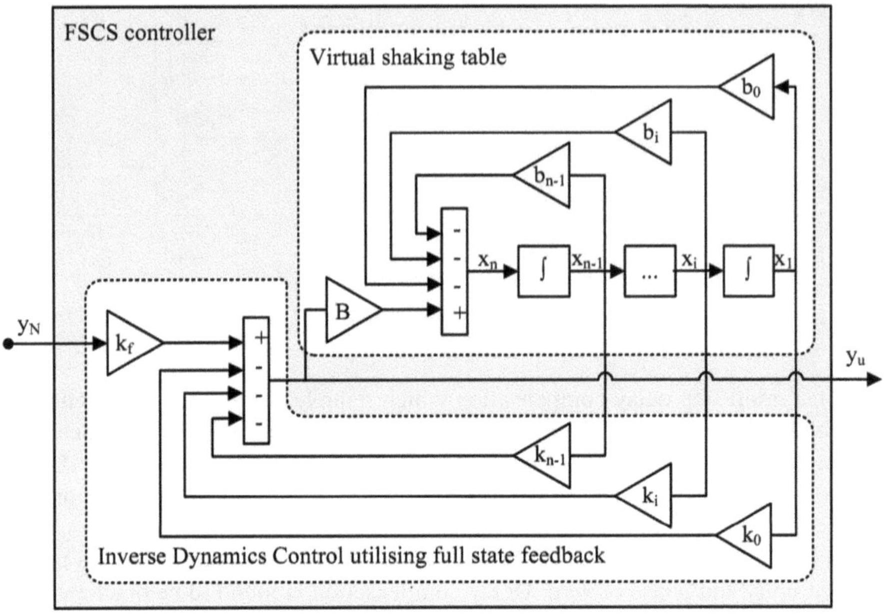

Fig. 11.9 Full state control via simulation

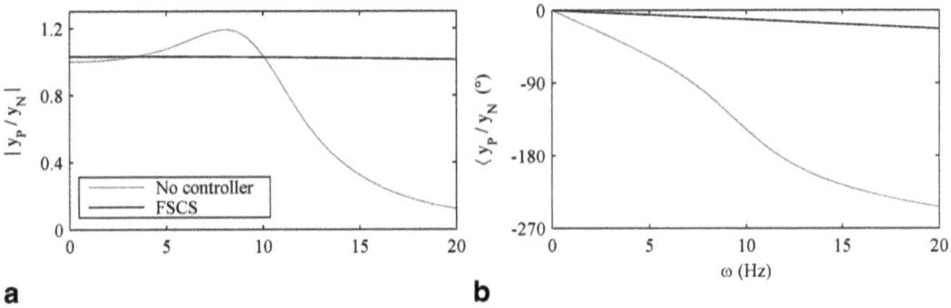

Fig. 11.10 FSCS performance assessment

11.3.2 RTDS for SSI

Linear modeling of the benchmark test has previously provided a reasonable correlation with the benchmark test results (Pitilakis et al. 2008). Due to the elapsed time since the benchmark test was conducted, the physical substructure was no longer available for use in the substructuring SSI tests. As a result, the 'physical' and numerical parts of this system are both modelled numerically. Note, however, that in the following discussion the terminology and the subscripts *N* and *P* are retained in order to distinguish between the different components of the system.

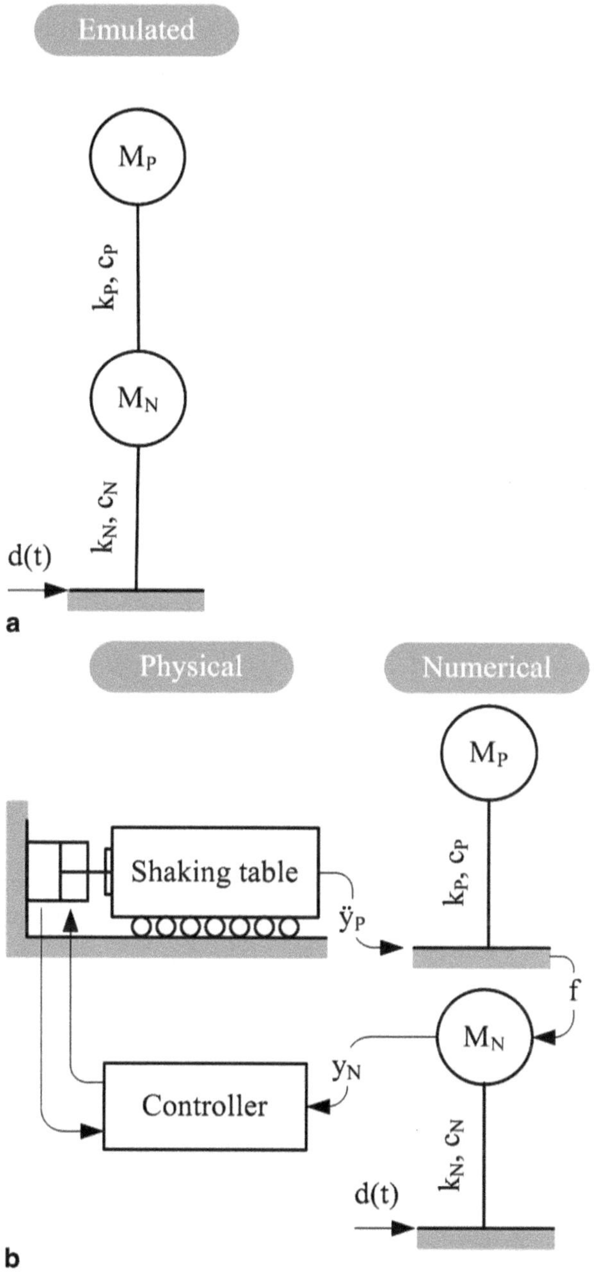

Fig. 11.11 Test system **a** emulated, **b** RTDS

Lumped mass models are adopted for both soil and structure. A schematic of the substructuring system used in the validation tests is that presented in Fig. 11.11. The superstructure of the benchmark test is represented by a single degree of freedom oscillator with the following parameters: $m_p = 50$ kg, $\omega_p = 22.5$ Hz, $\xi_p = 5.8\,\%$.

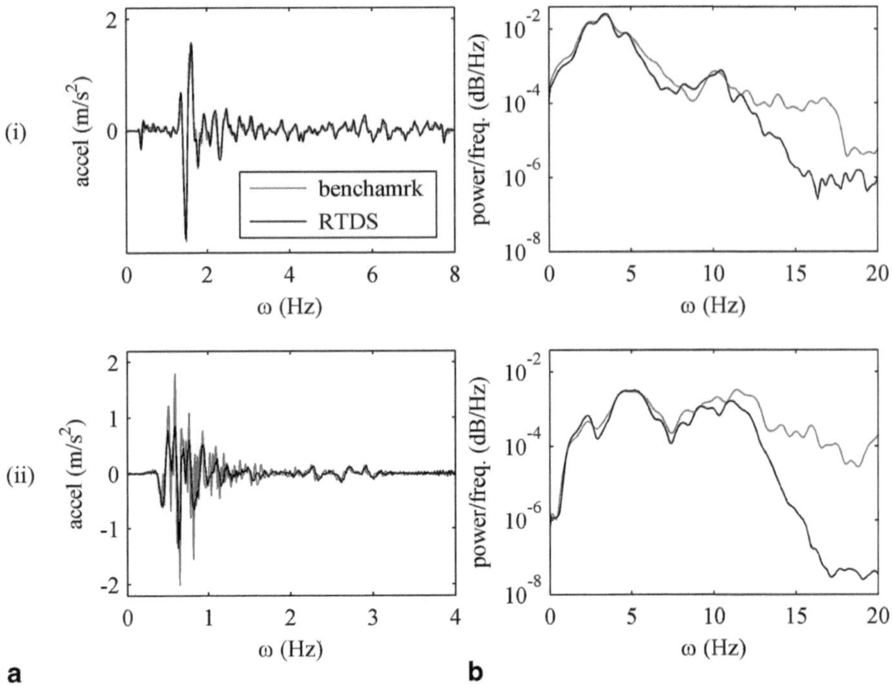

Fig. 11.12 Acceleration **a** time history and **b** power spectra response to the (*i*) Vrancea and (*ii*) Friuli excitation motions

The frequency and damping of the superstructure are those that were recorded for the benchmark test structure in the 'fixed-base' condition (i.e. by securing the foundation of the structure directly to the platform of the shaking table). The soil-foundation system is similarly represented as a single degree of freedom oscillator but with parameters: $m_N = 120.5$ kg, $\omega_N = 15.96$ Hz, $\xi_N = 14.1\,\%$. This parameter set was derived following a frequency domain system identification analysis and with reference to Fig. 11.4.

The frequencies exhibited by the benchmark test exceed the current substructuring capability. Some time scaling is consequently required. A factor of four is chosen which reduces the resonant frequencies of the physical and numerical substructures to 5.6 and 4.0 Hz respectively, and extends the duration of the seismic excitation signals accordingly.

The response of the system is assessed in terms of the acceleration measured on the shaking table ($\ddot{y}_P$) which, in the benchmark test, corresponds to the horizontal accelerations recorded on the foundation of the model structure. A comparison between the recorded acceleration time histories is presented at benchmark-test scale in Fig. 11.12a. The SSI RTDS test appears to perform reasonably well for the Vrancea excitation but poorly for the Friuli motion. The power spectra of the acceleration time histories presented in Fig. 11.12b unify the picture and reveals that for both excitation motions the RTDS test performs adequately up to 10 Hz benchmark scale.

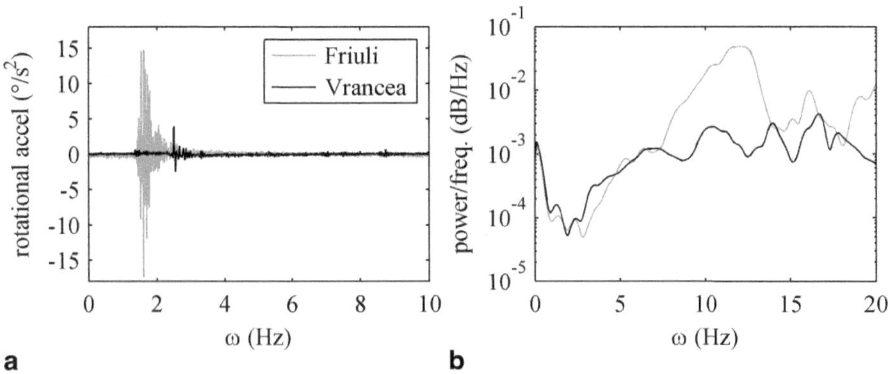

Fig. 11.13 Rocking acceleration of the benchmark test foundation **a** time histories, **b** power spectra

The reason for the malfunction of the RTDS test above 10 Hz is thought to be associated with two phenomena: firstly, the increasing nonlinearity of the soil-foundation system as the resonant frequency (measured at 12.6 Hz in Fig. 11.4) is approached; second, dynamic behaviour that is not taken into account by the RTDS test but is exhibited during the benchmark test. Figure 11.13 shows the rocking acceleration measured on the benchmark test foundation in both the time and frequency domain. Unlike the Vrancea input motion, the Friuli excitation motion induces significant rocking of the benchmark test structure. Furthermore, the frequency band in which the rocking occurs coincides with the frequency band where the RTDS and benchmark test data diverge. A better correspondence between benchmark and RTDS data will be achieved by addressing these issues either through the application of a nonlinear numerical substructure and by the development of the multi-axis shaking table RTDS method.

11.4 RTDH Test Method

The response of soils when subjected to dynamic loading is anisotropic and nonlinear at strain levels well below its failure condition. The stiffness of the material is stress-level and strain-level dependant and its behaviour may change during an earthquake, as primary loading/unloading evolves to reloading and volumetric deformations occur. As a result soil is a complicated material to model numerically. This section takes these considerations into account and considers an inverted Soil Structure Interaction (SSI) test methodology wherein the physical substructure is the soil—tested in a soil container on the shaking table—and the numerical model is the structure. Such testing involves mounting the substructuring transfer system (i.e. an actuator or system of actuators) on the shaking table above a soil container filled with the test soil. The use of an actuator within a substructuring control loop in conjunction with a shaking table pushes the methodology into the category of Real Time Dynamic Hybrid (RTDH) testing.

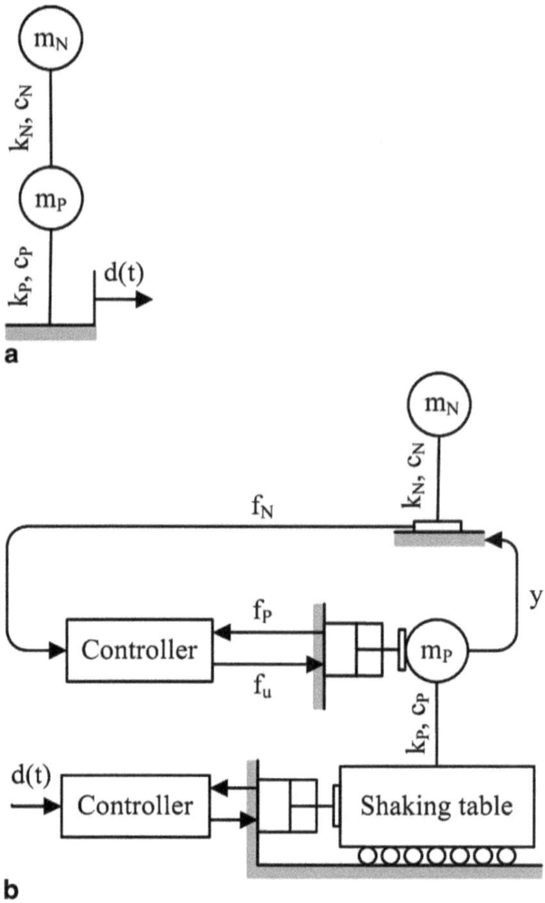

Fig. 11.14 Simulated system **a** emulated, **b** RTDH

RTDH testing combines shaking tables and reactive systems (actuators, reaction walls, strong floors, etc) with substructures and real-time processing hardware for full system simulation (Reinhorn and Shao 2004). Unlike substructure testing conducted away from a shaking table, hybrid testing imposes the true inertial forces in the specimen since the dynamic excitation is applied directly to the physical substructure. The physical and numerical substructures interact through the application of interface equilibrium and compatibility conditions. This is achieved by measuring the displacement imposed on the numerical substructure at the numerical-physical interface and imposing this displacement on the numerical substructure, hence satisfying the equilibrium condition. Then the resultant interface force computed from the numerical model is imposed on the physical system, satisfying the compatibility condition. The transfer system (consisting of a hydraulic or electrical actuator, for example) is required to impose the interface force calculated from the numerical substructure on the physical substructure. Compared to the RTDS described in § 11.3, the loop is reversed and the transfer system is operated in force, not displacement control.

To investigate the performance of a hypothetical RTDH SSI system, a complete model of a test system is developed and used to obtain a simulated RTDH response.

11.4.1 Generalised Hybrid Modelling

The hybrid test system is depicted in Fig. 11.14. Figure 11.14a) presents the emulated test system, Fig. 11.14b the RTDH model. It is the uppermost degree of freedom, modeled numerically in order to calculate the numerical reaction force f_N at the interface between the numerical and physical substructure that is commensurate with the feedback displacement y. f_p is the reaction force at the physical-numerical interface as imposed by the actuator with a desired value f_N. Two instances of FSCS are implemented to compensate for the dynamics of both the shaking table and the actuator.

The physical components of the RTDH test system include the soil (here the lowermost degree of freedom), the shaking table, and the actuator used to impose the interface force. Subjected to reference excitation $d(t)$, the physical substructure responds and the displacement is measured. This displacement is fed to the numerical substructure, producing a reaction force that is applied via the actuator. In essence, the shaking table applies the reference excitation to the physical substructure, the actuator applies the reaction force to the physical substructure.

The block diagram of the feedback control loop pertinent to Fig. 11.14 is depicted in Fig. 11.15 from which the following linear relationships for the different system components can be derived (Eqs. (11.7)–(11.9)).

$$GP = \frac{(c_P s + k_P)}{m_P s^2 + c_P s + k_P} \tag{11.7}$$

$$G_{Nf} = \frac{1}{m_P s^2 + c_P s + k_P} \tag{11.8}$$

$$G_{Ny} = \frac{m_N s^2 (c_N s + k_N)}{m_N s^2 + c_N s + k_N} \tag{11.9}$$

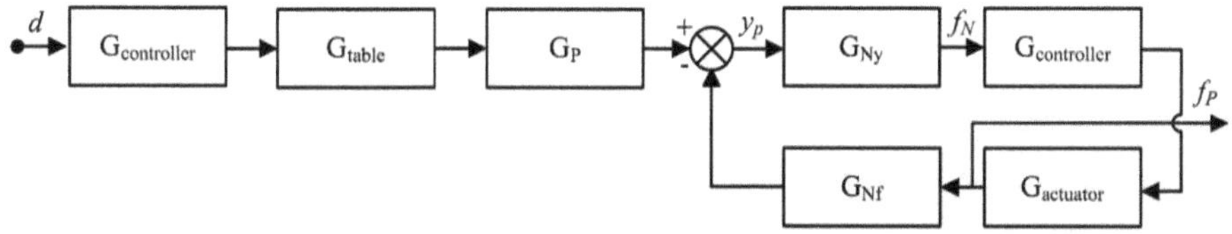

Fig. 11.15 Block diagram of a typical control strategy for a hybrid substructure testing

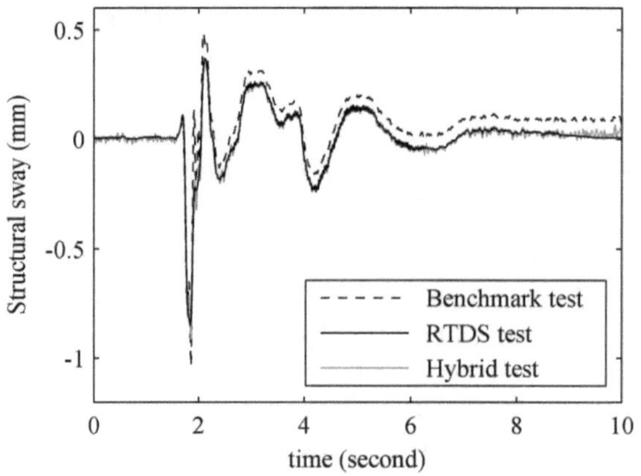

Fig. 11.16 Structural displacement recorded using three alternative test methods

The closed-loop transfer function of the RTDH test can be written:

$$\frac{f_P}{d} = \frac{G^2_{controller} G_{table} G_P G_{Ny}}{1 + G_{Nf} G_{Ny} G_{controller} G_{actuator}} \tag{11.10}$$

An expression for $G_{actuator}$ (from the command force to the applied force) can be obtained by applying Newton's second law to the forces on the actuator piston and following the modeling techniques presented by Carrion and Spencer (2007).

11.4.2 Hybrid Simulation of the Benchmark

The simulated displacement of the RTDH superstructure relative to the foundation is compared with that recorded during the benchmark test and during the RTDS test in Fig. 11.16. The input motion utilized is the N-S component of the Vrancea earthquake. The agreement between the different test methodologies is satisfactory, validating the RTDH test method for SSI studies on the shaking table. However before the method can be adopted by practice, there are a number of issues that require further attention.

As with all shaking table testing, the size of the specimen (the physical substructure) is limited by the capacity of the earthquake simulator. Ideally, for securing fast, accurate control of actuation, the capacity of the shaking table should be far in excess of the model mass. However, shaking table tests of geodynamic models generally involve large volumes of soil in an attempt to reduce the effects of the specimen boundaries. Filled with dry sand, the large and small shear stacks at the University of Bristol are 200 % and 33 % of the mass of the shaking table platform. It is reasonable to expect that such levels of mass will have significant effects on the performance of the shaking table, modifying the dynamic characteristics. The effects of specimen-table interaction would need to be compensated for in order to guarantee the performance of a RTDH test. Furthermore, thought needs to be given to the practicalities of mounting servo hydraulic actuators above soil containers on shaking tables. Seismically rigid reaction frames to support the actuator are required to prevent feedback signals becoming corrupted with components associated with reaction frame resonance. Due to FSCS being an inverse dynamics controller, there exists the possibility that small high frequency components within signals will be amplified and adversely influence test performance. Finally, as mentioned in the introduction to this section, soil is a highly nonlinear material whose stiffness and damping characteristics are time variant. The properties of a physical substructure have a domineering influence on the dynamic stability of a substructuring system. Thus, before RTDH testing of authentic (i.e. non-simulated) SSI systems can occur in practice, a stability analysis method for nonlinear hybrid test substructures needs to be developed.

11.5 Summary

This chapter has considered two implementations of the substructuring test technique for SSI studies using a shaking table. The first implementation separates the SSI system into a numerical soil model which is subjected to excitation at the base and acted upon from above by a physical model of a superstructure tested on a shaking table within the laboratory. The second implementation inverts the numerical-physical divide. The system component tested in the laboratory becomes the soil; the numerical model, the superstructure.

Adequate performance of substructuring tests requires the use of a controller to negate the dynamical attributes inherent within the physical test apparatus. Swept sinusoidal tests reveal that the characteristics of the shaking table significantly depart from the pure-delay model that underlies the delay compensation algorithms of the prevalent substructuring controllers. As such, delay compensation is ineffectual for shaking table substructuring. In response, a new control strategy called Full State Control via Simulation (FSCS) is introduced that conjoins inverse dynamic control and full state feedback to provide a noise free, inverse dynamics control signal that effectively nullifies shaking table dynamics.

The performance of FSCS controlled substructuring using a shaking table was verified through a comparison of the results with those of a benchmark test. The benchmark test of an entirely physical SSI system comprised of a shaking table, a soil container, a granular geomaterial, and a small scale model of a superstructure-foundation system. When the input motion fails to excite neither significant soil nonlinearity nor superstructure rocking, FSCS controlled substructuring test data align with those recorded during the benchmark test, verifying both variants of the test methodology. With more demanding input motions, nonlinear soil models and multi-axis substructuring (combining translational and rotational degrees of freedom) are required. The first of these requirements is addressed in Chap. 11, the second is a topic for future work.

Acknowledgments The research leading to these results has received funding from the European Community's Seventh Framework Programme [FP7/2007–2013] for access to the laboratory of the University of Bristol, UK under grant agreement n° 227887.

References

Carrion JE, Spencer BF (2007) Model-based strategies for real-time hybrid testing. Report No. NSEL-006, Newmark Structural Engineering Laboratory Report Series, University of Illinois at Urbana-Champaign, Urbana, Illinois

Carrion JE, Spencer BF Jr, Phillips BM (2009) Real-time hybrid simulation for structural control performance assessment. Earthq Eng Eng Vib 8(4):481–492

Chen Y, Lü X, Li P, Chen B (2006) Comparative study on the dynamic soil-structure interaction system with various soils by using shaking table model tests. China Civi Eng J 39(5):57–64 (in Chinese)

Chi F, Wang J, Wang Q, Jin F (2011) Delay-dependent stability analysis of mdof real-time dynamic hybrid testing considering compensation. Eng Mech 28(4):200–207 (in Chinese)

Crewe AJ, Lings ML, Taylor CA, Yeung AK, Andrighetto R (1995) Development of a large flexible shear stack for testing dry sand and simple direct foundations on a shaking table. In Elnashai AS (ed) European seismic design practice. Balkema, Rotterdam, pp 163–168

Dar AR (1993) Development of a flexible shear-stack for shaking table testing of geotechnical problems. PhD thesis, University of Bristol, UK

Darby AP, Williams MS, Blakeborough A (2002) Stability and delay compensation for real-time substructure testing. J Eng Mech ASCE 128:1276–1284

Heath A, Darby AP, Bawcombe J (2008) Substructure testing for dynamic soil-structure interaction. In: Proceedings of the second British Geotechnical Association International Conference on Foundations—ICOF Garston, Watford, UK: 1213–1224

Horiuchi T, Inoue M, Konno T, Namita Y (1999) Real-time hybrid experimental system with actuator delay compensation and its application to a piping system with energy absorber. J Earthq Eng Struct Dyn 28:1121–1141

Konagai K, Ahsan R (2002) Simulation of nonlinear soil-structure interaction on a shaking table. J. Earthq Eng 6(1):31–51

Mylonakis G, Gazetas G (2000) Seismic soil-structure interaction: beneficial or detrimental. J. Earthq Eng 4(3):277–301

Nakashima M, Kato H, Takaoka E (1992) Development of real-time pseudo dynamic testing. J Earthq Eng Struct Dyn 21:79–92

Pitilakis D, Dietz M, Wood DM, Clouteau D, Modaressi A (2008) Numerical simulation of dynamic soil–structure interaction in shaking table testing. Soil Dyn Earthq Eng 28:453–467

Pitilakis K, Raptakis D, Lontzetidis K, Tika-Vassilikou T, Jongmans D (1999) Geotechnical and geophysical description of EURO-SEISTEST, using field, laboratory tests and moderate strong motion recordings. J Earthq Eng 3(3):381–409

Reinhorn A, Shao X (2004) Advanced dynamic testing techniques in structural engineering. CIE616, Dept. of Civil, Structural and Environmental Engineering, University of Buffalo

Tagawa Y, Fukui K (1994) Inverse dynamics calculation of nonlinear model using low sensitivity compensator. Proceedings of dynamics and design conference, Akita, 185–188

Wallace MI, Sieber J, Nield SA, Wagg DJ, Krauskopf B (2005) Stability analysis of real-time dynamic substructuring using delay differential equations. J Earthq Eng Struct Dyn 34(15):1817–1832

Williams DM, Williams MS, Blakeborough A (2001) Numerical modeling of a servo-hydraulic testing system for structures. J Eng Mech ASCE 127(8):816–827

Wolf JP (1985) Dynamic soil-structure interaction. Prentice-Hall, Englewood Cliffs

Zhao J, French C, Shield C, Posbergh T (2003) Considerations for the development of real-time dynamic testing using servo-hydraulic actuation. J Earthq Eng Struct Dyn 32(11):1773–1794

Chapter 12
On the Control of Shaking Tables in Acceleration Mode: An Adaptive Signal Processing Framework

Vasileios K. Dertimanis, Harris P. Mouzakis and Ioannis N. Psycharis

12.1 Introduction

A central issue in structural testing is the reproduction of time–series for the application of forced excitation to structural components. This emerges for a series of different experimental studies that include reaction wall tests, seismic tests via shaking tables (Ji et al. 2009; Psycharis and Mouzakis 2012), hybrid testing (Saouma and Sivaselvan 2008), as well as combinations of these (Bursi and Wagg 2008).

The time–series selected as the excitation signal, henceforth referred to as the *reference signal*, may be as simple as a pure sinusoid, or as complex as a strong ground motion. In any case, a requirement of primary importance for the testing engineer is to be able to reproduce this signal with the least possible loss of frequency information. This is not a trivial task, as the mechanical facility[1] that transfers the reference signal to the specimen is characterized by its own dynamics. As a result, the actual signal (henceforth referred to as the *achieved signal*) that finally excites the specimen does not match the reference one.

Focusing on shaking tables, several researchers have tried to introduce further control actions, in addition to the three–variable–controller (TVC) (Tagawa and Kajiwara 2007; Plummer 2007) concept that is usually used to drive the actuators. Among others, a minimal control synthesis (MCS) framework is proposed in Stoten and Gómez (2001), which adaptively controls a shaking table in both displacement and acceleration modes, mainly by formulating an outer loop strategy (other alternatives also exist). Results of this strategy show actual performance improvement (Gómez and Stoten 2000), although the main recommendation of the authors is the

[1] The mechanical facility contains actuators, controller(s), data acquisition devices and possibly inertial elements, such as a table. Since it transfers the reference signal to the specimen, it will be henceforth referred to as the *transfer system*.

V. K. Dertimanis (✉) · H. P. Mouzakis · I. N. Psycharis
Laboratory for Earthquake Engineering, School of Civil Engineering, National Technical University of Athens, Athens, Greece
e-mail: bullit@mail.ntua.gr

F. Taucer, R. Apostolska (eds.), *Experimental Research in Earthquake Engineering*,
Geotechnical, Geological and Earthquake Engineering 35,
DOI 10.1007/978-3-319-10136-1_12

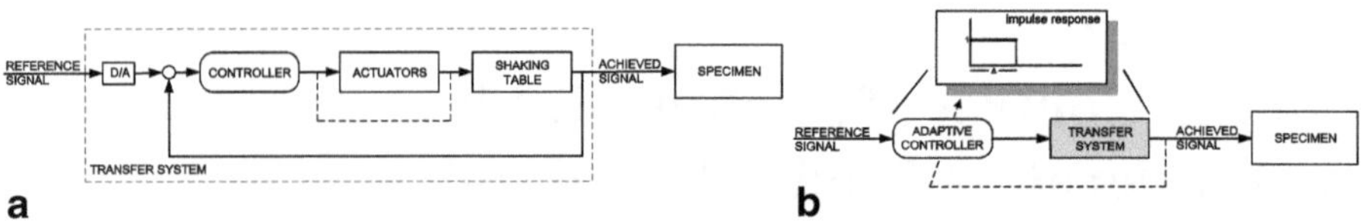

a b

Fig. 12.1 **a** Sketch that shows the fundamental configuration of a shaking table facility, **b** The proposed configuration, at which the reference signal corresponds to acceleration time–series

use of MCS in displacement mode, which retains certain limitations. More recently, an adaptive inverse controller that is based on the recursive least squares algorithm is presented in Shen et al. (2011), again as an outer loop control strategy that is used in conjunction to a TVC. A thorough review on control issues for structural testing is given by Plummer (2007), while the same author has attempted to analytically model a $5m \times 5m$ shaking table with 6 degrees-of-freedom (Plummer 2008).

As a general remark it is observed that (i) most of the proposed algorithms and methods seem to qualify the operation of shaking tables in displacement mode, rather than acceleration mode, and (ii) reported applications involving testing of large–scale specimens (with complex geometry and mass comparable to that of the table) in acceleration mode are limited. This Chapter aims at filling these gaps, by developing a novel framework for the conduction of shaking table tests in acceleration mode. The envisioned scheme takes advantage of established adaptive signal processing methods (Widrow and Wallach 2007) and generates a form of adaptive inverse control that compensates the dynamics of the transfer system, while providing an estimate of the total transfer delay. In this way, the cascade of the adaptive controller and the transfer system tends to become a delayed unit impulse response, so that the specimen gets excited by the reference signal unaltered. To verify the proposed methodology, a structure of sufficiently high mass (comparable to the mass of the table) and complex geometry, with $0.5s$ $(2Hz)$ fundamental period is used as the specimen and two waveform replication tests (sinusoidal and seismic) are performed with satisfying results.

12.2 Description of the Method

The idea behind any test that utilizes a shaking table facility is to force the latter to replicate a *reference signal*. Schematically, this is displayed in Fig. 12.1a: the reference signal, which is first converted into an analogue format by an appropriate digital–to–analogue (D/A) hardware, is driven to the actuators through the controller and it is finally transferred to the table, where a specimen has been installed. While this is a straightforward procedure, several issues arise with respect to

- the type of the reference signal: testing under displacement mode differs considerably from testing under acceleration mode, as in the former case the frequency band is significantly narrower. In addition, many recorded time–series that serve as reference signals (e.g. ground motions) come in the form of acceleration and their integration to displacement almost always results in some loss of information;

- the dynamics of the transfer system: these result in alteration of the actual reference signal in (i) its frequency content, due to the inherent dynamics of the converters, the controller, the actuators and the sensors, and (ii) its temporal realization, as the transfer system dynamics include a delay; and
- the dynamics of the specimen: these cause a serious alteration to the performance of the shake table. This alteration is a function of the specimen's mass and geometry.

To overcome the aforementioned issues, the configuration depicted in Fig. 12.1b is proposed. The idea is simple: place an adaptive controller in series to the transfer system, in order to produce a (delayed) unit impulse response from the reference to the achieved signal. To this end, the theory of adaptive inverse control is utilized. Specifically, two steps are employed and analysed in the rest of the Section: Step 1: An adaptive identification framework is formulated, both off–line and on–line, and the dynamics of the transfer system are identified, including the transfer delay; Step 2: An adaptive inverse identification scheme is performed and an inverse controller is identified. This controller filters the reference signal in a way that cancels the dynamics of the transfer system.

It is noted that, while the method is applicable to any type of reference signal, in the following it is assumed that all the indicated signals refer to acceleration.

12.2.1 Adaptive Identification

The adaptive procedure for specifying the unknown transfer dynamics is illustrated in Fig. 12.2. The (discrete–time) adaptive modeling system samples the reference ($x[t]$) and the achieved ($y[t]$) accelerations and adjusts its internal parameters, in

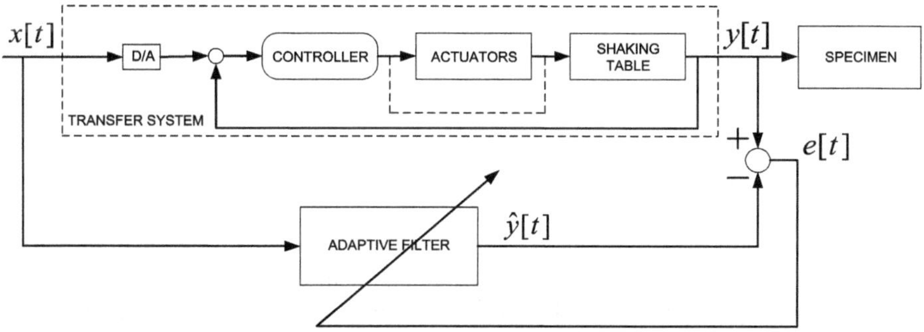

Fig. 12.2 Adaptive identification of the transfer system (having an impulse response $h[t]$). $x[t]$: reference acceleration signal; $y[t]$: achieved acceleration signal. The adaptive filter adjust its weights $\hat{h}[t]$, so that its output $\hat{y}[t]$ matches the achieved acceleration of the shaking table ($y[t]$). Notice that identification is performed with the specimen installed

Table 12.1 Steps of the decorrelated LMS algorithm. Parameter ρ is a trimming factor. Notation follows Fig. 12.2. (Glentis et al. 1999)

Step	Description	Equation
1	Calculate filter's output	$\hat{y}[t] = \hat{\mathbf{h}}^T[t]\mathbf{x}[t]$[a]
2	Calculate error	$e[t] = y[t] - \hat{y}[t]$
3	Calculate decorrelation coefficient	$a[t] = \dfrac{\mathbf{x}^T[t]\mathbf{x}[t-1]}{\mathbf{x}^T[t-1]\mathbf{x}[t-1]}$
4	Update gradient	$\mathbf{g}[t] = \mathbf{x}[t] - a[t]\mathbf{x}[t-1]$
5	Update step size	$\mu[t] = \dfrac{\rho e[t]}{\mathbf{x}^T[t]\mathbf{g}[t]}$
6	Update filter's weights	$\hat{\mathbf{h}}[t+1] = \hat{\mathbf{h}}[t] + \mu[t]\mathbf{g}[t]$

[a] $\hat{\mathbf{h}}^T[t] = [\hat{h}[t,0]\hat{h}[t,1] \ldots \hat{h}[t,n]]^T$ and $\mathbf{x}[t] = [x[t]x[t-1] \ldots x[t-n]]$

order to produce a sampled output ($\hat{y}[t]$) that closely resembles $y[t]$. Once the adaptation has converged, the adaptive filter provides sufficient information about the transfer system dynamics and can be used for further processing (Widrow and Wallach 2007).

Several issues arise with respect to the adaptive identification procedure diagrammed in Fig. 12.2, such as the choice of an appropriate input signal and the internal structure of the adaptive filter. It must be emphasized that the adaptation must be accomplished with the specimen installed on the shaking table, which imposes significant difficulty, since the specimen introduces a strong disturbance to the output of the transfer system. In fact, as the mass of the specimen approaches the mass of the shaking table, more disturbance is added to the output of the transfer system.

This feature limits the range of adaptive algorithms that can be effectively applied to those that retain the ability of very fast convergence. While several other alternatives are possible (and are currently being investigated) the applied method selected for the adaptive identification of the transfer system is the decorrelated LMS algorithm that belongs to a family of adaptive instrumental variable methods (Glentis et al. 1999) and it is characterized by improved convergence properties and low computational complexity. The internal steps of the algorithm are summarized in Table 12.1.

12.2.1.1 Delay Estimation

The application of the decorrelated LMS algorithm to the identification of the transfer system leads to the availability of a mathematical description which describes the unknown dynamics. Normally, any delay terms of the transfer system are estimated as leading zeros in the weights of the adaptive filter. However, due to the presence of disturbance from the specimen, of unavoidable instrumentation noise, as well as of non–Gaussian excitation, these leading terms may not converge to zero at all. To this, an off–line procedure is additionally employed for the verification of transfer delay, which applies a Gaussian white noise acceleration to the shaking table and performs two distinct non–parametric tests for the verification of the delay. The former test estimates the sample cross–correlation function between the reference and the achieved acceleration of the table, while the latter uses the same data to estimate the impulse response of the transfer system from the inverse discrete Fourier transform (DFT) of the frequency response function (FRF).

In detail, assume that the transfer system is described by a finite impulse response (FIR) model of the form

$$y[t] = \sum_{k=0}^{n} h[k]x[t-k-\tau_1] = \mathbf{h}^T[t]\mathbf{x}[t] \tag{12.1}$$

where $x[t]$ and $y[t]$ are the reference and the achieved accelerations of the transfer system, respectively, and τ_1 is the transfer delay. Their cross–correlation function is given by

$$R_{xy}[\tau] = E\{x[t]y[t+\tau]\} \tag{12.2}$$

where τ is the time lag. Substituting Eq. (12.1) to Eq. (12.2) and performing simple algebraic manipulations implies

$$R_{xy}[\tau] = \sum_{k=0}^{n} h[k]R_{xx}[\tau-k-\tau_1] \tag{12.3}$$

with R_{xx} denoting the autocorrelation of the reference signal, which is given by $\sigma_{xx}^2\delta[\tau]$, when the latter is zero mean Gaussian white noise of variance σ_{xx}^2. It follows that

$$R_{xy}[\tau] = \sigma_{xx}^2 \sum_{k=0}^{n} h[k]\delta[\tau-k-\tau_1] = \sigma_{xx}^2 h[\tau-\tau_1] \tag{12.4}$$

where the discrete–time equivalent of the shifting theorem has been applied. Since $h[k]$ is zero outside the $[0,n]$ interval, $R_{xy}[\tau]$ must be zero outside the $\left[\tau_1, n+\tau_1\right]$ interval. Thus, by estimating the sample cross–correlation function between the reference and the achieved acceleration, it is possible to obtain reasonable estimates for the transfer delay.

Remark 1 The previous analysis is valid only when the reference acceleration consists of an unfiltered Gaussian white noise time–series. In any other case, Eq. (12.4) does not hold, so the cross–correlations $R_{xy}[\tau]$ are not zero outside the prescribed interval. This includes the case of bandlimited white noise (Bendat and Piersol 2000).

Remark 2 In practice, the following steps can be utilized to the estimation of the transfer delay using the cross–correlation approach:

1. Excite the transfer system with Gaussian white noise acceleration (reference signal) and sample the achieved acceleration of the shaking table.
2. Estimate the normalized sample cross–correlation $\bar{R}_{xy}[\tau]$ for $\tau = 0, 1, \ldots, N/4$, where N is the length of data, as

$$\bar{R}_{xy}[\tau] = \frac{R_{xy}[\tau]}{\sqrt{R_{xx}[0] R_{yy}[0]}} \tag{12.5}$$

3. Under the assumption of normality, it can be shown (Brockwell and Davis 2002) that $\bar{R}_{xy}[\tau]$ can be considered zero at 95% level of significance, when is contained within the $\pm 1.96/\sqrt{N}$ bounds.
4. Estimate the transfer delay as the leading values (normally τ_1) that fall within the bounds of Step 3, before a non–zero one appears.
5. If required, combine Eqs. (12.4), (12.5) and obtain a rough estimate of $h[t]$ as

$$\hat{h}[t] = \frac{\sigma_{xx}}{\sigma_{yy}} \bar{R}_{xy}[\tau] \tag{12.6}$$

As noted above, the impulse response of the transfer system can be alternatively estimated by the inverse DFT of the estimated FRF

$$\hat{h}[t] = \mathrm{idft}\left(\frac{S_{xy}(f)}{S_{xx}(f)}\right) \tag{12.7}$$

where $S_{xy}(f)$ and $S_{xx}(f)$ are the cross–spectral and autospectral densities of the indicated quantities, respectively. In order to avoid circularly biased estimates (Bendat and Piersol 2000), it is recommended to use the two–sided estimates of $S_{xy}(f)$ and $S_{xx}(f)$ and to avoid using tapering. Naturally, the obtained impulse response estimate provides additional insight about the delay of the transfer system.

12.2.2 Adaptive Inverse Identification

The impulse response of the adaptive inverse controller of Fig. 12.2 is just the reciprocal of the one that describes the transfer system (including its delay). However, since the transfer system is dominated by internal delay, the inverse controller may have difficulty in overcoming it, as it must be a predictor. In addition, if the transfer system is nonminimum phase the inverse controller results unstable, while any disturbances bias the converged solution and prevent the formation of a proper inverse (Widrow and Wallach 2007).

To cope with these problems, two configurations are employed for the adaptive inverse identification of the transfer system, which are shown in Fig. 12.3. The first configuration (Fig. 12.3a) adapts the weight of the inverse model on–line, whereas the second configuration (Fig. 12.3b) performs the same operation off–line. The

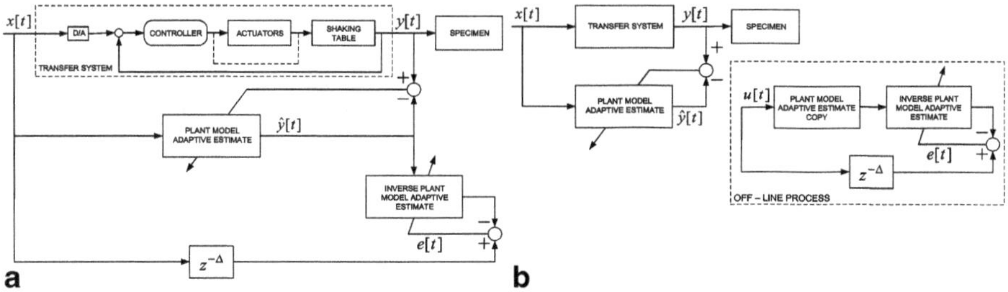

Fig. 12.3 Adaptive inverse identification of the transfer system **a** on–line, **b** off–line

critical point in both configurations is the use of the transfer system's adaptive estimate to the estimation of the inverse. This is justified by the fact that the estimate has essentially the same dynamic performance as the transfer system but is free of disturbance. Therefore, a closer approximation to the desired delayed inverse can be obtained, which is unbiased from plant disturbance.

The main disadvantage of the configuration of Fig. 12.3a is that the convergence of the on–line adaptive algorithm may be very slow, especially when conventional LMS algorithms are utilized, in order to cope with the high disturbance levels that appear to the transfer system due to the presence of the specimen (Widrow and Wallach 2007). To cope with this problem, the decorrelated LMS algorithm described in Table 12.1 is also applied to the inverse identification process, causing a dramatic decrease in the convergence rate. Alternatively, having an estimate of the transfer system's impulse response available, the configuration of Fig. 12.3b can be applied, which utilizes the same functionality. It consists of the on–line scheme of Fig. 12.2 that estimates the impulse response of the transfer system, followed by an off–line simulation that realizes a copy of that estimate to the estimation of the inverse controller. This configuration can be run much faster than real time and, at the same time, it permits the selection of a properly set simulation environment, in which several critical adaptation parameters can be tested and controlled, such as the choice of the input signal, the length of the inverse model and the step size.

In any case, the successful estimation of the transfer system and the adaptive inverse controller is a critical step towards the design of the process depicted in Fig. 12.1b and leads to a cascaded system with impulse response of the form $g[t] = \delta[t-\Delta]$ approximately, where Δ is is the total delay of the cascade of the transfer system and the adaptive inverse controller.

12.3 Application

The method is now applied and assessed through the testing of the specimen shown in Fig. 12.4a. Span length and height limitations imposed by the shaking table dimensions ($4.00m \times 4.00m$), as well as force and overturning moment capacity, required the construction of only one half of a frame. In specific, the specimen con-

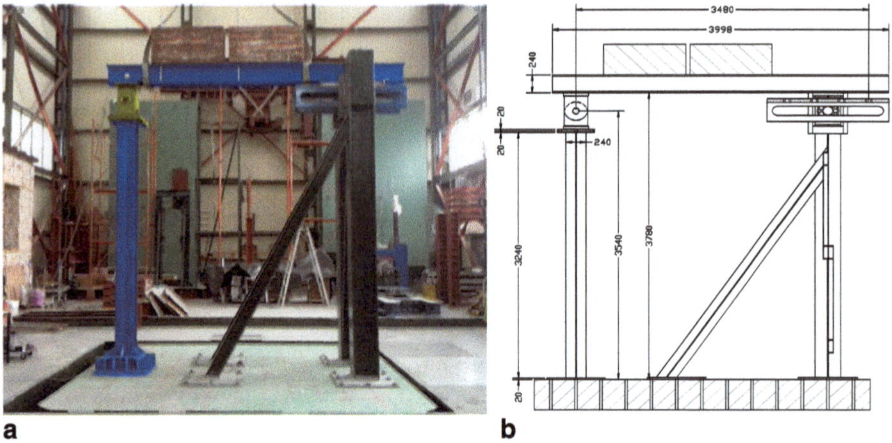

Fig. 12.4 The specimen was used for the tests **a** Installation, **b** Sketch

sists of a fixed-base column that is interconnected to a beam by a hinge. The boundary conditions applied to the free end of the beam pertain to free rotation around the horizontal axis and free sliding in the horizontal direction. This is achieved using a special device that provides a sliding pinned support. Both the column and the beam have grade S275 HEB240 cross sections, leading to a self weight of the specimen of 1102 kg, including the roller and its support. The weak axis of the column is placed parallel to the X axis of the shaking table and a lumped mass (i.e., dead weight) of 2350 kg is distributed on the beam, leading to a total mass of the specimen of 3452 kg. Under this configuration, the specimen is characterized by a single DOF with period at $0.5s$ ($2Hz$). A sketch of the specimen is shown in Fig. 12.4b. It must be stretched out that the conduction of initial tests after specimen installation revealed a strong alteration of the shaking table's performance, as a result of specimen's response. This is attributed to its mass (comparable to the mass of the shaking table) and geometry, which cause significant disturbance to the behavior of the shaking table. As a result, the corresponding reproduction of the reference accelerations prove extremely demanding, as a lot of effort of the controller must be paid to the compensation of any undesired performance.

The experimental procedure consists of (i) sine sweep (four–octave logarithmic, frequency range at [1 16] Hz) and Gaussian white noise ($rms = 0.014g$) tests for the specimen structural identification, (ii) transfer system adaptive identification and delay estimation, (iii) transfer system adaptive inverse identification (online and offline) and derivation of the overall transfer delay, (iv) waveform replication tests.

Figure 12.5 shows the achieved absolute acceleration of the specimen under the four–octave logarithmic sine sweep test. There's a clear pick around $105s$, followed by a smaller one at around $112s$. The corresponding FRF of the specimen, estimated using the recorded acceleration of the shaking table and the specimen acceleration of Fig. 12.5, is shown at Fig. 12.6 for the same frequency range. Estimation is based on Eq. 12.7 using Welch's method with Hanning windowing (number of

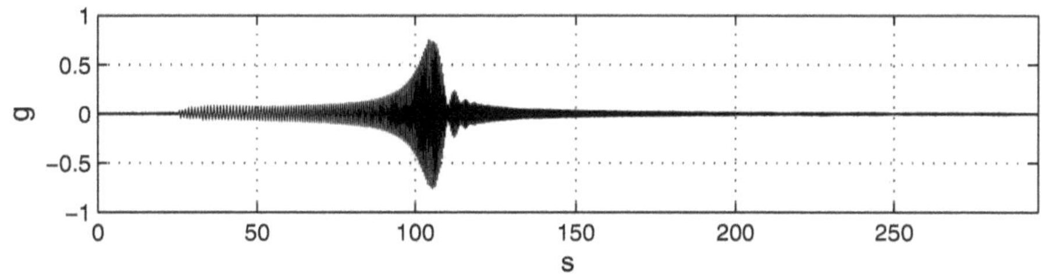

Fig. 12.5 Achieved acceleration of the shaking table at the sine sweep test

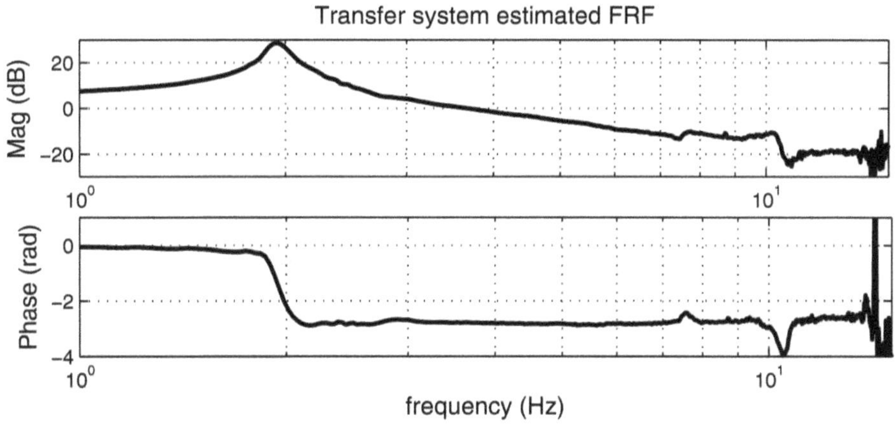

Fig. 12.6 Estimated FRF using sine sweep test data

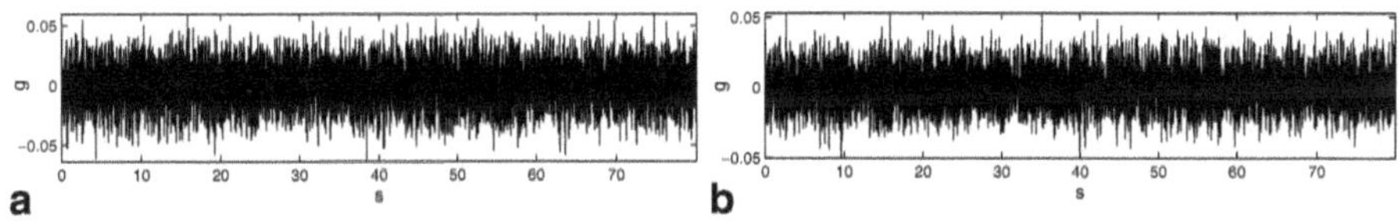

Fig. 12.7 Reference (**a**) and achieved (**b**) acceleration of the shaking table during the Gaussian white noise test

FFT points $\Delta = 250$, 50% overlapping). There's a clear peak in the area of $2Hz$, at which the phase changes from zero to almost $\pi rads$. Using the peak picking method (PPM), the natural frequency and the damping ratio of the specimen are estimated $\hat{f}_n = 1.953Hz$ and $\hat{\zeta}_n = 2.690\%$, respectively, showing that the natural frequency is very close to its theoretical counterpart, while the lightly damped nature of the specimen is confirmed.

The reference and achieved acceleration of the shaking table during the Gaussian white noise test are shown in Fig. 12.7. It is emphasized that the excitation signal is unfiltered, which means that it is characterized by a flat spectrum over the entire

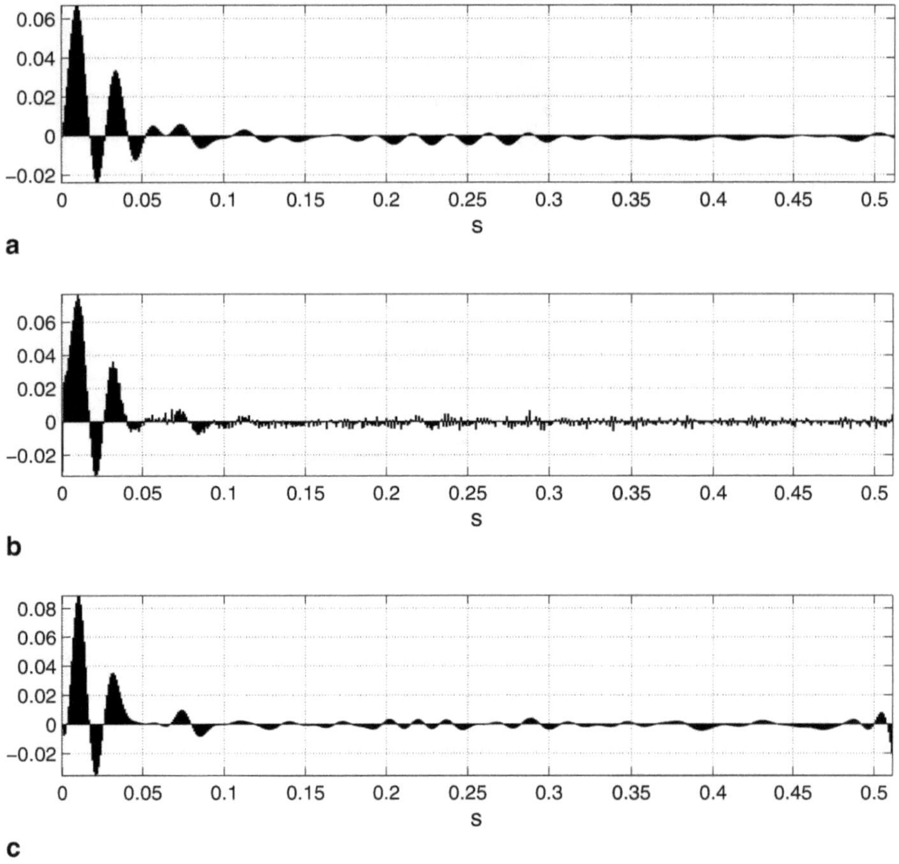

Fig. 12.8 Impulse response of the transfer system (off–line) **a** normalized sample cross–correlation estimate, **b** inverse DFT of the FRF estimate, **c** decorrelated LMS estimate

frequency range considered. Based on these data, Fig. 12.8 displays the impulse response of the transfer system, estimated using the normalized sample cross–correlation function, the inverse DFT of the FRF and the decorrelated LMS algorithm (off–line), respectively, as discussed in Sect. 12.2. All three estimates are characterized by a damped oscillating behavior, as a result of the specimen's presence. While the inverse FRF estimate (Fig. 12.8) has not identified any lag, the other two estimates have done so and this is very evident in Fig. 12.8, where the decorrelated LMS filter has four trailing "zeros". Moreover, as expected, all estimates are nonminimum phase, which means that their direct inverses are unstable and cannot be realized as inverse controllers. Thus, at a first glance, the impulse response of the transfer system is a nonminimum phase system with delay.

Having in mind this significant observation, an FIR model of the transfer system is identified on–line using the scheme of Fig. 12.2. Various FIR models are tested, using Gaussian white noise excitation of variance $\sigma_{xx}^2 = 0.05g$ as input, both unfiltered

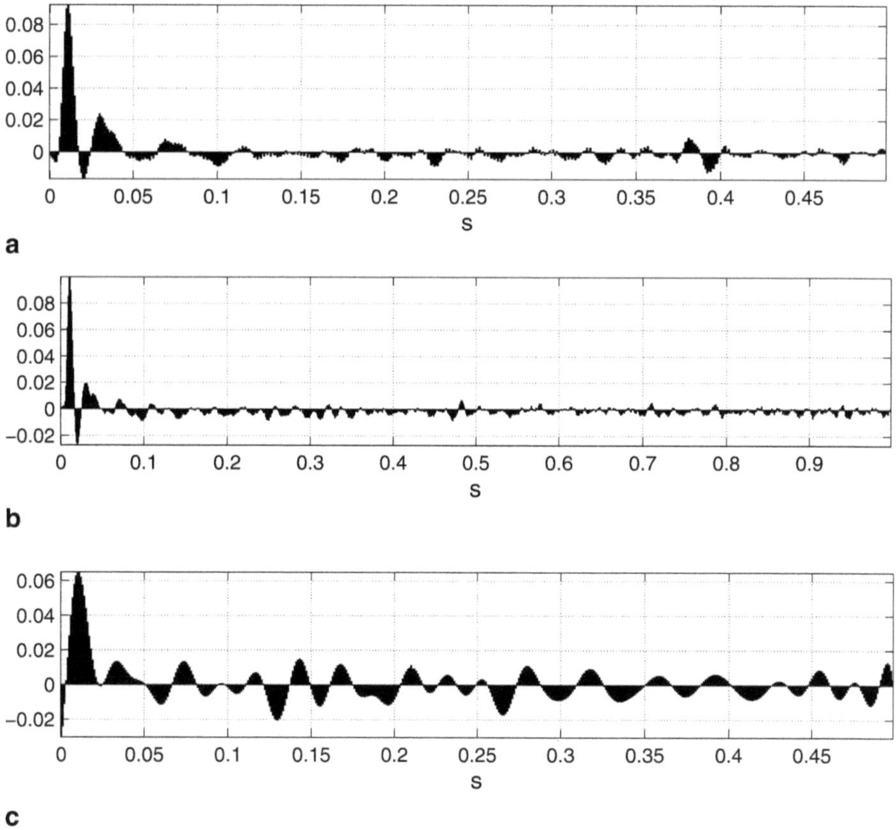

Fig. 12.9 Impulse response of the transfer system (on–line) **a** unfiltered data, FIR order 500, **b** unfiltered data, FIR order 1000, **c** filtered data, FIR order 500

and filtered with cut–off frequency equal to the Nyquist one ($F_s / 2 = 500Hz$). It is noted that, where applicable, the same filter (direct–form FIR transposed of 40*th* order) is applied to the measured shaking table acceleration.

Figure 12.9 shows the estimated impulse responses for three distinct tests. The first two impulse responses (Fig. 12.9a, b with filter lengths 500 and 1000, respectively) retain the same form as the off–line estimate presented in Fig. 12.8, since these two estimates where obtained using unfiltered input–output data. In addition, it seems that an order of 500 taps is sufficient, as higher weights do not seem to contribute with more significant information. On the contrary, the impulse response of Fig. 12.9 has the same damped oscillating structure, but differs substantially both in the leading and the trailing terms (observe that the delay has vanished). This is the result of the filtering action to the data. However, during the subsequent tests for the identification of the inverse controller, this estimate presented the best performance among the three and it was thus chosen as the FIR filter that describes the transfer system. It is noted that the selected FIR filter is also nonminimum phase.

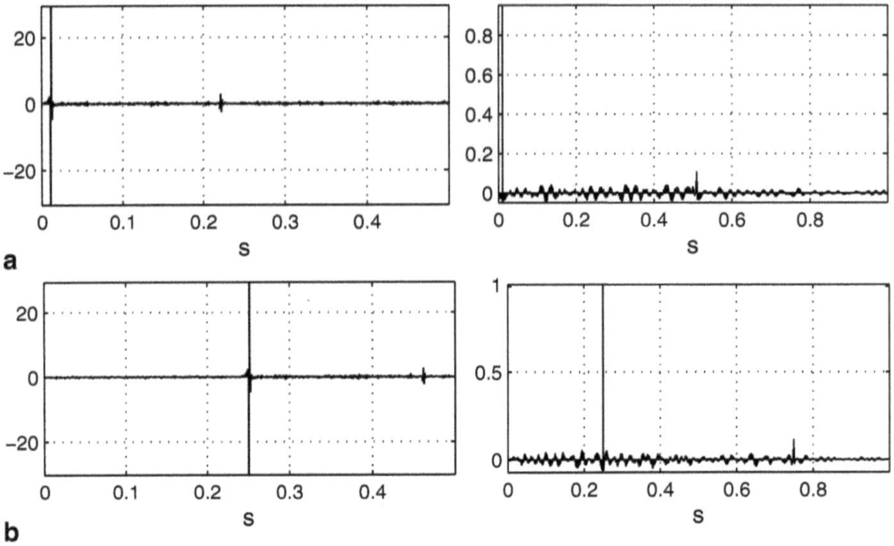

Fig. 12.10 Impulse response (*left part*) of the inverse controller and convolution (*right part*) with the estimated FIR model of the transfer system (Fig. 12.8) **a** FIR order 500, delay $\Delta = 10$, **b** FIR order 500, delay $\Delta = 250$

Due to time and safety reasons, the tests for the identification of the inverse controller are carried out off–line, by realizing the scheme of Fig. 12.3b. Two delay cases are investigated, namely a small delay of $\Delta = 10$ and a large delay of $\Delta = 250$ time steps, respectively, with the latter corresponding to half the FIR filter order of the transfer system. Figure 12.10 presents the results of the adaptation process, including the convolution between the estimated forward filter (impulse response of the transfer system) and estimated inverse filter (adaptive inverse controller). It is apparent that the controller with the large delay (Fig. 12.10b left) has many leading terms close to zero, which can be discarded, leading to a controller with much less delay. However, using the small delay controller in cascade with the transfer system leads to insufficient performance, as shown by Fig. 12.10a (right), where the convolution of these two controllers with the estimated FIR model of the transfer system is displayed. The convolution of the large delay controller with the estimated FIR model of the transfer system (Fig. 12.10b, right) leads to an impulse that is close to unity.

Having estimated the inverse controller and the total delay of its cascade with the transfer system ($\Delta = 250$), the next step is to perform a waveform replication test. Figure 12.11 illustrates the results of these tests. In the former (Fig. 12.11a), a pure sinusoidal acceleration of amplitude $0.1g$ and frequency $2Hz$ is applied to the system, forwarded at 250 time steps, and the achieved acceleration of the shaking table is recorded. As indicated in the figure, the matching of the two signals (reference and achieved, left) is quite good, except for some sharp spikes which are observed at the achieved acceleration. These spikes are the result of some detected inefficiencies in the current hardware of the facility related to analog filtering, and they are

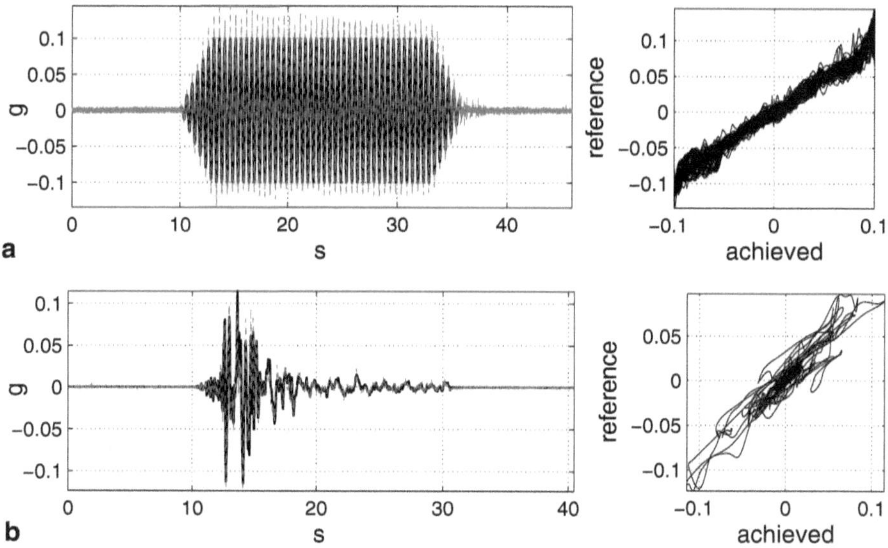

Fig. 12.11 Results of the waveform replication tests **a** sinusoidal input, **b** Kalamata earthquake (50 % scale). *Continuous black line*: reference acceleration, *dashed grey line*: achieved acceleration

not inherent inconsistencies of the proposed method. In the latter case (Fig. 12.11b) the procedure is repeated for the Kalamata earthquake (at 50% scale). Results are again satisfying, yet the amplitude of the achieved acceleration does not coincide to that of the reference and this is clearly reflected in the synchronization plot. This observation necessitates the implementation of additional measures at the adaptation process, as, for example, the realization of automatic gain controllers at both the transfer system estimation and the inverse controller adaptation stages.

12.4 Conclusion

The main objective of this work was to develop a framework for accurate waveform replication in shaking tables, when the latter operate in acceleration mode. To achieve this, the theory of adaptive signal processing was implemented. The main features of the proposed design are summarized as follows:

- Identification of the transfer system is carried out either on–line, or off-line, providing an estimate of its impulse response, as well as of the transfer delay.
- An adaptive inverse controller is subsequently identified either on–line, or off–line (preferable). The result of this stage is a controller that filters the reference signal in a way that cancels the dynamics of the transfer system.
- The process is designed to work in acceleration mode, dropping down the limitations of operation under the displacement mode.

The conducted tests on the specially design specimen were satisfying. Two issues that must be further investigated pertain to (i) the adoption or exclusion of low–pass filtering, as the latter may alter the adaptation process, and (ii) the introduction of automatic gain controllers, so as to assure that the cascade of the estimated FIR filters is indeed a delayed version of the Kronecker's delta function. Currently, more sophisticated adaptive inverse control methods are investigated, as it seems that advanced signal processing and data driven modeling are sufficient pathways, not only for conventional shaking table testing, but also for more advanced experimental procedures, such as real–time hybrid testing.

References

Bendat JS, Piersol AG (2000) Random data: analysis and measurement procedures, 3rd edn. Wiley, Chichester

Brockwell PJ, Davis RA (2002) Introduction to time series and forecasting, 2nd edn. Springer, New York

Bursi OS, Wagg DJ (eds) (2008) Modern testing techniques for structural systems. CISM courses and lectures, vol 502. Springer, New York

Glentis GO, Berberidis K, Theodoridis S (1999) Efficient least squares adaptive algorithms for fir transversal filtering. IEEE Signal Process Mag 16(4):13–41

Gómez EG, Stoten DP (2000) A comparative study of the adaptive mcs control algorithm on european shaking tables. In: Proceedings of the 12 WCEE, Auckland, New Zealand, paper 2626

Ji X, Kajiwara K, Nagae T, Enokida R, Nakashima M (2009) A substructure shaking table test for reproduction of earthquake responses of high-rise buildings. Earthq Eng Struct Dyn 38(12):1381–1399

Plummer AR (2007) Control techniques for structural testing; a review. Proc Inst Mech EngPart I: J Syst Control Eng 221(2):139–169

Plummer AR (2008) A detailed dynamic model of a six-axis shaking table. J Earthq Eng 12(4):631–662

Psycharis IN, Mouzakis HP (2012) Assessment of the seismic design of precast frames with pinned connections from shaking table tests. Bull Earthq Eng 10(6):1795–1817

Saouma V, Sivaselvan MV (eds) (2008) Hybrid simulation: theory, implementation and applications. Taylor & Francis, London

Shen G, Zheng ST, Ye ZM, Yang ZD, Zhao Y, Han JW (2011) Tracking control of an electro-hydraulic shaking table system using a combined feedforward inverse model and adaptive inverse control for real-time testing. Proc Inst Mech Eng Part I: J Syst Control Eng 225(5):647–666

Stoten DP, Gómez EG (2001) Adaptive control of shaking tables using the minimal control synthesis algorithm. Philos Trans Royal Soc A: Math, Phys Eng Sci 359(1786):1697–1723

Tagawa Y, Kajiwara K (2007) Controller development for the e-defense shaking table. Proc Inst Mech EngPart I: J Syst Control Eng 221(2):171–181

Widrow B, Wallach E (2007) Adaptive inverse control: a signal processing approach. Wiley, New York (reissue Edition)

Chapter 13
Refined and Simplified Numerical Models of an Isolated Old Highway Bridge for PsD Testing

Fabrizio Paolacci, Luigi Di Sarno, Raffaele De Risi, Giuseppe Abbiati, Arkam Mohammad Zeki Mohamad, Marialaura Malena and Daniele Corritore

13.1 Introduction

The project RETRO deals with the assessment of the seismic vulnerability of an old reinforced concrete viaduct with portal frame piers and studies the effectiveness of different isolation systems through pseudo-dynamic tests on a large scale model. The laboratory tests are carried out within the research program of the Seismic Engineering Research Infrastructures for European Synergies (SERIES), financially supported by the Seventh Framework Programme of the European Commission. The prototype structure is a 1:2.5 scale reinforced concrete (RC) existing bridge system tested with hybrid sub-structuring techniques at the ELSA laboratory in Ispra, Italy.

F. Paolacci (✉) · A. M. Z. Mohamad · M. Malena · D. Corritore
Department of Engineering, Roma Tre University, Rome, Italy
e-mail: fabrizio.paolacci@uniroma3.it

A. M. Z. Mohamad
e-mail: arkam.mohamad@uniroma3.it

M. Malena
e-mail: marialaura.malena@uniroma3.it

D. Corritore
e-mail: daniele.corritore@uniroma3.it

L. Di Sarno
Department of Engineering, University of Sannio, Benevento, Italy

R. De Risi
Department of Structures for Engineering & Architecture,
University of Naples Federico II, Naples, Italy
e-mail: raffaele.derisi@unina.it

G. Abbiati
Department of Civil Engineering, University of Trento, Trento, Italy
e-mail: giuseppe.abbiati@unitn.it

F. Taucer, R. Apostolska (eds.), *Experimental Research in Earthquake Engineering*,
Geotechnical, Geological and Earthquake Engineering 35,
DOI 10.1007/978-3-319-10136-1_13

The present chapter, which is complementary to the companion chapter focusing on the outcomes of a PsD test campaign on the Rio-Torto bridge, discusses the non-linear response of the bridge in the "as-built" and "isolated" configurations (Paolacci et al. 2014a). The seismic performance assessment is carried out by using refined non-linear structural models implemented in an advanced and reliable computer platform. The earthquake behaviour of the refined models used for the prototype structure accounts for non-linear phenomena of the viaduct, e.g. strain penetration of plain bars, shear deformation of transverse beams, flexural deformations in columns and beams. The finite element models are calibrated on the basis of the experimental tests results. The assessment of the seismic response system is investigated in terms of local and global response parameters. In addition, the effectiveness of the isolation systems used as retrofitting is also investigated numerically. The outcomes of the comprehensive nonlinear analyses are used to simulate the seismic response of the viaduct in the as-built and isolated configurations for Pseudo-dynamic testing, which is illustrated in a companion chapter.

13.2 Description of the Case Study

The case study structure is the Rio-Torto viaduct, an old reinforced concrete (RC) bridge consisting of a thirteen-span deck with two independent roadways sustained by 12 couples of portal frame piers (Fig. 13.1). The piers comprise two solid or hollow circular columns with diameters of either 120 or 160 cm, respectively. Such piers are connected at the top by a cap-beam and at various heights by transverse beams with rectangular cross-sections. The transverse beams may experience fragile collapse under earthquake lateral loadings.

Fig. 13.1 General view of the viaduct Rio-Torto

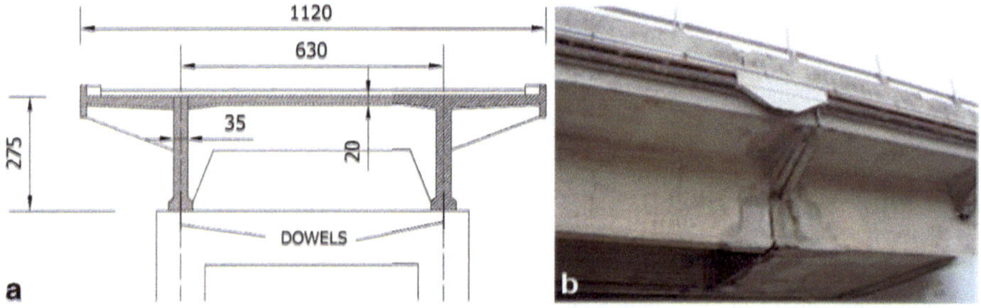

Fig. 13.2 **a** Cross-section of the deck **b** Gerber saddles view

The height of the piers varies between 13.8 m, near the abutments, to 41 m, at the center of the bridge. The deck includes two Π-shaped 2.75 m high reinforced concrete beams (Fig. 13.2a), which are interrupted by Gerber saddles (Fig. 13.2b) placed at the second, seventh and twelfth span, respectively. The deck is connected to the piers by two steel bars inserted in the concrete. The bearings at the abutments consist of fixed devices.

The length of the spans varies between 33 and 29 m. The columns have two types of cross-sections: a solid circular one with diameter of 120 cm and a hollow section with external and internal diameters equal to 160 cm and 100 cm respectively. Details of reinforcement of each portal frame and the geometry of all the piers can be found in (Paolacci et al. 2014b).

The uniformly distributed weight of the deck is approximately 170 kN/m for each road way. Thus, each pier has a vertical load varying between 5600 and 5300 kN.

13.3 Development of a Refined Nonlinear Model in OpenSEES

13.3.1 Non-Linear Phenomena in the As-Built System

A preliminary investigation aimed at studying the cyclic behaviour of the piers was carried out both numerically and experimentally. A test campaign performed at the structural laboratory of Roma Tre University (Italy) consisted of quasi-static cyclic displacements imposed to three 1:4 scale specimens of pier #12. Complete details on the experimental results can be found in Paolacci and Giannini 2012 The outcomes of the aforementioned experimental tests proved that the response of the pier is greatly affected by the behaviour of local details such as non-linear shear deformability of the transverse beam or strain-penetration of the plain steel bars. In Fig. 13.3a the numerical and experimental cyclic force-deflection responses of pier #12 are shown, whereas in Fig. 13.3b the experimental shear crack pattern of a transverse beam is shown.

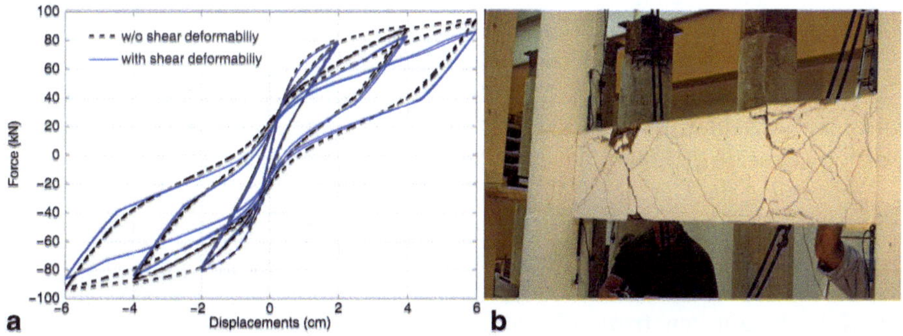

Fig. 13.3 Pier #12: **a** Force-deflection cycle, **b** Shear damage in the transverse beam

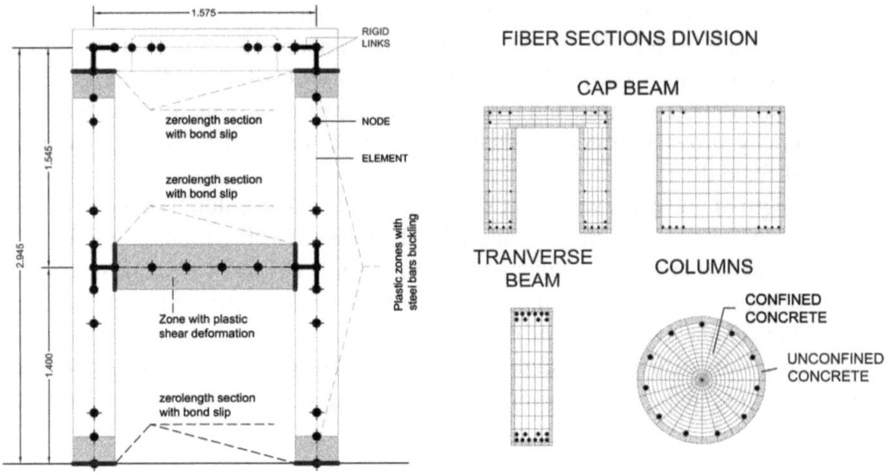

Fig. 13.4 Numerical model of pier #12

13.3.2 The FE Model of the "As Built" Viaduct

A non-linear finite element (FE) model of a single bridge lane was created in OpenSEES, a well-known software for seismic analysis of structures (McKenna et al. 2009). The model analyzed herein refers to the full scale system. The FE model comprises fiber beam elements simulating the actual layout of the reinforcing bars of beams and columns of the piers. The structural elements are modeled by nonlinear beam elements with a flexibility formulation. All degrees of freedom are fixed at the base of the finite element model. The section of each element is subdivided into fibers, reproducing the exact position and dimension of the reinforcing bars and their corresponding constitutive law.

For example, the cross sections of the pier #12 are shown in Fig. 13.4. The Kent-Scott-Park model is adopted for the behavior of concrete. The constitutive law has a first parabolic branch up to a compression peak stress of 26 MPa at a strain equal to 0.25 %, and a decreasing linear branch to 22 MPa at a strain of 0.6 %. According to results reported in the literature, especially from experimental tests, the contribution of the concrete tensile strength in modeling structures with plain steel bars and poor seismic details can be neglected (Arani et al. 2013).

The reinforcing steel bars are modeled according to the Menegotto-Pinto constitutive law. A yield stress equal to 360 MPa is assumed here, along with a modulus of elasticity equal to 205,000 MPa and a hardening parameter equal to 0.025.

Concerning the Gerber saddles, the saddles were modeled as hinges, with the possibility of transferring shear in the longitudinal and transversal directions.

13.3.2.1 Strain Penetration Effect of Plain Steel Bars

It is of primary importance in the FE model to consider the bond slip effect taking place at the bottom and top of the columns. This is due to the difference between the

Fig. 13.5 Crack opening for bar slippage

deformations of the bars and concrete, which yields the typical crack pattern shown in Fig. 13.5. In literature, the bond-slip problem and its contribution to the lateral flexibility of structures for horizontal forces has been widely investigated. Such effect may be significant for plain bars due to the low bond between concrete and steel. Following the approach proposed by Zhao and Sritharan (1996), the bond-slip effect may be accounted as a concentrated rotation representing the slippage of the bars in a section.

The bond-slip phenomenon was simulated by using a fiber zeroLengthSection element with a unit length. As a result, the element deformations correspond to section deformations and the moment-curvature is equivalent to the moment-rotation relationship. The rotation due to bond slip may be accounted for by defining a proper stress-slip relationship for steel, describing the interaction between concrete and the steel bar. The bond-slip phenomenon is accounted for when modelling the sections at the top and bottom of the columns and the end sections of the transverse beam. The bond-slip parameters are chosen according to experimental results of pull-out tests carried out in the laboratory of the Roma Tre University. The value of the slip s_y, corresponding to yielding of the bars, is equal to 0.5 mm, whereas the ultimate slip s_u is assumed equal to $40s_y$. For further details see Paolacci and Giannini (2012).

13.3.2.2 Modeling of Non-Linear Shear Behaviour

In order to calibrate the numerical model of the piers, shear behaviour of the transverse beam should be implemented. Shear response plays an important role especially for existing structures that do not meet seismic engineering design criteria. Several studies concerning the shear behaviour of RC beams or walls and their interaction with flexural response can be found in literature (Ceresa et al. 2007; Hidalgo et al. 2002).

Considering the formulations in the literature and the relatively scarce information for experimental results dealing with shear behaviour in the presence of plain longitudinal bars a phenomenological shear-strain hysteretic relationship for

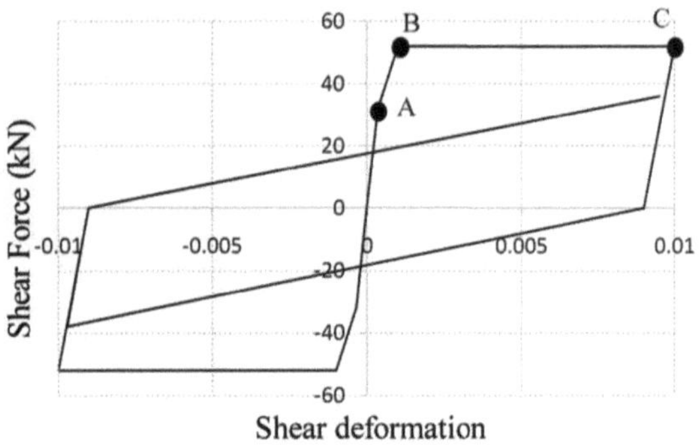

Fig. 13.6 Shear force-deformation relationship

shear behaviour of the transverse beam was assumed in the present investigations. It consists of a tri-linear envelope curve with stiffness and strength degradation with pinching response which is always observable in the RC elements subjected to shear forces. The model is similar to the one proposed by D'Ambrisi and Filippou (1999) and Lee et al. (2005) except for both the influence of axial force on the shear relationship, here neglected, and the use of a tri-linear backbone curve. The force deformation relationship for shear is implemented by using the OpenSEES command "Section Aggregator", which groups the behaviour of different materials into a single section force-deformation model. The shear and flexural behaviour are correlated by means of equilibrium equations, even though their mechanical formulations are uncoupled. A uniaxial material is chosen to represent the sectional shear behavior, defined through three points of the envelope curve (Fig. 13.6) (Paolacci and Giannini 2012).

The first point (A), corresponds to the onset of shear cracking, the second one (B) is referred to the maximum shear strength V_B of the beam. The third point has an ordinate $V_C = V_B$ which implies that the overall shear force remains constant for the last cycles. The force V_B was obtained according to the formula proposed by Priestly et al. 1988, and confirmed using the software Response 2000 (Bentz 2000), based on the Modified Compression Field theory (Vecchio and Collins 1988). The shear deformation corresponding to points A, B and C was obtained from the experimental data of Roma Tre test campaign on pier #12. In particular, points (A), (B) and (C) were characterized respectively by a shear deformation $\gamma_a = 3.5 \times 10^{-4}$, $\gamma_b = 1.0 \times 10^{-3}$ and $\gamma_c = 10\gamma_b$.

13.3.3 The FE Model of the "Isolated" Viaduct

The three-dimensional (3D) FE model was developed to simulate accurately the seismic response of the Rio Torto bridge. The numerical OpenSEES model is similar to the FE system used for the non-isolated case: nodes and elements numbering, modeling of the piers (considering all nonlinear sources, such as shear behaviour and fixed end rotation), deck simulated with elastic beam-elements, piers assumed fixed at the base, with transverse beams at different heights. Two differences were, however, included for the base isolated system: the Gerber-saddles were removed (i.e. a continuous beam was utilized) and isolation devices between the piers and the deck were introduced, representing the seismic retrofitting scheme.

The 3D-FE model implemented in OpenSEES was used to verify the design of the isolation system. In doing so, a simplified novel, yet robust procedure was developed for the design of the base isolators. Two assumptions were employed: the system is assumed partially isolated (i.e. isolation devices at the top of piers and pinned supports at the abutments), and nonlinear (bilinear) devices were used as isolators. The design procedure adopted is illustrated in (Della Corte et al. 2013).

13.3.3.1 Non Linear Response of Isolators

Three basic types of friction pendulum (FP) devices are generally used for new and existing constructions: (i) isolators with one spherical sliding surface, which may be at the top or at the bottom of the device, connected to a spherical hinge (Fig. 13.7); (ii) isolators with two main spherical surfaces and an interposed point rocker articulation that allows relative rotations (Fenz and Costantinou 2006); (iii) devices with two perpendicular cylindrical surfaces and two perpendicular cylindrical articulations allowing relative rotations (Marin 2006). The choice of the type of FP device depends on the structure to be retrofitted, and on the allowable displacement of the structural system (Priestley et al. 1988). Such displacements generally control the design of the isolators. FPs with two-spherical surfaces are often used to minimize

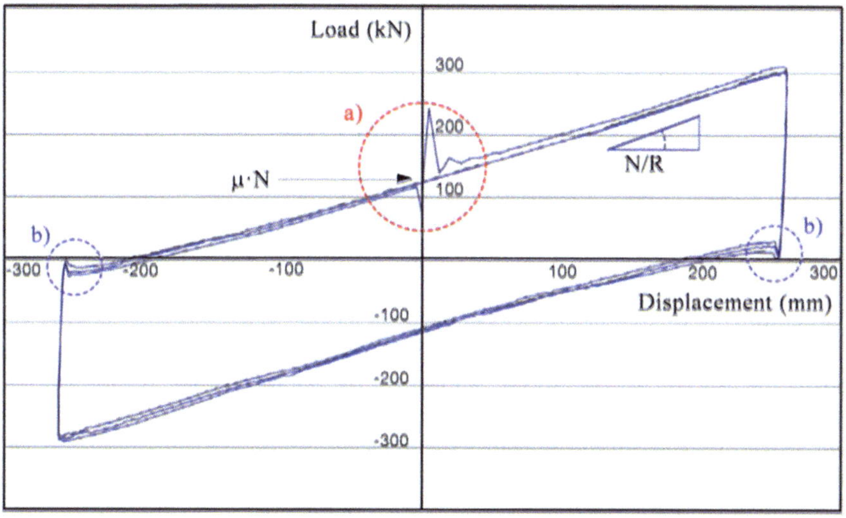

Fig. 13.7 Hysteretic behaviour obtained during dynamic tests on single concave surface sliding pendulum

the plan dimensions of the isolator and to limit the vertical load eccentricity caused by the horizontal displacement. The third type of FP isolator is used when a different behaviour is required in two perpendicular directions.

The first type of device is the most commonly adopted due to its simplicity; such devices were adopted to seismically isolate the Rio Torto bridge, which is characterized by relatively low displacements and similar response along the lateral and transverse directions. The basic elements of the single-surface FP include: an upper anchor plate, a sliding surface, a sliding material interface, a rotation element, a rotation sliding surface and a lower anchor plate.

The dynamic response of FP devices can be simulated by a bilinear force-displacement relationship:

$$V_{FPS} = \mu_f \cdot N + \frac{N}{R} \cdot \Delta_{iso} \tag{13.1}$$

where μ_f is the friction coefficient, N is the normal force, R is the device curvature radius and Δ_{iso} is the sliding displacement in the isolator. Figure 13.7 provides a typical hysteretic behaviour obtained during dynamic tests on a sliding pendulum with one sliding surface. The Figure also shows the variation of the friction coefficient associated to the breakaway of the motion (a) and the change in sign of velocity (b).

13.4 Earthquake Response of the Bridge Structure

13.4.1 Performance Criteria

The structural performance of the bridge was assessed at different earthquake levels. Elastic and inelastic response was considered in the analyses as discussed hereafter.

The Serviceability Limit State (SLS) can be identified with the formation of hairline cracks in the transverse beams and at the bottom sections of columns (Slight damage condition). SLS corresponds to elastic behaviour of the members so that the bridge system remains in service in the aftermath of the earthquake; structural safety is not impaired.

According to this definition and to the results of experimental tests conducted at Roma Tre University (Paolacci and Giannini 2012), it was possible to identify the displacements at the top of piers #9 and #11 corresponding to slight damage conditions. Threshold displacements for SLS can be assumed equal to about 7 cm and 4 cm, respectively. Figure 13.8 shows the shear-deformation cycle of the transverse beams and the moment-curvature cycle of the bottom section of the columns of pier #9 and #11.

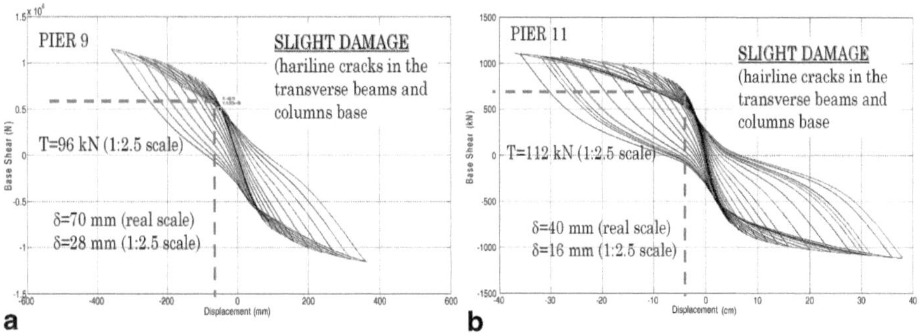

Fig. 13.8 Slight damage condition (SDC) for piers #9 (*left*) and #11 (*right*)

The performance at the ultimate limit state (ULS) is monitored through deformation and resistance parameters. Interstorey drifts corresponding to 2% were assumed as the maximum lateral drifts of the bridge piers (Paolacci and Giannini 2012). Significant drops, e.g. reduction larger than 20%, in the force-deformation curves were also assumed as structural response parameters used to check the ULS condition. Bar buckling at the beam-to-column connections, rupture of the stirrups, extensive shear failure and widespread concrete crushing were also used as performance criteria. The latter damage pattern was monitored visually at each load step during PsD testing. The load-deformation hysteretic curves were plotted during the PsD tests, allowing for detection of stiffness deterioration and strength degradation.

13.4.2 Earthquake Record Selection

Given the geographical position of the bridge and the recent earthquake swarms occurred in the region (especially the earthquake records of May 20th and 29th, 2012), the strong motions of the 2012 Emilia (Italy) earthquakes were employed. The Mirandola records (MRN station) were used because of their significant seismological characteristics, i.e. PGAs and duration of the accelerograms. The East-West May 29th record was used for the SLS; the North-South component was used to assess the seismic performance at the ULS.

Figure 13.9 shows the response spectrum of the record corresponding to the North-South component of the May 29th seismic event, as recorded in Mirandola. Values of spectral accelerations around 0.40 g are found in a period range 1–1.5 s, i.e. similar to the important natural periods of the part of the “as built bridge” in which pier #9 and #11 are located. The length of the signal was reduced assuming an Arias intensity of 99.4%. The input was chosen to be compatible with the ULS design spectra of Eurocode 8 (CEN 2005 (1994)). The filtered signal used for the ULS case is shown in Fig 13.11.

Figure 13.10 shows the response spectrum of the East-West component. In this case, values of PGA=0.2 g are attained within a range of periods 1–1.5 s, fully compatible with the SLS condition (see Sect. 13.5).

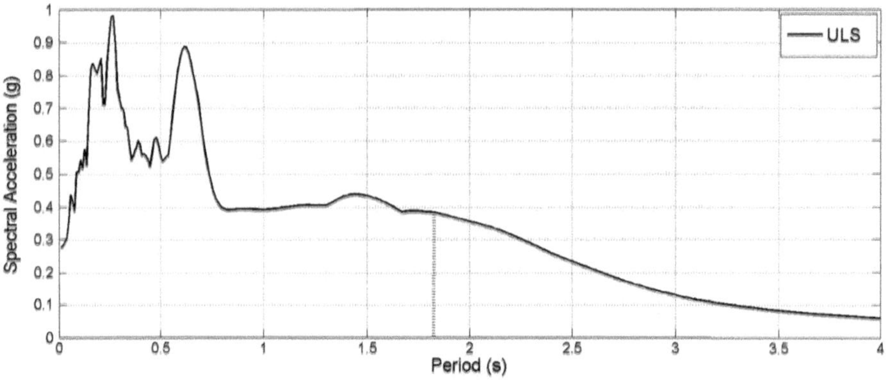

Fig. 13.9 Response spectrum acceleration of the 29 May 2012 record (North-South component)

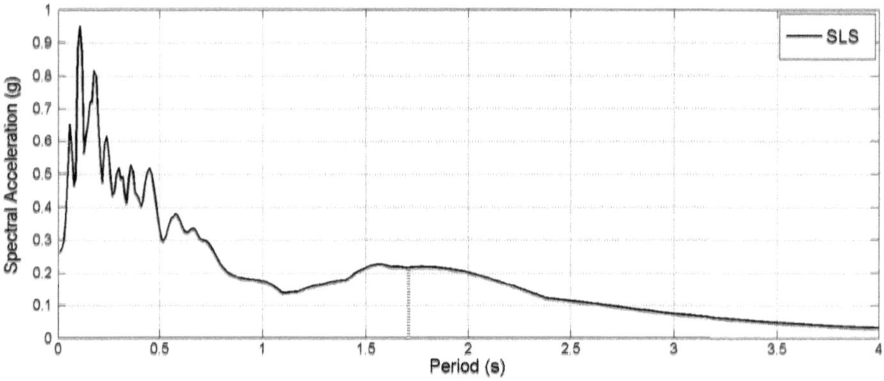

Fig. 13.10 Response spectrum acceleration of the 29 May 2012 record (East-West component)

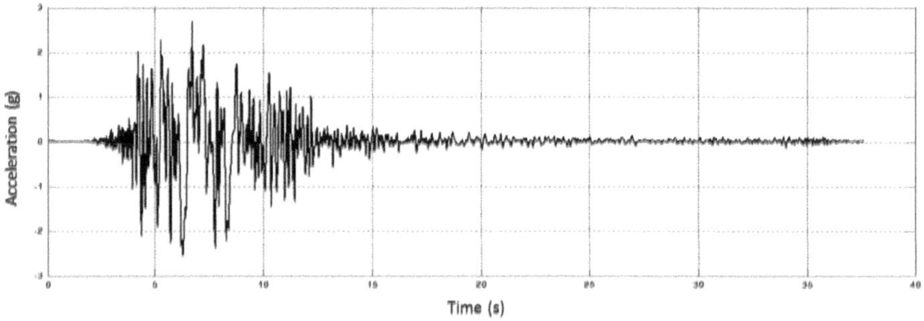

Fig. 13.11 Filtered signal used for ULS analysis

Table 13.1 Modal characteristics of the Rio-Torto viaduct in the transversal direction

Mode	Frequency [Hz]	Modal mass fraction	Cumulated modal mass fraction
1	0.67	0.0269	0.0269
2	0.71	0.4912	0.5181
3	0.78	0.1167	0.6348
4	1.24	0.2069	0.8417
5	1.27	0.0457	0.8874

13.4.3 Modal Analysis of the Viaduct

The reported modal characteristics of the FE model of the bridge refer to seismic loading, applied in the transversal direction. Accordingly, Table 13.1 summarizes frequencies, modal mass fractions and cumulated modal mass fractions up to Mode #5 of the non-isolated bridge.

As shown in Table 13.1 the first five eigenmodes contain 88.74 % of the modal mass in the transverse direction. Figure 13.12 shows deformed shapes of Modes #2 and #4 of the viaduct, which are characterized by the largest modal mass fractions in that direction, i.e. 49.12 and 20.69 %, respectively.

The influence of the distributed mass of piers on the dynamic response of the bridge is negligible, as confirmed by an additional model ran without the masses of piers. The influence of the rotational mass of the deck on the seismic response of the viaduct was also very limited. When the bridge is subjected to lateral displacements the piers rotate, but the excited mass involved in this movement is very limited, resulting in a variation of the normal forces in the pier columns that can be considered to depend mainly on the horizontal force applied at the top of the piers. Moreover, the elastic torsional modes of the deck are also negligible and therefore their contribution can be disregarded. Few modifications were introduced on the present model to simulate the isolated bridge. In greater detail, Gerber saddles were removed and a pair of OpenSEES *single surface FP bearing* isolator elements -one per column- were interposed between each pier portal frame and the deck.

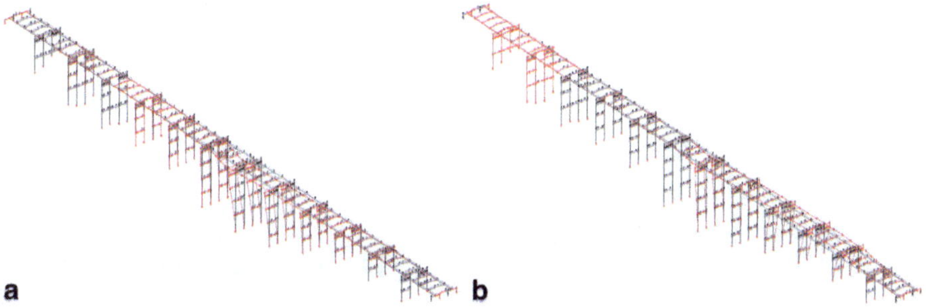

Fig. 13.12 **a** 2nd Eigenmode, **b** 4th Eigenmode

13.5 Numerical Analysis of the As-Built Model

13.5.1 Simulation of the Response for SLS

Figure 13.13a, b show the maximum lateral displacement and the maximum base shear of each pier respectively. The maximum absolute displacement (lower than 20 cm) occurs in the taller pier (#7), whereas pier #9 and #11 present maximum displacement of 65 and 55 mm, respectively. The SLS is thus fully satisfied for pier #9, whereas for pier #11 is exceeded (by nearly 35 %). Therefore, during the PsD tests carried out at the ELSA laboratory in Ispra, it was suggested to increase the PGA starting from a value of 50 % of the maximum PGA for SLS (0.25 g). The maximum base shear was approximately equal to 870 kN for pier #9 and 1000 kN for pier #11. These values are greater than the values predicted in the static cyclic analysis. This is due to the dynamic amplification of the motion within the actual modes of vibration of the test specimen.

The global and local cyclic response of the piers indicates that their behaviour is primarily elastic, as shown in Fig. 13.14. Figure 13.15 and Fig. 13.16 show the moment-curvature response at the bottom section of the right column of pier #9 and #11 respectively.

The transverse beams also responded in the linear range, at least for shear actions, as shown in Figs. 13.17 and 13.18. This is also demonstrated by the maximum drift level, below 0.6 %. The transverse beam shows non-negligible hysteresis loops in flexure, even if the number of cycles is very limited to (~ 1) and the plastic excursion is restricted to a maximum ductility of 2, and not for all piers.

It may thus be argued that the application of the input signal recorded during the Emilia earthquake of 29 May 2011 (Component WE) may result into slight damage conditions for the pier.

13.5.2 Simulation of the Seismic Test for ULS

The results are presented in terms of the maximum lateral displacement of piers, maximum base shear of piers, and the cyclic response of each pier (Figs. 13.19 and 13.20).

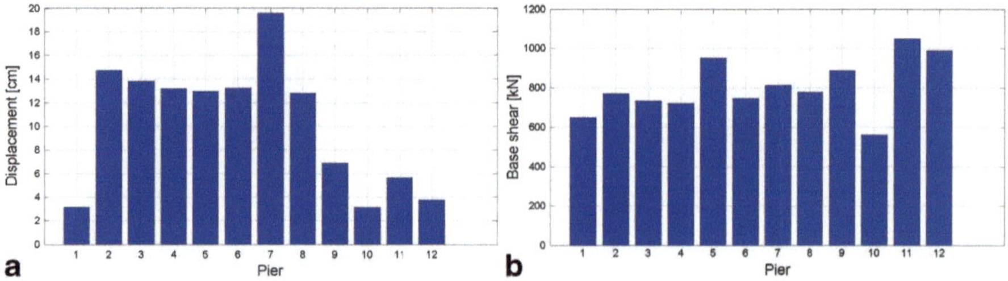

Fig. 13.13 **a** Maximum lateral displacement of piers for SLS **b** Maximum base shear of piers for SLS

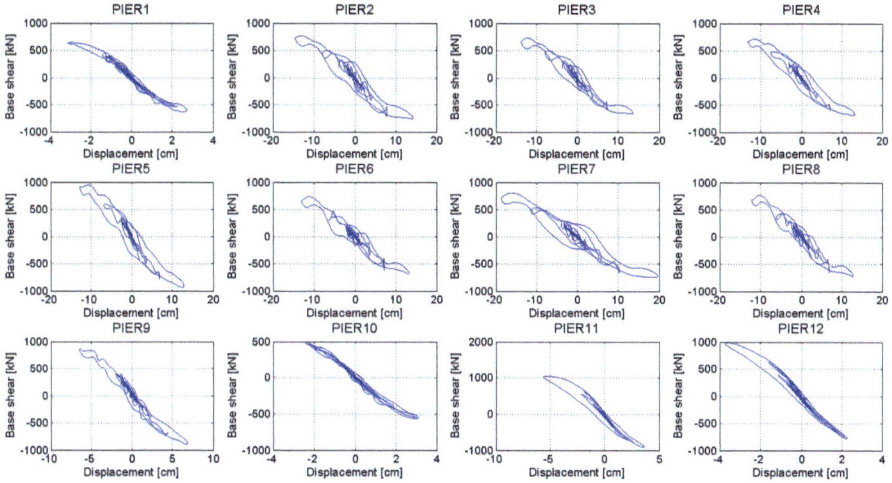

Fig. 13.14 Hysteretic cycle response of piers for SLS

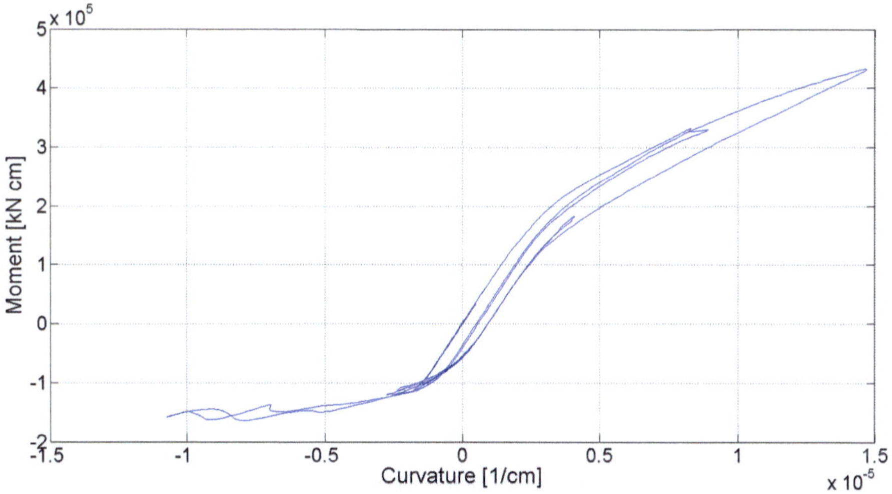

Fig. 13.15 Moment-curvature cycles at the bottom section of the *right* column of pier #9 (SLS)

Figure 13.19a shows that the maximum absolute displacement (<32 cm) occurs in the taller pier #7, whereas pier #9 and #11 exhibit maximum displacements of 17 cm and 15 cm, respectively. The maximum base shear is about 1200 kN, both for pier #9 and #11 (Fig. 13.19b).

The analysis of the cyclic response (Fig. 13.20) shows that plastic deformations occur in all piers. A pronounced pinching effect can be noticed in several piers, including pier #11. This is due to the effect of shear and bond. The expected level of crack width at the column base due to bar slippage is on the order of 1.5–2 mm,

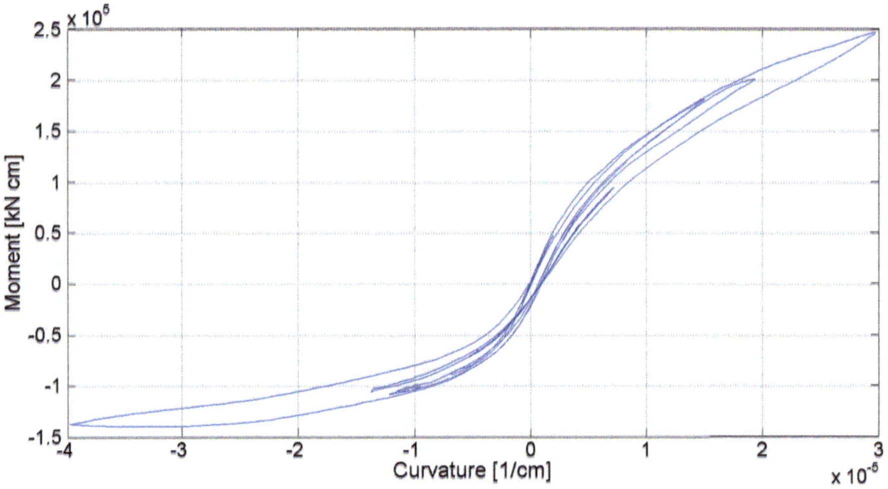

Fig. 13.16 Moment-curvature cycles at the bottom section of the *right* column of pier #11 (SLS)

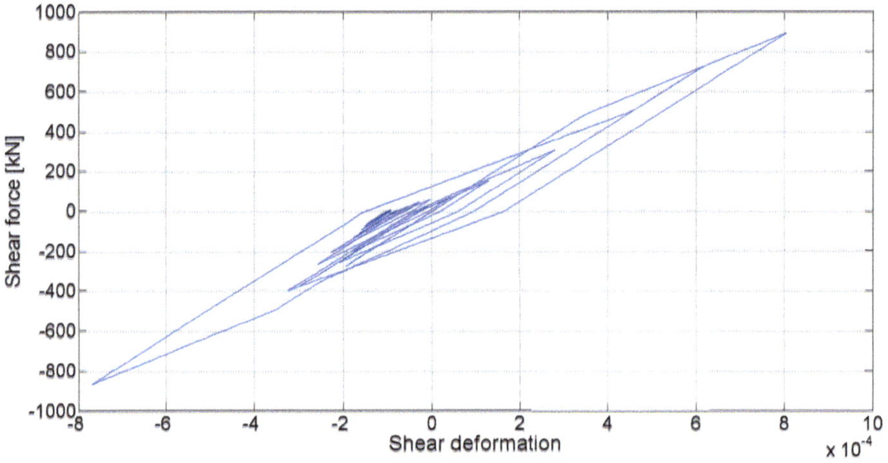

Fig. 13.17 Shear force-deformation cycles of the transverse beam at the 1st level of pier #9 (SLS)

as confirmed by the numerical model. The slippage is not enough to avoid flexural damage in the columns as shown by the moment-curvature cycles at the bottom section of the left column of piers #9 and #11 (Figs. 13.21 and 13.22); the maximum ductility is about 3.

The level of shear damage in the transverse beam is also high as confirmed by the hysteretic behaviour shown in Figs. 13.23 and 13.24. The maximum shear deformation is 3×10^{-3} and 6×10^{-3} for pier #9 and #11 respectively, which correspond to an extensive shear cracking pattern as already shown in (Paolacci and Giannini 2012). For pier #12, a 1 % drift was obtained, corresponding to shear failure of the

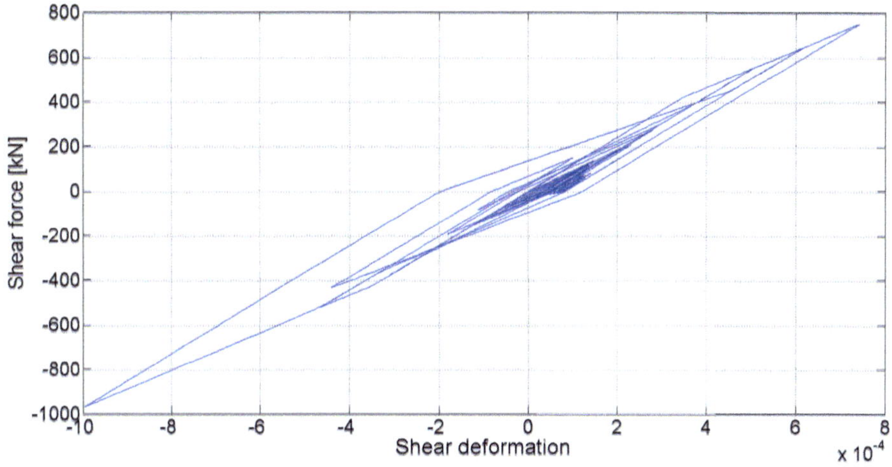

Fig. 13.18 Shear force-deformation cycles of the transverse beam at the 1st level of pier #11 (SLS)

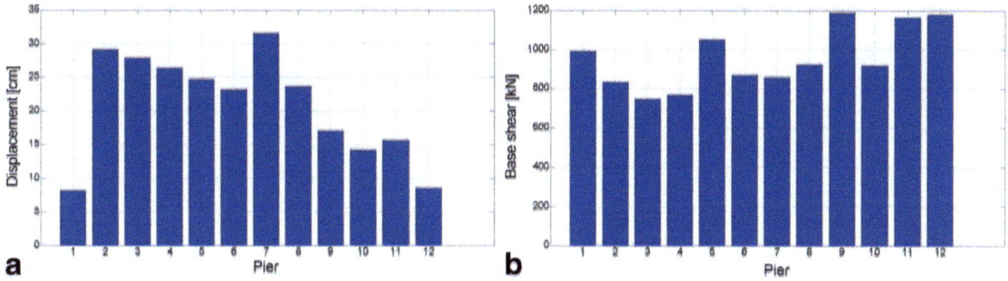

Fig. 13.19 **a** Maximum lateral displacement of piers for ULS **b** Maximum base shear of piers for ULS

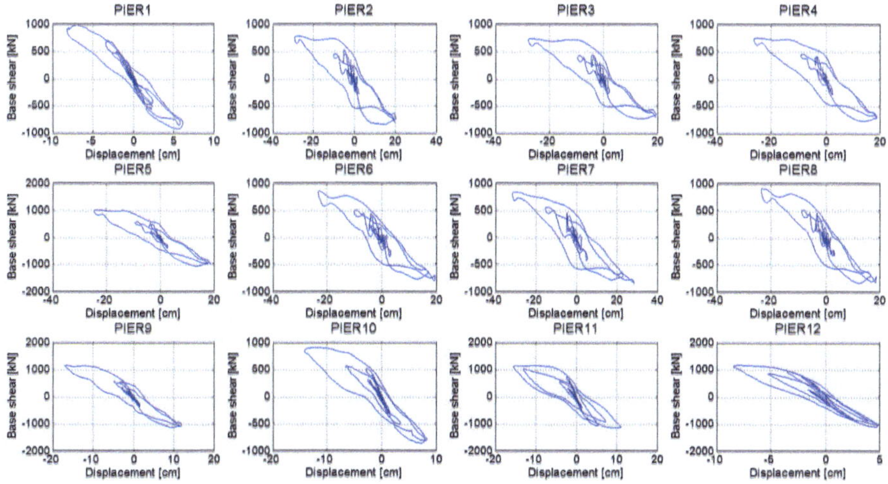

Fig. 13.20 Hysteretic cycle response of piers for ULS

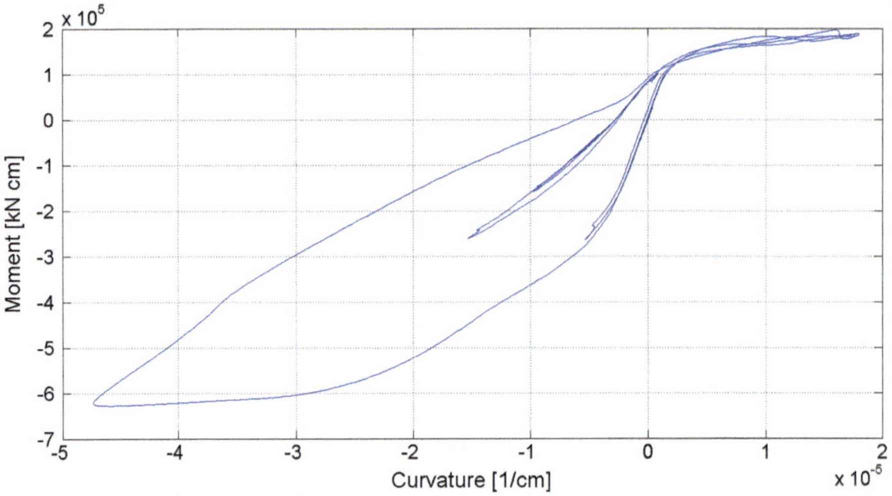

Fig. 13.21 Moment-curvature cycles at the bottom section of the left column of pier #9 (ULS)

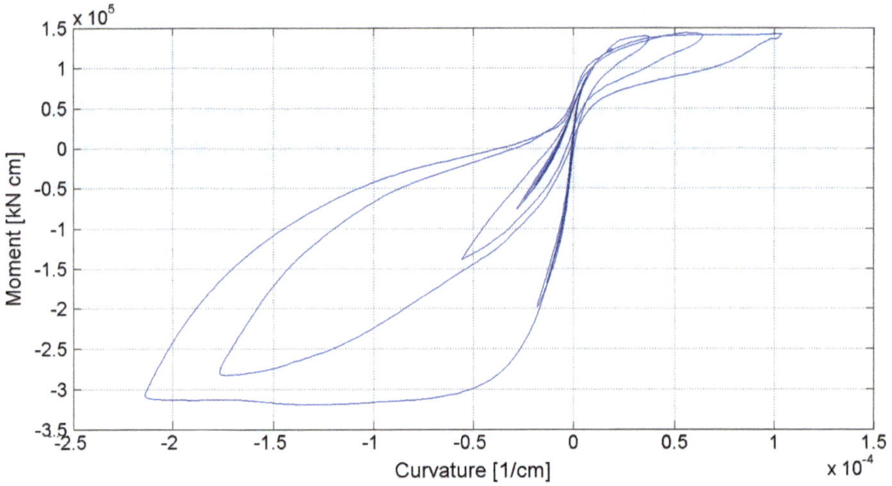

Fig. 13.22 Moment-curvature cycles at the bottom section of the left column of pier #11 (ULS)

transverse beam. Because the level of drift reached during the analysis is about 1 % for pier #11 and 0.8 % for pier #9, extensive shear damage is expected, at least at the 1st and 2nd level of the piers. The plastic flexural deformation of the transverse beam appears limited with a few number of cycles. Therefore, limited flexural damage is expected.

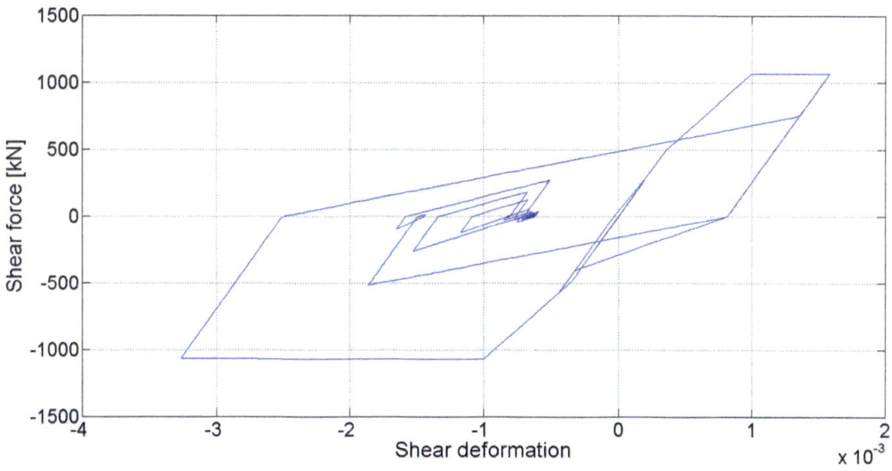

Fig. 13.23 Shear force-deformation cycles of the transverse beam at the 1st level of pier #9 (ULS)

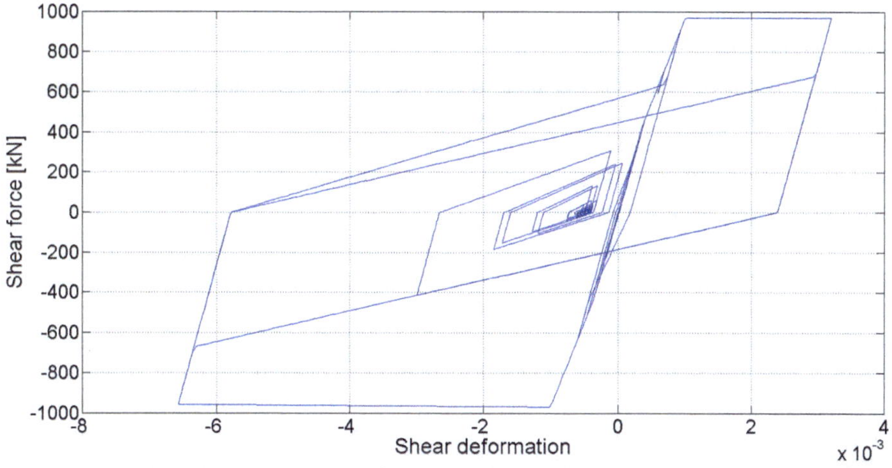

Fig. 13.24 Shear force-deformation cycles of the transverse beam at the 1st level of pier #11 (ULS)

13.6 Numerical Analysis of the Isolated Case

13.6.1 Simulation of the Response for Serviceability Limit State

The maximum transversal displacement of deck and piers is displayed in red and blue in Fig. 13.25, respectively. The maximum transverse displacement occurs at pier #6 and is on the order of 18 cm. The maximum lateral displacement of all piers is about 14 cm (pier #7). The deformed shape of the deck exhibits the maximum

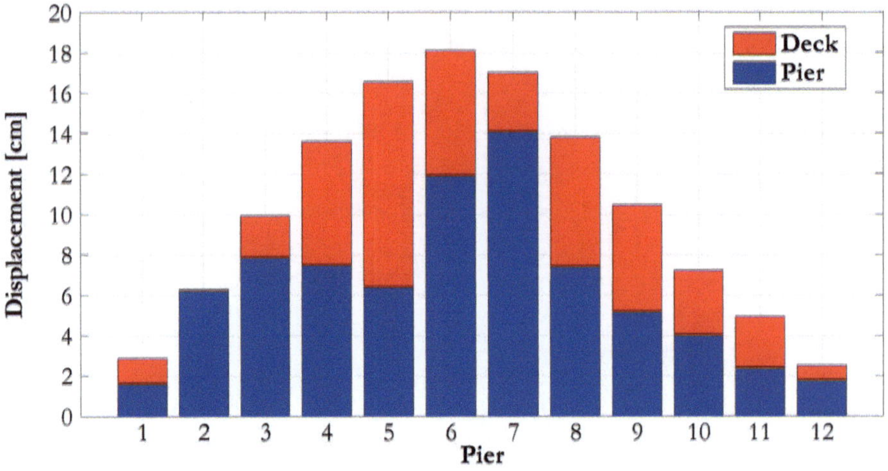

Fig. 13.25 Hysteretic cycle response of piers for SLS

at mid-span, and the maximum relative displacements occur for the deck between piers #4 and #9.

The difference between the displacements of the piers and the bridge deck is due to the activation of the FP devices. The devices are activated for a first sliding shear of about 100 kN, as expected from a simplified analysis (i.e. V equal to $\mu N \approx 0.04 * 2500 = 100$ kN). It is also found that the highest energy dissipation for the devices installed on the top of piers between piers #4 and #9 confirm the maximum relative displacements of the deck. Additionally, the activation of the device causes the limitation of the shear transferred from the deck to the piers; the latter behaving elastically, thus inhibiting the onset of structural and non-structural earthquake induced damage. Figure 13.26 illustrates the maximum base shear in the piers. It was found in (De Risi et al. 2011) that the first yielding of the bridge system is obtained for a base shear of about 500 kN. It is worth noting that Fig. 13.26 shows that the maximum base shear is lower than 500 kN, thus preventing the occurrence of inelasticity. As observed in (Di Sarno et al. 2011), the drift corresponding to shear failure of the transverse beam and to failure of the beam column joints is about 1 %. The values of drift in all piers are lower than 0.5 %, thus damage is prevented. It is also found from the hysteretic response of the single piers that piers #9 and #11 behave linearly. The insignificant irregular response that can be observed in the behavior of pier #9 is caused primarily by the shear behaviour of the transverse beams.

13.6.2 Simulation of the Seismic Test for the Ultimate Limit State

Figure 13.27 shows the maximum transversal displacement of the deck and piers in red and blue, respectively.

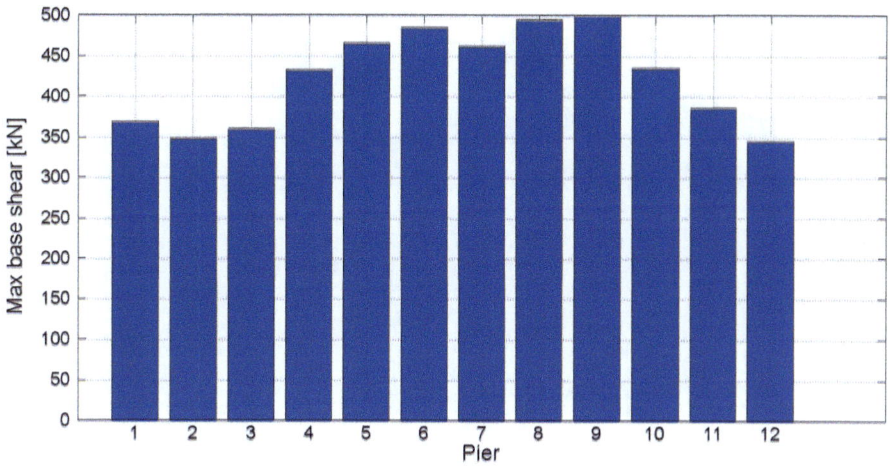

Fig. 13.26 Maximum base shear for SLS in the isolated case

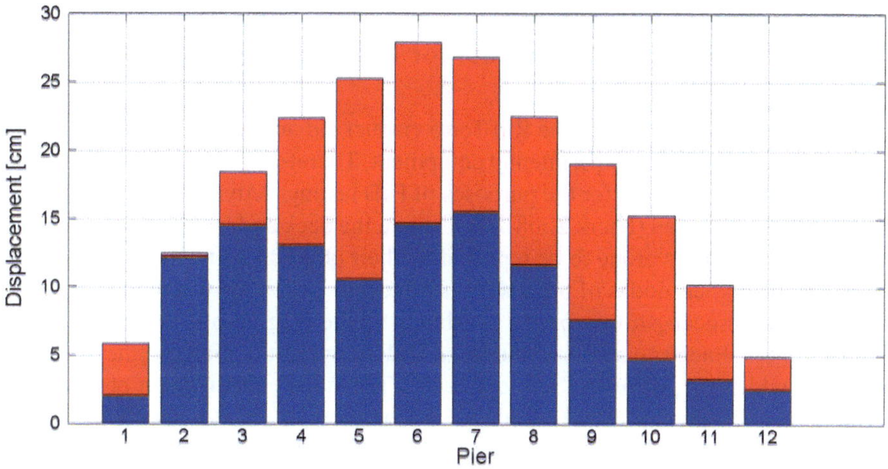

Fig. 13.27 Deck and piers maximum transversal displacement for ULS in Isolated case

The variations between the two depicted displacements are due to the activation of the devices. The devices are activated for a first sliding shear of about 100 kN, as expected from a simplified analysis (i.e. V equal to $\mu \cdot N \approx 0.04 * 2500 = 100$ kN). It is worth noting that the activation of the device causes the limitation of the shear transferred from the deck to the piers, which remain elastic. Figure 13.28 displays the maximum base shear in the piers and shows that the maximum strength is obtained for a base shear higher than 700 kN.

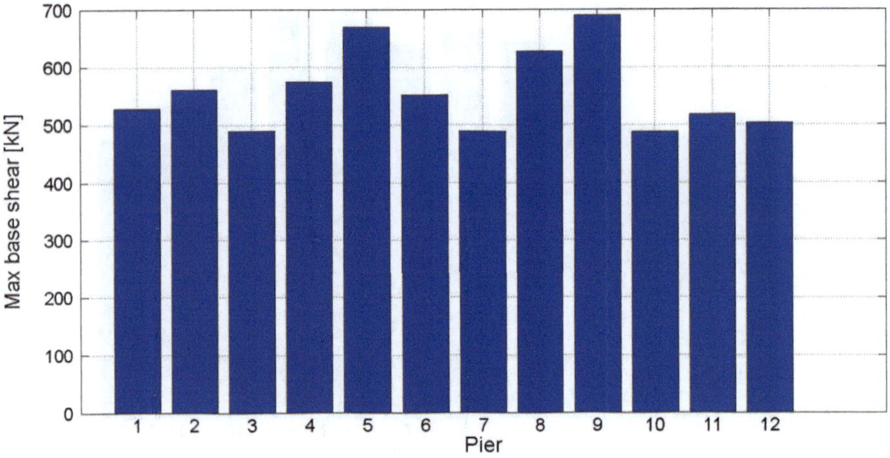

Fig. 13.28 Maximum base shear for ULS in isolated case

13.7 Dynamic Substructuring of the OpenSEES FE Model of the Viaduct for PsD Testing Purposes

The full OpenSEES model of the Rio Torto viaduct did not comply with the computational resources of the experimental equipment. Therefore, a rigorous reduction of the model was developed for the purposes of PsD testing. A dynamic substructuring approach was selected (Klerk 2008). As a result, the deck and isolators were considered as substructures prone to be replaced by their physical counterparts.

Since the higher modes of piers were not excited, a static Guyan reduction (Guyan 1965) was successfully applied to obtain Single-Degree-of-Freedom (S-DoF) reduced piers. Each translational interface DoF between the pier and the deck was retained, whilst the remainder was discarded. The static condensation procedure is detailed in (Abbiati et al. 2014). Reduced S-DoF piers were extended to the nonlinear range through Modified Bouc-Wen springs. They allowed for reproducing the hysteretic behaviour of OpenSEES piers. The resulting differential model of each piers reads as:

$$m\ddot{x} + c\dot{x} + r = -f \cdot a_g(t) + g(t) \tag{13.2}$$

$$\dot{r} = \left[\frac{\rho \cdot A}{\left(1 + \alpha \cdot x^2\right)} - (\beta \cdot \operatorname{sgn}(\dot{x} \cdot r) + \gamma) \, | \, r \, |^n \right] \cdot \dot{x} \tag{13.3}$$

where m, c and f are the linear parameters of the reduced pier obtained from the Guyan reduction, whilst ρ, A, β, γ and n are the parameters of the Modified Bouc-Wen

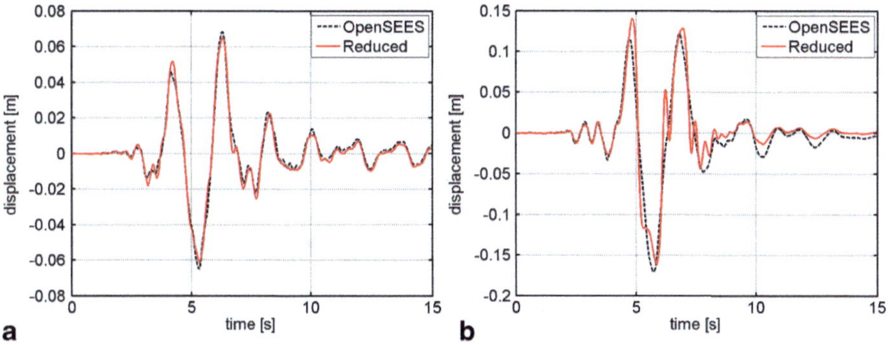

Fig. 13.29 Validation of S-DoF reduced models of Pier 9: displacement responses in the non-isolated case at **a** SLS and **b** ULS

spring. A was assumed equal to the linear stiffness of the S- DoF reduced pier, whilst ρ was introduced to represent its average degradation. The γ and n parameters were assumed equal to 0 and 1, respectively. In order to replicate the softening behaviour of the OpenSEES models, the softening factor α was introduced. Each single reduced S-DoF pier was considered as a stand-alone Single-Input-Single-Output (SISO) system for the purpose of the identification of its nonlinear parameters. OpenSEES time histories were taken as reference signals. In detail, the transversal force measured at the cap beam level $g(t)$ and the seismic accelerogram $a_g(t)$ were considered as inputs, whilst the displacement response at the same location was considered as output (Abbiati et al. 2014).

Figure 13.29 compares displacement responses of the reduced models of Pier 9 to OpenSEES reference solutions at the SLS and ULS limit states.

As shown in Fig. 13.29, displacement responses of the reduced models of Pier 9 were in good agreement with the OpenSEES reference solutions.

Concerning the isolator pairs, the OpenSEES *single surface FP bearing* elements were substructured through the bilinear state space model proposed by Mostaghel (1999). Figure 13.30 depicts both a schematic of the bilinear oscillator of Mostaghel and the entailing hysteretic loop.

The system of differential equations governing the behaviour of the bilinear system depicted in Fig. 13.30b reads:

$$\begin{cases} m\cdot\ddot{x}+c\cdot\dot{x}+\alpha kx+(1-\alpha)ku=\overline{P}_0\cdot p(t) \\ \dot{u}=\dot{x}\left(\overline{N}(\dot{x})\overline{M}(u-\delta)+M(\dot{x})N(u+\delta)\right) \end{cases} \tag{13.4}$$

where $N, M, \overline{N}\, and\, \overline{M}$ are polynomials of the signum function. The state space variable u represents the slip displacement. The integration of Eq. 13.3 defines the response of any non-degenerating hysteretic bilinear system under a given load.

The α, δ and k parameters were identified with respect to the OpenSEES recorded output histories considering the isolator as a SISO system. Figure 13.31 compares

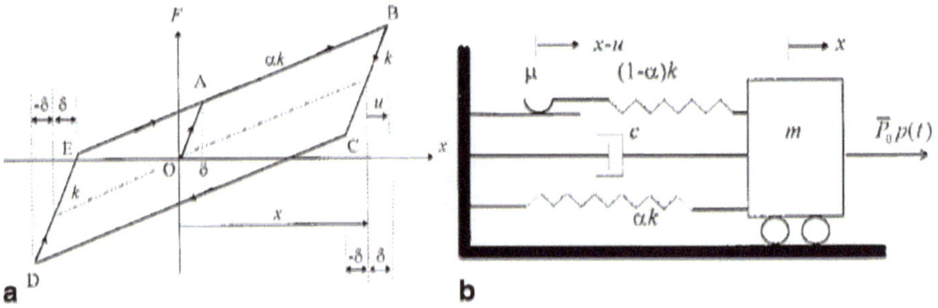

Fig. 13.30 **a** Hysteretic loop for the bilinear system, **b** bilinear S-DoF system after Mosthagel (1999)

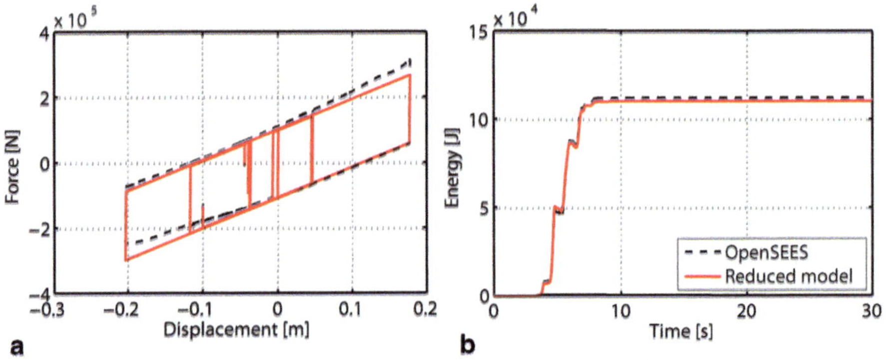

Fig. 13.31 Dynamic response of the *right* isolator of Pier 9 at ULS: **a** Hysteretic loop; **b** dissipated energy

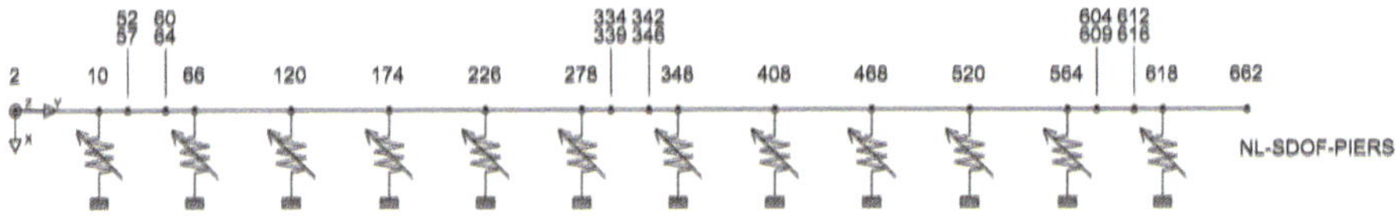

Fig. 13.32 Plan view of the reduced model of the bridge in the non-isolated case

the hysteretic loops and dissipated energy histories of the right OpenSEES isolator installed on Pier #9 to its reduced counterpart at ULS.

The nonlinear parameters of both piers and isolator pairs were identified with the MATLAB pattern search algorithm (Paolacci et al. 2014b). For testing purposes the deck was further condensed to its transversal DoFs.

For validation purposes reduced models of the bridge in both the non-isolated and the isolated configurations were implemented by assembling the substructured parts. Figure 13.32 shows the reduced model of the non-isolated bridge provided with nonlinear S-DoF piers and node numbering.

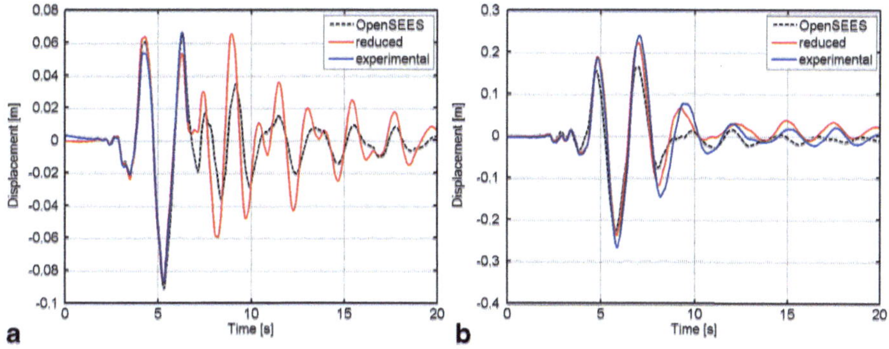

Fig. 13.33 Cap beam level transversal displacement responses of Pier 9 in the non-isolated case at **a** SLS, **b** ULS

Nonlinear time history analyses at both limit states proved the effectiveness of the reduced model in reproducing the dynamic response of the OpenSEES reference model. As a result, sub-structured component of the model were successfully implemented in the Cast3M FE code (Millard 1993) as Numerical Substructures (NSs) for the PsD tests at the ELSA laboratory of the Joint Research Centre of Ispra (Italy). The last upgrade of the ELSA controller software embeds the parallel partitioned time integration algorithm developed by (Pegon and Magonette 2002), i.e. the PM method. Its subcycling features allowed for interfacing and synchronizing the Cast3m FE code and the controller to perform continuous PsD testing with nonlinear NSs. Figure 13.33 compares cap beam level transversal displacement responses of Pier #9 of OpenSEES and reduced models to measure the experimental response of the non-isolated bridge at the SLS and ULS limit states.

As can be appreciated in Fig. 13.33, the experimental responses of Pier #9 agree with the predictive of the OpenSEES simulations, confirming the effectiveness of the tailored NSs.

In order to simulate a consistent degradation of the physical and numerical piers, a novel testing protocol was conceived and applied. It was based on recursive model identification and updating sessions aimed at propagating damage experienced by the specimens to the numerical piers (Paolacci et al. 2014a, b; Alessandri et al. 2013). A careful description of this procedure can be found in (Abbiati et al. 2014).

13.8 Conclusions

The main issues regarding the numerical simulation of the seismic response of the Rio-Torto reinforced concrete (RC) viaduct for the PsD test campaign ran at the ELSA Laboratory have been presented and discussed. Both the “as built” and “isolated” configurations have been considered.

The refined three-dimensional finite element (FE) non-linear model for the non-isolated case, developed in OpenSEES, is discussed. Such model encompasses non-linear flexural deformations of the members, non-linear shear deformations of the transverse beams and strain-penetration effects, which can be typically found in RC structures with plain steel bars. The model was calibrated on previous experimental test results carried out on scaled specimens.

The refined FE model was also adapted to the isolated case in which friction pendulum bearings were used as a retrofitting scheme of the as-built system. A special element specifically developed in OpenSEES was used (*single FP Bearing*).

Earthquake response analyses were carried out for the as-built and isolated configurations. Two different limit states were considered: serviceability (slight damage) and ultimate limit state. The serviceability condition corresponds to the onset of hairline cracks formation in the transverse beam and at the base of columns. At the ultimate limit state, shear failure of the transverse beam and large crack opening at the base of columns occured. Consequently, two different accelerograms were selected and used in the simulations. The records include the natural strong motions recorded during the Emilia earthquake event of 29 May 2012 in Italy.

Detailed local and global responses of the "as built" viaduct, including moment-curvature and shear force-deformation relationships, were used as a tool to check the response of the viaduct during the PsD test.

The simulations of the seismic response of the isolated viaduct showed the effectiveness of the isolators in reducing stresses, and consequently the damage levels in all the members of the piers.

The rational dynamic substructuring of the OpenSEES refined model of the bridge is presented for both the non-isolated and the isolated conditions. Reduced S-DoF models of the piers based on Modified Bouc-Wen springs replaced the OpenSEES model of the frames, while keeping their hysteretic responses. The isolators were represented by bilinear state space substructured models, whilst the deck was condensed to its transversal displacement DoFs. Time history analyses of reduced models of the bridge made of substructured parts confirmed the effectiveness of tailored NSs used for PsD testing purposes.

As result, PsD tests were successfully carried out, as further discussed in a companion chapter.

Acknowledgements The research leading to these results has received funding from the European Community's Seventh Framework Programme [FP7/2007–2013] for access to the EUROPEAN LABORATORY FOR STRUCTURAL ASSESSMENT of the EUROPEAN COMMISSION—JOINT RESEARCH CENTRE under grant agreement n° 227887. Opinions, findings and conclusions expressed in this paper are those of the authors and do not necessarily reflect those of SERIES.

References

Abbiati G, Bursi OS, Cazzador E, Ceravolo R, Mei Z, Paolacci F, Pegon P (2014) Pseudo-dynamic testing based on non-linear dynamic substructuring of a reinforced concrete bridge. SERIES Concluding Workshop—Joint with US-NEES "Earthquake Engineering Research Infrastructures", JRC-Ispra, May 28–30, 2013

Alessandri S, Giannini R, Paolacci F (2013) Aftershock risk assessment and the decision to open traffic on bridges. Earthq Eng Struct Dyn 42:2255–2275. doi:10.1002/eqe.2324

Arani KK, Marefat MS, Di Ludovico M, Prota A, Manfredi G (2013) Hysteretic cyclic response of concrete columns reinforced with smooth bars. Bull Earthq Eng 11(6):2033–2053

Bentz EC (2000) Sectional analysis of reinforced concrete members. PhD Thesis, Department of Civil Engineering, University of Toronto, 2000, 310

CEN 2005 (1994) Eurocode 8—Structures in seismic regions—Design—Part 2: Bridges. Comitè e Europaèen de Normalisation, Brussels

Ceresa P, Petrini L, Pinho R (2007) Flexure-shear fiber beam-column elements for modeling frame structures under seismic loading—state of art. J Earthq Eng 11:46–88

D'Ambrisi A, Filippou FC (1999) Modeling of cyclic shear behavior in RC members. J Struct Eng ASCE 125(3):1143–1150

De Risi R, Di Sarno L, Della Corte G, Paolacci F, Marioni A, Taucer F (2011) Seismic analysis and retrofitting of an existing RC highway bridge Part II: analysis of the isolated bridge. In: Proceedings of the 14th Italian National Conference on Earthquake Engineering, Bari, CD-Rom

Della Corte G, De Risi R, Di Sarno L (2013) Approximate method for transverse response analysis of partially isolated bridges. J Bridge Eng 18(11):1121–1130

Di Sarno L, Giannini R, Paolacci F, Taucer F (2011) Seismic analysis and retrofitting of an existing RC highway bridge part I: assessment of the "as-built" configuration. In: Proceedings of the 14th Italian National Conference on Earthquake Engineering, Bari, CD-Rom

Fenz DM, Constantinou MC (2006) Behaviour of the double concave friction pendulum bearing. J Earthq Eng Struct Dyn 35:1403–24

Guyan J (1965) Reduction of stiffness and mass matrices. AIAA J 3(2):380

Hidalgo PA, Jordan RM, Martinez MP (2002) An analytical model to predict the inelastic seismic behavior of shear-wall reinforced concrete structures. Eng Struct 24:85–98

Lee DH, Choi E, Zi G (2005) Evaluation of earthquake deformation and performance for RC bridge piers. Eng Struct 27:1451–1464

Marin CC (2006) Experimental and analytical study of the XY-friction pendulum (XY-FP) bearing for bridge applications. Ph.D thesis, State University of New York at Buffalo, Buffalo, New York

Millard A (1993) CASTEM 2000, Guide d'utilisation. Saclay, France. Rapport CEA 93/007

Mostaghel N (1999) Analytical description of pinching, degrading hysteretic systems. J Eng Mech 125(2):216–224. doi:10.1061/(ASCE)0733–9399(1999)125:2(216)

McKenna F, Mazzoni S, Scott MH, Fenves GL (2009) Open System for Earthquake Engineering Simulation (OpenSEES), http://OpenSEES.berkeley.edu

Paolacci F, Giannini R (2012) An experimental and numerical investigation on the cyclic response of a portal frame pier belonging to an old reinforced concrete viaduct. Earthq Eng Struct Dyn 41:1109–1127. doi:10.1002/eqe.1175

Paolacci F, Di Sarno L, Pegon P, Molina FJ, Poljansek M, Bursi OS, Abbiati G, Ceravolo R, Erdik M, De Risi R, Mohamad A (2014a) Assessment of the seismic behaviour of a retrofitted old R.C. highway bridge through PsD testing. SERIES Concluding Workshop—Joint with US-NEES "Earthquake Engineering Research Infrastructures", JRC-Ispra, May 28–30, 2013

Paolacci F, Pegon P, Molina FJ, Poljansek M, Giannini R, Di Sarno L, Abbiati G, Mohamad A, Bursi OS, Taucer F, Ceravolo R, Zanotti Fragonara L, De Risi R, Sartori M, Alessandri S, Yenidogan C (2014b) Assessment of the seismic vulnerability of an old RC viaduct with frame piers and study of the effectiveness of base isolation through PsD testing on a large scale model, JRC scientific and policy report, ISBN 978-92-79-35271-3, doi:10.2788/63472

Pegon P, Magonette G (2002) Continuous PSD testing with nonlinear substructuring: presentation of a stable parallel inter-field procedure. Technical Report 1.02.167, E.C., JRC, ELSA, Ispra, Italy

Priestley MJN, Seible F, Calvi GM (1988) Seismic design and retrofit of bridges. Wiley, New York

Vecchio FJ, Collins MP (1988) Predicting the response of reinforced concrete beams subjected to shear using the modified compression field theory. ACI Struct J 85(3):258–268

Zhao J, Sritharan S (1996) Modeling of strain penetration effects in fiber-based analysis of reinforced concrete structures. ACI Struct J (2007) 104(2):133–141

Chapter 14
Assessment of the Seismic Behaviour of a Retrofitted Old R.C. Highway Bridge Through PsD Testing

Fabrizio Paolacci, Luigi Di Sarno, Pierre Pegon, Francisco Javier Molina, Martin Poljansek, Oreste Salvatore Bursi, Giuseppe Abbiati, Rosario Ceravolo, Mustafa Erdik, Raffaele De Risi and Arkam Mohammad Zeki Mohamad

14.1 Introduction

The seismic vulnerability assessment of existing and new network systems, especially transportation systems, is becoming of paramount importance in resilient social communities. Bridge systems and highways are primary elements that can be utilized for rescue operations in the aftermath of moderate-to-major earthquake ground motions. The existing transportation systems were built chiefly in the late 1960s and early 1970s; they were designed primarily for gravity loads. As a result, such systems do not employ seismic detailing and hence their structural performance is inadequate under strong motions, as surveyed in the aftermath of major earthquakes worldwide.

F. Paolacci (✉) · A. M. Zeki Mohamad
Department of Engineering, University Roma Tre, Rome, Italy
e-mail: fabrizio.paolacci@uniroma3.it

A. M. Zeki Mohamad
e-mail: arkam.mohamad@uniroma3.it

L. Di Sarno
Department of Engineering, University of Sannio, Benevento, Italy
e-mail: disarno@unina.it

P. Pegon · F. Javier Molina · M. Poljansek
European Laboratory for Structural Assessment Unit, European Commission, Joint Research Centre (JRC), Institute for the Protection and Security of the Citizen (IPSC), Via Enrico Fermi 2749, 21027 Ispra VA, Italy
e-mail: pierre.pegon@jrc.ec.europa.eu

O. S. Bursi · G. Abbiati
Department of Civil Environmental and Mechanical Engineering, University of Trento, Trento, Italy
e-mail: oreste.bursi@unitn.it

F. Taucer, R. Apostolska (eds.), *Experimental Research in Earthquake Engineering*, Geotechnical, Geological and Earthquake Engineering 35,
DOI 10.1007/978-3-319-10136-1_14

Recently, a comprehensive research program was funded in Italy by the National Consortium of the University Laboratories Network for Earthquake Engineering (ReLUIS) to formulate pre-normative European guidelines for the assessment of existing bridges. The outcomes of the afore-mentioned research were summarized in a detailed technical report by Pinto and Mancini (2009). However, a number of design issues still remain unsolved and require further developments. The implementation of comprehensive guidelines for the seismic assessment and retrofit of existing bridges is time consuming as it involves the thorough understanding of complex local and global response mechanisms. Full scale testing programs should be employed to validate numerical models and assess their reliability for parametric analyses. In the US it took more than 10 years, dating back to the failures of modern bridges during the San Fernando of 1971 earthquake, for the Federal Highway Administration (FHWA) to issue the "Retrofitting guidelines for Highway Bridges" (FHWA-ATC 1983). Since 1992, a refined research program funded by FHWA has been undertaken to further investigate the seismic assessment and retrofit of bridges (e.g. FHWA 1995; FHWA-MCEER 2006). To date, the European standards, i.e. Eurocodes, include a document for the seismic design of new bridges, namely Eurocode 8 Part 2 (CEN-a 2006). The latter document, however, does not deal with existing bridge structures. Conversely, for existing building systems, Eurocode 8 Part 3 was issued (CEN-b 2006). The assessment of the seismic vulnerability of existing bridges is a complex and interdisciplinary process. The assessment scheme requires a comprehensive dataset, which is often unavailable due to the lack of information, especially for reinforced concrete (RC) structures. Moreover, adequate knowledge of the mechanical and geometrical characteristics of the structure and materials, information on structural modifications occurred during the life of the bridge, and a reliable estimation of the gravity loads are essential for sound estimates. The lack of sufficient and adequate level of knowledge affects detrimentally the seismic response evaluation; in turn, the structural vulnerability is not reliably evaluated.

The above discussion demonstrates that experimental test programs on RC bridge systems are deemed necessary to enhance the level of knowledge of

G. Abbiati
e-mail: giuseppe.abbiati@unitn.it

R. Ceravolo
Department of Civil Engineering, Politecnico di Torino, Turin, Italy
e-mail: rosario.ceravolo@polito.it

M. Erdik
Department of Earthquake Engineering, KOERI, Bogazici University, Istanbul, Turkey
e-mail: erdik@boun.edu.tr

R. De Risi
Department of Structure for Engineering and Architecture,
University of Naples Federico II, Naples, Italy
e-mail: raffaele.derisi@unina.it

existing structures and to investigate the efficiency of alternative seismic retrofitting schemes, especially base isolation systems. Moreover, it is crucial to formulate sound provisions for the seismic assessment of existing non-conforming structures. Towards this aim, a Pseudo-dynamic (PsD) testing program on a 1:2.5 scale RC existing bridge system was initiated within the European project "RETRO", which is the acronym for "Assessment of the seismic vulnerability of an old Reinforced concrete viaduct with portal frame piers and study of the Effectiveness of different isolation systems through Pseudo-dynamic Tests on a laRge scale mOdel". RETRO is a research program of the Seismic Engineering Research Infrastructures for European Synergies (SERIES) project, financially supported by the Seventh Framework Programme of the European Commission.

The present paper illustrates the results of the PsD experimental research program carried out at the ELSA Laboratory of JRC (Ispra, Italy) on the RC as-built and retrofitted bridge systems.

14.2 Description of the Case Study

The case study bridge is the Rio-Torto bridge, which is a RC viaduct built during the 1960s and located between Florence and Bologna in Italy. The structural system layout consists of a thirteen-span deck with two independent roadways, supported by 12 couples of portal frame piers (Fig. 14.1). The piers have either solid or hollow circular cross sections with diameters ranging from 120 to 160 cm. The piers are connected at the top by a cap-beam and by one or more transverse beams of rectangular sections which are located at different heights along the piers. The RC structural members were reinforced with plain steel bars. The deck consists of two RC Π-shaped 2.75 m high beams, which are interrupted by Gerber saddles placed at the second, seventh and twelfth bay respectively (Fig. 14.1). The deck is connected to the piers by two steel bars inserted in the concrete; the deck is simply supported at the abutments. The bridge foundations include large RC blocks resting on rock, hence soil-structure interaction was neglected in the earthquake response assessment. The uniformly distributed weight of the deck is approximately 170 kN/m for each roadway.

A global view of the sample bridge is provided in Fig. 14.2a. More details on geometrical and mechanical characteristics of the viaduct can be found in Paolacci and Giannini (2012).

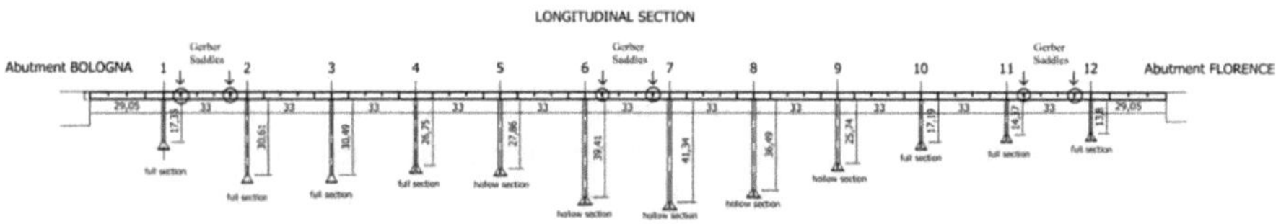

Fig. 14.1 Longitudinal view of the viaduct Rio-Torto

a b

Fig. 14.2 **a** General view of the bridge, **b** Section and reinforcement details of piers 9 (*tall*) and 11 (*short*)

Preliminary numerical simulations of the earthquake response of the as-built bridge, which will be presented in a companion paper, has shown for piers 9 and 11 (see also Fig. 14.2b) flexural failure of the columns, especially at beam-to-column and base-column sections, and widespread shear brittle ruptures in the transverse beams. Piers 9 and 11 were thus selected to be tested experimentally with the PsD method at the ELSA Laboratory, as further discussed in the next paragraphs.

14.3 Pseudo-Dynamic Test Design

A brief description of the large scale bridge pier specimens used for the PsD testing is provided hereafter. The set-up, loading protocol and testing method used during the laboratory experiments are also presented to set the ground for the discussion of the outcomes relative to the dynamic response of the seismic isolation devices and the sub-structured as-built and retrofitted RC bridge framed piers.

14.3.1 Testing Methods

The Pseudo-Dynamic (PsD) test method combines the numerical time integration of the equations of motion of a structure, properly condensed on a limited number of degrees-of-freedom (DoF), with the experimental measurement of the reaction forces resulting from this motion, applied by means of actuators. This hybrid analytical/experimental character is taken into account when introducing the substructuring technique (Dermitzakis and Mahin 1985), allowing to obtain the dynamic response of a structure with only a part, usually the most vulnerable one, present in the laboratory.

The ELSA laboratory has been involved for many years in the development of the continuous PsD testing method. Continuous PsD testing (PsD testing without hold period, (Magonette 2001) avoids load relaxation problems and increase the signal/noise ratio, consequently improving the quality of results. It allows also a considerable reduction of the test duration and provides much cleaner measurements.

This testing method, which is nowadays the only one used at ELSA, has been implemented by means of a synchronous process with a short control period (2 ms) and small time step for the time integration.

Within this framework, it is straightforward to add DoFs in the process controlling the structure, provided that the total number of DoFs remains small and the numerical structure elastic. This monolithic approach cannot be followed when the number of DoFs increases or when the numerical structure is nonlinear, as for RETRO. In this case two processes run in parallel, the first one responsible for the numerical structure and the other one responsible for the motion of the physical structure in the lab, preserving the smooth character of the continuous method. These two process had different time steps (It is virtually impossible to ensure that, within the effective clock duration—2 ms—of one laboratory time step, the analytical process is able to successively perform all the tasks required for advancing its solution, in particular when non-linearities are present) and ran different time integration schemes (conditional stability of the laboratory explicit central difference scheme is easy to fulfill, but the numerical structure should use the implicit scheme, to manage the larger time step and the larger number of DoFs).

Starting from a domain decomposition scheme presented in (Gravouil and Combescure 2001) it was shown that it is possible to transform a staggered asynchronous procedure in an inter-field parallel procedure suitable to work with one synchronous process, keeping most of the original characteristics of the scheme (Pegon and Magonette 2002). The non-linear case was also investigated and a combination of a non-iterative Operator Splitting strategy, (Nakashima et al. 1993; Combescure and Pegon 1997), and of a non-linear condensation technique was proposed to handle the analytical structure (Pegon and Magonette 2005). This distributed scheme was described in Pegon (2008), and a convergence analysis made in Bonelli et al. (2008).

RETRO includes a much larger specimen, some of them (the piers) accumulating damage in an irreversible way. It was thus needed to substantially upgrade the substructuring implementation of 2005, in order to have during the tests the same standards for error and alarm management, the same input (plus the additional requested information specific of substructuring), the same output definition (adding few substructuring related variables) as what it was the state-of-the art in continuous PsD testing at ELSA to handle conveniently and safely large-scale structures.

As with all substructured tests performed in the past at ELSA, the software used for the numerical part is Cast3M (Millard 1993), a multi-purpose finite element code which was slightly modified in order to introduce the non-linear models used during the tests.

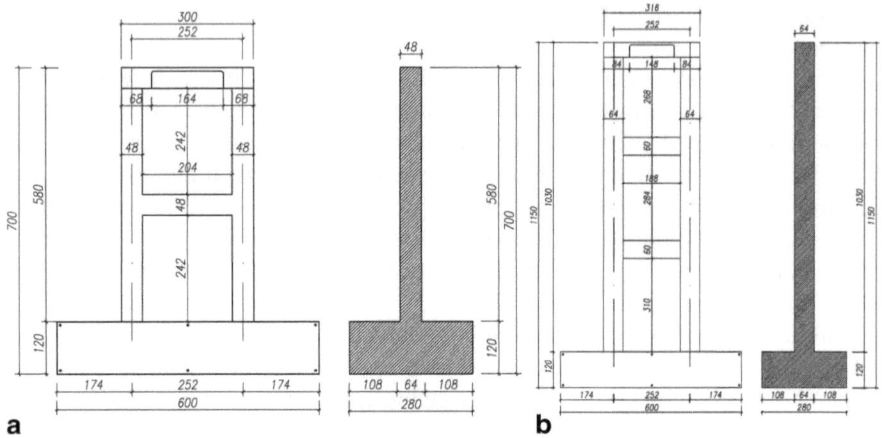

Fig. 14.3 Layout of Piers 11 (*left*) and 9 (*right*)

14.3.2 Tests Specimens

The laboratory tests carried out within RETRO include two RC portal frame piers of the Rio-Torto viaduct which were tested by using PsD techniques with substructuring. The specimens are 1:2.5 scaled models of piers 9 and 11 of the prototype bridge configuration (see also Fig. 14.2b). Pier 11 is a 2 level 1-bay portal frame; the frame total height is H = 580 cm. The foundation is a rigid RC 600 cm long and 280 m width block, as displayed in Fig. 14.3.

The columns of the portal frame specimen are constructed of solid circular sections with diameter D = 48 cm. The longitudinal steel reinforcement includes 20 plain ϕ8 mm bars and spiral smooth stirrups ϕ3 mm with a spacing of 5.6 cm. Pier 9 specimen consists of a 3 level 1-bay frame with a height H = 1030 cm. Further details of the sample pier specimens are provided in Fig. 14.3. The RC foundation blocks of piers 9 and 11 are similar.

The pier specimens were tested in the as-built and seismic isolated configurations. The isolators were sub-structured and connected to the foundation blocks of the piers as described.

14.3.3 Test Rig Configuration

The test rig configuration used for the two scaled piers in the non-isolated configuration is shown in Fig. 14.4. The base of each of the two models is fixed to the reaction floor by means of 16–36 mm diameter Dywidag bars pre-stressed to a load of 500 kN in the vertical direction. To prevent cracking and excessive deformation of the base during testing, 16 and 10-ϕ36 mm Dywidag bars were pre-stressed to a load of 500 kN in the tangent and normal directions of testing, respectively. The

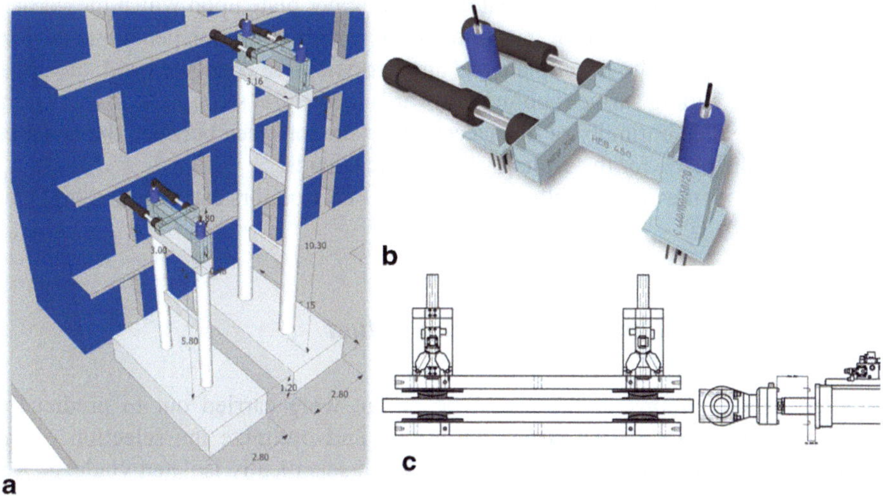

Fig. 14.4 Test rig configuration: **a** the two piers to be tested, **b** the horizontal and vertical load systems and **c** seismic isolation setup

two pier models were placed close to the reaction wall of ELSA laboratory as shown in Fig. 14.4a. Horizontal and vertical actuators are connected to the cap beam by means of a steel rig as displayed in Fig. 14.4b. The steel rig consists of HEB and C-shaped steel sections. Stiff rigid plates are used to connect the rig to the cap beams in each pier; the plates use 6-M27 class 8.7 bolts at each support for the connection of the rig to the cap beams.

Two 500 kN actuators with a displacement capacity of ± 50 cm and spaced at 1.34 m in the horizontal plane were connected to the steel rig, with a lever arm of 0.8 m with respect to the top of the cap beam. Such lever arm accounts for the distance from the centre of mass of the deck to the top of the cap beam of the prototype. The horizontal actuators are displacement-controlled so that during testing rotation in the horizontal plane is inhibited. Additionally, a metallic system to restrain unexpected out-of-plane displacements of the piers was installed for safety requirements. Such system did not interact with the direction of testing. Vertical gravity loads were applied to the piers by means of two 500 kN capacity vertical jacks. The jacks were connected to a 36 mm Dywidag bar running through the centre of the columns and connected to the base by means of a nut (the vertical load is self-equilibrated). For the taller columns, couplers were used to allow incremental extension of the Dywidag bars during construction.

PsD tests on the prototype bridge retrofitted with seismic isolation located between the top of the piers and the viaduct deck were carried out by means of the sub-structuring technique. The isolation system consisted of four friction pendulum devices per pier; during the tests, the devices were located at the ground level. They were connected to the pier specimen foundation blocks. The mock-up used for sub-structuring testing of the isolators is shown in Fig. 14.4c and consisted of a set of three rigid steel plates: a lower plate fixed to the laboratory strong floor, a central plate connected to a horizontal actuator, and an upper plate upon which the vertical loads were applied by means of four actuators. One pair of friction pendulum devices were connected with the lower plate and another pair were connected with the upper plate, with the sliding surfaces pointing towards the central steel plate. It was essential to impose the continuity of displacements at the interface between the isolation system and the piers, for the same level of vertical loads.

Details about the design method used for the FPS isolators can be found in (Della Corte et al. 2013).

14.3.4 Numerical Models of the Viaduct

Numerical simulations of the prototype bridge were carried out to predict the earthquake response of the structural system and optimize the selection of the experimental tests to perform on the as-built and seismically isolated viaducts. Refined and simplified finite element (FE) models were thus implemented.

In the refined model, the bridge response was simulated numerically by using the non-linear code OpenSees (McKenna et al. 2007). The FE scheme is illustrated in Fig. 14.5. The structural elements are modelled by nonlinear beam elements with a flexibility formulation. The model takes into account: (1) non-linear flexural behaviour of the element using a fibre discretization of the sections, (2) non-linear shear behaviour of transverse beams using a global model, calibrated by using experimental results and analytical models (Priestley et al. 1994) (3) strain-penetration effect of the reinforcing bars at the column-foundation joints using the Zhao and Sritharan (2007) model. The P-delta effect was also considered. For the behaviour of concrete the Kent-Scott-Park model was adopted. The contribution of the tensile strength of concrete was neglected. The rebars were modelled with the Menegotto-Pinto law. A yield stress of steel reinforcement equal to 350 MPa is assumed, along with a modulus of elasticity equal to 205,000 MPa and a hardening parameter equal to 0.025. The translational mass was defined on small portions (6 m) along the longitudinal, transversal and vertical directions (m_x, m_y and m_z), while the rotational mass was defined only along the longitudinal direction (global y-axis). The supports of the piers were considered fully fixed in all directions while the abutments at both sides

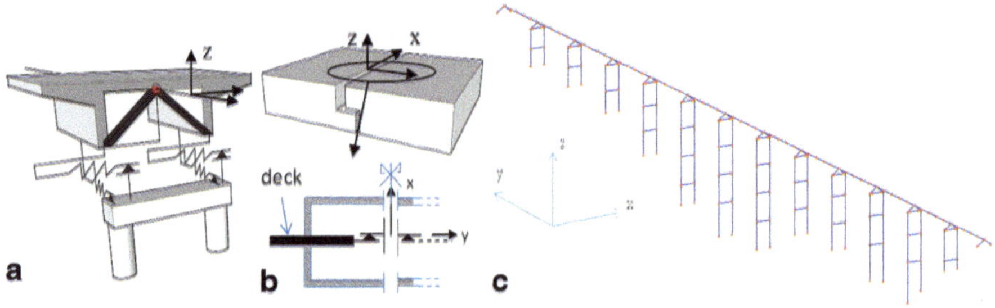

Fig. 14.5 **a** Pier-deck connection model **b** Gerber Saddle model **c** OpenSees model

of the bridge were assumed as simply supported in the longitudinal direction (global y), but restrained in the x and z directions. The Gerber saddles were modelled using rigid elements with a gap in the longitudinal direction and rigid rotational gap elements along the vertical direction. The relative displacements along the x-direction were considered restrained, thus including the possibility to transfer shear in the transverse direction (Fig. 14.5b).The pier-deck connection consists of two steel 34 mm diameter bars (dowels) at each column, modelled using elasto-plastic elements with shear strength and elastic stiffness representing the pair of steel bars (Fig. 14.5a). Moreover, the vertical relative displacements were restrained. The deck was modelled using elastic elements placed at its centre of gravity, connected to each pier using a couple of rigid beams.

For the isolated case the numerical model is similar to the FE system used for the non-isolated case: joints and elements labelling, modelling of the piers, deck represented as an elastic beam, piers fixed to the base and uniform transverse supported assumption. However, the base isolated system includes two retrofitting measures: the Gerber-saddle was removed (i.e. a continuous deck was obtained) and isolation devices between the piers and the deck were introduced. The FP devices used to seismically isolate the bridge deck of the Rio Torto were modelled using the "single-FPBearing" element object (Schellenberg 2010), defined by two nodes. The i-node represents the concave sliding surface and the j-node represents the articulated slider. The element can have zero length or the appropriate bearing height. The bearing has unidirectional (2D) or coupled (3D) friction properties (with post-yield stiffening due to the concave sliding surface) for shear deformations, and force-deformation behaviour defined by an Uniaxial Material in the remaining two (2D) or four (3D) directions. To capture the uplift behaviour of the bearing, the user-specified Uniaxial Material in the axial direction was modified for no-tension behaviour. By default P-Delta moments are entirely transferred to the concave sliding surface (i-node). Rotations of the concave sliding surface (rotations at the i-node) affect the shear behaviour of the bearing. To prevent the occurrence of artificial viscous damping in the isolation system, the bearing element does not contribute to Rayleigh damping. If the element has non-zero length, as in this case, the local x-axis is determined from the nodal geometry unless the optional x-axis vector is specified, in which case the nodal geometry is ignored and the user-defined orientation is utilized.

The Coulomb approach was used for the simulation of the devices for the Rio Torto bridge: kinetic friction does not vary with the sliding velocity. The latter assumption is compliant with the PsD test procedure. Simplified models, used during the PsD tests, were also implemented and calibrated on the outcomes of the refined FE-3D nonlinear bridge model. Since the higher modes of piers were not excited, the static Guyan reduction (Guyan 1965) was successfully applied to obtain single-degree-of freedom (S-DoF) reduced piers. Each translational interface DoF between the pier and the deck was retained, whilst the remainder was discarded. The static condensation is given by:

$$\mathbf{u} = \begin{bmatrix} \mathbf{u}_R \\ \mathbf{u}_L \end{bmatrix} = \begin{bmatrix} \mathbf{I} \\ \mathbf{\Phi}_R \end{bmatrix} \cdot \mathbf{u}_R = \mathbf{T} \cdot \mathbf{u}_R \tag{14.1}$$

where $\mathbf{u}_R$ and $\mathbf{u}_L$ are retained and deleted DoFs, respectively, whilst matrix $\mathbf{\Phi}_R$ collects constraint mode vectors projected on deleted DoFs, according to the Guyan formulation (Guyan 1965). Linear parameters of reduced S-DoF piers were obtained by:

$$k = \mathbf{T}^T\mathbf{KT}, m = \mathbf{T}^T\mathbf{MT}, f = \mathbf{T}^T\mathbf{ML} \tag{14.2}$$

where **K**,**M** and **L** are the stiffness, mass and Boolean lead matrices characterizing each single pier.

Proposed reduced S-DoF piers were not sufficient to describe accurately the nonlinear response. Modified Bouc-Wen springs capable of reproducing the typical hysteretic energy dissipation of piers were thus devised. Linear parameters of S-DoF reduced piers were assumed as references The nonlinear model can thus be expressed as follows:

$$m\ddot{x} + c\dot{x} + r = -fa_g(t) \tag{14.3}$$

$$r = \left[\frac{\rho A}{1+\alpha x^2} - \left(\beta \operatorname{sgn}(\dot{x}r) + \gamma\right)|r|^n\right]\dot{x} \tag{14.4}$$

where A, β, γ and n are parameters of the Bouc-Wen model. A was assumed equal to the linear stiffness element of the S-DoF reduced pier, whilst ρ was introduced to represent its average degradation. To reduce the number of tuning parameters, it was assumed $\gamma = 0$ and $n = 1$. The softening factor α was introduced according to the material properties of the refined OpenSEES model (RM). Interface forces recorded from OpenSEES RM time history analyses were considered as input applied to the coupling DoF of the substructured pier, whilst cap beam level displacements were considered as output. A penalty function was defined as the Normalized Root Mean Square Error (NRMSE) between the displacement response of the reduced pier and the displacement response calculated on the OpenSEES RM.

$$\{\hat{\rho}, \hat{\alpha}, \hat{\beta}\} = \min_{\rho,\alpha,\beta} NRMSE\left(x_{ref}, x_{red}(\rho,\alpha,\beta)\right) = \frac{\sqrt{\sum_{i=1}^{n}\frac{\left(x_{ref,i} - x_{red,i}(\rho,\alpha,\beta)\right)^2}{n}}}{\max\left(x_{ref}\right) - \min\left(x_{ref}\right)} \tag{14.5}$$

where x_{red} is the displacement response history of a generic reduced pier; x_{ref} is the displacement response history of a generic pier from the OpenSEES RM; and n is the sample number of displacement response histories.

The MATLAB pattern search algorithm was used to minimize the penalty function. Simplified models were used as efficient numerical models during the PsD tests.

14.3.5 Selection of Earthquake Input

An appropriate ground motion selection can reduce the bias and variance of structural response (Baker and Cornell 2006). Towards this end, it is advisable to perform an adequate seismic hazard assessment (SHA), which is a key step for seismic risk evaluation. In the present experimental study, earthquake record selection was carried out by means of both SHA and uniform hazard spectra (UHS) as implemented in the Italian seismic code of practice (NTC08 2008). The reference seismic hazard maps of Italy were developed by the Instituto Nazionale di Geofisica e Vulcanologia (INGV). Such maps proposed by INGV are implemented in the current Italian seismic code. The Rio Torto viaduct is located in the Emilia Romagna region, North-East of Italy. Geological investigations indicate that it is constructed on an extensive zone of argillite calcareous together with sandstones. The return period for the life safety condition is about 2000 years. The expected peak ground acceleration (PGA) ranges between 0.23 and 0.25 g, whereas for the collapse prevention condition (probability of 2% in 50 years) PGA ranges between 0.30 and 0.35 g. According to the current Italian seismic code (NTC08 2008) and Reluis guidelines (Pinto and Mancini 2009), assuming soil condition B, nominal life = 100 years and class of construction = IV, the maximum PGA for several limit states are: PGA = 0.174, 0.308 and 0.352 g for the damage, life safety and collapse prevention limit state conditions, respectively.

Two limit states are considered for the seismic performance assessment of the "as built" Rio-Torto bridge: Serviceability Limit State (SLS) and Ultimate Limit State (ULS). Given the geographical position of the bridge and the recent earthquake swarms occurred in the region (especially the earthquake records of the 20th and 29 May 2012), it was decided to use the seismic records of the 2012 Emilia (Italy) earthquakes. The Mirandola records (MRN station) were utilized because of their seismological characteristics: the PGAs and duration of the accelerograms match the outcomes of the INGV shake maps. The record of 29 May East-West was used for the SLS and the North-South component was used to assess the seismic performance at the ULS.

The response spectra of the sample record at the ULS shows a significant amplification for low-periods (between 0.5 and 1.0 s); values of spectral accelerations around 0.40 g are also found in the range 1–1.5 s, i.e. similar to the important natural periods of the portion of the "as built bridge" in which pier 9 and 11 are placed. The length of the waveform was reduced assuming a significant length corresponding to an Arias intensity is up to 99.4%. The input was selected to conform with the compatibility criteria with the ULS design spectra dictated by EN1998:1 and the Italian seismic code. For the SLS conditions the maximum spectral acceleration corresponding to the fundamental period of the bridge is about 0.2 g.

14.3.6 Testing Program

The experimental program included two sets of tests: the first group relates to the assessment of the seismic behaviour of the Rio-Torto viaduct; the second group

Table 14.1 List of significant tests performed during the RETRO test campaign

Name	Description	Parameters	Type of test	Physical part
d03	Test on isolator_P9	V=450 kN D=50,40,30,20,10 mm	Cyclic displacements	Isolator_P9
f03	Preliminary test for the stiffness of pier 11	V=450 kN D=1.5 mm	Cyclic displacements	Pier 11
f04	Preliminary test for the stiffness of pier 9	V=450 kN D=2 mm	Cyclic displacements	Pier 9
k06	Not isolated bridge 10% SLS	PGA=10% SLS	PsD test	Pier 9 & 11
k07	Not isolated bridge 100% SLS	PGA=100% SLS	PsD test	Pier 9 & 11
m01	Only physical isolators	V=450, 225, 175 kN H=30 mm, 1.88 mm/s	Cyclic displacements	Isolator_P9 & P11
l01	Isolated piers case 100% SLS	PGA=100% SLS, μ=4% (design value)	PsD test	Pier 9 & 11+isolator_P9
p02	Isolated piers case 70% ULS	PGA=70% ULS, μ=4% (design value)	PsD test	Pier 9 & 11+isolator_P9
k09	Not isolated bridge 100% ULS	PGA=100% ULS,	PsD test	Pier 9 & 11
r01	Pier 9—isolator case 65% ULS	PGA=65% ULS, μ=7% (actual value)	PsD test	Pier 9+Isolator_P9
r02	Pier 9—isolator case 80% ULS	PGA=80% ULS, μ=7% (actual value)	PsD test	Pier 9+Isolator_P9
r03	Pier 9—isolator case 90% ULS	PGA=90% ULS, μ=7% (actual value)	PsD test	Pier 9+Isolator_P9
k10	Aftershock—not isolated 100% ULS	PGA=100% ULS after a first sequence of 100% ULS (k09)	PsD test	Pier 9 & 11
k12	Afterhock—not isolated 200% ULS	PGA=200% ULS after a first sequence of 100% ULS (k09)	PsD test	Pier 9 & 11

deals with the evaluation of the effectiveness of FP bearings to augment the seismic performance of the retrofitted bridge system. The testing sequence includes the experiments summarized in Table 14.1.

The control system was preliminary checked and then it was used to provide the dynamic characteristics of the prototype structures to be tested, e.g. fundamental frequencies and the equivalent damping ratios of the viaduct. The test was carried out in the linear range. A random signal with a limited intensity level, i.e. maximum ground acceleration equal to 0.01 g, was used.

The cyclic behaviour of FP bearings provided by ALGA Spa was evaluated in tests *d03* and *m01*. The latter tests consist of a harmonic imposed displacement history with different amplitudes, frequencies and axial loads applied to the devices.

Three values of axial loads, frequencies and amplitudes were considered, for a total number of tests equal to 27.

The initial lateral stiffness of the two sample piers was determined experimentally, followed by cyclic displacements applied at the top of the piers (tests *f03, f04*). This phase aimed at providing the elastic stiffness of the two framed piers. A monotonic static displacement, including a corresponding unloading phase, was applied to each pier and the restoring force was measured. A maximum displacement of 1 and 0.5 cm was applied to piers 9 and 11, respectively.

A dynamic characterization of the bridge using the SLS accelerogram with a low intensity level was performed (test *k06*). The dynamic characteristics of the bridge, namely the fundamental frequency and damping, were computed experimentally. The selected strong motion was the record of 29 May East-West with an intensity level scaled to 0.01 g. A PsD test was then performed on the entire non-isolated bridge subjected to the 29 May East-West earthquake with an intensity level corresponding to a PGA of 0.25 g. This test (*k07*) was carried out to induce slight damage on the non-isolated piers, as it may occur for low return period earthquakes. Slight damage included hairline cracks on the transverse beams and at the bottom section of columns.

Seismic tests at SLS and ULS were performed on the isolated bridge (test *l01, p02*). The cyclic tests on the FP bearings showed a friction coefficient ($\mu=7\%$) greater than the designed value ($\mu=4\%$); the isolators of pier 11 exhibited irregular hysteresis loops. As a result, the tests were carried out by considering the physical isolators of pier 9 with a reduced vertical load to obtain $\mu=4\%$; the latter value of friction was also used for pier 11 and the numerical piers of the bridge system. For pier 11 the isolator was numerically simulated. The effectiveness of the physical FP bearings in protecting the bridge was demonstrated at different levels of PGAs corresponding to a given set of percentages of the ULS (*tests r01, r02, r03*).

To investigate the seismic vulnerability of the non-isolated bridge a seismic test of the non-isolated bridge using the record 29 May North-East component with PGA=0.27 g (ULS) was carried out (*tests k09*). Significant damage was caused to piers 9 and 11. Extensive crack patterns in the beams, crack opening at the base and at the top section of the columns were detected at the end of the PsD tests. Seismic tests of the non-isolated case for 100% ULS was also repeated to estimate the aftershock effects on the damaged bridge (test *k10*). Finally a seismic test of the non-isolated bridge for beyond-design condition using a 200% ULS accelerogram was conducted (test *k12*). The results of the experimental tests are illustrated.

14.4 Discussion of the Experimental Results

The preliminary results of the PsD experimental tests carried out for the prototype RETRO bridge are discussed in the next paragraphs. The outcomes of the characterization of the isolators and the single specimens are firstly introduced. Emphasis is placed on the results of the PsD tests the on isolated and isolated configurations. The high vulnerability of the viaduct is shown with respect to the as-built configuration. The effectiveness of the FP isolation system is stressed and critical issues related to the hysteretic response of FP devices are also pointed out.

14.4.1 Cyclic Characterization of the FPS Isolator

In order to characterize the cyclic behaviour of the isolators, several tests were performed. A first group of tests (test *b22, b23*) was carried with reference to the procedure proposed by Lomiento et al. (2013) for rubber-steel devices. Then, a series of cyclic tests with variable velocity and amplitudes was carried out to assess the sensitivity of the response of the devices to the variation of the above input parameters.

14.4.1.1 Cyclic Tests Procedure

The isolators used for the seismic retrofitting of the Rio-Torto bridge are different from the those used in the work by Lomiento et al. (2013), notwithstanding similar experiments were carried out within the range of the capacity of the ELSA laboratory. In Fig. 14.6, the force-displacement loop at 0.38 mm/s is compared with those at 0.63, 1.9, 3.8 and 6.3 mm/s and it is observed that the velocity effect has an increase of the friction coefficient at higher velocities. However, the cycling effect is not significantly observed for these velocities. Regarding the breakaway effect, it was not significant for 6.3 mm/s, but became more pronounced for slower speeds, even though it did not seem to affect the dissipated energy. The breakaway effect is observed not only when accelerating from zero, but also when decelerating to zero. This effect may, however, depend on the control characteristics.

14.4.1.2 Tests for Different Cycling Velocities and Amplitudes

Two types of cyclic tests were repeated under the same loading conditions for both isolator setups with a constant vertical load of 450 kN per isolator.

Tests *b16* and *d03* included the isolators subjected to 5 sine cycles at a constant frequency of 1/100 Hz. Amplitudes of 50, 40, 30, 20 and 10 mm were selected. The maximum velocity varied between 3.14 and 0.63 mm/sec.

Tests on isolators at the short (b22) and tall (b02) piers were also carried out by considering a series of 2 sine cycles at frequencies ranging between 1/500 and 1/5 Hz; a 30 mm constant amplitude was used. The maximum velocity varied from 0.63 to 37.7 mm/sec. However, due to the limitations of the testing setup at higher velocities, only the runs 1–5 with frequencies 1/500–1/30 Hz should be assumed reliable.

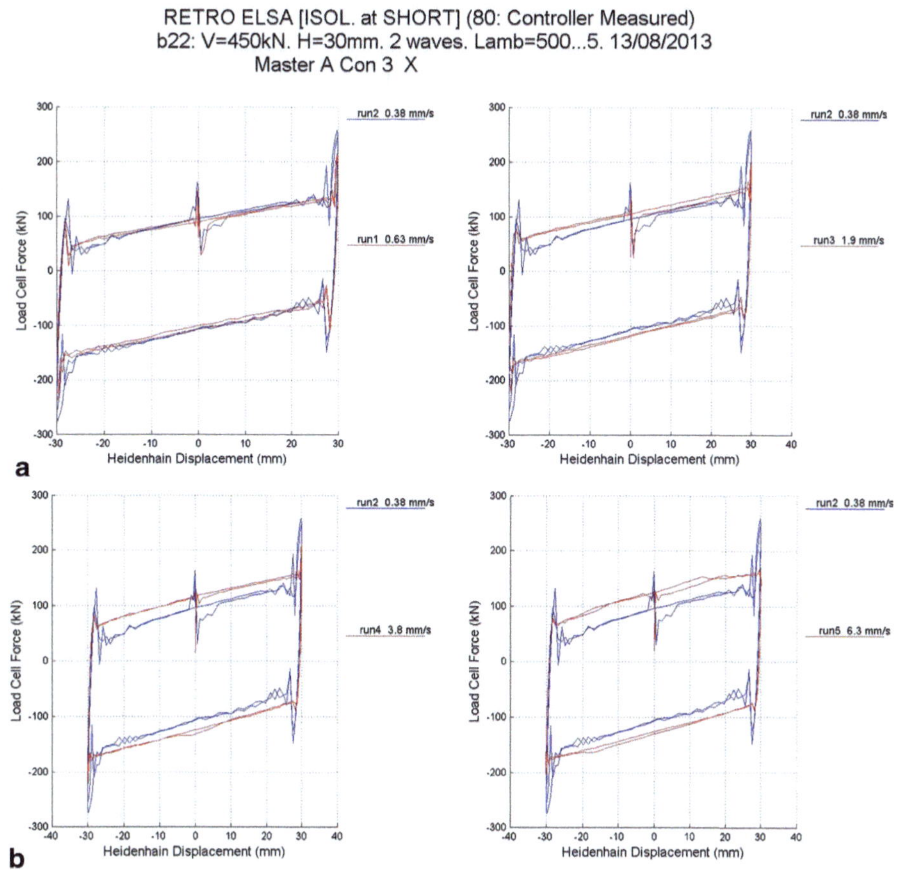

Fig. 14.6 Test results for the friction pendulum devices: sinusoidal input at different velocities

14.4.2 Static Characterization of the Specimens

Preliminary tests performed on the piers served for identification of the initial stiffness. Such tests allowed also tuning of the PID parameters of the control of their actuators. In test *f03*, sine cycles of 1.5 mm were applied to the short pier, whereas in test *f04*, sine cycles of 2 mm were applied to the tall pier.

At the beginning of each test, the vertical load was gradually applied by vertical force-controlled actuators. The applied vertical force remained constant in each column of the piers; the value of the axial load was 450 kN. Several runs of two consecutive horizontal cycles for the specified constant amplitude were then imposed at the head rig of the pier. The parameters of the PID controller for the horizontal actuator were calibrated during the tests to detect possible effects in the quality of control. The equivalent linear stiffness and equivalent-viscous damping ratios was

Table 14.2 Results of the static characterization of specimens

Pier	Test	Run	Amplitude (mm)	Stiffness (kN/mm)	Damping ratio
Short	f03	3	1.5	11.03	0.0366
Short	f03	4	1.5	11.01	0.0364
Short	f03	5	1.5	11.00	0.0363
Tall	f04	1	2	7.398	0.0393
Tall	f04	2	2	7.341	0.0327
Tall	f04	3	2	7.315	0.0286

estimated by a standard procedure in ELSA. The results of the structural identification are outlined in Table 14.2.

14.4.3 Test Results on the Entire Viaduct

In this section the results of the PsD tests on the entire isolated and not isolated bridge are illustrated. Local and global structural response quantities are used. Local response is assessed primarily in terms of curvatures and shear deformations, and the level of bond slip of bars. Global response is estimated by means of the displacements at the top of each pier and the base shear. The global displacement of the pier is computed as the average of the two horizontal actuators. The net total force on the pier, which is equal to the total force after removing the re-centring horizontal force given by the tensioned Dywidag bars inside of the columns of the pier, is considered. The latter net force can be computed as follows:

$$F_{Total}^{Net} = F_{Total} + K_{Geom} D \tag{14.6}$$

where a geometric stiffness K_{Geom} acts as an ad-hoc correction term.

For tests with large displacements, namely the tests labelled *k09, k10* and *k12*, the same approach, i.e. introduction of the correction term K_{Geom}, was used within the PsD algorithm before inserting the forces as restoring forces within the equation of motion.

14.4.3.1 Test Results on the Non-Isolated Viaduct

Test *k06* was carried out to check all instruments, LVDT sensors and the hydraulic actuators. The test was also used to estimate the displacements in each floor and determine the initial stiffness of piers. The force-deflection cycles for test *k07*, which corresponds to the non-isolated case with 100 % SLS input motion, show that he global behaviour remains mainly elastic. Limited hairline cracks were detected at the transverse beams in pier 11 and 9 due to shear damage. To avoid excessive and widespread damage in the short pier, the test was stopped at 6.6 s. The drift

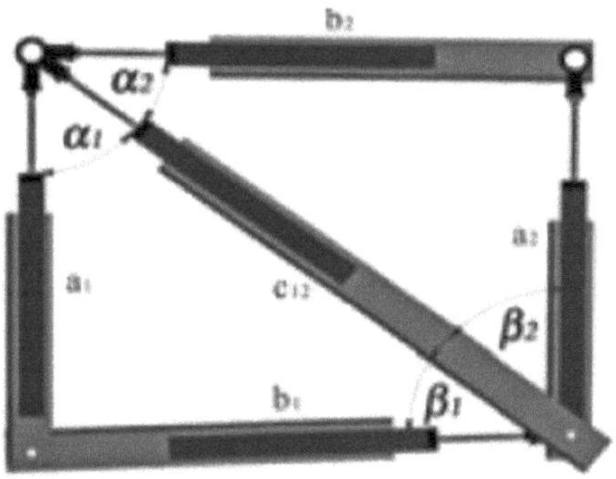

Fig. 14.7 Parameters for the evaluation of shear deformations of the transverse beams of the framed piers

corresponding to a slight damage estimated with the numerical models presented in Sect. 14.3.4 was about 3 cm for the tall pier and 1.6 cm for the short pier; the drift ratio was about 0.3 %. During the PsD test, the short pier reached a displacement of about 3 cm, which corresponds to a drift ratio of about 0.6 %. Such drift was mainly due to the higher horizontal deformability of pier 11, with respect to the lateral flexural stiffness assumed in design.

The crack pattern observed experimentally matched the damage predicted by the refined FE model of the prototype bridge structure. The maximum expected level of shear deformation γ was about 1×10^{-3} rad, which is close to the values predicted by the numerical simulations. Evaluation of the mean values of the shear deformation γ was carried out by using the following relationship:

$$\gamma = (\Delta\alpha + \Delta\beta)/2 \tag{14.7}$$

where α and β are the angles indicated schematically in Fig. 14.7, where the arrangement of the sensors used for measurements is also shown.

The variations of the angles α and β were computed as:

$$\Delta\alpha = \alpha_0(\alpha_2 + \alpha_1) \tag{14.8}$$

$$\Delta\beta = \beta_0(\beta_2 + \beta_1) \tag{14.9}$$

where $\alpha = \alpha_2 + \alpha_1$, $\beta = \beta_2 + \beta_1$, while α_0 and β_0 denote the angles of the initial undeformed configuration.

The time-history of the shear deformation of the transverse beam at the first storey of pier 11 is provided in Fig. 14.8.

Limited cracks were also observed at the bottom section of piers 9 and 11. Such cracks were determined by the level of bond slip measured by the LVDT sensors placed at the bottom of columns as depicted in Fig. 14.9. The time-history of the LVDT located at the base of pier 9 is shown in Fig. 14.10. The maximum level of displacement is 0.2 mm, which corresponds to a slight crack opening in the column of the piers.

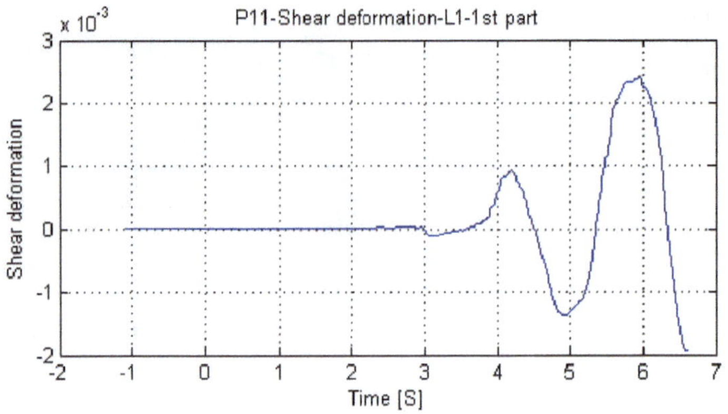

Fig. 14.8 Time history of shear deformations of the transverse beams of pier 11

Fig. 14.9 Parameters for the evaluation of the curvature of columns of the framed piers

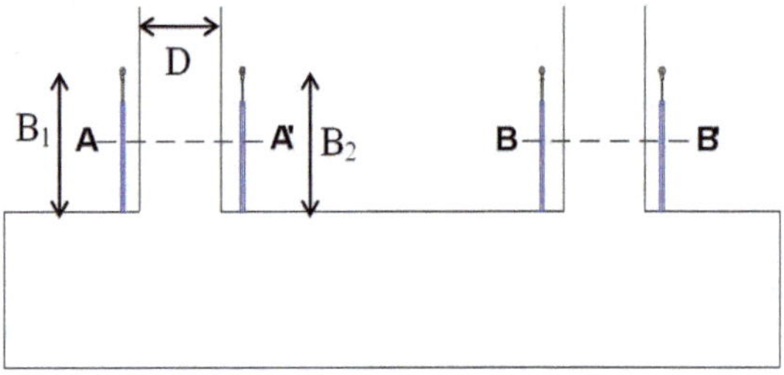

Fig. 14.10 Time history of flexural deformations at the column base of pier 11

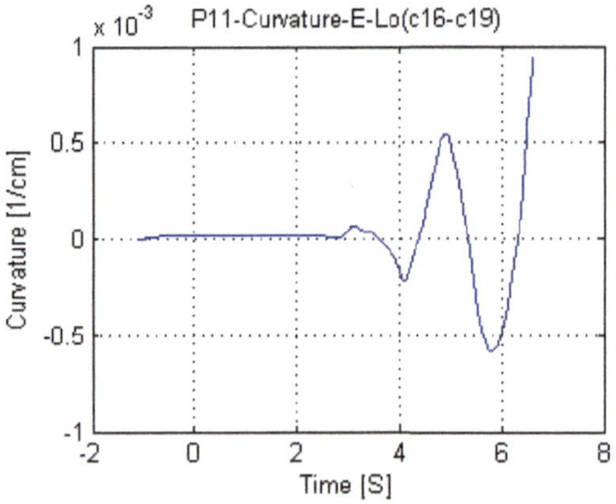

The experimental curvature X was evaluated as a function of displacements δ_1, δ_2 and includes the bond slip effect of rebars. The curvature was computed by using the following expression:

$$\chi = (\delta_1/B_1 + \delta_2/B_2)/D \tag{14.10}$$

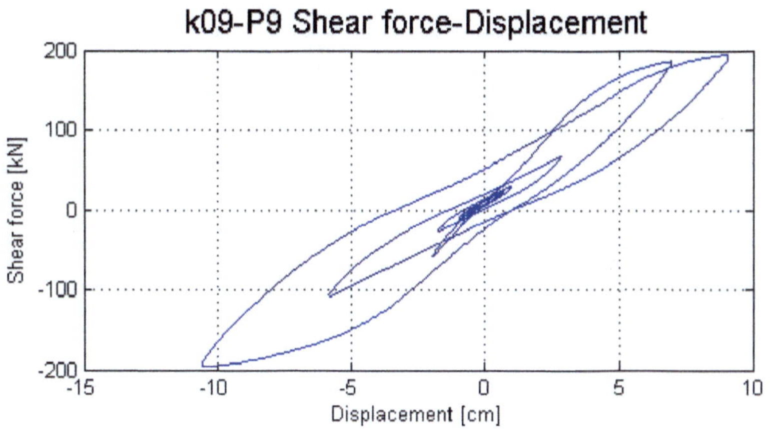

Fig. 14.11 Force-Deflection cycle of Pier 9 for test *k09* (ULS PGA = 100 %)

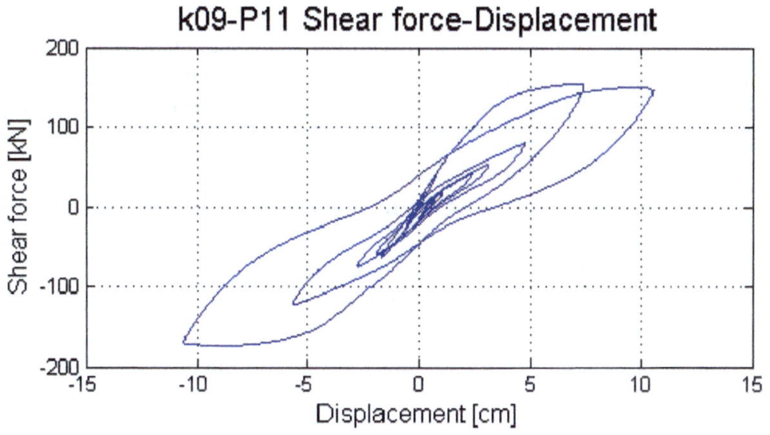

Fig. 14.12 Force-Deflection cycle of Pier 11 for test *k09* (ULS PGA = 100 %)

in which δ_1, δ_2 were measured by pairs of LVDTs with lengths B_1 and B_2 measured from the base (see also Fig. 14.9).

The curvature at the bottom section of one of the columns of pier 11, shown in Fig. 14.10, revealed a slight amount of non-linearity in the columns during the test.

The configuration of the non-isolated 100 % ULS was analysed with the *k09* test. During this test a significant shear crack pattern was observed in the transverse beams of both piers. Few cracks were observed between the cap beam and column joints. Base-column opening was also quite evident during the test cycles. Figures 14.11 and 14.12 show the force-deflection response; high non-linearities can be identified in both the physical piers.

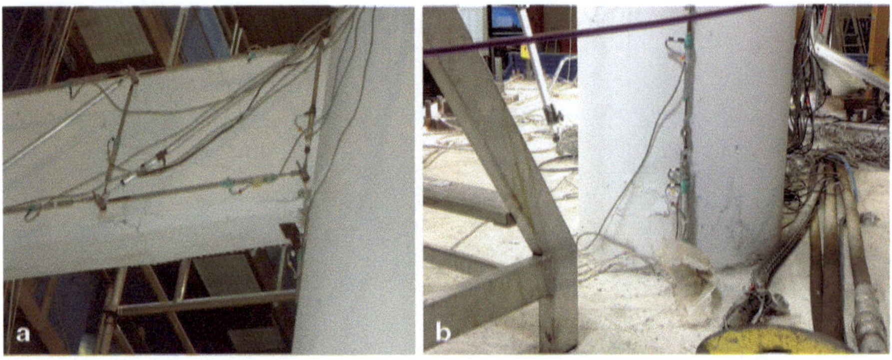

Fig. 14.13 Shear cracks pattern in the transverse beam (*left*) an column (*right*) of pier 11 after test *k09*

The high level of shear damage in the beams and the large amount of crack opening at the column base is shown in Fig. 14.13.

To assess the effects of aftershocks on the already damaged bridge, the input loading used for test *k09* was repeated with the same sequence as in test *k10*. The test aimed at verifying the level of degradation of non-linear behaviour and the level of damage of both piers. This test will also be used to assess the reliability of the adopted non-linear models in predicting the seismic behaviour of the damaged bridge during aftershocks. The test results show a decrease of stiffness due to an increase of the fix-end rotation effect, given by the high slippage of reinforcing bars at the top and bottom sections of columns. This effect was magnified by the increased openings of shear cracks in the transverse beams.

To test the piers beyond-design conditions and quantify global failure in terms of local and global response, a test with PGA=0.54 g, i.e. 200 % ULS, was performed. The hysteretic curves for the input strong motion at 200 %ULS are shown in Figs. 14.14 and 14.15. The threshold limit for the drift causing shear failure of the transverse beams is 2 % (Alessandri et al. 2013). A 2.4 % drift ratio was reached in the short pier, thus shear failure occurred in the transverse beams.

Figures 14.16, 14.17, 14.18, and 14.19 provide the damage pattern in the columns and transverse beams of the physical piers 9 and 11. An extensive damage is evident in both piers. Nevertheless, the damage is more pronounced and widespread in the short pier, especially in the transverse beam, where large typical shear cracks are found (see also Fig. 14.16). A large zone affected by spalling and buckling of the steel longitudinal bars was observed in the transverse beam of the short pier (Fig. 14.18).

Similar damage failure was detected in the transverse beam at the first level of the tall pier; the damage in such slender pier appears less widespread. The total drift ratio in pier 9 was 1.2 %, which may cause severe damage in the transverse beam, but far from the ultimate condition. Steel bar buckling occurred also at the base of the pier.

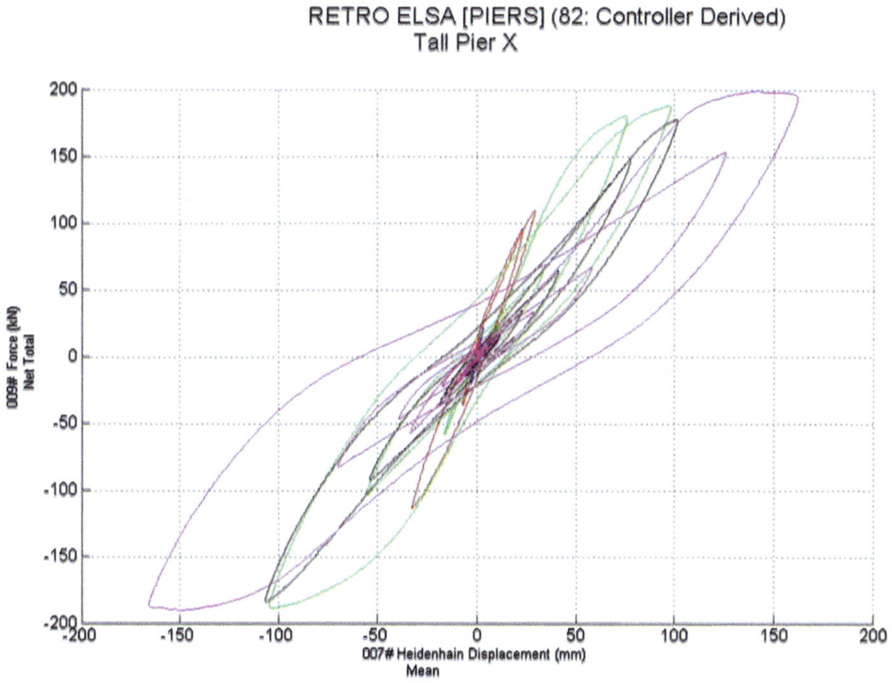

Fig. 14.14 Force-Deflection cycle of tall Pier for test *k12* (PGA = 200 % ULS)

14.4.3.2 Test Results on the Isolated Viaduct

Comprehensive PsD tests were carried out on the isolated viaduct at SLS and ULS. Design and actual values of friction μ for the friction pendulum devices, namely $\mu = 4\%$ and $\mu = 7\%$, were considered along with the physical and numerical piers. During the tests, only the isolators connected to the tall pier 9 were used as physical devices, as the set of devices connected to pier 11 exhibited an irregular hysteretic response, as discussed in Sect. 14.4.1. Physical piers and isolators were considered for tests *l01* and *p02*, which correspond to the 100 % SLS and 70 % ULS, respectively. These latter tests use the friction design value, i.e. 4 %. The bridge deck was assumed continuous, i.e. the saddles were eliminated, when the isolation system was introduced at the base of the RC deck as discussed in Sect. 14.3.4.

The bridge responded elastically when subjected to the 100 % SLS earthquake input. The effectiveness of the FP bearings in protecting the bridge was demonstrated by the nearly linear behaviour of the physical as well as of the numerical piers at SLS. Transverse beams also exhibited an elastic response at SLS; shear deformations in the lower transverse beam of pier 11 during test *l01* vary linearly. The linear elastic response of the piers stems also from the curvature at the bottom section of the left column of pier 11 during test *l01*, as depicted in Figs. 14.20 and 14.21. The hairline cracks on the transverse beams and at the bottom section of columns which appeared in the non-isolated configuration, namely in test *k07*, were not detected when base isolation devices were used, i.e. for test *l01*.

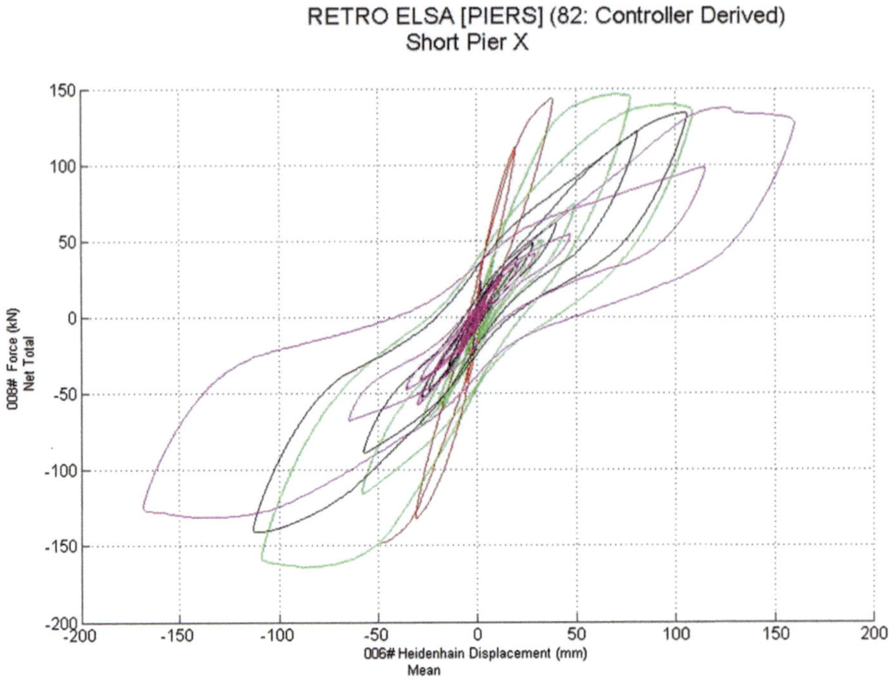

Fig. 14.15 Force-Deflection cycle of Pier 11 for test *k12* (PGA = 200 % ULS)

Fig. 14.16 Transverse beam damage on pier 11 after test *k12* (PGA = 200 % ULS)

The bridge configuration comprising physical piers (9 and 11) and physical isolators for piers 9 and 11 with $\mu = 4\,\%$ was tested with a PGA equal to 70 % ULS. The retrofitting system, i.e. base isolation, was used on a bridge possessing slight damage. The response was nearly elastic, as shown in Figs. 14.22 and 14.23, where force-deformation curves are plotted for the short and tall piers, respectively.

Fig. 14.17 Crack opening at the top of a column in the short pier after test *k12* (PGA = 200 % ULS)

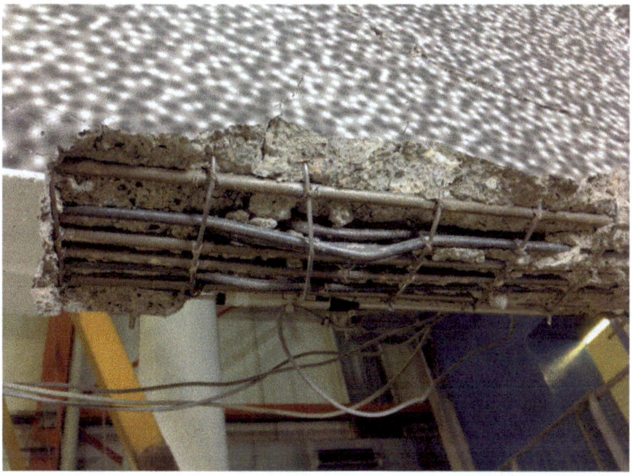

Fig. 14.18 Blow up of the transverse beam damage on short pier after test *k12* (PGA = 200 % ULS)

Fig. 14.19 Crack pattern in the first transverse beam of tall pier after test *k12* (PGA = 200 % ULS)

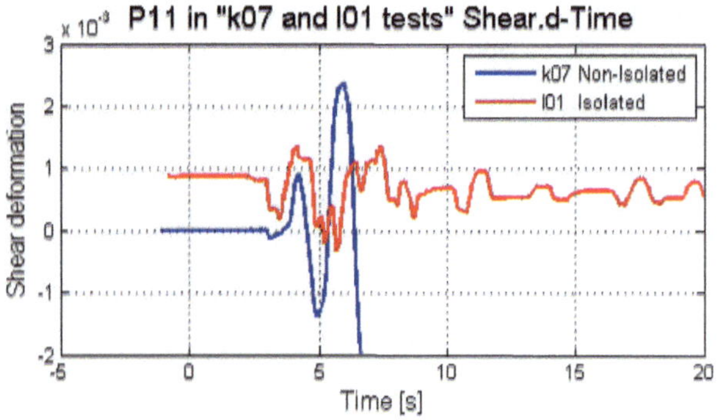

Fig. 14.20 Shear deformation in the lower transverse beam of pier 11 during tests *k07* and *l01* (SLS PGA = 100 %)

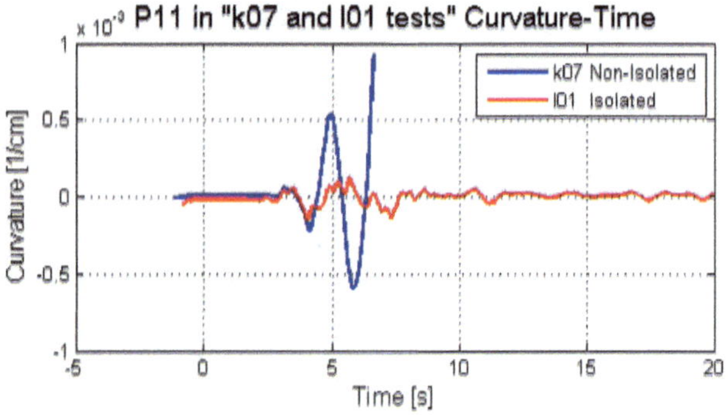

Fig. 14.21 Curvature at the bottom section of the *left* column of pier 11 during tests *k07* and *l01* (SLS PGA = 100 %)

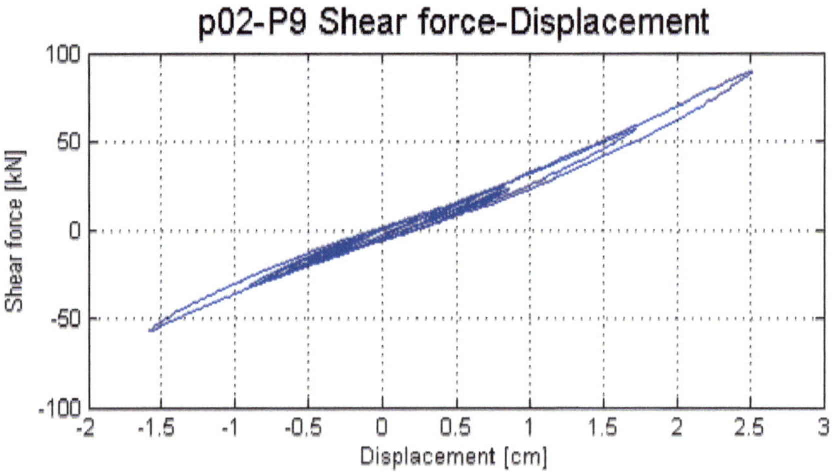

Fig. 14.22 Force-Deflection cycle of Pier 9 during test *p02* (ULS PGA = 70 %)

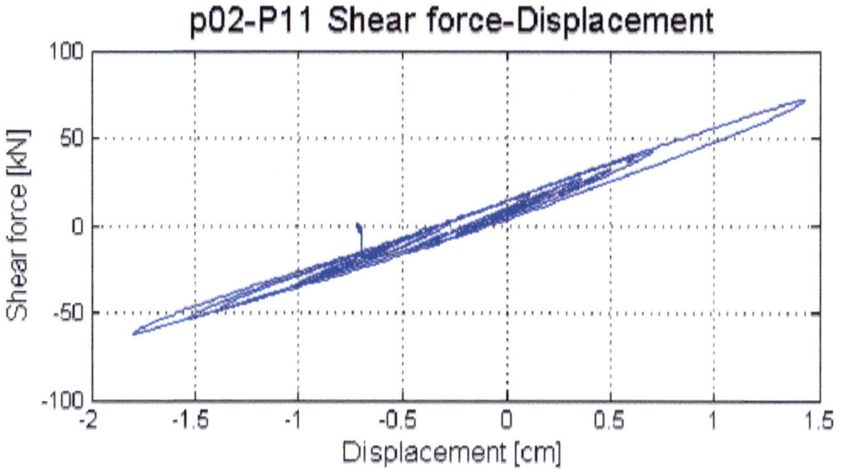

Fig. 14.23 Force-Deflection cycle of Pier 11 during test *p02* (ULS PGA=70%)

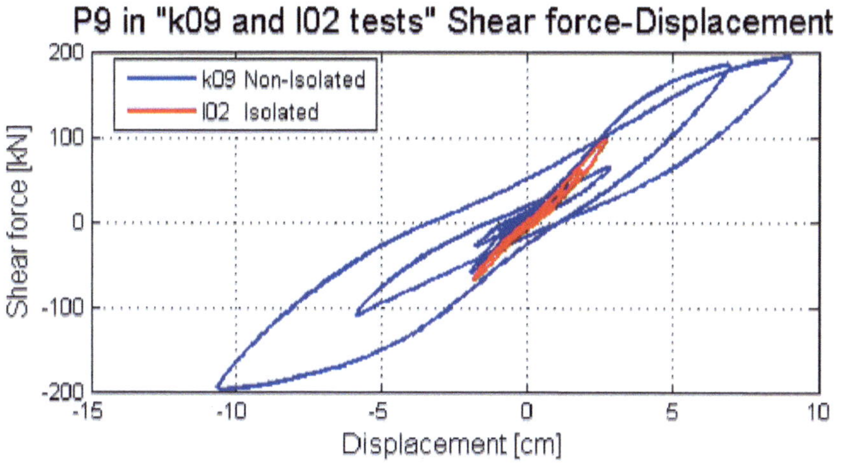

Fig. 14.24 Force-Deflection cycle of Pier 9 during test *k09* and *l02* (ULS 100%)

The effectiveness of the FP isolators may be fully observed by comparing the isolated and non-isolated case at 100% ULS (*k09* in the as-built and *l02* in the isolated case); Figs. 14.24 and 14.25 show the experimental force-deflection of Pier 9 and 11 in both cases. It is also clear in these figures that the isolators reduced the high vulnerability in the bridge.

The effectiveness of the physical FP bearings in protecting the bridge was further demonstrated when the strong motion at ULS is considered. Tests *r01, r02, r03* correspond to PGA accelerations equal to 65%, 80% and 90% of ULS, i.e. 0.25 g. For

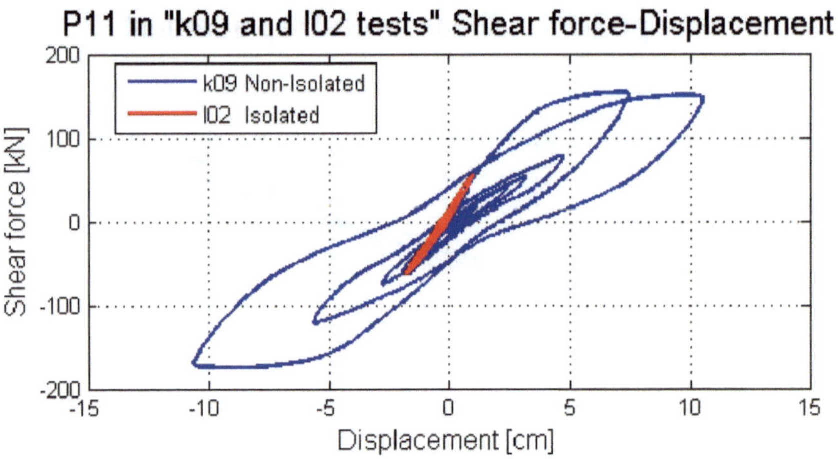

Fig. 14.25 Force-Deflection cycle of Pier 11 during test *k09* and *l02* (ULS 100 %).

Fig. 14.26 Force-Deflection cycle of Pier 9 during tests *r01, r02* and *r03* (ULS PGA = 65, 80 and 90 %)

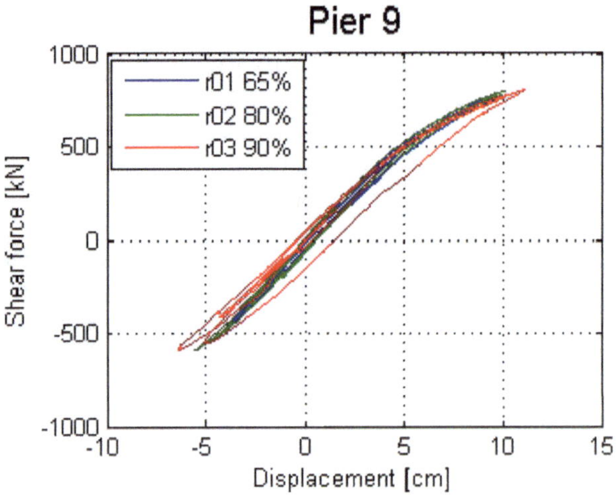

such tests, the physical pier 9 and the isolators connected to that pier; all remaining piers and isolators were simulated numerically. The force-deformation relationships of piers 9 and 11 are linear (Figs. 14.26 and 14.27).

The friction of the isolators was $\mu = 7\,\%$; the latter friction value, which was derived experimentally, was significantly higher than the design value, i.e. 4 %. From a practical standpoint, it is generally assumed that the higher friction may be caused, for instance, by inadequate selection of the devices, ageing phenomena, defects in manufacturing, etc. The latter are realistic conditions that may occur in practice.

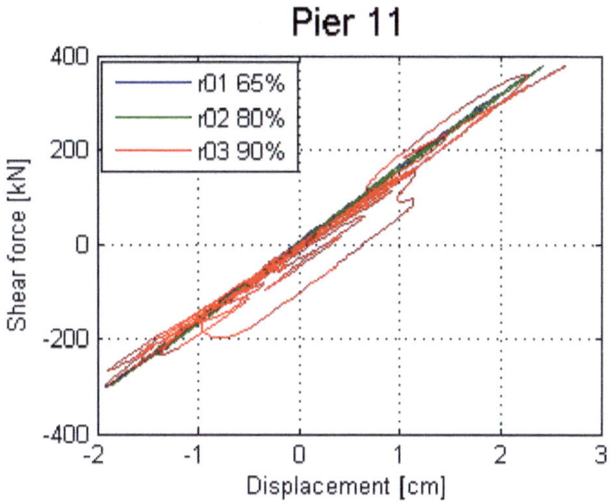

Fig. 14.27 Force-Deflection cycle of Pier 11 during test *r01, r02* and *r03* (ULS PGA=65, 80 and 90%)

14.5 Conclusions

The "Retro" TA Project deals with the assessment of the seismic vulnerability of an old reinforced concrete viaduct with portal frame piers and the study of the effectiveness of different isolation systems through pseudo-dynamic tests. The project was funded by the European commission within the SERIES project. The research activity focused on experimental and numerical investigations of old bridges, designed mainl for gravity loads. Towards this aim, the seismic vulnerability of an existing Italian viaduct with portal frame piers (Rio Torto Viaduct) was evaluated and an isolation system was designed based on friction-pendulum (FP) bearings.

A comprehensive experimental test campaign was performed at the ELSA Laboratory of the JRC (Ispra, Italy). Two one-bay RC frame piers (scale 1:2.5) with two (total height of 7.0 m) and three levels (total height of 11.5 m), were built and tested using the PsD technique with sub-structuring. Two configurations of the prototype bridge were considered: (1) an "as-built" viaduct imposing several levels of damage and (2) a retrofitted viaduct using friction bearings. Natural records, based on a specific hazard analysis, were selected as earthquake input for the experimental seismic tests.

The PsD campaign emphasized the high structural vulnerability of the as-built bridge, thus requiring an appropriate seismic retrofitting intervention. Friction pendulum devices were utilized for such an intervention. An extensive damage pattern was detected, especially in the short physical pier, where the transverse beam was subjected to severe cracking damage due to shear. Additionally, significant fix-end-rotations occurred at the base of the pier columns. The latter rotations were caused by the high bond slip effect typically found in plain steel bars.

Finally, the effectiveness of FPs was verified experimentally by performing tests at different amplitudes, frequency and axial load. It was found that friction coefficients may vary significantly during the tests; the actual friction values can be twice the nominal values used the design.

Acknowledgements The research leading to these results has received funding from the European Community's Seventh Framework Programme [FP7/2007–2013] for access to the EUROPEAN LABORATORY FOR STRUCTURAL ASSESSMENT of the EUROPEAN COMMISSION—JOINT RESEARCH CENTRE under grant agreement n° 227887. Any opinions, findings and conclusions or recommendations expressed in this paper are those of the authors and do not necessarily reflect those of SERIES. The authors would like to acknowledge the important contribution of Mauro Sartori and Agostino Marioni of Alga SpA Milan, for having kindly provided the FPS isolators and for their support in their design.

References

Alessandri S, Giannini R, Paolacci F (2013) Aftershock risk assessment and the decision to open traffic on bridges. Earthq Eng Struct Dyn 42:2255–2275. doi:10.1002/eqe.2324

Baker JW, Cornell CA (2006) Spectral shape, epsilon and record selection. Earthq Eng Struct Dyn 35(9):1077–1095

Bonelli, A., Bursi, O.S., He, L., Magonette, G., Pegon, P., Convergence analysis of a parallel interfield method for heterogeneous simulations with substructuring, 2008, International Journal for Numerical Methods in Engineering, 75, 7, 800-825.

CEN-a 2006 (1994) Eurocode 8—Structures in seismic regions—Design—Part 2: Bridges. Comitè e Europaèen de Normalisation: Brussels

CEN-b 2006 (2006) Eurocode 8: design of structures for earthquake resistance—Part 3: assessment and retrofitting of buildings

Combescure D, Pegon P (1997) α-operator splitting time integration technique for pseudodynamic testing: error propagation analysis. Soil Dyn Earthq Eng 16:427–443

Della Corte G, De Risi R, Di Sarno L (2013) Approximate method for transverse response analysis of partially isolated bridges. J Bridge Eng 18(11):1121–1130

Dermitzakis SN, Mahin SA (1985) Development of substructuring techniques for on-line computer controlled seismic performance testing. Technical report, Earth.Eng. Research Center, Univ. of California, Berkeley, CA, USA, 1985. Report No. UCB/EERC-85/04

FHWA (1995) Seismic retrofitting manual for highway bridges. Publ. FHWA-RD-94-052, Federal Highway Administration

FHWA-ATC (1983) Retrofitting guidelines for highway bridges. Report ATC-06-2, Applied Technology Council, Redwood City, California

FHWA-MCEER (2006) Seismic retrofitting manual for Highway Structures. Part 1- Bridges

Gravouil A, Combescure A (2001) Multi-time-step explicit-implicit method for nonlinear structural dynamics. Int J Numer Methods Eng 50:199–225

Guyan, R. J. (1965) Reduction of stiffness and mass matrices. American Institute of Aeronautics and Astronautics Journal, 3(2), 380

Lomiento G, Bonessio N, Benzoni G (2013) Friction model for sliding bearings under seismic excitation. J Earthq Eng. doi:10.1080/13632469.2013.814611

Magonette G (2001) Development and application of large-scale continuous pseudodynamic testing technique. Phil Trans R Soc Lond A 359:1771–1799

McKenna F, Fenves GL, Scott MH (2007) OpenSees: open system for earthquake engineering simulation. PEER, University of California: Berkeley, CA, 2007

Millard A (1993) CASTEM 2000. Guide dutilisation. Saclay, France. Rapport CEA 93/007

Nakashima M, Akazawa T, Sakaguchi O (1993) Integration method capable of controlling experimental error growth in substructure pseudo dynamic test. AIJ J Struct Constr Eng 454:61–71

NTC08 (2008) Norme tecniche per le costruzioni. Decreto Ministeriale 14.01.2008

Paolacci F, Giannini R (2012) An experimental and numerical investigation on the cyclic response of a portal frame pier belonging to an old reinforced concrete viaduct. Earthq Eng Struct Dyn 41:1109–1127. doi:10.1002/eqe.1175
Pegon P (2008) Continuous PsD testing with substructuring. In: Bursi OS, Wagg DJ (eds) Modern testing techniques for structural systems, dynamics and control. CISM, Italy
Pegon P, Magonette G (2002) Continuous PsD testing with non-linear substructuring: presentation of a stable parallel inter-field procedure. JRC-Special publication No. SPI.02.167
Pegon P, Magonette G (2005) Continuous PsD testing with non-linear substructuring: using the operator splitting technique to avoid iterative procedures. JRC-Special publication No. SPI.05.30
Pinto PE, Mancini G (2009) Seismic assessment and retrofit of existing bridges. The state of earthquake engineering research in Italy: the ReLUIS-DPC 2005–2008 Project, 111–140, © 2009 Doppiavoce, Napoli, Italy
Priestley MJN, Verma R, Xiao Y (1994) Seismic shear strength of reinforced, concrete columns. ASCE J Struct Eng 120(8):2310–2329
Schellenberg A (2010) Single concave friction pendulum bearing element. opensees Wiki. http://opensees.berkeley.edu/wiki/index.php/Single_Concave_Friction_Pendulum_Bearing_Element. Accessed 31 Dec 2012
Zhao J, Sritharan S (2007) Modeling of strain penetration effects in fiber-based analysis of reinforced concrete structures. ACI Struct J 104(2):133–141

Chapter 15
Full-scale Testing of Modern Unreinforced Thermal Insulation Clay Block Masonry Houses

Andreas Jäger, Suikai Lu, Hervé Degée, Christophe Mordant, Ambra Chioccariello, Zoran T. Rakicevic, Veronika Sendova, Luís Mendes, Paulo Candeias, Alfredo Campos Costa, António A. Correia and Ema Coelho

15.1 Introduction

The requirements of clay block masonry building constructions changed drastically during the last decades. Besides mechanical strength, also thermal insulation became more and more important in order to fulfil the increased legal requirements

A. Jäger (✉)
Wienerberger AG, Wienerbergstraße 11, 1100 Vienna, Austria
e-mail: andreas.jaeger@wienerberger.com

S. Lu
Consulting Engineer, Versorgungsheimstrasse 6, 1130 Vienna, Austria

C. Mordant · H. Degée
University of Liège, Place du 20-Août 7, 4000 Liège, Belgium
e-mail: cmordant@ulg.ac.be

A. Chioccariello
University of Genova, Genova, Italy

Z. T. Rakicevic · V. Sendova
Institute of Earthquake Engineering and Engineering Seismology-IZIIS, Ss Cyril and Methodius, Todor Aleksandrov 165, PO Box 101, 1000 Skopje, Republic of Macedonia
e-mail: zoran_r@pluto.iziis.ukim.edu.mk

L. Mendes · P. Candeias · A. Campos Costa · A. A. Correia · E. Coelho
Structures Department, Laboratório Nacional de Engenharia Civil, Lisbon, Portugal
e-mail: luismendes77@gmail.com

A. Campos Costa
e-mail: alf@lnec.pt

P. Candeias
e-mail: pcandeias@lnec.pt

A. A. Correia
e-mail: aacorreia@lnec.pt

E. Coelho
e-mail: ema.coelho@lnec.pt

F. Taucer, R. Apostolska (eds.), *Experimental Research in Earthquake Engineering,*
Geotechnical, Geological and Earthquake Engineering 35,
DOI 10.1007/978-3-319-10136-1_15

for heating/cooling energy demand. To ensure that monolithic clay wall constructions are able to fulfil these requirements still in the future, a new generation of high thermal insulating clay blocks has been developed and successfully launched by Wienerberger, the world's largest brick producing company.

Monolithic clay wall constructions represent a very common construction method in Europe. In these regions, seismic hazard is qualified by current standards as being low-to-moderate and requires a specific consideration.

In the context of modern masonry, the available test results mainly concern cyclic shear behaviour of wallets in order to assess the hysteretic behaviour of single structural members (Magenes and Calvi 1997, Magenes 2006, Tomazevic and Klemenc 1986, Tomazevic and Lutman 1988, Tomazevic 1999) but few tests have been performed on full-scale specimens to include the global structural behaviour.

The aim of this study is to validate experimentally the effective three-dimensional dynamic response of unreinforced masonry buildings by means of full-scale shaking table tests. For this purpose, two real scale mock-ups have been constructed and tested on LNEC's 3D shaking table with six degrees of freedom using a series of seismic records with increasing intensity. These mock-ups were designed as 2-storey structures with the main difference on the plan regularity: the first one was regular while the second included significant irregularities.

In this paper, the main results from the experimental campaign and the initial structural analysis are presented. The experimental setup, the dynamic behaviour tested using impulse excitation and the preliminary test results are also addressed in some detail. In addition, the most probable collapse mechanism is identified and behaviour factors (q-values) are estimated using the measured maximum acceleration on the storeys that reached a near collapse situation.

Nevertheless, further analysis is still needed in particular using numerical models to analyze and extend the results from the experimental campaign.

15.2 Experimental Setup

15.2.1 Mock-up Idealization and Geometry

Two full-scale mock-ups were designed. These mock-ups had 2 storeys and had plan dimensions of 3.7×4.2 m^2 at the ground floor and a height of 5.4 m. They were built on specially designed steel foundations (see Fig. 15.1). Mock-up A is regular in plan while Mock-up B includes significant irregularities. A plan view of both mock-ups is presented in Fig. 15.2 and an elevation view is shown in Fig. 15.6 and in Fig. 15.7 The clear height of the walls is 2.5 m and the storey slabs are 0.20 m thick. All door openings were conceived with a height of 1.9 m and the window openings were 1.15 m tall. The length of the individual shear walls ranges from only 0.8 m to 2.1 m. In addition, the walls are 20 cm thick, which is a relatively low value for this type of structural solutions. As a comparison, according to EN 1998-1—Table 9.2 (EN 1998-1, CEN 2004), the minimum thickness of shear walls for unreinforced

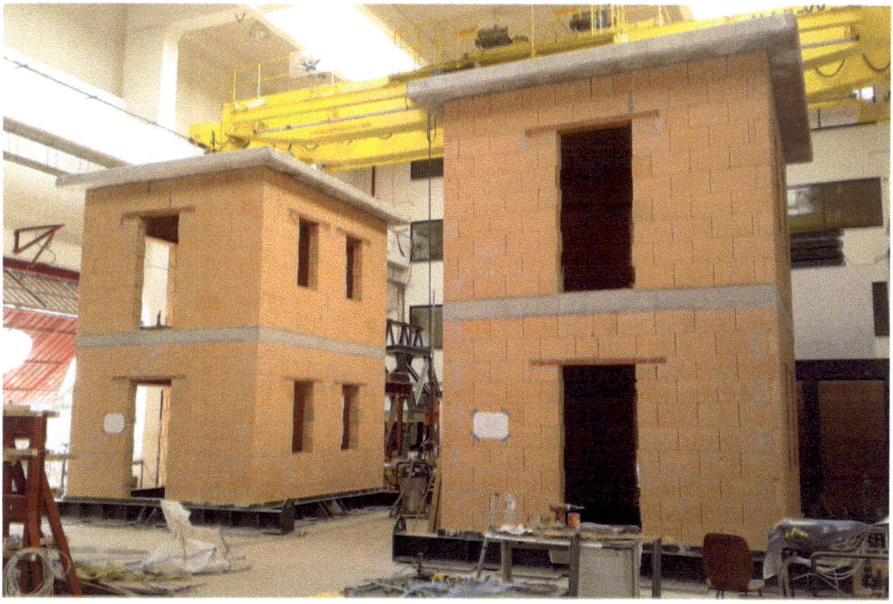

Fig. 15.1 Mock-up A and B in the test lab before testing

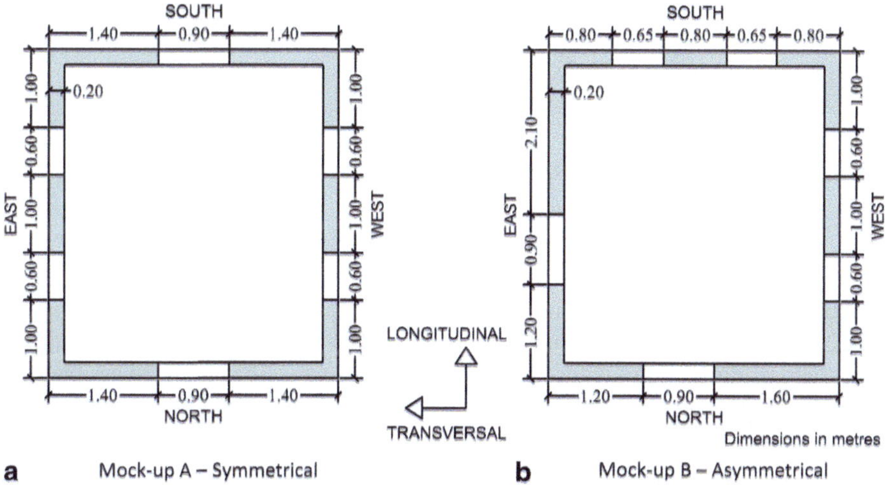

Fig. 15.2 Plan view of the mock-ups

masonry is generally 24 cm and this value can be reduced to 17 cm only in low seismicity regions (low seismicity is defined as zones with $\gamma_I \times a_{gR} \times S \leq 0{,}10$ g).

The first and second storey slabs were prefabricated reinforced concrete slabs with a thickness of 20 cm. The first floor slabs are rectangular with $3.70 \times 4{,}20$ m^2 in plan, whereas the ones used on the second floor are 4.40×4.90 m^2.

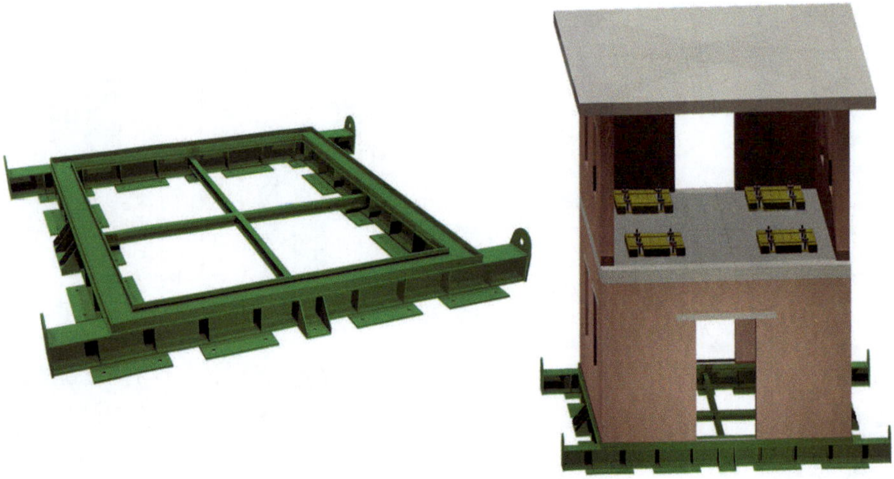

Fig. 15.3 Mock-up details

In order to simulate a live load of 2 kN/m^2, four additional masses weighting 6 kN each were fixed on top of the first floor slab. Their positions were chosen in order to have similar inertial properties to the corresponding distributed load (see Fig. 15.3b). The total mass including walls, ceilings, steel foundation, and additional masses amounts to 31.7 t for each mock-up.

15.2.2 Steel Foundations

Specially designed steel foundations were built to attach the mock-ups to the shaking table (see Fig. 15.3a). The design was made to limit deflections at midspan to L/1000 during transportation on to the shaking table and to avoid damaging the mock-ups as much as possible. The transportation was made by suspending the mock-ups using a bridge crane. In the end, HEB300 steel beams were adopted for the main beams and UNP260 steel beams were chosen to lay a conventional mortar bed to support the masonry.

15.2.3 Construction

The construction of both mock-ups was executed in the following steps:

First floor:

- Placing the first layer of clay blocks in the U-shaped steel frames of the steel foundation on top of a conventional mortar

- Erecting the walls using thin-layer mortar up to the wall height of 2.5 m. Above door and window openings, prefabricated lintels with a height of 6.5 cm were used
- On top of the last layer, sanded bitumen sheeting was glued using thin-layer mortar. On top of this bitumen sheeting the height was levelled out using conventional mortar. This represents an approved construction detail which ensures proper transmission of shear forces (see e.g. German national annex of Eurocode 6, DIN EN 1996-2/NA, 2011)
- Placing of prefabricated reinforced concrete slabs in wet (not hardened) mortar

Second floor:

- Placing the first layer of clay blocks on top of the RC slab using conventional mortar
- Following the same steps adopted for the first floor

15.2.4 Steel Ties

Six steel ties (two in each WE façade and one in each NS façade) were used to mitigate cracking during transportation by reducing the base deformation and by creating a favourable compression state. Load cells were used to control the axial force installed in the steel ties, which was about 50 kN in each NS tie and 30 kN in each WE tie.

15.2.5 Material Parameters

Premium clay blocks from Wienerberger (Porotherm 20–40 W.i. Plan) with excellent mechanical and thermal performance were used for building the mock-ups (see Fig. 15.4a). These special blocks are characterized by relatively large voids in which additional insulation material, either mineral wool or perlite, is placed. The mechanical and geometrical parameters of the blocks are given in Table 15.1.

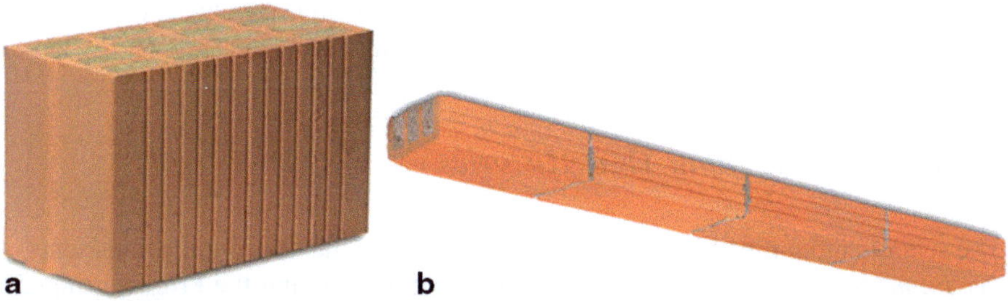

Fig. 15.4 **a** High thermal insulating clay block, Wienerberger Porotherm 20–40 W.i. Plan and **b** prefabricated lintel, Wienerberger Porotherm Sturz 9 cm

Table 15.1 Geometrical and mechanical clay block characteristics (Wienerberger Porotherm 20–40 W.i. Plan)

Material characteristic	Value
Mean compressive strength $\overline{f_b}$ (EN 772-1) [N/mm²]	10.7
Normalized compressive strength f_b (EN 772-1) [N/mm²]	13.4
Voids ratio (EN 772-3) [%]	48
Gross dry density (EN 772-3) [kg/m³]	755
Dimensions L × H × W (EN 772-16) [mm]	400 × 249 × 200

Table 15.2 Characteristics of thin-layer mortar

Material characteristic	Value
Bending strength (EN 1015-11) [N/mm²]	3.1
Compressive strength (EN 1015-11) [N/mm²]	12.1
Density (EN 1015-6) [g/cm³]	1.39

Table 15.3 Mechanical masonry characteristics

Material characteristic	Value
Mean masonry strength f_i (EN 1052-1) [N/mm²]	5.6
Characteristic masonry strength f_k (EN 1052-1) [N/mm²]	5.3
Young modulus (EN 1052-1) [N/mm²]	4500
Characteristic shear strength f_{vk0} (EN 1052-3) [N/mm²]	0.32

The bed joints of these blocks have an accurate geometry, which is achieved by grinding, thus allowing the construction using a thin mortar layer of approximately 1 mm thickness, compared to conventional mortar with a thickness of 12 mm. The mechanical parameters of the thin-layer mortar are listed in Table 15.2.

The mechanical properties of the masonry were tested according to relevant testing standards. The material parameters obtained are given in Table 15.3.

The lintels for doors and windows were executed using two prefabricated lintel beams (Wienerberger Porotherm Sturz 9 cm) as shown in Fig. 15.4b. The dimensions of the beams were 9 cm in width and 6.5 cm in height. The length was adjusted to the window/door size and is given in Figs. 15.6 and 15.7.

15.2.6 *Seismic Input Time Histories*

The seismic input time histories were generated from the horizontal components of the *Tolmezzo-Diga Ambiesta* station records acquired during the Friuli earthquake that occurred in 1976 in the northeast of Italy. These records have 15 s duration and were adapted to match the EC8 elastic response spectrum (EN 1998-1, CEN 2004), 5 % damping, type 1 and ground type C ($S=1.15$; $T_B=0.2$ s; $TC=0.6$ s;

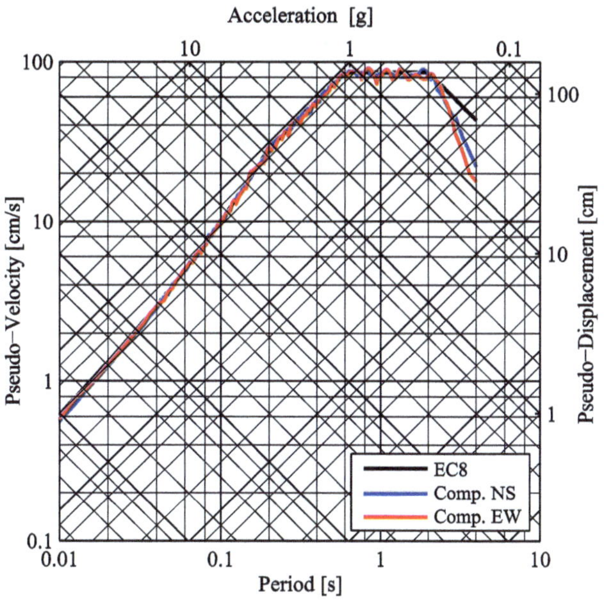

Fig. 15.5 Reference vs. EC8 response spectrum of both horizontal components for 5 % damping (all scaled to PGA = 0.32 g)

TD = 2.0 s), as shown in Fig. 15.5. The resulting records were processed to be used on the shaking table, in particular the time-history begin and end segments were smoothened and the components with higher periods were filtered out in order to reduce peak displacements and velocities and cope with shaking table limitations, namely a maximum stroke of ± 20 cm and a peak velocity of 70 cm/s. The effect of such processing can be seen in the constant displacement zone of the response spectra presented in Fig. 15.5.

The reference signals (REF) were obtained by scaling the N-S (E-W) component to PGA = 0.36 g (0.32 g) and used in the shaking table transverse (longitudinal) direction (see Fig. 15.2). These signals were scaled down to 12.5, 25.0, 37.5, 50, 62.5, 75 and 87.5 % resulting in a total of eight seismic stages with PGA values listed in Table 15.4.

15.2.7 Instrumentation Plan

The adopted instrumentation setup comprised the shaking table displacements and accelerations, four biaxial absolute displacements for the first and second storey slabs, 26 acceleration records (steel foundation, slabs and out-of-plane wall movements) and 18 relative displacements measured by LVDTs (rocking, sliding and diagonal relative displacements). A schematic representation of the sensor layout is presented in Fig. 15.6 for Mock-up A (symmetric) and in Fig. 15.7 for Mock-up B (asymmetric). A high speed data acquisition system was used to register the tests and the data was stored in an appropriate data format.

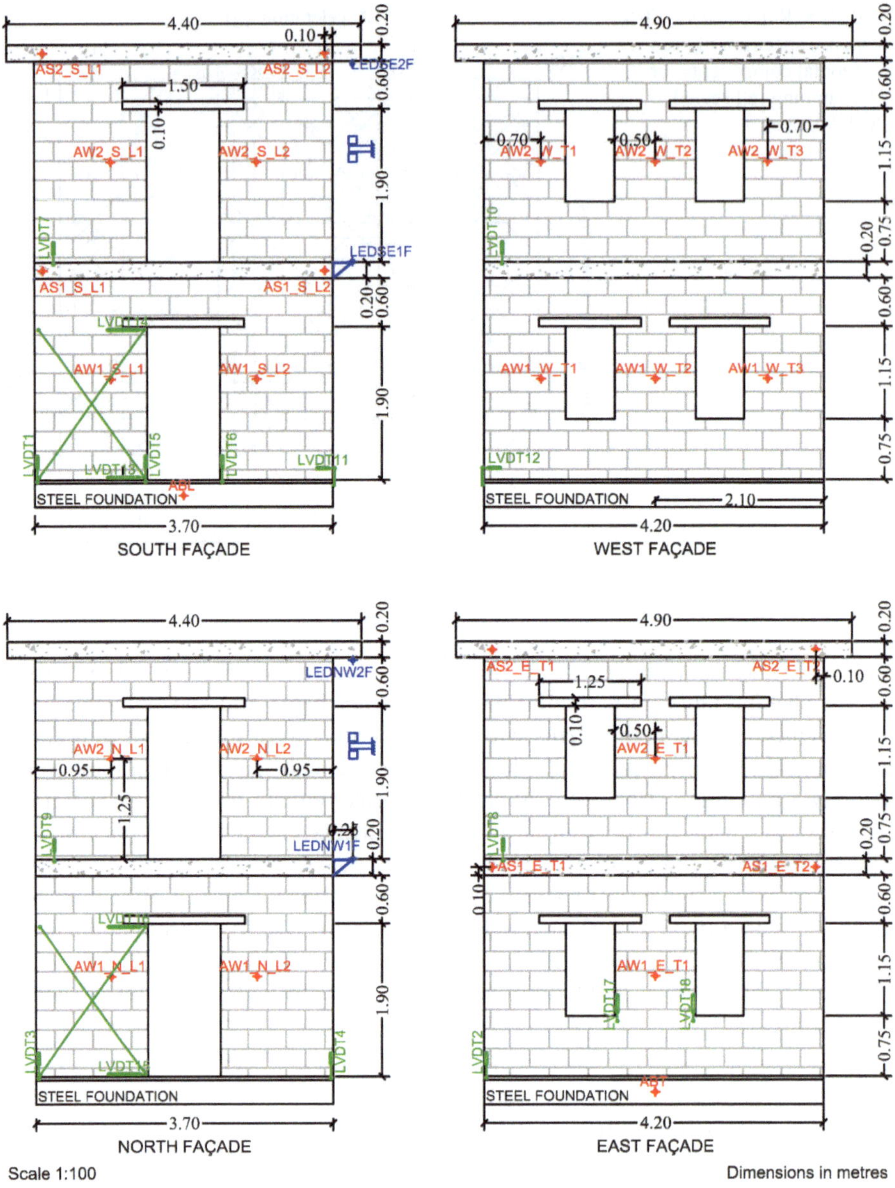

Fig. 15.6 Instrumentation setup adopted for Mock-up A (symmetric)

15.2.8 Testing Procedure

As mentioned before, eight seismic stages were generated from the REF signals. To enhance the comparison with numerical results, it was decided to adopt alternating uniaxial and biaxial test stages. Consequently, the odd test stages were divided into two tests for each horizontal direction acting separately. For the other stages, both

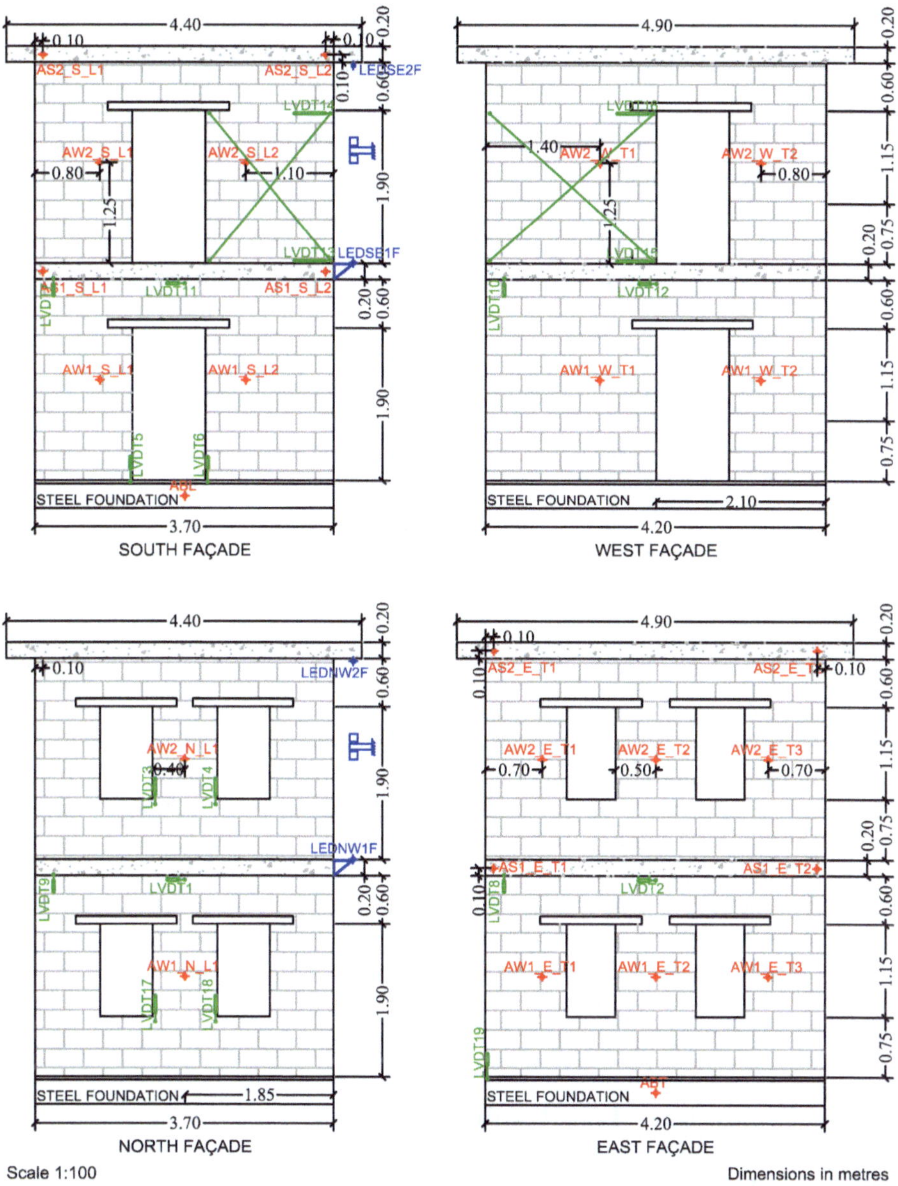

Fig. 15.7 Instrumentation setup adopted for Mock-up B (asymmetric)

horizontal components were used at the same time. The test sequence is presented in Table 15.4.

It is a well-known fact that shaking tables reproduce the target signals with a certain approximation due to several aspects of control systems' engineering that are beyond the scope of this paper, *e.g.*, table-specimen interaction or oil-column resonance. An adaptive technique is often used to minimize the differences between

Table 15.4 Measured peak accelerations for load sequences applied to Mock-up A and Mock-up B

Mock-up A (symmetric)				Mock-up B (asymmetric)			
Stage	NS (Long.) [m/s²]	EW (Trans.) [m/s²]	No. of shakes [-]	Stage	NS (Long.) [m/s²]	EW (Trans.) [m/s²]	No. of shakes [-]
01T	0.096	0.433	6	01T	0.092	0.428	5
01L	0.491	0.110	5	01L	0.636	0.084	5
02	1.013	0.913	5	02	1.000	0.949	6
03T	0.280	1.388	5	03T	0.141	1.249	5
03L	1.419	0.616	5	03L	1.505	0.620	4
04	3.734	2.143	5	04	2.016	1.882	5
05T	0.486	2.857	5	05T	0.664	2.616	4
05L	2.526	0.844	6	05L	3.193	1.148	5
06	3.099	2.684	6	06	3.918	2.105	6
07T	0.646	3.068	7	07T	1.008	4.415	4
07L	3.541	0.830	5	07L	3.639	1.141	3
08	3.718	5.362	2	08	3.685	4.583	2
Total No. of shakes			62	Total No. of shakes			54

target and effective motions of the shaking table. This technique consists in progressively incrementing the shaking table drive motions and by minimizing in each step the error time series using the information available from the dynamic properties of the global system (mock-up and shaking table). This adaptive technique led to an increase of the total number of shakes experienced by the mock-ups, which are also listed in Table 15.4.

During the tests and as a safety measure in case of global collapse, both slabs were connected by cables to the bridge crane. These cables were loose for typical testing displacements and would hold the slabs and part of the masonry walls if a collapse would have occurred.

15.3 Preliminary Test Results

15.3.1 Qualitative Observations and Collapse Modes

This section summarizes the direct observations made during the tests or based on the analysis of pictures and movies from the experiments. It describes the structural global behaviour of the mock-ups and identifies the most probable collapse mechanisms. These preliminary observations will be later crossed with the outcomes of the measurement devices in order to validate and quantify some of the conclusions drawn.

Regarding Mock-up A, the collapse occurred on the second floor and Fig. 15.8 illustrates the failure mode, in which the façades with door openings are the most

Fig. 15.8 Observed collapse mechanisms of symmetric Mock-up A

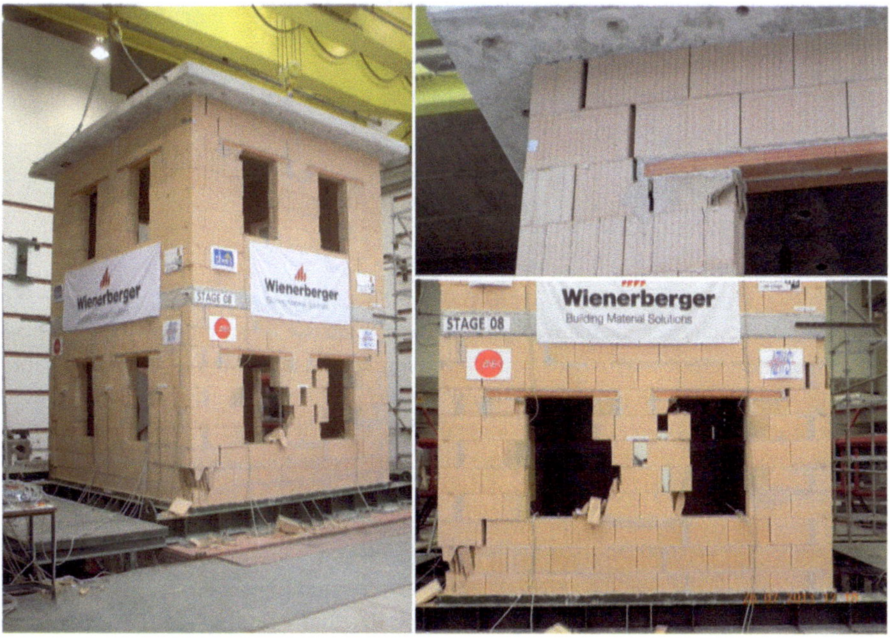

Fig. 15.9 Observed collapse mechanisms of asymmetric Mock-up B

cracked. The observed failure mechanisms are a combination of shear failure, sliding and local crushing.

The main damages observed on Mock-up B are concentrated in one of the façades with window openings. The main collapse mechanism identified is the shear failure of the central wall (see Fig. 15.9, left), although some localized crushing of units is also observed (see Fig. 15.10). The total collapse of the mock-up was avoided thanks to the fact that the floor slab was initially four-side supported. Therefore, even after totally losing the load-bearing capacity of the front façade, the slab is still supported on three sides and the structure remains stable after the seismic tests.

Fig. 15.10 Local failure of asymmetric Mock-up B

15.3.2 Dynamic Characterization

Dynamic characterization tests were performed before, in-between and after the seismic stages. These tests were executed to estimate the frequency evolution of the main vibration modes. This was achieved using impulse excitation introduced into the mock-ups by the shaking table using low amplitude square wave displacement time histories. This type of tests was executed in between each seismic test stage and the shaking table and mock-up accelerations were measured using high sensitivity piezoelectric accelerometers.

The results extracted for the two main structural vibration modes are listed in Table 15.5 and presented graphically in Fig. 15.11 It should be noted that during these tests the mock-ups are placed on top of the shaking table, meaning that the dynamic interaction between the mock-up and the shaking table must be taken into account. As a result, the apparent structural mode frequency values are known to be changed (often a decrease), due to coupling between specimen and shaking table dynamic characteristics (mass and stiffness) and even their relative order can be modified. Further analysis is required using a numerical model to analyze the absolute values of the measured mode frequencies and also their order of appearance (longitudinal *vs.* transverse direction). Nevertheless, the relative decrease of the main mode frequencies gives an idea of the dynamic characteristics, evolution.

For what concerns the symmetric Mock-up A, the frequency values gradually decrease with the acceleration level increase, which is associated with the test stages. This behaviour is expected and such a frequency drop has been already

Table 15.5 Results from dynamic characterization tests

Mock-up A (symmetric)			Mock-up B (asymmetric)		
BEFORE STAGE	ST-IMP-1 (Long.) [Hz]	ST-IMP-2 (Trans.) [Hz]	BEFORE STAGE	ST-IMP-1 (Long.) [Hz]	ST-IMP-2 (Trans.) [Hz]
STAGE 1T	–	6.9	STAGE 1T	5.4	5.8
STAGE 1L	5.7	6.9	STAGE 1L	5.4	5.7
STAGE 2	5.5	6.7	STAGE 2	5.4	5.8
STAGE 3T	5.5	6.7	STAGE 3T	5.3	5.8
STAGE 3L	5.2	6.3	STAGE 3L	5.3	5.8
STAGE 4	5.2	6.1	STAGE 4	5.2	5.8
STAGE 5T	5.2	5.9	STAGE 5T	5.2	5.2
STAGE 5L	5.2	5.9	STAGE 5L	5.2	5.2
STAGE 6	5.2	6.0	STAGE 6	5.2	5.2
STAGE 7T	5.0	5.5	STAGE 7T	5.1	4.3
STAGE 7L	5.1	5.2	STAGE 7L	5.0	4.1
STAGE 8	4.8	5.3	STAGE 08	4.9	4.1

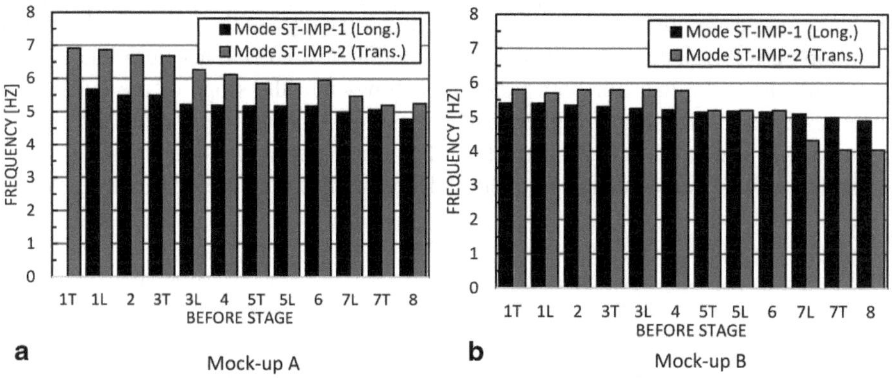

Fig. 15.11 Results from dynamic characterization

observed in other test campaigns within the SERIES project (Mordant et al. 2013) as in other projects (Michel et al. 2011). In this case, the frequency decrease is more pronounced along the transverse direction (W-E), showing that a higher degradation occurred on the façades with door openings. These observations are in agreement with the qualitative observations presented before, where it was shown that collapse occurred in a transverse façade.

The same analysis for the asymmetric Mock-up B leads to similar observations and conclusions. A frequency decrease is also observed in both directions, however in this case it is more significant in the transverse one which is expected because the collapse mechanism occurred along this direction.

15.3.3 Seismic Response

The main results available so far concern the evolution of the maximum acceleration at the first and second floor level with respect to the measured maximum table acceleration (see Table 15.4) and the maximum measured inter-storey drift.

The first analysis allows the identification of the damaging effects. The mock-up behaviour can be considered as elastic as long as the increase of the table acceleration leads to a proportional increase of the accelerations measured at the slabs levels. If this relation becomes non-proportional, this may be due to a loss of stiffness and, therefore, to lower natural frequencies and/or to an increase of the damping, leading hence to lower spectral accelerations. From the difference between accelerations for theoretically ideal elastic behaviour, extrapolated from low load intensities, and measured accelerations at failure, the behaviour factors *q* can be estimated (Degée et al. 2007).

In this work, only the plots of the storey where failure occurred are presented in Fig. 15.12 for Mock-up A and in Fig. 15.13 for Mock-up B. The graphs on the left show the transverse accelerations while those on the right show the longitudinal accelerations. In the upper graphs, the accelerations measured by two accelerometers on the slabs are plotted, in the graphs below, the mean values and the corresponding idealized elastic behaviour are plotted. As mentioned above, the idealized elastic behaviour is estimated as a linear fit to the low intensity loading stages, with linear relation between shaking table acceleration and measured slab acceleration.

Analysis of the data for both mock-ups shows that the slope of the curves are progressively decreasing at higher load intensities, highlighting the ductile behaviour of this structural system. The start of the non-linear range has a good correspondence with the identified frequency decrease. For the case of Mock-up A, the

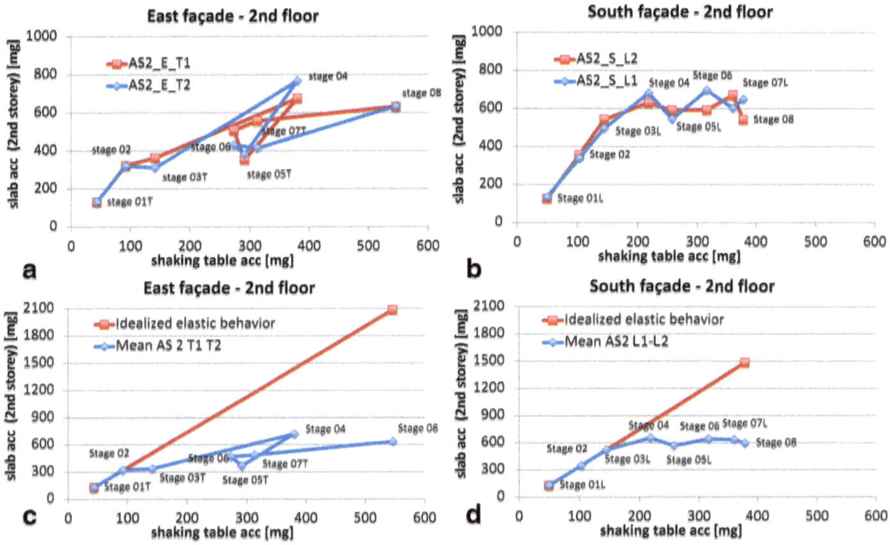

Fig. 15.12 Maximum acceleration measured on Mock-up A on the second floor slab: **a** and **b** measured values, **c** and **d** mean values and idealized elastic behaviour

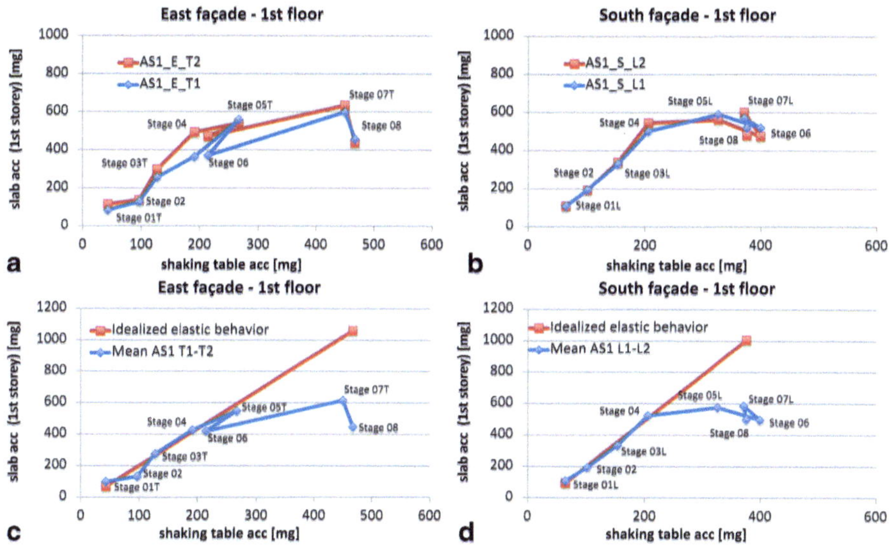

Fig. 15.13 Maximum acceleration measured on Mock-up B on the first floor slab: **a** and **b** measured values, **c** and **d** mean values and idealized elastic behaviour

first damage in the transverse direction is visible after Stage 02 (see Fig. 15.12). This fits well to the small frequency drop visible at this stage in Fig. 15.11. On the other hand, for Mock-up B the first severe damage occurred at Stage 04, indicated by a frequency drop (see Fig. 15.11) and by the beginning of the non-linear global response, as shown in Fig. 15.13.

The ratio between the elastic extrapolation and the actual measured maximum acceleration for the last point of the curves (Stage 08) gives an estimation of the behaviour factor q of this structural system. On the basis of the results related to the storey where the failure occurred (2nd for Mock-up A and 1st for Mock-up B) in the transverse direction a $q=3.3$ is obtained for Mock-up A and of $q=2.4$ for Mock-up B. In the longitudinal direction, the q-values were determined as 2.5 for Mock-up A and 2.0 for Mock-up B. These observed values are similar or even higher than the ones recommended in Eurocode 8 (EN 1998-1, CEN 2004), which indicates q-values between 1.5 and 2.5 for unreinforced masonry.

The second set of results consists in the acceleration-drift curves which are plotted in Fig. 15.14 and 15.15 for Mock-up A and Mock-up B, respectively. For what concerns Mock-up A, the maximum drift values are higher in the second storey (see Fig. 15.14), which complies with the qualitative observations made and with the collapse mechanism observed. The near-collapse state was considered to happen during Stage 08 (the maximum drift of about 0.3 % in Stage 07 and about 0.8 % in Stage 08). With respect to the recommendations given by Eurocode 8-Part 3 (EN 1998-3, CEN 2005), this corresponds to the range of maximum value suggested for walls failing due to shear (0.4 %).

Regarding Mock-up B, the collapse occurred at the first storey and the corresponding drift values are given in Fig. 15.15 The most significant damage is

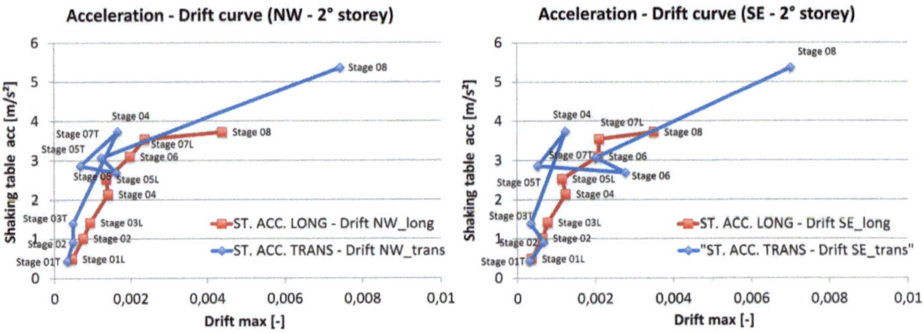

Fig. 15.14 Maximum inter-storey drift (2nd Storey) of Mock-up A as a function of the measured shaking table acceleration

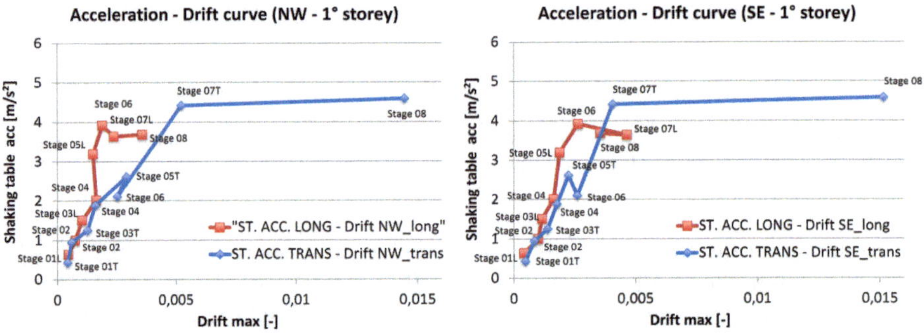

Fig. 15.15 Maximum inter-storey drift (1st Storey) of Mock-up B as a function of the measured shaking table acceleration

observed during Stage 05 and for the transverse direction, translating the loss of shear resistance of the middle wall of the North façade. At this stage, the maximum drift is about 0.4 %, which is again in agreement with the values recommended by the standards. The overall limit state is reached during test Stage 08, for which the maximum drift increases from 0.5 to 1.5 %. It is also interesting to note that, if the relative drift is calculated with respect to the height of the middle wall instead of the story height, it exceeds largely a value of 2 %, corresponding to an advanced damage state.

All the preliminary observations reported here will be further complemented by additional analysis of the results and with comparisons with numerical models.

15.4 Conclusions

This paper presented the first results of the seismic vulnerability assessment carried out on two 2-storey full-scale mock-ups built with modern unreinforced masonry made of highly thermal insulating clay blocks.

The mock-ups were tested in LNEC's 3D shaking table and were loaded with a sequence of uniaxial and biaxial seismic excitations with increasing intensity. After each test, the natural frequencies were identified by dynamic characterization tests. The tests were continued until a near-collapse state was reached. During the tests all relevant displacements and accelerations were recorded by the test instrumentation. Based on these results the following preliminary conclusions can be drawn:

- Evolution of natural frequencies: For each test stage, a continuous decrease in the main mode frequencies was observed for both mock-ups. For interpretation of the absolute values further analyses are required to assess the interaction between the shaking table and the mock-up.
- Type of failure: Both mock-ups reached a near-collapse state due to failure of the walls along the transverse direction. The highest frequency drop was also identified for the transverse direction. The observed failure mechanisms are a combination of shear failure, sliding and local crushing.
- Behaviour factors: Both mock-ups show a significant amount of ductility. The behaviour factors (q-values) evaluated from the slab vs. shaking table accelerations ranged between 2.5 and 3.3 for Mock-up A and between 2.0 and 2.4 for Mock-up B. These values are higher than those prescribed in Eurocode 8 for unreinforced masonry, ranging between 1.5 and 2.5.
- Drift values: The observed drift values are equal or even higher than the ones defined in Eurocode 8-Part 3(EN 1998-3, CEN 2005).

Acknowledgements The research leading to these results has received funding from the European Union Seventh Framework Programme [FP7/2007–2013] for access to the Laboratório Nacional de Engenharia Civil (LNEC), Portugal under grant agreement no. 227887. The authors are grateful for the support.

References

Degée H, Denoël V, Candeias P, Campos Costa A, Coelho E (2007) Experimental investigation on the seismic behaviour of north European masonry houses. In: Proceedings of SISMICA 07, Congresso Nacional de Sismologia e Engenharia Sísmica, Aveiro, Portugal

DIN EN 1996-2/NA: 2011-10—National Annex—Nationally determined parameters—Eurocode 6: design of masonry structures—Part 2/NA: design considerations, selection of materials and execution of masonry

EN 1998-1 (2004) Eurocode 8: design of structures for earthquake resistance—Part 1: general rules, seismic actions and rules for buildings. CEN-EUROPEAN COMMITTEE FOR STANDARDIZATION

EN 1998-3 (2005) Eurocode 8: Design of structures for earthquake resistance—Part 3: assessment and retrofitting of buildings. CEN-EUROPEAN COMMITTEE FOR STANDARDIZATION

Magenes G (2006) Masonry building design in seismic areas: recent experiences and prospects from a European Standpoint. In Proc. 13th European Conference on Earthquake Engineering, Geneva, Switzerland, Paper No K 9

Magenes G, Calvi GM (1997) In-Plane seismic response of brick masonry walls. Earthq Eng Struct Dyn 26:1091–1112

Michel C, Zapico B, Lestuzzi P, Molina FJ, Weber F (2011) Quantification of fundamental frequency drop for unreinforced masonry buildings from dynamic tests. Earthq Eng Struct Dyn 40(11):1283–1296)

Mordant C, Dietz M, Degée H (2013) Seismic behaviour of thin-bed layered unreinforced clay masonry shear walls including soundproofing elements. In: Ilki A, Fardis MN (eds) Proc SERIES Workshop. Geotechnical, Geological and Earthquake Engineering series (Chap. 6)

Tomazevic M (1999) Earthquake-Resistant design of masonry buildings. Imperial College Press, London

Tomazevic M, Klemenc I (1986) The behaviour of horizontally reinforced masonry walls subjected to cyclic lateral in-plane load reversals. Proc. 8th European Conf. on Earthq. Eng. 4:7.6/1–8

Tomazevic M, Lutman M (1988) Seismic resistance of reinforced masonry walls. In: Proc. 9th World Conf. on Earthq. Eng. 6:VI/97–102

Chapter 16
Assessment of Innovative Solutions for Non-Load Bearing Masonry Enclosures

João Leite, António A. Correia, Paulo B. Lourenço, Elizabeth Vintzileou, Vasiliki Palieraki, Paulo Candeias, Alfredo Campos Costa and Ema Coelho

16.1 Introduction

This paper presents some of the results of the research project "Masonry Enclosures" developed within the framework of the transnational access (TA) to LNEC's triaxial shake table of the FP7 project SERIES. The TA project addressed the seismic performance of masonry enclosures in European countries with moderate to high seismicity and how they affect global structural response. It consisted in the experimental evaluation of the seismic response of reinforced concrete (RC) frames with innovative solutions for masonry infill walls, considering both the in-plane and out-of-plane behaviour of the enclosures.

The recent L'Aquila earthquake of 2009 has reminded seismic engineers that current masonry infill solutions are not effective, as illustrated by the considerable in-plane damage and out-of-plane collapses verified throughout the affected areas. Eurocode 8 (2010) addresses this issue by imposing the use of reinforced infill solutions but fails to give design and detailing methodologies. With this in mind, a shake table experimental research programme was devised using two different approaches.

The first phase of the research activity involved seismic testing of a two-storey RC frame building designed according to Eurocode 2 (2010) and Eurocode 8 (2010) and built at a 1:1.5 scale, schematically shown in Fig. 16.1a. As a follow-up of previous tests performed at LNEC, using RC buildings either with double leaf unreinforced masonry infill walls or single leaf masonry plus bed joint reinforcement,

A. A. Correia (✉) · P. Candeias · A. Campos Costa · E. Coelho
Department of Structures, National Laboratory for Civil Engineering (LNEC), Lisbon, Portugal
e-mail: aacorreia@lnec.pt

J. Leite · P. B. Lourenço
Department of Civil Engineering, University of Minho and ISISE, Guimarães, Portugal

E. Vintzileou · V. Palieraki
Department of Structural Engineering, National Technical University of Athens (NTUA), Athens, Greece

F. Taucer, R. Apostolska (eds.), *Experimental Research in Earthquake Engineering,*
Geotechnical, Geological and Earthquake Engineering 35,
DOI 10.1007/978-3-319-10136-1_16

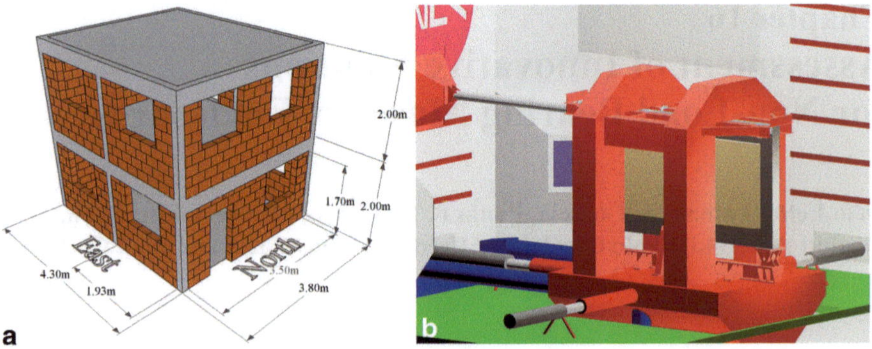

Fig. 16.1 Schematic representation of **a** building model, **b** wall panels' test setup

the reduced scale model was built with single leaf clay bricks and reinforced mortar coating: wire mesh reinforcement was placed within the mortar coating on both sides of the infill walls and anchored to the RC frame and masonry units. From these tests it was possible to assess the evolution of the seismic behaviour of infills and their influence on the RC structure through several acceleration inputs of increasing amplitude, associated to cumulative damage limit states.

The second phase of the TA project comprised three full-scale tests of closed RC plane frames infilled with masonry, simultaneously for in-plane and out-of-plane dynamic actions: (i) one unreinforced masonry infill wall; (ii) one infill wall specimen with horizontal reinforcement layers between masonry units; (iii) one infill wall with wire mesh reinforcement in the mortar coating. The test setup simultaneously uses the shake table, one reaction wall and the Testing device for Innovative Masonry infills (TIM), as represented in Fig. 16.1b. This unique testing setup was specifically designed for these tests and is mainly composed of a stiff steel caisson three-dimensional frame which moves rigidly with the shaking table. It is fixed to the specimen's upper beam in the transverse direction, while a system of rollers allows for an independent motion in the longitudinal direction. This phase of the TA project is presented herein conceptually, since no results are available yet.

The geometry of the reduced scale building model is shown in Fig. 16.1a. The dimensions of all models respect the typical dimensions for RC frames according to the survey in Pereira (2013). Moreover, traditional clay brick units, horizontally perforated, were used in all models.

16.2 Building Model Tests

16.2.1 Building Specimen and Test Setup

The materials adopted for RC elements were C30/37 and S500 for concrete and steel reinforcing bars, respectively. All four façades of the model had infill walls:

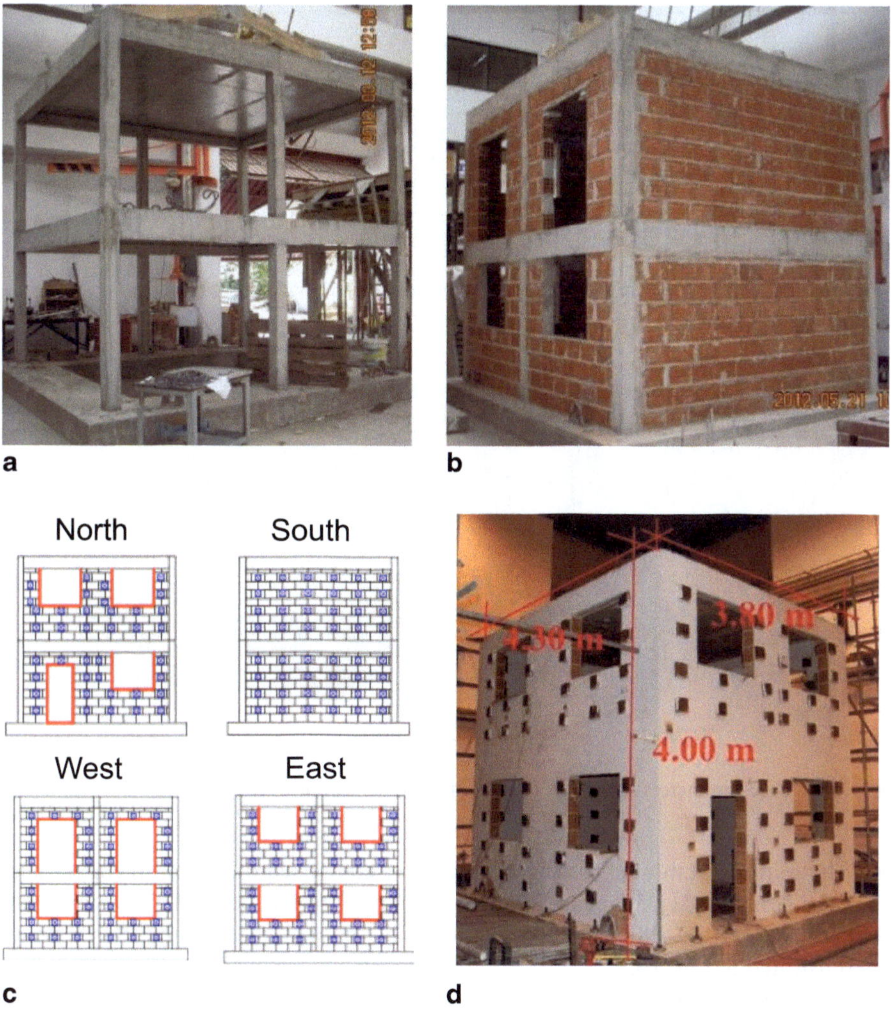

Fig. 16.2 Construction views and geometry of the reduced scale (1:1.5) building model

one façade without openings and around 20 % of the area corresponding to openings in all other façades (see Fig. 16.2). The clay brick units used were only scaled in the thickness, while the other dimensions were kept at a 1:1 scale.

Cauchy-Froude similitude requirements between the prototype and the reduced scale model were duly respected during these tests. Namely, the specific mass of the model was modified by using two types of additional steel masses: 12 large masses in total (around 1.2 t each) attached to the floor slabs and a set of 334 small masses (around 7 kg each) distributed throughout the walls of the building. The total weight of the mock-up (building model, RC foundation and additional masses) was nearly 434 kN.

Fig. 16.3 Details of the wire mesh reinforcement application

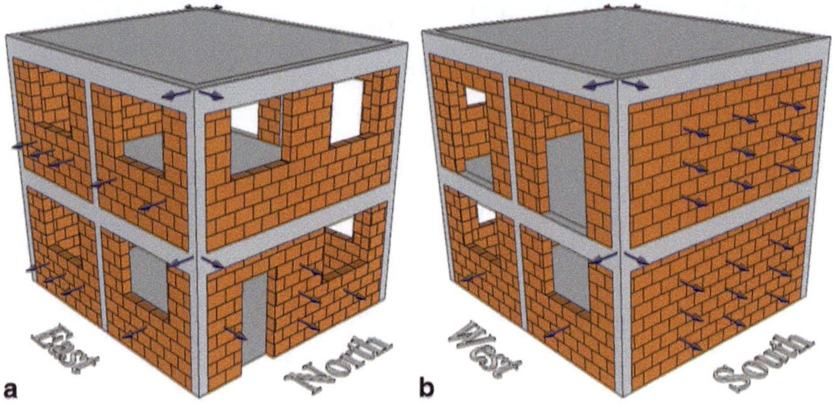

Fig. 16.4 Accelerometers setup for out-of-plane wall behaviour

The reinforced mortar coating adopted was composed of a reinforcement mesh Bekaert Armanet $\phi 1.05$ mm 12.7 mm × 12.7 mm, a pre-batched M5 class mortar and Hilti X-M8H10–37-P8 connectors for attaching the wire mesh to the RC elements, as illustrated in Fig. 16.3.

The abovementioned masses were used for connecting the wire mesh to the masonry units instead of the Hilti connectors, as shown in Fig. 16.2, since the former had to be used anyway in order to comply with the Cauchy-Froude similitude requirements (Carvalho 1998).

After the RC foundation was attached to the shake table, the model was thoroughly instrumented in order to characterize its behaviour during the tests. The instrumentation setup was based on the expected response of the model, obtained from preliminary studies (Leite 2009), and the objectives of the test. The instrumentation can thus be divided in two groups: (i) setup to acquire the out-of-plane behaviour of the infill walls; (ii) setup to monitor the global response of the RC building. The out-of-plane behaviour of the infill walls was captured by a set of 34 accelerometers distributed on the surface of the walls according to Fig. 16.4. On the other hand, the global response of the RC frame was captured using a total of eight accelerometers placed orthogonally at the Northeast and Southwest corners of each RC slab.

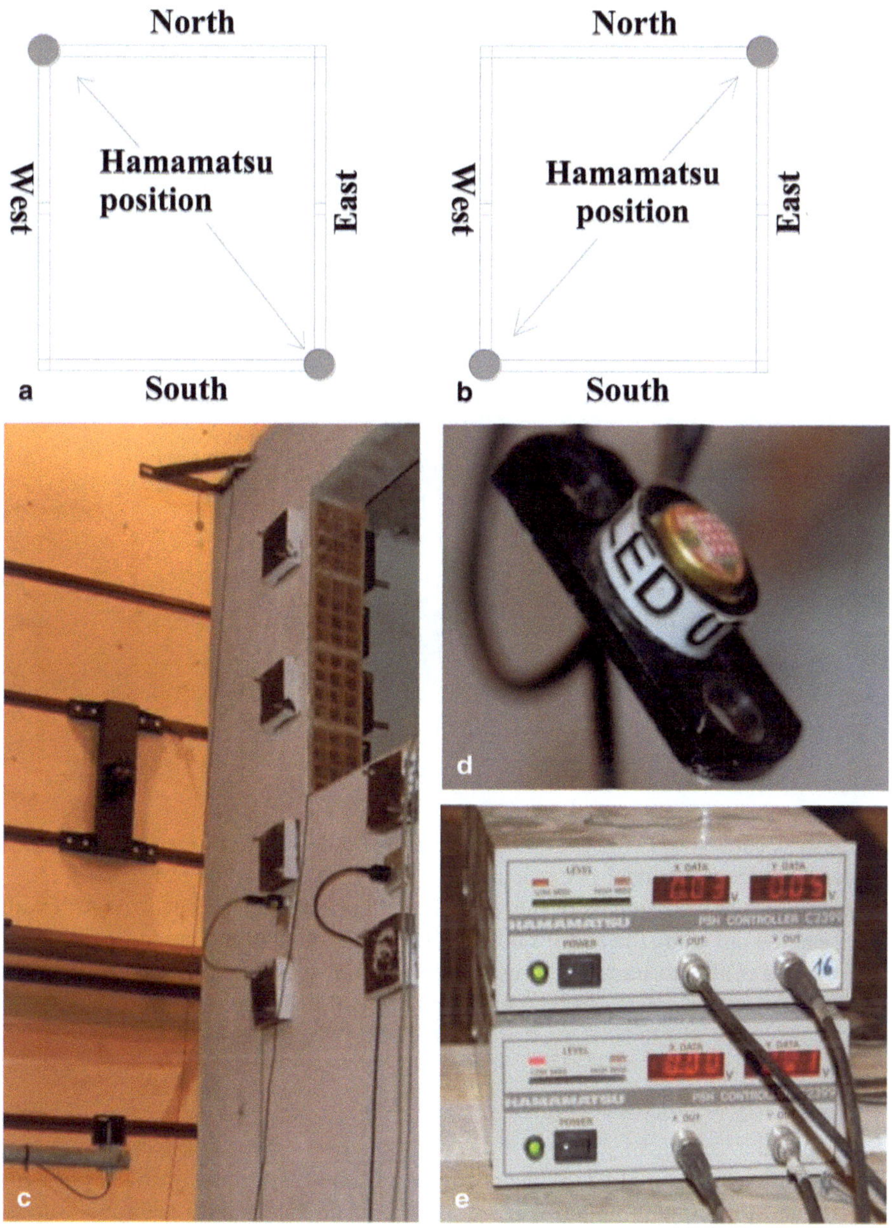

Fig. 16.5 Motion tracking cameras **a** position at first storey slab, **b** position at roof slab, **c** camera and led setup, **d** infrared led, **e** controller

Four motion tracking cameras were also used at the corners of the slabs, as depicted in Fig. 16.5 (Hamamatsu 2013). They are capable of determining the planar motion of an infrared LED each. Data acquisition from all the equipment mentioned

Fig. 16.6 K600 camera system by (Krypton 2003) **a** CCD cameras, **b** calibration equipment, **c** acquisition system, **d** LEDs distributed on the upper floor infill wall of the North façade

above was carried out using one PXI-1052 chassis connected to a PXI-8106 controller, both from National Instruments.

Additional measurements of out-of-plane motions at the North façade infill wall were carried out using the K600 camera system (Krypton 2003), composed of three CCD cameras able to track the three-dimensional displacement of a set of infrared LEDs. The latter were spread on the upper floor wall (see Fig. 16.6).

16.2.2 Input Signal and Test Sequence

The seismic loading applied to the specimen was a bi-directional artificial horizontal ground motion, with uncorrelated components and around 30 s duration, compatible with the Eurocode 8 (2010) response spectra for Lisbon, Portugal, soil type A and seismic action type 1 (large magnitude, long distance). It was applied in three stages of increasing amplitude associated to cumulative damage limit states.

Six artificial accelerograms were generated (Mendes and Campos Costa 2008), resulting in a pair of accelerograms for each stage of the shake table tests. The accelerograms were fitted to the 5 % damped elastic response spectra of Eurocode 8 (2010), scaled according to the damage state considered for each stage: Damage

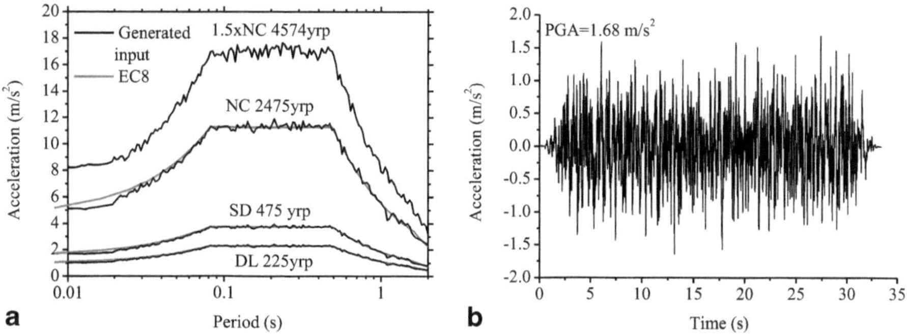

Fig. 16.7 Longitudinal (North-South) component of horizontal ground motion **a** pseudo-acceleration response spectra, **b** target accelerogram for stage 2 (SD)

Table 16.1 Shake table test sequence for the building specimen

Stage	Identification	Description
1	DI 0	Initial dynamic identification test
	DL	Seismic test for Damage Limitation limit state
	DI 1	Dynamic identification at the end of stage 1
2	SD	Seismic test for Significant Damage limit state
	DI 2	Dynamic identification at the end of stage 2
3	NC	Seismic test for Near Collapse limit state
	DI 3	Dynamic identification at the end of stage 3

Limitation (DL—225 years of return period (YRP)); Significant Damage (SD—475 YRP); Near Collapse (NC—2475 YRP). Figure 16.7 represents the longitudinal component of such ground motions, which comply with the Cauchy-Froude similitude requirements, thus evidencing a time contraction with respect to the code spectra values, by a factor of $1.5^{1/2}$, with no scaling in the acceleration values. The test sequence consisted on the application of these three stages, while dynamic identification tests have been performed in between them, as shown in Table 16.1.

16.2.3 Results and Discussion

16.2.3.1 Overall Response and Damage Evolution

The specimen did not show any visual signs of damage after stage 1. However, during stage 2, cracks developed in the mortar coating at all four corners of the building, starting at its base, and between the windows on the intermediate columns of the East and West façades. Small cracks starting at the corners of some of the ground floor openings and moving towards the RC frame were also visible, whilst the first floor presented almost no damage. Figure 16.8 presents the crack patterns

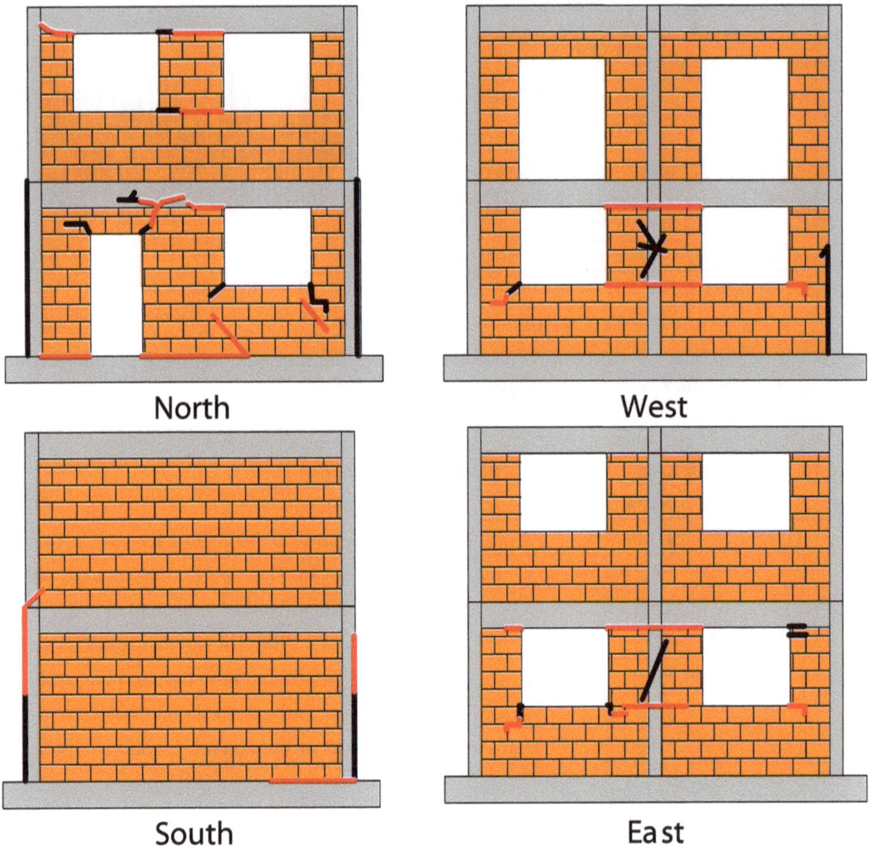

Fig. 16.8 Crack patterns after stages 2 (*black lines*) and 3 (*red lines*). Note: the lines on the RC frame represent damage on the mortar coating, not necessarily on the RC frame

after stages 2 and 3. These results are in agreement with the dynamic data, which shows a frequency decrease of 3.1 % on average for all relevant modes and with respect to the initial state. Such stiffness loss can be related to the separation of the infill walls from the RC frame and to micro-cracking in the RC structure. This damage is either difficult to detect or hidden by the mortar coating.

After the third stage, the building presented wider and longer cracks and some mortar coating detachment at all corners of the ground floor, as shown in Fig. 16.9b and d. Considerable cracking was present between the windows at the ground level of the East and West façades (Fig. 16.9c), as well as at the corners of the North façade openings (Fig. 16.9a). Globally, the structure presented light damage, given the amplitude of the first three stages of the test, and concentrated at the ground level. However, analysis of the dynamic data showed a frequency decrease of 22.1 % on average for all relevant modes and with respect to the initial state, suggesting a more extensive damage than the observed one.

It is pointed out that during the application of the seismic input of stage 3 there were some difficulties on the shake table control, leading to a larger than desired

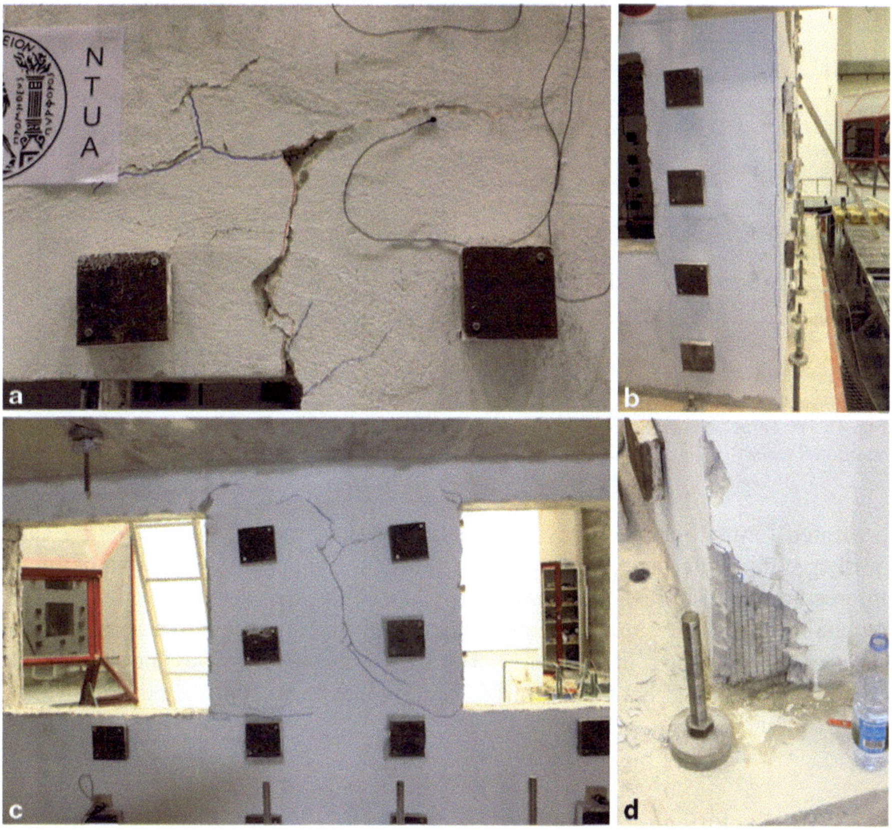

Fig. 16.9 Damage after stage 3 **a** extensive cracking at the corner of the ground floor window of the North façade, **b** cracking in the coating along the height of the Northwest corner, **c** cracks at the West façade window jambs of the ground floor, **d** coating detachment at the base of the Southwest corner

input motion for frequencies above 4 Hz. For this reason, and given the light overall damage in the specimen, the model was submitted to the first three stages of loading again. As shown in Fig. 16.10, this attempt also failed to fulfil the stage 3 target motion.

After the additional three test stages, the specimen did not develop new cracking patterns but the existing ones widened considerably, (see Fig. 16.11). On the other hand, the mortar coating became increasingly detached from the infill walls, most probably due to the connection provided by the additional masses. These results highlight the importance of establishing a good connection of the reinforced coating to the infill walls. Regarding the upper storey, it presented no significant damage, with only small cracks forming at the corners of the openings.

Upon removal of the additional masses, it was observed that the reinforced coating was completely detached from the infill walls and that the supports of the additional masses worked as connectors, thus preventing the mortar coating from collapsing. Moreover, a careful analysis of the infill walls showed that, although these

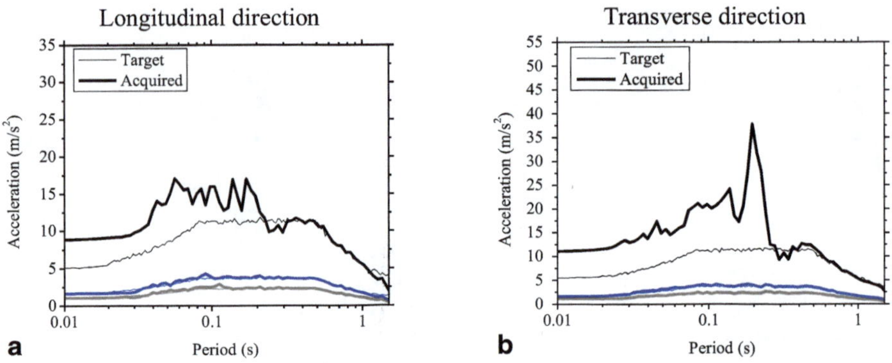

Fig. 16.10 Target and measured pseudo-acceleration response spectra of the input motion for the additional tests

presented limited damage, they were detached from the RC frame along a large portion of their interface. This implies that there was a major contribution of the reinforced coating in preventing the out-of-plane collapse of the infill walls. Regarding the RC structure, extensive cracking was detected at some of the beam-column connections, as shown in Fig. 16.12. The RC frame was very flexible under these conditions, meaning that the reinforced coating may have contributed significantly to prevent the collapse of the entire specimen.

16.2.3.2 Evolution of Modal Properties

The evolution of the five lower vibration frequencies of the building model, detected during the dynamic identification tests (DI0–DI3), are herein used for illustrating the structural degradation of the specimen. The initial modes of vibration (DI0), with frequencies ranging from 6.3 to 35.6 Hz, are plotted in Fig. 16.13: the first two longitudinal and transverse modes and the first torsional one.

Figure 16.14a presents the evolution of the frequencies with the test stages for each of the five modes. Additionally, Fig. 16.14b presents a similar evolution for the out-of-plane vibration modes of the infill walls.

After stage 2, the longitudinal direction presented no frequency decrease, the transverse direction presented an average 5.1 % decrease and the torsional mode presented a 5.5 % decrease, in comparison to DI0. After stage 3, the average decrease in the longitudinal direction was 15.8 % and the average frequency loss in the transverse direction was 24.0 %. This larger stiffness loss in the transverse direction is related to the larger dependence of the transverse vibration modes on the behaviour of the infill walls. As these get detached from the RC frame, due to in-plane damage, the overall stiffness decreases and so does the frequency. The infill walls contributing to the stiffness in the longitudinal direction also detached from the RC frame; however, the response in this direction is more dependent on the RC frame

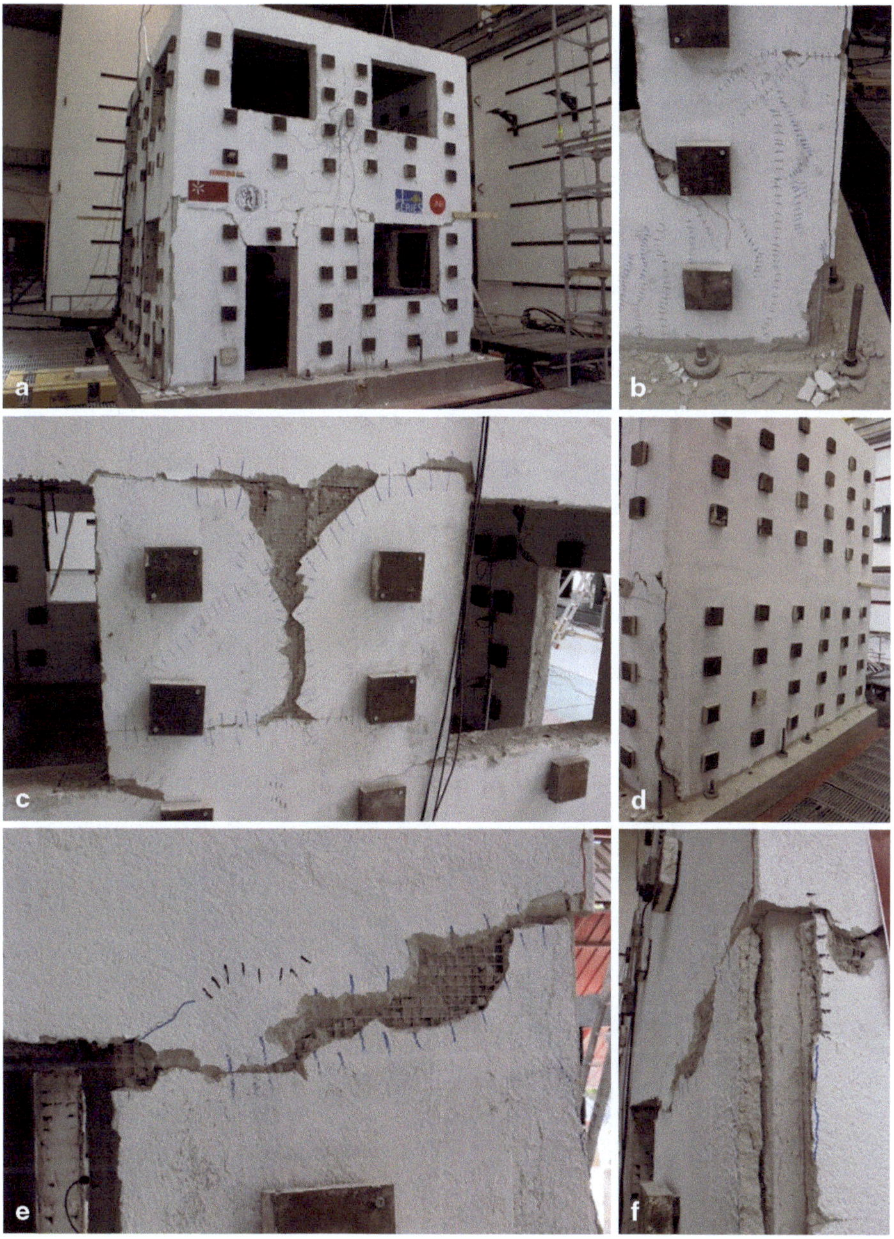

Fig. 16.11 Damage after additional seismic tests **a** North façade, **b** crack and mortar coating loss at the Northwest corner, **c** crack at the interior jambs in the infill wall at the ground floor of the East façade, **d** South façade. **e** crack at the a lateral jamb in the infill wall at the ground floor of the East façade, **f** crack and mortar coating loss at Northeast corner

Fig. 16.12 Extensive cracking observed upon reinforced coating removal at the beam-column connection in the Northwest corner

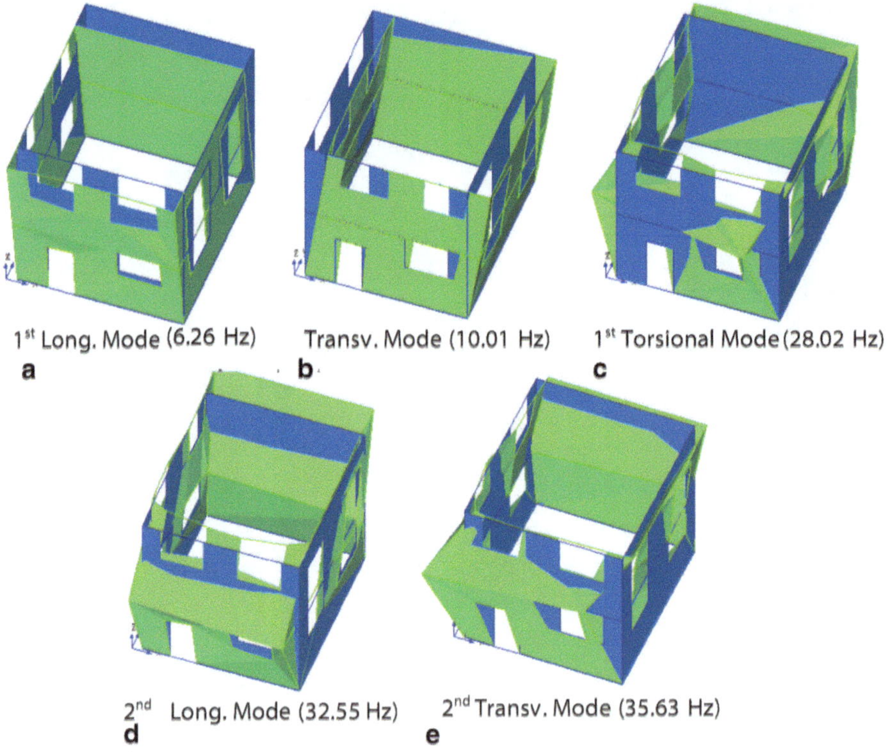

Fig. 16.13 Initial vibration modes and frequencies of the building model on the shake table

and less dependent on the infill walls. Finally, the torsional mode presented a 31.1 % frequency decrease, possibly associated to a loss of connection between the infill walls and the RC frame since the high frequency of the torsional mode is very dependent on the response of the infill walls (Leite 2009). This evolution of dynamic properties is in agreement with the observed overall damage.

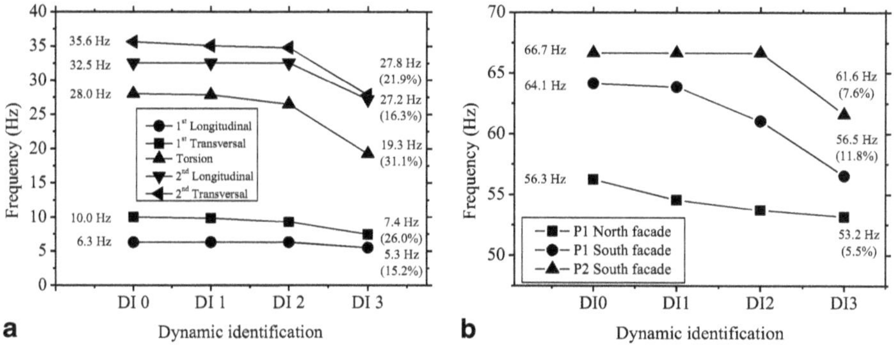

Fig. 16.14 Evolution of the vibration frequencies along the test stages and final reduction with respect to the initial state **a** initial five global vibration modes of the building, **b** out-of-plane vibration modes of the infill walls

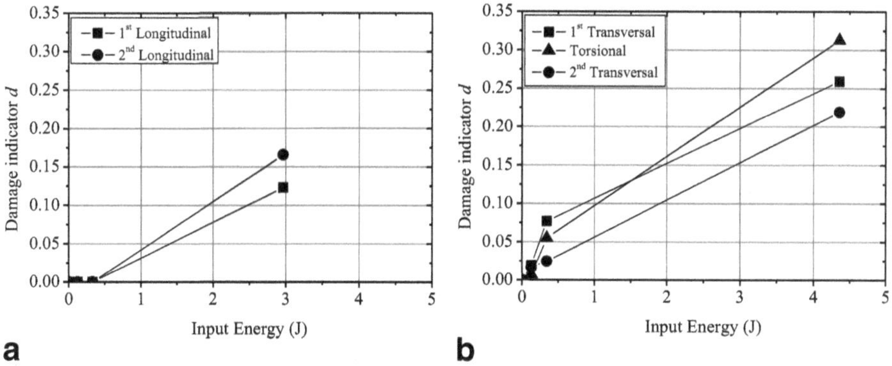

Fig. 16.15 Seismic vulnerability curves using input energy as intensity measure

Figure 16.15 depicts the seismic vulnerability curves for each vibration mode of the building specimen, in terms of relative frequency decrease versus the seismic input energy as intensity measure. These confirm that the longitudinal direction presented no considerable damage until stage 3. After this stage, the transverse direction presented considerably more damage when compared to the longitudinal one, which is clearly associated to a larger energy input.

16.2.3.3 Displacement Demand

The maximum values of the floor displacements and interstorey drifts measured in the building model during the three test stages increase progressively with the seismic amplitude, as shown in Fig. 16.16. The first floor displacements and its drift with respect to the ground floor reached similar values in both directions for all three stages, while roof displacements were slightly larger in the longitudinal direction,

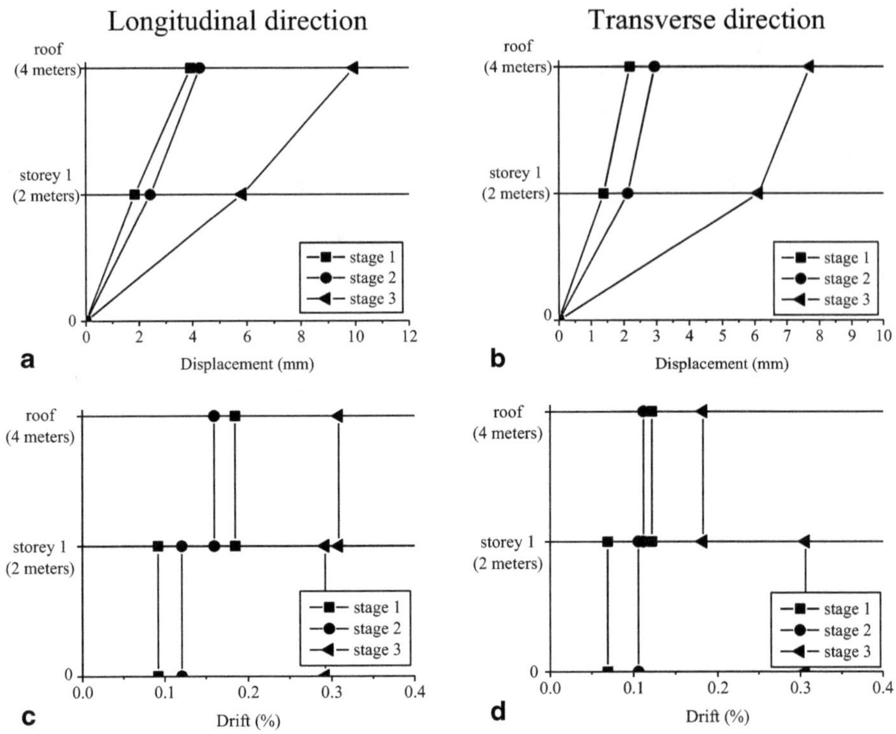

Fig. 16.16 Maximum interstorey displacements and drifts

according to the first mode shape, even though the second mode presented a larger frequency decrease. These results are in agreement with the observed damage and modal frequencies variation, since both directions show similar responses in the first two stages, with no damage of the model, while in the third stage both directions register large displacements and drift increments.

The maximum values of interstorey drift were, in general, larger on the upper floor than on the lower floor. Nevertheless, their difference decreased as the seismic intensity increased. The exception to the former trends occurred during stage 3 in the transverse direction: the interstorey drift at the upper floor was smaller than the drift at the lower floor and the drift in the longitudinal direction.

Similar comments can be made on the basis of the results obtained in the additional seismic tests presented in Fig. 16.17. Nevertheless, the maximum values of floor displacements and interstorey drifts were significantly larger than those obtained on the first set of three test stages.

16.2.3.4 Comparison with Previous Tests

Two building models, similar to the specimen analysed within this project, were previously tested at the Earthquake Engineering and Structural Dynamics Division

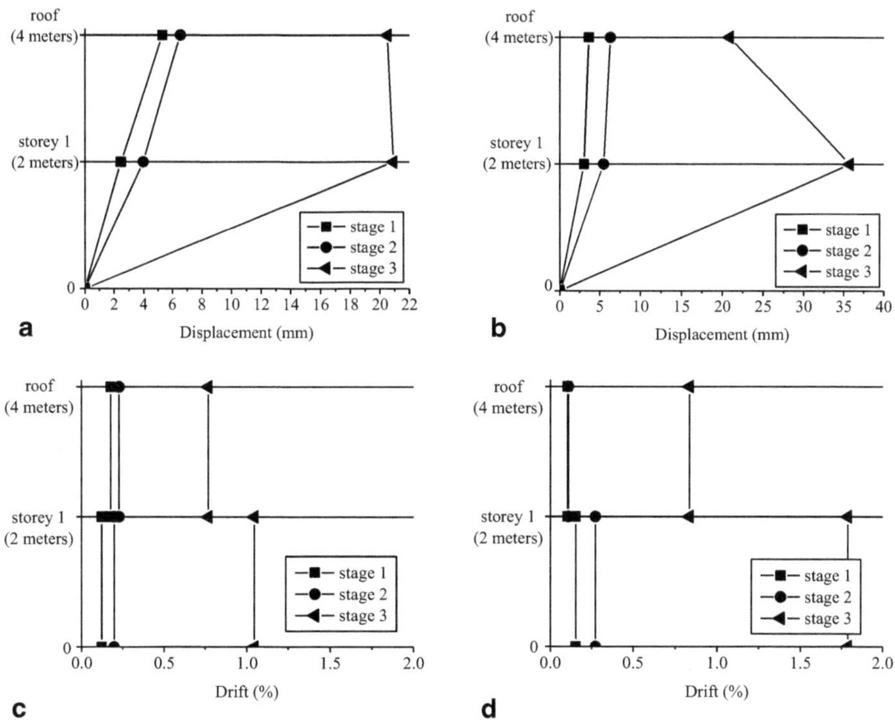

Fig. 16.17 Maximum interstorey displacements and drifts for the additional tests

(NESDE) of LNEC. The first one, model 1, was designed following the Portuguese pre-Eurocode normative, RSA (1983) and REBAP (1983), and built with the most commonly used classes of concrete and steel reinforcing bars (C20/25 and S400, respectively). The masonry infill walls were built with double leaf, unreinforced, clay brick units. Hence, model 1 represents the built heritage in Portugal for the last three decades. The second specimen, model 2, was designed according to Eurocode 2 (2010) and Eurocode 8 (2010) and built with concrete and steel reinforcing bars of higher classes (C30/37 and S500, respectively), together with single leaf clay brick infill walls with bed joint reinforcement. The latter was connected to the RC frame with steel dowels at every second bed joint. Single leaf infills are preferred as they allow for an external thermal insulation system to be placed. Therefore, model 2 and the specimen built within this project (model 3) represent possible solutions to be adopted in the construction of future RC frames with masonry infills.

All models were subjected to the same three seismic input stages described previously. Additionally, models 1 and 2 were subjected to a fourth stage with an input motion corresponding to 1.5 times the input motion from stage 3 (return period of 4574 years), as shown in Fig. 16.7.

Model 1 had a good performance until the end of stage 2, with no visible damage. Nevertheless, both the RC frame and the infill walls clearly showed a frequency loss. During stage 3, some of the thin masonry units, positioned on the external

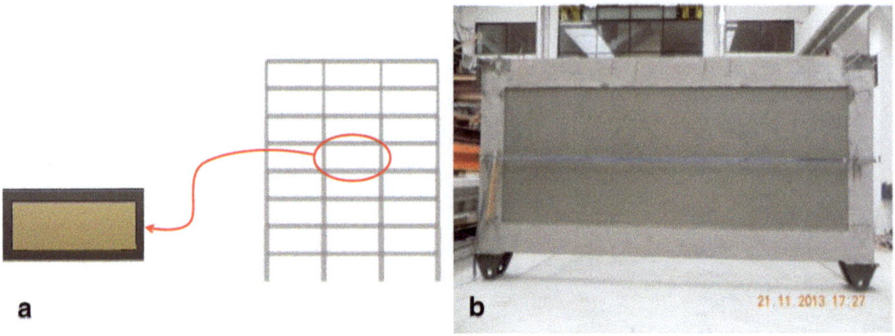

Fig. 16.18 Wall panel of typical RC frame building **a** schematic view, **b** test specimen

leaf outside the RC columns and beams in order to avoid thermal bridges (a very common solution in the Portuguese building stock), cracked and fell. During stage 4 the double leaf unreinforced infill walls of the ground floor performed poorly, collapsing out-of-plane by a rigid-body rotation around the base of the model. A highly undesirable *soft-storey* collapse mechanism developed, with low energy dissipation capacity and brittle behaviour. Both the interior and exterior leaves presented a similar seismic behaviour.

Regarding model 2, the single leaf infill walls with bed joint reinforcement connected to the RC frame presented a very good seismic performance until the end of stage 2, with no visible damage and only a small decrease in the modal frequencies of the structure. After stage 3, the unreinforced mortar coating was severely damaged, especially at the corners of the model. The model did not collapse during stage 4 but presented severe, and most likely irreparable, damage. Moreover, despite Eurocode 8 (2010) imposing structural detailing in order to ensure a *beam-sway* mechanism, the RC structure developed a *soft-storey* mechanism. Taking into account the mid-height cracks that developed at all RC columns, it is possible to assume that the infill walls and their openings influenced negatively the seismic response of the RC structure. None of the infill walls collapsed out-of-plane, but it was clear that those with openings at the ground floor would have collapsed if there was no bed joint reinforcement and a connection to the RC frame.

16.3 Wall Panels Tests

16.3.1 Wall Panels Specimens and Test Setup

The second part of the transnational access activity comprised the dynamic testing of a closed RC plane frame with 6.40×3.25 m external dimensions. The beams and column of the frame had cross-sectional dimensions of 0.5×0.4 m and 0.4×0.4 m, respectively. These specimens were tested simultaneously for in-plane and out-of-

Fig. 16.19 Masonry infill construction with bed joint reinforcement

plane dynamic actions, representing the response of a 4th floor frame panel of an 8-storey RC building (see Fig. 16.18). The columns have 360 kN of centred pre-stress, representing the vertical load from the floors above.

The first two masonry enclosure solutions tested were the unreinforced masonry and the masonry with horizontal reinforcement between masonry units (Bekaert Murfor RND/Z −5–200). Afterwards, the unreinforced masonry infill was demolished and rebuilt inside the same RC frame using a reinforced mortar coating. Figure 16.19 illustrates the construction process of the models.

As mentioned in the introduction, the test setup for the wall panels simultaneously uses the shake table, one reaction wall and the Testing device for Innovative Masonry infills (TIM), as represented in Fig. 16.1b.

The in-plane motion enforces an inter-storey drift time-history on the frame by restraining the upper beam and by imposing the displacement of the shake table on the lower beam. The upper beam, which also has 360 kN of centred prestress for withstanding push-pull actions, is prevented from moving in the longitudinal direction through a strut connection to the reaction wall. This connection between the strut and the reaction wall is performed via a pyramidal support, as depicted in Fig. 16.20a, which distributes the strut reaction on the wall. A long rod then links the pyramidal support to the RC frame upper beam through hinged connectors. All beam-column joints are free to rotate in the plane of the infill, through special hinged base supports, visible in Fig. 16.20a.

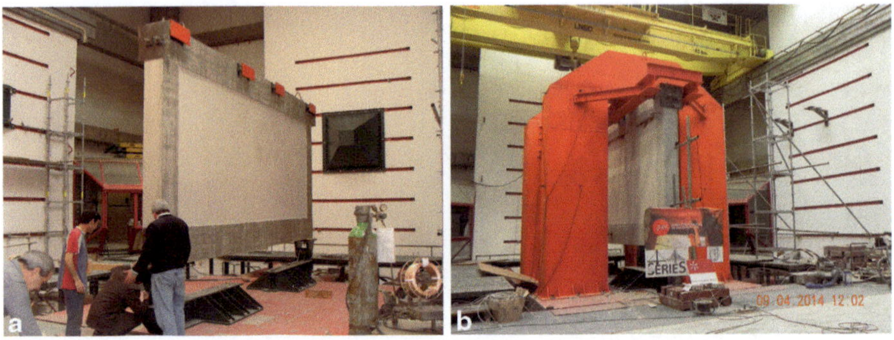

Fig. 16.20 Wall panels test setup **a** Hinged base and pyramidal support, **b** complete setup

Fig. 16.21 Out-of-plane wall panel deformation measurement system. (Krypton 2003)

On the other hand, the out-of-plane motion consists on a rigid-body vibration of both the upper and lower beams, reproducing storey absolute accelerations and thus inducing high-frequency inertia forces perpendicular to the masonry panel and leading to a local vibration of the infill wall. Note that this shake table motion is transmitted to the top beam through the rigid steel caisson structure of TIM. The design of TIM was thus controlled by the requirement of a very stiff behaviour in the transverse direction, which was ensured by a vibration frequency for the first mode of vibration above 20 Hz. The entire test setup is shown in Fig. 16.20b.

The instrumentation of the wall panels comprised several different sensors:

(i) K600 camera system by Krypton (2003) for measuring the out-of-plane deformations of the wall panel using 16 leds (Fig. 16.21);
(ii) Video camera for measuring the RC node in-plane deformation using data image correlation methods (Fig. 16.22);
(iii) Optical system by Hamamatsu (2013) for measuring horizontal motions of the bottom and top corners of the RC frame (Fig. 16.23);
(iv) 36 accelerometers for vibration monitoring of the shake table, TIM, RC frame and masonry wall panel (Fig. 16.23);
(v) Load cells for measuring the dynamic reaction on the strut connecting the reaction wall and the top beam of the RC frame (Fig. 16.24).

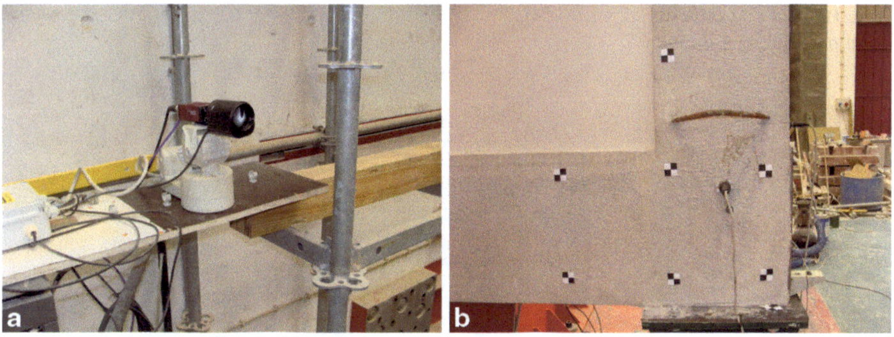

Fig. 16.22 Video camera and target points for data image correlation measurement of in-plane deformations at one RC frame node

Fig. 16.23 **a** Hamamatsu (2013) setup for measuring horizontal translations of the RC frame nodes. **b** accelerometer setup for out-of-plane vibration measurements

16.3.2 Input Signal and Test Sequence

The same seismic inputs used for the building specimen were adopted in the wall panels' tests at full scale. These accelerograms were applied at the base of a representative building model, shown in Fig. 16.25, in order to obtain the time-history of the wall panels' motions.

The response time-history in terms of interstorey drift at the 4th floor for the seismic input considered corresponds to the shake table motion to be applied in the

Fig. 16.24 Load cells for strut reaction measurement

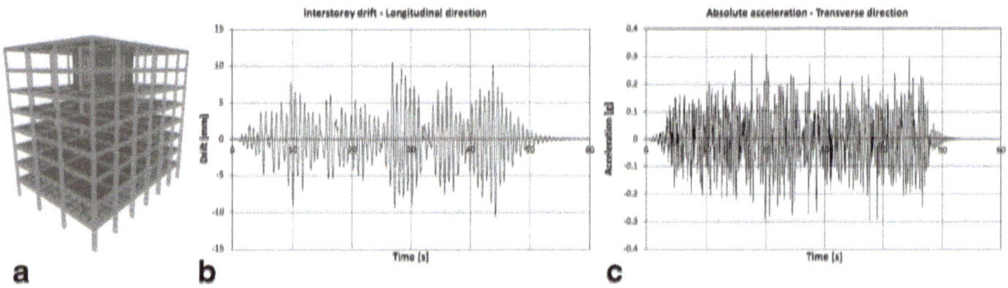

Fig. 16.25 Representative building model for wall panel input time-histories definition and resulting in-plane (longitudinal) and out-of-plane (transverse) time-histories

longitudinal direction. On the other hand, the absolute acceleration observed in the out-of-plane direction corresponds to the shake table motion transmitted by TIM to the masonry panel inside the RC frame. Both longitudinal and transverse motions are represented in Fig. 16.25.

16.4 Conclusions

Reinforced masonry infill wall solutions able to comply with the performance levels' requirements of Eurocode 8 (2010) were tested using LNEC's triaxial shake table. The building model, thoroughly described in this work, used single leaf clay brick infill walls with a reinforced mortar coating. This solution presented only minor damage for a seismic input corresponding to the Significant Damage limit state (475 years of return period) and light damage for a seismic input corresponding to the Near Collapse limit state (2475 years of return period).

After additional tests were performed, with a repetition of the three test stages, the building still presented light damage for a visual observation. Nevertheless, there was a considerable stiffness loss of the specimen. After the removal of the reinforced coating, it was possible to verify that the RC columns did not present plastic hinges at the extremities neither cracks at mid-height, despite considerable cracking at some beam-column joints, hence no undesirable collapse mechanism developed. It may be concluded that the infill walls had no negative influence on the seismic performance of the RC structure, unlike what was observed in a previous test using bed joint reinforcement.

The infill walls were not significantly damaged during the tests, even though the dynamic identification presented a clear stiffness loss. This was due to the detachment observed between the reinforced coating and the infill walls and between these and the RC frame, which allowed for relative motions of the RC frame with respect to the infill walls. Finally, although the reinforced coating concealed the damage on the infill walls, it prevented their out-of-plane collapse.

The second phase of the project included the development of an innovative testing setup and methodology for conducting full-scale seismic tests on wall panels. These were presented only conceptually, but preliminary results seem promising.

Acknowledgements The research leading to these results has received funding from the European Union Seventh Framework Programme [FP7/2007–2013] for access to the Laboratório Nacional de Engenharia Civil (LNEC), Portugal under grant agreement n° 227887. Opinions, findings and conclusions expressed in this paper are those of the authors and do not necessarily reflect those of SERIES.

References

Carvalho EC (1998) Seismic testing of structures. Proceeding of the 11th European Conference on Earthquake Engineering, Paris, France

Eurocode 2 (2010) NP EN 1992-1-1. Eurocódigo 2-Projecto de estruturas de betão—Parte 1–1: Regras gerais e regras para edifícios. Instituto Português da Qualidade (IPQ), Lisbon, Portugal

Eurocode 8 (2010) NP EN 1998-1. Eurocódigo 8-Projecto de estruturas para resistência aos sismos—Parte 1: Regras gerais, acções sísmicas e regras para edifícios. Instituto Português da Qualidade (IPQ), Lisbon, Portugal

Hamamatsu (2013) C5949 characteristics. http://pdf.datasheetcatalog.com/datasheet/hamama-tsu/C5949.pdf. Accessed: 30 June 2013

Krypton (2003) K400/K600-Hardware & Software Guide (Krypton Help Pages), Leuven, Belgium

Leite J (2009) Avaliação da segurança baseada em deslocamentos: aplicação a edifícios de betão armado sem e com paredes de preenchimento (in Portuguese). MSc Dissertation, Dept. of Civil Engineering, University of Minho, Guimarães, Portugal

Mendes L, Campos Costa A (2008) LNEC-SPA: Signal Processing and Analysis tools for civil engineers. Earthquake Engineering and Structural Dynamics Division. Dept. of Structures, National Laboratory for Civil Engineering (LNEC), Lisbon, Portugal

Pereira MFP (2013) Avaliação do desempenho das envolventes dos edifícios face à acção dos sismos (in Portuguese). PhD Thesis, Dept. of Civil Engineering, University of Minho, Guimarães, Portugal

REBAP (1983) Regulamento de Estruturas de Betão Armado e Pré-Esforçado. Diário da República, Portugal

RSA (1983) Regulamento de Segurança e Acções para Estruturas de Edifícios e Pontes. Diário da República, Portugal

Chapter 17
Seismic Behaviour of Thin-Bed Layered Unreinforced Clay Masonry Frames with T- or L-Shaped Piers

Christophe Mordant, Matthew Dietz, Colin Taylor and Hervé Degée

17.1 Introduction

Among traditional construction methods, masonry is one of the main and oldest used for centuries for several types of buildings, like private dwellings, town halls, churches, etc. Historically, the design of masonry structures has relied on good practice methods. The contribution of engineers was often limited, usually leaving the task of designing the buildings to architects and mason-workers. Nevertheless, civil engineers are nowadays more frequently involved and required in the design phase. Indeed, masonry structures are progressively used closer to the limit of the material capacity, for example with compression ratio under service loads higher than even in a recent past (e.g. through the spreading of multi-storey apartment buildings or of lightweight concrete houses). Moreover, ecological and economical considerations require a more rational use of materials to reduce cost and consumption. All these considerations are at the base of Eurocode 6—Design of masonry structures (EN 1996, CEN 2004), where modern design methodologies and verification rules are proposed.

The earthquake impacts cannot be adequately considered without an additional proper understanding of the structural behaviour under these specific horizontal dynamic actions. The basic principles of this characterization and the consequent analysis and design methodologies are presented in (Tomazevic 1999) and are the base of the European standard dedicated to the design of buildings in seismic regions (EN 1998-1, CEN 2004a). These general considerations need however to be

C. Mordant (✉)
University of Liege, Chemin des Chevreuils 1, 4000 Liège, Belgium
e-mail: cmordant@ulg.ac.be; h.degee@ulg.ac.be

H. Degée
University of Hasselt, Agoralaan Gebouw H 3590, Hasselt (Belgium)
e-mail: herve.degee@uhasselt.be

M. Dietz · C. Taylor
University of Bristol, University Walk, Bristol BS8 1TR, UK
e-mail: m.dietz@bristol.ac.uk; colin.taylor@bristol.ac.uk

F. Taucer, R. Apostolska (eds.), *Experimental Research in Earthquake Engineering*,
Geotechnical, Geological and Earthquake Engineering 35,
DOI 10.1007/978-3-319-10136-1_17

specifically transposed to each particular type of masonry structure. In the past 15 years several research works have been carried out in this perspective (Kazemi et al. 2010; Nakagawa et al. 2012; Schermer 2005; Calio et al. 2012; Milani et al. 2009). None of the identified references is however specifically dedicated to the most common type of masonry structural configuration used in North-Western European areas, namely relatively thin bearing walls (from 10 to 20 cm) with high strength units (compression resistance up to 15 N/mm^2 or even more) working at a very high compression ratio under service loads and more and more implemented using horizontal thin-bed layered jointing and open "tongue and groove" vertical joints for constructional efficiency purpose.

Furthermore, some researchers have pointed out the over-conservatism and the mismatch with common construction habits of current standards design rules (Degée et al. 2007; Karantoni and Lirantzaki 2009) when comparing the resistance predicted based on code design rules with experimental results on full scale houses. One of the reasons for this discrepancy is that the codes are commonly not considering the contribution of the walls perpendicular to the seismic action. Another point concerns horizontal elements, such as lintels or spandrels, which are often neglected or whose modelling is questionable.

Consequently, tests on single-storey unreinforced masonry frames with T- or L-shaped piers have been carried out in the Earthquake and Large Structures Laboratory (EQUALS) at the University of Bristol. The test specimens have been designed to investigate the dynamic response of unreinforced clay masonry frames targeting several aims. First, it is expected to develop a better understanding of the behaviour of masonry frames and of the influence of the spanning structural elements. Then, a specimen with T-shaped piers was specifically designed in order to trigger torsion effects, while a second specimen with L-shaped piers aims at comparing different connection methods between the shear wall and the perpendicular flange wall constituting the piers. Finally, the influence of the gravity loading case is investigated, in particular the case of a frame with piers partially loaded shaken in the direction perpendicular to the plan of loaded walls.

This paper summarizes the observations derived from the direct experimental measurements, followed by preliminary conclusions. The detailed assessment of the test results and the corresponding model identification and calibration are planned to be presented in an additional contribution.

17.2 Description of the Tested Specimens

17.2.1 Mechanical and Geometrical Characteristics of the Units

The units used to construct the walls are "*Zonnebeke POROTHERM*" produced by Wienerberger. The walls are built with thin-bed layered masonry with "*PORO+*" glue-mortar (Wienerberger 2014). Vertical joints are empty tongue and groove systems. Unit size is:

$$(Length \times Width \times Heigth \equiv 300\ mm \times 138\ mm \times 188\ mm)$$

Mechanical characteristics of the units and masonry are:

- Normalised compressive strength of units (EN 772−1 Annex A)

$$f_b = 13.0\ N/mm^2$$

- Measured characteristic masonry compressive strength (EN 1052−1)

$$f_k = 5.6\ N/mm^2$$

- Characteristic compressive strength (EN 1996-1-1)

$$f_k = 4.2\ N/mm^2$$

- Characteristic compressive strength (NBN-EN 1996-1-1)

$$f_k = 3.9\ N/mm^2$$

17.2.2 Description of the Specimens

The geometry of the specimens is chosen in such a way to fit with practical configurations. In masonry houses, the total inner space is divided by a set of simple walls perpendicular to each other create different rooms. As it can be seen in Fig. 17.1, most of these walls are actually T- or L- shaped walls, separated by openings necessary for the circulation of the inhabitants. As a consequence, the chosen geometry of the present test specimens consists in frames with two T- or L-shaped piers connected by a RC lintel (see Fig. 17.2 and 17.3).

The dimensions of the piers and openings are governed by the shaking table size limitation as follows:

- Shear walls: 0.74 m × 2.0 m × 0.14 m (Length × Height × Width)
- Flanges: 0.74 m × 2.0 m × 0.14 m (Length × Height × Width)
- Opening: 0.9 m × 1.8 m (Length × Height)
- Lintel: 1.8 m × 0.2 m × 0.14 m (Length × Height × Width)

Note that the term "Shear wall" refers to the part of the pier along the frame plan while the "Flange" refers to the part of the pier perpendicular to the frame plan.

The frame with T-shaped piers was designed with piers oriented differently in order to trigger a global torsional behaviour. The frame with L-shaped piers is

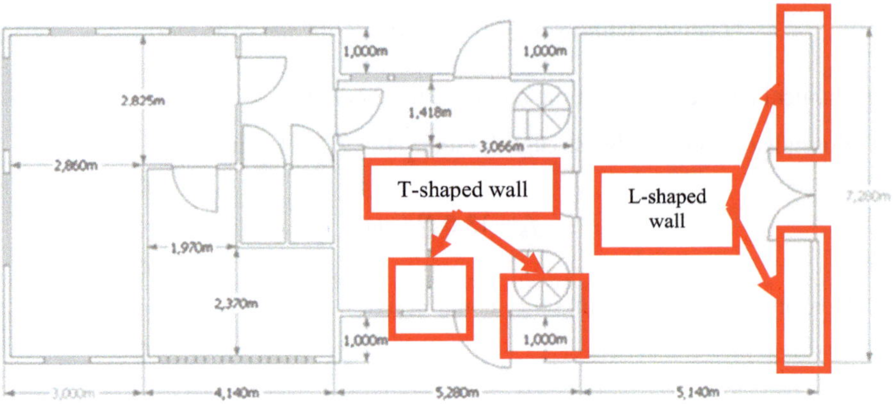

Fig. 17.1 Example of building

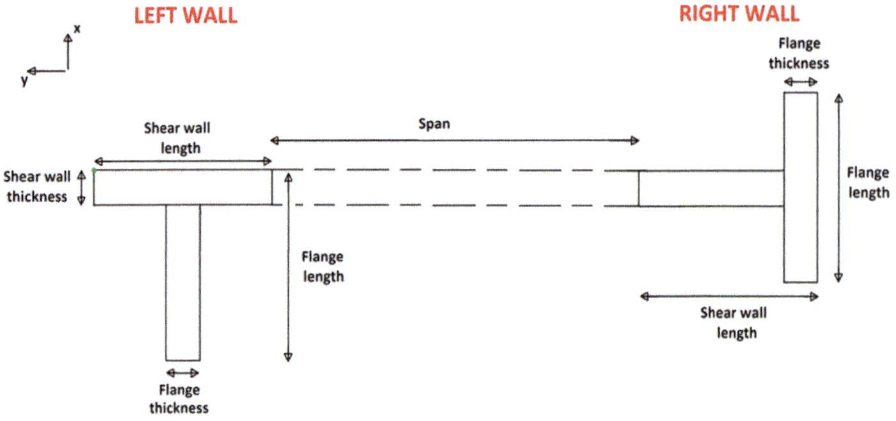

Fig.17.2 Geometry of the T-shaped frame

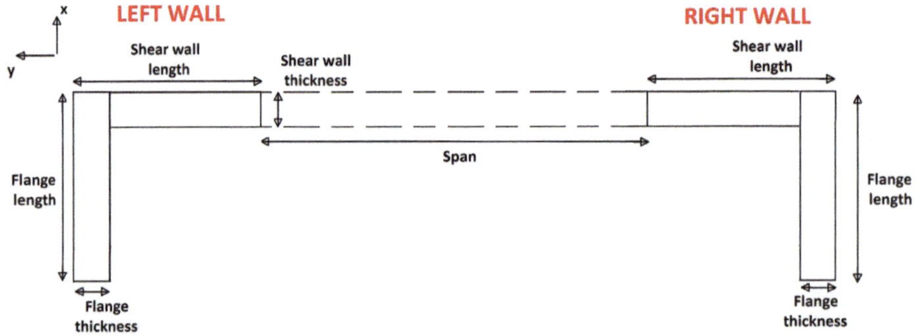

Fig. 17.3 Geometry of the L-shaped frame

Fig. 17.4 Additional mass

symmetrical from a pure geometric point of view, but the shear wall and the flange of each pier are connected in two different ways. For the first one, the flange is glued to the shear wall with a high performance mortar while, for the second one, the flange and shear wall are classically connected by an alternated mason work.

The structural floor load is simulated by an additional 5 t mass fixed on a reinforced concrete slab lying on top of the specimen (see Fig. 17.4). The connection between the slab and the masonry frame is ensured by steel elements which can be moved to simulate different loading cases (Fig. 17.5).

The resulting average compressive stress in the loaded walls is about 0.135 MPa for fully loaded configurations and about 0.25 MPa for the partially loaded case. These values lie in the usual range of compression ratio for this type of masonry.

17.2.3 Preliminary Assessment Design

The main parameter characterizing the input action on the experimental specimens is obviously the acceleration of the table. In order to estimate the maximum value

Fig. 17.5 Steel connectors

that could be sustained by these specimens, a theoretical assessment based on structural models commonly used for seismic analysis of masonry structures was performed before the experimental tests.

The model is based on an equivalent static loading similar to the one used in Mordant et al. (2013a), with an equivalent static horizontal load V applied to the specimen at the location of its gravity centre, namely in the present case in the middle of the RC slab. Once the internal forces are derived for both piers, the procedure consists in the calculation of the compressive length for each pier through equilibrium equations considering no tensile strength at the interface between the base of the pier and the foundation. At this interface, a linear distribution of the compressive stresses is assumed. Then, final checks are made with respect to shear forces, compression of the units and global overturning of the piers. This procedure results in an estimate of the maximum allowable equivalent horizontal load V_{max}.

Unlike the specimens studied in Mordant et al. (2013a), the present mock-ups can hardly be considered as SDOF systems and moreover exhibit a significant non-linear behaviour associated with the load redistribution from one pier to the other. Therefore, the methodology used to convert the maximum allowable equivalent horizontal load into acceleration is based on the derivation of the full force-displacement pushover curve using the N2-method. The general methodology and its particular application are detailed in EN (1998), Annex B, CEN (2004). The ultimate limit state is assumed being reached when the maximum total displacement reaches a specific threshold value defined from recommendations given by Oropeza (2011) and depending on the wall geometry and compression ratio.

Details of the procedure and calculation are given in Mordant (2012) and final outcomes are summarized in Table 17.1 and 17.2 for the frame with T-shaped piers, respectively for an earthquake along the frame plan and perpendicular to it. The results for the frame with L-shaped piers are tabulated in Table 17.3 and 17.4. These tables also compare the results obtained according to different assumptions on the boundary conditions of the lintel (built-in or simply supported) and on the support conditions of the slab (resting resp. on the shear wall, on the flanges or on both).

17.3 Test Description

17.3.1 Axis Convention

For all upcoming information regarding instrumentation and results, the following convention is followed according to the usual practice at EQUALS laboratory:

- x-axis is positive from the near end to the far end of the table;
- y-axis is positive from right to left;
- z-axis is positive upward

The reference for notions of "near end", "far end", "left" and "right" of the shaking table is the control mezzanine, from where the picture in Fig. 17.6 has been taken.

Table 17.1 T-shaped frame: preliminary assessment results along the frame plane

Support conditions of the lintel	Load case	Horizontal shear (N)	Acceleration (g)
Left and right built-in	Shear wall	13,745	0.939
	Flange	4148	0.795
	Both	8912	0.857
Left and right simply supported	Shear wall	10,808	0.780
	Flange	3527	0.640
	Both	6797	0.763

Table 17.2 T-shaped frame: preliminary assessment results perpendicular to the frame plane

Support conditions of the walls	Load case	Horizontal shear (N)	Acceleration (g)
Top wall free	Shear wall	1548	0.013
	Flange	6800	0.055
	Both	5652	0.046

Table 17.3 L shaped frame: preliminary assessment results along the frame plane

Support conditions of the lintel	Load case	Horizontal shear (N)	Acceleration (g)
Left and right built-in	Shear wall	13,795	1.200
	Flange	8220	0.752
	Both	9610	0.997
Left and right simply supported	Shear wall	10,850	0.854
	Flange	7120	0.654
	Both	8430	0.829

Table 17.4 L shaped frame: preliminary assessment results perpendicular to the frame plane

Support conditions of the walls	Load case	Horizontal shear (N)	Acceleration (g)
Top wall free	Shear wall	7632	0.057
	Flange	10,960	0.081
	Both	8516	0.063

Fig. 17.6 Axis convention

17.3.2 Instrumentation of the Specimens

Besides the internal instrumentation of the table and additional accelerometers measuring the horizontal acceleration in both directions and fixed on the table, the instrumentation layout includes three types of devices:

- Acceleration measurements (SETRA type 141A accelerometers). Relative displacement measurements: displacements of the tested specimens with respect to the table or relative displacements between parts of the specimens are monitored (using LVDTs). Global displacement measurements: an Imetrum Vision System monitors the horizontal and vertical displacements of the RC slab and of the top of the specimen thanks to targets placed at the right place. The instrumentation layout is slightly different for the two mock-ups. It is only illustrated for the frame with L-shaped piers, complete information is given in Mordant (2012). A legend and an overview, given in Fig. 17.7 and 17.8, refer to the cross sections drawn in Fig. 17.9, 17.10, 17.11. The position and the digit of the targets for the vision system are illustrated in Fig. 17.12. The left picture is on right side on the table, the middle one is on the near end and the right one is on the left of the table.

17.3.3 Testing Procedure

The testing procedure alternates two types of tests. The first type is a low level white noise test, performed alternatively along and perpendicular to the frame plan. This type of test is performed to derive the main dynamic characteristics of the wall (natural modes and frequencies, damping ratio). Each test lasts about 3 min

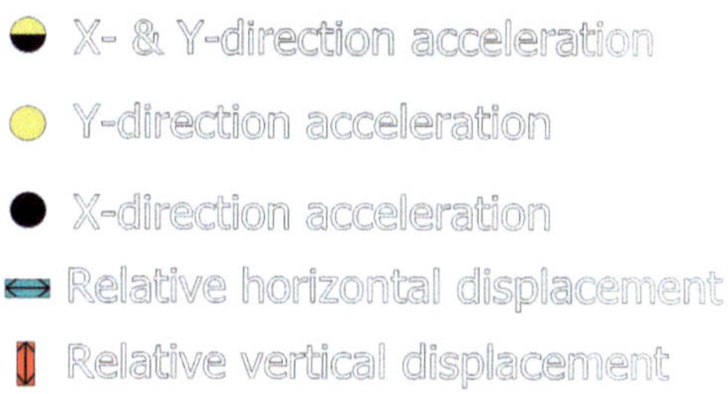

Fig. 17.7 Legend of the instrumentation layout

so as to reach a proper steady vibration state. A random white noise excitation with frequency content between 1 and 100 Hz at level of about 0.1 g RMS is generated using an Advantest R9211C Spectrum Analyser.

The second type of tests corresponds to real earthquake action. These tests consist in sending a waveform to the table to simulate seismic ground motions. The characteristics of the waveform are given in Fig. 17.13. Successive input signals with increasing values of the peak acceleration are applied to the table (see Table 17.5 for the considered acceleration levels).

17.3.4 Excitation Waveforms for Seismic Tests

Specimens were pre-assessed according to Eurocodes 6 and 8. The waveform of the input signal used for the seismic phases of the testing procedure is chosen consistently as an artificial input compatible with a Eurocode 8—type 2 spectrum. The excitation waveform and the corresponding response spectrum are given in Fig. 17.13 where TRS is the target spectrum and RRS the spectrum corresponding to the generated signal.

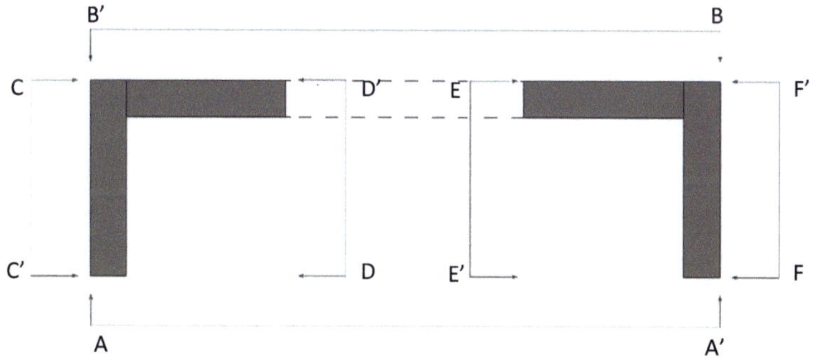

Fig. 17.8 Plan view (reference for the instrumentation layout)

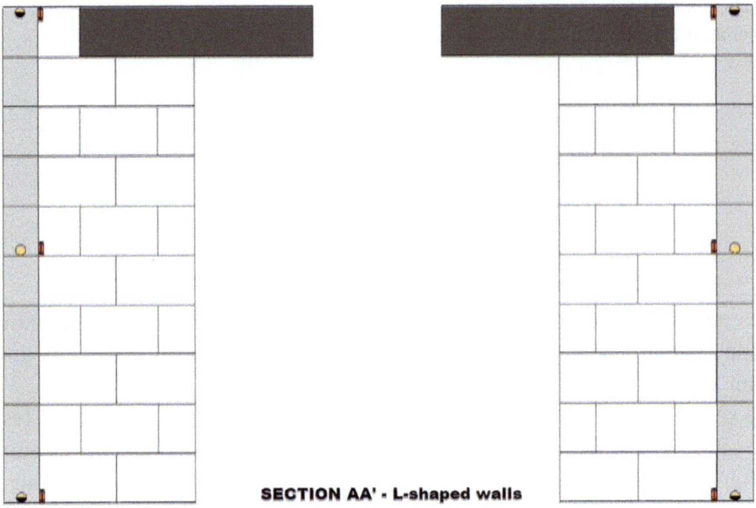

Fig. 17.9 Back view

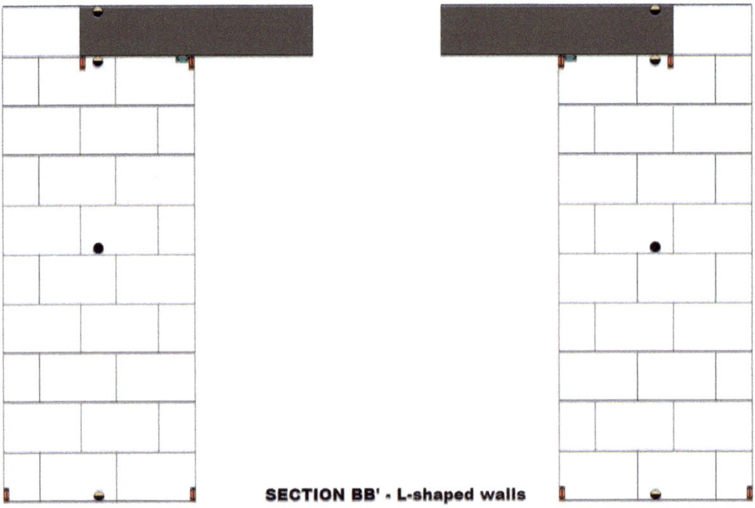

Fig. 17.10 Front view

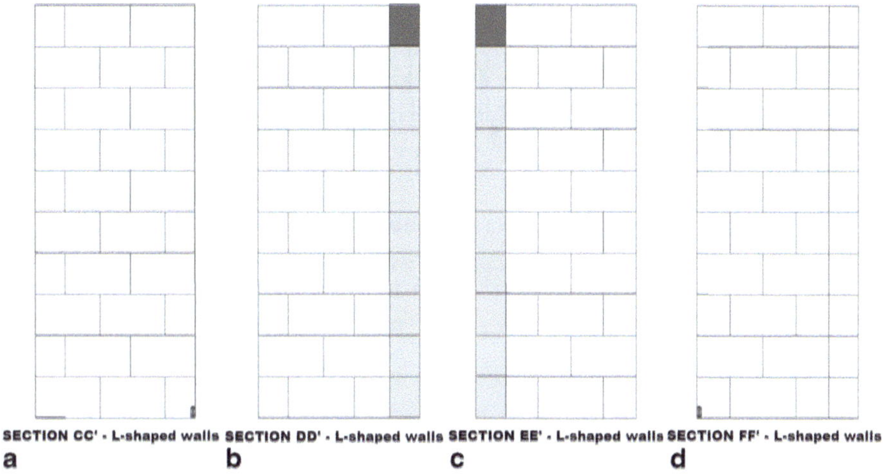

Fig. 17.11 Side views

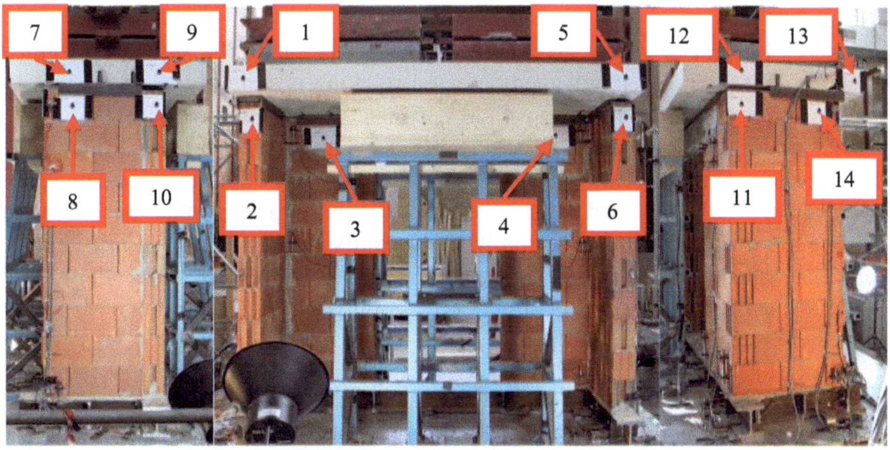

Fig. 17.12 Targets of the imetrum vision system

17.4 Test Results

17.4.1 Qualitative Observations

As already observed for tests on simple unreinforced clay masonry shear walls performed in a first phase of this experimental program (Mordant et al. 2013a), a major observation is that the two mock-ups exhibit significant rocking behaviour, involving progressive degradation of the mortar base joint.

It is also observed that the connection of the lintel to the piers failed during the first steps of the tests on both specimens (i.e. at low level accelerations). As shown

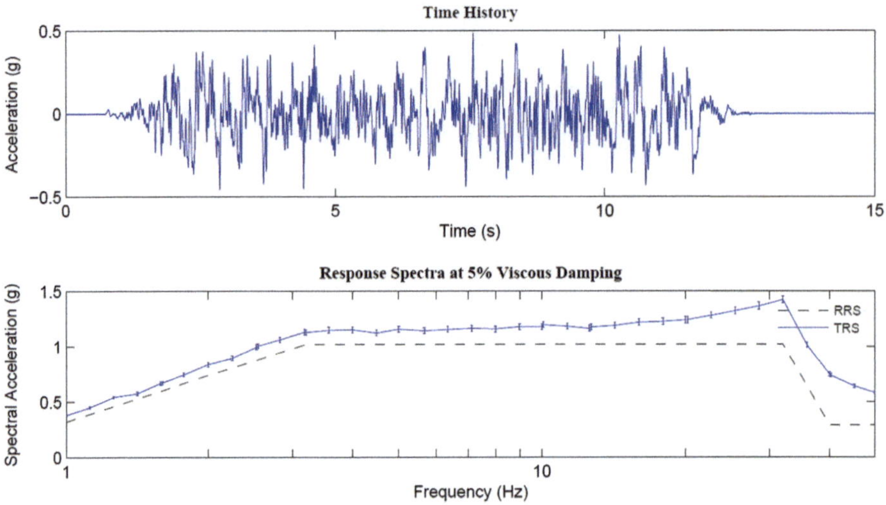

Fig. 17.13 Excitation waveforms

in Fig. 17.14, cracks appear and, therefore, the piers cannot be anymore assumed as clamped at their tops. Such an anticipated cracking is certainly related to the absence of units above the lintel as well as between the lintel and the slab.

The loading configuration is also identified as having a significant influence on the behaviour of the specimen. By removing the influence of progressive damage of the specimen with L-shaped piers for tests with increasing amplitude, it is observed that the top displacements are actually significantly larger when the piers are only partially loaded, especially when the seismic action is perpendicular to the loaded part of the piers. The stabilizing effect of the gravity load is indeed much smaller in this latter situation.

Fig. 17.14 Cracking of the lintel supports

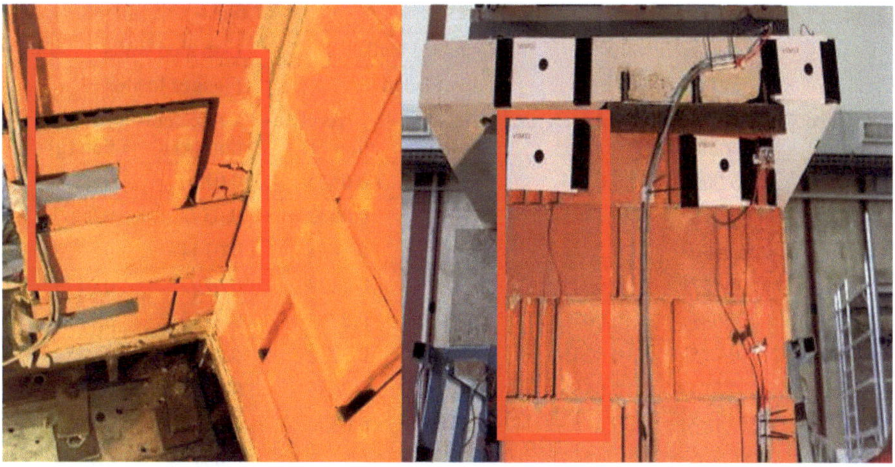

Fig. 17.15 Collapse of the specimens (*left*: T-shaped frame—*right*: L-shaped frame)

A final direct observation deals with the failure mode of the specimens. Collapse of the mock-up with T-shaped piers was induced by local torsional effects on the piers. As it can be seen on the left hand side of Fig. 17.15, some units shifted from their initial position in a combined translation/rotation motion.

For the specimen with L-shaped piers, the collapse was triggered by failure of the vertical connection of the shear wall with the flange for a seismic input along the frame plan (Fig. 17.15 right).

These qualitative observations lead to two main comments on the design procedure. On the one hand, the verification models proposed by the current standards do not provide any verifications or rules for the torsional behaviour of a pier, while this effect is here identified as possibly the most critical. On the other hand, it is emphasized that a local verification of the capacity to transfer longitudinal shear at the connections between perpendicular walls may be required to prevent the risk of early collapse.

Measured PGAs of each seismic test are summarized in Table 17.5. The failure of both specimens was reached for acceleration levels lower than estimated by the preliminary assessment, because the observed failure modes were actually not explicitly considered in this preliminary stage, as explained above. Quantitative results are provided in the following sections in terms of natural frequencies, modal shapes and seismic response of the system. More comprehensive information on the tests results is available in Mordant (2012).

In the perspective of further commenting the results, it is relevant to remind that the seismic tests on the L-shaped walls with gravity loads only acting on flanges are carried out on the same specimen as the tests with piers fully loaded and are thus starting from a pre-damaged situation.

When compared to the theoretical uniaxial testing procedure, Table 17.5 shows that the seismic input is actually bidirectional, although one component (residual

Table 17.5 Measured seismic input

	T-shaped walls		L-shaped walls		L-shaped walls	
	Fully loaded		Fully loaded		Flanges loaded	
After seismic test at	x Acceleration (g)	y Acceleration (g)	x Acceleration (g)	y Acceleration (g)	x Acceleration (g)	y Acceleration (g)
S1	0.0069	*0.0735*	0.0059	*0.0655*	0.0110	*0.0630*
S2	0.0182	*0.1475*	*0.0375*	0.0042	*0.0393*	0.0067
S3	0.0462	*0.2846*	0.0143	*0.1438*	0.0144	*0.0828*
S4	*0.0379*	0.0049	*0.0871*	0.0128	*0.0797*	0.0179
S5	*0.0831*	0.0079	0.0364	*0.2205*	0.0259	*0.1334*
S6	*0.1759*	0.0245	*0.1347*	0.0232	*0.1246*	0.0352
S7	*0.2763*	0.0565	0.0771	*0.2686*	0.0390	*0.1946*
S8	0.1074	*0.4765*	*0.1798*	0.0424	*0.1697*	0.0365
S9	/	/	/	/	0.0577	*0.1974*

component) is significantly lower than the other one (main component, in bold in Table 17.5). The residual component induced by an imperfect control of the table motion ranges from 8 to 29 % of the main component.

17.4.2 Natural Frequencies Identification

The modal characterisation is carried out under low-level white noise excitation, during which the 'input' and 'response' channels of a spectrum analyser are connected to the appropriate instrumentation (i.e. table x- or y-accelerometer and top mass x- or y-accelerometer, respectively). The spectrum analyser acquires 32 segments of time data, converts to the frequency domain and averages results to produce a transfer function. Natural frequency values are determined for all coherent resonances using a curve-fitting algorithm running directly on the spectrum analyser. Identification is carried out before and after any seismic shake. This allows the characterisation of possible deterioration of the walls.

Three main peaks are observed in both directions for the two specimens, except in the frame plan of the last configuration (namely the frame with L-shaped piers only loaded on the flanges). The evolution of the frequencies is shown in Fig. 17.16 for the x-direction and Fig. 17.17 for the y-direction.

A first comment about these results is that the first natural frequency of the mock-up with T-shaped piers is higher than the frequency of the mock-up with L-shaped piers, which is in contradiction with the theoretical expectation (see below).

The natural frequencies in the x-direction, shown in Fig. 17.16, are decreasing with the increase of the acceleration level, in general. This decrease is faster for the specimen with T-shaped piers because the acceleration levels are actually increased at a faster rate from one test to the next. The sequence of acceleration increase for the specimen with L-shaped piers is slower because of a wish to allow testing the

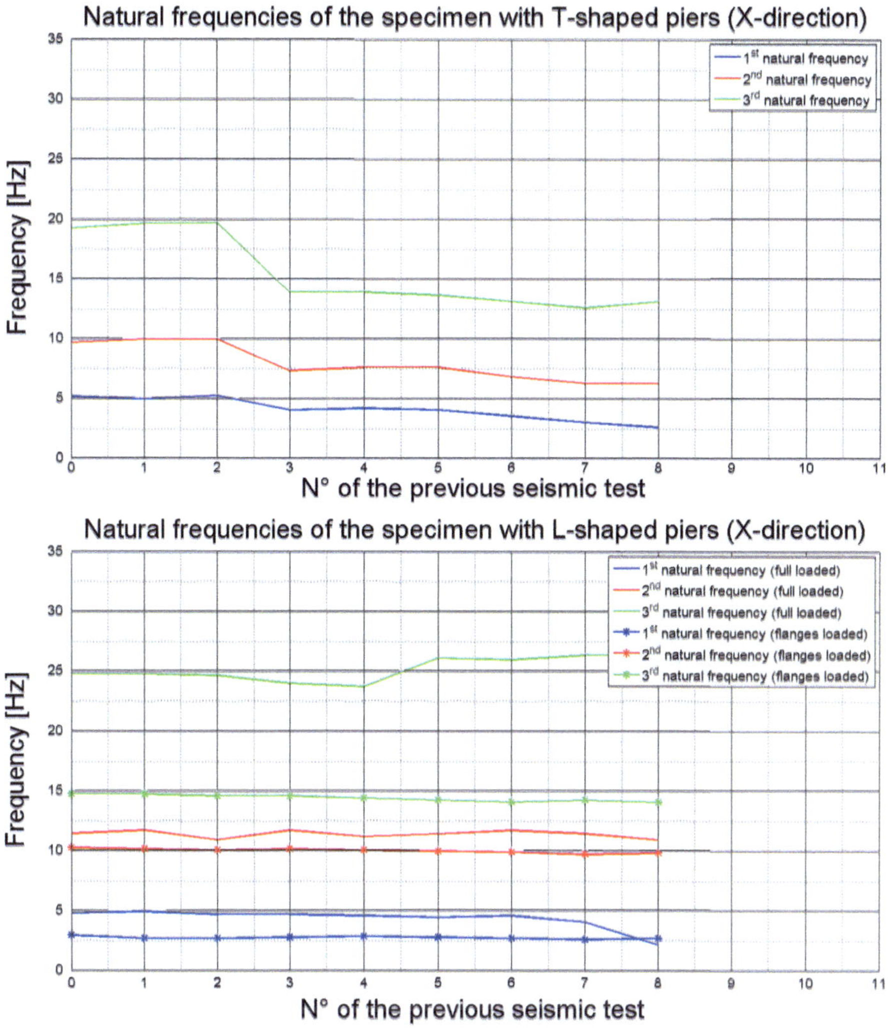

Fig. 17.16 Natural frequencies in x-direction (*top*: T-shaped frame—*bottom*: L-shaped frame)

same specimen for different loading cases without too significant damages (see also Table 17.5). The main frequency drop is observed for the fundamental frequency, whereas the third natural frequency is more constant or even increasing.

These observations are also valid in the y-direction, illustrated in Fig. 17.17. The frequency values are however higher, as it can be expected given that the direction of the frame plan is clearly stiffer. The numerical values drawn in the Fig. 17.16 and 17.17 are tabulated in the Tables 17.6 and 17.7 for the x and y direction respectively.

A comparison of the measured first natural frequency in the undamaged state with the value obtained from the theoretical model used in the preliminary assess-

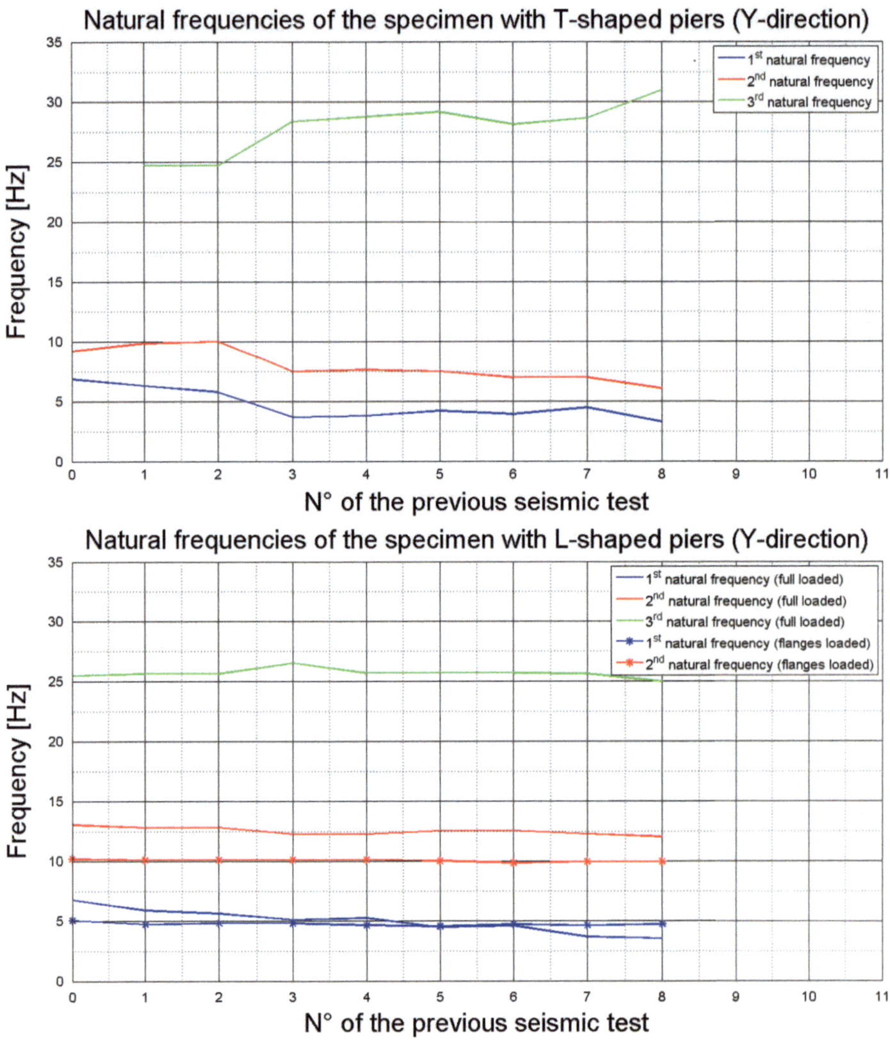

Fig. 17.17 Natural frequencies in y-direction (*top*: T-shaped frame—*bottom*: L-shaped frame)

ment with material properties according to Eurocode 6 is given in Table 17.8 for the x-direction and in Table 17.9 for the y-direction.

As said above, the expected fundamental frequency of the frame with T-shaped piers was expected to be lower than the frequency of the frame with L-shaped piers, which is not in good agreement with the measurements.

The prediction based on standard models yields frequencies higher than measured, except for the case of the specimen with T-shaped piers in the x-direction. According to the authors opinion, this discrepancy can be partially explained on

Table 17.6 Natural frequencies in x-direction

	T-shaped walls			L-shaped walls			L-shaped walls		
	Fully loaded			Fully loaded			Flanges loaded		
After	1st peak frequency. [Hz]	2nd peak frequency [Hz]	3rd peak frequency [Hz]	1st peak frequency. [Hz]	2nd peak frequency [Hz]	3rd peak frequency [Hz]	1st peak frequency. [Hz]	2nd peak frequency [Hz]	3rd peak frequency [Hz]
Before	5.18	9.71	19.28	5.02	11.49	24.82	2.96	10.28	14.80
S1	4.99	9.98	19.70	4.96	11.76	24.82	2.70	10.19	14.80
S2	5.25	9.98	19.70	4.70	10.97	24.69	2.70	10.1	14.63
S3	4.07	7.35	13.92	4.70	11.76	24.03	2.79	10.19	14.63
S4	4.20	7.62	13.92	4.57	11.23	23.77	2.87	10.1	14.45
S5	4.07	7.62	13.66	4.44	11.46	26.12	2.79	10.01	14.28
S6	3.55	6.83	13.13	4.57	11.76	25.99	2.70	9.93	14.11
S7	3.02	6.30	12.61	4.05	11.49	26.38	2.61	9.75	14.28
S8	2.63	6.30	13.13	2.17	10.97	26.38	2.70	9.84	14.11
S9	/	/	/	/	/	/	/	/	/

one hand, by the assumption on the elastic modulus value given by Eurocode 6 that suggests a value of E equal 1000 f_k, with f_k being the characteristic compressive strength of the masonry wall, and by Eurocode 8 that recommends to divide this value by two to account for cracking, which seems actually too rigid. On the other hand, the assumption of piers clamped at their top seems to be quite optimistic. Indeed, the in-plane stiffness of a frame can be expressed by Eq. 17.1:

$$K = \frac{1}{\frac{H^3}{\alpha EI} + \frac{H}{kGA}} \tag{17.1}$$

where

H	is the specimen height
E and G	are the elastic and shear moduli
I	is the sum of the moment of inertia of the piers
A	is the sum of the cross sectional area of the piers
α	is a parameter depending of the support condition

If the piers are clamped at both their extremities, the parameter α is equal to 12. For the specimen with L-shaped piers, the best frequency fitting is obtained if the α parameter is equal to 6.3 and the elastic modulus is equal to $400 f_k$. It can therefore be concluded that the recommended value for the elastic modulus in EC 6 and EC 8 is too stiff for the type of masonry considered in the present study, where the vertical joints are not filled with mortar.

Table 17.7 Natural frequencies in y-direction

	T-shaped walls			L-shaped walls			L-shaped walls		
	Fully loaded			Fully loaded			Flanges loaded		
After	1st peak frequency. (Hz)	2nd peak frequency (Hz)	3rd peak frequency (Hz)	1st peak frequency. (Hz)	2nd peak frequency (Hz)	3rd peak frequency (Hz)	1st peak frequency. (Hz)	2nd peak frequency (Hz)	3rd peak frequency (Hz)
Before	6.86	9.19	/	6.79	13.06	25.47	5.05	10.19	/
S1	6.30	9.85	24.68	5.88	12.80	25.60	4.71	10.10	/
S2	5.78	9.98	24.68	5.62	12.80	25.60	4.79	10.10	/
S3	3.68	7.48	28.36	5.09	12.28	26.52	4.79	10.10	/
S4	3.80	7.62	/	5.22	12.28	25.67	4.62	10.10	/
S5	4.20	7.48	29.15	4.44	12.54	25.67	4.53	10.01	/
S6	3.94	6.96	28.10	4.57	12.54	25.67	4.70	9.84	/
S7	4.46	6.96	28.62	3.66	12.28	25.60	4.61	9.93	/
S8	3.28	6.04	30.99	3.53	12.02	24.95	4.70	9.93	/
S9	/	/	/	/	/	/	/	/	/

Table 17.8 Comparison with the theoretical predictions (x-direction)

	Measured frequency [Hz]	Frequency based on EC6 [Hz]
Specimen with T-shaped piers	5.18	5.03
Specimen with L-shaped piers	5.02	5.65

Table 17.9 Comparison with the theoretical predictions (y-direction)

	Measured frequency [Hz]	Frequency based on EC6 [Hz]
Specimen with T-shaped piers	6.86	8.72
Specimen with L-shaped piers	6.79	9.48

17.4.3 Modal Shapes

Natural vibration modes were identified assuming that the slab is behaving as a rigid body with a motion fully described by 3 DOFs. Modal shapes are identified from the low-level white noise tests results using a square matrix frequency-dependant of cross PSD of the slab based on the measurements obtained from the 4 accelerometers located on the slab. The identification procedure is detailed in Mordant (2012). It is also relevant to note that, for the low level of ground acceleration used for the identification stages, no significant rocking is induced in the structure and the system can be assumed to behave as fixed in the foundation. The analysis focuses on the first three vibration modes of the structure, consistently with outcomes of Sect. 17.4.2 related to the natural frequencies.

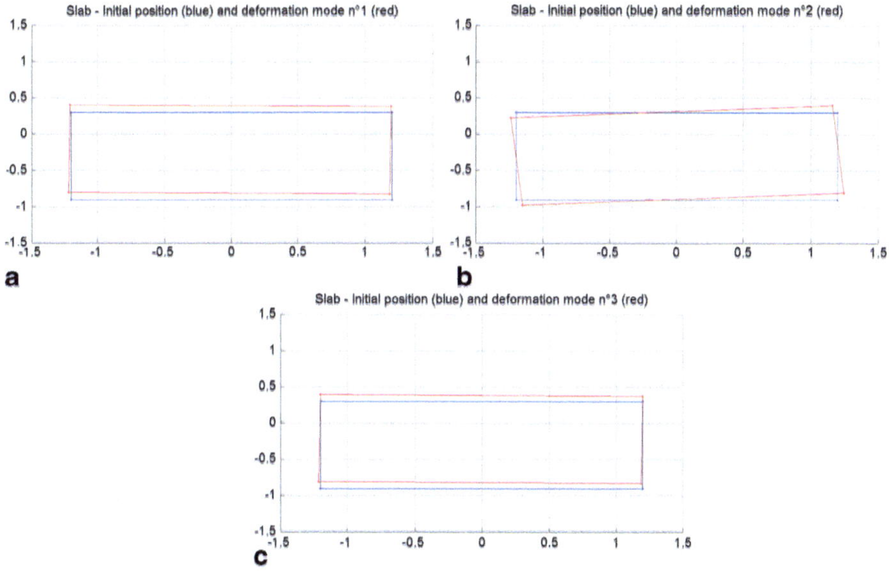

Fig. 17.18 Three first modal shapes (T-shaped frame—x-direction)

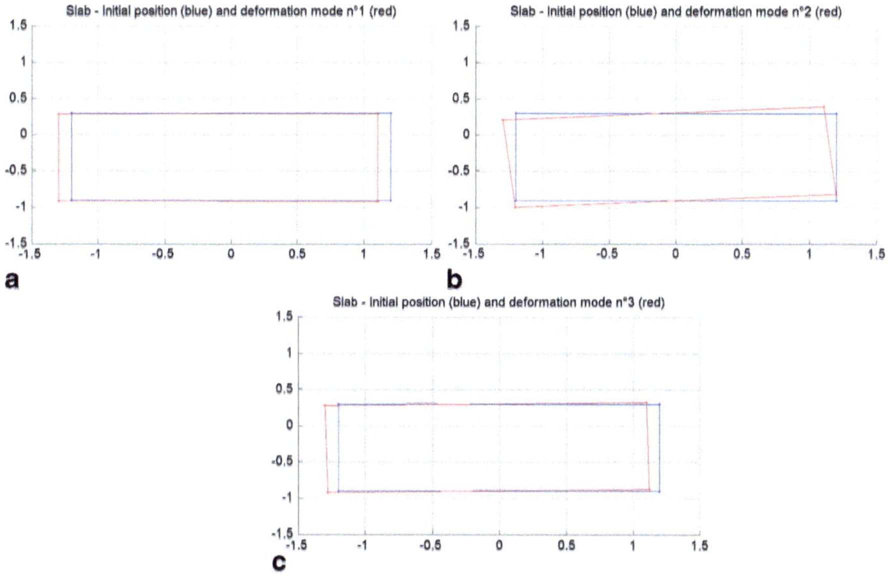

Fig. 17.19 Three first modal shapes (T-shaped frame—y-direction)

Modal shapes are depicted in Fig. 17.18 and 17.19 for the T-shaped walls, respectively for white noises in the x and y direction. Observation of the identified modal shapes lead to the following comments, also valid for L-shaped walls. The first and third modes are characterized by a combination of translation and rotation,

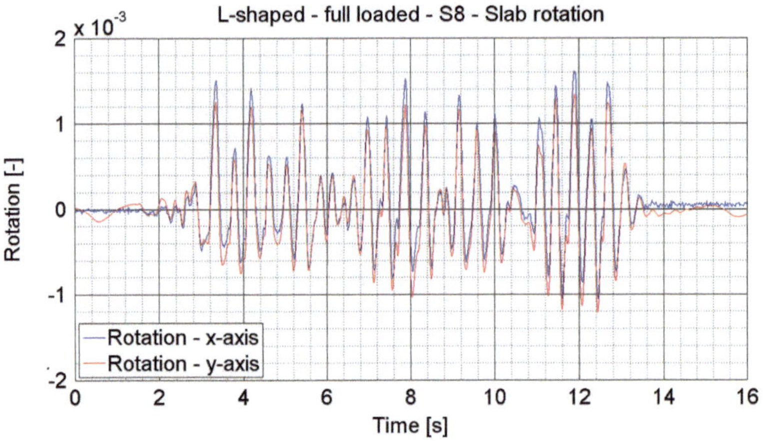

Fig. 17.20 Slab rotation—full loaded L-shaped frame

the importance of the latter depending on the specimen studied and different for both modes. The second mode can be qualified as a quasi-pure rotation mode, with a negligible translation component.

17.4.4 Seismic Behaviour

The exploitation of the seismic tests results as presented in the present contribution is essentially limited to the analysis of the motion of the slab with respect to the ground.

First of all, it is necessary to specify that the measurements (displacements) taken by the Imetrum Vision system (VS) are only relevant when the main seismic input is perpendicular to the plan of the target, i.e. the use of the measurements recorded by the targets 1–6 (see Fig. 17.12) is for instance only possible for the seismic tests along the y-axis. However, the displacements at given points can be recovered by integrating the accelerometric signals measured at the same locations.

Figures 17.20 and 17.21 plot the time history of the slab torsional rotation during the tests with high amplitude derived from the displacement measurements described above. Although the configuration of the frame with L-shaped piers is symmetric regarding the x-direction, and thus no rotation should be theoretically observed, such a motion is however triggered by accidental eccentricities, the position of the slab, the reliability of the connexion system and the mass distribution being not exactly controlled. The torsional motion remains nevertheless less important than for the T-shaped specimen. Indeed, Fig. 17.22 shows that the amplitude of the rotation is higher than 2.0 mrad, which is more or less the maximum value reached in the case of the L-shaped specimen.

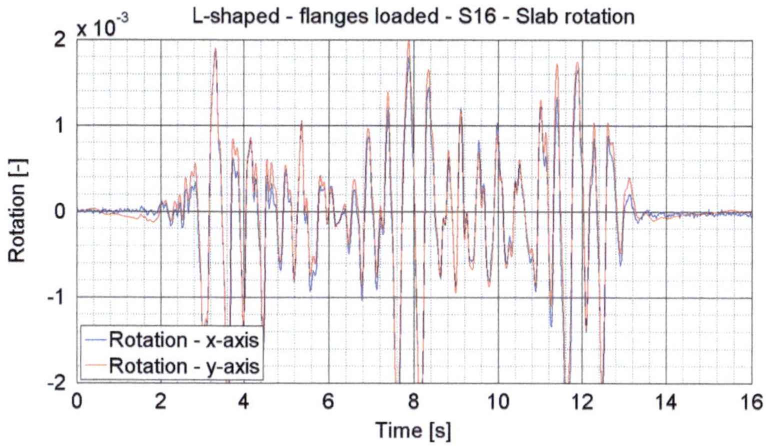

Fig. 17.21 Slab rotation—L-shaped frame with flanges loaded only

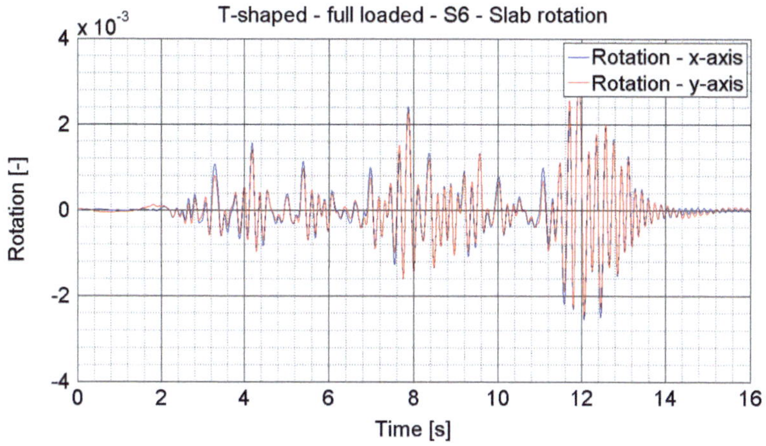

Fig. 17.22 Slab rotation—full loaded T-shaped frame

The influence of the gravity load configuration can also be highlighted. In this perspective, a comparison of the slab displacement along the earthquake direction is done in the Fig. 17.23 for a seismic input in the y-direction (PGA around 0.14 g) and in Fig. 17.24 for a seismic input in the x-direction (PGA around 0.13 g).

The comparison shows larger displacements for the same acceleration level, when only the flanges are loaded. The difference is also more important when the input is perpendicular to the loaded walls. It must also be pointed out that the specimen is actually able to sustain a higher acceleration level than the one predicted by a

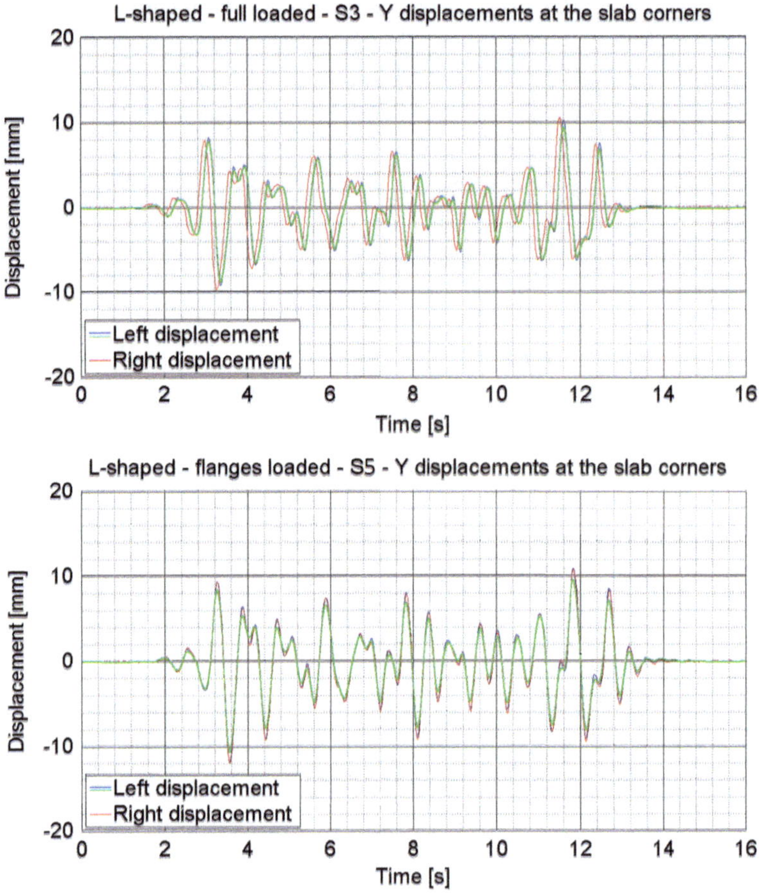

Fig. 17.23 y-displacements—full loaded L-shaped frame (*top*: full loaded—*bottom*:partially loaded)

theoretical model assuming that the walls are considered separately or that they are represented by an equivalent simple wall with the same total length, namely 1.5 m. According to these two assumptions, the maximum sustainable acceleration expected is respectively about 1.55 and 0.149 g, as shown in Mordant (2012). The interest of considering the frame effect and the perpendicular walls is therefore obvious.

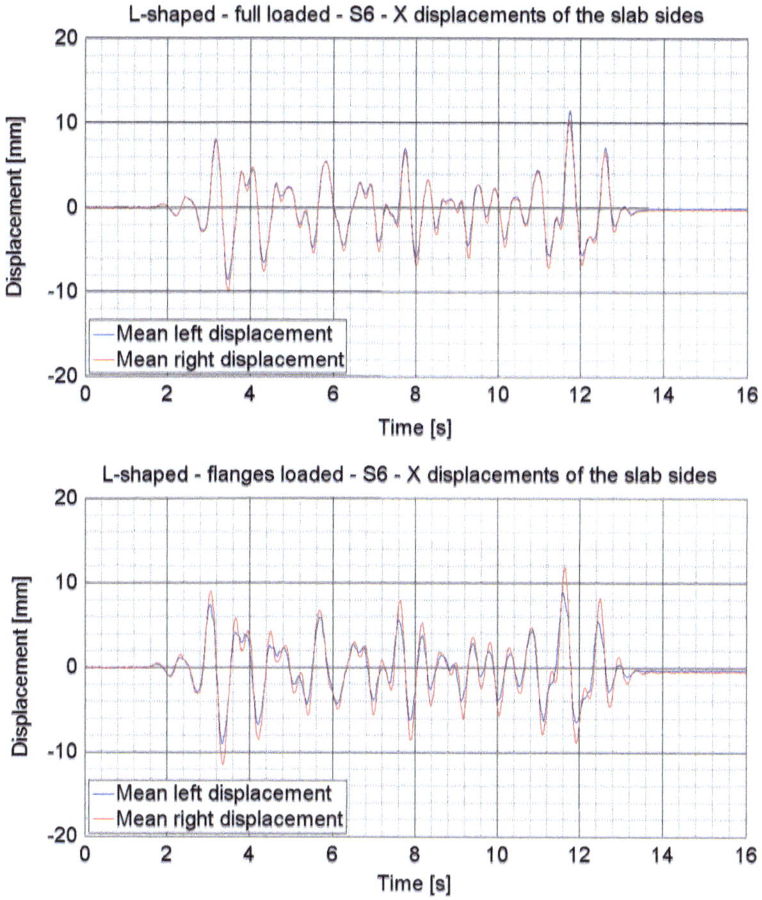

Fig. 17.24 x-displacements—full loaded L-shaped frame (*top*: full loaded—*bottom*: partially loaded)

17.5 Conclusions

This paper has presented preliminary results of experimental shake table tests carried out with the main objective of characterizing the dynamic rocking behaviour of high strength clay masonry sub-structures with glued horizontal joints and empty vertical joints subjected to seismic action. The geometries of the specimens are defined such that the effects of torsion, the influence of the gravity loading configuration and the consequences of the connection method can be studied.

The main observations that can be directly derived from the observations done during the testing campaign are the following:

- A significant rocking motion is observed for both tested frames, with L- or T-shaped piers.

- The identification of the natural vibration modes under low level white noise excitation raises the question of the accuracy of the values of the elastic modulus usually suggested in the main reference documents, this modulus being apparently overestimated for the studied type of masonry. Similar questions also rise regarding the boundary conditions to be considered in theoretical models.
- The effects of local torsion of masonry piers can currently not be properly assessed with the available design models. The control of this behaviour needs to be improved in order to take the associated stress state appropriately into account, likely to induce anticipated unexpected failures such as those observed in the present tests.
- The connection line at the intersection of perpendicular vertical walls is also a critical point and requires appropriate local verifications.
- The importance of the gravity loading configuration and, under certain conditions, of the consideration for the walls perpendicular to the seismic action has been highlighted and is shown as influencing significantly the maximum sustainable acceleration.

At this stage, additional processing and developments are however still required to improve the exploitation and the interpretation of the tests results. Detailed assessment of the test results and corresponding model identification and calibration are planned to be presented in upcoming contributions currently in preparation.

Ackonwledgements The research leading to these results has received funding from the European Union Seventh Framework Programme (FP7/2007–2013) under grant agreement n° 227887. H. Degée also acknowledges the direct support received from F.R.S.-FNRS (Belgian Fund for Research).

References

Caliò I, Marletta M, Pantò B (2012) A new discrete element model for evaluation of the seismic behaviour of unreinforced masonry buildings. Eng Struct 40:327–338

Degée H, Denoël V, Candeias P, Campos Costa A, Coelho E (2007) Experimental investigation on the seismic behaviour of North-European masonry houses. Proceeding of the congress SISMICA, p 12

EN 1996 (2004) Seismic design of masonry structures—Part 1.1: common rules for reinforced and unreinforced masonry structures. CEN—European committee for standardization

EN 1998 (2004a) Eurocode 8: Design of structures for earthquake resistance—Part 1: general rules, seismic actions and rules for buildings. CEN—European committee for standardization

EN 1998 (2004b) EurcoCode8: Annex B(Informative): determination of the target displacement for nonlinear static (pushover) analysis. CEN—European committee for standardization

Karantoni FV, Lirantzaki FN (2009) Seismic behaviour of simple masonry buildings according to EN 1998. 7th international conference on earthquake resistant engineering structures

Kazemi MT, Hoseinzadeh AM, Bakhshi A, Rahimzadeh RF (2010) Shaking table study of a full-scale signle storey confined brick masonry building. Sci Iranica 17(3):184–193

Milani G, Beyer K, Dazio A (2009) Upper bound limit analysis of meso-mechanical spandrel models for the pushover analysis of 2D masonry frames. Eng Struct 31:2696–2710

Mordant C (2012) Contribution to experimental tests on the seismic behaviour of masonry structural elements. Master dissertation, University of Liège

Mordant C, Dietz M, Degée H (2013a) Seismic behaviour of thin-bed layered unreinforced clay masonry shear walls including soundproofing elements. In: Alper I, Fardis Michael N (eds) Proc SERIES Workshop. Geotechnical, geological and earthquake engineering series (Chap. 6)
Nakagawa T, Narafu T, Imai H et al (2012) Collapse behaviour of a brick masonry house using a shaking table and numerical simulation based on the extended distinct element method. Bull Earthq Eng 10:269–283
Oropeza Ancieta M (2011) Fragility functions for seismic risk in regions with moderate seismicity. EPFL, Lausanne
Schermer DC (September 2005) Autoclaved aerated concrete innovation and development. 4th International conference on autoclaved aerated concrete, AAC, London (UK)
Tomazevic M (1999) Earthquake-resistant design of masonry buildings. Imperial College Press, London
Wienerberger (2014) Rouge. La Force. www.wienerberger.be. Accessed 25 Nov 2014

Chapter 18
Shake Table Testing of a Half-Scaled RC-URM Wall Structure

Marco Tondelli, Sarah Petry, Igor Lanese, Simone Peloso and Katrin Beyer

18.1 Introduction

In Switzerland many residential buildings are constructed with reinforced concrete (RC) and unreinforced masonry (URM) walls where the two types of structural systems are coupled by RC slabs (Fig. 18.1). The seismic behaviour of these mixed structures strongly depends on both types of structural elements (RC and URM) since the stiffness of the two types of walls are comparable. Despite being very common in design and retrofit practice these structures have not been tested in the past under seismic loading and their seismic behaviour is still not well understood. Furthermore design codes do not provide specific provisions for mixed RC-URM structures; Eurocodes and Swiss design codes (CEN 2004, 2005; SIA 2003) do not provide indications regarding the choice of the force reduction factor q and the selection of appropriate deformation limits. Consequently, Swiss design practice neglects the effect of URM walls on the seismic behaviour of the structure assigning all forces to RC walls and adopting q-factors applicable to RC walls. An additional problem related to URM structures is the out-of-plane vulnerability of URM piers. Recent studies (Dazio 2008) showed that most Swiss buildings do not fulfil the slenderness criteria provided by Eurocode 8.

The CoMa WallS project, based on the results of uni-directional shake table tests on a half-scale structure, has as main goal the assessment of the uni-directional shake table of mixed RC-URM structures. The results from the tests will allow to

K. Beyer (✉) · M. Tondelli · S. Petry
Earthquake Engineering and Structural Dynamics Laboratory (EESD),
School of Architecture, Civil and Environmental Engineering (ENAC),
École Polytechnique Fédérale de Lausanne (EPFL),
Switzerland Earthquake Engineering, Lausanne, Switzerland
e-mail: katrin.beyer@epfl.ch

I. Lanese · S. Peloso
EUCENTRE—European Centre for Training and Research in Earthquake Engineering,
Pavia, PV, Italy

F. Taucer, R. Apostolska (eds.), *Experimental Research in Earthquake Engineering*,
Geotechnical, Geological and Earthquake Engineering 35,
DOI 10.1007/978-3-319-10136-1_18

Fig. 18.1 3- and 8-storey residential buildings with RC walls and URM walls. (Reproduced by permission of Thomas Wenk)

address several issues such as: understanding the stiffness degradation and damage evolution with increasing levels of shaking, definition of performance limits for RC-URM mixed structures, insights into displacement and acceleration profiles during seismic loading, information on the internal force distribution among the different structural elements, effect of boundary conditions on the seismic behaviour of out-of-plane loaded URM walls.

18.2 Test Unit

The test unit, which was built at half-scale, represented a typical Swiss residential building where most of the walls are constructed as URM walls and some as RC walls, the two different vertical systems are coupled by RC slabs. The structure, which is shown in Fig. 18.2, was composed of four in-plane loaded URM walls, two out-of-plane loaded URM walls and two RC walls.

The URM walls were constructed using clay-brick units which were produced for the project by the Swiss brick manufacturer "Morandi Frères SA". This half-scale brick unit was 9.5 cm wide, 15 cm long and 9.5 cm high. The masonry was assembled using a standard mortar of class M15. The brick-units were developed within a test campaign investigating the effects of scaling on the properties of masonry which will be reported in the following section.

RC walls were cast using class C28/35 concrete. The reinforcement consisted of a double layer of longitudinal steel bars and shear reinforcement of class B450C. RC slabs where constructed using the same concrete as for the RC walls; the thickness of the slab was 15 cm and they were reinforced with a double layer of steel net. More details on the reinforcement layout and the reinforcement properties can be found in Tondelli et al. (2014).

Since the structure was built at half-scale it was necessary to load the structure with additional masses, which were added in the form of concrete blocks and cast directly on top of each slab. A plastic sheet between slab and concrete block ensured that the concrete blocks did not add to the stiffness of the slab. A summary of the structural masses is listed in Table 18.1.

Fig. 18.2 Transversal (North) and longitudinal (West) views of the test unit, dimensions are in mm

Table 18.1 Summary of structural masses

Foundation	13.1 t
Mass of the structure	34.9 t
Additional masses	34.9 t
Total mass	82.9 t
Total mass without foundation	69.8 t
Total mass during transportation	48.0 t

Fig. 18.3 Test specimen during transportation

The test unit was built outside the testing facility and was then jacked up and pulled inside the TREES laboratory by means of hydraulic jacks and a slider system. In order to prevent the structure from being damaged during the transportation phase the URM walls were pre-tensioned (Fig. 18.3).

18.3 Tests Conducted at EPFL in Preparation of the Shake Table Test

The masonry walls of the shake table test unit were constructed using purpose-manufactured model bricks at scale 1:2. To develop these, numerous types of model bricks were tested at EPFL. Figure 18.4 shows the finally selected model brick and the corresponding prototype brick. For the shake table test it was particularly important that the behaviour of the masonry walls under horizontal loading was well reproduced. For this reason, quasi-static cyclic tests on masonry piers were conducted in addition to standard material tests on masonry wallettes (vertical compression tests, diagonal compression tests, shear triplet tests). All tests were conducted at scale 1:2 and 1:1 in order to compare the behaviour of the model masonry to that of the prototype masonry (Petry and Beyer 2014a).

Figure 18.5 shows the experimental setups for the quasi-static cyclic tests at full- and half-scale. In these setups two vertical hydraulic actuators applied the normal force and the bending moment at the top of the masonry pier and a third actuator applied the horizontal displacement. The three actuators were operated in a completely coupled mode and controlled in such a manner that the normal force and the height of zero moment remained constant throughout the test. For each scale, five quasi-static cyclic tests were carried out. The tests differed with regard to the applied normal force and the height of zero moment. The results of these tests showed that the model masonry reproduces the prototype masonry very well with regard to the horizontal stiffness, the strength and the failure mechanism of the masonry piers.

a b

Fig. 18.4 **a** Prototype of clay brick, and **b** model brick for masonry at scale 1:2

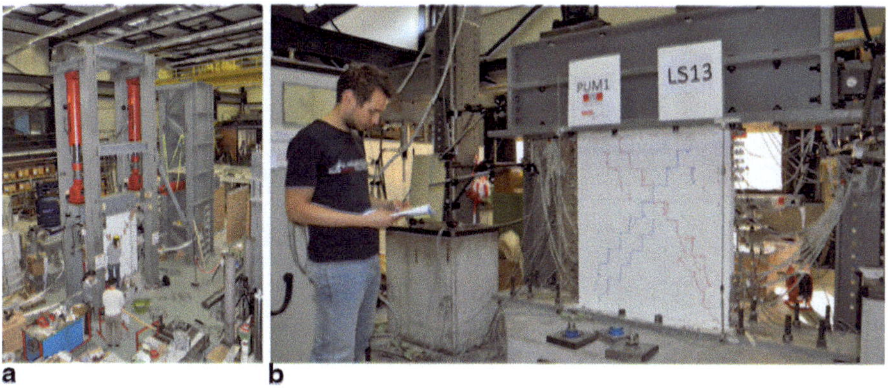

Fig. 18.5 Test setups for quasi-static cyclic tests on URM piers at scale 1:1 (**a**) and at scale 1:2 (**b**) (Petry and Beyer 2014a)

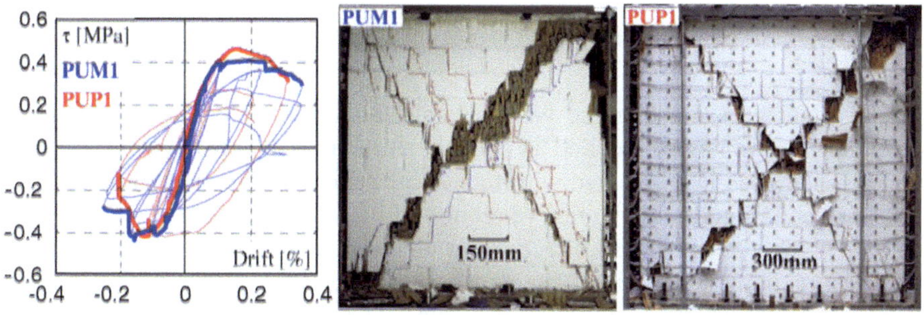

Fig. 18.6 Results of quasi-static cyclic tests on URM piers: Shear stress vs. drift relationships for piers at scale 1:2 (*PUM1*) and piers at scale 1:1 (*PUP1*) as well as the crack patterns after axial load failure (Petry and Beyer 2014b)

The deformation capacity of the model masonry was on average slightly larger than that of the prototype masonry. As an example, Fig. 18.6 shows the shear stress vs. drift relationship of the first test units of the model (PUM) and the prototype (PUP) series as well as the corresponding crack patterns after axial load failure of the test units.

To better understand the interaction of URM and RC walls in mixed structural systems, quasi-static tests on a two-storey structure were conducted (Fig. 18.7a, Paparo and Beyer 2014). A specially designed experimental setup allowed measuring the reaction forces at the base of the URM wall. From these forces and the applied horizontal and vertical loads, the axial force, shear force and bending moment at the base of the RC wall could be back-calculated. The shear force-displacement hysteresis in Fig. 18.7b shows that the RC wall can prevent a premature softening of the overall system. Furthermore, the fracture pattern shows that, in a mixed system—unlike in buildings with URM walls only—the horizontal deformation is distributed over a greater height and not concentrated in the first storey only.

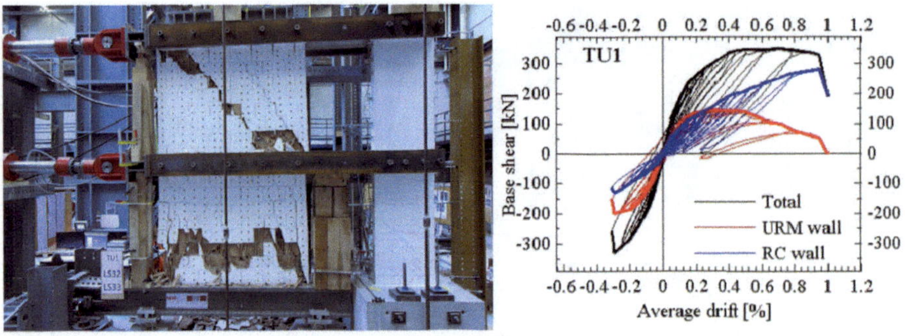

a b

Fig. 18.7 Quasi-static cyclic tests on system with a RC wall and an URM wall: Crack pattern after axial load failure of the unreinforced masonry wall (**a**) and horizontal force-displacement hysteresis (**b**) (Paparo and Beyer 2014)

18.3.1 Instrumentation

The test unit was instrumented with in total 20 accelerometers, 49 displacement transducers, 24 omega gages and an optical measurement system.

Each slab was instrumented with four accelerometers, two recording accelerations in the longitudinal direction (direction of the motion) and two recording accelerations in the transverse direction. In addition two accelerometers recording acceleration in the longitudinal direction and one recording in the transverse direction were installed on the foundation of the structure. Finally one accelerometer was installed directly on the shake table recording accelerations in the longitudinal direction.

Since the most severe deformations, and therefore damages, were expected to take place at the first storey of the structure, a dense network of displacement transducers monitored the deformations of RC walls and URM piers. The East face RC wall was instrumented with eight displacement transducers measuring the elongation of the wall edges; in addition wire potentiometers were installed along the diagonals of the wall in order to record shear and rocking deformations. Wire potentiometers were also installed on the East side URM walls along the two diagonals. The diagonals were connected to the slab in order to record both shear and rocking deformations.

Deformations of the of the out-of-plane loaded URM piers were recorded using displacement transducers; each pier was instrumented with one potentiometer measuring the out-of-plane deformations at mid-height and the internal and external vertical displacement of the top and bottom rows of bricks. The omega gages were used to measure the out-of-plane deformations of the first-storey slab; two different locations of the slab were instrumented with six gages both on the top and bottom face of the slab.

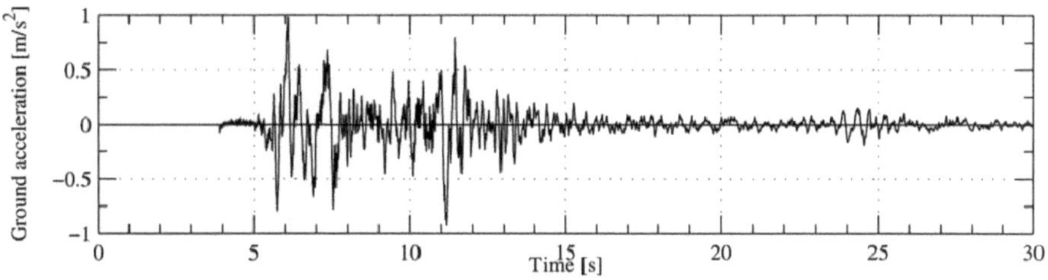

Fig. 18.8 Input signal for shake table test for a PGA of 1 m/s²

The displacements of the West façade of the building during the motion were measured by means of an optical measurement system. The motion of the structure was detected through reflecting markers glued onto the specimen whose movement was acquired by a high definition camera system.

18.3.2 Input Ground Motion

The input ground motion that was used for the uni-directional shake table test was the Ulcinj-Hotel Albatros station record of the 1979 Montenegro earthquake. The accelerogram, which was applied in the longitudinal direction of the building, was scaled in time reducing the duration by the square root of two to account for the fact that the structure was built at half-scale. For the scaling of the test specimen the Artificial Mass Simulation scaling law was applied (Krawinkler 1979). The acceleration time history of the record is shown in Fig. 18.8 for a maximum PGA of 1 m/s², scaled to match different levels of peak ground acceleration ranging from 0.05 to 0.9 g.

18.3.3 Shake Table Test

The shake table test was performed at the TREES laboratory of the European Centre for Training and Research in Earthquake Engineering (EUCENTRE, Pavia, Italy). The actual testing lasted four days and included nine runs with shakings of different levels of intensity (PGA). Table 18.2 summarizes the testing chronology. Each single shake table test was followed by a survey of the structure intended to identify and report the evolution of the structural damage. Each test was also followed by a dynamic identification of the structure performed to investigate the evolution of the fundamental period of vibration of the structure after each shaking.

The structure developed very limited levels of damage after the first three runs, which were characterized by a low intensity level of shaking. Hairline cracks formed in the URM piers of the first storey, mainly concentrated at the corner of the masonry panels.

Table 18.2 Summary of the nominal PGAs and testing dates

Test number	Nominal PGA (g)	Testing date
1	0.05 g	18.12.2012
2	0.1 g	18.12.2012
3	0.2 g	19.12.2012
4	0.3 g	19.12.2012
5	0.4 g	19.12.2012
6	0.6 g	20.12.2012
7	0.4 g	20.12.2012
8	0.7 g	20.12.2012
9	0.9 g	21.12.2012

Fig. 18.9 Cracks in the first storey West side URM piers after test number 6

Tests number 4 and 5 led to the formation of the first hairline cracks in the second storey URM piers and in the first storey RC walls and slab. The diagonal cracks in the URM walls of the first storey grew in length and extended at the end of test 5 over the entire length of the diagonals. The residual crack widths were, however, still very limited.

Test number 6 was actually the first test which damaged the structure more severely. The pre-existing cracks in the URM piers of the first storey started widening up to a crack width up to 0.8 mm, even if in the majority of cases the crack width was still smaller than 0.2 mm (Fig. 18.9). Moreover, at this point the formation of cracks in the URM piers extended to all storeys of the structure.

The seventh test was performed with a lower level of intensity with respect to test number 6 in order to simulate the evolution of structural damage induced by an after-shock during a real seismic event. The survey of cracks widths showed, however, that no severe additional damage was induced during this run.

Fig. 18.10 View of the East, North and West face of the structure after test number 9

The structure reached a significant level of damage during test number 8 as the deformation in the URM piers of the first and second storey started concentrating in a single diagonal crack with an average crack width equal to 0.8 mm. Moreover, for the first time cracks passing through masonry bricks were identified. Until this point all cracks had followed the mortar joints.

During the ninth and final test the structure came rather close to collapse since the in-plane loaded URM walls of the first and second storey lost their axial load bearing capacity (Fig. 18.10). The axial loads were finally carried by the RC walls and the out-of-plane loaded URM walls. The failure mechanism of the in-plane loaded URM piers of the first and second storey was the crushing of the compression diagonal of the piers at the top and bottom corners and at the middle height where two diagonals were intersecting. In the outer URM piers failure occurred for loading towards the North. For this loading direction, the axial force in the outer URM piers increased and therefore the deformation capacity of the URM piers was reduced (Petry and Beyer 2014b)

A striking feature of the test specimen was the equal level of damage experienced by the first two storeys (Fig. 18.11). Generally, in masonry structures all the damage is concentrated in the first storey. The addition of RC walls to the URM structure led therefore to a spread of the damage over the first two storeys.

The behaviour shown by the RC walls depended on the loading direction. When the structure was pushed towards North the walls were in tension and developed horizontal flexural cracks along the South edge over the entire height of the first storey. These cracks showed a residual deformation of approximately 1 mm indicating that the longitudinal reinforcement had yielded (Fig. 18.12). When the structure was instead pushed toward the South, the RC walls were in compression and remained largely uncracked; the few cracks that developed for this loading direction showed a residual deformation close to zero indicating that at this end of the walls the longitudinal reinforcement had not yielded.

During the last test, for the first time in the experimental test campaign, a clear out-of-plane deformation of an URM pier was observed. The URM panel at the

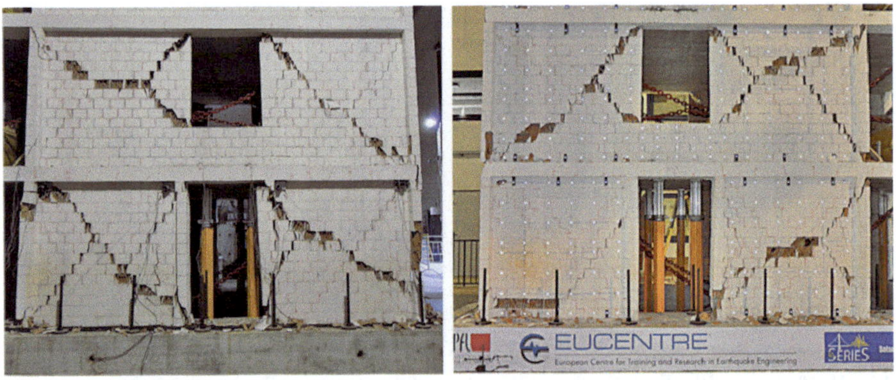

Fig. 18.11 Damage in the URM piers of the first two storeys after test number 9

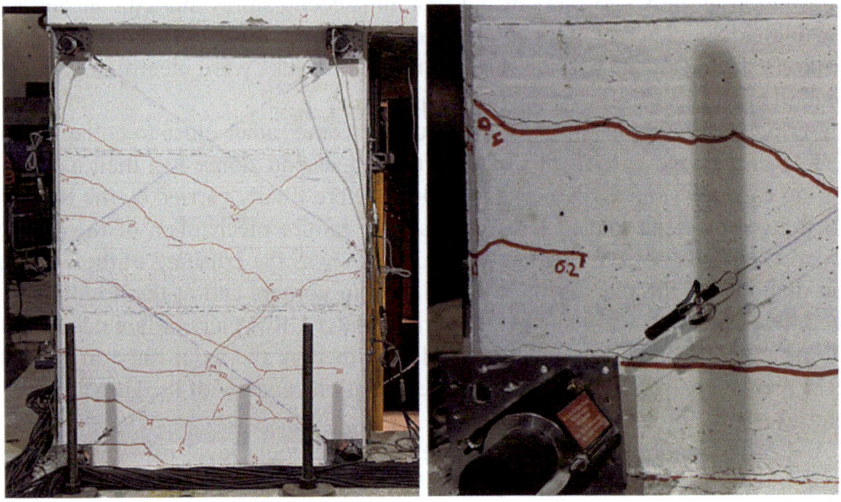

Fig. 18.12 Cracks in the East face RC wall after test number 9

fourth storey of the North façade showed a large out-of-plane deformation during shaking and touched the support structure. The support structure for the out-of-plane elements consisted of steel and timber elements that were placed both inside and outside of the structure to prevent that URM piers would fall onto the shake-table. The survey after the test showed that horizontal cracks in the mortar joint at the top, bottom and mid-height of the pier had formed (Fig. 18.13). As expected the boundary conditions played an important role on the out-of-plane behaviour of the URM walls. While the top pier on the North face of the building reached collapse the corresponding pier of the South face of the building did not show any large deformation during the test and also did not present any cracks at the end of the test.

Fig. 18.13 Out-of-plane failure of URM wall: horizontal cracks at mid height and at the top of the wall

18.4 Conclusions

The shake table test on the four-storey mixed RC-URM structure allowed the investigation of the seismic performance of such structures for a large range of ground motion intensities. The first shaking, with a low level of intensity, induced only the formation of small and few cracks in the structure. As the intensity increased increasing damages were observed in the structure; finally the ninth and last test brought the structure very close to collapse since the in-plane loaded URM walls of the first two storeys lost their axial load bearing capacity and the gravity loads were carried only by the RC walls and the out-of-plane loaded URM walls.

During the shake table tests about 90 hard-wired instruments recorded local deformations; additionally an optical measurement system allowed tracking the position of approximately 300 markers on the West face of the structure.

The shake table test showed how the presence of RC walls influences the performances of URM buildings. While in a masonry building damage and drifts are concentrated in the first storey, in the mixed structure the first two storeys were subjected to similar levels of drift and damage. The two upper storeys experienced significantly lower levels of drift, which were however sufficiently large to cause cracking.

During the last test one URM pier failed due to out-of-plane accelerations. The observed behaviour pier underlined how the different boundary conditions can influence the seismic behaviour of out-of-plane loaded piers. The data collected during the test will allow to address different open issues related to the seismic behaviour of mixed RC-URM structure.

Acknowledgements The research leading to these results received founding from the European Community's Seventh Framework Programme [FP7/2007–2013] for access to TREES laboratory of EUCENTRE under grant agreement n° 227887 for the SERIES project. Additional financial support was received from the Office Fédéral de l'Environement (OFEV) in Switzerland. The authors appreciate and gratefully acknowledge both financial contributors. The bricks were purpose-manufactured by Morandi Frères SA, Switzerland. The authors would like to thank all members of the TREES laboratory and in particular the head of the laboratory, Prof. Alberto Pavese, for their invaluable support during the entire duration of the project as well as the staff of the structural engineering laboratory at EPFL who helped with the quasi-static cyclic tests and material tests.

References

CEN (2004) Eurocode 8: design of structures for earthquake resistance—part 1: general rules, seismic actions and rules for buildings. European Code EN 1998-1, European committee for standardization. Brussels, Belgium

CEN (2005) Eurocode 8: design of structures for earthquake resistance—part 3: assessment and retrofitting of buildings. European code EN 1998-3, European committee for standardization. Brussels, Belgium

Dazio A (2008) The effect of the boundary conditions on the out-of-plane behaviour of unreinforced masonry walls. 14th world conference on earthquake engineering (WCEE 2008), 12–17 October, 2008, Beijing, China

Krawinkler H (1979) Possibilities and limitations of scale-model testing in earthquake engineering. In: Proceedings of the second U.S. national conference on earthquake engineering. Stanford, pp 283–292

Paparo A, Beyer K (2014) Quasi-static cyclic tests of two mixed reinforced concrete–unreinforced masonry wall structures, Engineering Structures 71:201–211

Petry S, Beyer K (2014a) Scaling unreinforced masonry for reduced-scale testing, Bulletin of Earthquake Engineering 12(6):2557–2581, 2014

Petry S, Beyer K (2014b) Influence of boundary conditions and size effect on the drift capacity of URM walls, Engineering Structures 65(C):76–88

SIA (2003) Actions on structures, SIA 261:2003. Swiss design code. Swiss society of engineers and architects, Zurich, Switzerland

Tondelli M, Petry S, Peloso S and Beyer K (2014) Dynamic testing of a four-storey building with reinforced concrete and unreinforced masonry walls: Data set. Submitted to Bulletin of Earthquake Engineering

Chapter 19
Experimental and Numerical Investigation of Torsionally Irregular RC Shear Wall Buildings with Rutherma Breakers

Ahmet Yakut, Alain Le Maoult, Benjamin Richard, Gabriela M. Atanasiu, Frederic Ragueneau, Stefen Scheer and Seref Diler

19.1 Introduction

Seismic response of reinforced concrete shear wall buildings with irregularities has not been studied in detail and thus requires detailed experimental as well as numerical investigations. Therefore, a challenging prototype building having irregularly placed shear walls in plan was designed and tested on the AZALEE shaking table at the SEMT/EMSI laboratory in CEA/Saclay. The research project, called ENISTAT, was funded by yhe SERIES project via Transnational Access to the CEA facility in France. The project has three main objectives: (1) Study the behaviour of the mock-up under increasing bi-directional horizontal synthetic earthquake records; (2) Attempt to evaluate the experimental results using modern experimental techniques for data acquisition; (3) Implement & monitor performance of a new structural element that allows for wall-slab connection to reduce thermal energy losses. A half scale reinforced concrete shear wall mock-up with an irregular plan and irregular distribution of shear walls having similar geometry with respect to the SMART specimen

A. Yakut (✉)
Department of Civil Engineering, Middle East Technical University, Ankara, Turkey
e-mail: ayakut@metu.edu.tr

A. Le Maoult · B. Richard
CEA, DEN, DANS, DM2S, SEMT, Laboratoire d'Etudes de Mécanique Sismique, Gif-sur-Yvette, France

G. M. Atanasiu
Faculty of Civil Engineering and Building Services, Technical University Gheorghe Asachi of Iasi, Iasi, Romania

F. Ragueneau
LMT-Cachan, ENS-Cachan/CNRS, Cachan, France

S. Scheer
SCHOECK France SARL, Entzheim, France

S. Diler
SCHÖCK Company, Baden-Baden, Germany

F. Taucer, R. Apostolska (eds.), *Experimental Research in Earthquake Engineering*, Geotechnical, Geological and Earthquake Engineering 35,
DOI 10.1007/978-3-319-10136-1_19

(SMART 2013) was designed according to EN-1998-1 (Eurocode 8 2004) requirements. The specimen was constructed at the SEMT/EMSI Laboratory in CEA.

Reinforced concrete shear walls are among the most important structural members providing earthquake resistance in RC buildings. Although a comprehensive knowledge and practice is available on design of shear walls, their complete nonlinear response and structural performance under seismic loadings has not been fully understood and identified. In order to bring up more information on the strength and deformation capabilities of structural walls, extensive experimental studies were carried out. Shear walls were tested mostly as isolated members under the action of horizontal and axial loads and in a few tests as representative subassemblies of structural buildings taking into account actual loading conditions. Various seismic loadings, ranging from monotonically increasing static loads to dynamic loading on shaking tables were applied on the experimental specimens (Oesterle et al. 1976; Vallenas et al. 1979; Palermo and Vecchio 2002; Lu 2002). Despite comprehensive earlier research generally focusing on individual reinforced concrete walls, there is still need to investigate their behavior in regular buildings for the purpose of understanding the relationship between the damage and commonly used response parameters. Limited numbers of previous studies employing 3D systems including shear walls generally focused on overall structural response and have not revealed enough data to investigate the effects of some important parameters of the response.

Previous experimental work conducted on real time aimed to study unidirectional excitation (Kotronis et al. 2005), bi-directional excitation (Kotronis et al. 2005; Reynouard and Fardis 2001) and irregular buildings (Juster-Lermitte et al. 2009).

However, due to the design geometry and detailing, few of those tests allowed the investigation of strong torsional effects. Only the cases of walls or part of a real structure have been treated. Concerning instrumentation, real time dynamic experiments on structures need classical instrumentation techniques such as acceleration and displacement measurement devices to apprehend the global behaviour of the specimen. More recent advances using digital images by tracking techniques allow obtaining displacements of targets. The ENISTAT specimen and test program aimed to combine all these techniques to study the behaviour of irregular RC buildings with shear walls.

19.2 Description of ENISTAT Specimen

19.2.1 Geometry

The ENISTAT specimen was chosen to represent a typical office building at half scale. Thus, the specimen dimensions, material properties, member detailing and masses were determined consistent with the scaling factor. The model is 3-story high with an irregular plan shown in Fig. 19.1. The structural system is composed of three shear walls, one column and beams. Two walls (Wall 2 and Wall 3) have openings whereas one wall (Wall 1) is solid. The dimensions of structural members are given in Table 19.1. Thermal break elements are placed at the second floor between the slab and the walls.

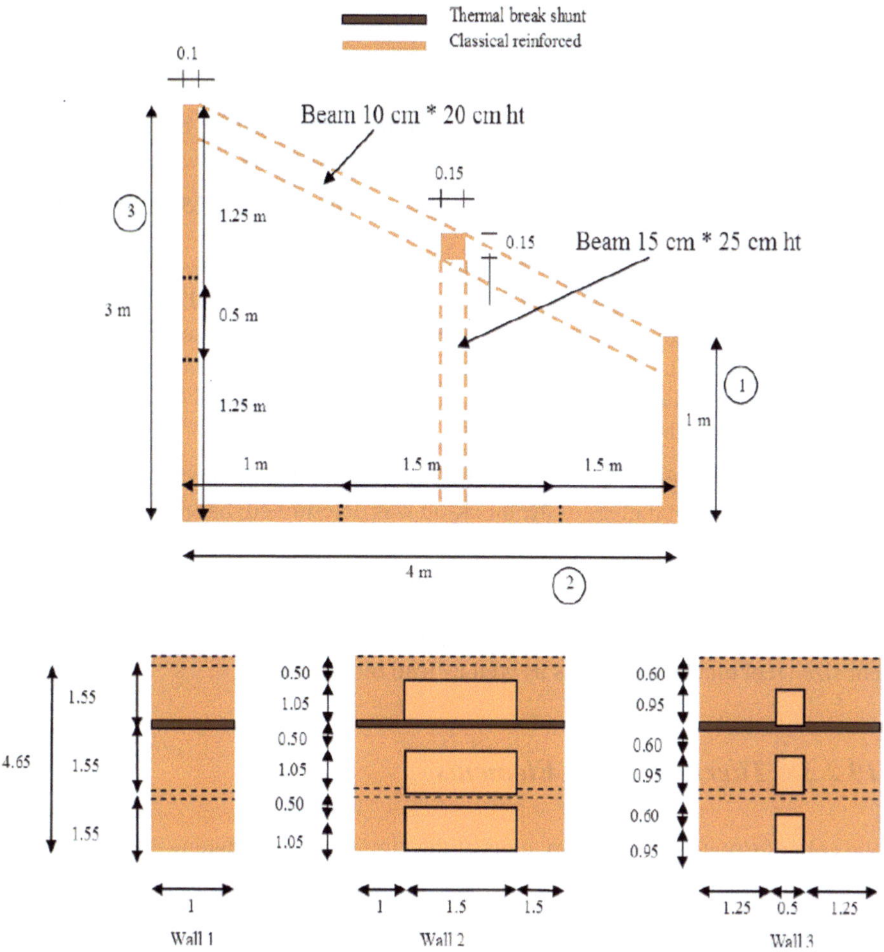

Fig. 19.1 Plan and elevation of ENISTAT specimen (wall dimensions are in meters and the *solid part* in the floor represents slabs with thermal break elements)

Table 19.1 Main dimensions of structural members

	Length (m)	Thickness (m)	Height (m)
Wall #W1	1.00	0.10	4.65
Wall #W2	4.00	0.10	4.65
Wall #W3	3.00	0.10	4.65
Beam 1	4.30	0.10	0.20
Beam 2	1.65	0.15	0.25
Column	0.15	0.15	4.93

19.2.2 Design of Specimen

The target in design of the model building was to use requirements of EC8 and to have the model building respond in the nonlinear range under increasing seismic excitations. The following criteria have been considered in the design:

- The concrete and reinforcement grades were selected as C25/30 and Fe500-3, respectively
- The structure is composed of uncoupled walls and thus according to EC8 it is considered in the DCM ductility class and classified as a torsionally flexible system
- The seismic parameters were chosen to represent a moderate seismic area (Mediterranean seismic area): The design spectrum was represented by Type I with soil type C. The nominal acceleration value was assumed to be 0.3 g. Target period of the model was aimed at 0.08 s.

The design of the specimen was done by a professional company in order to reflect current practice. Total mass of the mock-up was determined, based on the geometry shown in Fig. 19.1, as 18.7 t. In order to achieve the desired stresses consistent with the scaling factor additional masses, determined based on the similitude law, were placed at each floor. The amount of additional masses per floor was 6.45 t adding to a total of 19.45 t for the mock-up. The additional masses were applied using different size steel blocks and plates as well as lead blocks.

19.2.3 Thermal Break Elements

Recent construction technology aims at developing and producing energy efficient buildings. One of the efficient ways of reducing energy losses is to prevent thermal bridges, especially at wall-slab connections. The thermal-break elements are innovative technological components under study in several countries around Europe. While the thermal benefits have already been confirmed, the mechanical effects of such a wall-slab connection in a building in the context of earthquake loading have not been assessed. To evaluate the seismic vulnerability modification due to these thermal break elements, the connections at the second floor level of the mock-up were made using these structural elements (Figs. 19.2 and 19.3). In such a way, the global behaviour of the specimen is not be disturbed. The objective is to show the capability of such a sustainable design to resist strong earthquakes. The thermal breaker is a seismically designed module able to manage shear forces of the slab during the seismic excitation. All modules have been specifically designed by Schöck Company (Schock 2010) for ENISTAT mock up.

19.2.4 Construction of the Specimen

The ENISTAT specimen was constructed by a professional company at the TAMARIS laboratory and transported to the shaking table for testing. The construction

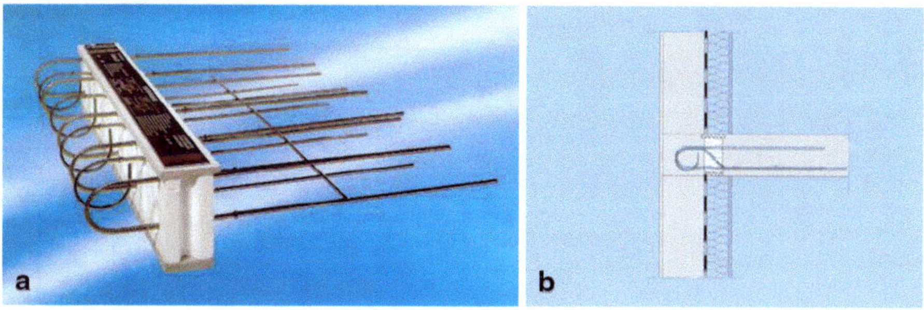

Fig. 19.2 Rutherma thermal break elements

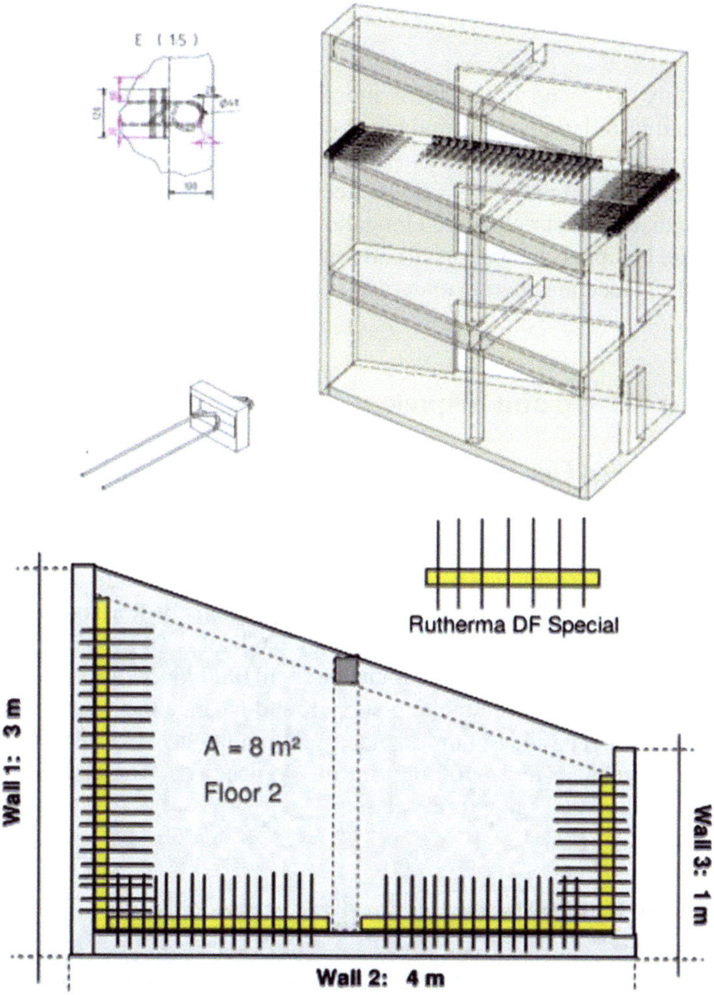

Fig. 19.3 Layout of thermal break elements on ENSITAT specimen

Fig. 19.4 View of ENISTAT specimen and installation of thermal break elements at the SEMT/EMSI laboratory in CEA

started in July 2012 and lasted for 3 months. Figure 19.4 shows the completed specimen and installation of thermal break elements at the second floor.

Material properties of concrete and steel used in the construction of the specimen were determined through testing. B500B steel of 6, 8, 10 and 12 mm bars were used. Yield strength, tensile strength and elongation of these specimens were determined. For concrete, compressive and tensile strengths were determined. Test results indicated that the steel and concrete materials fulfill design.

19.3 Test Set-Up and Sequence

19.3.1 Test Set-Up

All tests were performed on the six degrees of freedom (6 DOF) AZALEE shaking table (Fig. 19.5). The table is fixed in a pit, located in the middle of a 2700 t concrete reaction massif, by eight hydraulic actuators (four for horizontal excitations and four for vertical excitations). Its maximum payload is 100 t. Its square plate is 6 m × 6 m. Each actuator has a maximum force of 1000 kN. Four static pneumatic supports are placed under the table to support and balance the weight of the table and the specimen. The maximum displacement amplitude range is ± 125 mm for the two horizontal axes and ± 100 mm for the vertical axis. The table was built by welding aluminum plates in a rigid caisson.

19.3.2 Instrumentation

The instrumentation of the ENISTAT specimen was designed based on the following considerations:

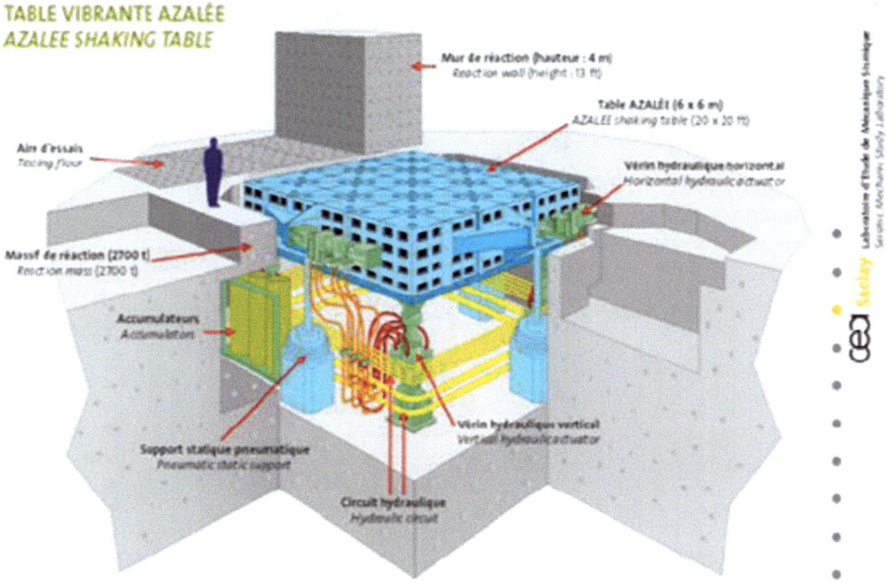

Fig. 19.5 AZALEE shaking table

- Global measurements of the shaking table and the mock up
- Local measurements of the second floor with thermal break elements
- Local measurements of Wall 1 and Wall 3 at level 1

The instrumentation for global measurements of the shaking table and the mock-up consisted of conventional sensors including accelerometers and displacement transducers. To achieve this objective, 21 accelerometers and 12 displacement transducers were installed at the mockup, whereas 19 accelerometers and eight displacement transducers were used on the shaking table. These sensors were placed at different points to take measurements at global X, Y and Z directions, including also measurements at actuators. The instrumentation points on the mock-up are shown in Fig. 19.6.

The main objective behind local measurements was to monitor thermal break elements installed in the second floor. Strain gages were used to monitor thermal break element deformations. Inductive displacement sensors with short strokes were used to measure displacements between the second floor slab and the walls. Finally, accelerometers were used to compare the vertical motion of each floor. The instrumentation to meet this objective consisted of 15 relative displacement sensors in X, Y and Z directions, and 34 strain gages at the steel reinforcement of each thermal break element. Additionally, 21 accelerometers (seven in the second floor) were installed on the floors.

To measure local deformations of wall 1 and wall 3 digital techniques were employed. A 3D optical device with 48 targets was used to measure displacements in the X, Y and Z directions of wall 1 and the foundation. A 3D optical device was used to measure the strain inside a 500 mm × 500 mm area on Wall 3.

Examples for some of the sensors used are shown in Fig. 19.7.

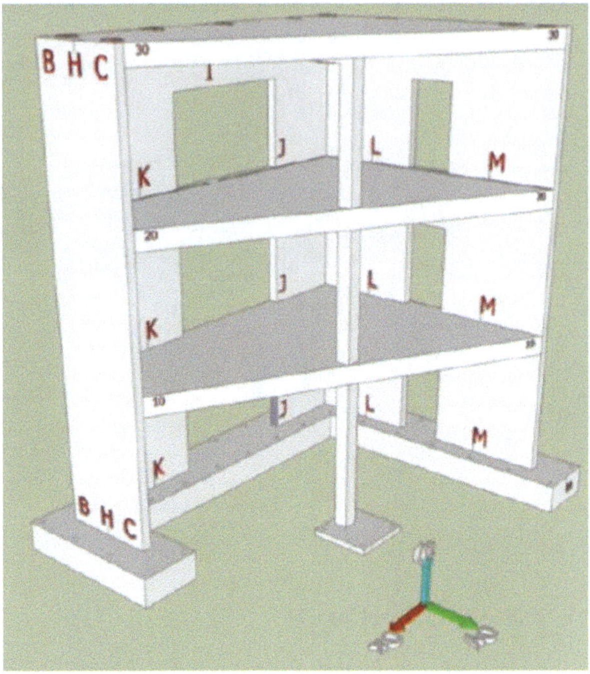

Fig. 19.6 Instrumentation points on the mock-up of ENISTAT

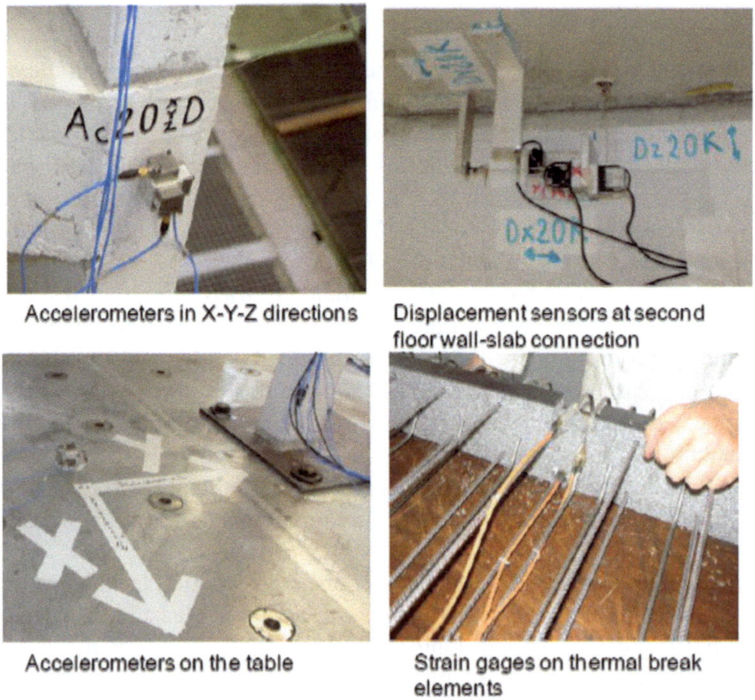

Fig. 19.7 Location and type of sensors used for measurements

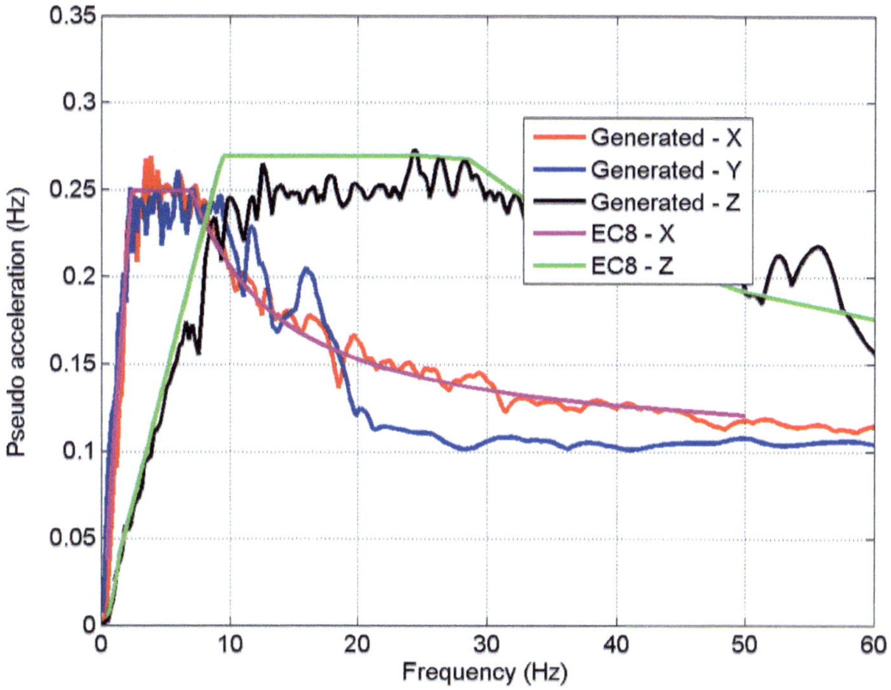

Fig. 19.8 Comparison of spectra for artificial inputs with EC 8 spectra

19.3.3 Test Sequence

Three artificial time histories were generated and used as the seismic excitation. These ground motion records were obtained to be compatible with the type 1, soil Class C, EC8 vertical and horizontal spectra (Fig. 19.8). These ground motions were then scaled to yield different levels of PGA. The horizontal spectrum was normalized to the lowest seismic level of 0.1 g PGA level and the vertical spectrum was then normalized to 0.09 g PGA level. The test sequence applied is summarized in Table 19.2. All the tests were completed in 2 months. Following some pre-tests conducted for control and frequency determination purposes (runs 1–15 not shown in Table 19.2), three tri-axial seismic tests at 0.024, 0.05 and 0.075 g were carried out. Afterwards, uniaxial tests at 0.2 g in the two horizontal tests were followed by 0.4, 0.6 and 0.8 g bi-axial seismic tests to conclude the test program. White noise tests were performed at the end of the seismic tests to determine frequencies of the mock-up. After completing the intended test sequence, two post tests at 0.2 and 0.3 g were performed to study the effect of aftershocks.

Table 19.2 Test sequence

Run	Type	Excitation direction			PGA of reference signals		
		X	Y	Z	X	Y	Z
16	Seism	✓	✓	✓	0.025 g	0.025 g	0.022 g
17	Seism	✓	✓	✓	0.05 g	0.05 g	0.045 g
18	Seism	✓	✓	✓	0.075 g	0.075 g	0.068 g
19	Seism	✓	✓		0.1 g	0.1 g	
20	Seism	✓	✓		0.2 g	0.2 g	
21	White noise	✓			0.09 g		
22	White noise		✓			0.08 g	
23	Seism	✓	✓		0.2 g		
24	Seism		✓			0.2 g	
25	Seism	✓	✓		0.4 g	0.4 g	
26	White noise	✓			0.1 g		
27	White noise		✓			0.08 g	
28	Seism	✓	✓		0.6 g	0.6 g	
29	White noise	✓			0.1 g		
30	White noise		✓			0.08 g	
31	Seism	✓	✓		0.8 g	0.8 g	
32	White noise	✓			0.1 g		
33	White noise		✓			0.08 g	
34	Seism	✓	✓		0.2 g	0.2 g	
35	Seism	✓	✓		0.3 g	0.3 g	
36	Seism	✓	✓		0.3 g	0.3 g	

19.4 Results and Observations

19.4.1 Test Results

As explained before, a comprehensive instrumentation was used to measure global as well as local response of the mock-up. In this paper, only some of these measurements that include modal frequencies, deformations at the thermal break elements

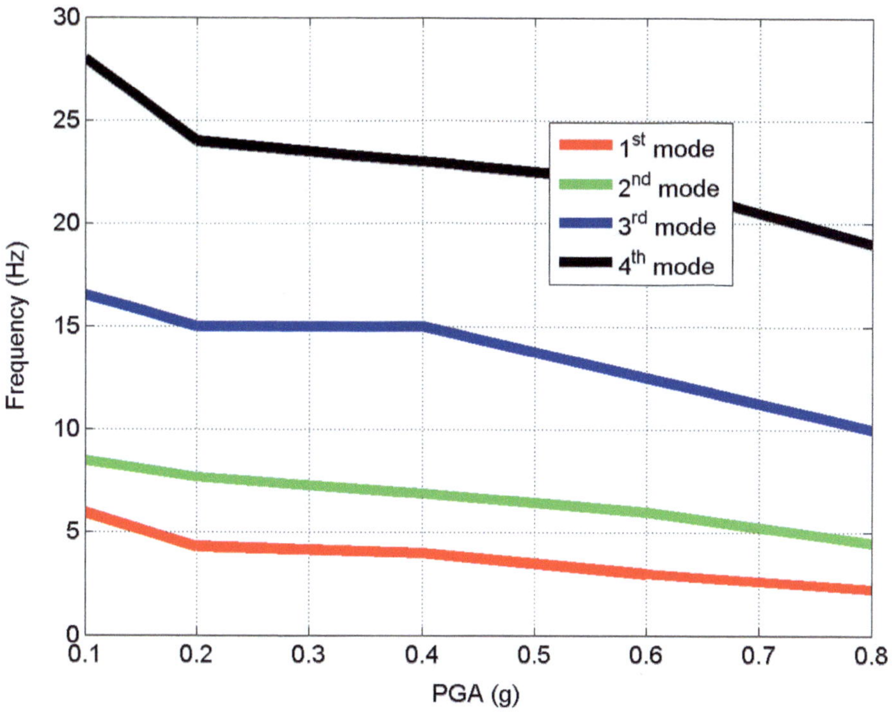

Fig. 19.9 ENISTAT specimen eigen-frequencies in the first four modes after each seismic test

Table 19.3 Experimental eigen-frequencies of mock-up in the first four modes after each test

Level of the previous seismic test (g)	0	0.2	0.4	0.6	0.8
Mode 1 (Hz)	5.9	4.3	4	3	2.3
Mode 2 (Hz)	8.5	7.7	6.9	6	4.5
Mode 3 (Hz)	16.5	15	15	12.5	10
Mode 4 (Hz)	28	24	23	22	19

and accelerations over the mock-up are presented. The change of the specimen's frequency with the seismic tests applied is shown in Fig. 19.9 and tabulated in Table 19.3 for four modes of vibration. The vibration mode shapes, as determined from the numerical analyses, are coupled modes as shown in Fig. 19.10. The damage on the specimen is apparent in terms of softening in the specimen at each mode.

One of the important points to note is that in AZALEE shaking table there is interaction between actuators and the specimen leading to the applied signals being generally different from the ones measured on the shaking table. Due to this control fact that is common to all shaking tables, the intended ground motion could not be exactly applied as can be seen in Fig. 19.11, where both signals are compared. The difference is larger at small excitation levels but diminishes as the excitation level increases.

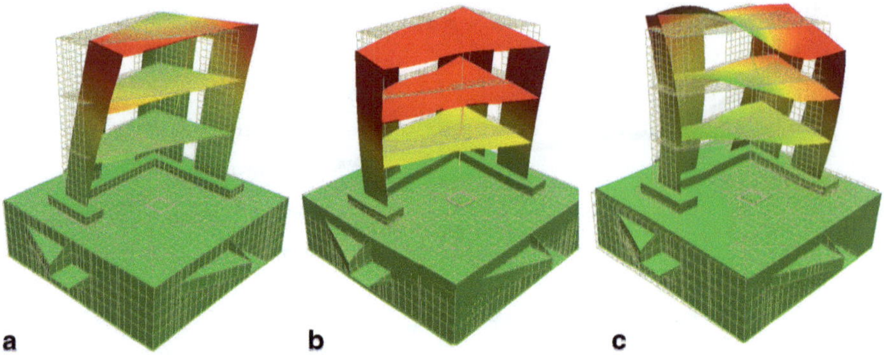

Fig. 19.10 First three mode shapes from numerical analyses

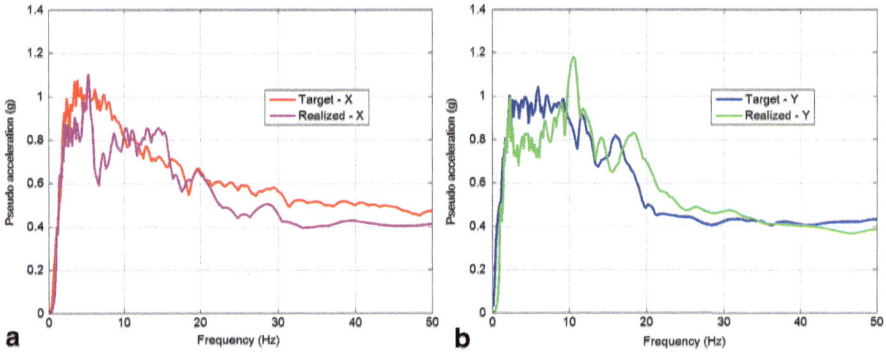

Fig. 19.11 Comparisons of applied and measured signals on the shaking table

The steel bars of some of thermal break elements were equipped with strain gages. Strain measurements in thermal break elements indicated that the strains were still in the linear range. The maximum absolute value of the strains recorded for each seismic test is presented in Fig. 19.12. Although there is a significant increase in strain with the seismic excitation level, these strains are below yield level. The first vertical resonant frequency of each floor was also measured using the transfer function between accelerometers on the floor and accelerometers on walls closed to the floor. This frequency is driven by damage in the area of the connection between the floor and the wall. Comparison of the decrease of this frequency for each floor provides insight into the performance of the thermal elements. These frequencies are shown in Fig. 19.13. The thermal break elements cut the mechanical continuity of concrete at the wall/slab connection of the second floor. This is the main difference between the first and the second floor, as they have the same design and the same additional masses. This explains the close value of the initial frequency of the first and the second floor (41 and 40 Hz). After the 0.2 g seismic test, the measured frequency of the first floor did not change significantly, while the frequency of the

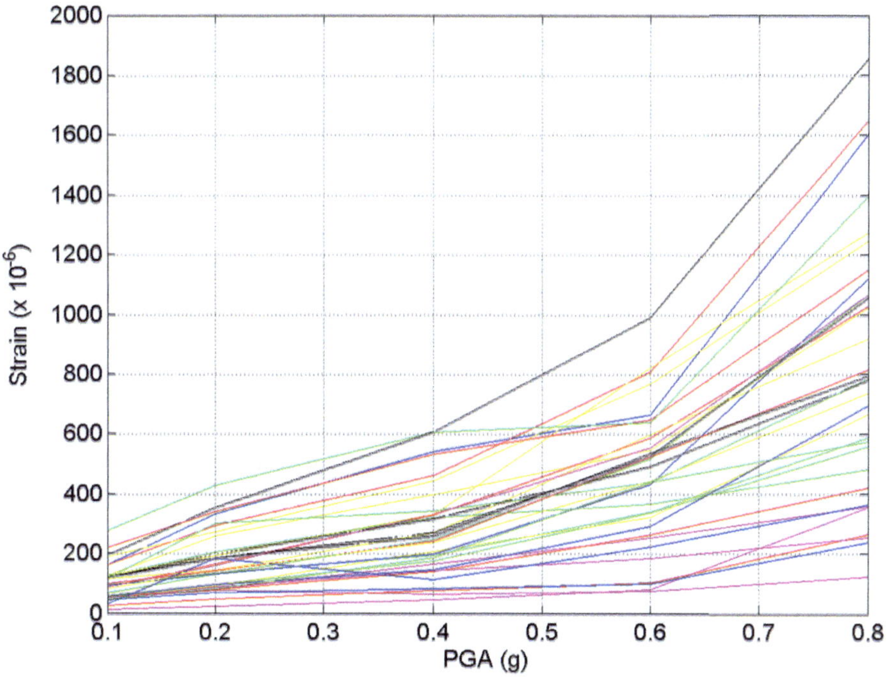

Fig. 19.12 Maximum strains measured on thermal break elements during the tests

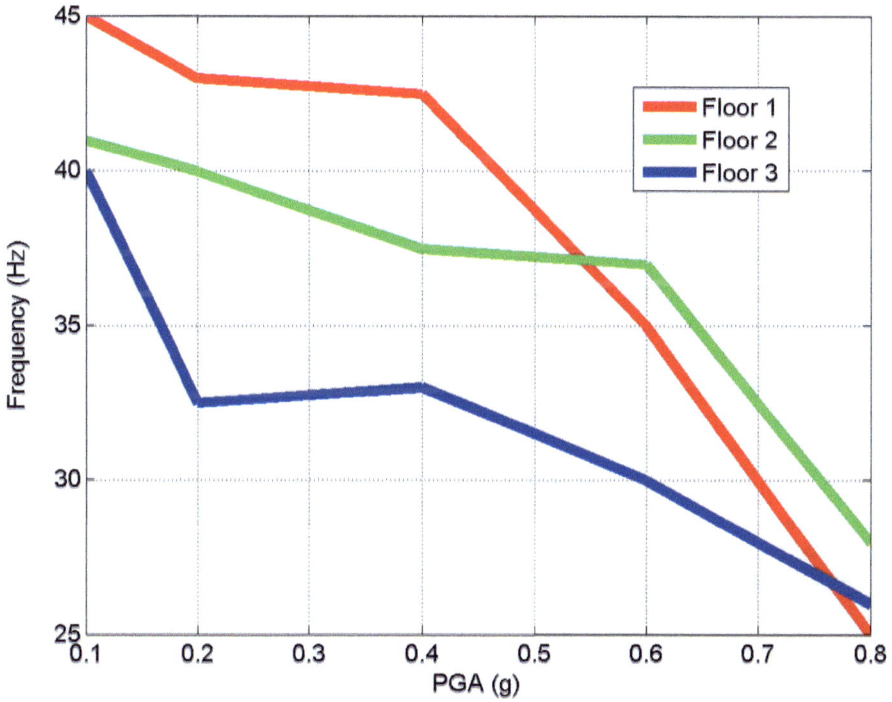

Fig. 19.13 Fundamental frequency of floors obtained after each seismic test

second floor decreased from 40 to 32 Hz (even if the acceleration was higher at the first floor during the test). It is likely that the frequency change is due to thermal break elements flexibility. Nevertheless, in higher seismic level tests, measured frequencies of the first and the second floor converged to the same value (26 Hz). Thermal break elements seem to create an initial decrease of frequency but do not affect the floor frequency at higher seismic levels.

The maximum and minimum values of the measured acceleration and displacement are given at different points on the first and second floors in Table 19.4. Figure 19.14 displays time histories at some selected points for 0.1 and 0.8 g runs. Points A, B, C and D are located at each corner of the floor, whereas points K and L show intermediate points (Fig. 19.6). Aside from the significant increase in the acceleration and displacement with the increase in seismic excitation, the torsional behavior is evident. The maximum displacement is measured at point B in the Y direction and at point D in the X direction, indicating the flexible sides on the floor. Point A, being closer to the center of rigidity, has smaller displacements. The maximum relative displacement between the wall and the slab at the second floor is around 0.6 mm at 0.8 g, which is not significant.

19.4.2 Damage Observations

Observations were made by assessing the mock-up after every seismic test. During the first three tests, i.e. up to 0.4 g, no significant damage was observed in the structural members, except for minor hairline cracks on the spandrel beams. At 0.6 g test, more cracks in the beams were observed without any major crack in the walls (Fig. 19.15). Strain measurements in Rutherma steel sections indicated that the strains were within the elastic range (Fig. 19.12). During the 0.8 g test, separation of the shear wall member (wall 1) on the flexible side from the foundation was observed (Fig. 19.16). All cracks were marked to illustrate their geometry and extent. The damage was observed to concentrate mostly on the first floor.

After the test, major cracks that are mainly confined on wall 1 in the first floor were observed. The most significant damage occurred in the corners where longitudinal reinforcements were exposed (Fig. 19.16). No damage on the Rutherma breakers was observed.

Similar but less significant damage occurred on Wall 2 at the first floor level, where cracks extending towards the foundation were also observed (Fig. 19.17). The connection between the mock up foundation and wall 1, and part of wall 2, was made with steel bars.

Regarding floors, most of cracks were observed in the first floor despite very low damage in the second floor (Fig. 19.18). Some damage in terms of cracks was observed between wall 2 and the beam connecting to it, these cracks were generally observed in the first floor level (Fig. 19.18).

Table 19.4 Maximum and minimum values measured during the tests (accelerations are in g, displacements are in mm)

Name	Max.	Min.	Max.	Min.	Max.	Min.
	0.1 g		0.2 g		0.8 g	
Axtab	0.10	−0.12	0.21	−0.20	0.96	−0.79
Aytab	0.12	−0.12	0.18	−0.19	0.82	−0.72
Aztab	0.02	−0.03	0.04	−0.07	0.30	−0.19
Ac10xA	0.16	−0.16	0.25	−0.26	1.13	−0.85
Ac10yA	0.14	−0.13	0.25	−0.24	1.02	−1.07
Ac10zA	0.11	−0.13	0.17	−0.18	0.42	−0.89
Ac20xA	0.27	−0.24	0.41	−0.37	1.27	−1.21
Ac20yA	0.21	−0.22	0.41	−0.38	1.36	−1.34
Ac20zA	0.11	−0.13	0.18	−0.17	0.47	−0.71
Ac30xA	0.36	−0.32	0.55	−0.48	1.46	−1.48
Ac30yA	0.31	−0.34	0.60	−0.52	1.92	−1.61
Ac30zA	0.11	−0.13	0.17	−0.17	0.52	−0.71
Ac10yD	0.14	−0.13	0.26	−0.23	1.07	−1.08
Ac10xD	0.20	−0.20	0.41	−0.34	1.40	−1.21
Ac10zD	0.07	−0.07	0.15	−0.16	0.54	−0.69
Ac20yD	0.21	−0.22	0.39	−0.38	1.38	−1.33
Ac20xD	0.37	−0.30	0.69	−0.62	1.59	−1.96
Ac20zD	0.08	−0.09	0.16	−0.21	0.67	−0.80
Ac30yD	0.28	−0.33	0.55	−0.53	1.70	−1.58
Ac30xD	0.50	−0.48	0.92	−0.94	2.74	−3.25
Ac30zD	0.09	−0.08	0.16	−0.20	0.71	−0.86
Dx10A	5.69	−4.59	10.45	−8.39	43.52	−32.90
Dx20A	5.43	−4.95	10.47	−9.87	48.28	−36.93
Dx30A	5.96	−4.91	11.84	−10.57	52.12	−41.90
Dy10A	5.58	−4.40	10.83	−8.86	43.99	−39.02
Dy20A	6.02	−4.43	11.15	−9.07	45.34	−42.43
Dy30A	6.01	−4.52	11.34	−9.22	47.38	−45.69
Dy10B	6.63	−4.15	12.12	−10.11	44.59	−48.36
Dy20B	8.68	−7.33	15.31	−13.84	57.85	−62.01
Dy30B	10.57	−9.45	18.25	−17.41	77.38	−78.06
Dx10D	5.81	−4.87	11.15	−10.34	49.99	−41.47
Dx20D	7.17	−6.07	13.21	−13.99	61.81	−58.47
Dx30D	9.12	−6.82	16.03	−17.64	74.75	−73.61

Axtab, *Aytab*, *Aztab* Table acceleration in X, Y and Z directions respectively
Ac10xA Acceleration of point A at level 1 in X direction
Dx10 A Displacement of point A at level 1 in X direction

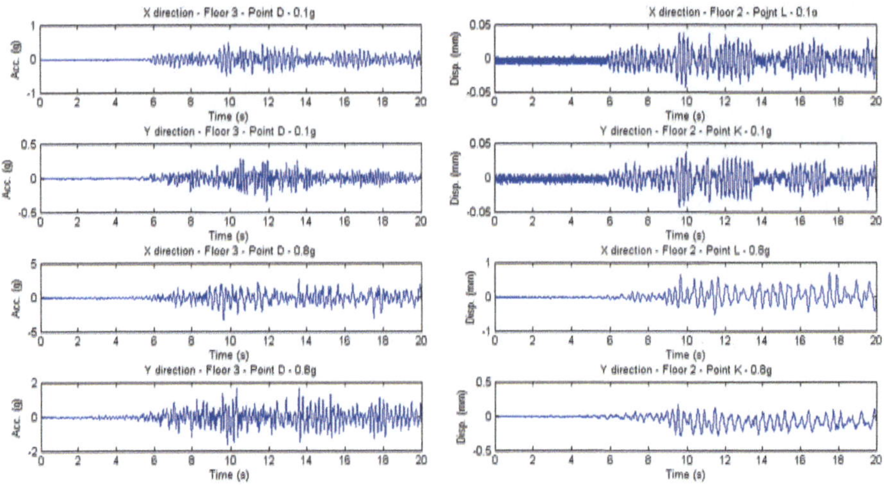

Fig. 19.14 Acceleration and displacements measured at various points on the mock-up

Fig. 19.15 Damage observations after 0.6 g tests

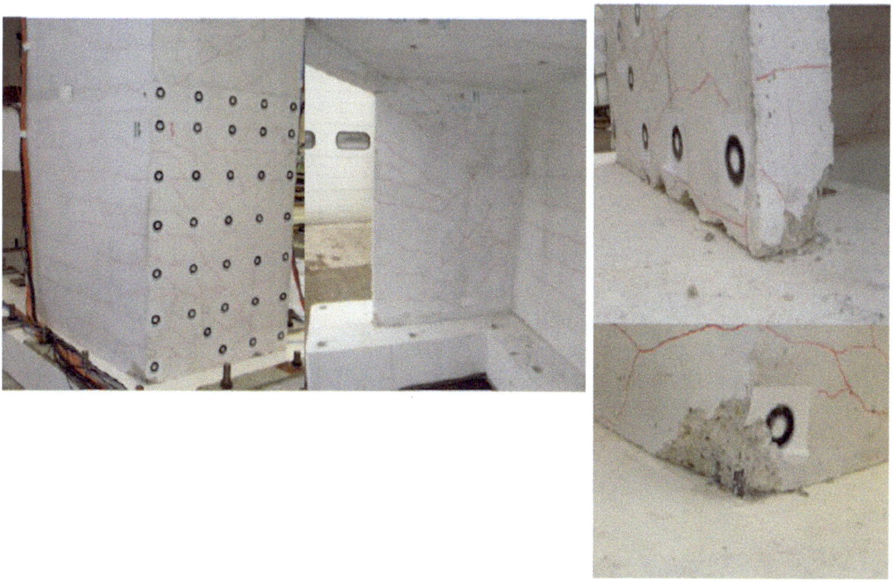

Fig. 19.16 Damage observed on wall 1 after 0.8 g tests

Fig. 19.17 Damage observed on wall 2 after the tests

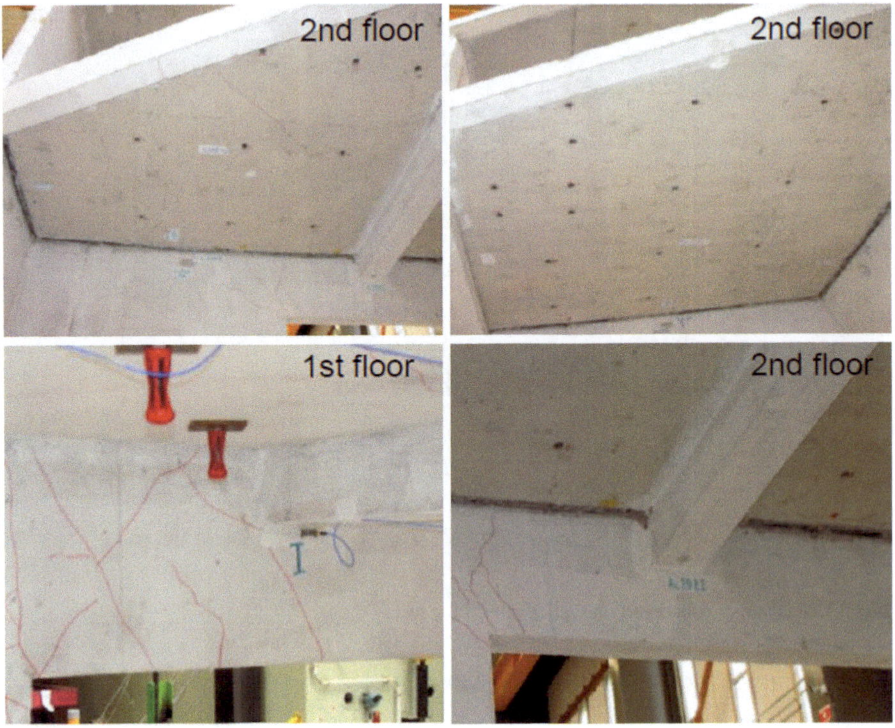

Fig. 19.18 Damage on slabs and beams

19.5 Conclusions

A torsionally irregular RC building model with RC walls (ENISTAT specimen) was tested on the AZALEE shaking table at the SEMT/EMSI laboratory in CEA. The mock-up, representative of an office building, was designed according to EC8. One of the main objectives of the study was to test Rutherma members (thermal break elements) under seismic excitation. The Ruthermas, manufactured specifically for this project, were placed between the shear walls and the slab at the second floor level, and were instrumented for monitoring their performance. Three synthetic ground motions compatible with EC8 type 1 spectrum for horizontal and vertical motions were generated. These ground motions were scaled at PGA levels of 0.1, 0.2, 0.4, 0.6 and 0.8 g and were applied to the mock-up consecutively. Before each test, low level white noise tests were carried out to determine frequencies of the mock-up. The mock-up responded almost linearly at the seismic levels up to 0.4 g, where no significant damage was observed in the structural members, except for minor hairline cracks on the spandrel beams. At the 0.6 g test, more cracks in beams were observed without any major cracks in walls. Strain measurements in Rutherma steel sections indicated that strains were within the elastic range.

Heavy damage occurred during the 0.8 g test, separation between the shear walls and the foundation was observed. The foundation almost fully disconnected from the shear wall.

No damage on Rutherma elements was observed. Thermal break elements seem to create an initial frequency decrease of the floor but did not affect the floor frequency at higher seismic levels. The results based on the ENISTAT specimen and the test protocol indicated that the thermal break elements can resist strong bi-axial seismic excitations (0.8 g in this case) without increasing the vulnerability of the building.

The use of digital image correlation techniques gives a large amount of data to quantify at the structural scale. Experimental data on crack propagation and growth are crucial for quantification of damage explaining nonlinear effects in 3D response of wall elements. The processing of the stereo correlation digital data is underway.

Numerical analyses are ongoing and are expected to provide tools for the assessment of the damage observed and for validation of simplified nonlinear analysis tools based on beams, shells and macro-elements so as to handle the parametric analysis needed for the building vulnerability assessment. A special emphasis is given to the modeling of thermal break elements that are being tested numerically.

Acknowledgements The research leading to these results received founding from the European Community's Seventh Framework Programme [FP7/2007–2013] for access to the TAMARIS laboratory in CEA/Saclay, France under grant agreement n 227887.

References

European Committee for Standardization (2004) Eurocode 8: design of structures for earthquake resistance Part 1: general rules, seismic actions and rules for buildings. Eurocode 8, Brussels

Juster-Lermitte S, Chaudat T, Courtois A (2009) SMART 2008 project experimental tests on a reinforced concrete building subject to torsion. 20th SMIRT conf., Finland

Kotronis P, Ragueneau F, Mazars J (2005) A simplified modelling strategy for R/C walls satisfying PS92 and EC8 design. Eng Struct 27:1197–1208

Lu Y (2002) Seismic behaviour of multistorey RC wall-frame system versus bare ductile frame system. Earthq Eng Struct Dyn 31(1):79–97

Oesterle RG, Fiorato AE, Johal LS, Carpenter JE, Russell HG, Corley WG (1976) Earthquake resistant structural walls—tests of isolated walls. Report to National Science Foundation, Portland Cement Association, Skokie, Illinois, p 315

Palermo D, Vecchio FJ (2002) Behavior of three-dimensional reinforced concrete shear walls. ACI Struct J 99(1):81–89

Reynouard JM, Fardis MN (2001) Shear walls structures. ECOEST—ICONS Report No 5

Schöck (2010) Design of the test specimen for ENISTAT—RUTHERMA DF Thermal break elements

SMART (2013) Project web site. http://www.smart2013.eu. Accessed May 2012

Vallenas MV, Bertero VV, Popov EP (1979) Hysteretic behaviour of reinforced concrete structural walls. EERC Report 79/20. Earthquake Engineering Research Center, University of California, Berkeley

Chapter 20
Assessment of the Seismic Response of Concentrically-Braced Steel Frames

Brian M. Broderick, Jamie Goggins, Darko Beg, Ahmed Y. Elghazouli, Philippe Mongabure, Alain Le Maoult, Alan Hunt, Suhaib Salawdeh, Primoz Moze, Gerard O'Reilly and Franc Sinur

20.1 Introduction

Concentrically braced frames (CBFs) offer an economical and efficient form of earthquake resistance for steel structures. For small, more frequent earthquakes, they provide sufficient stiffness and strength to meet serviceability requirements. For larger earthquakes, adequate seismic design and detailing can ensure a dissipative response that limits structural response. Diagonal bracing members CBFs are critical elements that during severe seismic loading experience repeated cycles involving yielding in tension and member buckling in compression. The performance of these members depends on various factors, including local slenderness, global slenderness, material yield strength, section shape and end restraint (Elghazouli 2003).

Due to the difficulty in accurately modelling their complex seismic response, numerous experimental studies have been carried out to study the cyclic inelastic behaviour of bracing members. Early studies examined the load-displacement

B. M. Broderick (✉) · A. Hunt
Department of Civil, Structural and Environmental Engineering,
Trinity College, Dublin, Ireland
e-mail: bbrodrck@tcd.ie

J. Goggins · S. Salawdeh · G. O'Reilly
College of Engineering and Informatics, National University of Ireland, Galway, Ireland

D. Beg · P. Moze · F. Sinur
Faculty of Civil and Geodetic Engineering, University of Ljubljana, Ljubljana, Slovenia

A. Y. Elghazouli
Department of Civil and Environmental Engineering, Imperial College, London SW7 2AZ , UK

P. Mongabure · A. Le Maoult
CEA Saclay, DEN/DANS/DM2S/SEMT/EMSI, 91191 Gif sur Yvette Cedex, France

F. Taucer, R. Apostolska (eds.), *Experimental Research in Earthquake Engineering*,
Geotechnical, Geological and Earthquake Engineering 35,
DOI 10.1007/978-3-319-10136-1_20

hysteretic response which was shown to be most strongly influenced by global slenderness (Popov and Black 1981). Slender members lost compressive resistance more rapidly than stocky members, resulting in fewer inelastic response cycles and less energy dissipation. More recently, attention has shifted to examination of the factors influencing the fracture life of bracing members. Through experimental testing, both global and local slenderness were found to influence fracture life (Tremblay 2002), and empirical expressions for the fracture life and ductility capacity of hollow section bracing members have been proposed (Goggins et al. 2006; Nip et al. 2010).

Gusset-plate connections employed in CBFs in which out-of-plane brace buckling is envisaged must be designed to accommodate the large brace end-rotations experienced at large storey drifts. This implies the formation of a stable ductile plastic hinge within the gusset plate. The design details must also prevent gusset plate buckling in compression or yielding in tension (AISC 2005). However, current design guidance and practice can lead to the use of over-sized gusset plates which reduce the seismic performance of the brace members themselves. More recently, balanced gusset plate detailing rules have been recommended, which result in more efficient connection designs while improving the seismic performance of the CBF overall (Lehman et al. 2008).

20.2 Experimental Aims and Methodology

20.2.1 Research Objectives

The overall aim of the BRACED project was to improve the understanding of, and numerical modelling methods for, steel CBFs subjected to seismic loading; and to assess the implications for design methods and guidance. This is achieved by addressing three principal objectives:

- A large body of research has generated theoretical and empirical formulae to predict key brace member performance parameters for earthquake response. These include the stiffness, resistance and ductility of individual brace members and whole CBFs. The BRACED project examines the validity of these predictive formulae under realistic dynamic response conditions using shake table testing of model CBFs, supported by complementary quasi-static cyclic testing and numerical modelling.
- More recent research has shown that standard practice in brace connection design can be improved by using alternative geometrical details. This project assesses the influence of different gusset plate designs on the dynamic response of CBFs to earthquake ground motion.
- Finite element software has been used in engineering research and practice to model the global response of braced frames and the local response of brace members and brace connections. In this project, OpenSees modelling techniques

proposed for this form of structure will be investigated, and the software's ability to model the earthquake response of CBFs will be validated through correlation with test results.

20.2.2 Methodology

The BRACED Project was initiated as part of the Transnational Access programme offered by the European Commission's Seventh Framework Programme (FP7) project SERIES (Seismic Engineering Research Infrastructures for European Synergies). The main experimental phase was carried out at the TAMARIS Laboratory in the Laboratoire d'Etudes de Mécanique Sismique (EMSI) at CEA Saclay, France. The tests were carried out on the AZALEE shake table. The AZALEE platform has an area of 6×6 m and can accommodate test masses up to 100 t. It is capable of triaxial excitations up to 1.0 g, offering six degrees of freedom and maximum longitudinal and lateral displacement of ± 125 mm.

These series of shake table tests investigated the ultimate behaviour of CBFs through shake table testing of a model test frame incorporating pairs of brace specimens with different brace member and gusset plate characteristics. The seismic performance of such structures during strong earthquakes can be affected by the limited ductility capacity of brace members under low cycle fatigue conditions. The experimental programme was designed to assess whether existing models for evaluating the ductility capacity of hollow section bracing members (which were generally developed using quasi-static cyclic test results) capture observed behaviour during dynamic response conditions. Hence the test programme examines brace members with different cross-section sizes, and therefore different global and local slendernesses. Recent proposals for improved design and detailing guidance for gusset plate connections in dissipative CBFs may extend the fatigue life of hollow brace section members. Thus, the brace-gusset plate test specimens compare the performance of specimens designed using conventional methods with the proposed balanced design method.

The test frame (or 'mock-up') used for the BRACED experiments on the Azalee shake table was designed to facilitate the testing of multiple pairs of brace-gusset plate specimens, by allowing the specimens to be exchanged between experiments. The brace member and connection details were varied between experiments to investigate the range of global and local member slenderness found in European design practice, and to assess the effect of conventional and novel gusset plate designs. In each experiment, three separate earthquake tests were performed with table excitations scaled to produce elastic response, brace buckling and/or yielding and brace fracture. The principal outcomes included measurements of the displacement ductility capacity of the brace specimens; an evaluation of the influence of gusset plate detailing on connection ductility; observations on the contributions of brace and connection yielding to overall inelastic deformation of CBFs; measurements of equivalent viscous damping in CBFs; assessment and improvement of Eurocode 8 design guidance for CBFs; and validation of numerical models.

20.2.3 Shake Table Experimental Programme

The shake table experimental programme investigated the performance of CBFs at different levels of seismic excitation, including ultimate behaviour, through shake table testing of a model frame incorporating a pair of brace-gusset plate specimens. To address project objectives, three different test parameters were examined in these tests: brace cross-section size; brace connection configuration and gusset plate design. The following notation is used to identify the properties of the brace-gusset plate specimens examined in individual experiments:

Brace Section Size:
S1 80 × 80 × 3.0 SHS
S2 100 × 50 × 3.0 RHS
S3 80 × 40 × 3.0 RHS
S4 60 × 60 × 3.0 SHS
Connection Configuration:
CA Gusset connection to beam and column flange
CB Gusset connection to beam flange only
Gusset Plate Design:
G1 Conventional design with Standard Linear Clearance (SLC)
G2 Balanced design with Elliptical Clearance (EC)

The schedule of tests presented in Tables 20.1 and 20.2 was designed to address all of the above test parameters. In this table, the values for non-dimensional slenderness, λ_{nom}, are calculated assuming actual member cross-section areas, pinned-pinned boundary conditions with bending about the minor axis (K = 1.0) and both nominal (f_y = 235 MPa for braces; f_y = 275 MPa for gusset plates) and actual material yield strengths. The λ_{nom} values shown were selected to cover the range of brace slenderness allowed by Eurocode 8. The b/t values imply Class 1 cross-sections, as required for dissipative brace members, but the higher values are close to the boundary with Class 2 to capture the influence of local buckling on brace ductility. The β_{ww} parameter (Lehman et al. 2008) represents the ratio of the plastic tension resistances of the brace member and gusset plates. Specimens with conventionally designed gusset plates have low β_{ww} values, implying that gusset yielding will not occur, whereas those designed using the balanced design approach have higher β_{ww} values, implying that gusset yielding may occur.

A pair of identical brace-gusset plate specimens is tested in each experiment. A total of 12 experiments are included in the programme, and the test frame is designed to allow the brace-gusset plate specimens to be exchanged between experiments. To this end, the gusset plates are welded to flange plates which are bolted to the flanges of the beam and column members. All specimens were tested under uniaxial seismic excitation using the same earthquake record scaled to three different levels. The signal employed is a natural ground record from the PEER database, recorded in Imperial Valley (California, USA) during the 1940 earthquake. In each test, three levels of earthquake were examined: (i) low-level with elastic response, (ii) medium-level with brace buckling and yielding, and (iii) high-level with brace fracture. These are represented by earthquake events with

Table 20.1 Specimen properties: brace member

Test	Specimen	Section size	Brace length (mm)	Characteristic strength		Yield capacity		Non-dimensional slenderness		Buckling capacity	
				R_{eH} (MPa)	R_m (MPa)	$N_{pl,Rd}$ (kN)	$N_{pl,ck}$ (kN)	λ_{nom}	λ_{ck}	$N_{b,Rd}$ (kN)	$N_{b,ck}$ (kN)
1	S1-CA-G1	80 × 80 × 3.0	2413	372.5	437.0	214.8	340.8	0.83	1.04	167.6	217.3
2	S3-CA-G1	80 × 40 × 3.0	2427	384.3	430.5	158.4	259.0	1.59	2.03	53.6	56.3
3	S4-CA-G1	60 × 60 × 3.0	2425	347.5	404.0	158.4	234.2	1.11	1.35	93.0	103.4
4	S2-CA-G1	100 × 50 × 3.0	2413	341.5	377.7	200.7	291.6	1.24	1.49	102.0	110.0
5	S1-CA-G2	80 × 80 × 3.0	2502	337.8	389.0	214.8	308.7	0.86	1.03	163.6	199.7
6	S2-CA-G2	100 × 50 × 3.0	2509	341.7	380.2	200.7	291.8	1.28	1.55	96.2	102.9
7	S3-CA-G2	80 × 40 × 3.0	2504	370.5	429.7	158.4	249.7	1.64	2.05	50.8	53.0
8	S1-CB-G1	80 × 80 × 3.0	2395	336.5	388.8	214.8	307.6	0.82	0.98	168.4	208.7
9	S2-CB-G1	100 × 50 × 3.0	2395	340.0	377.8	200.7	290.4	1.23	1.48	103.1	111.3
10	S4-CB-G2	60 × 60 × 3.0	2437	347.5	404.0	158.4	234.2	1.12	1.36	92.4	102.6
11	S2-CB-G2	100 × 50 × 3.0	2433	341.5	377.7	200.7	291.6	1.25	1.50	100.7	108.4
12	S3-CB-G2	80 × 40 × 3.0	2420	370.5	429.7	158.4	249.7	1.58	1.99	53.9	56.4

Table 20.2 Specimen properties: gusset plate connection

Test	Specimen	Gusset type	t_p $\mathrm{t_p}$ (mm)	Characteristic strength		Yield capacity		Balance factor		Buckling capacity	
				R_{eH} (MPa)	R_m (MPa)	$N_{pl,\,Rd}$ (kN)	$N_{pl,\,ck}$ (kN)	$\beta_{ww,\,Rd}$	$\beta_{ww,\,ck}$	$N_{b,\,Rd}$ (kN)	$N_{b,\,ck}$ (kN)
1	S1-CA-G1	SLC	12	369.0	506.7	835.6	1121.2	0.33	0.28	737.5	941.4
2	S3-CA-G1	SLC	8	337.0	490.0	557.1	682.6	0.36	0.35	412.6	461.6
3	S4-CA-G1	SLC	8	337.0	490.0	513.1	628.7	0.39	0.36	370.5	411.4
4	S2-CA-G1	SLC	12	369.0	506.7	901.6	1209.8	0.28	0.23	804.8	1032.8
5	S1-CA-G2	EC	5	336.0	444.7	348.2	425.4	0.79	0.73	302.6	356.3
6	S2-CA-G2	EC	4	388.0	488.0	300.5	424.0	0.85	0.68	260.3	341.5
7	S3-CA-G2	EC	4	388.0	488.0	278.5	393.0	0.72	0.62	233.5	299.9
8	S1-CB-G1	SLC	12	369.0	506.7	835.6	1121.2	0.33	0.27	747.6	960.6
9	S2-CB-G1	SLC	12	369.0	506.7	901.6	1209.8	0.28	0.24	798.1	1020.2
10	S4-CB-G2	EC	4	388.0	488.0	256.5	361.9	0.79	0.64	126.0	134.6
11	S2-CB-G2	EC	4	388.0	488.0	300.5	424.0	0.85	0.69	140.7	149.5
12	S3-CB-G2	EC	4	388.0	488.0	278.5	393.0	0.72	0.62	109.1	114.1

50, 10 and 2% probability of exceedance in 50 years respectively. Low-level white noise excitation was also applied before and after each earthquake level to monitor the evolution of elastic properties with brace member damage.

The spectrum of the input seismic signal was normalized to the lowest seismic level expected in the test sequence, namely a PGA of 0.1 g. This was considered to be the 0 dB level, and the signal was then amplified for different excitation levels. The original signal was filtered at low frequency to limit the maximum displacement under the ± 100 mm limit value for the AZALEE table. The high pass filter eliminated frequencies under 0.7 Hz. This filtering was also necessary to obtain a null value of table displacement at the end of the test. The signal duration was 40 s with five seconds at null value added at the beginning to provide time to check the correct triggering of the data acquisition system.

20.3 Test Frame and Specimens

20.3.1 Test Frame

The test frame was designed as a dedicated single-storey model CBF structure capable of accommodating the full range brace and gusset-plate connection specimens set out in Tables 20.1 and 20.2. These specimens were designed so that they could be tested to failure within the capacity of the shaking table. The following specific requirements drove the primary test frame geometry and strength specimen design:

- The test frame should have a realistic storey height and natural period
- The mass supported by the test frame should not exceed 50 t
- The PGA required to fracture the brace specimens should not exceed 1.0 g
- The displacement ductility demand required for brace fracture ($\mu_\Delta > 4$) should be accommodated by the test frame
- Brace lengths shall should be realistic; non-dimensional slenderness should be within or close to Eurocode 8 limits
- Braces members should possess class 1 cross-sections, but small b/t and d/t ratios should be avoided to ensure that local buckling and fracture is observed
- The brace- and gusset plate specimens should be easily exchanged between tests
- Brace connections to beam and column (CA) and beam only (CB) should be accommodated
- Two gusset plate designs should be included: conventional design (G1) with standard linear clearance (SLC) plastic hinge detailing, and balanced design (G2) with elliptical clearance (EC) plastic hinge detailing

These objectives were completed whilst remaining within the AZALEE shake table constraints described in the previous section. Figures 20.1 and 20.2 show the resulting design of the test frame. The lateral resistance of the frame was provided by a pair of brace specimens in Frame B which were positioned in the same plane

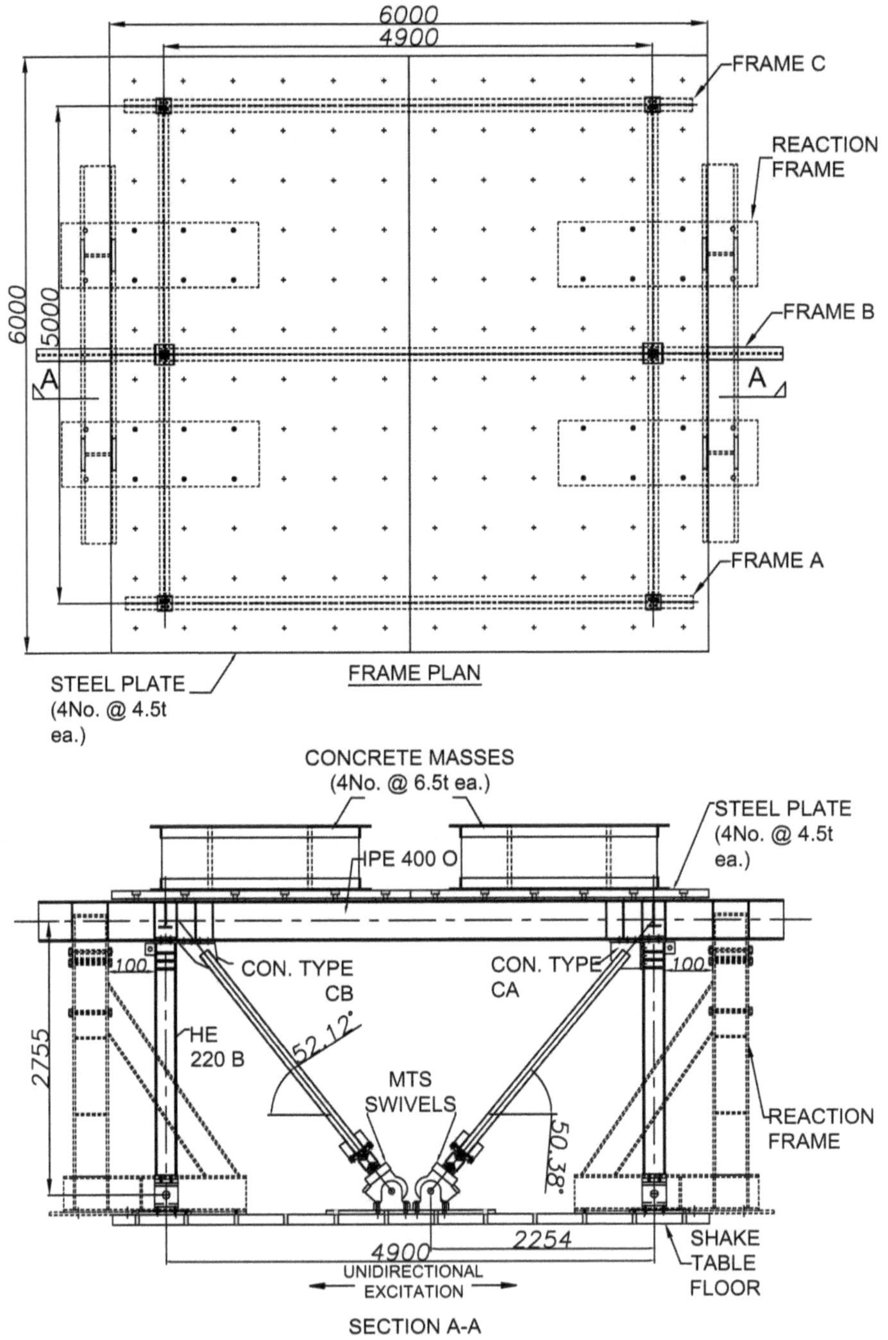

Fig. 20.1 Plan and elevation of BRACED test frame on Azalee platform. *CA* and *CB* connections shown for illustration, identical brace specimen pairs were used in all tests

Fig. 20.2 Test frame without added masses and MTS swivel bearing with load cells

to prevent any significant torsional response. The test frame is symmetrical at either side of Frame B. Two additional unbraced frames (Frame A and C) were located on either side of the CBF model to provide lateral stability; lateral beams support the added mass of the frame. All column members in Frames A and C were pinned at the top and bottom ends. Columns in Frame B were pinned at their bottom ends and bolted connected at their top ends to the primary beam by a flush end plate bolted connection.

The principal elements of the test frame are:

- a main beam in Frame B (IPE 400), length 7500 mm
- two columns in Frame B (HE 220 B) supporting the IPE 400
- two columns each on Frames A and C (HE 120 A)
- six beams (IPE 270), forming a square horizontal roof grid, supported by the outer columns and fixed to the main IPE 400 beam in Frame B
- four transverse braces (100 × 20 mm solid cross-section) to provide lateral stability to the frame in the direction perpendicular to Frames A–C
- two MTS swivel bearings (described below) with load cells assemblies
- two brace members, which are the elements tested, mounted in the main plane between the swivels and the IPE 400/HE 220 angle
- two mechanical devices to return the frame to vertical plumb after each test

20.3.2 Brace-Gusset Plate Specimens

Twenty-four identical brace tube pairs were designed using four different cross-section sizes for two connection types (CA and CB) and two gusset plate types (G1 and G2). The yield and buckling capacities (based on nominal and measured steel strengths) are presented in Tables 20.1 and 20.2. The corresponding non-dimensional slenderness is also presented. The area used for the calculation of brace resistance is the product of the Whitmore width and the gusset plate thickness t_p. Two connection design approaches were applied to each of the four brace cross-section sizes and the two connection types, CA and CB (Fig. 20.3). These are conventional designs using

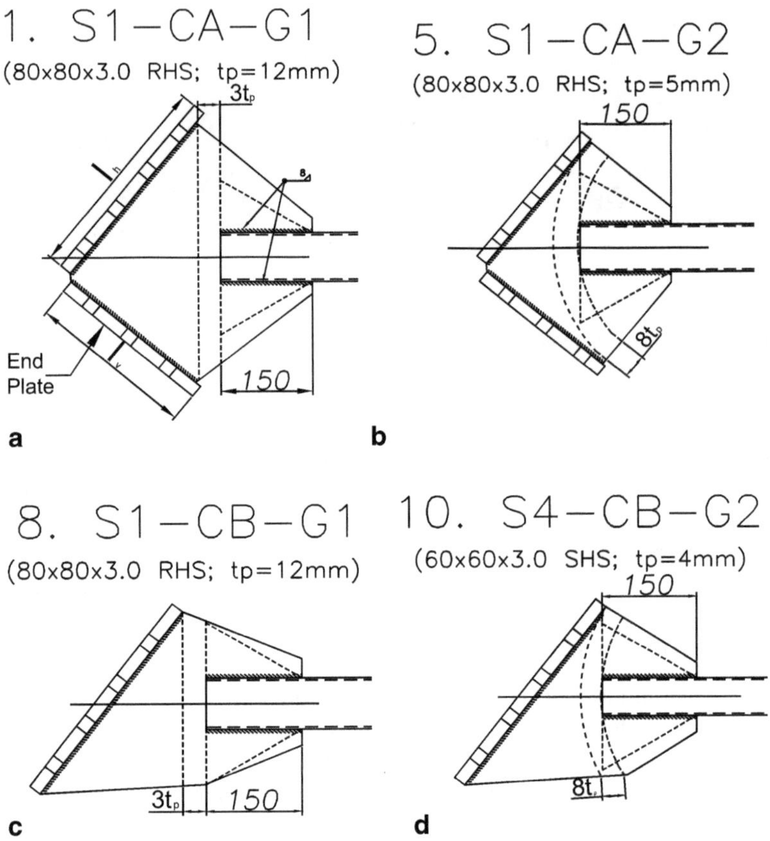

Fig. 20.3 Sample gusset plate connection designs **a** CA-G1 **b** CA-G2 **c** CB-G1 and **d** CB-G2

the standard linear clearance detailing rule for the gusset plate and balanced design using the elliptical clearance detailing rule (Lehman et al. 2008).

The concept of balancing the brace tensile yielding and gusset yielding mechanisms has been encapsulated using the balance factor β_{ww}. The conservative nature of the G1 gusset designs resulted in low (~0.2–0.3) β_{ww} values. A higher range of β_{ww} values (~0.6–0.75) was achieved for the G2 designs by specifying thinner gusset plates and employing the more compact EC detailing rather than the SLC detailing used in the G1 specimens. For all G1 specimens a linear plastic hinge clearance length of $3t_p$ was used while an elliptical clearance zone of thickness $8t_p$ was used for the G2 plastic hinge. Values for expected yield stress ratio R_y were used to calculate $\beta_{ww,\,Rd}$ values, while characteristic material strengths were used to calculate values of $\beta_{ww,\,ck.}$

20.4 Experimental Results

A wide range of response variables were measured in each test, including table and response (roof) accelerations and displacements, brace elongation and axial force, and strains in the brace member and gusset plate. Figure 20.4 presents a sample of some of the recorded results from the final run in Test 4, where brace fracture occurred after approximately 30 s after the test had been completed.

Brace fracture was observed in all tests, either in the third or fourth earthquake exciation run. During the low level test excitations the frame remained elastic with no brace buckling. Brace buckling and yielding occurred in the intermediate level runs, sometimes with large out-of-plane brace buckling deformations, but always limited plastic deformation demand. A fully inelastic response was observed in all high level excitation tests, usually causing fracture in one or both braces. In some tests, an additional failure level earthquake excitation run was added to cause brace fracture. A similar pattern of failure was displayed in most cases: brace buckling in compression led to large out-of-plane brace bending and the formation of a plastic

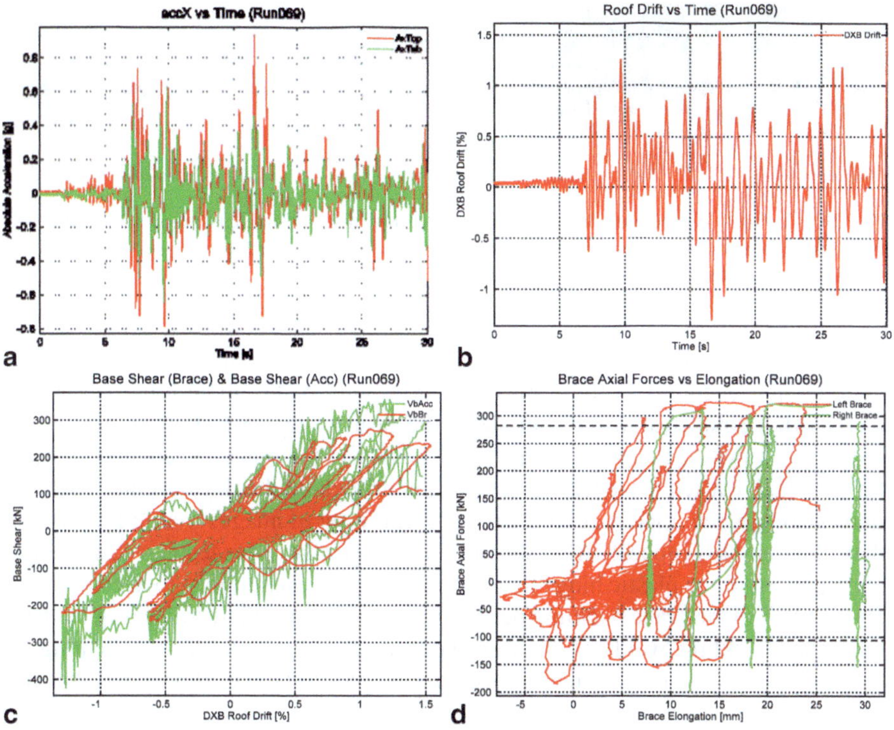

Fig. 20.4 Measured response in Test 4: **a** table and response accelerations; **b** Frame storey drift; **c** Base shear-drift hysteresis; **d** brace force-elongation hysteresis

Fig. 20.5 Images of specimen deformation and fracture. **a** local buckle at brace mid-length with initial tear; **b** incomplete fracture of brace; **c** deformed shape of brace cross-section after local buckling and fracture; **d** yield pattern in gusset plate with SLC detail

hinge close to brace mid-length. During large amplitude displacement cycles, local buckling occurred in these plastic hinges, and as the hinge rotation demand increased, a small tear would initiate at the peak of the local buckle. Upon subsequent reversal of the direction of frame response the brace experienced tension forces which caused these tears to propagate throughout the depth of the cross section causing brace fracture. Figure 20.5 presents a set of images from different tests that illustrate this process. Also shown is a gusset plate after testing, displaying a yield pattern in the plastic hinge that must form in the gusset plate to accommodate large out-of-plane brace buckling deformations. No gusset plate failures (plate fracture, plate buckling, weld or bolt failure) occurred in any of the tests, validating the capacity design and overstrength procedures employed (AISC 2005; CEN 2004).

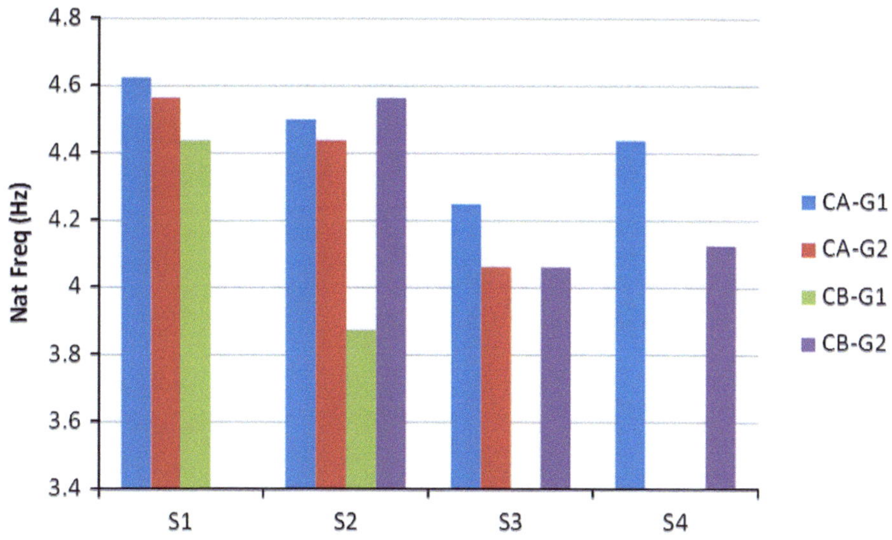

Fig. 20.6 Initial natural frequency of test frames by specimen characteristics

20.4.1 Frame Stiffness

Figure 20.6 compares the initial natural frequency of each test by brace specimen cross-section size and gusset plate connection configuration. The values vary in a relatively narrow range between 3.8 and 4.6 Hz approximately. The larger cross-sections (S1 and S2) tend to display higher natural frequencies than the smaller ones (S3 and S4), but not in every case, and the differences are less than the differences in brace cross-section area. For a given cross-section size, the highest natural frequency (and therefore frame stiffness) tends to be displayed by specimens with the conventional CA-G1 connection configuration, but the reductions in natural frequency observed with other configurations are small.

Figure 20.7 examines the evolution of frame natural frequency in each individual test, grouped by brace-gusset specimen cross-section size. In Fig. 20.7a the natural frequency of the frame with S1 cross-sections does not reduce much between the initial and final runs because the maximum drift demand (in the 10 % probability of exceedance in 50 years earthquake test run) remained less than 1 %, and the brace had not experienced large out-of-plane buckling deformations. In contrast, Fig. 20.7b–d display large reductions in natural frequency after subsequent runs (10 and 2 % probability of exceedance in 50 years where additional failure level earthquake runs were executed). The S2, S3 and S4 cross-section sizes in these specimens lead to larger out-of-plane brace buckling slendernesses, and larger brace buckling deformations.

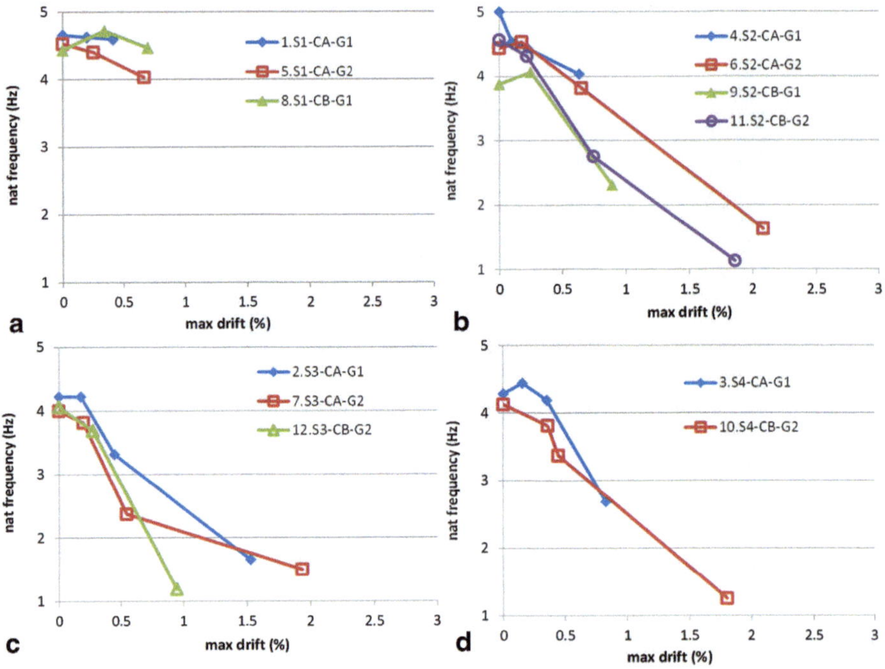

Fig. 20.7 Evolution of natural frequency with previous maximum drift demand in individual tests by brace-gusset plate specimen cross section. **a** S1 brace cross-sections, **b** S2 brace cross-sections, **c** S3 brace cross-sections, **d** S4 brace cross-sections

20.4.2 Frame Drift and Brace Ductility

Figure 20.8 compares the variation in maximum drift demand with the PGA displayed in each test. The results are grouped by brace-gusset plate specimen cross-section size. The larger cross sections (S1 and S2) display a mostly linear relationship between drift and PGA, while the smaller cross sections (S3 and S4) exhibit increasing drift values for higher PGA. This behavior may be expected in short period structures that are subjected to ground excitations substantially greater than those required for initial yield.

Measured maximum frame drift and brace force data can be combined to give a high-level indication of the influence of brace-gusset plate specimen connection type on the global ductility capacity of the test frame. The design of the experimental programme provided pairs of tests in which the specimens differ in only one of the main test variables (brace cross-section, connection type and gusset plate design). Figure 20.9 compares the response of pairs of tests which both employed the same brace cross-section, but different connection details. The plots shown compare the variation in the maximum normalized brace force observed in each run with the maximum drift experienced by the test frame in that run. Three plots compare the

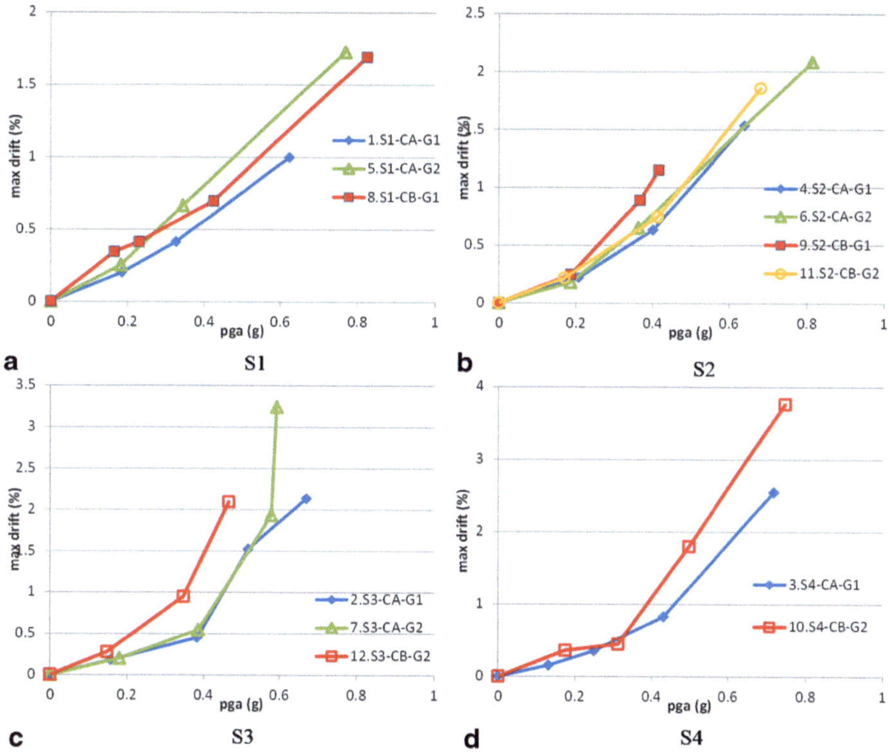

Fig. 20.8 Variation of maximum storey drift demand with PGA by specimen cross section

application of the conventional and balanced design methods to CA-type connections. In each case, the balanced design reaches a larger drift before brace fracture. This is especially noticeable with the 80 × 80 specimens in which the conventional design experienced brace fracture at a drift of only 1 %. The maximum brace forces are also greater in the balanced design cases. Overall, the comparisons presented in Fig. 20.9 support the hypothesis that the use of the balanced gusset plate design method leads to a more ductile and dissipative response in CBFs without loss of brace resistance.

Figure 20.10 presents the observed displacement ductility capacity of the brace-gusset plate specimens. The brace ductility capacity values shown are obtained by normalizing the brace fracture elongation by the brace yield displacement. The brace fracture elongation is the maximum measured change in overall brace length in a fractured brace during the earthquake test run in which that brace fractured. This change in length may be an increase in length (elongation under tension) or a reduction in length (shortening under compression) and includes the effects of axial deformations in the tube length and gusset plate strains. The brace yield displacement is obtained by multiplying the length of the unstiffened brace tube by its char-

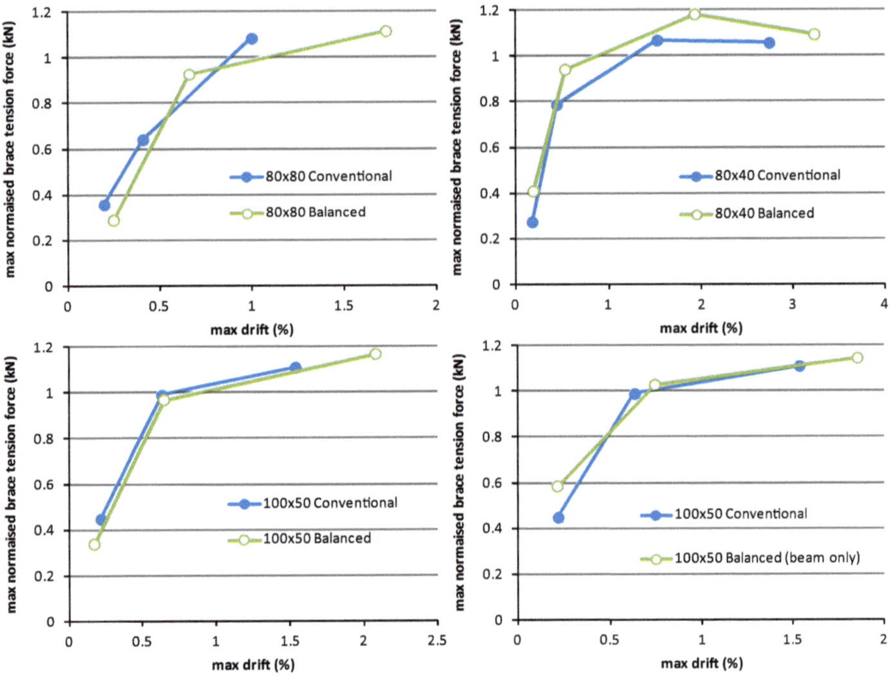

Fig. 20.9 Variation of maximum storey drift demand with PGA by specimen cross section

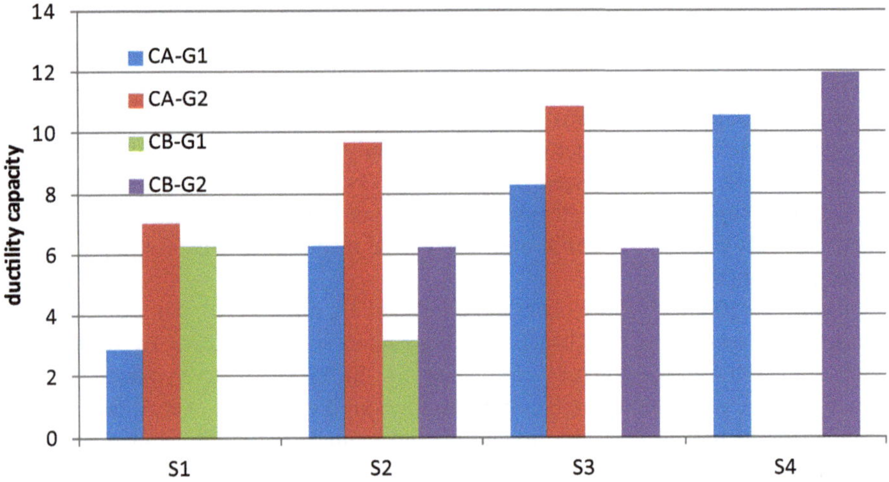

Fig. 20.10 Brace displacement ductility capacity by specimen characteristics

acteristic yield strain, identified from the results of the characteristic steel strengths presented in Tables 20.1 and 20.2.

The measured brace displacement ductility capacities vary between 2.9 and 12.0, with a mean value of 7.5. The variation between the values identified in each test is attributable to the main test specimen parameters: member slenderness, cross-section slenderness, connection type and gusset plate design method. Larger ductility capacities were displayed by more slender specimens with smaller cross sections (S3 and S4), and the use of the balanced gusset design approach (G2) leads to larger brace ductility capacity than the conventional approach (G1). In the four cases where direct comparison between specimens designed using these two approaches can be made, the improvement ranges from 30–140 %, with a mean improvement of 80 %.

20.5 Conclusions

The BRACED project completed a series of shake table tests on a model CBF employing various brace members with different cross-section and gusset plate connection details. Amongst other properties, the test results identified the evolution of frame stiffness with drift level, the sensitivity of frame drift to PGA level, and the brace displacement ductility capacity displayed with different brace member-gusset plate combinations. In particular, the tests confirmed that the use of a balanced design approach in which gusset plate and brace member resistances are designed to ensure a more uniform distribution of plastic strains can lead to higher brace ductility capacities.

The experimental results can also be used to validate and improve empirical models for the ductility capacity of hollow section bracing members, identify active yield mechanisms and failure modes in different brace member/connection configurations, develop and validate numerical models for simulating the inelastic seismic response of CBFs, and provide essential data on the earthquake response of European CBFs.

Acknowledgements The research leading to these results received founding from the European Community's Seventh Framework Programme [FP7/2007–2013] for access to the TAMARIS laboratory in CEA/Saclay, France under grant agreement n 227887.

References

AISC (2005) ANSI/AISC 341-05 Seismic provisions for structural steel buildings. American institute of steel construction, Chicago, Illinois

CEN (2004) EN 1998-1:2004, Eurocode 8: Design of structures for earthquake resistance—Part 1: general rules, seismic actions and rules for buildings. European committee for standardization, Brussels, Belgium

Elghazouli AY (2003) Seismic design procedures for concentrically braced frames. Struct Build 156:381–394

Goggins JM, Broderick BM, Elghazouli AY, Lucas AS (2006) Behaviour of tubular steel members under cyclic axial loading. J Constr Steel Res 62:121–131

Lehman DE, Roeder CW, Herman D, Johnson S, Kotulka B (2008) Improved seismic performance of gusset plate connections. J Struct Eng 134:890–901

Nip KH, Gardner L, Elghazouli AY (2010) Cyclic testing and numerical modelling of carbon steel and stainless steel tubular bracing members. Eng Struct 32:424–441

Popov EP, Black RG (1981) Steel struts under severe cyclic loadings. J Struct Div, ASCE 107(7):1857–1881

Tremblay R (2002) Inelastic seismic response of steel bracing members. J Constr Steel Res 58:665–701

Chapter 21
Shaking Table Test Design to Evaluate Earthquake Capacity of a 3-Storey Building Specimen Composed of Cast-In-Situ Concrete Walls

Salvador Ivorra, Dora Foti, Ilaria Ricci, Giada Gasparini, Stefano Silvestri and Tomaso Trombetti

21.1 Introduction

Several different construction techniques characterized by low cost, limited installation time, great flexibility and high energy/acoustic efficiency have been proposed in recent years for the accomplishment for the construction of low-rise residential buildings at worldwide level (Vanderwerf et al. 2005). Among these, structural systems composed of cast-in-situ squat concrete walls, which make use of lightweight material (e.g. polystyrene) as a support for structural concrete (as in case of Shotcrete, i.e. prefabricated modular pre-reinforced polystyrene support panels) appear to be extremely promising. Indeed these systems allow to obtain high structural, thermal and acoustic efficiency. Concrete guarantees the load bearing capacity, while the lightweight material is left in place (once the construction is completed) to ensure thermal and acoustic insulation. When applied to low-rise residential buildings, these innovative technologies allow to obtain a cellular structure (i.e. structure characterized by bundled-tube behaviour) composed of cast-in-situ sandwich squat concrete walls. As far as the seismic behaviour of concrete walls are concerned, most of the research work accomplished up to date has been focused on slender cantilever walls (e.g. to Paulay and Priestley 1992 and to Coull and Stafford Smith 1991) with limited attention devoted to squat walls, despite hav-

S. Ivorra (✉)
Department of Civil Engineering, Universidad de Alicante,
Apartado 99, 03080 Alicante, Spain
e-mail: Sivorra@ua.es

D. Foti
Department of Civil and Environmental Engineering (DICA),
Technical University of Bari, Via Orabona 4, 70125 Bari, Italy

I. Ricci · G. Gasparini · S. Silvestri · T. Trombetti
Department of Civil, Chemical, Environmental and Materials Engineering (DICAM), University of Bologna, Viale Risorgimento 2, 40136 Bologna, Italy

F. Taucer, R. Apostolska (eds.), *Experimental Research in Earthquake Engineering*,
Geotechnical, Geological and Earthquake Engineering 35,
DOI 10.1007/978-3-319-10136-1_21

ing shown good performances during high intensity earthquakes (e.g. Wood 1991). Squat sandwich walls need specific studies not only due to their global geometrical squat configuration, but also because they have basic characteristics—like thickness, amount of vertical reinforcement, constructive details—differing from those of the traditional ones.

In recent years an exhaustive experimental campaign has been carried out by the University of Bologna and the EUCENTRE labs in Pavia (Trombetti et al. 2010a, 2010b, 2010c). The effort was devoted to the assessment of the structural and seismic performances of single panels composed of cast-in-situ sandwich squat concrete walls. In order to obtain a correct characterization of the seismic behavior (stiffness, strength, ductility, energy dissipation) of such structural elements, a number of pseudo-static tests with cyclic loading were performed on two-dimensional (3.0 m by 3.0 m) cast-in-situ sandwich squat concrete walls (with and without openings) and on a 2-storey H-shaped portion of the structure. A number of horizontal in-plane loading cycles were imposed to the specimens, while the vertical load was kept constant. The results obtained have shown that the tested elements are characterized by: (i) absence of a real failure to the applied loads; (ii) high values of the maximum horizontal load applied to the structural systems (higher than the applied vertical load); (iii) residual bearing capacity with respect to the vertical loads, also when large lateral deformations were developed; (iv) a good degree of kinematic ductility.

To validate the theoretically and partially-experimentally anticipated (through cyclic tests under horizontal loads) good seismic behaviour of cellular structures composed of cast-in-situ squat sandwich concrete walls, shaking table tests were designed and performed at the laboratory facilities of the EUCENTRE in Pavia. The tests comprised a full-scale 3-storey building structure composed of cast-in-situ squat sandwich concrete walls and characterized by 5.50×4.10 m in plan and 8.25 m in height. The shaking -table tests, performed on 6–7 December 2011, represent the main objective of the research project "SEismic behaviour of structural SYstems composed of cast-in-situ COncrete WAlls" (SE.SY.CO.WA) and are part of the SERIES Project (Seismic Engineering Research Infrastructures for European Synergies).

21.2 The Construction System

21.2.1 The Modular Panels

The construction system is based on the production and use of prefabricated modular pre-reinforced polystyrene panels (simply referred as modular panels) that act as support for the casting the structural concrete. These modular panels (Fig. 21.1) have a length of 1120 mm and an adjustable height equal to the interstorey height. They are composed of a single expanded polystyrene sheet (this sheet can be produced with thickness varying between $60 \div 160$ mm in order to suit the specific

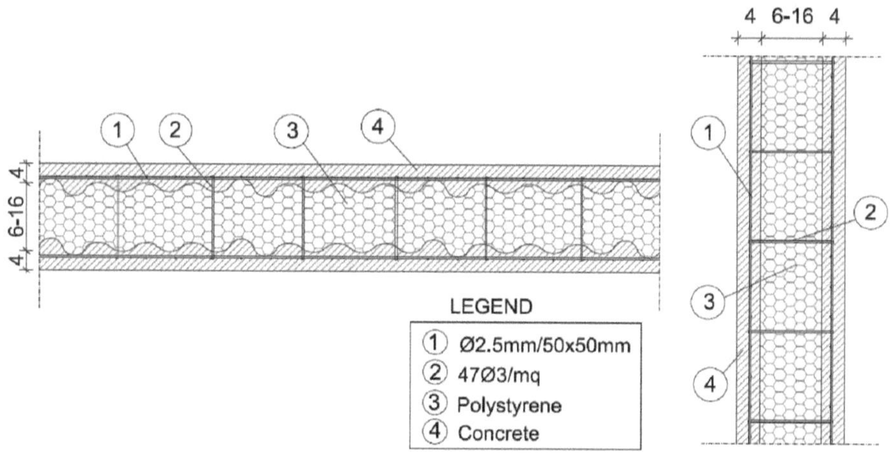

Fig. 21.1 Modular panels: polystyrene edges and wire meshes

needs for thermal and acoustic insulation) featuring two grids of galvanized and electro-welded steel wire mesh on the two external faces. Notice that the expanded polystyrene sheet is shaped with a waved profile along the horizontal direction. The two grids of wire meshes are mechanically linked together with metallic ties (having diameter of 3 mm in quantity of 40 ÷ 50 ties per m^2) placed during production at the factory and embedded within the polystyrene. The modular panels display a peculiar design of the polystyrene edges and the wire meshes (represented in Fig. 21.1) so that when the modular panels are positioned one beside each other (the reason of this positioning will be provided in the following paragraph) the meshes are overlapped to about 100 mm to guarantee the continuity of the horizontal reinforcement. The polystyrene used for the panel is typically characterized by a density of 15 kg/m^3. Both the meshes and the ties are typically constructed using galvanized steel with low carbon content and breakage tension of $f_{tk} = 700$ MPa, with a diameter of 2.5 mm and a mesh of 50 × 50 mm.

The connections between (i) walls and foundation, (ii) orthogonal walls, and (iii) walls and floors, are designed to ensure complete transmission of the actions (i.e. shear, bending, and eventually any axial force) exerted through the adjacent orthogonal walls. In this way it is possible to achieve a cellular behaviour with respect to the horizontal loads. Moreover, to ensure the complete transmission of actions during rare seismic events, the connections are designed following a capacity design approach. These connections are created using special panels and placing the appropriate amount of horizontal and vertical reinforcement.

21.2.2 The Structural System Obtained and Its Features

The construction system allows to obtain a structure characterized by the following specific features: (i) the sandwich walls are characterized by a light amount

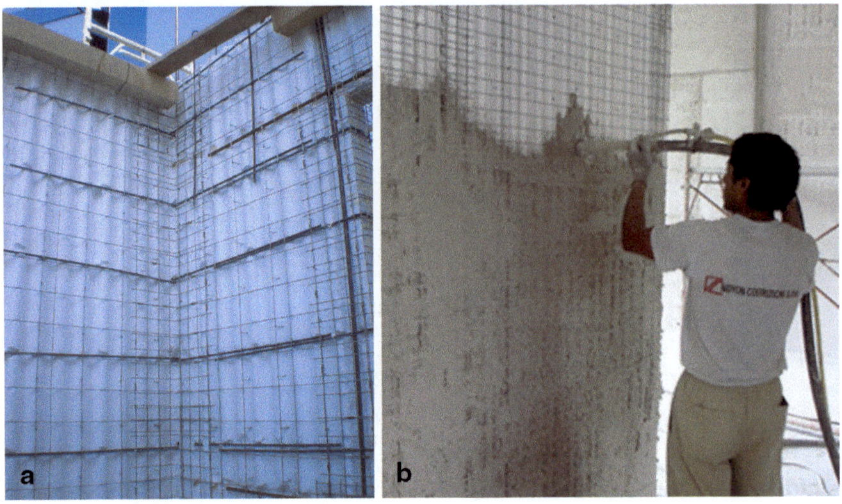

Fig. 21.2 **a** Vertical wall at the inside corner of the building. **b** Shotcrete on the panels

reinforcement, equal in the vertical and horizontal directions (Fig. 21.2a and b); (ii) the system is characterized by a bundled-tube behaviour (i.e. cellular behaviour) (Coull and Stafford Smith 1991). Moreover, given that this structural system is typically used for the realization of low-rise residential buildings, the sandwich walls are characterized by: (iii) a squat configuration; (iv) low values of vertical stresses.

21.3 Shaking Table Test: Design of the Structure

The objective of the shaking table tests is to validate the theoretically and partially-experimentally anticipated (as obtained with pseudo-static cyclic tests carried out on single panels) good seismic behaviour of cellular structures composed of cast-in-situ squat sandwich concrete walls. In more detail, the tests are aimed at identifying the possible differences in the behaviour of the structural system at hand between the cases of (i) pseudo-static cyclic loading and (ii) dynamic seismic loading. To satisfy the above mentioned objectives, the specimen tested on the shaking table test was designed to: (i) be representative of common real built structures (squat configuration of the walls); (ii) be easily transportable from the construction site to the table (this involves the design of specific lifting and lowering systems); (iii) lead to significant results (carried out up to the collapse of the walls, in order to capture also the post-yielding behavior and the ductility resources); (iv) prevent damage of the shaking table (analytical and numerical prediction of the specimen behavior should be developed). Various solutions have been proposed for both the specimen and the foundation systems. The final setup consists in a full-scale 3-storey structural system composed of cast-in-situ squat sandwich concrete walls of 5.50×4.10 m in plan and 8.25 m in height adopting the construction system previ-

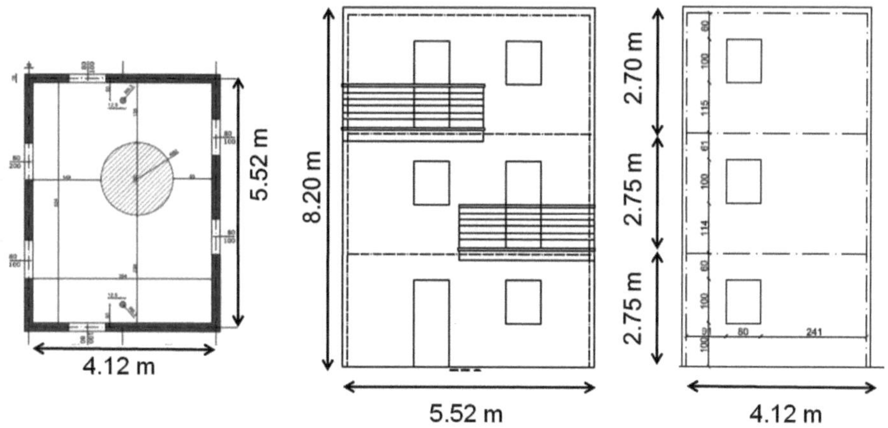

Fig. 21.3 Geometry of the designed building

ously described but halving the amount of vertical and horizontal reinforcement in the walls (φ2.5/100 × 100 mm instead of φ2.5/50 × 50 mm).

The foundation is composed of four beams with the following dimensions: 6.40 × 5.10 m. The two long sides of the foundation have a 70 cm × 40 cm rectangular section whereas the two short sides have a 65 cm × 40 cm rectangular section.

The shaking table of the EUCENTRE Lab in Pavia is a one degree of freedom facility; the seismic input was applied in the direction parallel to the long walls (called in the following "parallel walls", with a length of $l_{//}$ = 5.52 m). The other two walls, the short ones, with a length of $l_{\perp}$ = 4.12 m, will be indicated as "perpendicular walls" (Fig. 21.3). In the following, (i) the stresses (i.e. demand) in the walls (perpendicular and parallel) due to the spectral acceleration, Sa, equal to 1 g, will be evaluated together with (ii) the corresponding strength (i.e. capacity). The strength of the walls will be evaluated applying the analytical formulas developed by the authors on the basis of the classical hypotheses for traditional reinforced concrete elements (Trombetti et al. 2010a, b, c). The validity of these formulas have been validated through experimental results obtained from pseudo-static cyclic tests performed on two-dimensional (3.0 m by 3.0 m) cast-in-situ sandwich squat concrete walls (with and without openings) and on a 2-storey H-shaped portion of structure. Comparing the stresses due to 1 g spectral acceleration with the corresponding strength, the sequence of the possible collapse mechanism of the structure has been identified.

21.3.1 Comparison Between Demand Due To 1 g Spectral Acceleration and Capacity

The total shear, T_{ED}, and the total bending moment, M_{ED}, at the base of the structure, due to a spectral acceleration Sa = 1 g, are equal to 66 t and 360 tm, respectively. For the evaluation of the total bending moment at the base of the structure a triangular distribution of the horizontal forces was assumed. The moment distribution under the walls was evaluated assuming (i) linear elastic behavior and (ii) conservation of plane sections.

The bending moment of the parallel and perpendicular walls results in:

$$M_{Ed,\,//} = 111 \text{ tm and } M_{Ed,\,\perp} = 249 \text{ tm}$$

The parallel bending moment on the single parallel wall (ignoring the presence of the openings) is:

$$M_{\text{Ed, single // wall}} = M_{Ed,\,//}/2 = 56 \text{ tm}$$

The perpendicular bending moment results in stresses of compression and tension for the perpendicular walls equal to 45 t. The total shear at the base of the structure is taken by the single parallel walls and ignoring the presence of the openings is equal to 33 t.

The in plane first yielding bending moment of the single parallel wall is given by (Trombetti et al. 2010a, b, c) $M_{y1} = 149$ tm .

The in plane ultimate bending moment of the single parallel wall is given by (Trombetti et al. 2010a, b, c) $M_u = 181$ tm

The shear strength of the single parallel wall is given by (Trombetti et al. 2010a, b, c):

$$T_{Rd} = \min(T_{Rcd}, T_{Scd}) = 60 \text{ t}$$

where: T_{Rsd} in the steel shear strength, T_{Rcd} is the concrete shear strength.

The tensile strength of the single perpendicular wall (ignoring the concrete tensile strength) is given by $N_{Rd,\perp\text{wall}} \simeq 20$ t

The sliding base strength of the structure is given by (Trombetti et al. 2010a, b, c) $S_{Rd,//\text{ wall}} = 75$ t

A more detailed description of the analytical analysis developed is presented in (Ricci et al. 2012)

21.3.2 *Synthesis of the Predicted Behaviour*

Considering a triangular distribution of the horizontal forces along the height of the structure and an amplification factor equal to 2.5, the sequence of the collapse mechanism is hereafter presented:

1. Tension-Yielding of the perpendicular walls, for Sa = 0.72 g and PGA = 0.29 g
2. Bending-Yielding, in-plane parallel walls, for Sa = 1.32 g and PGA = 0.53 g

3. Ultimate bending moment of the parallel walls, for Sa = 1.50 g and PGA = 0.60 g
4. In-plane shear crisis of parallel walls, for Sa = 1.82 g and PGA = 0.73 g
5. Sliding shear at base, for Sa = 2.28 g and PGA = 0.91 g

21.3.3 Transportation Phases

The construction of the specimen was carried outside of the laboratory; it was lifted and pulled inside using hydraulic jacks and a roller system that reduces the friction coefficient on the ground surface to less than 0.1. This process produces several changes on the final boundary conditions; in fact a prestress system was designed for the building and the foundation to assure the integrity of the structure during the transportation process.

The transportation of the "structure-foundation" specimen was as followes: (1) the specimen is uploaded in position 1 with four actuators; (2) the specimen is positioned on sliders and pulled with chains up to position 2; (3) in position 2 the specimen is lowered and re-uploaded; (4) the specimen is positioned on sliders and pulled with chains up to the shaking table (position 3). Figure 21.4a presents a scheme of the procedure. Figure 21.4b, c present two positions of this phase, outside and inside the laboratory.

To design the reinforcing system a linear elastic 3D FEM was designed. The structure was analyzed to consider the arch effects when the structure and foundation are lifted using four jacks at the corners. To keep all vertical walls under compression during the transportation phase, the whole structure was postensioned under a load of 400 kN. Similarly, the four beams that constitute the foundation were postensioned to guarantee the strength capacity of these beams during this phase. An exterior steel structure was used to stiffen the concrete foundation and to connect the jacks at the four lifting points.

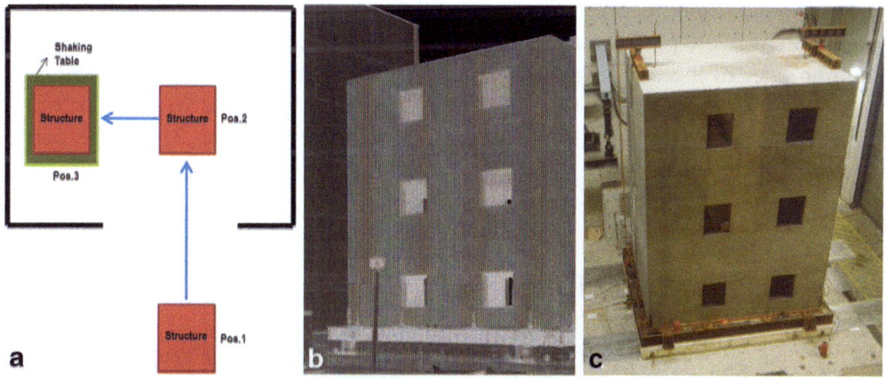

Fig. 21.4 **a** Scheme of the trasportation phase. **b** General view of the building with the reinforcing system for the transportation phase. **c** General view of the building inside the laboratory

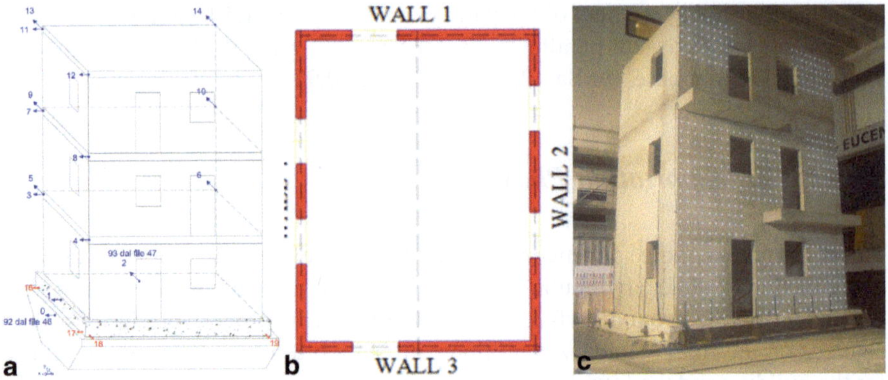

Fig. 21.5 **a** Location of the accelerometers on the building; **b** Wall reference; **c** Position of the reflectors for the high resolution cameras

21.4 Instrumentation

In order to provide measures of the accelerations at different locations of the structure, 9 mono-directional accelerometers and 3 bi-directional accelerometers were located at the shaking table, at the foundation and at the three floors of the structure, as displayed in Fig. 21.5a.

Four potentiometers were installed in order to monitor the relative displacement between the shaking table and the foundation. In detail: (i) Potentiometer n. 16 and 17 were oriented along the x direction; (ii) Potentiometer n. 18 and 19 were oriented along the y direction.

All the external walls (referred to as wall 1, 2, 3 and 4, see Fig. 21.5b) were instrumented with extensometers in order to provide measures of deformation. In total 64 extensometers were installed: 20 extensometers (100 mm base length); 2 extensometers (200 mm base length); 42 extensometers (50 mm base length); and 2 extensometers (200 mm base length).

During the tests the structure was monitored by seven high resolution cameras. Four cameras were placed outside of the building and each of them recorded one of the four sides of the building. The remaining three cameras were placed inside the building at the ground floor (Fig. 21.5c) in order to record specific portions of the walls: between the openings of walls 2 and 4 (cameras n. 2 and 3) and the corner between walls 1 and 4.

21.5 Shaking-Table Tests

21.5.1 The Reference Seismic Input and the Test Program

The reference seismic input (Fig. 21.6a) was the accelerogram recorded at the Ulcinj station (Hotel Albatros, x component) during the Montenegro Earthquake of

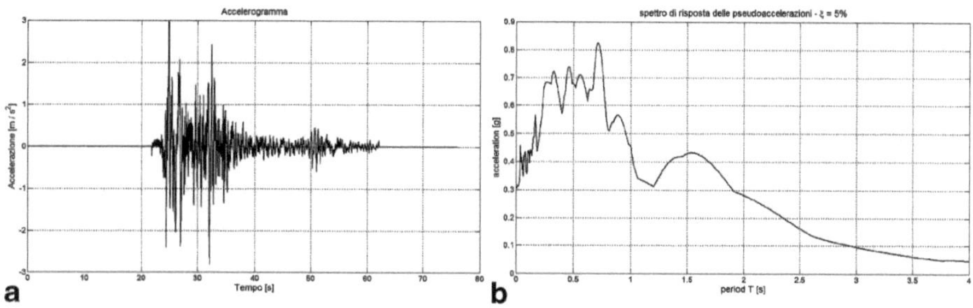

Fig. 21.6 **a** Accelerogram of the reference seismic input. **b** Spectrum of the reference input

Table 21.1 Test program

N.	Test description
WN1	White noise 0.05 g
ST1	Seismic test 0.05 g
WN2	White noise 0.10 g
ST2	Seismic test 0.15 g
WN3	White noise 0.10 g
ST3	Seismic test 0.50 g
WN4	White noise 0.10 g
ST4	Seismic test 1.00 g
WN5	White noise, amplitude increasing from 0.05 to 0.30 g
WN6	White noise a 0.50 g
ST5	Seismic test 1.20 g
WN7	White noise 0.30 g
ST6	Seismic test 1.20 g
WN 8	White noise 0.30 g

1979 (PGA=0.305 g). In Table 21.1 the test program is presented. For the specific characteristics of the control system of the shaking table, a long calibration was performed. For this reason, the structure was subjected to a limited number of seismic tests. The reference input was scaled at increasing levels of PGA and, among the different seismic tests, important sessions of white noise were performed to keep monitored the changes in natural frequency (Fig. 21.6b).

21.5.2 *Results*

21.5.2.1 Experimental Frequencies

In Table 21.2, the values of the experimental frequencies observed during the tests are presented. Looking at Table 21.2, it is possible to observe that (i) up to the 1 g seismic test:

Table 21.2 Main frequencies of the building

Test	Experimental frequencies (Hz)
Before 0.05 g test	10.0
Between 0.05 g and 0.15 g tests	10.0
Between 0.15 g and 0.50 g tests	10.0
Between 0.50 g and 1.00 g tests	10.0
Between 1.00 g test and 0.30 g white noise	10.0
Between 0.30 g white noise and 1.2 g first test	8.6
Between 1.20 g first test and 0.50 g white noise	–
Between 0.50 g white noise and 1.20 g second test	8.2
After 1.20 g second test	–

- the experimental frequencies remained constant and equal to 10 Hz
- the structure did not show any damage

whereas (ii) after the 0.30 g white noise (performed with the objective of cracking the concrete with high frequencies vibrations) a decrease of frequency (even though modest) was observed from 10 to 8.6 Hz. Another decrease of frequency was observed after the 0.5 g white noise.

From these results in terms of frequencies, it is possible to note that:

- there were no substantial changes in frequencies with the increase of the seismic intensity; the structure showed a stable behavior during the whole test
- this lack of changes in frequencies proved that the structure developed a linear elastic behavior, and the predicted sequence of the collapse mechanisms was not observed.

Moreover, it is worth noting that to obtain the recorded frequency values with a FEM model which makes use of linear shell elements, a Young Modulus of about $E_{concrete}/2$ (roughly 15,000 MPa) should be introduced to capture the experimental results. This indicates that the structure basically remained in the linear elastic field. This result is different from those that could be expected starting from the results of the previously experimental (cyclic) tests, where the specimens behaved in totally cracked conditions from the beginning of the tests; in such conditions the reduction of the Young Modulus to match the experimental stiffness was found to be equal to $E_{concrete}/10$ (roughly 3000 MPa), both for simple 3 m × 3 m walls and for the 2-storey H-shaped structure made of the same identical construction system. On the contrary, during the actual (dynamic) shaking-test, the structure behaved between the uncracked and cracked conditions.

21.5.2.2 Cracking Pattern

As far as the cracking pattern is concerned, it is worth to point out that up to the 1 g test no visible cracks were developed. Then, after the 0.5 g white noise (in

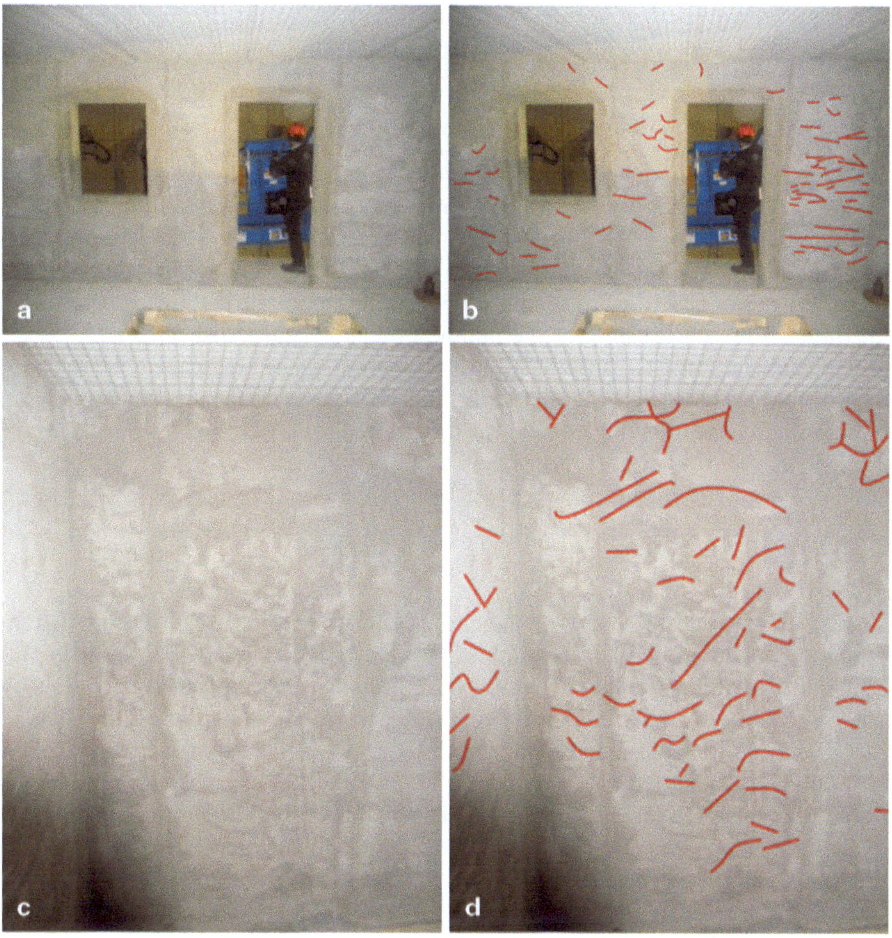

Fig. 21.7 **a** and **c** Cracking pattern after the 0.5 g white noise (inside third floor) **b** and **d** same walls with cracks underlined in red

conjunction with decrease of frequency), a very limited number of hairline and not spread cracks appeared especially on the top floor walls. From the 0.5 g white noise to the final 1.2 g seismic test, the cracking pattern remained stable and almost unchanged. Figure 21.7a and c show the picture of the walls of the top floor after the 0.5 g white noise, whereas, Fig. 21.7b and d show the same pictures but with the cracks underlined.

21.5.2.3 Overstrengths

From the data registered by the accelerometers placed on the structure, the following information has been evaluated:

1. experimental total base shear, $T_{base}(t) = \sum_{i=1}^{3} m_i a_i(t)$

2. experimental total base bending moment, $M_{base}(t) = \sum_{i=1}^{3} m_i a_i(t) h_i$

In the previous equations m_i, $a_i(t)$ and h_i are respectively the mass, the acceleration and the height of the i-th floor, with $i = 1, 2, 3$. Comparing the maximum experimental value of the total base shear and the total base bending moment with the predicted values described in Section 21.3, another important difference regards the maximum strength capacity of the building under seismic loads compared to the previous cyclic tests: the shaking table tests were designed to achieve the collapse of the building with lower accelerations than those used in the shaking table tests, however, after the tests no damage was observed.

21.6 Conclusions

In this paper the design phase of shaking-table tests on a full-scale 3-storey structural system composed of cast-in-situ squat sandwich concrete walls has been presented. Some of the preliminary results have been synthetically anticipated, as they will be accurately discussed in following research works. In particular, the shaking - table tests performed on 6–7 December 2011 allowed to verify the cellular behavior of this kind of structures, highlighting important differences between the results of the previous pseudo-static cyclic experimental tests and the results of the actual dynamic shaking-tests of the building. One of the main differences regards the stiffness: this parameter presents significant variations from the experimental results carried out previously on both simple 3 m × 3 m walls and the 2-storey H-shaped structure made of the same identical construction system.

Another important difference regards the maximum strength capacity of the building under seismic loads compared to previous tests: the tests were designed to have the collapse of the building with lower accelerations that those used in the shaking table tests, however, following the tests no damage was observed on the structure. This difference could be explained by the different loading modalities (pseudo-static cyclic load vs. dynamic seismic ones). The preliminary interpretation of the experimental results leads to the following fundamental observations:

- the prototype building exhibited a high seismic performance: it was able to sustain increasing levels of seismic input up to 1.2 g PGA without visible damage;
- the prototype building seismic response was very close to that of the "theoretical" response of an uncracked structure up to approximately 1.0 g PGA (in terms of both natural frequencies and deformation response);
- the prototype building exhibited unexpected "overstrengths" which did not allow to observe the expected mechanisms of failure, whose analytical predic-

tions were partially-based on the results of the pseudo-static reversed cyclic tests performed on both full scale single panels (with and without openings) and on a simple H-shaped building structure;

- the results of the previous pseudo-static cyclic tests did not show any "concrete contribution" to the shear strength of the panels (i.e. the contribution due to the concrete tensile strength). Therefore this contribution was neglected in the design phase;
- if the shear strength of the wall is evaluated according to the Modified Compression Field theory (i.e. accounting for the tensile strength contribution) a significantly higher strength (superior than the maximum shear applied during all the seismic tests) is obtained prior to concrete cracking;
- the comparison between the time-history of the base moment (i.e. the base moment due to the applied "external" loads, obtained from the recorded accelerations) and the internal base moment (i.e. the base moment due to the internal forces, obtained from the recorded deformation) also confirmed that the response of the prototype building is close to that of the "theoretical" uncracked model.

Based on all the above observations it clearly appears that a structural system composed of a r/c sandwich panel, which is designed according to current seismic design requirements, is able to withstand very high seismic shaking (PGA larger than 1 g) without exhibiting visible damages, thanks to the development of the so called "box behaviour" which basically induces in plane actions (shear and moments) which are well sustained by the sandwich structural r/c walls.

Furthermore, it should be noted that the specific prototype building showed "unexpected" overstrengths, with respect to those exhibited during previous pseudo-static cyclic tests. A possible explanation of the observed unexpected response is related to the different behaviour of the concrete observed during the pseudostatic tests and the dynamic tests: while during the pesudostatic tests the concrete behaves essentially in cracked conditions (from the beginning of the tests), during the shaking table tests no visible cracks were observed, suggesting a concrete uncracked behaviour. On this regard, if the strengths of the walls are computed accounting for the concrete tensile contribution the observed "overstrengths" can be explained.

Acknowledgements The research leading to these results received founding from the European Community's Seventh Framework Programme [FP7/2007–2013] for access to TREES laboratory of EUCENTRE under grant agreement n° 227887. The authors thank Prof. Alberto Pavese and Dr. Simone Peloso of the EUCENTRE.

References

Coull A, Stafford Smith B (1991) Tall buildings structures: analysis and design. Wiley

Paulay T, Priestley MJN (1992) Seismic design of reinforced concrete and masonry buildings. Wiley Interscience Press publication, Wiley

Ricci I, Gasparini G, Silvestri S, Trombetti T, Foti D, Ivorra S (2012) Design of a shaking table test on a 3-storey building composed of cast-in-situ concrete walls. In: Proceeding of world conference on earthquake engineering, at Lisbon, Portugal, Vol 1

Trombetti T Silvestri S, Gasparini G, Ricci I (2010a) Results of pseudo-static tests with cyclic horizontal load on concrete/polystyrene sandwich bearing panels without openings. Sustainable development strategies for construction in Europe and China, Roma, 19–20 April 2010
Trombetti T, Silvestri S, Gasparini G, Ricci I (2010b) Results of pseudo-static tests with cyclic horizontal load on H shaped substructure composed of concrete/polystyrene sandwich bearing panels without openings. Sustainable development strategies for construction in Europe and China, Roma, 19–20 April 2010
Trombetti T, Silvestri S, Gasparini G, Ricci I (2010c) Correlations between the experimental results of pseudo-static tests with cyclic horizontal load on concrete/polystyrene sandwich bearing panels and their analytical counterparts. Sustainable development strategies for construction in Europe and China, Roma, 19–20 April 2010
Vanderwerf PA, Panushev IS, Nicholson M, Kokonowski D (2005) Concrete systems for gomes and low-rise construction: a Portland cement association's guide for homes and lo-rise buildings. Mc Graw Hill, New York
Wood SL (1991) Performance of reinforced concrete buildings during the 1985 chile earthquake: implications for the design of structural walls. Earthq Spectra 7(4):607–637

Chapter 22
High-Performance Composite-Reinforced Earthquake Resistant Buildings with Self-Aligning Capabilities

Bohumil Kasal, Tiberiu Polocoser, Pablo Guindos, Shota Urushadze, Stanislav Pospisil, Andreas Heiduschke, Norbert Rüther and Zbigniew Zembaty

22.1 Introduction

The use of laminated wood in medium-rise buildings is increasingly being used due to numerous advantages. Wood structures commonly provide ease of manipulation and construction, important light-weight properties for transportation and low ecological footprint. Wood structural elements have high strength-to-weight and stiffness-to-weight ratios, which allow for good performance in the event of dynamic loads such as earthquake excitations, Kasal et al. (2004). Despite the high stiffness-to-weight ratio, stiffness of wood is still low when compared to other classical structural construction materials such as steel or reinforced concrete and relatively large drifts are often experienced (Yasumura 1998; Kasal et al. 2004). Indeed, the inter-story drifts tend to be the limiting engineering design parameter (EDP) for these types of structures. The beam to column connection (BTC) is typically the least stiff part of the system and small rotations between beams and columns cause large drifts in the system. It appears that developing a stiff BTC can help to reduce

T. Polocoser (✉)
Technical University of Braunschweig, Braunschweig, Germany
e-mail: t.polocoser@tu-braunschweig.de

B. Kasal · P. Guindos · N. Rüther
Fraunhofer Wilhelm Klauditz Institut (WKI), Braunschweig, Germany

S. Urushadze · S. Pospisil
Institute of Theoretical and Applied Mechanics, Academy of Science of the Czech Republic, Prague, Czech Republic

A. Heiduschke
HESS TIMBER GmbH & Co. KG, Kleinheubach, Germany

Z. Zembaty
University of Technology, Opole, Poland

F. Taucer, R. Apostolska (eds.), *Experimental Research in Earthquake Engineering*,
Geotechnical, Geological and Earthquake Engineering 35,
DOI 10.1007/978-3-319-10136-1_22

excessive drift but cannot control it. In addition, the design of an appropriately stiff BTC is a challenging task since local over-stresses may arise. Capacity design was used to avoid brittle failures of the BTC. The behavior of the BTC connections is frequently governed by plastic deformations of the steel fasteners and is well described in previous research as a non-conservative and non-linear system with pinching (Heiduschke et al. 2009). When designed properly, stiffness degradation of the BTC can be a positive aspect that changes resonant frequencies during the loading process. Also, this feature facilitated the opportunity to design self-aligning frames by providing a reservoir of elastic energy stored in the wooden members. The challenge was in balancing the dissipated and stored energies with stiffness degradation while providing sufficient fixity of column-to-foundation connections relative to the BTC connection.

Buchanan and Fairweather (1993) analyzed the most beneficial ductility for various BTC connections of low rise frames in order to create a balance between drift control and brittle failure of structural members. Ohashi et al. (1994) obtained large drifts in three-story timber buildings and therefore suggested the design of higher stiffness BTC connections. Miyazawa (1994) concluded that timber frames can experience large sways without failure after testing two story frames with shear walls. Ceccotti et al. (1994) measured maximum rotations from 0.03 to 0.04 radians in a full-scale single story frame. Kikuchi (1994) considered the pinned BTC as unfavorable and reported quite large drifts in frames up to four stories, consequently the frames did not meet the Japanese standards requirements. Palermo et al. (2006) presented considerable hysteretic dissipation of energy with flag-shaped loops, good self-aligning capacity, and no damage of the BTCs consisting of glued-in mild-steel bars designed to yield in tension and compression. Frenette et al. (1996) used embedded steel plates in the BTC and concluded that the connections were responsible for about 90 % of the total frame deformation. They also noted the resonant frequencies of the building decreased when increasing the BTC damage. Yasumura (1998) mounted single portal frames with inserted steel plates, and observed good self-aligning capabilities of the frame up to an inter-story drift ratio (IDR) of approximately 1/24. Kasal et al. (2004) tested conventional, densified and glass-reinforced frames. Good self-aligning capabilities were obtained and it was concluded that BTCs can provide significant energy dissipation. Another frame method employed by Newcombe et al. (2010) using post tensioned beams showed very good self aligning capabilities at 2% drift. Ponzo et al (2012) also designed the BTCs based on yielding of steel members which were mounted in a scaled, post-tensioned, three-story frame. The test revealed no significant damage even after high seismic excitations.

This paper describes the experimental results and main conclusions of the international project entitled "High-Performance Composite-Reinforced Earthquake Resistant Buildings with Self-Aligning Capabilities" developed within the seventh framework-SERIES program, University of Patras (2011). In this project, the dynamic performance of two different timber frames was assessed. First, a full-scale one-story frame was tested on a shake table under several dynamic excitations. This structure used relatively rigid BTCs, which were constructed by means of long self-tapping screws and hardwood blocks. Second, the response of a one-to-three scaled three-story frame was also evaluated when subjected to a series of distinct dynamic excitations. The joints of this second structure were based on the dissipation of energy through friction rather than yielding of the steel elements. In both frames, several preliminary cyclic-loading tests were performed in order to assess the mechanical capabilities of the two distinct BTCs design alternatives.

22.2 Rigid BTC with Long Self-Tapping Screws and Beech Blocks

A series of BTC connections were tested, which allowed choosing one prototype for implementation in a full scale mock-up test. The connection described was chosen from two dimensional testing, and then used in a three dimensional frame. The beams and columns of the moment-resisting connection were made of laminated spruce (*Picea*) in accordance to standard tests (DIN 1052 2004–2008). The manufacturer of the beams and columns was HESS TIMBER GmbH & Co. KG, located in Kleinheubach, Germany. The assembly of the beams and columns was performed in the Institute of Theoretical and Applied Mechanics (ITAM) laboratory, Prague, Czech Republic. The column width, depth, and length were 223 × 220 × 2200 mm, while for the beam it was 160 × 360 × 1500 mm. The moisture content (MC) of each specimen was measured and was on average 8.6 %. The timber density ranged from 415 to 450 kg/m^3. Between the beam and column there was intermediary hardwood blocks made of laminated beech wood (*Fagus*). The dimensions of the blocks were 65 × 170 × 160 mm and where used to transfer forces through self-tapping screws.

Self-tapping screws were used for their high pull out strength and their effectiveness as wood connectors. They were manufactured by SFS Holding AG, Heerbrugg, Switzerland (2011) in standard lengths of 120–350 mm with diameters ranging from 6.5 to 12 mm. The same screws were used for both the 2-D and 3-D moment connections. The fasteners used to connect the beams to column were star-drive SSK 10 × 350 mm long self tapping screws. There were eight screws that went through the hardwood blocks and into the beam at a 45° angle to the beam, while two screws were placed alternating between them at a 45° angle into the column in toe-nail like fashion. Figure 22.1 shows the specimen on the laboratory floor before testing.

The connection was tested using the sinusoidal load function described in DIN EN 12512 (2002). It consists of applying pseudo-static loading with increasing displacement amplitude. The amplitudes were derived from a static monotonic test. The moment-rotation behavior of the connection was linear throughout the entire test with a high moment to rotation ratio, thus indicating a stiff connection. The connection showed a maximum moment capacity of 105.8 kNm with a maximum rotation of 0.0075 radians. The cumulative dissipated energy reached 1.45 kNm radians.

floor before testing.

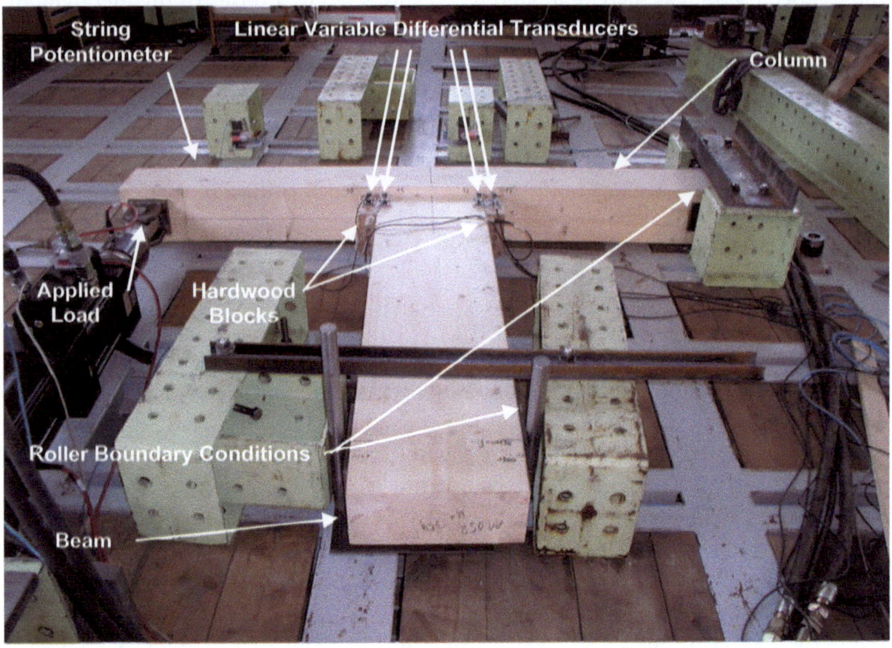

Fig. 22.1 Two dimensional connection test setup. The testing was displacement controlled with the string-potentiometer located opposite to the applied load. Load was measured with a load cell in the hydraulic cylinder. Four linear variable differential transducers were placed next to the connection to measure the displacements used for calculating rotation

22.3 One Story Mock-Up

The BTC with self-tapping screws was incorporated into a full-scale, one-story frame in order to test its performance on a shaking table. The materials of this frame were identical to those used for the connections. The columns were 2560 mm in length with a cross section of 220 × 220 mm. The beams measured 3600 mm in length, 340 mm in height and 220 mm in width. The cross laminated timber (CLT) deck had a footprint of 4400 × 4400 mm, with a density of 500 kg/m^3 and 200 mm thick, and was mounted on top of the beams. Two layers of the CLT, 100 mm each were glued and screws were used to apply the necessary pressure. Screws were used to attach the decks to the beams. The screws were placed 100 mm on center. The full live load was simulated by placing four steel masses of 1000 kg each on the deck. The total mass of the structure was about 6100 kg—see Fig. 22.2.

The dynamic tests were performed at the shaking table of the Earthquake Engineering Research Centre (EERC), at the Department of Civil Engineering of the University of Bristol, Bristol, UK. The shaking table's footprint was 3000 × 3000 mm, and allowed for a payload capacity of 17 t with a maximum testing frequency of 100 Hz. The maximum displacements, velocities and accelerations of this table

Fig. 22.2 Main components of the one-story full-scale frame tests including: 2560 × 220 × 220 mm columns, 3600 × 340 × 220 mm beams, 4400 × 4400 × 200 mm deck, 3000 × 3000 mm shaking table, four added steel masses of 1000 kg each and stiff steel reference frame for measuring the drifts of the frame respect to the shaking table

range respectively from ± 150 mm, ± 1100 mm/s, and ± 60 mm/s^2. The series of dynamic tests included noise excitation, snapback, sinusoidal sweep and seismic loads. The synthetic seismic spectrum contained significant amount of energy close to the frame's first natural frequency to ensure the most unfavorable loading conditions. The seismic excitation included X, Y and Z directions. The damage of the joints was accounted via the progressively decreasing natural frequencies of the frame which were measured following each seismic test.

The accelerations, drifts, strains and rotations of the connections were measured at multiple points. The acceleration recording set up was arranged by 14 accelerometers: three accelerometers on the shake table, four on the deck, six at the connections, and one attached to a beam. The drift of the frame was recorded by placing four string potentiometers on an extremely stiff steel structure placed beneath the timber deck, as seen in Fig. 22.2. The relative rotation between beam and columns was measured in two connections at different directions with four LVDTs. The strain recording arrangement was composed by 16 strain gauges, four of them at the deck, four at the bottom of the beams, and eight at the columns.

Three different seismic tests of increasing peak ground acceleration (PGA) were performed from 0.08 g through 0.78 and 1.43 g. The maximum rotation during the first, second and third seismic tests were respectively 0.015, 0.73 and 1.83. During the first seismic test the maximum strain was 0.06 mm/m; this value increased to 0.48 and 1.08 mm/m, in the second and third test respectively. The maximum recorded drifts ranged from 2 mm through 55 and 95 mm along the first, second and

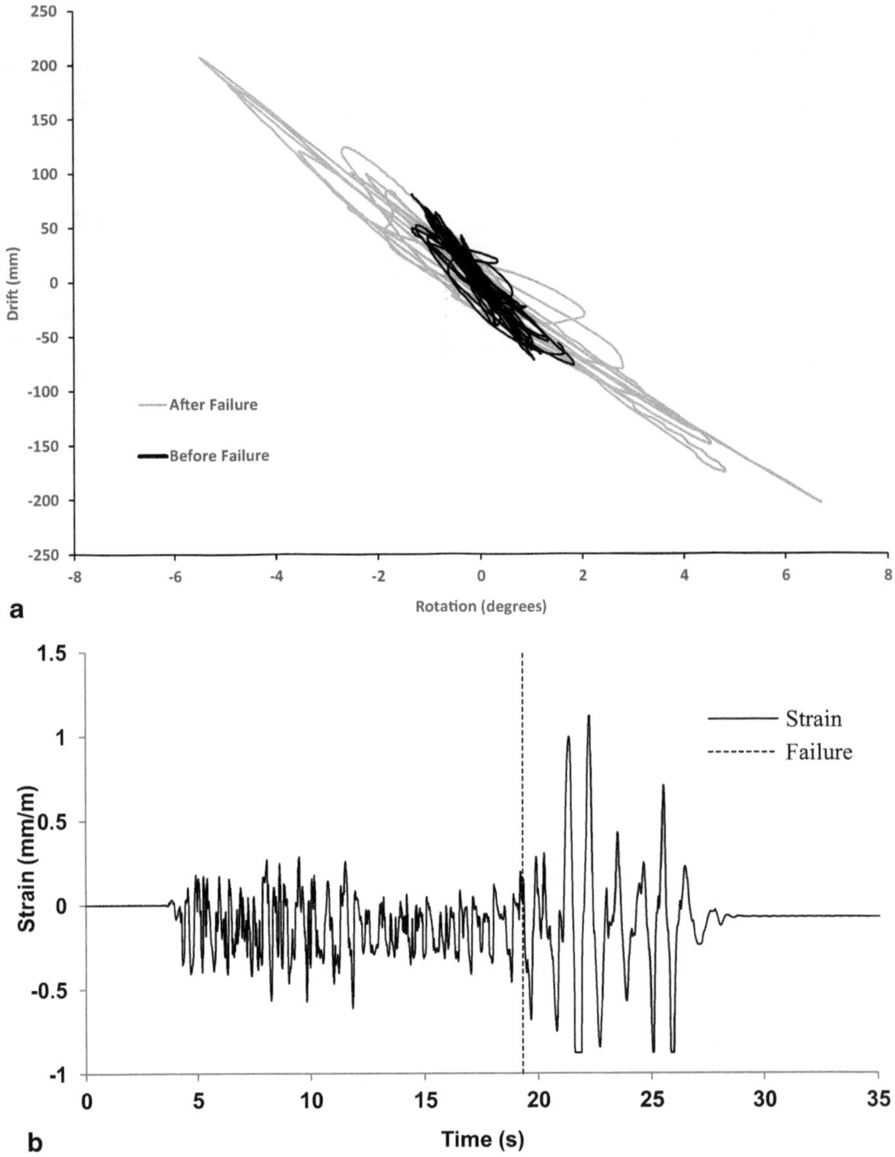

Fig. 22.3 **a** Rotations versus drifts, and **b** strains in the one-story mockup during the third seismic test (1.43 g PGA). The rotations and drifts were measured at the BTCs of columns 3 and 4 respectively. The strains were measured at the bottom of column 3 (see Fig. 22.2 for further details about measuring devices' locations)

thirds seismic test respectively. The first natural frequency of the frame decreased from 2.42 to 2.02 Hz and then to 2.00 Hz following the second and third test accordingly—due to BTC degradation. The frame was able to resist a PGA of 1.25 g without significant damage. Failure occurred during an excitation of 1.5 g (PGA) by shearing off one of the columns in the direction perpendicular to grain. At the time of failure, the maximum rotation, drift, and strain were up to 6.69, 225 mm and 1.08 mm/m respectively. Yet, the wood members were far from yield limits (about 2 mm/m). The rupture occurred in the column at the BTC connection. After inspection it was found that the frame failed in a brittle manner because of the reduced column cross-sectional area due to the multitude of screws in the BTC connection. Figure 22.3 shows the rotations, drifts and strains recorded during the third seismic test (1.43 g PGA)

22.4 Frictional BTC

Similarly to the rigid BTC with long self-tapping screws and beech blocks, these joints were made with laminated timber of strength class GL24h, made from Spruce (*Picea*) according to DIN current standards (DIN 1052 2004–2008). The complete manufacturing process was developed at the Department of Structural Engineering and Construction of the Fraunhofer WKI-Institute for Wood Research, Braunschweig, Germany. Subsequently, the testing of their performance was carried out at the ITAM laboratory in Prague, Czech Republic.

Unlike former BTCs where the main energy dissipation mechanism is provided by yielding of the steel screws, the BTC was designed to dissipate most of the energy through frictional energy between the steel parts while wood members perform within the elastic range. The joint was constructed and tested as a three-dimensional connection. The geometry of the connection and test positioning is shown in Fig. 22.4. The BTC consisted of a $3000 \times 140 \times 140$ mm column and four $1500 \times 120 \times 240$ mm beams. All these beams were oriented perpendicular to the face of columns at half of their height. The column and one of the beams were reinforced with unidirectional glass fiber fabric material type HP-U600E of 660 g/m^2, produced by HP-Textiles GmbH, Shapen, Germany (2010). This local reinforcement was adhered at the external surfaces of the wooden members which were close to the BTC by means of an Epoxy Resin system type HP-E45KL, also manufactured by HP-Textiles GmbH (2010).

The boundary conditions and measuring devices of the test are presented in Fig. 22.4 of the two adjacent non-reinforced beams were constrained at their ends with roller-type supports, i.e. the vertical displacements were inhibited while horizontal displacements were freely permitted. The other two beams, one reinforced and the other unreinforced, were cyclically loaded in the vertical direction by means of two hydraulic actuators which allowed for displacement control. Similarly to the former rigid BTCs, the testing protocol followed by the actuators was that commonly prescribed for the testing of timber connections with fasteners (DIN 12512 2002–2008). In addition to the cyclic loading, a compressive load of 15 kN was axially applied by a third actuator at the column end, pre-stressing the structure before the experiment and guaranteeing its stability during testing. Since the geometry and the

Fig. 22.4 Main components of the frictional beam to column connections (BTC), including: one 3000 × 140 × 140 mm column, four 1500 × 120 × 240 mm beams, three hydraulic actuators (one of them constantly compresses the column), two roller supports (located opposite to the hydraulic actuators) and the glass fiber (GF) reinforcement. Only the column and beam subjected to hydraulic actuator 2 were reinforced

boundary conditions of the connection were symmetric (with the exception of the reinforced beam), this test also allowed for assessing the dynamic performance of reinforced versus unreinforced BTC. During testing, the displacement of the beams relative to the columns was recorded by means of eight string potentiometers, which were located above and below the cross-section mid-span of each beam. A metal filament was used to precisely connect each of these potentiometers to the column. Also the horizontal displacement of each beam was recorded with four additional potentiometers placed perpendicularly to the beam's ends (see Fig. 22.5). The hydraulic actuators were moving in phase.

The details of the BTC are shown in Fig. 22.6. The BTC consisted of two metal plates. The first was a trapezoidal-shaped kerf metal plate with a base of 335 mm, a length of 450 mm and a thickness of 4 mm, which was embedded at the mid-width of the beams. This plate was fastened to the beam by means of 36.7 mm self-tapping screws type WS-T-7, manufactured by SFS Holding AG, Heerbrugg, Switzerland (2011). The second was a T-shaped steel plate with a base of 340 mm, a length of 100 mm and a thickness of 10 mm, which was mounted at the column's face with four standard S235, 9 mm diameter bolts. The moments transmitted by these bolts to the column may generate perpendicular-to-grain failure, which is very critical

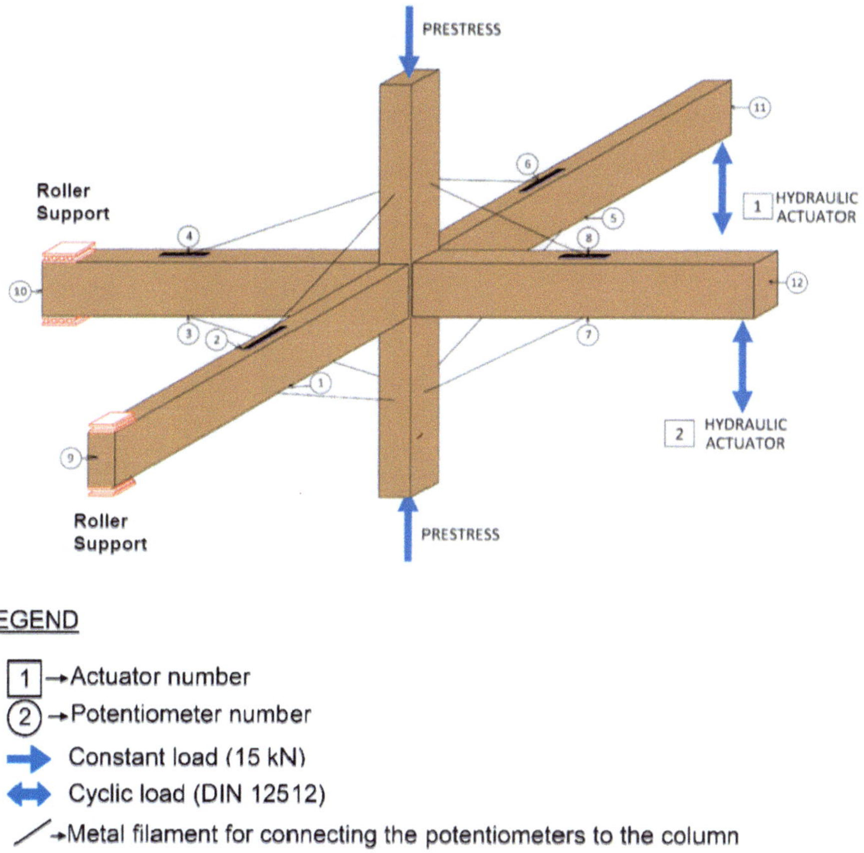

Fig. 22.5 Measuring devices and boundary conditions of the frictional BTC. The potentiometers from 1 to 8 measured the relative displacement between beams and columns, meanwhile the potentiometers from 9 to 12 measured the axial displacement of the beams

for wood. In order to prevent this premature failure, six 9 mm WRT-9 screws, also manufactured by SFS Holding AG, were inserted with an inclination of 30° around the column's bolts. Two standard S235, 27 mm diameter bolts were used to connect the kerf plate to the T-shaped plate. These two bolts were tightened with a torque of 500 Nm, which allowed for considerable friction between metal plates while permitting the wooden members to remain below their elastic limits. Furthermore, to increase the frictional energy, two additional 90×90 mm, 10 mm thick plates were placed in-between the metal plates and the bolts' nuts, one at each bolt end.

The connection showed a flagged-shaped moment-rotation curve with good energy dissipation. The maximum moment capacity was 23.28 kNm with a maximum rotation of 0.049 radians. The cumulative dissipated energy reached a value of 21.06 kNm-radians.

Fig. 22.6 Detail of the frictional BTC during testing, including: potentiometer and metal filament to measure the relative displacement between beam and column. Beam's kerf plate with a base of 335 mm, a length of 450 mm and a thickness of 4 mm fixed with WS-T-7, 7 mm self-tapping screws. Column's T-shaped metal plate with a base of 340 mm, a length of 100 mm and a thickness of 10 mm fixed with four S235, 9 mm bolts. Two S235, 27 mm frictional bolts tightened with a torque of 500 Nm and four additional 90 × 90 × 10 mm friction plates governed the friction of the BTC

22.5 Scaled Three-Story Frame

The size of the frame was scaled 1:3. The materials used in the construction of the scaled frame were the same as those used in the testing of the connections and the previous mock-up testing. The columns were 1680 mm in length with a cross section of 140 × 140 mm for the first story. The lengths of the column were then decreased to 1200 mm for the following two stories. The beams measured 2160 mm in length, 200 mm in depth and 100 mm in width. The length of the overhang beams were 825 mm and were located only on one corner of the structure. The purpose of these overhangs was to simulate adjacent bays. The beams and columns were fiber reinforced at each of the BTC connections in the same manner described for the individual connection test. Each floor was made with a CLT deck with a 3000 × 3000 mm footprint. The deck was made of plywood, with a thickness of 50 mm and a density of 500 kg/m^3. Figure 22.7a shows the completed structure ready for testing. Steel masses for a total of 150 kg per floor, were placed on the deck next to one column to simulate the combination of dead and live loads. The masses were placed only

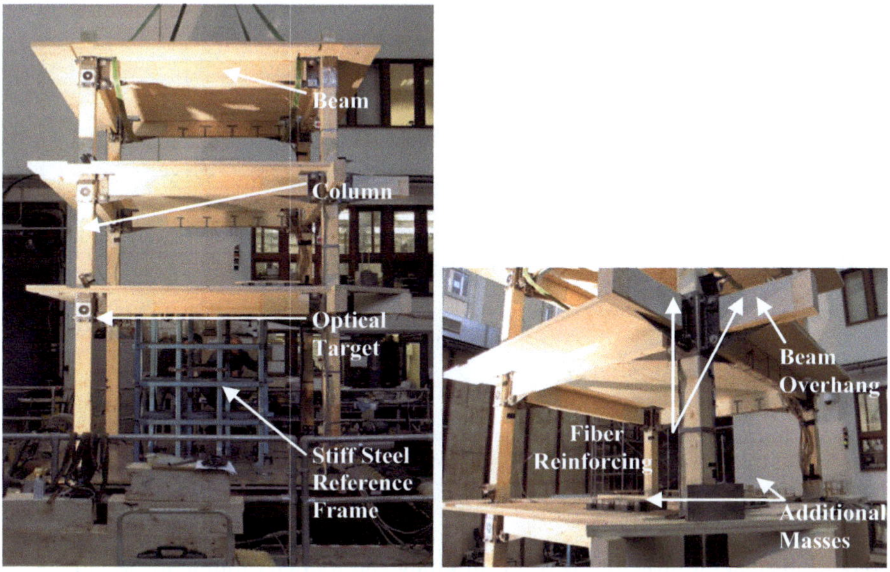

Fig. 22.7 **a** Main components of the three-story full-scale frame tests including: 1680 × 140 × 140 mm columns, 2160 × 200 × 100 mm beams, 3000 × 3000 × 50 mm deck, the 3000 × 3000 mm shaking table, 12 added steel masses of 38 kg each and a stiff steel structure below the 1st story for measuring the drifts of the frame with respect to the shaking table. **b** The locations of the added masses and overhanging beams connected to only one corner of the structure. Visible here is the fiber reinforcement placed on the beams and columns at the connections

on the first floor for one static load test and then on all floors for the full seismic tests. The total mass of the structure with loads was approximately 1750 kg. See Fig. 22.7b for additional clarification of the mass locations.

Figure 22.8 provides an overview of the locations of the measurement devices used. Channels 1 through 8 are string potentiometers used for determining joint rotations, while channels 9 through 12 are LVDTs used for determining sway (drift) of the first floor only. These channels were attached to a stiff steel reference frame located beneath the first floor. Using these instruments, displacements were measured directly. These measurements can be used to double check the results of double integration of accelerometer channels placed on the first floor, which will determine the accuracy of displacements gathered from accelerometer channels on the higher floors. Channels 13 through 18 were LVDTs used for measuring cross-bay sway of the next two stories. All of these channel locations are seen in Fig. 22.8a. In Fig. 22.8b accelerometers Channels 19 through 42 are shown. It was important to place accelerometers on each floor at the floor boundaries to be able to capture the torsional effects during testing. This was important because the structure was not symmetric in mass distribution and geometry. Channels 43 through 62 were unidirectional strain gauges placed on both beams and columns. Their locations are shown in Fig. 22.8c. These measurements were important for validating the

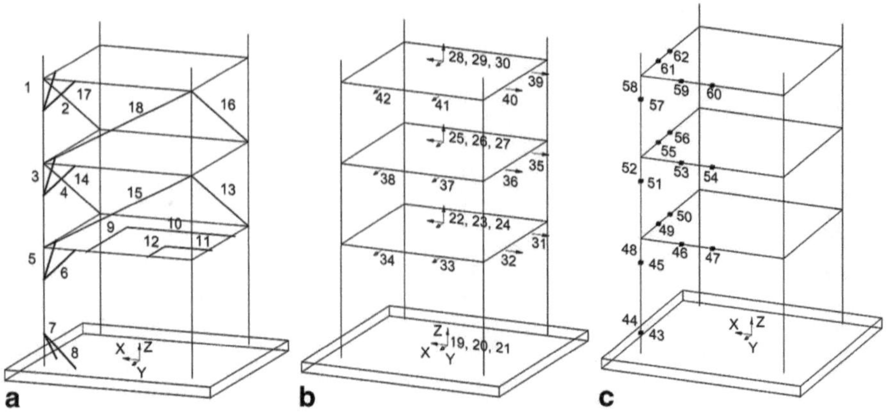

Fig. 22.8 Instrumentation locations and device types. **a** Channel numbering for the string potentiometers and LVDTs. **b** Channel numbering and locations for the accelerometers. **c** Channel numbering and locations of the strain gauges

hypothesis that the structure was self-aligning and for determining the strain levels. The last system of measurement was the optical measurements taken using targets on the corners of the structure. The locations of the targets are seen in Fig. 22.7a.

A variety of testing types were implemented in the testing protocol in order to assess the structure's properties in every direction during differing conditions. A total of 20 different tests were performed as documented in Table 22.1. The impact test was the first and done in both horizontal directions. It is important to note the directions of testing because of structure asymmetry. The impact testing was performed using a large mass against one of the protruding beams of the adjacent bays, for determining the structure's response to a single impulse. The next significant test was a snapback test which was essential for determining the system's damping parameters and initial stiffness. This was done by pulling the top of the structure to a certain deflection, cutting the chord which released the structure to oscillate and allow for damping measurements. The sine sweep testing was conducted to determine the resonant frequencies of the structure, which would allow for calibration of the synthetic seismic spectrum used for testing. The structure's natural frequencies were then evaluated using low level white noise tests, so as not to harm or degrade the structure any further after each seismic test.

There were five seismic tests conducted; the structure failed during the last test at approximately 2 g (PGA). Following each of the first four tests, a white noise test was done to measure the degradation of the structures stiffness and BTC connections via a transfer function. The transfer function of the white noise data indicates how each natural frequency decreased following each individual seismic test. The benefit of a degrading system during a seismic event is that it prevents the structure from continuously vibrating at resonant frequencies. It was found that the self-aligning capability of the wooden frames was achieved. However the drift was not sufficiently controlled.

Table 22.1 Testing protocol

No.	Excitation type	Excitation level (PGA/RMS/Amplitude)	Excitation direction
1	Impact test	–	X
2	Impact test	–	Y
3	Static load test	No dead load	–
4	Static load test	150 kg on 1st floor (Column 1)	–
5	Snapback-pullover	2300 N	Y
6	White noise	~ 0.1 g (RMS*)	X
7	White noise	~0.1 g (RMS)	Y
8	White noise	~0.1 g (RMS)	X+Y
9	Sine sweep	~0.1 g (Amplitude)	X
10	Sine sweep	~0.03 g (Amplitude)	X
11	Sine sweep	~0.03 g (Amplitude)	Y
12	First seismic test	~0.1 g (RMS)/0.02 g**(PGA***)	X+Y+Z
13	White noise	~0.1 g (RMS)	X+Y
14	Second seismic test	~0.3 g (RMS)/0.06 g*(PGA)	X+Y+Z
15	White noise	~0.1 g (RMS)	X+Y
16	Third seismic test	~0.5 g (RMS)/0.1g*(PGA)	X+Y+Z
17	White noise	~0.1 g (RMS)	X+Y
18	Fourth seismic test	~1.0 g (RMS)/0.2g*(PGA)	X+Y+Z
19	White noise	~0.1 g (RMS)	X+Y
20	Fifth seismic test	~2 g (PGA)	X+Y

*Root Mean Square (RMS)
**Vertical excitation (Z)
***Peak Ground Acceleration (PGA)

22.6 Conclusion

The results of these sets of experiments revealed that it is very difficult to control the drifts of timber frame structures. Relatively rigid moment connections are possible but cannot sufficiently control the drift. There were relatively large drifts even though the rotations of the BTC connections were quite small. The frames exhibited self-aligning properties as desired. The self-correcting property of the wood-laminated frames can be achieved by balancing the stiffness and degradation parameters of the BTC and connections between the frame and the ground.

The stiffness degradation can be viewed as a positive aspect which, at large amplitudes, prevents the frame from vibrating at resonant frequencies. The frames resisted considerable loads without exceeding the theoretical ultimate strains, so that beams and columns remained in the elastic range during large excitations.

Acknowledgements The research leading to these results has received funding from the European Union Seventh Framework Programme [FP7/2007–2013] for the access to the laboratory of the University of Bristol, UK under grant agreement n° 227887. A number of individuals were involved in this project and helped at various stages of the experiments: Matt Dietz and Luisa Dihoru, University of Bristol, UK; Piotr Bobra and Andrzej Marynowicz, University of Technology Opole, Poland. Help of the staff of the Fraunhofer WKI Braunschweig and University of Bristol is gratefully acknowledged.

References

Buchanan AH, Fairweather RH (1993) Seismic design of glulam structures. Bull N Z Natl Soc Earthq Eng 26:415–436

Ceccotti A, Vignoli A, Goirdana C (1994) Seismic tests on full-scale timber structures. In: Proceedings from pacific timber engineering conference, Gold Coast 1:232

DIN 1052 (2004–2008) Design of timber structures—General rules and rules for buildings. Deutsches Institut für Normung, Berlin

DIN EN 12512 (2002–2008) Cyclic testing of joints made with mechanical fasteners. Deutsches Institut für Normung, Berlin

Frenette C, Foschi RO, Prion HG (1996) Dynamic behaviour of timber frame with dowel type connections. In: Proceedings from the international wood engineering conference, New Orleans, USA 4:89

Heiduschke A, Kasal B, Haller P (2009) Shake table tests of small- and full-scale laminated timber frames with moment connections. Bull Earthq Eng 7:323–339

HP-Textiles (2010) Epoxi-System HP-E45KL—Laminierharzsystem. Technisches Datenblatt, Schapen

Kasal B, Pospíšil S, Jirovsky I, Drdacky M, Heiduschke A, Haller P (2004) Seismic performance of laminated timber frames with fiber reinforced connections. Earthq Eng Struct D 33:633–646

Kasal B, Guindos P, Polocoser T, Heiduschke A, Urushadze S, and Pospisil S (2014) Heavy Laminated Timber Frames with Rigid Three-Dimensional Beam-to-Column Connections. J. Perform. Constr. Facil. 28, SPECIAL ISSUE: Performance of Timber and Hybrid Structures, A4014014

Kikuchi S (1994) Earthquake resistance of multi-storey timber frame structures. In: Proceedings from pacific timber engineering conference, Gold Coast, Australia 1:205

Miyazawa K (1994) Timber shear walls and skeleton structure analyses and experiments. In: Proceedings from pacific timber engineering conference, Gold Coast, Australia 1:241

Newcombe MP, Pampanin S, Buchanan AH (2010) Global response of a two storey Pres-Lam timber building. In: New Zealand society for earthquake engineering conference, Wellington, New Zealand, pap 28

Ohashi Y, Sakamoto I, Isoda H (1994) Experiments and response analyses on three storeyed timber frame structures. Proceedings from the Pacific Timber Engineering Conference, Gold Coast, Australia, 11-15 July, Volume 1, pp. 222–231

Palermo A, Pampanin S, Buchanan A, Fragiacomo M, Deam MB (2006) Code provisions for seismic design of multi-storey post-tensioned timber buildings. In: Proceedings of CIB W18, Florence, Italy

Ponzo CF, Smith T, Di Cesare A, Pampanin S, Carradine D, Nigro D (2012) Shaking table test of a multistorey posts tensioned glulam building: design and construction. In: proceedings of world conference on timber engineering, Auckland, New Zealand 15–19 July 1:44

SFS intec GmbH (2011) Bemessungsgrundlagen System WR. Datenblatt No. 1, Heerbrugg, Switzerland

University of Patras (2011) TA project: CERBSAC. High-performance composite-reinforced earthquake resistant buildings with self-aligning capabilities. http://www.series.upatras.gr/TA_project-ITAM. Accessed 7 May 2013

Yasumura M (1998) Structural behavior of timber joints under earthquake loading. In: Proceedings of the COST C1 international conference, Liege, Belgium, p 337

Chapter 23
Experimental Study on Seismic Performance of Precast Concrete Shear Wall with Joint Connecting Beam Under Cyclic Loadings

Xilin Lu, Dun Wang and Bin Zhao

23.1 Introduction

Since the advent of the use of precast concrete structural systems, for many decades numerous investigations on the performance and development of novel precast concrete systems have been conducted, such as unbonded post-tensioned precast concrete frame structures (El-Sheikh et al. 1999), unbonded post-tensioned precast concrete walls (Kurama et al. 1999), self-centering precast concrete wall systems with energy dissipators (Restrepo and Rahman 2007) and partially prefabricated laminated composite RC walls (Zhang et al. 2011). Structural connections are a major feature of precast concrete shear walls—distinguished from monolithic shear walls—as they play an important role on the seismic performance and safety of precast concrete shear wall structures, including shear capacity, stiffness and ductility, energy dissipation, and force-transfer (fib Bulletin No. 43 2008). Plenty of studies have been conducted on the performance of precast concrete shear walls with various joint connections (Pekau and Hum 1991; Schultz et al. 1994; Henry et al. 2010). In China, Jiang (Jiang et al. 2011b) first conducted a test on 81 plug-in filled holes for lap-joint steel bar samples, considering factors such as reinforcement bar diameter, concrete strength and anchorage length. These tests were followed by elastic quasi-static tests on a three-story full-scale model using the same connecting method for the steel bars (Jiang et al. 2011b). Qian (Qian et al. 2011) conducted an ex-

X. Lu (✉) · B. Zhao
State Key Laboratory of Disaster Reduction in Civil Engineering, Tongji University, Shanghai, China
e-mail: lxlst@tongji.edu.cn

B. Zhao
e-mail: binzh@tongji.edu.cn

D. Wang
Research institute of Structural Engineering and Disaster Reduction, Tongji University, Shanghai, China
e-mail: waltdon77@126.com

F. Taucer, R. Apostolska (eds.), *Experimental Research in Earthquake Engineering*, Geotechnical, Geological and Earthquake Engineering 35,
DOI 10.1007/978-3-319-10136-1_23

perimental study of five wall specimens with an aspect ratio of 2.25, connecting the vertical reinforcement with various splicing techniques such as a single connecting rebar, a mechanic sleeve connection and a sleeve-mortar splicing connection. Ong (Ong et al. 2006) reported the flexural behavior of two precast concrete specimens with one-to-one horizontal loop connections and concluded that an increase in the loop overlapping length, a decrease in the internal diameter, or an overlap spacing of the loop resulted in an increase in the flexural strength of the precast specimen tested. Ma (Ma et al. 2012) conducted an investigation of longitudinal joint details with tight bend diameter U-bars for accelerated bridge construction.

In this paper the concept of a Joint Connecting Beam (JCB) is developed to connect the vertical reinforcement of precast concrete shear walls. The JCB is composed of staggered splicing rectangular steel loops protruding from the wall panel, an assembly of longitudinal steel bars and stirrups, and casted concrete. The influence of the location and height of the JCB on the performance of precast concrete shear walls is studied and comparisons are made with the results of quasi-static tests on monolithic shear walls. The failure mode, the top lateral force-displacement hysteresis curve, lateral strength capacity, deformation, stiffness degradation, reinforcement strains and energy dissipation are presented in detail. The work indicated that the concept and detailing of the JCB are feasible and applicable to precast concrete shear wall structures. Some findings are presented based on the research and suggestions are made for further investigations and engineering applications.

23.2 Experimental Programme

23.2.1 Specimen Design

Seven full-scale wall specimens were designed according to the Chinese code for design of reinforced concrete structures (GB 50010 2010). This included six precast concrete shear walls and a monolithic shear wall. All the wall specimens have the same rectangular section of 1500 × 200 mm and same concrete strength grade C40. The height of the wall specimens is 2.60 m with a top beam of 400 × 400 × 1800 mm (height and width and thickness) for lateral loading and vertical loading, resulting in an aspect ratio of 2.8 (measured from the bottom of the wall to the centerline of the top beam). The main difference between the precast wall specimens and the monolithic conventional wall is the existence of the joint connecting beam (JCB). The height of the JCB varies with values of 150, 200, 300 mm, chosen to study the effect of the lapped length of steel bars according to the Chinese code (GB 50010 2010). The locations of the JCB are in the middle and at the bottom of the precast wall, with the purpose of checking its role in determining the performance of precast wall. The JCB has the same concrete strength grade C40 of the wall. Detailed information of wall specimens can be found in Table 23.1. All the wall specimens were constructed in a batch at the State Key Laboratory for Disaster Reduction in Civil Engineering of Tongji University. The characteristic values of concrete and steel bars obtained from material property tests are listed in Table 23.2 and 23.3.

Table 23.1 Design parameters of wall specimens

Speci-mens	Joint connecting beam				Boundary element		Distrib-uted bars	Concrete
	Height (mm)	Location	longi-tudinal bars	Stirrups	longi-tudinal bars	Stirrups	Horizon-tal and vertical	Strength
SW-2	–	–	–	–	6ϕ10	ϕ6@75	ϕ8@150	C40
PCW-7	150	Bottom	4ϕ12	ϕ6@100	6ϕ10	ϕ6@75	ϕ8@150	C40
PCW-8	200	Bottom	6ϕ10	ϕ8@100	6ϕ10	ϕ6@75	ϕ8@150	C40
PCW-9	200	Middle	6ϕ10	ϕ8@100	6ϕ10	ϕ6@75	ϕ8@150	C40
PCW-10	300	Middle	6ϕ10	ϕ10@100	6ϕ10	ϕ6@75	ϕ8@150	C40
PCW-11	200	Bottom & Middle	6ϕ10	ϕ8@100	6ϕ10	ϕ6@75	ϕ8@150	C40
PCW-12	300	Bottom & Middle	6ϕ10	ϕ8@150	6ϕ10	ϕ6@75	ϕ8@150	C40

Table 23.2 Mean value of concrete properties

Location of Concrete	Cubic Strength (MPa)	Prism Strength (MPa)	Modulus of Elasticity (N/mm^2)	Design Strength Grade
Precast concrete wall	39.1	26.15	3.33×10^4	C40
Joint connecting beam	38.5	25.81	3.05×10^4	C40

Table 23.3 Mean value of steel bar strength

Diameters	Yielding strength	Ultimate ten-sile strength	Position of reinforcement	Strength grade
(mm)	(Mpa)	(Mpa)		
6	467.5	607.5	Stirrups A, Ties	HPB300
6	510	620	Stirrups B	HRB400
8	557.5	630	Distributed reinforce-ment, stirrups B	HRB400
10	552.5	657.5	Reinforcement A and B, stirrups B	HRB400
12	497.5	632.5	Reinforcement B	HRB400

A denotes boundary element; B denotes Joint Connecting Beam (JCB)

23.2.2 Measurement and Test Procedure

Figure 23.1 illustrates the test set-up and the loading history of the wall specimens. A vertical load was first applied to the fixed wall specimens by four hydraulic jacks on top of the loading beam, resulting in an axial force of 1200 kN that was kept constant throughout the entire testing process. A horizontal load was then applied by a Schenck actuator with a 200 mm maximum stroke and 1000 kN capacity, as shown in Fig. 23.1a. A displacement-controlled loading history was exerted on the specimens throughout the entire testing, as shown in Fig. 23.1b. The displacement amplitude of the loading history was applied at 1 mm cycle increments in the elastic range; the yield displacement was determined from the strain of the outermost reinforcement in the boundary element of wall specimen. In the inelastic range displacements were applied at increments equal to the measured yield displacement, with three cycles repeated at each displacement value. The yield displacement for the wall specimens was equal to 8 mm (for PCW-11), 10 and 11 mm (for PCW-12).

Before the start of testing, cracks due to temperature shrinkage during the construction stage along the top interface of the JCB were monitored and marked to distinguish them from cracks due to the damage of specimens during testing. During the entire process of tests, the top displacement, horizontal force, longitudinal and transverse rebar strains, and concrete strains were measured. Displacement of the anchoring beam at the bottom of each wall specimen, which is usually negligible, was also monitored by a displacement transducer, in order to derive the net drift of the wall specimen. In order to monitor the slip displacement between the top interface of the JCB and the precast wall panel above, some additional steel bars with a diameter of 6 mm were installed horizontally in the specimen along the interface of the joint connecting beam and the wall panel above.

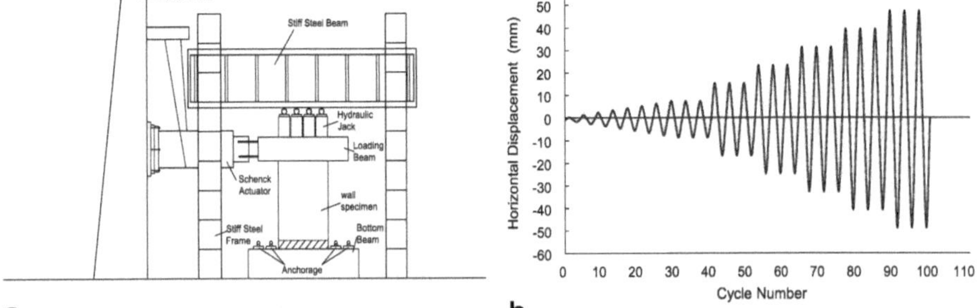

a b

Fig. 23.1 Test set-up and loading history

23.3 Test Results

23.3.1 Overview

The sequence of damage of the wall specimens was observed as follows: at first cracks emerged along the top interface of the JCB and merged with cracks formed during the construction stage, gradually becoming a line. Minor cracks then appeared at the end side and at the bottom of the wall. Vertical cracks occurred at the end and at the middle of the JCB. Meantime, minor slant cracks developed at the corner of the wall and developed upwards. With the increase of lateral displacement, minor spalling of concrete occurred at the top interface of the joint connecting beam, with main diagonal inclined cracks developing both in the pull and push directions. With the increase of vertical cracks and new horizontal cracks in the JCB, concrete at the corner of the wall began to crack severely, resulting in initial spalling of the concrete cover. After reaching the maximum load capacity of the wall specimens, complete spalling of the concrete cover, buckling of the longitudinal rebar and/or fracture of the longitudinal rebars was observed. The test was stopped when the lateral capacity decreased to at least 15 % of the maximum load capacity of the wall specimen or when there was a loss of the vertical load bearing capacity. After completion of all tests it was concluded that the failure mode of all wall specimens was a combination of flexure and shear.

Figure 23.2 shows the damage patterns of some typical wall specimens. The drift ratio θ of the wall specimen is defined as the top net displacement divided by the height of the wall specimen. It is observed that when the JCB is located at the bottom of the wall specimen concrete was severely damaged, with complete spalling of the concrete cover of the JCB along a large region (Fig. 23.2a), gap-opening on the top interface of the JCB (Fig. 23.2b and e), minor buckling or no buckling with not fracture of the longitudinal rebar with (Fig. 23.2b and d). When the JCB is located in the middle of the wall specimen, concrete at the bottom damaged severely with spalling of the concrete cover, and buckling and fracture of the longitudinal rebar (Fig. 23.2c), similar to the damage observed for the monolithic wall specimen (Fig. 23.2f).

23.3.2 Hysteresis Behavior and Skeleton Curve

The hysteresis loops of some representative wall specimens are shown in Fig. 23.3. As shown in Fig. 23.3, the characteristics of the hysteresis curves of precast concrete wall specimens are similar to that of the monolithic wall specimen. Initial behavior is practically linear-elastic with few hysteretic cycles developing before appearance of the first cracks. However, with the increase of lateral horizontal displacement after cracks develop, the curve begins to slope down towards the displacement axis, with hysteretic loops of larger area and the pinching effect gradually emerging as

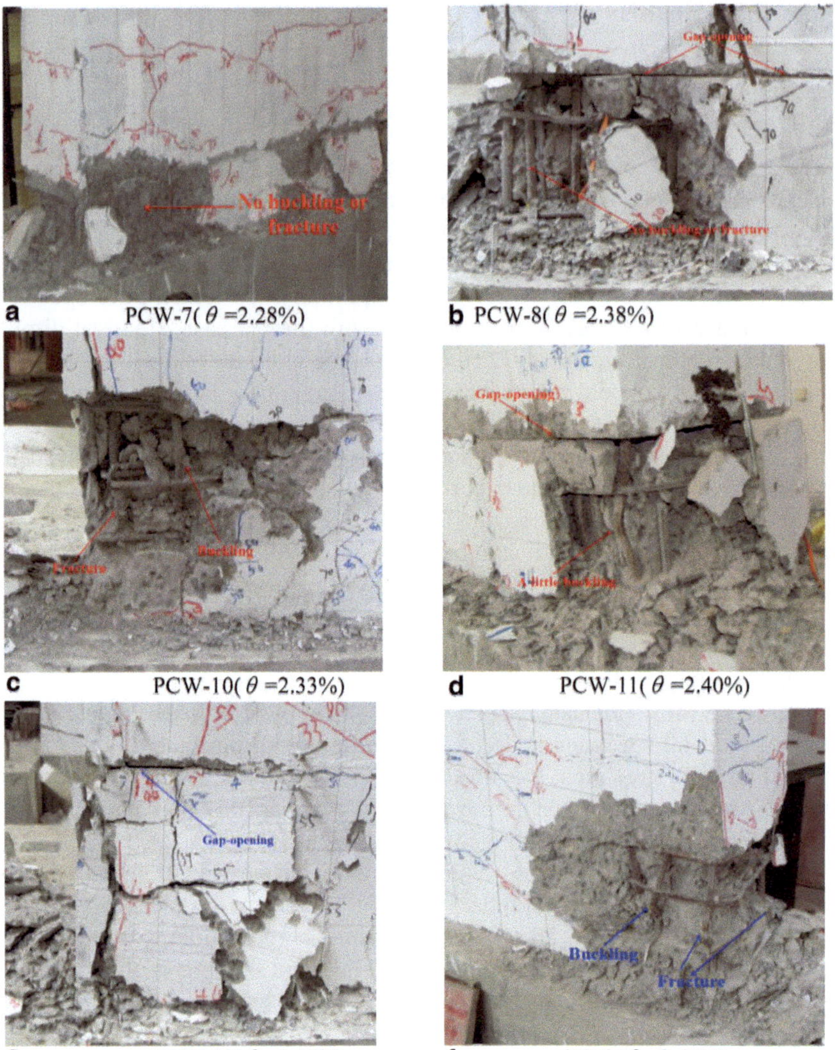

a PCW-7(θ =2.28%) b PCW-8(θ =2.38%)

c PCW-10(θ =2.33%) d PCW-11(θ =2.40%)

e PCW-12(θ =2.36%) f SW-2(θ =2.45%)

Fig. 23.2 Damage conditions of wall specimens

diagonal cracking develops. The lateral load capacity of wall specimen decreases after reaching the maximum load capacity. The test was stopped due to the loss of vertical load carrying capacity, also demonstrated by the skeleton curves.

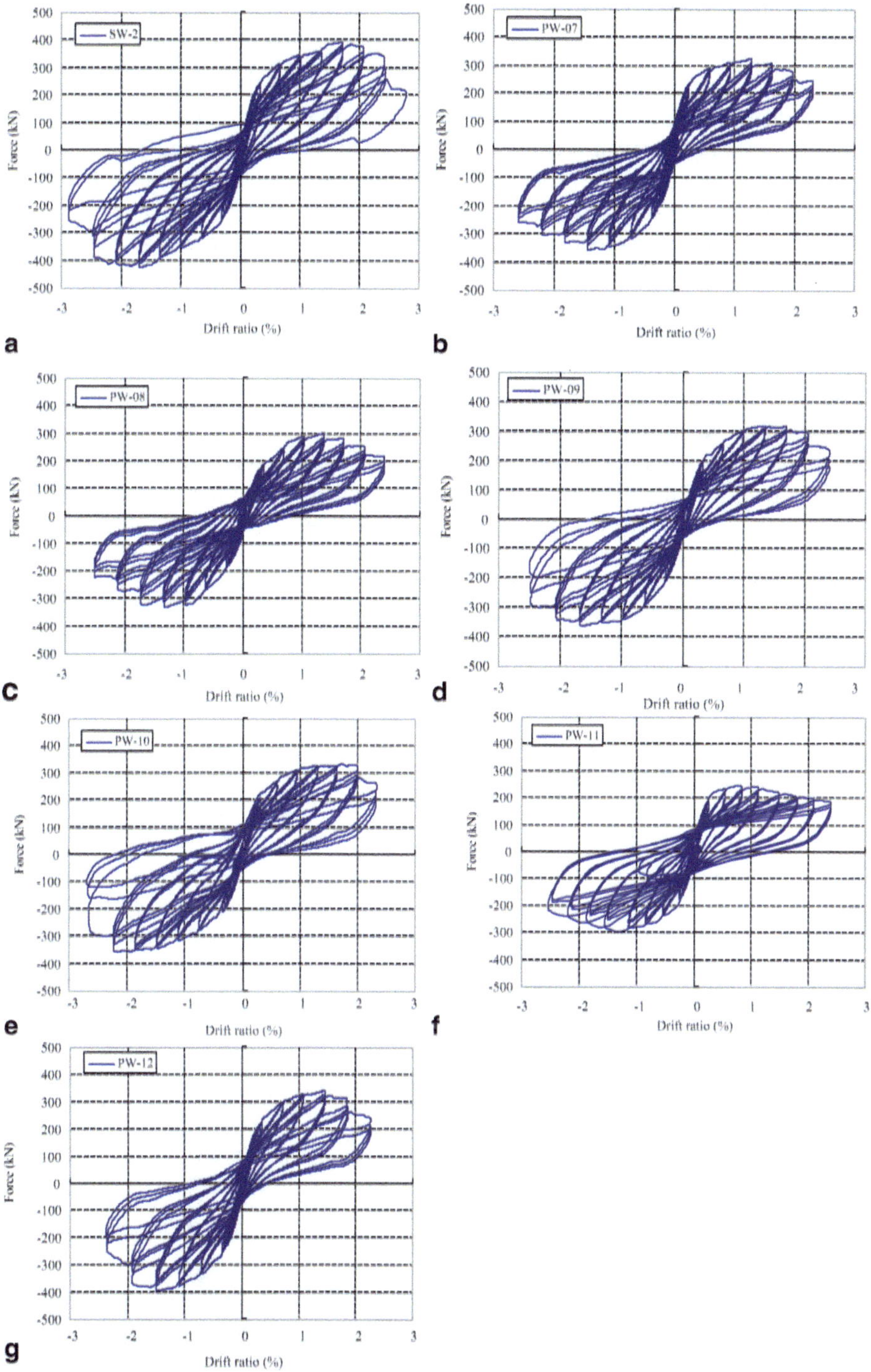

Fig. 23.3 Hysteresis loops of wall specimens

23.3.3 Lateral Strength

Table 23.4 presents the lateral forces of wall specimens in the directions of pull and push at different states, such as cracking, yielding and maximum load capacity. The skeleton curves of the wall specimens are illustrated in Fig. 23.4 and 23.5 and show the ratio of maximum lateral shear strength of precast wall specimens to that of the monolithic wall SW-2. As can be seen from Fig. 23.4 and 23.5, the lateral strength of the precast concrete shear wall with JCB is lower than the lateral strength of the monolithic shear wall, but not lower than 80 % (except for PCW-8 and PCW-11). Moreover, with the increase of the height of the JCB when located at the middle of the wall specimens, the lateral strength rises up, with the exception of the case when the JCB is located at the bottom (especially for specimens PCW-8). Further studies are needed on determining how the location of the JCB affects the lateral strength of the shear wall. With smaller heights of the JCB at the bottom of the wall, the precast concrete wall showed the lowest load capacity, such as in specimens PCW-11. This lower capacity of the PCW-11 specimen is attributed to the fact that the PCW-11 specimen was tested for the first time, and to double casting of the JCB. Another reason is a different loading history, with cycles of 8 mm increments applied in the inelastic range.

Table 23.4 Lateral force of wall specimens at various states

Specimens	F_{cr}/kN			F_y/kN			F_{max}/kN		
	Push	Pull	Average	Push	Pull	Average	Push	Pull	Average
SW-2	135.8	144.8	140.3	283.4	297.8	290.6	394.4	426.1	410.25
PCW-7	99	108.1	103.55	237.07	266.27	251.67	325.7	357.7	341.7
PCW-8	89.63	85.47	87.55	222.37	246.2	234.29	301.74	330.21	315.98
PCW-9	84.6	89.2	86.9	232.94	267.5	250.22	319.7	362.4	341.05
PCW-10	87.63	96.48	92.06	241.77	252.88	247.33	333.67	355.9	344.79
PCW-11	131.1	203.9	167.5	188.87	211.07	199.97	245.6	295.6	270.6
PCW-12	97.02	93.48	95.25	246.27	280.69	263.48	391.68	340.59	366.14

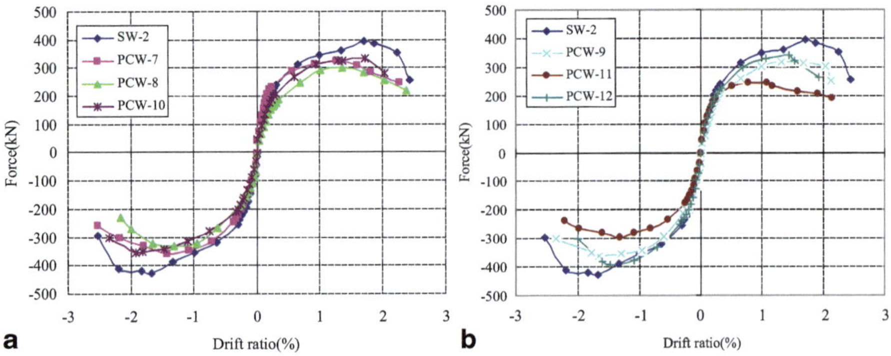

Fig. 23.4 Skeleton curves of wall specimens

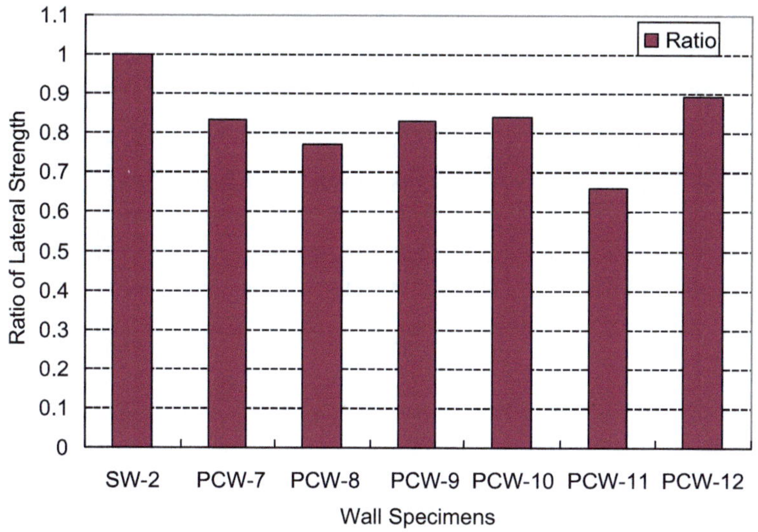

Fig. 23.5 Ratio of lateral strength of wall specimens

23.3.4 Ductility Evaluation

Ductility characterizes the capacity of a structure or member to deform without significant degradation of its loading capacity, and is generally defined by the so-called displacement ductility coefficient μ_Δ:

$$\mu_\Delta = \Delta_u / \Delta_y \tag{23.1}$$

where Δ_u and Δ_y are the ultimate and yield deformations. As mentioned by many researchers (Chen et al. 2009), different definitions may result in a remarkable difference in the displacement ductility coefficient. In this paper, the yield displacement during the test used for the loading history was determined by the first yield strain of the outermost longitudinal reinforcement in the boundary element of the wall specimen. The yield displacement for calculating the displacement ductility coefficient was determined by a combination of the Park-method (Park 1989) and the method of equivalent energy of ASTM ES2126-02a (ASTM 2003). It was found that there are some differences in the yielding displacement calculated according to these formulas, yet not listed here. That the displacement ductility coefficient μ_Δ was computed following the expression:

$$\mu_\Delta = \frac{|+\Delta_u| + |-\Delta_u|}{|+\Delta_y| + |-\Delta_y|} \tag{23.2}$$

Table 23.5 Deformation of wall specimens at various states and ductility

Specimens	Δ_{cr}/mm			Δ_y/mm			Δ_u/mm			Ductility	
	Push	Pull	Average	Push	Pull	Average	Push	Pull	Average	μ_Δ	$\xi(\Delta_u)$
SW-2	3.5	2.8	3.15	14.49	15.21	14.85	63.97	65.53	64.75	4.36	0.808
PCW-7	1.9	2.2	2.05	7.11	13.45	10.28	53.96	60.49	57.23	5.57	0.824
PCW-8	3.53	0.89	2.21	15.31	15.08	15.2	57.32	54.15	55.74	3.68	0.826
PCW-9	1.3	3.2	2.25	14.15	14.37	14.26	58.5	62.63	60.57	4.25	0.855
PCW-10	2.19	2.51	2.35	13.07	16.74	14.91	59.94	65.27	61.11	4.1	0.815
PCW-11	2.2	2.3	2.25	6.82	11.82	9.32	50.01	58.77	54.39	5.83	0.899
PCW-12	2.02	2.13	2.08	12.43	13.63	13.03	49.1	51.17	50.14	3.85	0.882

Considering the complexity and different definitions of the yield displacement, a new index of the ductility factor is adopted in this paper to evaluate the ductility characteristics of wall specimens. The relative strain energy $\xi(\Delta_u)$ is defined as follows (fib Bulletin 2008):

$$\xi(\Delta_u) = W_{\text{int}}(\Delta_u) / (F_{\max} \Delta_u) \tag{23.3}$$

where $W_{\text{int}}(\Delta_u)$ is the area defined by the load-displacement relationship of the wall specimen at ultimate displacement and $F_{\max}$ is the maximum load capacity of the wall specimen. Values of the ductility factor for different indices are listed in Table 23.5, showing that the ductility of precast wall specimens is slightly superior to that of a monolithic wall specimen.

23.3.5 *Strain Distribution of Reinforcements*

The strain distribution of reinforcement in the boundary element, 20 mm above the bottom cross section of the wall specimen is shown in Fig. 23.6 for the 2 mm, 10 mm (8 mm for PCW-11, 11 mm for PCW-12) and 30 mm (24 mm for PCW-11, 33 mm for PCW12) loading in the pull direction, which corresponds approximately to the cracking stage, the yield stage and occurrence of the main diagonal cracks, respectively. It is noted that the plane-section assumption holds for all the wall specimens before the yielding stage, thereafter, for the wall specimens with the JCB at the bottom, the plane-section assumption is no longer maintained when compared with the other wall specimens. Moreover, none of the walls (regardless of the location and height of the JCB and for the monolithic wall specimens SW-2) conform

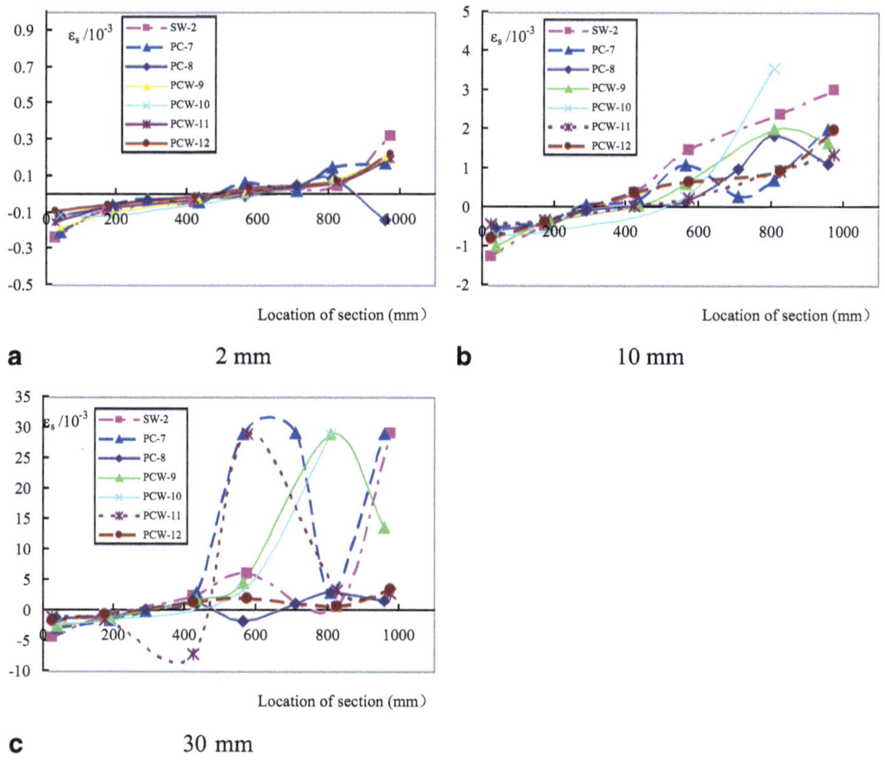

Fig. 23.6 Strain reinforcement distribution of boundary elements at different loading states

with the plane-section assumption at sustained levels of loading after yielding and formation of the main diagonal cracks.

23.3.6 Energy Dissipation

The energy dissipation of the wall specimens with the increase of lateral displacement is calculated by calculating the area under the force-displacement curves as shown in Fig. 23.7. The curves show that prior to yield (i.e. 10 mm), a very small amount of energy is dissipated. After that, a larger amount of energy is dissipated as displacements increase (except for PCW-12). When the JCB is located at the bottom, precast wall specimens dissipate less energy than a monolithic shear wall. When the JCB is located in the middle of the wall, the energy dissipation of wall all specimens is nearly the same. The height of the JCB seems not to have a primary influence on the energy dissipation of wall specimens.

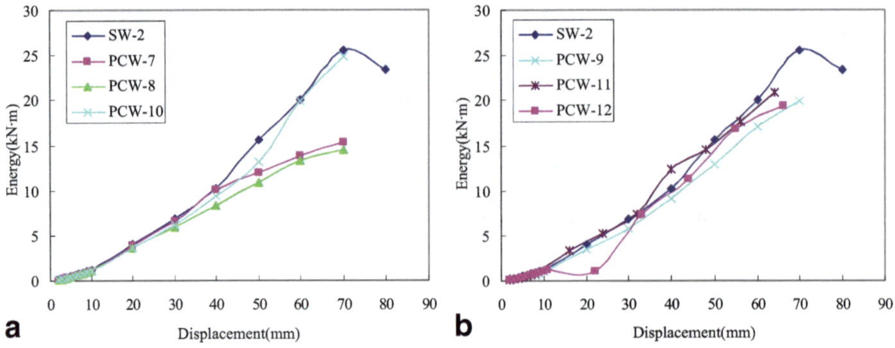

Fig. 23.7 Energy dissipation of all the wall specimens

23.3.7 Stiffness Degradation

Stiffness K_i is defined as the secant stiffness of a wall specimen at the maximum displacement of each cycle under cyclic lateral displacement(JGJ101-96 1996); the calculation formula is as follows:

$$K_i = \frac{|F_i| + |-F_i|}{|\Delta_i| + |-\Delta_i|} \tag{23.4}$$

where F_i is the maximum force during the ith cycle, and Δ_i is the horizontal displacement corresponding to the maximum force. The positive and negative signs represent the direction of pull and push, respectively. It is clearly demonstrated from Fig. 23.8 that stiffness degradation of all wall specimens is nearly the same, especially in the inelastic stage, with a few differences in the elastic stage, regardless of the location and the height of the JCB.

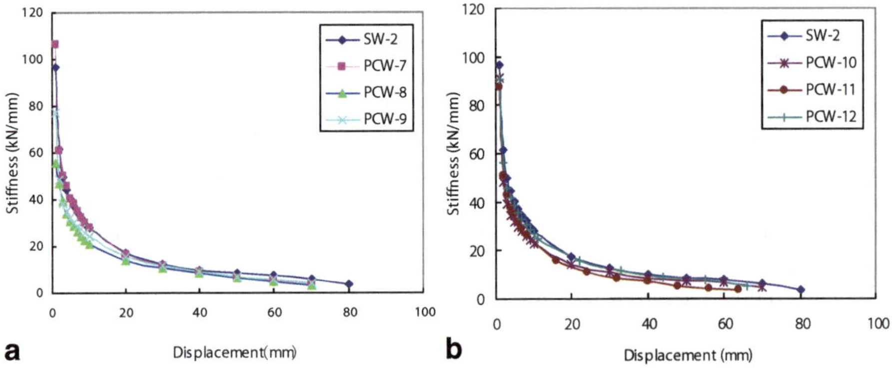

Fig. 23.8 Stiffness degradation of wall specimens

23.4 Conclusions

In this paper, six precast concrete shear wall members with a Joint Connecting Beam (JCB) were designed according to the current seismic Chinese design code. Cyclic tests were conducted at the State Key Laboratory of Disaster Reduction in Civil Engineering, Tongji University. The following conclusions are drawn after comparative analysis of the test results:

- The failure mode of the precast wall with the JCB specimen is similar to that of a monolithic wall specimen, which includes flexural-shear failure with crushing of concrete at the bottom, tensile yield or bucking, and in some cases fracture of vertical reinforcement in the boundary elements, all these observed when the JCB is located in the middle of the wall specimen.
- The lateral load capacity of precast wall specimens with the JCB is lower than that of a monolithic shear wall, not exceeding 20 %, except for the PCW-11 specimen. The height and location of the JCB have an important role in the strength of the specimen, the bigger the height of the JCB, the higher the strength of the wall specimen. For JCBs located at the bottom of the wall adverse conditions are found, especially for smaller heights due to the existence of a plastic hinge region. By using higher strength concrete and steel fiber concrete, this situation may be improved. The deformation of precast walls with JCB is slightly smaller and occurs earlier than for the monolithic case; however, the ductility of precast walls with JCB is superior to that of a monolithic wall, as confirmed by the index of relative strain energy. The higher the height of the JCB, the lower the ductility found in the precast walls. Energy dissipation and stiffness degradation of precast walls with JCB are similar to those of monolithic wall specimens, regardless of the location of JCBs.
- The concept of the joint connecting beam is feasible and effective. It can effectively transfer stresses in the reinforcement of the wall above and below the JCB. However, further studies are needed such as the use of high quality JCBs to improve the performance of precast concrete walls, so as to promote the application of precast concrete walls with joint connecting beams.

Acknowledgments The financial supports from the Key Technology R&D Program for the 12th 5-year Plan of China under Grants No. 211BAJ09B01-03 are gratefully acknowledged.

References

ASTM (2003) Standard test method for cyclic (Reversed) load test for shear resistance of walls for buildings, ES 2126-02a

Chen LZ, Lu XL, Jiang HJ, Zheng JB (2009) Experimental investigation of damage behavior of RC frame members including non-seismically designed columns. Earthq Eng Eng Vib 8(2):301–311

El-Sheikh M, Sause R, Pessiki S, Lu L-W (1999) Seismic behavior and design of unbonded post-tensioned precast concrete frames. PCI J 44(3):54–71

Fédération internationale du béton, Fédération internationale du béton (2008) Task Group 6.2, Structural Connections for Precast Concrete Buildings. fib Bulletin No. 43., the International Federation for Structural Concrete (*fib*), 2008

GB 50010 (2010) The code for design of concrete structures of China

Henry R, Aaleti S, Sritharan S, Ingham J (2010) Concept and finite-element modeling of new steel shear connectors for self-centering wall systems. J Eng Mech 136(2):220–229

Jiang HB, Chen ZX, Zhang JQ et al (2011a) Quasi-static test of precast reinforced concrete shear wall structure. J Build Struct 32(6):34–40 (in Chinese)

Jiang HB, Zhang HS, Liu WQ et al (2011b) Experimental study on plug-in filling hole for steel bar anchorage of the PC structure. J Harbin Inst Technol 43(4):28–36 (in Chinese)

JGJ101-96 (1996) Specification of test methods for earthquake resistant building of China

Kurama Y, Sause R, Pessiki S, Lu LW (1999) Lateral load behavior and seismic design of unbonded post-tensioned precast concrete walls. ACI Struct J 96(4):622–633

Ma ZGJ, Cao Q, Chapman CE, Burdette EG, French CEW (2012) Longitudinal joint details with tight bend diameter U-bars. ACI Struct J 109(6):815–824

Ong KCG, Hao JB, Paramasivam P (2006) Flexural Behavior of precast joints with horizontal loop connections. ACI Struct J 103(5):664–671

Park R (1989) Evaluation of ductility of structures and structural assemblages from laboratory testing. Bull New Zealand Natl Soc Earthq Eng 22(3):155–166

Pekau OA, Hum D (1991). Seismic response of friction jointed precast panel shear walls. PCI J, 36(2):56–71

Qian JR, Yang XK, Qin H (2011) Tests on seismic behavior of pre-cast shear walls with various methods of vertical reinforcement splicing. J Build Struct 32(6):51–59 (in Chinese)

Restrepo J, Rahman A (2007) Seismic performance of self-centering structural walls incorporating energy dissipators. J Struct Eng 133(11):1560–1570

Schultz AZ, Mafana RA, Tadros MK (1994) Experimental study of joint connections in precast concrete walls. Fifth U.S. National Conference on Earthquake Engineering, Vol II, 1994

Zhang HM, Lu X L, Duan YF (2011) Experimental study and numerical simulation of partially prefabricated laminated composite RC walls. Adv Struct Eng 14(5):967–979

Chapter 24
The Importance of connections in Seismic Regions: Full-Scale Testing of a 3-Storey Precast Concrete Building

Dionysios Bournas, Paolo Negro and Francisco Javier Molina

24.1 Introduction

A collaborative three-year research project called SAFECAST was undertaken by European national associations of precast concrete producers, along with universities and research centres, to study the behaviour of precast concrete structures under earthquake loading. The main objective of the project was to fill the gap in the knowledge of the seismic behaviour of precast concrete structures, with emphasis on the connections between precast members. A major part of the experimental phase of this programme consisted of pseudodynamic tests on a full-scale 3-storey precast concrete building, carried out at the European Laboratory for Structural Assessment (ELSA), Joint Research Centre (JRC) of the European Commission in Ispra.

24.2 The Mock Up

The structure of the specimen was a three-storey full-scale precast residential building, with two 7 m bays in each horizontal direction as shown in Fig. 24.1 The structure was 15 × 16.25 m in plan, with a height of 10.9 m (9.9 m above the foundation level) and floor-to-floor heights equal to 3.5, 3.2 and 3.2 m for the 1st, 2nd and 3rd level, respectively. The floor systems, which were of high interest in this research, were carefully selected to gather the largest possible useful information.

D. Bournas (✉)
Department of Civil Engineering, University of Nottingham, Nottingham, UK
e-mail: dionysios.bournas@nottingham.ac.uk

P. Negro · F. J. Molina
European Laboratory for Structural Assessment Unit, European Commission,
Joint Research Centre (JRC), Institute for the Protection and Security of the Citizen (IPSC),
Via Enrico Fermi 2749, Ispra VA 21027, Italy

F. Taucer, R. Apostolska (eds.), *Experimental Research in Earthquake Engineering*,
Geotechnical, Geological and Earthquake Engineering 35,
DOI 10.1007/978-3-319-10136-1_24

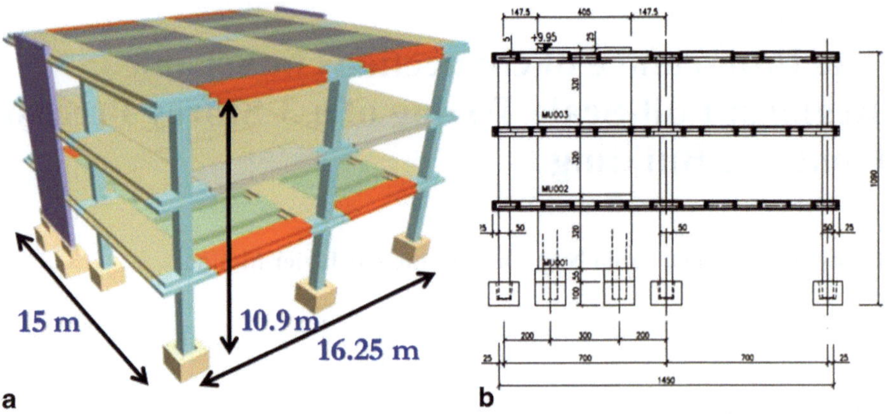

Fig. 24.1 **a** Three-dimensional representation of the test structure (dimensions in m). **b** Section view of the test structure in the loading direction (dimensions in cm)

To accomplish this, three different pretopped floor systems were adopted. As shown in Fig. 24.1, the 1st floor at 3.5 m was constructed with box-type elements welded to each other side by side; similarly the 2nd floor at 6.7 m comprised double-tee elements welded to each other side by side; and finally the 3rd floor at 9.9 m was constructed with the same box slab elements of the 1st floor, but spaced to simulate diaphragms with openings.

The precast three-storey columns had a cross-section of 0.5 × 0.5 m, kept constant along their height. The columns were embedded 0.75 m into 1-m-deep, 1.3 × 1.3 m in plan, pocket foundations. All columns were constructed with wide capitals at the level of each floor, with widths of 0.90 and 2.25 m in the loading and transverse directions, respectively, in order to allow testing the mechanical beam-to-column connection, the details of which are presented in a companion paper. The capitals of the columns were designed as cantilevers, fixed at the axis of the columns with bending and shear reinforcement. The longitudinal beams connected to the columns' capitals were precast box-type hollow core elements, with a cross-section 0.4 × 2.25 m and a length of 6.38 m.

As it can be seen in Fig. 24.1b two 4.05 m-long × 9.6 m-tall × 0.25 m-thick (4.05 m × 9.6 m × 250 mm) precast concrete walls were connected to the mock-up in order to compose with the columns a dual frame-wall precast system. Each wall comprised 3-wall hollow-core precast elements 3.2 m-tall (Fig. 24.2), which were joined among themselves by means of vertical reinforcement crossing their gaps at the level of each floor. Concrete was cast only at the two edge cores of the section, where the wall vertical reinforcement was concentrated in "boundary elements". The confinement of the concrete was also there foreseen (Fig. 24.2). The longitudinal reinforcement was lap-spliced at the mid-height of the second floor. Similarly to the columns, the walls-to-foundation connection was constructed through two pocket foundations in which the walls' longitudinal reinforcement was anchored (Fig. 24.2).

Fig. 24.2 **a** Assembling phase of the precast walls. **b** Pocket foundations used for the walls

All precast elements (columns, walls, beams, slabs) were cast using the same concrete class, namely C45/55, which turned out to have a 28-day strength, measured on 150 × 150 mm cubes, equal to 64.5 MPa. The steel reinforcement cast into the members had a yield stress of 527 MPa, a tensile strength of 673 MPa, and an ultimate strain equal to 10%. Table 24.1 summarizes the dimensions and percentages of steel reinforcement of all prefabricated structural bearing elements used for the construction of the mock-up.

The mock-up was constructed in such a way that the effectiveness of four different structural precast systems could be investigated experimentally. Therefore, the behaviour of a series of parameters, including several types of mechanical connections (traditional as well as innovative) and the presence or absence of shear walls along with the framed structure, could be assessed. The first layout (prototype 1) comprised a dual frame-wall precast system, where the two precast shear wall

Table 24.1 Dimensions and percentages of reinforcement of all prefabricated bearing elements

Precast element	Concrete Strength fc (MPa)	Type of the cross-section	Dimensions of the cross-section (m)	Amount of longitudinal reinforcement (mm²)	Geometrical ratio of longitudinal reinforcement ρs, %
Column	64.5	Solid	0.5 × 0.50	2513	1.00
1st floor beams	64.5	Hollow-core	0.4 × 2.25	1810	0.35
2nd floor beams	64.5	Hollow-core	0.4 × 2.25	1609	0.32
3rd floor beams	64.5	Hollow-core	0.4 × 2.25	1473	0.29
Wall	64.5	Hollow-core	2.4 × 0.25	8952	1.30
Wall end	55.0	Solid	0.8 × 0.25	2767	1.38
Wall web	64.5	Hollow-core	2.45 × 0.25	3418	1.20

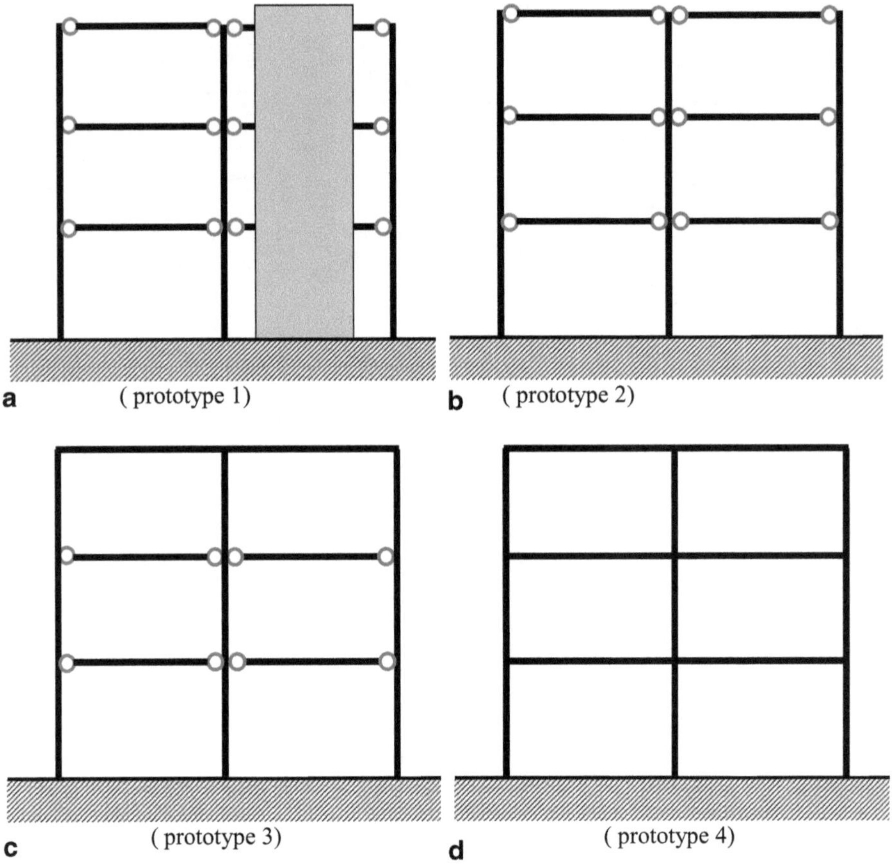

Fig. 24.3 Structural layouts of the four prototypes

units were connected to the mock-up (Fig. 24.3a). In this structural configuration, the effectiveness of precast shear walls in terms of increasing the stiffness of a relatively flexible three-storey precast building with hinged beam-to column joints was examined. In the second layout (prototype 2, Fig. 24.3b), the shear walls were disconnected from the structure, to test the building in its most typical configuration, namely with hinged beam-column connections by means of dowel bars (shear connectors). This configuration, which represents the most common connection system in construction practice in European countries, had been investigated only for industrial, typically single-storey, precast structures (Negro et al. 2007). The second layout investigates for the first time experimentally, the seismic behaviour of a flexible multi (three)-storey precast building with hinged beam-to-column connections, where the columns are expected to work principally as cantilevers. The possibility of achieving emulative moment resisting frames by means of a new connection system with dry connections was investigated in the third and fourth structural configurations. With the objective of providing continuity to the longitudinal reinforcement crossing the joint, an innovative connection system embedded in the precast elements was activated by means of bolts connecting the steel devices in the columns and beams. A special mortar was placed to fill the small gaps between beams and columns. In particular, the first solution examined was oriented to reduce the flexibility of such structures with hinged beam-to-column joints by restraining just the last floor of multi-storey buildings; and thus, in the third layout (prototype 3, Fig. 24.3c), the connectors were restrained only at the third floor. Finally, in the last and fourth layout, the connection system was activated in all beam-column joints (prototype 4, Fig. 24.3d).

24.3 Testing Programme

The continuous PsD method developed at the ELSA laboratory of the European Commission JRC (Pegon et al. 2008) was used for testing the mock-up. The PsD method couples the properties of the structure into a physical quasi-static model tested in the laboratory with a computer model representing inertia.

Translational masses of 186,857 kg at the first floor, 168,404 kg at the second floor and 132,316 kg at the top floor, were numerically represented in the PsD test of prototype 1. The corresponding masses for the layouts without shear walls (prototype 2, 3 and 4) were considered equal to 170,948, 157,978 and 127,013 kg, for the first, second and third floor, respectively. The above simulated masses in the PsD tests are larger than the actual masses of the specimen in order to reproduce the effect of additional loads beyond self-weights.

An overview of the experimental set up adopted is shown in Fig. 24.4. The horizontal displacements were applied on the mid axis of the two transversal bays by two hydraulic actuators with a capacity of 1000 kN each at the 2nd and 3rd floor

Fig. 24.4 General view of the experimental set-up

levels, while at the 1st floor level (due to the availability of these devices in the laboratory), four actuators with a capacity of 500 kN were used (two of which were controlled in force). Steel beams were placed along the two actuators axes to connect all the floor elements and distribute the applied forces.

The reference input motion used in the PsD tests is a unidirectional 12 s long time history, shown in Fig. 24.5a for a PGA of 1.0 g. The selected seismic action is represented by a real accelerogram (Tolmezzo 1976) modified to fit Eurocode 8 (CEN 2005) response spectrum type B for the considered frequency interval. Figure 24.5b illustrates the spectra of the modified EW component of Tolmezzo recording and the EC8 specification. The accelerogram was scaled to the chosen peak ground accelerations of 0.15 g for the serviceability limit state, and 0.30 g for the no-collapse limit state. Two PsD tests at PGAs of 0.15 g (Prot1_0.15 g) and 0.30 g (Prot1_0.30 g) were initially conducted on prototype 1, namely the dual frame-wall precast system. After the walls were disconnected from the structure, the same excitation sequence was repeated for prototype 2 (Prot2_0.15 and Prot2_0.30 g) which had hinged beam-column connections in all joints. Prototype 3, which had emulative beam-column connections only at the top floor, was subjected only to the higher intensity earthquake of 0.30 g (Prot3_0.30 g), whereas prototype 4, which had emulative connections in all beam-column joints, was tested at the PGAs of 0.30 g (Prot4_0.30 g) and 0.45 g (Prot4_0.45 g). A zero-acceleration signal was

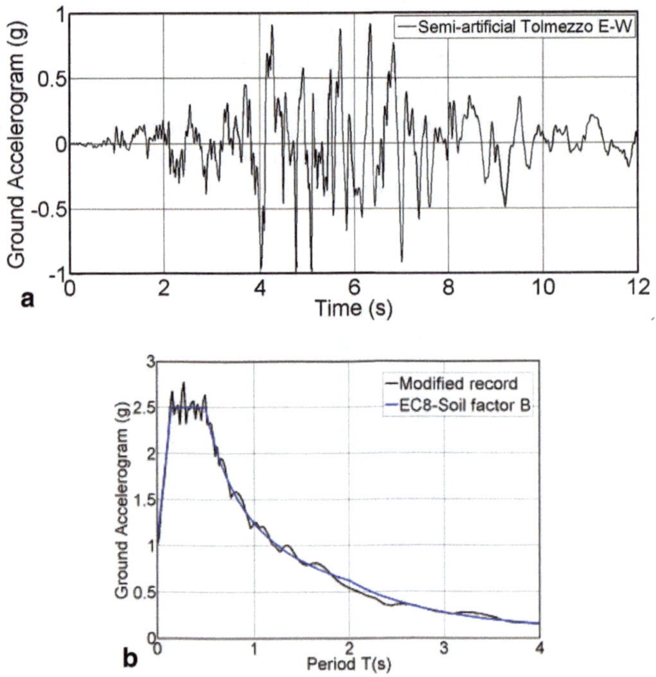

Fig. 24.5 **a** Input motion, scaled to PGA of 1 g. **b** Spectra of the modified EW component of the Tolmezzo record compared to EC8 spectrum

added after the end of the record, to allow for a free vibration of the test structures, giving total durations ranging between 15 to 19 s for the applied record. To approach the ultimate capacity of the structure, a final "funeral" sequence of cyclic tests was performed, controlling the top displacement of the structure and constraining the floor forces to an inverted triangular distribution, which is consistent with the assumptions of most seismic codes including EC8.

24.4 Results

Detailed results concerning the behaviour of the mechanical connections used between various precast elements are given in Negro et al. (2013) and Bournas et al. (2013). In the next sections of this chapter the global test results related to the overall pseudodynamic response of the four prototypes are presented in detail.

24.4.1 Prototype 1

Prototype 1 (Fig. 24.3a) was tested under two input motions scaled to a PGA of 0.15 and 0.30 g. The time histories of floor displacements and restoring forces measured in these two PsD tests are shown in Fig. 24.6. This dual wall-frame precast

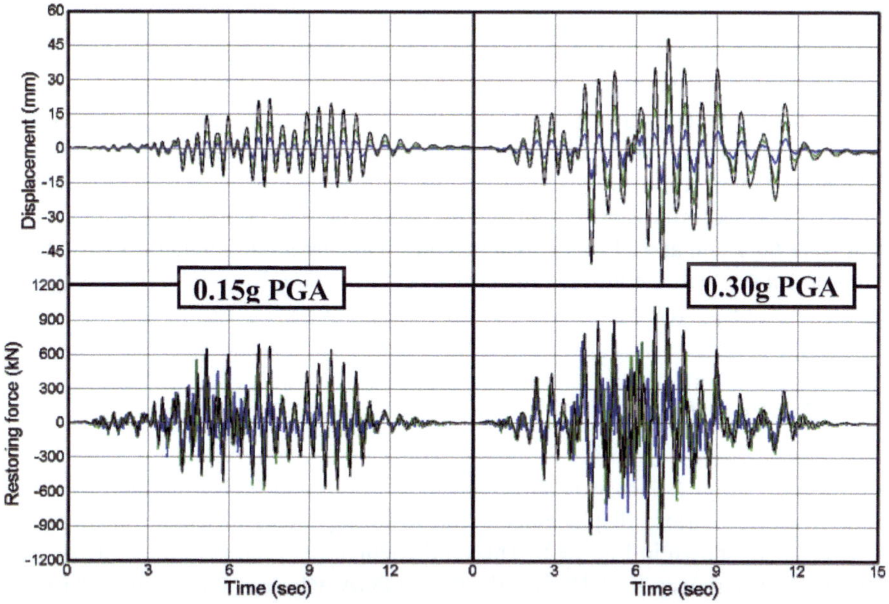

Fig. 24.6 Time histories of floor displacements and restoring forces of prototype 1 at the PsD tests

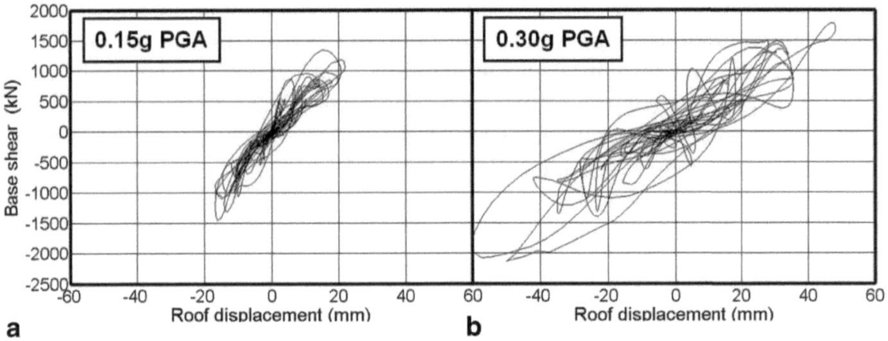

Fig. 24.7 Base shear versus roof displacement response of prototype 1 at PGAs of: **a** 0.15 g and **b** 0.30 g

system was as expected stiff, with an (experimental) fundamental natural vibration period of T=0.30 s for the 0.15 g PGA. At the higher intensity earthquake, namely 0.30 g PGA, the response curves were characterised by lower frequencies (natural vibration period shifted to T=0.46 s). These period estimations, as for all the tests, were obtained from the measured response by means of identification of equivalent linear models (Molina 2011). The global base shear force versus roof (3rd floor) horizontal displacement hysteretic response is plotted in Fig. 24.7 for the 0.15 and 0.30 g tests. At the 0.15 g PsD test, corresponding to the serviceability limit state (SLS) earthquake, the response remained practically within the elastic range as illustrated in Fig. 24.7a. This PsD 0.15 g test deformed the building to a maximum roof displacement equal to 21.9 mm, while the maximum base shear force was 1457 kN. The maximum interstorey drift ratio was recorded at the third floor, equal to 0.31 %. Figure 24.7b plots the base shear versus roof displacement hysteretic curves for the 0.30 g test. At this higher intensity earthquake corresponding to the ultimate design limit state (ULS), the response of the dual wall-frame system was characterized by some non linear effects with noticeably wider force-displacement loops. The peak roof displacement and maximum base shear force measured in this test were 60.3 mm and 2146 kN, respectively. The maximum interstorey drift ratios recorded at the first, second and third floor kept low, namely equal to 0.42 %, 0.71 % and 0.72 %, respectively.

24.4.2 Prototype 2

Prototype 2 (Fig. 24.3b) was subjected to the 0.15 and 0.30 g earthquakes. The time histories of floor displacements and restoring forces measured in these two PsD tests are shown in Fig. 24.8. The fundamental vibration period of this flexible structural system was T=1.10 s for the 0.15 g PGA, whereas at the 0.30 g PGA, the response curves were characterised by lower frequencies (natural vibration period shifted to T=1.40 s). The seismic response of prototype 2 was much influenced

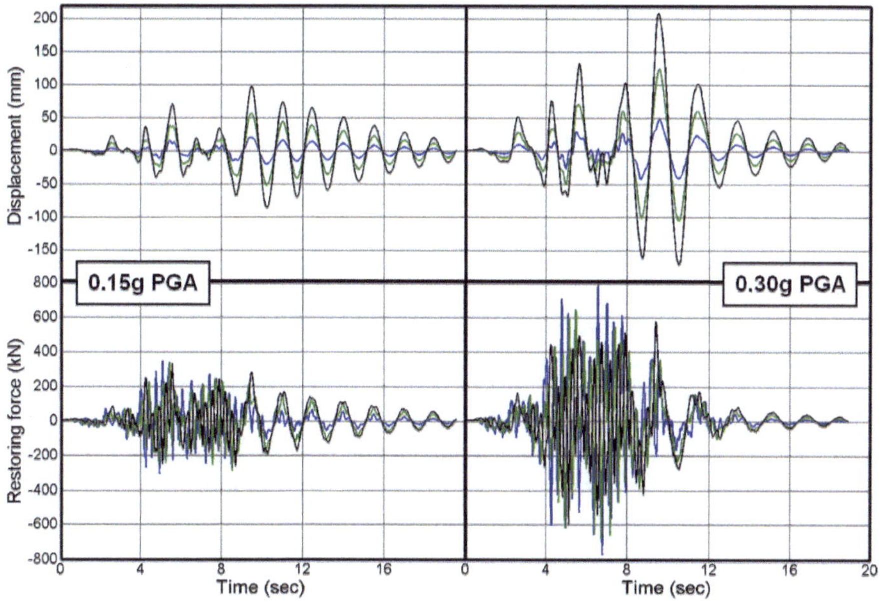

Fig. 24.8 Time histories of floor displacements and restoring forces of prototype 2 at PGAs of 0.15 and 0.30 g

by the effect of higher modes. As can be observed in Fig. 24.8, the floor displacements and restoring forces are out of phase for both earthquake intensities at some moments. In addition, from the shape of the base shear force—top displacement curves, illustrated in Fig. 24.9, it is evident that the higher modes significantly influenced the values of shear forces, for both the 0.15 and 0.30 g earthquakes. Moreover, there seems to be no upper limit for the storey forces when the structure enters into the nonlinear regime, an effect which was anticipated in preliminary nonlinear calculations (Olgiati et al. 2010). This effect, which is a direct consequence of the large higher modes contribution, results into large force demands in the connections

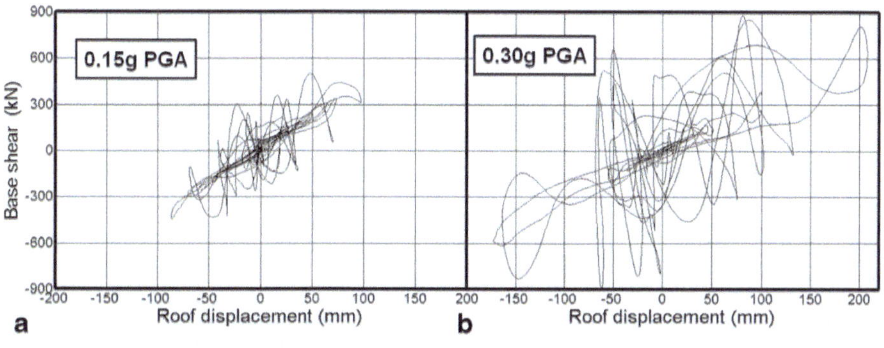

Fig. 24.9 Base shear versus roof displacement response of prototype 2 at PGAs of: **a** 0.15 g and **b** 0.30 g

(Bournas et al. 2013). It should be also pointed out here that, although the peaks of the input ground acceleration occurred during the first 7 s of the accelerogram (Fig. 24.5a), the corresponding peak floor displacements were recorded in the last 4 s, when the seismic excitation was considerably reduced. For this reason a zero-acceleration signal was added after the end of the record, to allow for a free vibration of the test structure, giving a total duration close to 19 s for the applied record.

At the 0.15 g PsD test, corresponding to the frequent (serviceability) seismic action, the overall response of prototype 2 remained practically within the elastic range as illustrated in Fig. 24.9a. However, the EC8 damage limitation requirement for buildings, which is simply expressed by an upper limit on the interstorey drift ratio, equal to 1 % for the serviceability limit state, was exceeded in the second and third floor. In particular, the maximum interstorey drifts were equal to 0.57, 1.06, and 1.18 %, at the first, second and third storey, respectively. Thus, to meet the requirements imposed by EC8 a multi-storey hinged frame structure would require larger cross-sections of the columns. Figure 24.9b plots the base shear versus roof displacement hysteretic curves for the 0.30 g test. At this higher intensity earthquake corresponding to the no-collapse design limit state, the response of this precast system with hinged beam-to-column joints was characterized by excessive deformability. The peak roof displacement and maximum base shear force measured in this test were 208 mm and 895 kN, respectively. Maximum interstorey drift ratios recorded at the first, second and third floor were 1.29, 2.18 and 2.37 %, respectively.

24.4.3 *Prototype 3*

In the third structural configuration, the mechanical connection system embedded in the beam-column joints was activated to create moment-resisting connections only at the top floor. Prototype 3 (Fig. 24.3c) was subjected to the earthquake of 0.30 g. The time histories of floor displacements and restoring forces measured in this PsD test are shown in Fig. 24.10a. The fundamental vibration period of the

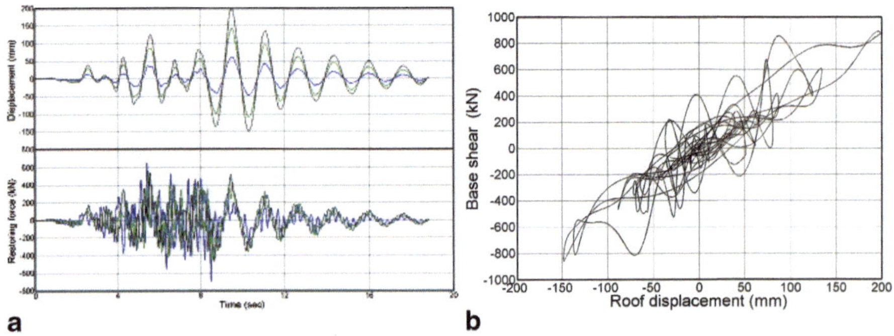

Fig. 24.10 **a** Time histories of floor displacements and restoring forces of prototype 3 at the 0.30 g PsD test. **b** Base shear versus roof displacement response of prototype 3 at the 0.30 PGA

structural system was reduced by 23 % in comparison with its counterpart with hinged beam-column connections (T = 1.08 s, as opposed to T = 1.40 s in prototype 2). However, it turned out that the concept of emulative beam-column joints at the top floor only was not quite effective as a means of reducing interstorey drifts and the overall displacements of the structure. In fact, as illustrated in Fig. 24.10a, the floor displacements and restoring forces are still out of phase for the design level earthquake, a fact which, as in the case of prototype 2, corresponds to a seismic response considerably influenced by the effects of higher modes.

Figure 24.10b plots the base shear versus roof displacement hysteretic curves for the 0.30 g test. The peak roof displacement and maximum base shear force measured in this test were 199 mm and 889 kN, respectively. The maximum interstorey drifts were equal to 1.56, 2.23, and 1.50 %, at the first, second and third storey, respectively. Comparing the above values to the corresponding maximum interstorey drifts marked in prot2_0.30 g, (namely 1.29, 2.18 and 2.37 %), it is evident that by constraining the beam-column joints of the top floor, the problem of large interstorey drifts was moved from the third to the second and first floors. Therefore, it appears that the damage limitation requirement for the serviceability limit state imposed by EC8, would not be met for the columns of lower stories by restraining only the top joints of a precast system with hinged beam-column connections.

24.4.4 Prototype 4

In the fourth and last layout, the mechanical connection system was activated in all beam-column joints with the aim of fully creating emulative moment-resisting frames. Prototype 4 (Fig. 24.3d) was tested under two input motions scaled to 0.30 and 0.45 g. Figure 24.11 illustrates the time histories of floor displacements and restoring forces measured during these two PsD test. The natural vibration period of Prototype 4 was 0.66 s, approximately half of the period measured in its counterpart with hinged beam-column joints (prot2_0.30 g). This stiffer precast system led as a consequence to higher inertia forces and lower storey drifts. Moreover, its vibration for 0.30 g PGA was not affected by the higher modes to the same extent, as was the case in the prototypes with hinged beam-to-column joints. This can be clearly noticed in Fig. 24.11, where the restoring forces at the three floors are in phase with the applied horizontal floor displacements. At the maximum considered earthquake, namely 0.45 g PGA, though, the response curves were characterised by much lower frequencies (natural vibration period shifted to T = 1.25 s) due to crack opening of the beam-column joints, caused by the previous 0.30 g test.

Figure 24.12 presents the base shear versus roof displacement loops of prototype 4 for both 0.30 and 0.45 g earthquakes. In the ULS seismic excitation of 0.30 g, the response of prototype 4 underwent reduced non-linear effects (Fig. 24.12a). The 0.30 g PsD test deformed the building to a maximum roof displacement of 132.5 mm, while the maximum base shear force was 1454 kN. The maximum interstorey drifts were significantly lower than the corresponding ones in prototypes

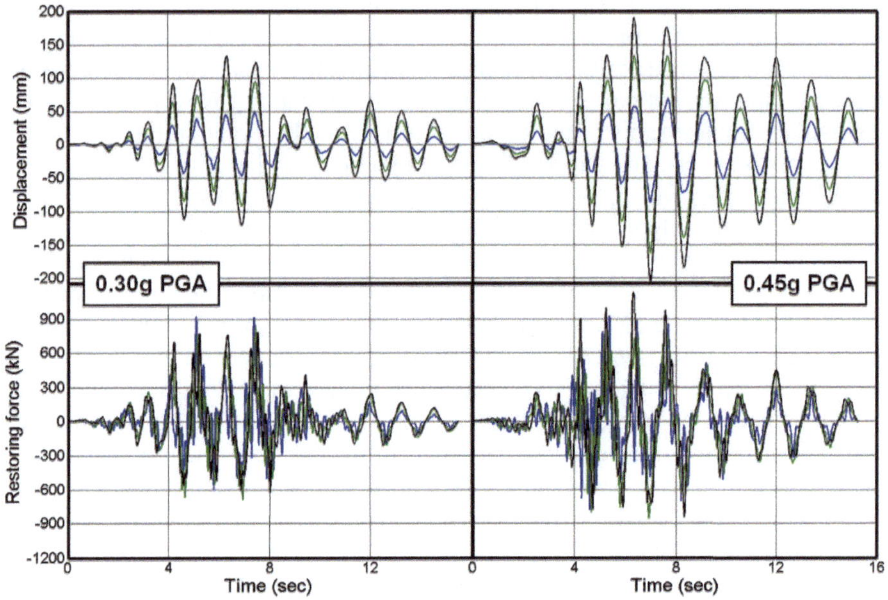

Fig. 24.11 Time histories of floor displacements and restoring forces of prototype 4 at PGAs of 0.30 g and 0.45 g

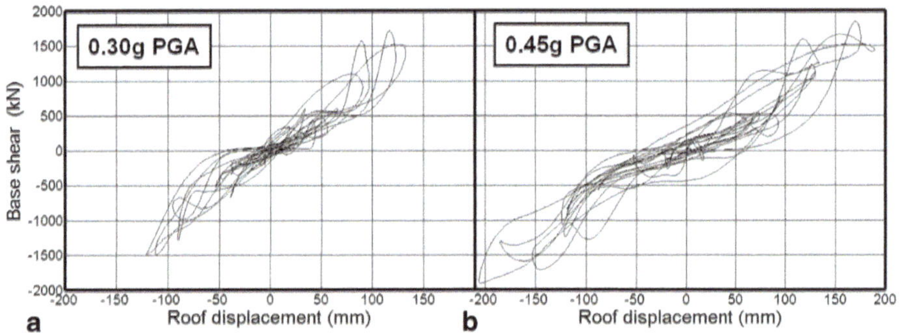

Fig. 24.12 Base shear-roof displacement response of prototype 4 at PGAs of 0.30 and 0.4 5 g

2 and 3, namely they were equal to 1.35 %, 1.51 %, and 1.05 %, at the first, second and third storey, respectively. A visual inspection made at the end of the 0.30 g earthquake at prototype 3 did not reveal any new damages to the specimen. Since prototype 4 survived the design level earthquake (0.30 g PGA) without significant damage, it was decided to proceed with the more intense seismic excitation of 0.45 g, which might be regarded as representative of the maximum considered earthquake. The peak roof displacement and maximum base shear force measured in this test were 206.5 mm and 1902 kN, respectively. Under the 0.45 g excitation,

the maximum interstorey drifts were increased to 2.21, 2.37, and 1.61 %, at the first, second and third storey, respectively. A visual inspection of the structure after the 0.45 g PsD test revealed dense flexural cracking at the base of the ground floor columns, but without considerable damage.

Peak ground accelerations of 0.30 or 0.45 g could be assumed to be the intensities for the no-collapse limit state for a zone with high seismicity, however, the seismic tests did not lead prototype 4 to heavy damage. To approach the ultimate capacity of the structure, a final "funeral" sequence of cyclic tests was performed, controlling the top displacement of the structure and constraining the floor forces to an inverted triangular distribution. The starting displacement amplitude was selected to approximately coincide with the maximum displacement recorded during the PsD tests, while the successive cycle(s) progressively increasing by 90 mm corresponded roughly to 1 % interstorey drift increases. Two cycles were repeated at increasing amplitudes of ±210 mm and ±300 mm. During the reverse cycle of the larger amplitude (300 mm), the fastenings of one of the actuator force distribution beams were detached at the third floor. For this reason the final loading cycle of the test was performed by applying the actions only at the first and second floors and controlling the displacement at the second floor at ±360 mm amplitude. The fact that the actuators of the third floor were disconnected during the last cycle, therefore applying zero force, did not drastically affect the building's maximum base shear force, nor its "collapse" mechanism, since the damage concentration and interstorey drifts measured already during the two cycles (amplitudes of ±210 mm and ±300 mm), were in the 1st and 2nd floor, substantially higher than in the 3rd one.

Figure 24.13a presents the base shear versus roof displacement hysteretic curves recorded for the cyclic test. The peak roof displacement and maximum base shear force measured in the cyclic test were 415.6 mm (4.2 % global drift) and 2237 kN, respectively. The maximum interstorey drifts were considerably high, namely equal to 5.97, 4.84, and 1.99 %, at the first, second and third storey, respectively. During

a b

Fig. 24.13 **a** Base shear versus roof displacement response of prototype 4 during the cyclic test. **b** Disintegration of concrete and bar buckling at columns' base in prototype 4 after the cyclic test

the cyclic test, the response of prototype 4 underwent extensive damages and approached the non-collapse limit state with very wide force-drift cycles. Nevertheless, even under the very high horizontal displacements that prototype 4 was subjected to, its peak recorded strength was not considerably reduced in any of the loading directions and thus the building's failure as it is conventionally defined (20 % drop in peak strength) was not reached. The cyclic test was terminated when the stroke of the actuators in the first floor (±250 mm) was about to be exhausted.

The major part of inelasticity and damage was concentrated at the base of all ground floor columns, namely at the cross-sections of maximum moment. There, the concrete cover and part of the core over the lower 300 mm of the columns disintegrated and bar buckling initiated after the concrete cover spalled off, as shown in Fig. 24.13b. The average maximum concrete tensile and compressive strain (for the 9 columns) at those sections were excessive (7.61 and 3.24 %, respectively). The ground floor columns attained a drift ratio of about 6 % (Fig. 24.13b), while their mean curvature ductility factor was rather high, in the order of 22.

24.5 Modal Decomposition of Prototype's Response

To further investigate the influence of the hinged and emulative beam-column connections on the seismic behaviour of the four prototypes, a modal decomposition was executed. The procedure in (Molina et al 2011; Molina 2011) was applied to compute how the effective period of the three modes of the test structures in the direction of testing evolved during the seismic response: a secant stiffness matrix, K, and a viscous damping matrix, C, based on the same DoFs as in the PsD test is fitted by least squares, to the measured restoring force vector, R, and the imposed vectors of displacement and velocity, d and v, respectively, as $R = Kd \pm Cv$. Note that the PsD equation of motion was formulated without viscous damping, but now this equivalent linear model with viscous damping is used for different purposes. The identification of the matrices was done repeatedly from the experimental data of a moving time-window on the PsD test results. On the basis of the fitted stiffness matrix, K, and damping matrix, C, and of the theoretical mass matrix M (companion paper), complex eigenvalues and modes were obtained for the central instant of every time-window.

Figure 24.14 illustrates the modal contribution response history to the applied ground motions of 0.15 and 0.30 g for prototype 2 at the second storey force. For each one of the three modes, its contribution is plotted (black line) together with the total response (blue line). This structural configuration with pinned beam-column joints, which was most influenced by the higher modes, has the highest interest for what concerns the storey forces and consequently the design shear forces for the beam-column connections. It is clear by comparing Fig. 24.14a–b that by doubling the earthquake intensity from 0.15 to 0.30 g, the second storey forces attributed to the first vibration mode are practically the same, whereas the corresponding forces attributed to mode 2 are more than double for the 0.30 g intensity. This explains the

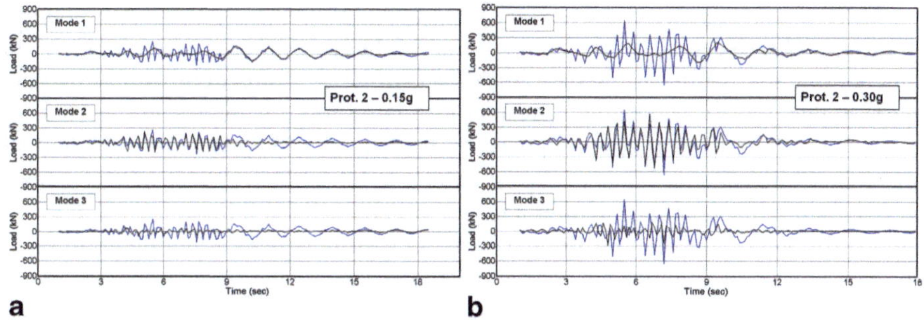

Fig. 24.14 Modal contribution response history of the second storey force of prototype 2 to the applied ground motions of: **a** 0.15 g, **b** 0.30 g

approximately double (total) storey forces for the 0.30 g in comparison with the 0.15 g seismic excitation (Bournas et al. 2013), a fact that should be carefully considered in the design of multi-storey systems, where the simplification of reducing the storey forces by q might not be adequate.

The problem of the large storey forces, and thus of the large actions on connections for the hinged multi-storey structure, had been anticipated by the preliminary numerical simulations and then verified experimentally and should be adequately reflected in design. The magnification factors for the storey forces, which determine the demand on connections, were very high in all stories of prototype 2. If the designer does not include shear walls in these flexible systems (i.e. solution of prototype 1), the large magnification of storey forces (determining the capacity design of connections) should be considered. A possible conservative simplification could be to multiply the design forces in all stories by q.

24.6 Conclusions

A full-scale three-storey precast building was subjected to a series of PsD tests at the European Laboratory for Structural Assessment. The mock-up was constructed in such a way that four different structural configurations were investigated experimentally. Therefore, the behaviour of various parameters like the types of mechanical connections (traditional as well as innovative) and the presence or absence of shear walls along the framed structure were assessed.

The presence of two stiff precast wall units in prototype 1 was quite effective in limiting the maximum interstorey drift ratios for both the serviceability and ultimate limit states. In such a dual frame-wall system the first vibration mode dominated the PsD response for both earthquake intensities. However the proper connection of stiff RC walls or cladding elements to precast diaphragms still remains a challenging task.

The seismic response of prototype 2 was highly influenced by the effects of higher modes. There seemed to be no upper limit for the storey forces when the structure entered into the nonlinear regime. This effect, which is a direct consequence of the large higher modes contribution, results into large force demands in the connections. The 1 % drift limitation imposed by EC8 for the SLS was exceeded and, at the higher intensity earthquake, corresponding to ULS, the response of this precast system with hinged beam-to-column joints was characterized by excessive deformability. However, despite the limited stiffness of this structural configuration and the fact that the maximum interstorey drifts were above the limits imposed by EC 8, prototype 2 did not suffer significant damage in its structural members during the 0.30 g PsD test. A visual inspection made at the end of the design level earthquake revealed only some very slight damages.

After the seismic test results of prototype 3, it turned out that the concept of emulative beam-column joints at the top floor only was not much effective as a means of reducing interstorey drifts and the overall displacements of the structure, as well as for controlling the effect of higher modes on the structure's seismic response.

Finally, when activated at all the floors, the proposed connection system is quite effective as a means of implementing dry precast (quasi) emulative moment-resisting frames. This was the case in prototype 4, where lower maximum interstorey drifts were recorded and the first vibration mode dominated the PsD response. In the design level test (0.30 g), prototype 4 underwent little non linear effects and thus it was subjected to a PGA of 0.45 g. In this MCE the structure revealed dense flexural cracking at the base of the ground floor columns, but again without considerable damage. In the final cyclic test, prototype 4 underwent extensive damages and approached the non-collapse limit state with very wide force-drift cycles. The major part of inelasticity and damage was concentrated at the base of all ground floor columns which attained a drift ratio of about 6 %.

References

Bournas DA, Negro P, Molina FJ (2013) Pseudodynamic tests on a full-scale 3-storey precast concrete building: behavior of the mechanical connections and floor diaphragms. Elsevier Eng Struct 57:609–627

CEN European Standard (2005) EN 1998-3: Eurocode 8: design of structures for earthquake resistance-part 3: assessment and retrofitting of buildings. European Committee for Standardization, Brussels

Molina FJ (2011) Spatial and filter models. MATLAB functions freely available at MATLAB CENTRAL FILE EXCHANGE. http://www.mathworks.com/matlabcentral/fileexchange/32634

Molina FJ, Magonette G, Pegon P, Zapico B (2011) Monitoring damping in pseudodynamic tests. J Earthq Eng 15(6):877–900

Negro P, Mola E, Ferrara L, Zhao B, Magonette G, Molina J (2007) Precast structures EC8: seismic behaviour of precast concrete structures with respect to EC8-contract N. G6RD-CT-2002-00857. Final report of the experimental activity of the Italo-Slovenian Group, p 190

Negro P, Bournas DA, Molina FJ (2013) Pseudodynamic tests on a full-scale 3-storey precast concrete building: global response. Elsevier Eng Struct 57:594–608

Olgiati M, Negro P, Colombo A (2010) Safecast project: definition and analysis of the three-storey precast buildings. In: Proc. of the 14th ECEE

Pegon P, Molina FJ, Magonette G (2008) Continuous pseudo-dynamic testing at ELSA. In Saouma VE, Sivaselvan MV (eds) Hybrid simulation: theory, implementation and applications. Taylor & Francis/Balkema Publishers, pp 79–88

Chapter 25
Caisson Foundations Subjected to Seismic Faulting: Reduced-Scale Physical Modeling

Ioannis Anastasopoulos, Orestis Zarzouras, Vasileios Drosos and George Gazetas

25.1 Introduction

Despite being the source of an earthquake, faulting has not been given the same attention as seismic shaking. Hence, while a substantial effort has been made to understand the mechanisms of soil-structure interaction for seismic shaking, little attention has been given to the effects of an outcropping fault rupture. Interest on the subject was refueled by the devastating 1999 earthquakes in Turkey and Taiwan, which offered ample evidence of the damage potential of tectonic faults, highlighting the need to develop design methods and guidelines against faulting-induced loading. During these two earthquakes (Kocaeli, Turkey; and Chi-Chi, Taiwan) various structures were crossed by outcropping faults, yielding a substantial number of case histories (Youd et al. 2000; Chang et al. 2000; Kawashima 2001; Dong et al. 2003; Pamuk et al. 2005; Faccioli et al. 2008).

Structures on top of the emerging surface fault scarp may undergo significant differential movements that could lead to failure. For instance, several bridges were severely damaged or collapsed during the Chi-Chi earthquake (Kawashima 2001; Pamuk et al. 2005). Similar bridge failures were observed during the Düzce and Kocaeli earthquakes in Turkey (Pamuk et al. 2005; Ulusay et al. 2002), such as the Arifiye bridge and the Kaynasli Viaduct. It is therefore understandable that seismic codes had in the past prohibited construction in the "immediate vicinity" of active faults. However, for long structures such as bridges, such a strict prohibition is difficult if not impossible to respect.

In fact, during the aforementioned earthquakes, there were several examples of successful performance (Ulusay et al. 2002; Youd et al. 2000; Erdik 2001; Bray 2001), indicating that for structures "on top of a major fault outcrop" survival is

I. Anastasopoulos (✉) · O. Zarzouras · V. Drosos · G. Gazetas
Laboratory of Soil Mechanics, National Technical University of Athens, Athens, Greece
e-mail: i.anastasopoulos@dundee.ac.uk

F. Taucer, R. Apostolska (eds.), *Experimental Research in Earthquake Engineering*,
Geotechnical, Geological and Earthquake Engineering 35,
DOI 10.1007/978-3-319-10136-1_25

not impossible (Anastasopoulos and Gazetas 2007a, b), confirming the belief that structures can be designed to withstand large tectonic displacements (Duncan and Lefebvre 1973; Niccum et al. 1976; Youd 1989; Berill 1983). The performance of bridges subjected to faulting-induced deformation has been studied analytically in (Anastasopoulos et al. 2008a), focusing on the response of bridges supported on deep foundations: piles and caissons. The latter were found to be quite advantageous compared to pile foundations, being capable to resist the deformation and divert the fault rupture.

This paper presents an experimental study on the interaction of embedded caisson foundations with normal and reverse fault ruptures, aiming to gain further insight into the interaction mechanisms and produce experimental data to be used for model calibration and validation. Besides being typical for bridge structures, the choice of embedded caisson foundations offers the possibility of observing strongly nonlinear phenomena, such as the diversion and bifurcation of the fault rupture path. A series of reduced-scale physical model tests were conducted at the Laboratory of Soil Mechanics of the National Technical University of Athens, focusing on the effect of foundation location relative to the fault rupture.

25.2 Physical Modeling Methodology

25.2.1 Problem Definition

As schematically illustrated in Fig. 25.1, the conceptual prototype refers to a typical massive caisson foundation of dimensions 10 × 5 × 5 m ($H \times B \times D$), fully embedded in a dense sand stratum to a depth of 15 m. In the tests reported herein, the relative density of the sand stratum was approximately 80 %, and the weight of the caisson foundation roughly 20 MN (always in prototype scale). The bedrock is subjected to

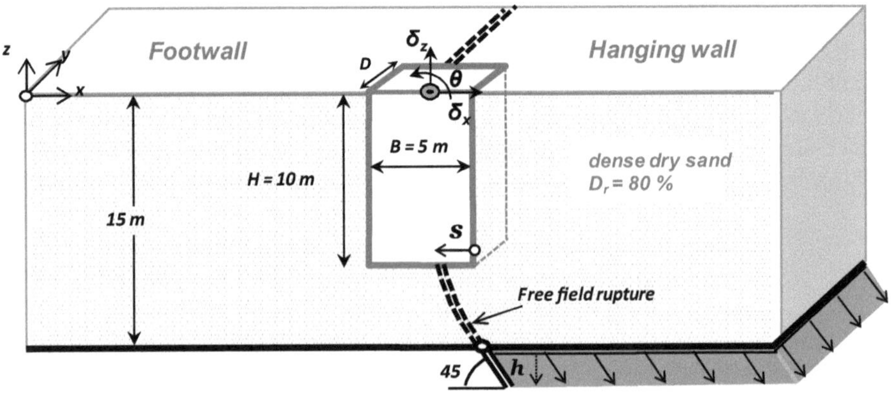

Fig. 25.1 Schematic illustration of the studied problem with key dimensions in prototype scale

Table 25.1 Summary of physical model tests: normal and reverse faulting at different locations

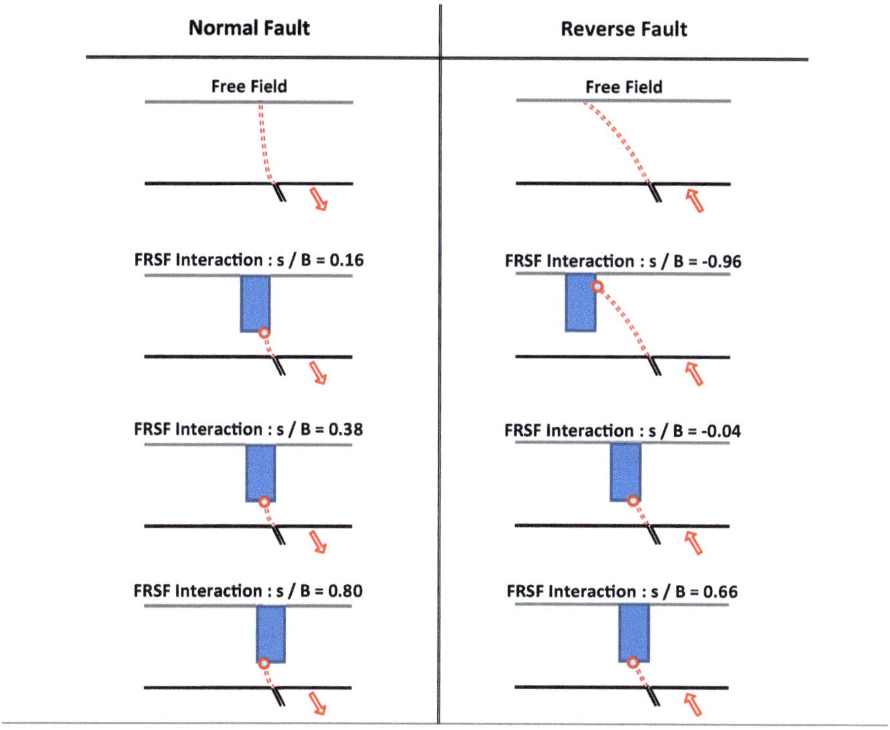

a 45° dip-slip fault rupture (normal or reverse) of vertical offset h, while the caisson is positioned at a distance s relative to the unperturbed fault rupture breakout (defined as the horizontal distance between the caisson right corner and the free field rupture outcrop at the caisson base), which is parametrically investigated. As summarized in Table 25.1, a total of 8 different configurations were modeled, parametrically investigating: (a) the style of dip-slip faulting (normal versus reverse), and (b) the exact position of the caisson foundation relative to unperturbed (i.e., free field) fault rupture.

As discussed in more detail in the sequel, the displacements of the caisson, Δx, Δz, and θ, and the associated displacements of the soil mass are recorded during the experiments through image analysis and laser scanning of the soil surface. In the first case, digital high-resolution cameras are utilized to capture photographs during the incrementally imposed fault rupture displacement h, which are then processed through image analysis. In the latter case, a novel technique was developed and applied through a custom system, designed and constructed in-house. After each displacement increment, the model surface was scanned with 8 laser displacement transducers, which travel along the specimen at a constant speed to produce a digital relief of the deformed surface.

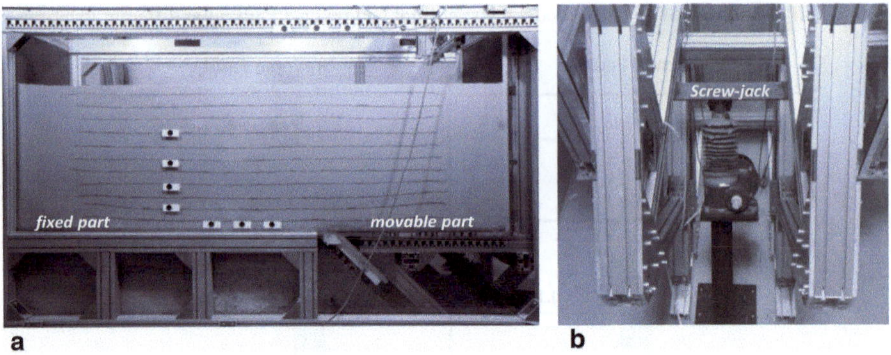

Fig. 25.2 The Fault Rupture Box (FRB) used for the experiments: **a** photo of the FRB filled with soil just before the test (shown for the free-field case) **b** photo of the screw-jack actuator

25.2.2 Experimental Setup

The experiments were conducted utilizing a custom-built Fault Rupture Box (FRB). Designed and constructed in-house, the FRB can simulate quasi-static fault rupture propagation through soil and its interaction with foundation-structure systems. As shown in Fig. 25.2, the FRB is essentially a split-box container with internal dimensions 2.6 × 1.1 × 0.9 m (length × height × width). It consists of a fixed and a movable part; the latter can move downwards or upwards to simulate normal or reverse faulting, respectively. Composite "windows" are installed at the two sides of the FRB, consisting of an outside perspex sheet for rigidity and durability, accompanied by a thinner interior glass sheet to minimize friction and prevent scratching of the (relatively) soft outer surface by the (quartz) sand grains. These windows allow observation of the evolution of fault rupture propagation, and direct optical monitoring of the deformation of the soil mass and the displacement of the structure. The movement of the split base is achieved by means of an electromechanical (screw-jack) actuator of 8 t capacity, capable of a maximum displacement of 20 cm at a controllable speed ranging from 0 to 5 cm/s.

Physical modelling requires *accuracy, reproducibility,* and *repeatability*. Considering dry sand, this reduces to the control of the relative density D_r. For this purpose, an automated sand raining system was designed and constructed in house, and is used for model preparation. The system consists of a sand hopper supported by a beam, which can travel back and forth at a controllable velocity, operated through an electric motor along a pair of parallel beams fixed on the ceiling. A second electric motor is used to adjust the vertical position of the sand hopper, sliding along two vertical support beams. The movement of the sand hopper in both directions is controlled by a digital controller, which allows remotely controlling the pluviation speed and height during sample preparation. During pluviation, the hopper moves automatically back and forth covering the entire length of the container. The density depends on the raining height, and the sand discharge rate. The height is controlled by the vertical position of the hopper, whereas the discharge rate is controlled by the aperture of the hopper and the pluviation speed. The raining system has been calibrated for the "Longstone" sand used in the experiments, as described in Anastasopoulos et al. (2010).

Taking account of the capacity of the FRB, a 1:20 model scale was chosen. Model dimensions and materials were selected on the basis of scaling laws for 1 g modeling (Gibson 1997). The model caisson was made of steel of density 7.8 Mg/m^3. A "white noise" pattern was glued on the visible side of the caisson to facilitate image analysis. The backfill consists of dry "Longstone" sand, a very fine and uniform quartz sand with $D_{50}=0.15$ mm and uniformity coefficient $D_{60}/D_{10}=1.42$. The void ratios at the loosest and densest state were measured, as discussed in Gibson (1997). Based on the procedure described in Kolbuszewski (1948): $e_{max}=0.995$, $e_{min}=0.614$, and $G_s=2.64$. Direct shear tests were carried out to measure the strength of the sand. Tests were performed for normal stresses ranging from 13 to 300 kPa. The very low normal stress is considered representative of the average stress level prevailing in 1 g testing. The angle of shearing resistance appears to depend strongly on stress level: for stresses higher than 120 kPa $\varphi' = 35°$, while for stresses lower than 100 kPa φ' increases from 36° up to 51° at normal stress $\sigma = 13$ kPa.

25.2.3 Model Preparation and Instrumentation

Model preparation begins with sand pluviation within the FRB using the automated sand raining system. The layering of the sand takes place in layers of approximately 5 cm. In the case of free field tests, this procedure was repeated until the total height of 75 cm was achieved. For the caisson tests, sand pluviation is first carried out for the first 25 cm of soil (5 m in prototype scale). At this height the foundation base level is reached and the caisson model is installed at its designated position. The foundation is placed very carefully, so as to achieve verticality and minimize initial soil deformations. A silicone "sausage" is placed in the perimeter of the caisson to prevent sand flowing between the caisson and the FRB window. Thereafter, the remaining 50 cm of soil are pluviated in the same way.

During the test, the FRB base displacement is imposed in small consecutive increments, each one of them being of the order of 2 mm. After each displacement increment, a high-resolution digital camera was used to take photos of the deformed physical model. The digital images were subsequently analyzed using the Geo-PIV software (White et al. 2003) to calculate caisson displacements and shear strains within the soil. In addition to the above, after each displacement increment the ground surface was scanned using a custom-built system. As shown in Fig. 25.3, the latter consists of a moving row of 8 laser displacement transducers, set above the model. The row of the laser transducers was placed perpendicularly to the axis of the model and vertically. The row is connected to an automated apparatus which moves horizontally from one end of the container to the other at constant velocity,

Fig. 25.3 Custom-build scanning system: "moving row" of laser displacement transducers, traveling back and forth at constant velocity, measuring the deformed relief of the ground surface

scanning the surface of the model. Applying this procedure, the deformed relief of the ground surface was measured for each increment of fault dislocation h.

25.3 Normal Faulting

25.3.1 Free-Field Normal Faulting

The experimental results for the free field case are presented first. Characteristic snapshots of the deformed physical model are presented in Fig. 25.4. Unless otherwise stated all of the results are presented in prototype scale. Just after 0.16 m of base fault offset, a steep rupture zone can be observed at the lower 3/4 of the soil deposit (Fig. 25.4b). The increase of h to 0.35 m leads to the development of a slightly more inclined rupture, indicating that the initial shear plane is not kinematically admissible any more (Fig. 25.4c). A similar response has been observed in trapdoor centrifuge tests of White et al. (1994). Indeed, it is this second rupture that outcrops, generating a sharp dislocation at the surface. For $h=0.4$ m, a secondary antithetic rupture emanates (Fig. 25.4d), outcropping for $h=1.5$ m (Fig. 25.4e). Further increase of h leads to development of a less steep antithetic rupture (Fig. 25.4f). Between the two shearing zones, a fault graben develops, which subsides due to loss of support (Anastasopoulos et al. 2010).

The primary fault rupture emerges on the surface at a distance of about 3 m from the base discontinuity, having an almost constant dip along its path, of the order of 75°. The greater propagation angle observed during the experiment is in agreement with former experimental results (Cole and Lade 1984; Bransby et al. 2008a; Anastasopoulos et al. 2007; Loli et al. 2012) and field observations (Bray et al. 1994a). Normal faults tend to refract on the soil–bedrock interface, propagating to the ground surface at increased dip angles; nonetheless, it is believed that the phenomenon is overestimated in the 1 g tests due to the increased soil strength and dilation at such small stresses (i.e., scale effects).

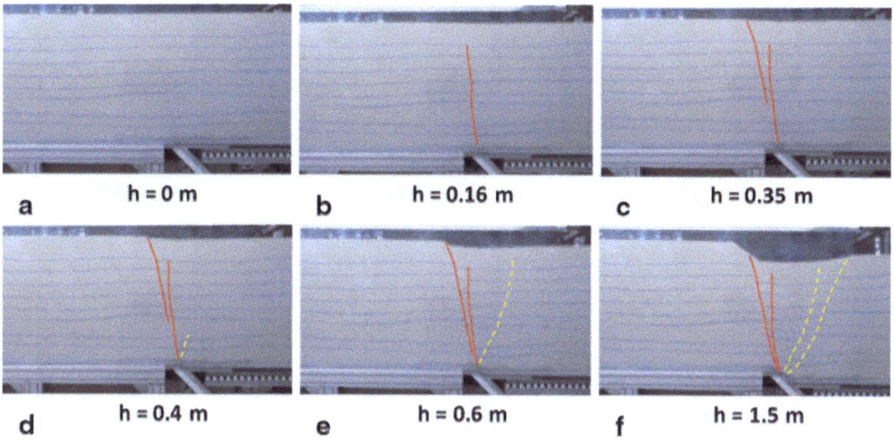

Fig. 25.4 Free-field normal faulting. Snapshots of the deformed soil for characteristic bedrock fault offsets: **a** $h=0$ m; **b** $h=0.16$ m; **c** $h=0.35$ m; **d** $h=0.4$ m; **e** $h=0.6$ m; and **f** $h=1.5$ m. Solid lines denote primary ruptures; *dashed lines* are used for secondary antithetic ones

25.3.2 Fault Rupture–Caisson Interaction: s/B = 0.16

In this test, the caisson was positioned so that the free field rupture would cross its base in the vicinity of its right corner, as sketched in Table 25.1. Due to space limitations, selected results are presented, focusing on key response mechanisms. The detailed results can be found in Psycharis et al. (2013). Characteristic snapshots captured during the experiment are shown in Fig. 25.5. The initial state (i.e., the un-deformed model) is shown in Fig. 25.5a. For $h=0.16$ m (Fig. 25.5b), significant deformation takes place on the hanging wall and underneath the right corner of the caisson. This is an indication that the already steep fault rupture plane becomes even steeper, due to the divergence of the fault rupture. The consequent movement of the caisson to the right causes a field of active pressures behind the foundation (at the footwall).

Further increase of the base displacement up to 35–40 cm leads to the development of a distinguishable active failure prism (Figs. 25.5c and d). This active wedge, which forms behind the foundation (i.e., towards the footwall, to the left), is directly related to the outward movement (i.e., towards the hanging wall, to the right) of the caisson, which due to its location is rotating substantially towards the hanging wall. In this way, the active shear plane that develops behind the caisson facilitates the development of another rupture zone, which becomes the primary shearing plane for $h>0.6$ m (Figs. 25.5e and f). The translational and rotational faulting-induced displacements of the caisson as a function of the bedrock fault offset h are illustrated in Fig. 25.6. Following the hanging wall, the caisson undergoes large horizontal and vertical displacements Δx and Δz, respectively, and rotation θ which reaches 14° for $h=2$ m.

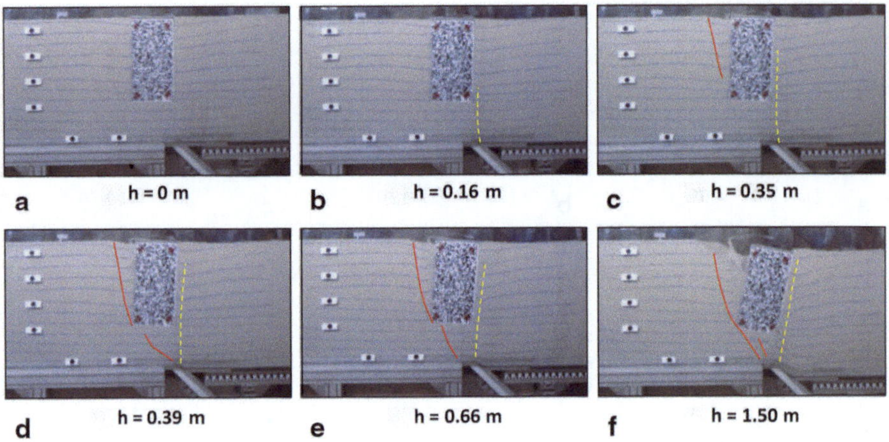

Fig. 25.5 Interaction of a normal fault with the caisson foundation positioned at $s/B=0.16$. Snapshots of the deformed model for characteristic bedrock fault offsets: **a** $h=0$ m; **b** $h=0.16$ m; **c** $h=0.35$ m; **d** $h=0.39$ m; **e** $h=0.66$ m; and **f** $h=1.5$ m. Solid lines denote primary ruptures; *dashed lines* are used for secondary antithetic ones

25.3.3 Fault Rupture–Caisson Interaction: s/B = 0.8

In this case, the caisson is positioned further to the right so that the free field rupture would cross its base closer to its left edge. The evolution of the interaction mechanisms is presented in Fig. 25.7. For $h=0.16$ m the rupture diverges to the left of the caisson (towards the footwall) leaving the foundation totally in the hanging wall (Fig. 25.7b). The rupture outcrops for $h=0.31$ m (Fig. 27.7c). Contrary to the previous case, the response is dominated by vertical displacement (Fig. 25.8), as the caisson practically moves with the hanging wall. Quite interestingly, a counterclockwise rotation of the caisson is now observed, which is mainly due to loss of support at the left edge of the caisson. Consequently, active conditions are now developed at the right wall of the caisson (for $h=0.66$ m), leading to the development of a secondary antithetic rupture and an active wedge towards the hanging wall (Figs. 25.7d, e and f).

25.4 Reverse Faulting

25.4.1 Free-Field Reverse Faulting

The results of the free field case are presented in Fig. 25.9. In agreement with former experimental results (Bransby et al. 2008b, Loli et al. 2011), the reverse fault rupture outcrops at the ground surface of the soil deposit for greater base offset compared to the corresponding value of the normal fault case. The initial (i.e.,

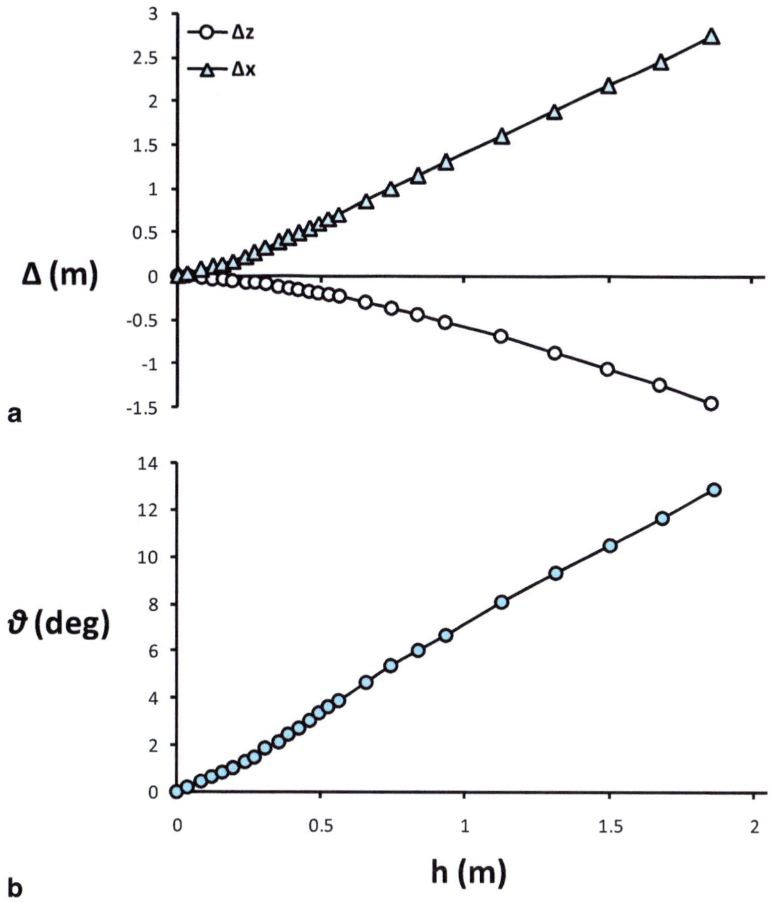

Fig. 25.6 Interaction of a normal fault with the caisson foundation positioned at $s/B=0.16$. Evolution with bedrock fault offset h of: **a** horizontal Δx and vertical Δz caisson displacements; and **b** rotation θ (all measured at the top of the caisson)

un-deformed) model is presented in Fig. 25.9a. The first rupture (yet not the primary one) appears for $h=0.46$ m (Fig. 25.9b). Although there is no visible strain localization, a wide failure zone can be identified in the image, which starts from the fault application point and propagates towards the soil surface within approximately 1/3 of the soil specimen.

After 0.7 m of additional fault offset ($h=1.18$ m), a displacement discontinuity (i.e., a fault scarp) appears on the ground surface at a horizontal distance of about 16 m from the fault initiation point, and the shear failure zone is visible throughout the soil layer. The shear failure is again distributed to a zone rather than a distinct surface. With further increase of the bedrock fault offset ($h>2$ m), the shear failure zone is enveloped by the two failure planes (Fig. 25.9e and 25.9) and the failure pattern does not change any more. The primary fault rupture emerges on the surface

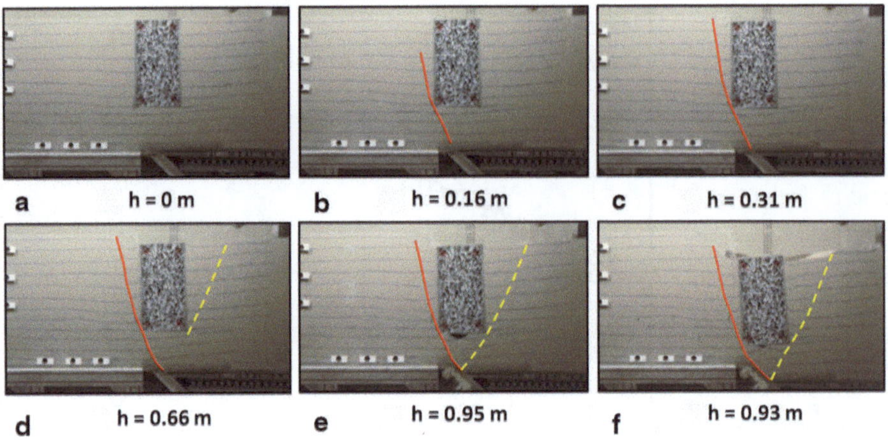

Fig. 25.7 Interaction of a normal fault with the caisson foundation positioned at $s/B=0.8$. Snapshots of the deformed model for characteristic bedrock fault offsets: **a** $h=0$ m; **b** $h=0.16$ m; **c** $h=0.31$ m; **d** $h=0.66$ m; **e** $h=0.95$ m; and **f** $h=0.93$ m. Solid lines denote primary ruptures; *dashed lines* are used for secondary antithetic ones

at a distance of about 17 m from the base discontinuity, clearly bending downwards (i.e., towards the hanging wall) closer to the surface. This is in agreement with former experimental results (Cole and Lade 1984; Bransby et al. 2008a; Anastasopoulos et al. 2007; Loli et al. 2012), especially taking account of the increased soil strength and dilation at such small stresses (i.e., scale effects).

25.4.2 Fault Rupture–Caisson Interaction: $s/B=-0.04$

As sketched in Table 25.1, in this test the caisson is positioned so that the free field rupture would cross its base in the vicinity of its right corner. The evolution of the faulting mechanisms is presented in Figs. 25.10 and 25.11. The stressing of the right part of the model is observable only after applying more than 0.4 m of vertical displacement at the base (Fig. 25.10b). The rupture propagates to the surface with a substantially larger dip angle, clearly indicating the intense interaction taking place between the rupture and the caisson. Indeed, for $h=0.75$ m, the rupture "hits" the right bottom corner of the caisson, being clearly diverted towards the footwall (Fig. 25.10c). This is attributed to the rigidity of the caisson, which poses a very strict kinematic constraint, effectively diverting the rupture path.

Further increase of the imposed bedrock offset leads to another interesting failure mechanism. The primary rupture grazes the right sidewall of the caisson, propagating almost vertically towards the ground surface. With the failure taking place along the caisson-soil interface, a vertical step is formed at the soil surface adjacent to the caisson for $h>1$ m (Fig. 25.10d). At the same time, a secondary rupture zone

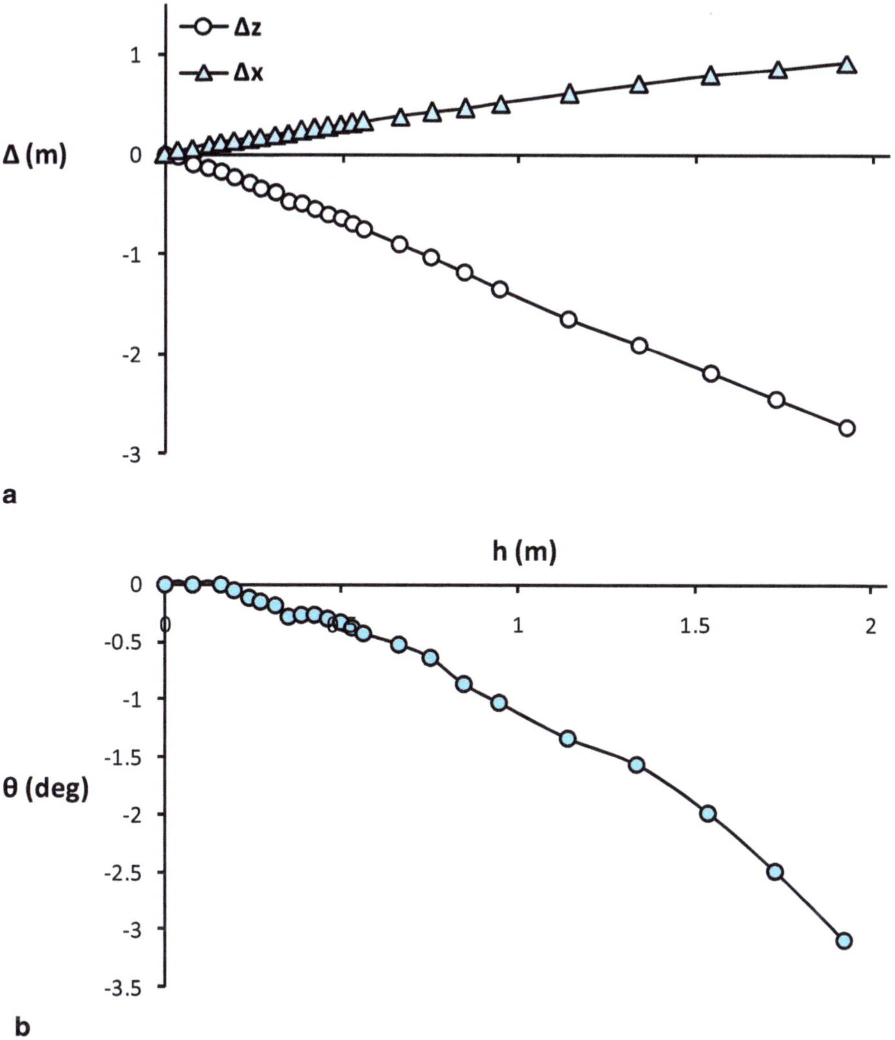

Fig. 25.8 Interaction of a normal fault with the caisson foundation positioned at $s/B = 0.8$. Evolution with bedrock fault offset h of: **a** horizontal Δx and vertical Δz caisson displacements; and **b** rotation θ (all measured at the top of the caisson)

develops and propagates with a smaller dip towards the left side of the caisson base. However, this zone does not go beyond the caisson foundation level, at least for the applied base dislocation ($h \leq 1.8$ m). The caisson undergoes major horizontal and vertical displacements and rotations, as shown in Fig. 25.11.

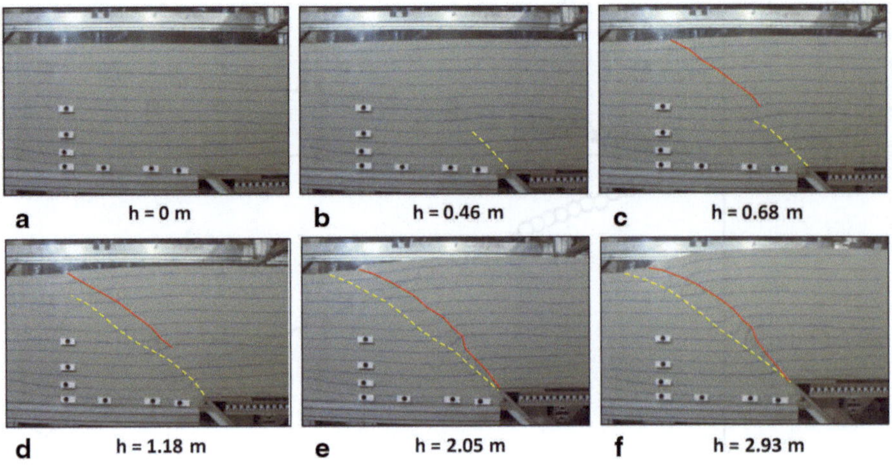

Fig. 25.9 Free-field reverse faulting. Snapshots of the deformed model for characteristic bedrock fault offsets: **a** $h=0$ m; **b** $h=0.46$ m; **c** $h=0.68$ m; **d** $h=1.18$ m; **e** $h=2.05$ m; and **f** $h=2.93$ m. Solid lines denote primary ruptures; *dashed lines* are used for secondary ones

25.4.3 Fault Rupture–Caisson Interaction: s/B=0.66

Finally, the caisson is positioned further towards the footwall (to the right) so that the free field rupture would cross its base at the middle. The evolution of the interaction mechanisms are presented in Fig. 25.12. Similarly to the previous case, the

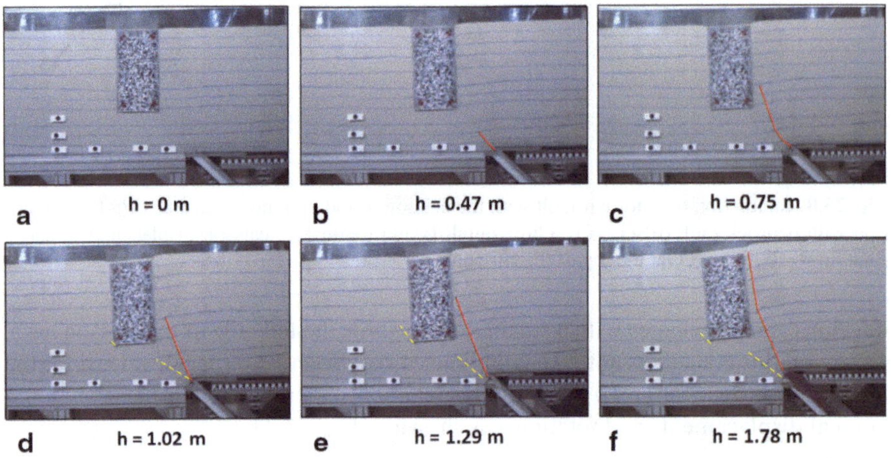

Fig. 25.10 Interaction of a reverse fault with the caisson foundation positioned at s/B=-0.04. Snapshots of the deformed model for characteristic bedrock fault offsets: **a** $h=0$ m; **b** $h=0.47$ m; **c** $h=0.75$ m; **d** $h=1.02$ m; **e** $h=1.29$ m; and **f** $h=1.78$ m. Solid lines denote primary ruptures; *dashed lines* are used for secondary ones

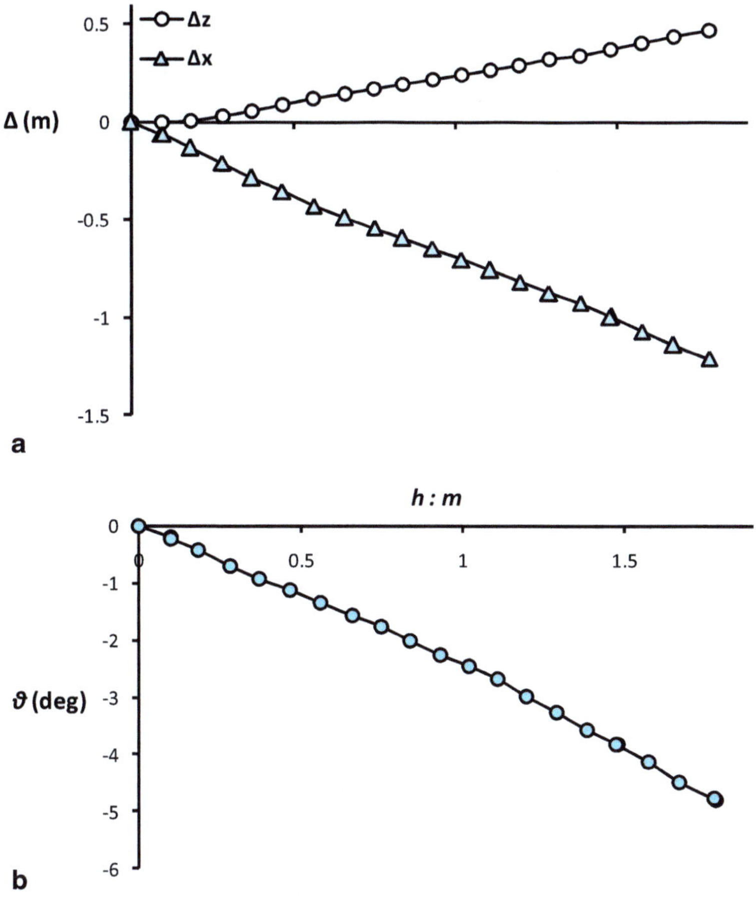

Fig. 25.11 Interaction of a reverse fault with the caisson foundation positioned at $s/B = -0.04$. Evolution with bedrock fault offset h of: **a** horizontal Δx and vertical Δz caisson displacements; and **b** rotation θ (all measured at the top of the caisson)

presence of the caisson leads to bifurcation of the fault rupture and substantial diffusion of shear deformation in the vicinity of the caisson base. The primary branch is diverted towards the hanging wall, just missing the left edge of the caisson base. The secondary branch grazes the right sidewall of the foundation, with shearing taking place at the caisson-soil interface. As a result, the caisson moves substantially (Fig. 25.13), following the movement of the hanging wall to a large extend. Yet, the rotations are substantially reduced compared to the previous case.

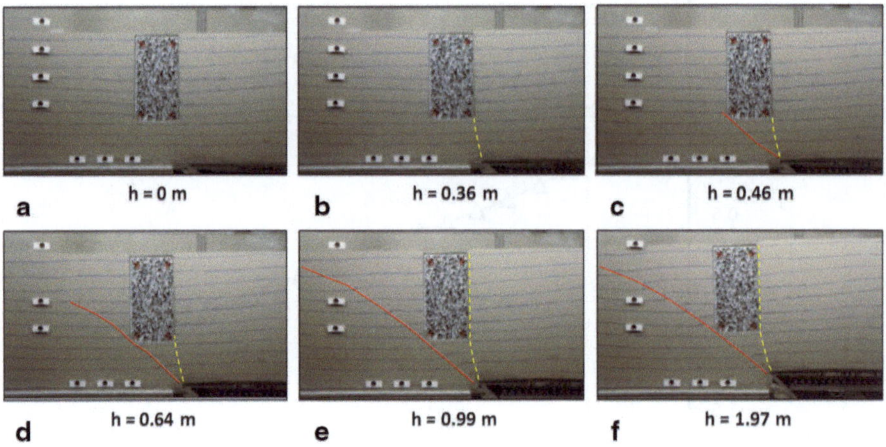

Fig. 25.12 Interaction of a reverse fault with the caisson foundation positioned at $s/B=0.66$. Snapshots of the deformed model for characteristic bedrock fault offsets: **a** $h=0$ m; **b** $h=0.36$ m; **c** $h=0.46$ m; **d** $h=0.64$ m; **e** $h=0.99$ m; and **f** $h=1.97$ m. Solid lines denote primary ruptures; *dashed lines* are used for secondary ones

25.5 Conclusions

The paper has presented the key results of a series of reduced-scale physical model tests on the interaction of a caisson foundation with normal and reverse faults. The massive caisson foundation is proven to act as a kinematic constraint, substantially altering the rupture path (compared to that in the free field). Its presence may either force the rupture to totally divert towards the footwall or the hanging wall (depending on the location of the caisson), or to bifurcate and diffuse around the foundation. The exact interaction mechanism and the corresponding response of the foundation are shown to be radically affected by the exact location of the caisson relatively to the rupture surface. Such massive foundations have an apparent advantage compared to piled foundations and spread footings.

Acknowledgments The research leading to these results received founding from the European Community's Seventh Framework Programme [FP7/2007–2013] for access to the Laboratory of Soil Mechanics, National Technical University of Athens under grant agreement n 227887.

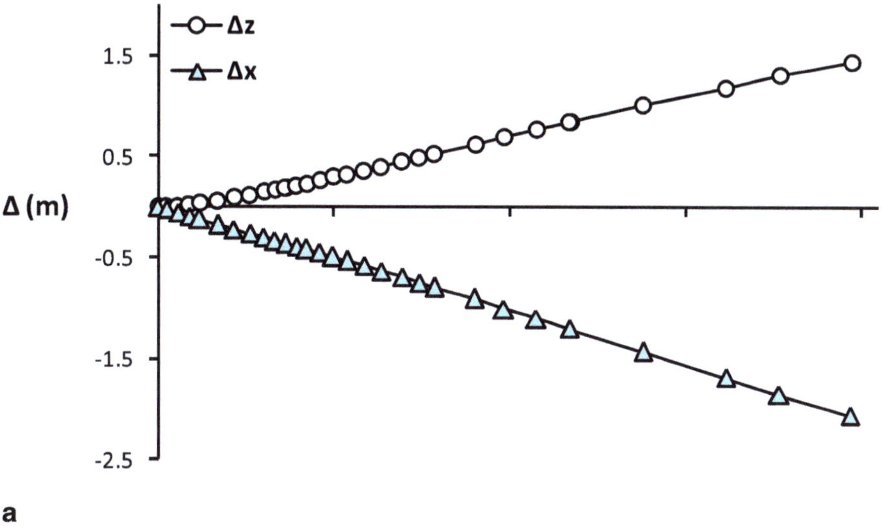

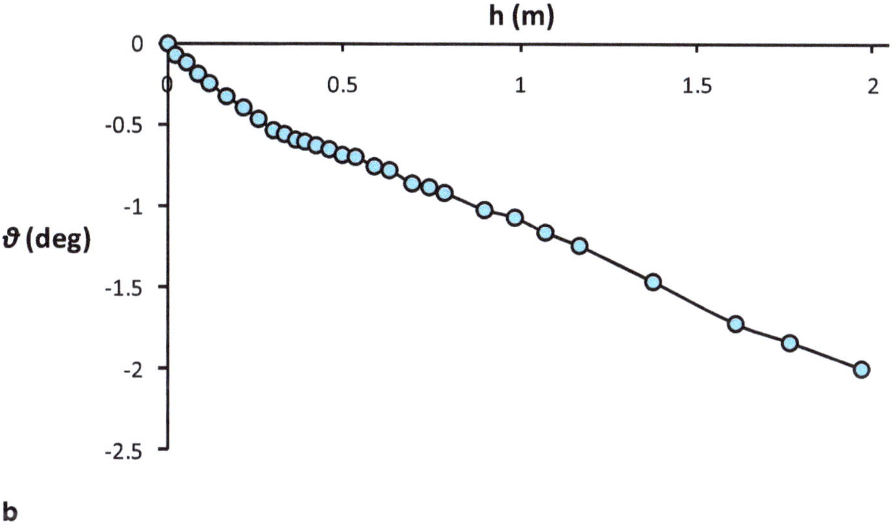

Fig. 25.13 Interaction of a reverse fault with the caisson foundation positioned at $s/B = 0.66$. Evolution with bedrock fault offset h of: **a** horizontal Δx and vertical Δz caisson displacements; and **b** rotation θ (all measured at the top of the caisson)

References

Anastasopoulos I, Gazetas G (2007a) Foundation-structure systems over a rupturing normal fault: Part I. Observations after the Kocaeli 1999 earthquake. Bull Earthq Eng 5(3):253–275

Anastasopoulos I, Gazetas G (2007b) Behaviour of structure-foundation systems over a rupturing normal fault: part II. Analysis of the Kocaeli case histories. Bull Earthq Eng 5(3):277–301

Anastasopoulos I, Gazetas G, Bransby MF, Davies MCR El Nahas A (2007) Fault rupture propagation through sand: finite element analysis and validation through centrifuge experiments. J Geotech Geoenviron Eng ASCE 133(8):943–958

Anastasopoulos I, Gazetas G, Drosos V, Georgarakos T, Kourkoulis R (2008a) Design of bridges against large tectonic deformation. Earthq Eng Eng Vib 7(4):345–368

Anastasopoulos I, Gerolymos N, Gazetas G, Bransby M F (2008b) Simplified approach for design of raft foundations against fault rupture. Part I: free-field. Earthq Eng Eng Vib 7:147–163

Anastasopoulos I Georgarakos T Georgiannou V Drosos V Kourkoulis R (2010) Seismic performance of bar-mat reinforced-soil retaining wall: shaking table testing versus numerical analysis with modified kinematic hardening constitutive model. Soil Dyn Earthq Eng 30(10):1089–1105

Berill, JB (1983) Two-dimensional analysis of the effect of fault rupture on buildings with shallow foundations. Soil Dyn Earthq Eng 2(3):156–160

Bransby MF, Davies MCR, El Nahas A, Nagaoka S (2008a) Centrifuge modelling of normal fault-foundation interaction. Bull Earthq Eng 6(4):585–605

Bransby MF, Davies MCR, El Nahas A, Nagaoka S (2008b) Centrifuge modelling of reverse fault-foundation interaction. Bull Earthq Eng 6(4):607–628

Bray JD, Seed RB, Cluff LS, Seed HB (1994a) Earthquake fault rupture propagation through soil. J Geotech Eng 120(3):543–561

Bray JD (2001) Developing mitigation measures for the hazards associated with earthquake surface fault rupture. Workshop on seismic fault-induced failures–possible remedies for damage to Urban facilities, University of Tokyo Press, pp 55–79

Cole DA Jr, Lade PV (1984) Influence zones in alluvium over dip-slip faults. J Geotech Eng 110(5):599–615

Chang KC, Chang DW, Tsai MH, Sung YC (2000) Seismic performance of highway bridges. Earthq Eng Eng Seismol 2(1):55–77

Dong JJ, Wang CD, Lee CT, Liao JJ, Pan YW (2003) The influence of surface ruptures on building damage in the 1999 Chi-Chi earthquake: a case study in Fengyuan city. Eng Geol 71:157–179

Duncan JM, Lefebvre G (1973) Earth pressure on structures due to fault movement. J Soil Mech Found Eng ASCE 99:1153–1163

Erdik M (2001) Report on 1999 Kocaeli and Düzce (Turkey) earthquakes. Structural control for civil and infrastructure engineering. World Scientific

Faccioli E, Anastasopoulos I, Callerio A, Gazetas G (2008) Case histories of fault–foundation interaction. Bull Earthq Eng 6(4):557–583

Gibson AD (1997) Physical scale modeling of geotechnical structures at one-G, Report No. SML 97-01. California Institute of Technology, Pasadena, pp 413

Kawashima K (2001) Damage of bridges resulting from fault rupture in the 1999 Kocaeli and DÏzce, Turkey, earthquakes and the 1999 Chi-Chi, Taiwan earthquake. Workshop on seismic fault–induced failures, University of Tokyo Press, pp 171–190 (Konagai K et al (eds))

Kolbuszewski J (1948) An experimental study of the maximum and minimum porosities of sands. Proceedings of the 2nd ICSMFE 1:158–165

Loli M, Anastasopoulos I, Bransby MF, Waqas A, Gazetas G (2011) Caisson foundations subjected to reverse fault rupture: centrifuge testing and numerical analysis. J Geotech Geoenv Eng ASCE 137(10):914–925 (J32)

Loli M, Bransby MF, Anastasopoulos I, Gazetas G (2012) Interaction of caisson foundations with a seismically rupturing normal fault: centrifuge testing versus numerical simulation. Géotechnique 62(1):29–43 (J37)

Niccum MR, Cluff LS, Chamoro F, Wylie L (1976) Banco central de nicaragua: A case history of a high-rise building that survived surface fault rupture. In: Humphrey CB (ed) Engineering geology and soils engineering symposium, No. 14, Idaho Transportation Department, Division of Highways, pp 133–144

Pamuk A, Kalkan E, Ling HI (2005) Structural and geotechnical impacts of surface rupture on highway structures during recent earthquakes in Turkey. Soil Dyn Earthq Eng 25(7–10):581–589

Psycharis I et al (2013) D14.4. Shaking table test techniques and fault rupture box testing for SSI, SERIES Research project Report

Ulusay R, Aydan O, Hamada M (2002) The behaviour of structures built on active fault zones: examples from the recent earthquakes of Turkey. Struct Eng Earthq Eng 19(2):149–167 Japan Society of Civil Engineers

White RJ, Stone KJL, Jewel RJ (1994) Effect of particle size on localization development in model tests on sand. In: Leung CF, Lee FH, Tan TS (eds) Proc Inst Centrifuge Conf, Balkema, Rotterdam, The Netherlands, pp 817–822

White DJ, Take WA, Bolton MD (2003) Soil deformation measurement using particle image velocimetry (PIV) and photogrammetry. Géotechnique 53(7):619–631

Youd T L (1989) Ground failure damage to buildings during earthquakes. Found Eng Curr Princ Pract 1:758–770. ASCE, New York

Youd T L, Bardet JP, Bray JD (2000) Kocaeli, Turkey, earthquake of August 17, 1999 reconnaissance report. Earthq Spectr 16(Suppl A):456

Chapter 26
Development of New Infinite Element for Numerical Simulation of Wave Propagation in Soil Media

Vlatko Sesov, Mihail Garevski, Kemal Edip and Julijana Bojadjieva

26.1 Introduction

Seismic soil-structure interaction analysis of massive engineering structures such as dams, power plants and high rise buildings is a very complex issue, which has gained an increasing importance for the last decades. The interaction effects considerably influence the seismic performance of these structures. Lessons from recent strong earthquakes (Christchurch 2010–2011; Great Tohoku 2011; Sichuan 2008, 2013, etc.) undoubtedly show that soil-foundation-structure interaction and local site effects can significantly increase total damage if they are neglected or not treated appropriately.

The present study is focused on the numerical simulation of underlying soil media which are unbounded and extend to infinity. In many engineering applications, the numerical treatment of unbounded domains is of a considerable interest. The first attempts to numerically treat infinite domains involved applying the finite element method directly by simply truncating the outer region. Although this method worked well for static cases, in dynamic analysis, the results diverged enormously due to the reflected waves on the artificially introduced boundaries. Artificial boundaries simulating energy radiation towards infinity were proposed by many researchers. In the work of Kausel et al. (1974), a layered half space was considered by introducing viscous stress boundaries. Tzong and Penzien (1983) developed the system identification method to calculate the dynamic stiffness matrix for the exterior soil region. In his work, Liou (1989) explained an analytical method without using the Green's function for layered media of rigid bedrock. Manolis and Beskos

V. Sesov (✉)
Institute of Earthquake Engineering and Engineering Seismology,
IZIIS, Ss Cyril and Methodius, Todor Aleksandrov No.165, 1000 Skopje,
Republic of Macedonia
e-mail: vlatko@pluto.iziis.ukim.edu.mk

M. Garevski · K. Edip · J. Bojadjieva
Institute of Earthquake Engineering and Engineering Seismology,
IZIIS, Ss Cyril and Methodius, Skopje, Republic of Macedonia

F. Taucer, R. Apostolska (eds.), *Experimental Research in Earthquake Engineering*,
Geotechnical, Geological and Earthquake Engineering 35,
DOI 10.1007/978-3-319-10136-1_26

(1983) used a boundary element method in the time domain enabling usage of non-linear material models in the finite elements domain.

An infinite element attempts to simulate the behaviour of the unbounded domain. The development of infinite elements dates back to a more recent period. The concept is very similar to that of finite elements including the concept of infinity to the element domain. The use of infinite elements together with the well known finite elements is a promising choice for the investigation of such unbounded domains. An infinite element is an element that represents the behaviour of unbounded domains. One of the first publications introducing mapping infinite elements dates back to Zienkiewicz and Bettess (1976).

There are mainly two types of infinite elements. The first type uses the decay function together with a shape function, which approaches zero at infinity. In the case of the second one, the geometry is mapped from a finite to an infinite domain. The mapping of infinite elements has the advantage that application of standard Gauss (Zienkiewicz et al. 1983) integration formulas is possible. Bettess (1992) showed that mapped infinite elements work very well for static analysis of elastic media. The application of infinite elements in wave propagation requires more attention to be paid to outwardly propagating waves. The application of infinite elements is extensive and can be used in many fields of engineering. Application of infinite elements in mass transport is explained in the work of Zhao and Valliappan (1994). Sullivan and O'Neill (1992) use infinite elements in mass transfer problems which require far boundary representations. In the work of Askar and Lynn (1984), ground freezing problems are considered using infinite elements. Soil-structure interaction problems considering infinite elements were the subject of investigation of many authors such as: Bettess and Zienkiewicz (1977), Zienkiewicz et al. (1985), Chow and Smith (1981), Rajapakse and Karasudhi (1986), Chuhan and Chongbin (1987), etc. A comprehensive overview for the acoustic case is given by Astley (2000). Medina and Penzien (1982) proposed a different type of infinite element considering both P-wave and S-wave propagation, although shape functions appeared to be extremely complicated for such applications. In the work of Haggblad and Nordgren (1987), static infinite elements are combined with an absorbing layer, leading to good results. In the work of Edip et al. (2011), infinite elements are upgraded by absorbing properties at each node and validation is done accordingly. In this work, the infinite element is further developed such that the number of nodes is increased and the mapping functions are arranged accordingly.

The basic idea of the newly developed infinite element is to consider the assumption that the displacement field approaches zero at infinity, absorbing the outward propagating waves. Application of this type of an infinite element in soil-structure interaction problems is preferable due to the formulation, which is similar to that for finite elements. Thus, the exterior domain is partitioned into a finite number of infinite elements, which are directly connected with the finite element mesh of the interior domain.

26.2 Governing Equations of the Newly Developed Infinite Elements

In the case of conventional infinite element formulations, two 'global to local' mappings are employed. One is for the coordinates and the other for the unknown function values. The coordinate transformation (mapping) is used to compute the Jacobian matrix, which is used to obtain the stiffness and mass matrices. As it is presented by Bettess (1992), the physics of infinite element formulations shows that the finite element is stretched to be an infinite element retaining the integration to be of a Gauss-Legendre type. The new element is based on a mapping function, which maps the global to the local coordinate system. The mapping used can be written as follows:

$$x = M^0(\xi)\cdot x_0 + M^2(\xi)\cdot x_2 \tag{26.1}$$

where x and ξ represent the global and local coordinates. Functions M^0 and M^2 stand for mapping functions and are given as follows:

$$M^0(\xi) = \frac{-\xi}{1-\xi};\ M^2(\xi) = 1 + \frac{\xi}{1+\xi} \tag{26.2}$$

Solving the above mappings for ξ, the inverse mapping containing the length of element ($2a$) is obtained as:

$$\xi = 1 - \frac{2a}{x - x_0} \tag{26.3}$$

In terms of polynomial mappings, the polynomial in local coordinates $P = k_0 + k_1\xi + k_2\xi^2 + \ldots\ldots$ is mapped into a polynomial in global coordinates such as $P = t_0 + {}^{t_1}\!/_{\xi} + {}^{t_2}\!/_{\xi^2} + \ldots$ The term t_0 implies the value of the function at infinity. If the function decays to zero at infinity, then equation $t_0 = 0$ holds. The extension of the mapping functions to two dimensional cases is straightforward as given in the work of Zienkiewicz et al. (1985). The domain mapping in the infinite elements is shown in Fig. 26.1.

The basic idea consists of mapping an infinite region onto a finite one. This technique needs appropriate mapping functions by which a domain, unbounded in one direction, is transformed into a bounded one. These mapping functions are obtained as a product of standard shape functions for the finite direction and special mapping functions for the infinite direction. In this element, the field variables are approximated using standard interpolation functions. Then, it is possible to obtain an expression with $1/r$ type decay of the field variables in the unlimited domain,

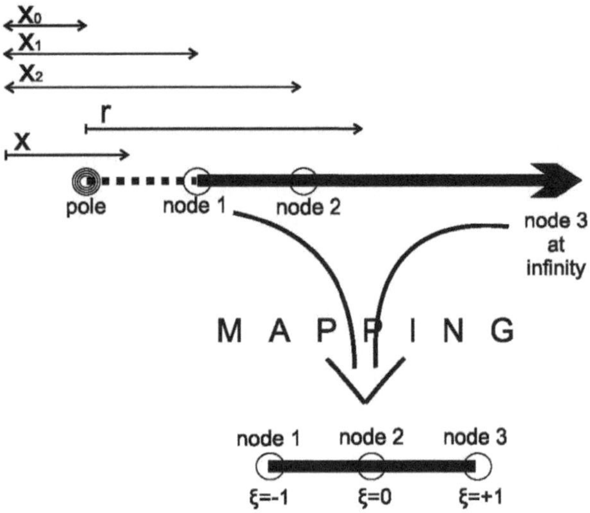

Fig. 26.1 Representation of domain *mapping* in infinite element

where r is the distance from the 'pole' of the transformation to a point belonging to the infinite domain. Some attention should be paid to positioning the pole(s) of infinite elements, as the geometry and the field variable expansion depend on their position Zienkiewicz et al. (1985).

Attention has also to be paid to ensuring uniqueness of the mapping and the continuity of the solution between elements with common sides. The number of nodes plays an important role for the effectiveness of infinite elements. In order to couple the finite and infinite elements, the number of nodes has to be considered carefully. In Fig. 26.2, coupling of finite and infinite elements is given.

Referring to Fig. 26.2, the coupling between the finite and infinite elements can be presented as follows. The finite element has eight nodes, with three nodes on the end side. The infinite element has three nodes on the side which allows complete coupling with the finite element.

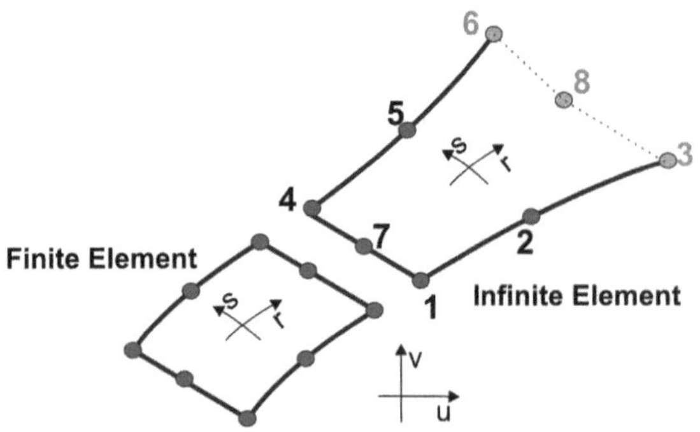

Fig. 26.2 Coupling of *finite* and *infinite* elements

The element displacement in the **u** and **v** directions is interpolated with the usual shape functions N^1, N^2, N^4, N^5 and N^7:

$$u = \begin{bmatrix} N^1 & N^2 & 0 & N^4 & N^5 & 0 & N^7 & 0 \end{bmatrix} \mathbf{u}$$
$$v = \begin{bmatrix} N^1 & N^2 & 0 & N^4 & N^5 & 0 & N^7 & 0 \end{bmatrix} \mathbf{v} \tag{26.4}$$

In expression (26.4), **u** and **v** are vectors with nodal point displacements in global coordinates. The shape functions are given as follows:

$$\begin{aligned} N^1 &= -(r-1)(-1+s)(s+1+r)/4 \\ N^2 &= (r-1)(1+r)(-1+s)/2 \\ N^4 &= -(r-1)(1+s)(s-1-r)/4 \\ N^5 &= -(r-1)(1+r)(1+s)/2 \\ N^7 &= (-1+s)(1+s)(r-1)/2 \end{aligned} \tag{26.5}$$

Based on the isoparametric concept, the infinite element in the global coordinates is mapped onto an element in the local coordinate system using the expression given in (26.6).

$$r = \begin{bmatrix} M^1 & M^2 & 0 & M^4 & M^5 & 0 & M^7 & 0 \end{bmatrix} \mathbf{r}$$
$$s = \begin{bmatrix} M^1 & M^2 & 0 & M^4 & M^5 & 0 & M^7 & 0 \end{bmatrix} \mathbf{s} \tag{26.6}$$

The mapping functions are given as follows:

$$\begin{aligned} M^1 &= -\frac{(1-s)rs}{1-r} \\ M^2 &= -\frac{(1-s)(1+r)}{2(1-r)} \\ M^4 &= \frac{(1+s)rs}{1-r} \\ M^5 &= -\frac{(1+s)(1+r)}{2(1-r)} \\ M^7 &= -\frac{2r(1+s)(1-s)}{(1-r)} \end{aligned} \tag{26.7}$$

In expression (26.7), r and s are vectors of nodal point displacements in local coordinates where it is to be pointed out that, on the side of infinity ($r=1$), no mappings are assigned to the nodes as it is taken that displacement decays at infinity. The number and location of the nodes connecting finite and infinite elements must coincide to guarantee a continuity condition between the elements. The main advantage of the proposed infinite elements is that the number of nodes on the infinite element allows coupling with finite elements with eight nodes which are used for displacement sensitive problems. Construction of element matrices is done by using the usual procedures as described in Bathe (1982). The new coordinate interpolation functions are taken into consideration in the Jacobian matrix as described in Bettess (1992). For the absorbing layer of the infinite element, the Lysmer-Kuhlmeyer approach (Plaxis 2D Dynamics Manual) is used. In all cases, a plane strain two dimensional case is studied. For impact of plane waves on element sides, normal and tangential stresses are derived as follows:

$$\begin{bmatrix} \sigma_n \\ \tau \end{bmatrix} = \begin{bmatrix} a\rho c^p & 0 \\ 0 & b\rho c^s \end{bmatrix} \begin{bmatrix} \dot{u}^n \\ \dot{u}^t \end{bmatrix} \tag{26.8}$$

where c^p and c^s indicate the wave velocities for the P wave (compressional) and S wave (shear), respectively. The term ρ stands for the density of the soil medium. In order to take into account the directions of the incident waves, coefficients a and b are used as multipliers. Transformation from local to global coordinates is done by ANSYS software (2006), such that there is no need for defining transformation matrices. Time derivatives are approximated by the Newmark's method. The programming part of the infinite element was performed by using the Programmable Features of the ANSYS software.

26.3 Verification of the Infinite Element

To study the accuracy of the proposed infinite element, several dynamic problems were considered. Firstly, wave propagation through a soil layer was considered. Then, propagation of body waves in the two dimensional domain were analyzed. Finally, a soil layer domain subjected to seismic input was considered in two different discretization scenarios.

26.3.1 Wave Propagation—One Dimensional Case

In order to check the wave absorbing properties of the infinite elements, a long soil layer with given properties (see table Table 26.1) was considered. In order to verify

Table 26.1 Soil properties

Young's modulus	E	18,000	kPa
Poisson's ratio	*n*	0.2	–
Density	ρ	2.04	t/m³

the absorbing properties of the newly programmed infinite elements, simulation of one dimensional wave propagation problem was performed using the soil layer properties as given in thePlaxis Validation Manual. The soil layer was horizontal, with a total length of 10 m. The soil layer was simulated in two alternative ways. First, only finite elements with fixed boundaries and then the same finite elements with absorbing infinite element boundaries. Figure 26.3 shows the soil domain with point A in the middle.

The soil domain was discretized by 40 elements. The properties of the soil layer are given in Table 26.1.

The compressional wave velocity V_p is:

$$V_p = \sqrt{(1-\nu)E/\rho((1+\nu)(1-2\nu))} \tag{26.9}$$

Using the values given in Table 26.1, $V_p = 100$ m/s was obtained. In order to give a more general insight into the 1D wave propagation, a horizontal displacement on the left of the domain was applied as two different types of functions:

- Heaviside step function
- Impulse function

The first type of displacement application is considered to be of Heaviside step type. In this case, a displacement $u_x = 0.001$ m is applied at approximately 0.05 s.

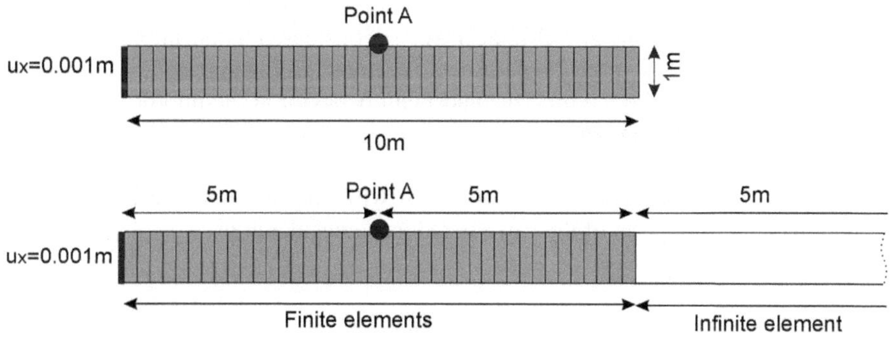

Fig. 26.3 Soil layer domain

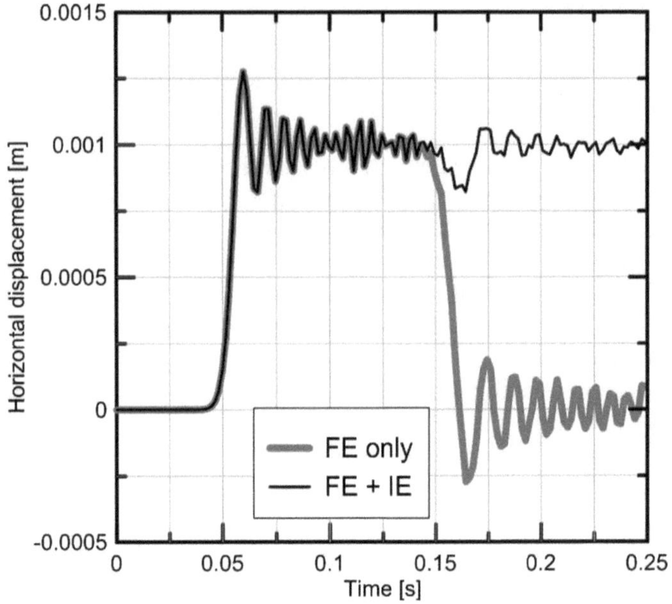

Fig. 26.4 Time displacement history—Heaviside step function

Figure 26.4 shows the time history of the traveling P wave at the middle point (point A).

Figure 26.4 clearly shows that, in the case of a domain composed of finite elements only, the wave is reflected, while in the case of a coupled approach when both finite and infinite elements are applied, the wave is absorbed at the boundary so that the value of the horizontal displacement oscillates around the initially applied displacement value ($u_x = 0.001$ m). The time required for the P-wave ($V_p = 100$ m/s) to reach point A is 5 m/100 m/s = 0.05 s, see Fig. 26.4. This proves the correctness of the numerical simulation using coupled finite and infinite elements. In the subsequent case, application of a displacement is considered as an impulse function for only 0.001 s and then the displacement is removed. The system is analyzed for 0.3 s in order to observe the time-displacement behavior at the middle point of the domain (Point A, see Fig. 26.5).

In Fig. 26.5, it is clearly shown that in the case of finite elements with fixed boundaries, the impulse is reflected at 0.15 s, while in the case of infinite element boundaries, the reflection of the wave is reduced considerably.

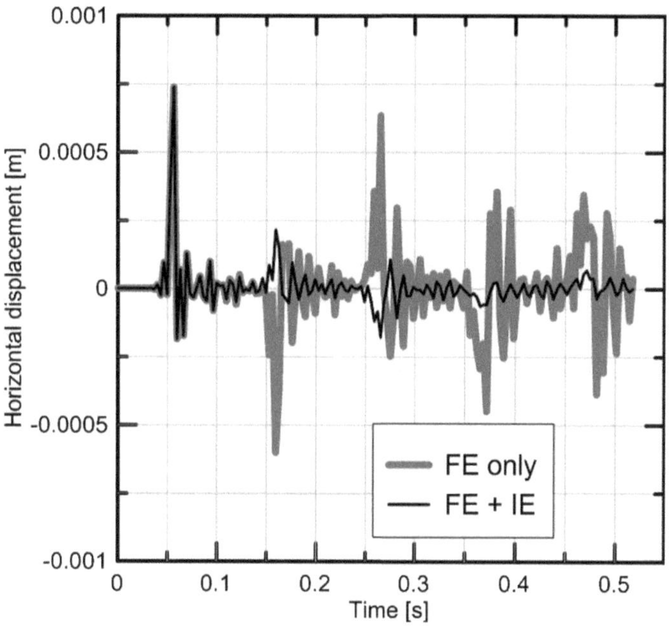

Fig. 26.5 Time displacement history—impulse function

26.3.2 Wave Propagation—Two Dimensional Case

In order to verify the absorbing properties of the newly programmed infinite elements, two dimensional wave propagation including both compressive and shear waves are simulated. The soil medium is considered in two alternative ways. First, only finite and infinite elements without absorbing properties are considered and then the same infinite elements with absorbing properties are considered. Figure 26.6 shows the soil domain with a total width of 40 m and a height of 20 m.

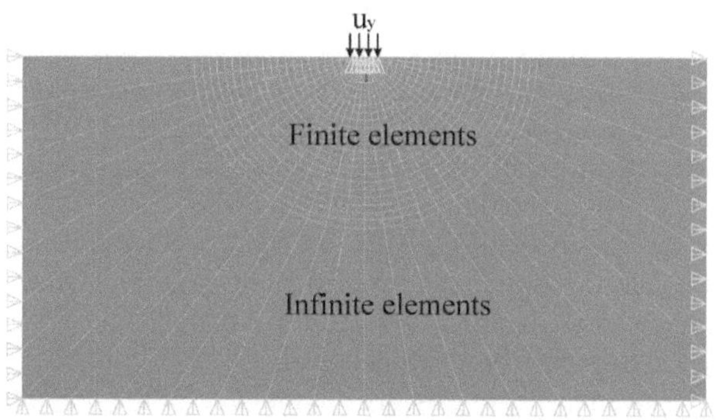

Fig. 26.6 Soil domain of *finite* and *infinite* elements

Table 26.2 Soil properties

Young's modulus	E	50,000	kPa
Poisson's ratio	*n*	0.35	–
Density	ρ	2.70	t/m³

The soil domain is discretized by finite and infinite elements as shown in Fig. 26.6. The properties of the soil are given in Table 26.2.

The applied displacement is considered to be an impulse type. In this case, a displacement of $u_y = 0.001$ m is applied as an impulse function at time 0.01 s.

As can be seen from Fig. 26.7, there are two waves propagating in the domain. The first one is the compressive P wave while the second one is the shear S wave. The infinite elements are able to absorb both types of waves although their velocities of propagation are not the same. It is to be mentioned that, at time $t = 0.07$ s, the P wave enters the infinite domain. In the case of absorbing infinite elements, the wave is almost completely absorbed while in the case of infinite elements without absorbing properties, the P wave reflects back to the finite element domain. At $t = 0.09$ s, in the case of a domain surrounded by absorbing infinite elements, the S-wave propagation at the finite element boundaries is observed. On the other hand, in the domain of infinite elements without absorbing properties, the outward propagating S-wave collides with the reflected P wave, as shown in the figure. At time $t = 0.15$ s, in the domain composed of absorbing infinite elements boundaries, there is no sign of wave propagation since the waves are absorbed in the infinite element region.

26.3.3 Soil Layer Simulation

In order to verify the applicability of infinite elements in the seismic response of a soil layer, two different models of a soil layer were formulated. In the first case, discretization was done by finite elements only in which case the number of elements was large and the analysis lasted quite long (Fig. 26.8a).

In the second case, the layer domain was discretized by both finite and infinite elements in which the number of elements was dramatically reduced, Fig. 26.8b. Soil properties of the soil layer are given in Table 26.3.

The acceleration time history of the Imperial Valley EQ, El Centro record, 1940-May-18 was applied to both models. The results obtained at the top of the domain are shown in the figure below, Fig. 26.9.

As can be seen from Fig. 26.9, both acceleration and displacement time histories obtained by finite and coupled finite-infinite elements are presented. In order to compare the numerical simulation of both domains, the input acceleration is applied at the boundary of the domains.

The analysis reveals that, in the case of the domain composed of finite and absorbing infinite elements, the results are not different from those obtained for the domain composed of a huge number of finite elements only. The difference is that,

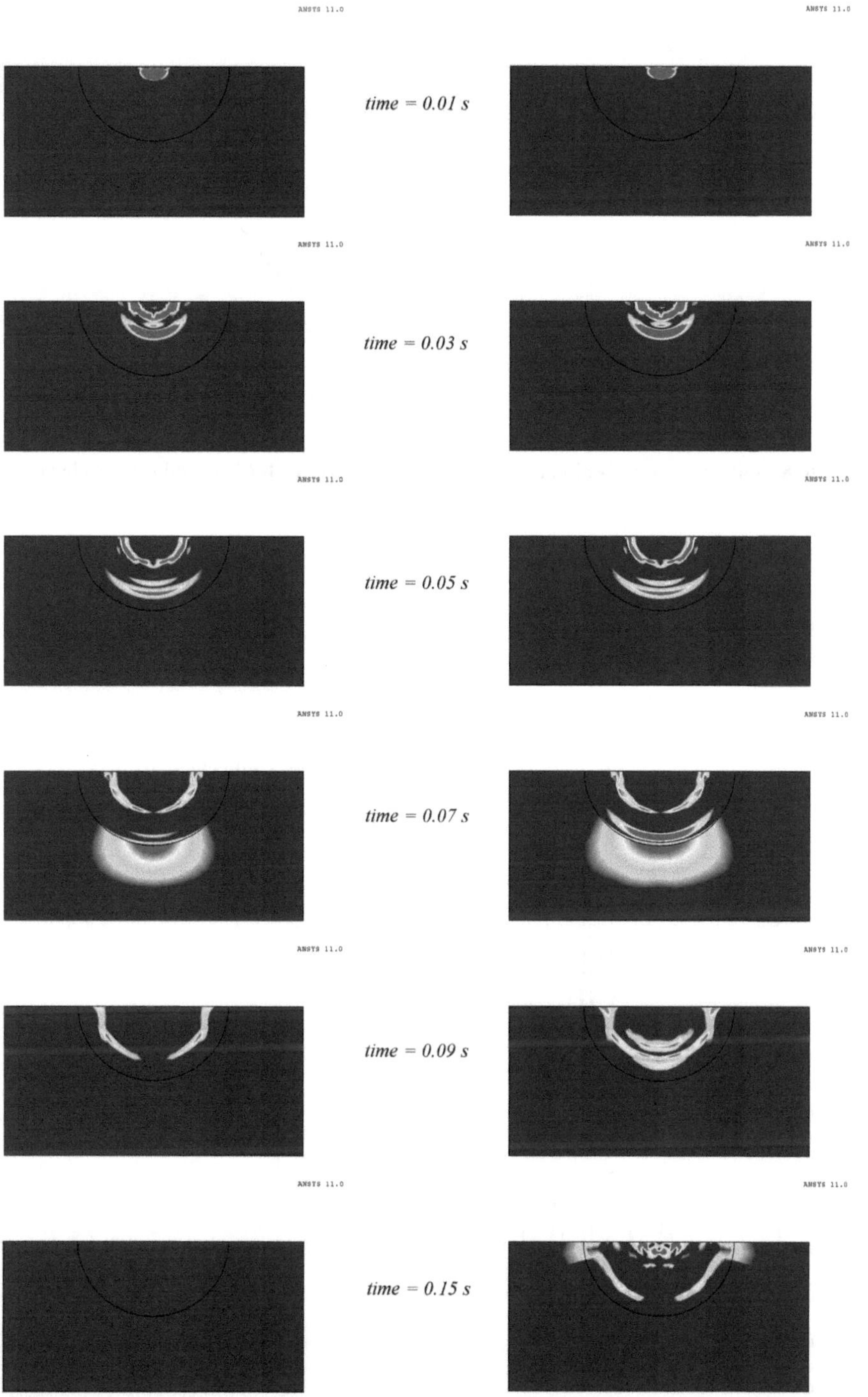

Fig. 26.7 Wave propagation at different time spots using (**a**) infinite elements and (**b**) infinite elements with absorbing layer

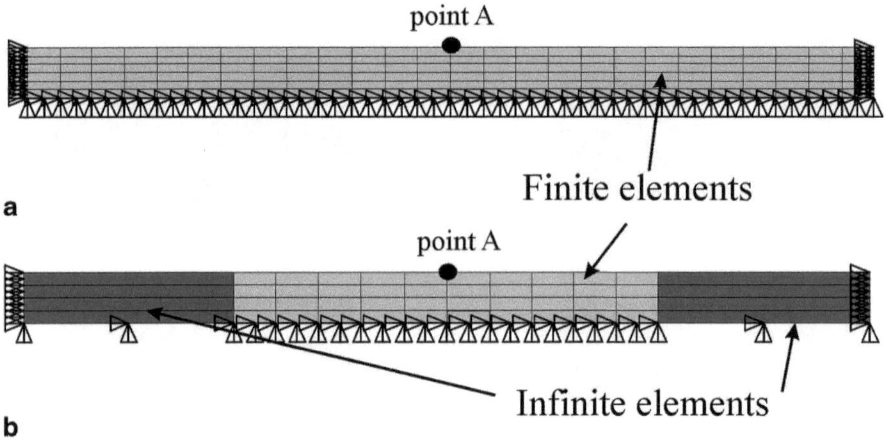

Fig. 26.8 Soil layer discretization cases. **a** *Finite* elements only. **b** *Finite* and *infinite* elements

Table 26.3 Soil properties

Young's modulus	E	50,000	kPa
Poisson's ratio	*N*	0.35	–
Density	P	2.70	t/m³

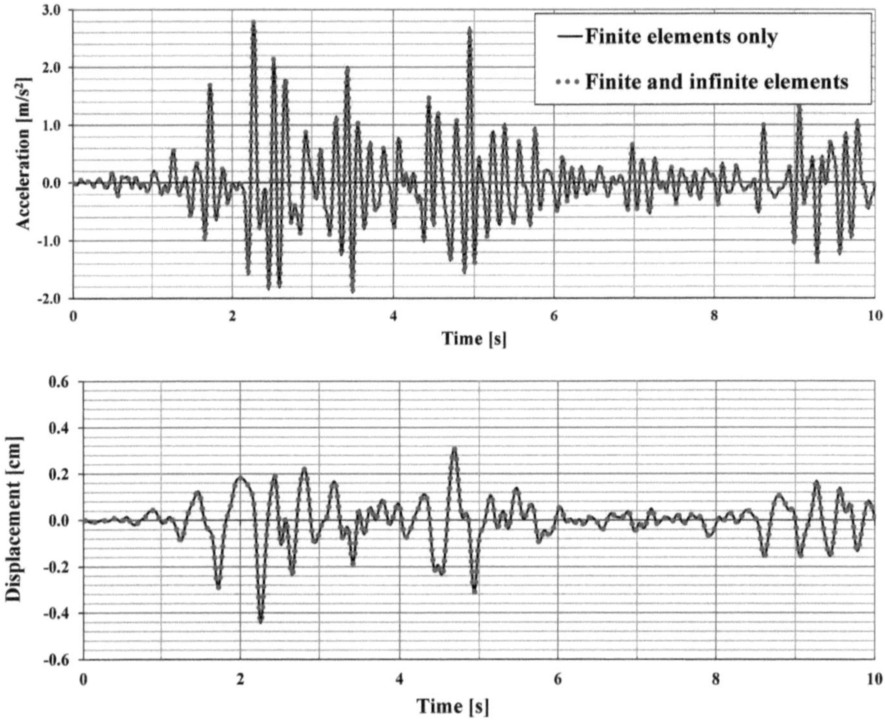

Fig. 26.9 Acceleration and displacement time histories at the top of the soil layer (point A)

in performing the analysis without infinite elements, the number of finite elements is much bigger and thus the analysis requires a longer time. The usage of infinite elements reduces the number of elements used while conserving the correctness of results.

26.4 Conclusion

A coupled computational method of finite and infinite elements was applied for analysis of selected geotechnical problems. The new developed model uses mapping infinite elements to approximate the stiffness of the unbounded exterior domain. Infinite elements were added with absorbing characteristics which were used in this study to provide a promising tool for coping with problems in dynamic analysis domain. The efficiency of the coupled finite and infinite elements model is shown by several numerical examples. One- and two-dimensional wave propagation cases were first considered considered. Then, a soil layer extending to infinity was subjected to seismic excitation. From the examples, it can be concluded that using the coupled finite-infinite element approach the number of elements and nodes is reduced without affecting the accuracy of the results in the interior domain.

Presented in this paper are efforts toward developing and applying new mapped infinite elements with absorbing properties to simulate the external boundaries of unbounded soil media. The proposed mapping functions together with the absorbing properties were applied to a couple of practical problems and the results were compared accordingly. The study is based on infinite elements of a horizontal type implemented in the general finite element software—ANSYS. Apart from mapping functions, these elements can be treated as usual finite elements. The proposed absorbing infinite elements give very good results and preserve the simplicity of the finite element method.

References

ANSYS (2006) FEM Software

Askar H, Lynn P (1984) Infinite elements for ground freezing problems. J Eng Mech 110(2):157–172

Astley RJ (2000) Infinite elements for wave problems: a review of current formulations and an assessment of accuracy. Int J Numer Methods Eng 49(7):951–976

Bathe KJ (1982) Finite element procedures in engineering analysis. Prentice-Hall, Englewood Cliffs

Bettess P (1992) Infinite elements. Penshaw, New Castle

Bettess P, Zienkiewicz O (1977) Diffraction and refraction of surface waves using finite and infinite elements. Int J Numer Methods Eng 11(8):1271–1290

Chow Y, Smith I (1981) Static and periodic infinite solid elements. Int J Numer Methods Eng 17(4):503–526

Chuhan Z, Chongbin Z (1987) Coupling method of finite and infinite elements for strip foundation wave problems. Earthq Eng Struct Dyn 15(7):839–851

Edip K et al (2011) Numerical simulation of wave propagation in soil media. In: Proceedings of the 21st European Young Geotechnical Engineers' Conference Rotterdam, 2011. Rotterdam, Netherland

Häggblad B, Nordgren G (1987) Modelling nonlinear soil-structure interaction using interface elements, elastic-plastic soil elements and absorbing infinite elements. Comput Struct 26(1–2):307–324

Kausel E, Roesset J, Waas G (1974) Forced vibrations of circular foundations on layered media. Department of Civil Engineering, School of Engineering, Massachusetts Institute of Technology

Liou GS (1989) Analytic Solutions for Soil-Structure Interaction in Layered Media. Earthq Eng Struct Dyn 18(5):667–686

Manolis G, Beskos D (1983) Dynamic response of lined tunnels by an isoparametric boundary element method. Comput Methods Appl Mech Eng 36(3):291–307

Medina F, Penzien J (1982) Infinite elements for elastodynamics. Earthq Eng Struct Dyn 10(5):699–709

Plaxis 2D Dynamics Manual Version 9.0, www.plaxis.nl

Rajapakse R, Karasudhi P (1986) An efficient elastodynamic infinite element. Int J Solids Struct 22(6):643–657

Sullivan JM, O'Neill K (1992) Application of infinite elements to phase change situations on deforming meshes. Int J Num Methods Eng 33(9):1861–1874

Tzong TJ, Penzien J (1983) Hybrid modeling of soil-structure interaction in layered media. NASA STI/Recon Technical Report N 84:30137

Zhao C, Valliappan S (1994) Transient infinite elements for contaminant transport problems. Int J Num Methods Eng 37(7):1143–1158

Zienkiewicz OC, Bettess P (1976) Infinite elements in the study of fluid-structure interaction problems. Computing methods in applied sciences (Second Internat. Sympos., Versailles, 1975), pp. 133–172. Lecture Notes in Phys., Vol. 58, Springer, Berlin, 1976

Zienkiewicz OC, Emson C, Bettess P (1983) A novel boundary infinite element. Int J Num Methods Eng 19(3):393–404

Zienkiewicz O et al (1985) Mapped infinite elements for exterior wave problems. Int J Num Methods Eng 21(7):1229–1251

Chapter 27
Analysis of the Dynamic Behaviour of Squat Silos Containing Grain-like Material Subjected to Shaking Table Tests—ASESGRAM Final Report

Dora Foti, Tomaso Trombetti, Stefano Silvestri, Laura Di Chiacchio, Salvador Ivorra, Colin Taylor and Matthew Dietz

27.1 Introduction

The assessment of horizontal interactions is of particular interest when evaluating the actions exerted by grain-like materials on the walls of flat-bottom silos during an earthquake. This interest is based on the possibility of providing more appropriate design rules closer to the effective seismic behaviour of silos. A careful evaluation of the forces produced by the material in the silos makes it possible to safely design silos in seismic areas without excessive redundancy.

In static design, the identification of the effective mass (Pozzati and Ceccoli 1972) follows the approach proposed by Janssen (1895) and Koenen (1896), in which the grain mass mobilized is completely supported by the walls. Also Eurocode 1 Part 4 (CEN 2004) stands on this procedure.

In a previous research work (Silvestri et al. 2012) the authors developed an analytical theory for the evaluation of the mobilized mass and the interaction between the ensiled material and the silo wall.

Since the analytical theory is focused on a simplified idealized system, the design issues of the silo specimen have to deal with the differences between the idealised model and the characteristics of an actual silo subjected to earthquake ground motion. It is clear that a real silo is characterized by compressible (soft) grain material, flex-

D. Foti (✉)
Department DICAR, Technical University of Bari, Bari, Italy
e-mail: d.foti@poliba.it

T. Trombetti · S. Silvestri · L. Di Chiacchio
Department DICAM, University of Bologna, Bologna, Italy

S. Ivorra
University of Alicante, Alicante, Spain
e-mail: Sivorra@ua.es

C. Taylor · M. Dietz
University of Bristol, Bristol, England

F. Taucer, R. Apostolska (eds.), *Experimental Research in Earthquake Engineering*,
Geotechnical, Geological and Earthquake Engineering 35,
DOI 10.1007/978-3-319-10136-1_27

ible walls and, during an earthquake it is subjected to seismic accelerations. On the other hand, the theory developed by Silvestri et al. (2012) has been developed with reference to the following assumptions (i.e. the ideal model deals with three fundamental hypotheses) of (i) incompressible material (in dry conditions, compact, without voids, as it were composed by a number of infinitely stiff and infinitely resistant spherical balls), (ii) infinitely stiff walls, (iii) time-constant acceleration. For these reasons, the shaking table tests and thus the test specimens have been designed both to meet the ideal conditions of the analytical theory and to investigate the influence of the type of input. Consequently, the two objectives of the experimental tests are:

- to verify the findings of the analytical theory developed by Silvestri et al. (2012);
- to evaluate the influence of the assumption made on the base input (earthquake vs. sinusoidal/constant) for the transition from the ideal model case to the real silo case.

27.2 Test Set-up

27.2.1 The Specimen: Geometry and Materials

Given that the dimensions of the EQUALS shaking-table is 3 × 3 m, a scaled (roughly 1/10) circular tank specimen was designed. The specimen consists of a 1.2 m diameter, 1.6 m high, 3 mm wall thickness polycarbonate container built for the project. Polycarbonate (E = 2.3 GPa) has been selected as the material for the walls in order to increase strains and thus to facilitate their measuring. The circular tank was obtained by folding two single sheets on themselves to create two U sections, which were interlocked and cemented with a 50 mm seam using epoxy and bolts. A ring is placed at the top of the specimen to avoid local deformations. Ballotini glass was chosen as the ensiled material to simulate incompressible grain-like material (Fig. 27.1). Furthermore, Ballottini glass is very close to the idealized ensiled ma-

a b c

Fig. 27.1 **a** The silo specimen. **b** The setup of the instrumentations. **c** The insertion of the Ballotini glass

terial and its weight (20 kN) leads to horizontal forces which are compatible with the capacity of the shaking table system (150 kN). A 0.4–0.6 mm diameter for the particles of the Ballotini glass content was selected for scaling reasons.

For this purpose, the following instrumentation was installed (Fig. 27.2):

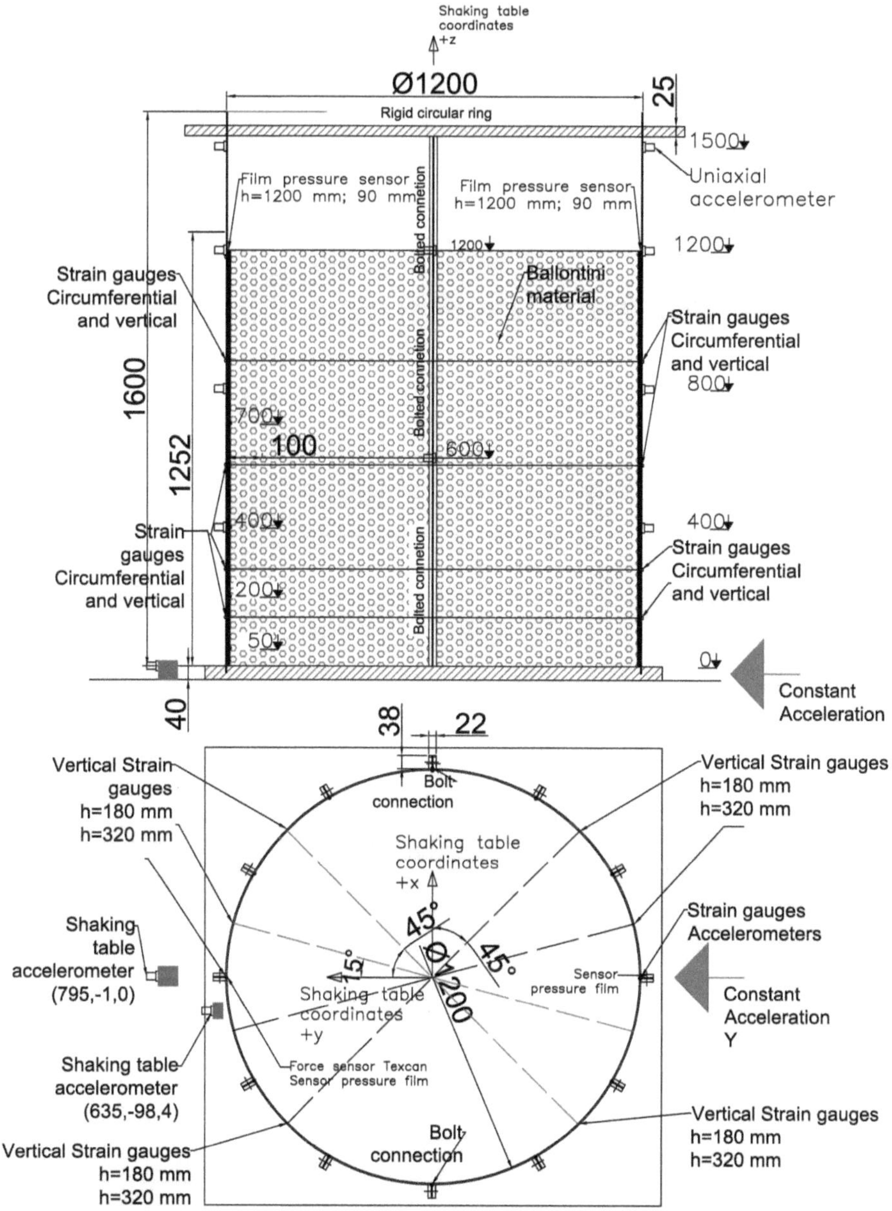

Fig. 27.2 The instrumentation set-up

The specimen tank was filled up to different heights h during the tests:

- $h = 1.2$ m, to obtain a slenderness ratio equal to 1 (Fig. 27.3);
- $h = 0.6$ m, to obtain a slenderness ratio equal to 0.5 (Fig. 27.4).

a b

Fig. 27.3 **a** The specimen filled with Ballotini glass in the First session. **b** The specimen filled with Ballotini glass in the Second and Third sessions

a b

Fig. 27.4 **a** The specimen filled with Ballotini glass up to 0.6 m with roughened walls. **b** The specimen filled with Ballotini glass up to 1.6 m with roughened walls

Table 27.1 Measured values of the parameters

Measured value					
$E^{polycarbonate}$	$\gamma^{Ballotini}$	φ	λ	$\mu^{G,\,B}$	$\mu^{G,\,W}$
2.3 GPa	1481 kg/m^3	26°	0.55	0.45	0.30

Table 27.1 summarizes the reasonable values that can be assumed for the parameters involved in the analytical formulation to be validated (Table 27.1).

27.2.2 *The Test Instrumentation*

The test setup was designed in order to provide measures of:

- table, structure and grain accelerations at different locations
- structure deformation at different locations
- displacement of the rigid circular ring at the top of the silo
- local pressures exerted by the grain on the walls

The following instrumentation was installed in the specimen:

- mono-directional accelerometers located at the shaking-table foundation, glued to the silo walls and placed inside the Ballotini glass
- vertical and horizontal strain gauges positioned on the exterior side of the walls at four different heights
- LVDTs placed at the top of the structure
- film and Flexi Force pressure sensors placed on the interior side of the walls

27.2.3 *The Test Sessions*

Three sessions of tests have been performed on the same specimen:

- First session: the silo is characterized by smooth walls (grain-wall friction coefficient: $\mu_{G,W} = 0.30$) and filled with Ballotini glass up to a height of 120 cm (Fig. 27.3a), tested under white noise, sinusoidal input and earthquake accelerograms applied along the horizontal direction only;
- Second session: the silo is characterized by roughened walls (grain-wall friction coefficient: $\mu_{G,W} = 0.45$) and filled with Ballotini glass up to a height of 60 cm (Fig. 27.4a), tested under white noise, sinusoidal input and earthquake accelerograms applied along both the horizontal and the vertical directions;

- Third session: the silo is characterized by roughened walls (grain-wall friction coefficient: $\mu_{G,W}$=0.45) and filled with Ballotini glass up to a height of 120 cm (Fig. 27.3b), tested under systematic sinusoidal input along the horizontal direction only.

The purpose of the first session of tests was to determine the dynamic properties of the silo specimen and to observe the variation in the dynamic properties of the prototype silo.

The objectives of the second session of tests were to analyse the repeatability and the initial conditions of the silo and to monitor the effect of a lower slenderness ratio (Ballotini up to 0.6 m). Three repetitions of the same input were carried out to evaluate how the compaction of the ensiled material could affect the results. Also, for this purpose, the initial tests were repeated at the end of the campaign.

The aim of the third session of tests was to analyse the same conditions presented in the first session of tests (Ballotini up to 1.2 m) with particular attention on the effects of the increased roughness of the silo wall and the consequent changed internal friction grain-wall coefficient on the behaviour of the flat-bottom silo.

27.2.4 *The Test Input*

The following three tables report the test input for each one of the three sessions of tests (Table 27.2, 27.3 and 27.4). The notation PGA is herafter used to indicate the Peak Table (Ground) Acceleration.

Table 27.2 Test input for the First session of tests

Input	Test no.	PGA (g)
White noise	N1–N5	0.05–0.30
1 HZ sinusoidal (Y)	S1–S8	0.05–0.40
White noise	N6	0.30
1 HZ sinusoidal (Y)	S9	0.03
0.5 HZ sinusoidal (Y)	S10–S13	0.01–0.15
White noise	N7	0.30
1 HZ sinusoidal (Y)	S14	0.5
White noise	N8	0.30
Earthquake input	E1–E18	0.04–0.40
White noise	N9	0.30

Table 27.3 Test input for the Second session of tests

Input	Test no.	PGA (g)
1 HZ sinusoidal (Y)	S1–S3	0.10
White noise	N1	0.20
1 HZ sinusoidal (Y)	S4–S36	0.10–0.58
2 HZ sinusoidal (Y)	S37–S48	0.30–0.76
2 HZ sinusoidal (Y&Z in phase)	S49–S54	0.30
2 HZ sinusoidal (Y)	S55–S67	0.10–0.70
1 HZ sinusoidal (Y&Z in and out of phase)	S68–S75	0.10–0.40
Earthquake input (Y)	E1–E3	0.30–0.60
White noise	N2	0.30
Earthquake input (Y)	E4	0.10
Earthquake input (Y&Z)	E5	0.30
2 HZ sinusoidal	S76–S78	0.30–0.70
Earthquake input (Y&Z)	E6	0.60
2 HZ sinusoidal (Y)	S79–S86	0.80–1.10
2 HZ sinusoidal (X)	S87–S89	0.30–0.50
2 HZ sinusoidal (Y&Z in phase)	S90–S105	0.10–0.85
Earthquake input (Y)	E7–E25	0.07–1.2
Earthquake input (Y&Z in phase)	E26–E44	0.08–1.11

Table 27.4 Test input for the Third session of tests

Input	Test no.	PGA (g)
1 HZ sinusoidal (Y)	S1	0.10
White noise	N1	0.20
1 HZ sinusoidal (Y)	S2–S8	0.10–0.60
2 HZ sinusoidal (Y)	S9–S30	0.10–0.95

27.3 Experimental Results

27.3.1 Frequency

At first a white noise input was used in order to evaluate the dynamic properties of the silo. The first initial frequency was about 12 ÷ 13 Hz. Table 27.5 reports the

Table 27.5 Own frequencies of the silo for each session of tests

White noise	Ensiled material	Frequency (Hz)
First Session (August 2012)	Full Ballotini	12.72–43.9
Second Session (January 2013)	Half Ballotini	28
Third Session (February 2013)	Full Ballotini	16–45.43

frequency of the silo for the different sessions using the white noise input at 0.3 g.

The grain inside the silo seems to act as a stiff material. No differences in the acceleration time-history were significantly observed during all tests performed under different base excitations (white noises, sinusoidal inputs and earthquake ground motions).

27.3.2 Compaction of the Ensiled Material

During the first session of tests, immediately after the repetitions of the first tests, a residual settlement was observed, probably due to the compaction of the ensiled material. The compaction was measured in 13 positions (Fig. 27.5 and Table 27.6, pp. 1–13) in terms of the distance to a reference ideal line on the top ring of the silo. The average compacted thickness of the ensiled material was around 3 ÷ 4 cm.

During the second session of tests, compaction due to dynamic excitation and its effects on the dynamic frequencies of the silo were systematically observed. The first tests were repeated three times to check the compaction phenomenon. The measurements were taken from the top of the surface of the ensiled material, at a central point, to the circular ring (Table 27.7).

Fig. 27.5 Measurement of the settlement of the surface of Ballotini glass in terms of distance from a reference ideal line on the top ring of the silo

Table 27.6 Measures of the settlement of the top-layer surface of Ballotini glass [mm]

Test type	Dia-meter	p. 1 [mm]	p. 2 [mm]	p. 3 [mm]	p. 4 [mm]	p. 5 [mm]	p. 6 [mm]	p. 7 [mm]	p. 8 [mm]	p. 9 [mm]	p. 10 [mm]	p. 11 [mm]	p. 12 [mm]	p. 13 [mm]
0	121.5	37	36.5	37	36.5	36.5	36.5	36.5	37	37	37.5	38	38	38
S1	121.5	37	37	37	36.5	36.5	36	37	37	37	37	37.5	38	38.5
S2	121.5	36.5	37	37	36.5	36	36	37	37	37	37	37.5	38	39
S3	121.5	39.5	38	38	37.5	37	37	37	37.5	37.5	38	38.5	39	40
S7	121.6	40	39	38.5	38	38	37.5	38.5	38.5	38.5	39	39	40	41
S8	121.6	39.5	39.5	40	39.5	39.5	40	40	39.5	39.5	40	40	40	40.5
E5	121.6	40	39.5	39.5	39.5	39.5	40	40	40	40	40	40	40	40.5
E17	121.6	40	39.5	39.5	39.5	40	40	40	40	40	40	40	40	41

Table 27.7 Measures of the settlement of the top-layer surface of Ballotini glass [mm]

Test type	p. 1 [m]
0	1.000
S1 (first repetition) 0.1 g 1 Hz	1.000
S2 (second repetition) 0.1 g 1 Hz	1.002
S3 (third repetition) 0.1 g 1 Hz	1.003
S4 (first repetition) 0.1 g 1 Hz	1.003
S5 (second repetition) 0.1 g 1 Hz	1.003
S6 (third repetition) 0.1 g 1 Hz	1.003

With the objective of understanding how much initial conditions affect the results, in the third session the tests were repeated three times and the initial test was repeated at the end of a set of tests. Then the settlement of Ballotini glass inside the silo was measured for white noise, sinusoidal input and earthquake simulation. An average of 3–4 cm was detected.

27.3.3 Acceleration Sinusoidal Input

The following Figures (Fig. 27.6, 27.7, 27.8) represent the horizontal acceleration versus time at various heights from which it is possible to distinguish the type of input used for the test (sinusoidal). Figure 27.9 represent the profiles of horizontal acceleration along the height of the specimen. In this case, from the analysis made, the silo's behaviour validates the hypothesis that assumes an infinitely stiff silo. No amplification has to be considered and the spectral acceleration coincides with the ground acceleration (i.e. the response of the acceleration of the silo does not vary along the height of the silo). Furthermore, the behaviour of the two parts (left and right) of the silo is identical.

The previous figures are representative not only of the reported test but of the 149 sinusoidal tests. The following conclusion can be formulated: the vertical distribution of horizontal accelerations under horizontal sinusoidal input is almost constant.

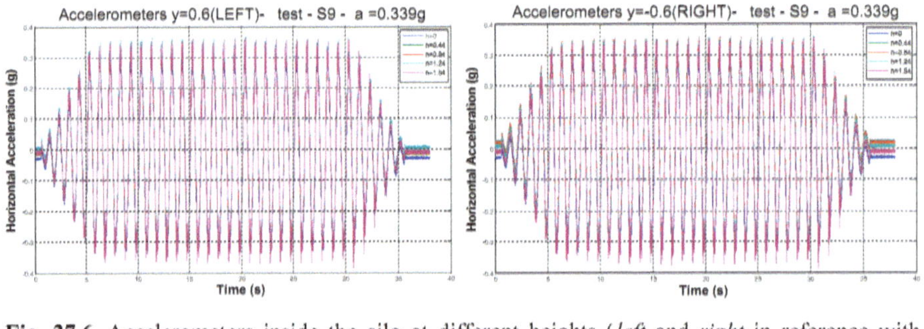

Fig. 27.6 Accelerometers inside the silo at different heights (*left* and *right* in reference with Fig. 27.3a)

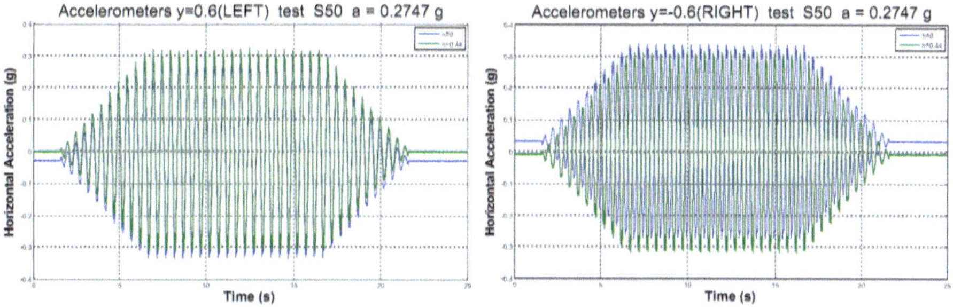

Fig. 27.7 Accelerometers inside the silo at different heights (*left* and *right* in reference with Fig. 27.3a)

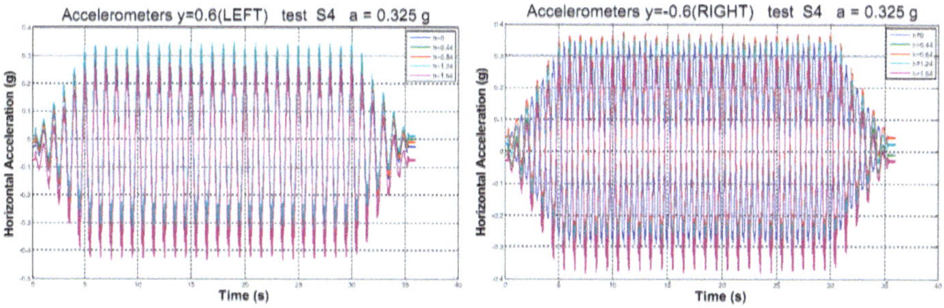

Fig. 27.8 Accelerometers inside the silo at different heights (*left* and *right* in reference with Fig. 27.3a)

27.3.4 Acceleration Earthquake Input

Figure 27.10 represents the profiles of the horizontal acceleration for the left and right sides of the silo (with respect to Fig. 27.3a) and it is representative not only of the reported tests but of most of the 62 earthquake tests performed. The following conclusion can be formulated: the vertical distribution of horizontal accelerations under a horizontal earthquake input (Fig. 27.11) is roughly linear. Therefore, there is an acceleration amplification given by the dynamic system “silo + grain” along the height of the silo.

The results indicate that the analytical theory briefly recalled in the introduction should be tailored to a given distribution of horizontal accelerations along the height of the silo.

27.3.5 Vertical Strains

Figures 27.12, 27.13 and 27.14 represent the vertical strains recorded for the right and left parts of the silo (see Fig. 27.3a). It may be observed that the maximum

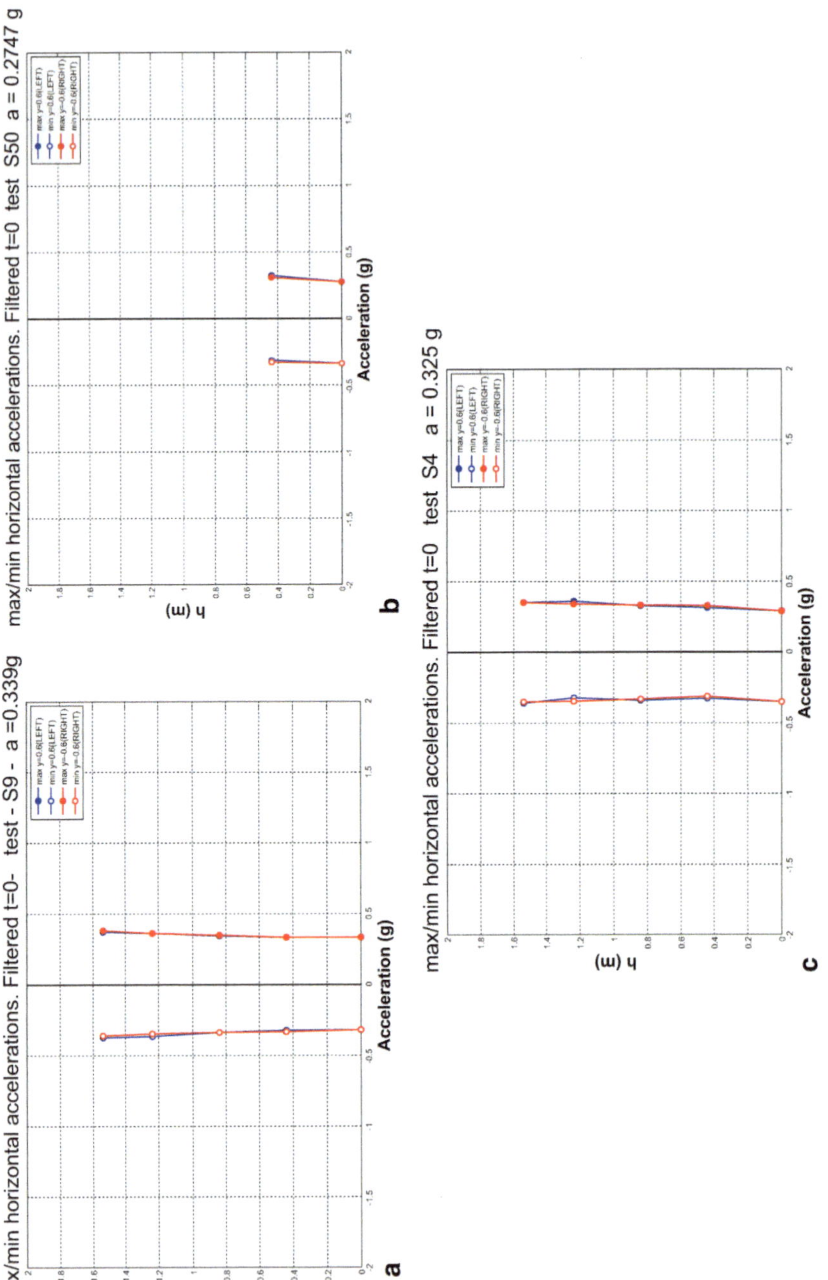

Fig. 27.9 Maximum and minimum *horizontal* acceleration profiles under sinusoidal input: **a** First session. **b** Second session. **c** Third session.

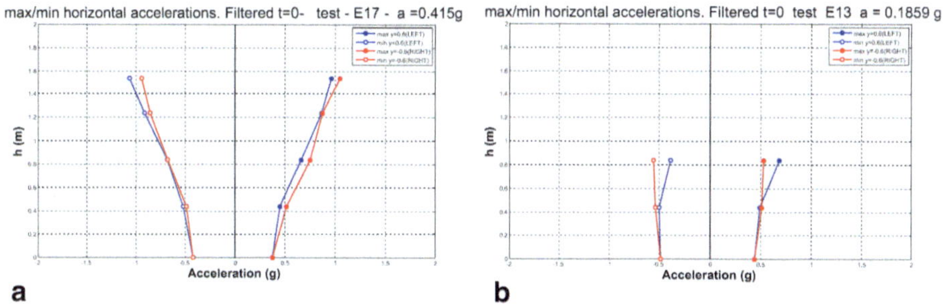

Fig. 27.10 Maximum and minimum *horizontal* acceleration profile under earthquake input: **a** First Session. **b** Second Session

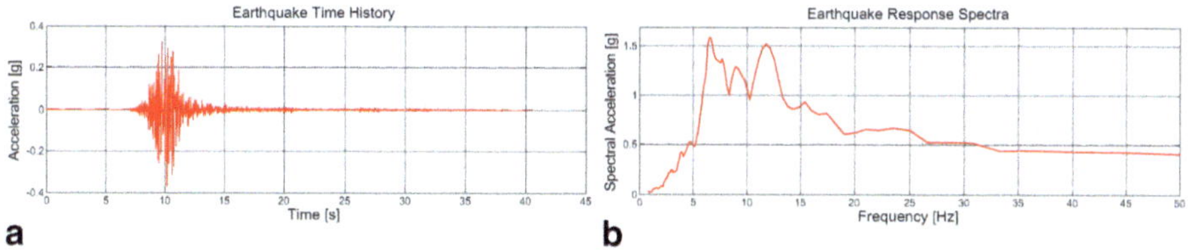

Fig. 27.11 FFT of the earthquake input "South Iceland"

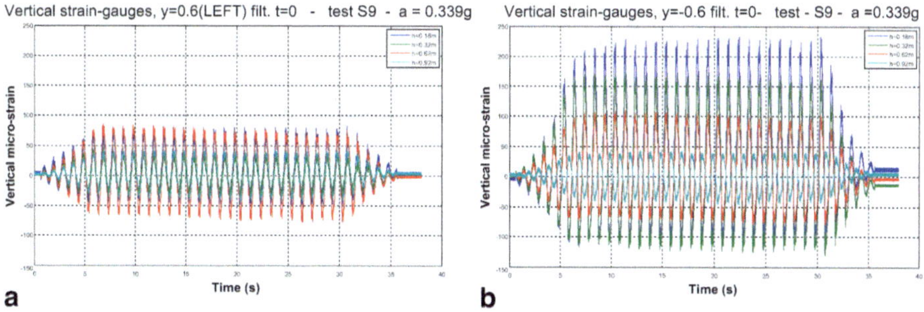

Fig. 27.12 First session of tests: *vertical* strain at different heights (*left* and *right*, Fig. 27.3a)

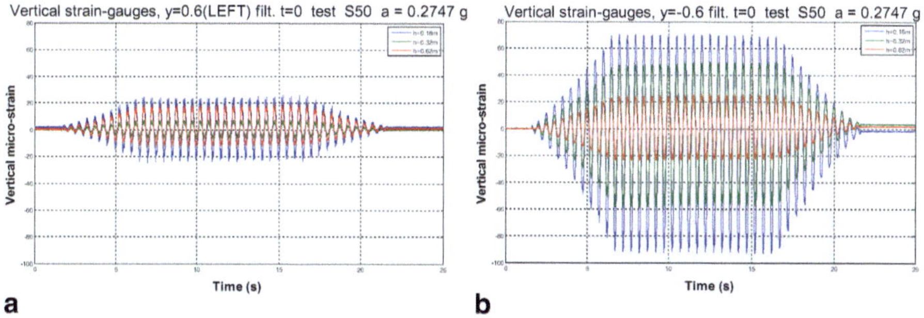

Fig. 27.13 Second session of tests: *vertica*l strain at different heights (*left* and *right*, Fig. 27.3a)

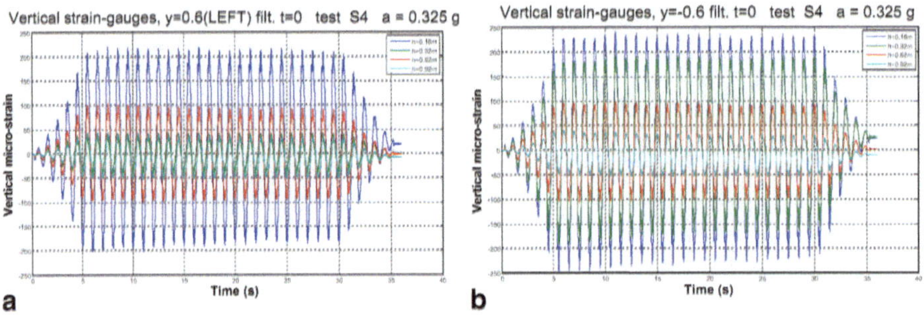

Fig. 27.14 Third session of tests: *vertical* strain at different heights (*left* and *right*, Fig. 27.3a)

vertical strain is reached at a height of 0.18 m (the lowest one), while the lowest vertical strain is obtained at a height 0.92 m.

These figures are representative not only of the reported tests but of most of the 149 sinusoidal tests. The residual strains were removed from the figures to eliminate the static condition. The following conclusion may be formulated: the response of the strain gauges placed at the base of the silo shows that the assumption of "plane sections remain plane" cannot be applied to such systems.

27.3.6 Horizontal Strains

Figure 27.15 represents the horizontal strains recorded for the right and left parts of the silo (see Fig. 27.3a). The asymmetry of the registered horizontal strains may be explained by the squeezing of Ballotini glass along the junction between the two polycarbonate semicircles (Fig. 27.16) that leads to a residual in the horizontal strain.

These figures are representative not only of the reported test but of most of the sinusoidal tests performed.

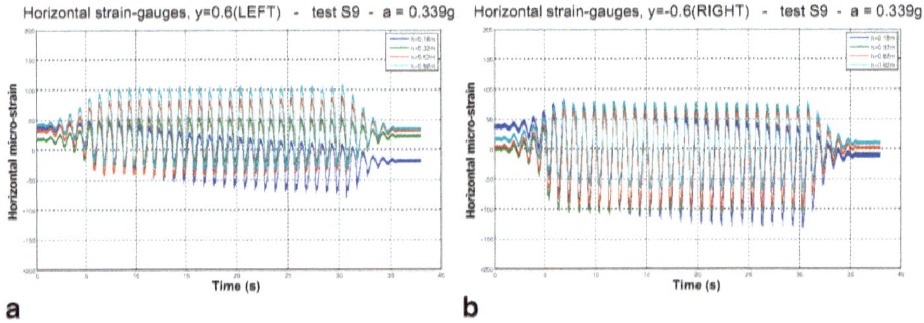

Fig. 27.15 *Horizontal* strain at different heights (*left* and *right* in reference with Fig. 27.3a)

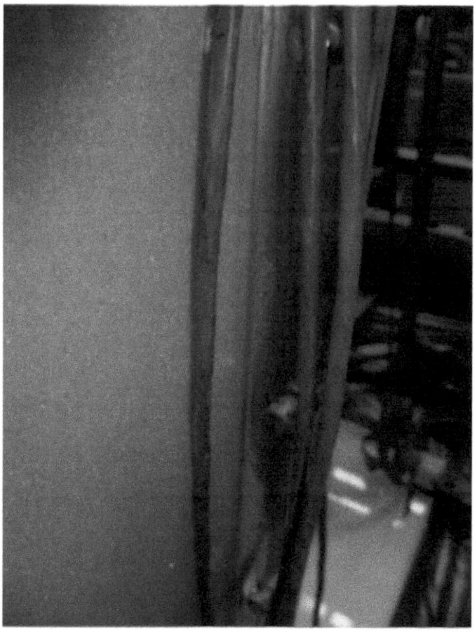

Fig. 27.16 Squeezing of Ballotini glass along the junction between the two polycarbonate semicircles

27.3.7 *Change in the Physical Behaviour/Response for Increasing Peak Table Acceleration*

Figure 27.17 presents the average values (over the maximum values obtained for each one of the 10–15 cycles of the sinusoidal input) in each one of the vertical strain gauges (two in the principal direction for the first session, and ten in the principal direction and at 45° for the second and the third sessions) placed at the base of the silo as a function of the actual measured acceleration on the shaking-table (PGA).

The following observations can be made:

- there is a clear change in the physical behaviour at a PGA of around 0.35 g (black line—Fig. 27.17a).
- Figure 27.17d shows that for rough walls the change in the physical behaviour seems to appear around 0.25 g.
- the same consideration can be made for the input in the Y and Z directions (Fig. 27.17b)

27.3.8 *Vertical and Horizontal Inputs in Phase*

Figure 27.18 shows that the presence of a vertical input in-phase with, and equal to the horizontal input, seems to increase the base strain values to about $+30 \div +50\,\%$.

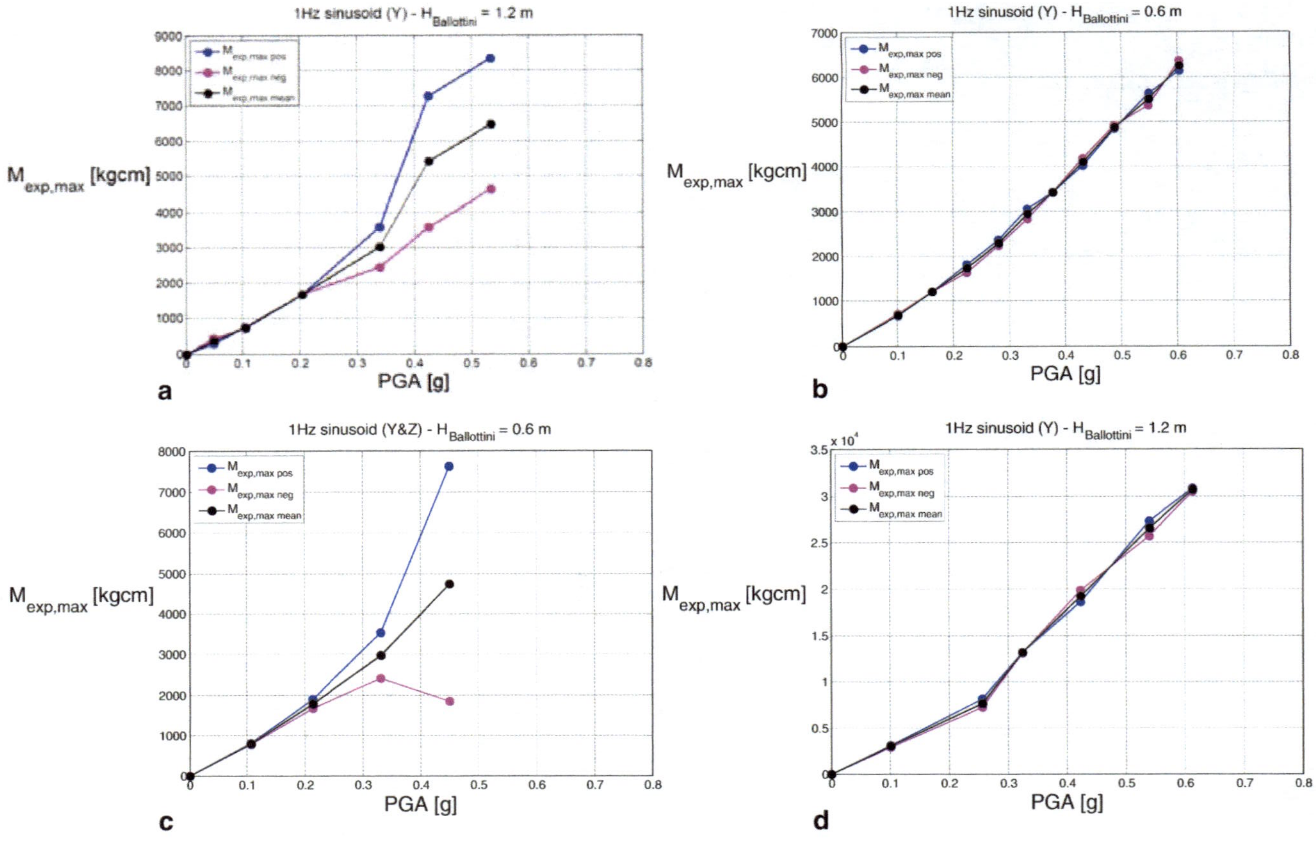

Fig. 27.17 **a** Strain Trend of First Session Test. **b** Strain Trend of Second Session Test, 1 Hz in direction *Y*. **c** Strain Trend of Second Session Test, 1 Hz in directions *Y&Z*. **d** Strain Trend of Third Session Test, 1 Hz in direction *Y*

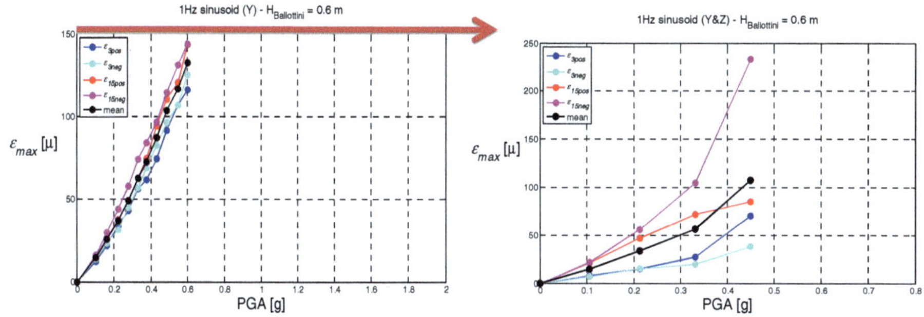

Fig. 27.18 **a** Strain Trend of Third session, 1 Hz *Y* direction; **b** Strain Trend of Third Session 1 Hz *Y&Z* directions

27.3.9 The Influence of the Wall-Grain Friction Coefficient

Figure 27.19 compares the bending moments at the base as reconstructed from the base strain values for the following sessions:

- First session: Ballotini height equal to 1.2 m and smooth walls (black line)
- Second session: Ballotini height equal to 0.6 m and roughened walls (blue line)
- Third session: Ballotini height equal to 1.2 m and roughened walls (red line)
- as obtained in the case of 1 Hz sinusoid input (only horizontal direction along the Y axis) at an acceleration of 0.33 g (first session test n. 9 vs. third session test n. 4).

  The results of the reconstruction and comparison of the base bending moment clearly indicate that the wall-grain friction coefficient strongly affects the experimental base bending moment.

Then the following conclusions can be drawn:

- The experimental results do not match Eurocode 8 (CEN 2006) prescriptions, which do not take into account wall-grain friction coefficient
- From a qualitative point of view, according to the analytical theory suggested by the authors (Silvestri et al. 2012), a higher wall-grain friction coefficient (roughened walls) leads to higher actions inside the walls

27.3.10 Bending Moments at the Base

Figure 27.20 presents the reconstructed bending moments at the base with respect to the provisions of Eurocode 8 (CEN 2006, simplified and accurate procedures) and the prediction of Silvestri et al. (2012). It is possible to note the significant difference in the magnitude of the base bending moment under the same base acceleration and low frequency equal to 1 Hz. At different input frequencies and accelerations, the value of the experimental bending moment at the base of the silo is much closer to the value given by the analytical theory (Silvestri et al. 2012) than to the two

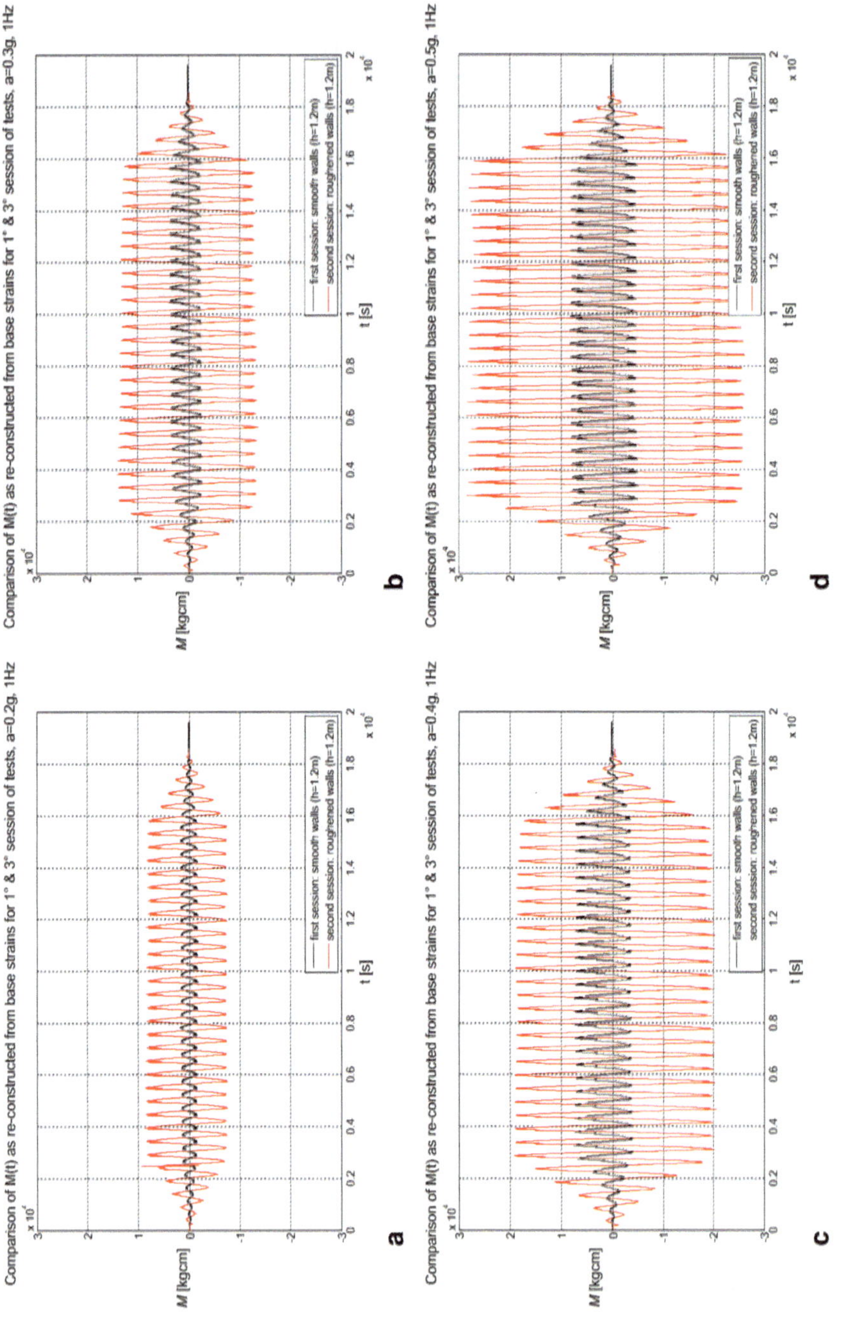

Fig. 27.19 Comparison between First, Second and Third session: bending moment at the base at 1 Hz and PGA equal to 0.2 g (**a**), 0.3 g (**b**), 0.4 g (**c**) and 0.5 g (**d**)

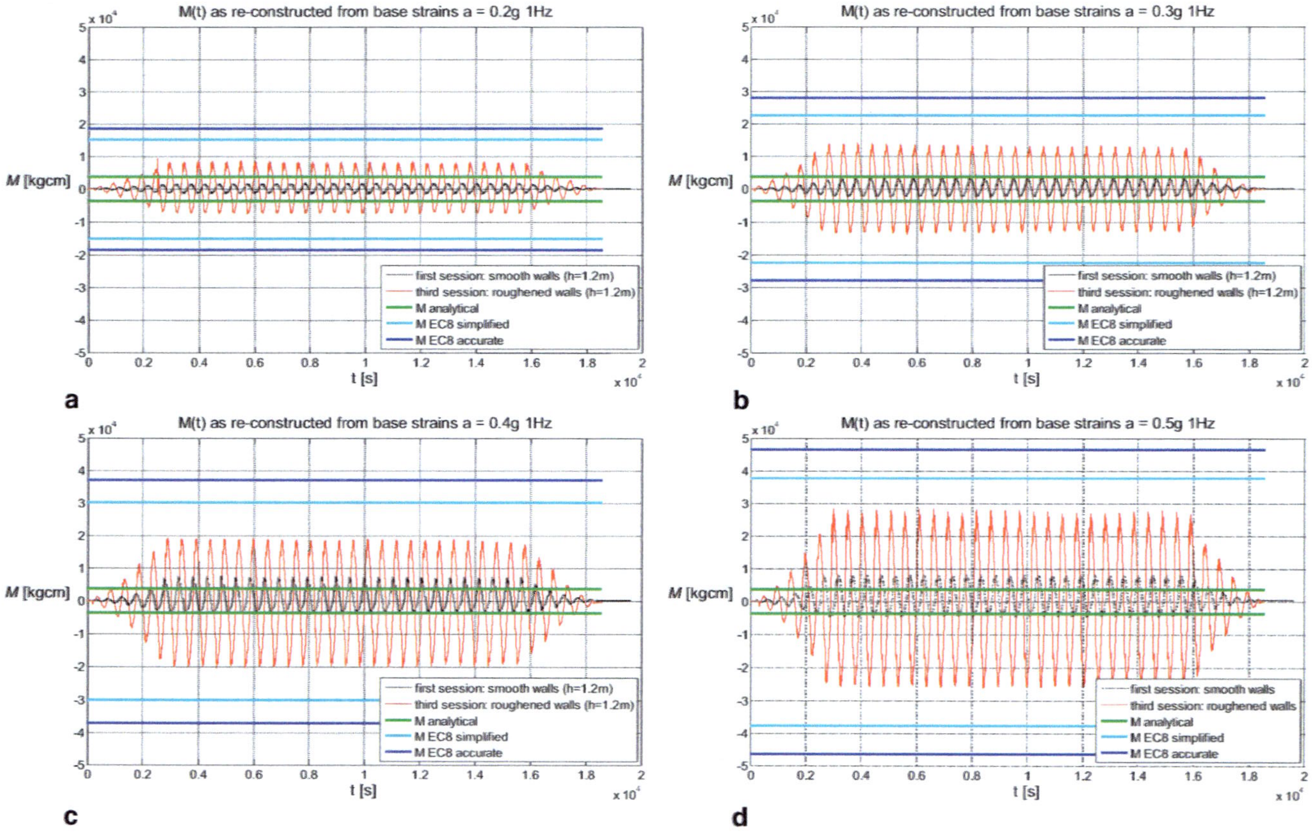

Fig. 27.20 Comparison of bending moments at the base in case of smooth and roughened surface of the silo wall, with respect to the expected value of the Eurocode 8 provisions (accurate and simplified procedures) and the analytical prediction by Silvestri et al. (2012)

values (simplified and accurate procedures) given by the Eurocode 8 (CEN 2006) prescriptions. For the tests for which the value of the acceleration does not exceed the value of 0.30 g, the experimental value of the bending moment is even smaller than the value given by the analytical theory. On the contrary, when the acceleration reaches values of about 0.4 ÷ 0.5 g, the experimental value of the bending moment exceeds the analytical one. This is maybe due to the fact that the analytical theory is valid for small values of acceleration, basically ascribable to the values of the grain-grain and grain-base friction coefficients (Silvestri et al. 2012).

27.4 Conclusions

This paper briefly presents the results of the ASESGRAM project, which was devoted to the experimental verification of the analytical findings obtained by the authors (Silvestri et al. 2012) regarding the actions induced by grain-like material upon the walls of flat-bottom silos. In more detail, the objective of the shaking-table tests was to investigate the pressures given by grain-like materials under constant horizontal acceleration confined in a stiff cylindrical element in order to compare them with the analytical formulation and with the Eurocode 8 (CEN 2006) provisions.

The following concluding remarks containing preliminary but yet clear indications can be drawn from the ASESGRAM experimental campaign:

- The grain inside the silo seems to act as a stiff material. No differences in the acceleration time-history were significantly observed during all tests performed under different base excitations (white noise, sinusoidal input and earthquake ground motions).
- Grain compaction occurs under dynamic excitation.
- The vertical distribution of horizontal accelerations under a horizontal sinusoidal input is almost constant. No acceleration amplification has to be considered for the dynamic system “silo + grain”.
- The vertical distribution of horizontal accelerations under a horizontal earthquake input is roughly linear. Thus, there is some acceleration amplification given by the dynamic system “silo + grain” along the height of the silo.
- By looking at the response of the strain gauges placed at the base of the silo, it seems that the assumption of “plane sections remain plane” cannot be applied to such systems.
- The experimental results clearly indicate that the wall-grain friction coefficient strongly affects the experimental bending moment at the base of the silo. This does not match with Eurocode 8 prescriptions which do not take into account the wall-grain friction coefficient.
- From a qualitative point of view, according to the analytical theory suggested by the authors (Silvestri et al. 2012), higher wall-grain friction coefficient leads to higher actions inside the walls.

- For a height h = 1.2 m of Ballotini glass inside the silo, the bending moment recorded during the tests is much closer to the predictions given by the analytical formulas provided by the authors (Silvestri et al. 2012) rather than to the values corresponding to Eurocode 8 (CEN 2006) provisions.
- For a height h = 0.6 m of Ballotini glass inside the silo, there is some interaction between the mass of grain and the stiff circular ring at the top of the silo, whose presence seems to be no more negligible.
- The presence of vertical input in-phase with, and equal to the horizontal input seems to increase the strain values at the base between 30 and 50 %.

Acknowledgements The research leading to these results received founding from the European Community's Seventh Framework Programme [FP7/2007–2013] for access to the Bristol Laboratory of Civil Engineering (EQUALS) under grant agreement n 227887. The authors thank Prof. Colin Taylor and Dr. Matt Dietz of the Bristol Laboratory of Civil Engineering (EQUALS). Also, the authors acknowledge the University of Alicante for the partial financial support by the grant n ACIE12-04.

References

CEN (2004) EN 1991-4: Eurocode 1. Basis of design and action on structures. Part 4: actions in silos and tanks. CEN, Brussels

CEN (2006) EN 1998-4: Eurocode 8. Design of structures for earthquake resistance, Part 4: silos, tanks and pipelines. CEN, Brussels

Janssen HA (1895) Versuche uber Getreidedruck in Silozellen. Z Ver Dt Ing 16:446–449

Koenen M (1896) Berechnung des Seiten und Bodendrucks in Silozellen. Centralblatt Bauverwalt 16:446–449

Pozzati P, Ceccoli C (1972) Teoria e Tecnica delle Strutture, Volume Primo, Preliminari e Fondamenti. UTET, Torino (in Italian)

Silvestri S, Gasparini G, Trombetti T, Foti D (2012) On the evaluation of the horizontal forces produced by grain-like material inside silos during earthquakes. Bull Earthq Eng 10:1535–1560

Chapter 28
Multi-Building Interactions and Site-City Effect: An Idealized Experimental Model

Logan Schwan, Claude Boutin, Matthew Dietz, Luis Padron, Pierre-Yves Bard, Erdin Ibraim, Orlando Maeso, Juan J. Aznárez and Colin Taylor

28.1 Introduction

The usual approach regarding seismic response of cities is to consider that the ground motion results from the seismic source and the characteristics of the substratum only. However, the observation of long signals and beatings in some records (Chavez-Gracian and Bard 1994) made in urban area during the Mexico earthquakes led Wirgin and Bard (1996) to suggest that the seismic response of a site might also depend on the resonant city.

Theoretical models and experiments on single or limited number of closed spaced buildings (Castellaro and Mulargia 2010; Kitada 1999; Lee and Wesley 1973; Luco and Contesse 1973) showed that (multiple) soil-structure interactions can have noticeable effects on the dynamic response of the structures and of the surrounding ground motion. Besides, numerical calculations based on various methods (finite/boundary elements methods, Green functions, etc) and performed on idealized cities (2D or 3D, periodic or randomly distributed) (Clouteau and Aubry 2001; Ghergu and Ionescu 2009; Guéguen et al. 2002; Tsogka and Wirgin 2003; Uenishi 2010) or on groups of nearby buildings (Padrón et al. 2009) support the idea that the city can interact significantly with its site especially at the buildings' resonance.

L. Schwan (✉) · C. Boutin
ENTPE/LGCB – LTDS - CNRS UMR 5513, Université de Lyon, Vaulx-en-Velin, France
e-mail: logan.schwan@gmail.com

M. Dietz
BLADE, Department of Civil Engineering, University of Bristol, Bristol, U.K.
e-mail: m.dietz@bristol.ac.uk

E. Ibraim · C. Taylor
Department of Civil Engineering, University of Bristol, Bristol, U.K.

L. Padron · O. Maeso · J. J. Aznárez
University Institute SIANI, Universidad de Las Palmas de Gran Canaria, Las Palmas, Spain

P.-Y. Bard
ISTerre/IFSTTAR, Université de Grenoble, Grenoble, France

F. Taucer, R. Apostolska (eds.), *Experimental Research in Earthquake Engineering,*
Geotechnical, Geological and Earthquake Engineering 35,
DOI 10.1007/978-3-319-10136-1_28

An alternative approach consists in finding an analytical model of the macroscopic equivalent behavior of the unbounded city (Boutin and Roussillon 2004; Boutin and Roussillon 2006). Provided that the representative surface element of the city is much smaller than the seismic wavelength, the homogenization of periodic systems (Auriault et al. 2009; Sanchez-Palencia 1980) provides a rigorous method for describing the city as a surface impedance that modifies the usual condition of free field (Schwan and Boutin 2013).

It appears that only sparse comparisons between these several approaches have been performed, probably due to the radically different assumptions and methodology required by each. This Chapter aims at building up a common study case that favors experimental, numerical and theoretical crossed-analyses. The multiple Structure-Soil-Structure interactions are investigated through an idealized experimental model and the resulting data is compared with two models: (i) the theoretical city-impedance model derived from homogenization (Boutin and Roussillon 2004; Boutin and Roussillon 2006; Schwan and Boutin 2013); and (ii) a numerical model consisting in a 2D hybrid BEM-FEM approach (Padrón et al. 2004).

Section 28.2 presents the physical principles leading to the city-impedance and Sect 28.3 the numerical model. The experimental set-up is designed in Sect 28.4 and experimental results are presented and compared to the numerical models in Sect 28.5.

28.2 Theoretical Model for Soil-City Interactions: The City-Impedance Analysis

28.2.1 Homogenization of the City into a Surface Impedance

The urban landscape, especially in modern cities, is usually made up of high and heavy buildings organized in grids. This observation leads us to idealize the city as a periodic distribution of identical resonant structures (Fig. 28.1).

Consequently, consider a Σ-periodic distribution of all identical linear oscillators lying on a top plane surface Γ of a homogeneous isotropic half-space (shear modulus μ, Poisson's ratio v, weak damping $\xi_L << 1$ and density ρ).

We analyze the propagation of harmonic waves of frequency $f=\omega/2\pi$ in this linear system (the factor $e^{-i\omega t}$ is implicit in the sequel) assuming the characteristic size ℓ of Σ is much smaller than the shear wavelength $\Lambda=c_S/f$ in the medium where $c_S=\sqrt{(\mu/\rho)}$ is the shear wave velocity. This scale separation condition is quantified by the scale ratio $\varepsilon=2\pi\ell/\Lambda<<1$.

Under this scale separation condition, the dynamic has two scales (see Fig. 28.1): at a macroscopic scale, seismic waves propagate with the characteristic length Λ; at a micro-scale, the oscillators on the surface, set in motion by the waves, induce a heterogeneous distribution of local forces **F**. The oscillators are all in mutual interactions and their heterogeneous local wavefields interfere with one another with a characteristic length ℓ. But far from the surface their interferences are constructive and produce a quasi-homogeneous wavefield: small scale variations vanish far from Γ.

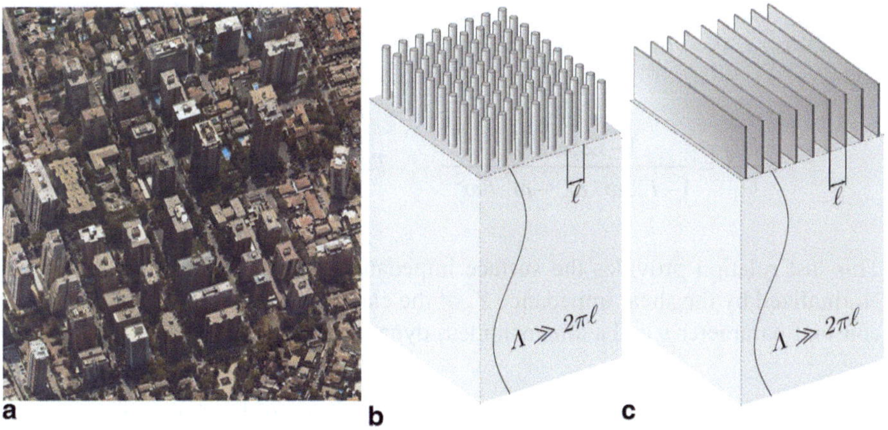

Fig. 28.1 **a** Satellite view of San Cristóbal, Santiago, Chile. **b**, **c** Examples of periodic distribution of oscillators on the uppermost surface of an elastic medium: **b** isotropic cylindrical oscillators, **c** anisotropic unidirectional oscillators

Such a situation corresponds to a boundary layer located in the vicinity of the surface, superimposed upon the macro-field. The local force distribution **F** exerted by the oscillators on Γ is balanced by both the macroscopic and the boundary layer fields. The a-priori-unknown surface condition for the macro-field is derived from the global equilibrium of the periodic boundary layer following the homogenization method (not presented here, see Boutin and Roussillon 2006). It shows that at dominant order, the macro-field balances the mean surface stress $\mathbf{T}=\mathbf{F}/|\Sigma|$ undergone by one period.

Furthermore, since the linear oscillator on the period Σ responds to the surface motion $\mathbf{U}_\Gamma$, the force **F** it exerts on Σ is given as the linear function $\mathbf{F}=-i\omega\boldsymbol{Z}\mathbf{U}_\Gamma$ where the frequency-dependent impedance matrix $\boldsymbol{Z}$ depends on the dynamic behavior of the oscillator. The surface stress $\mathbf{T}=\mathbf{F}/|\Sigma|$ is then linked to the surface motion by the surface impedance $\boldsymbol{Z}_\Gamma=\boldsymbol{Z}/|\Sigma|$ standing for the city-impedance.

28.2.2 Features of the Resonant City-Impedance

Conversely to the scalar shear impedance $Z_S=\rho c_S=\sqrt{(\mu\rho)}$ of the elastic medium, $\boldsymbol{Z}_\Gamma$ is a matrix that varies with frequency. To illustrate these key properties, consider first that the oscillator on the period Σ has a Single-Degree-Of-Freedom (SDOF) characterized by a stiffness k, a viscous damping coefficient c and a mass m_o in one horizontal direction. Focusing on this direction, the problem is scalar. The harmonic macroscopic surface motion U_Γ induces a motion U_m of the mass of the oscillator. The force F exerted by the oscillator on Σ balances the mass inertia:

$$F = m_o\omega^2 U_m = (k - i\omega c)(U_m - U_\Gamma) \tag{28.1}$$

Defining the eigenfrequency $f_o = \omega_o/2\pi = \sqrt{(k/m_o)}$ and the weak damping ratio $\xi = c/2\sqrt{(km_o)} << 1$, the transfer function of the oscillator and the macro-stress become

$$\frac{U_m}{U_\Gamma} = \frac{1 - i2\xi\omega/\omega_o}{1 - i2\xi\omega/\omega_o - \omega^2/\omega_o^2} \quad ; \quad T = \frac{m_o\omega^2 U_m}{|\Sigma|} = -i\omega Z_o U_\Gamma \tag{28.2}$$

This last relation provides the surface impedance Z_o in the oscillator's direction. Normalized by the shear impedance Z_S of the elastic medium, Z_o is the product of a constant parameter η and a dimensionless dynamical function:

$$\frac{Z_o}{Z_S} = \eta \frac{(i\omega/\omega_o)(1 - i2\xi\omega/\omega_o)}{1 - i2\xi\omega/\omega_o - \omega^2/\omega_o^2} \quad ; \quad \eta = \frac{\sqrt{k\, m_o}/|\Sigma|}{\sqrt{\mu\rho}} = \frac{\ell\omega_o}{c_S}\frac{m_o}{\rho|\Sigma|\ell} \tag{28.3}$$

As for the impedance parameter η, the first expression shows that it is the ratio between (i) the impedance $\sqrt{(km_o)}$ built from the stiffness and mass of the oscillator divided by the surface of the period Σ and (ii) the impedance of the elastic medium. The condition of free surface is recovered when $\eta = 0$. To have a substantial interaction, the oscillator must therefore be heavy and rigid compared to the soil. The second expression shows that η is at best of the first order (in $\varepsilon_o = \ell\omega_o/c_S << 1$ under scale separation at city resonance) if the resonating mass m_o is of the same order as the mass $\rho|\Sigma|\ell$ of the medium under one period on a depth ℓ. A value of η around 10 % or more leads to significant effects.

Considering the asymptotic behaviors of the dynamical function and the order of magnitude of η, the impedance ratio Z_o/Z_S can be assessed in the low, high and resonant frequency ranges and gives equivalent surface conditions to the oscillators. In quasi-static ($\omega << \omega_o$) or inertial ($\omega >> \omega_o$) regime, the impedance Z_o is much smaller than the impedance of the medium Z_S similarly to the usual free surface condition. Conversely, at the resonance $\omega \approx \omega_o$ of weakly-damped oscillators, the city impedance is significant compared to the soil impedance (the impedance ratio tends to $\eta/2\xi >> 1$) similarly to a rigid condition imposed in the oscillator's direction only.

Consider now that the oscillator on the period Σ has a SDOF in both horizontal directions denoted by X and Y. Focusing on these two directions, the problem is planar. The oscillator can have the same resonant behavior in both directions if it is isotropic (see Fig. 28.1), but we will rather consider an anisotropic oscillator which can resonate in X but remain inert in Y, so that we can compare the two behaviors. For more accuracy, we take into account the inert mass m_i of the oscillators in the resonating X direction. In the inert Y direction, the inert mass is therefore $m_i + m_o$. Consequently, the city surface impedance matrix reads

$$Z_\Gamma = \begin{bmatrix} Z_X & 0 \\ 0 & Z_Y \end{bmatrix} \quad ; \quad Z_X = Z_o + i\omega\frac{m_i}{\rho|\Sigma|c_S} \quad ; \quad Z_Y = i\omega\frac{m_o + m_i}{\rho|\Sigma|c_S} \tag{28.4}$$

28.2.3 Application to the Case of a City Lying on a Layer

Consider that a ℓ-periodic city lies on a h_L-high layer verifying the scale separation assumption $\Lambda >> \ell$ and define z as the upward vertical axis with the origin taken on the surface. The layer is shaken at its base horizontally in the X direction and harmonically at the circular frequency ω: it generates harmonic stationary SH waves with a horizontal displacement $U_\uparrow$ propagating upwards from the base to the surface of the layer and $U_\downarrow$ propagating downwards from the surface to the base.

At the base, the displacement U_b is imposed; at the surface, the displacement $U_\Gamma = U_\uparrow + U_\downarrow$ puts the oscillators in motion, which impose the city-impedance Z_X

$$\begin{aligned} z = 0: \quad & i\omega\, Z_S^*\left(U_\uparrow - U_\downarrow\right) = -i\omega\, Z_X\left(U_\uparrow + U_\downarrow\right) \\ z = h_L: \quad & U_\uparrow \exp\left(-i\omega\, h_L / c_S^*\right) + U_\downarrow \exp\left(+i\omega\, h_L / c_S^*\right) = U_b \end{aligned} \tag{28.5}$$

where $Z_S^* = \rho c_S^*$ is the complex impedance of the layer with $c_S^* = c_S\sqrt{(1 - i2\xi_L)}$. Denoting $f_L = \omega_L/2\pi = c_S/4h_L$ the eigenfrequency of the layer with a free surface, the system leads to the surface displacement for a unitary displacement of the layer's base

$$\frac{U_\Gamma}{U_b} = 1 \Big/ \left(\cos\left(\frac{\pi}{2}\frac{\omega}{\omega_L}\frac{1}{\sqrt{1 - i2\xi_L}} \right) + i\frac{Z_X}{Z_S^*}\sin\left(\frac{\pi}{2}\frac{\omega}{\omega_L}\frac{1}{\sqrt{1 - i2\xi_L}} \right) \right) \tag{28.6}$$

and to the displacement in the layer at the depth $z<0$

$$\frac{U}{U_b} = \frac{U_\Gamma}{U_b}\left(\cos\left(\frac{\pi\omega}{2\omega_L}\frac{|z|/h_L}{\sqrt{1 - i2\xi_L}} \right) + i\frac{Z_X}{Z_S^*}\sin\left(\frac{\pi\omega}{2\omega_L}\frac{|z|/h_L}{\sqrt{1 - i2\xi_L}} \right) \right) \tag{28.7}$$

Let us suppose that the resonance of the city in the X direction coincides with the resonance of the layer (Fig. 28.2). As a result of the soil-city interactions, the transfer function U_Γ/U_b is drastically modified compared to the one of the layer with the free surface: at the frequency at which the layer's resonance should be, there is a surface antiresonance due to the rigid-like condition imposed by city's resonance. The layer's resonance peak is split into two soil-city resonance peaks, one at a lower and one at a higher frequency. The amplitude of these peaks is substantially reduced compared to the layer's resonance peak and the phase is drastically modified too. Note that in the time domain, this two-peak-spectrum is favorable to beatings.

In the Y direction, the equations are the same as in the X direction but with the surface impedance Z_Y. Assuming the inert mass $m_i + m_o$ of the city is small compared to the mass $\rho|\Sigma|h_L$ of layer under one period, the transfer function U_Γ/U_b is then the

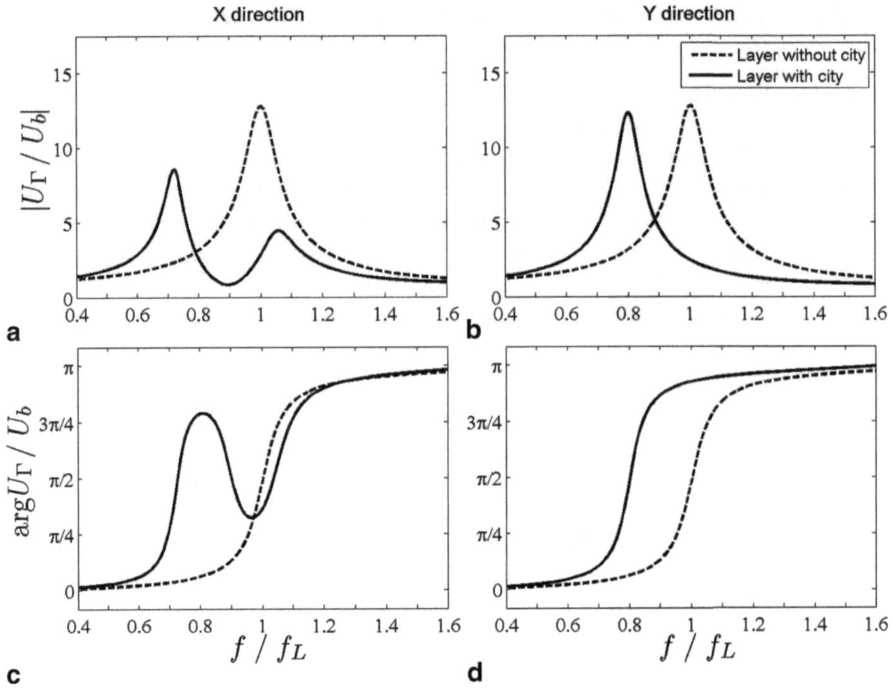

Fig. 28.2 Transfer function modulus (a and b) and phase (c and d) of layer's surface U_Γ to base motion U_b: a, c in X-direction; b, d in Y direction. Calculations performed with $\eta = 11\,\%$; $\xi = \xi_L = 5\,\%$; $f_o/f_L = 90\,\%$ and $m_i/\rho|\Sigma|h_L = 18\,\%$ in X and $26\,\%$ in Y)

same as the one of a layer with a free surface, but with a reduced eigenfrequency due to added inert mass on the surface.

$$Y : \frac{U_\Gamma}{U_b} \approx 1/\cos\left(\frac{\pi}{2}\frac{\omega}{\tilde{\omega}_L}\frac{1}{\sqrt{1-i2\xi_L}}\right) \quad ; \quad \frac{\tilde{\omega}_L}{\omega_L} \approx 1/\left(1+\frac{m_o+m_i}{\rho|\Sigma|h_L}\right) \tag{28.8}$$

In an oblique direction (e.g. 45°) the response of the layer is obtained by superposition in the X and Y directions. Combining the frequency-dependent free-like/rigid-like anisotropic conditions on the surface leads to unconventional depolarization effects whereby the system's motion may strongly depart from the direction of the input motion of the layer's base (see also Boutin and Roussillon 2006; Schwan and Boutin 2013). For instance, shaking harmonically the base of the layer in the 45° direction at the common city/layer resonance leads to (i) nearly no surface displacement in the resonating X direction due to city-resonance; and (ii) amplified displacement in the inert Y direction due to the layer's resonance. As a result, the surface should move mainly in the Y direction while the base moves in the 45° direction.

28.3 Numerical Model for Soil-City Interactions: Hybrid BEM-FEM Analysis

This section presents briefly the model and the numerical harmonic 2D boundary element-finite element (BEM-FEM) method that is used to describe the soil-city interactions in the resonant X direction. This numerical method, developed by Padrón et al. 2004 aims at fulfilling the basic assumptions of the theoretical city-impedance model while taking into account other phenomena that have been left aside by the city-impedance model as correctors to the dominant order description.

28.3.1 Model for the Layer and the Oscillators

The "layer" is modeled numerically as a 2D elastodynamic solid using Boundary Elements while an oscillator is modeled using Finite Elements as a horizontal mass/spring/dashpot SDOF system. As the numerical model has to deal with a limited number N of oscillators on the surface of the elastic medium, the elastic "layer" has to be of finite horizontal extension too and, as such, will be referred to as an elastic block in what follows.

The elastic block has the same height h_L as the infinite layer in the city-impedance model, but its width is the total width $N \times \ell$ of the periods on the surface. The block is made up of homogeneous, elastic, linear, isotropic, weakly-damped material with the same density ρ, damping ξ_L and Poisson's ratio ν as the infinite layer. Like the infinite layer, the damping is taken into account considering a complex Young's modulus $E^*=E(1-2i\xi_L)$. The Young's modulus E is chosen so that the finite-sized block with its boundary conditions (detailed hereafter) has the same eigenfrequency as the infinite layer in the city-impedance analytical model.

The horizontal harmonic displacement U_b is imposed at the base while the top surface (ignoring the oscillators) and the lateral surfaces are free. However, because of its limited width, the block may undergo a gyration of the top surface (i.e. rotation around the horizontal axis perpendicular to the horizontal base motion along X, like a bending beam) in addition to pure shear translation motion. To avoid this kinematic that does not occur in an infinite layer, any vertical displacement is prevented on the top free surface of the block.

The oscillators are distributed ℓ-periodically on the whole surface of the block. The oscillator's mass m_o, eigenfrequency f_o, damping ξ and inert mass m_i are the same as in the city-impedance model. The masses are expressed per unit length since the model is 2D. The oscillator is linked to the block surface by a perfectly-bonded rigid surface of length $2b$. This base stands for the foundation of the oscillator and allows the establishment of equilibrium and compatibility conditions between the boundary elements on the block and the finite element structures attached to them.

28.3.2 Brief Description of the Numerical Methodology

The elastodynamic block is analyzed using a harmonic 2D direct boundary element formulation proposed by Dominguez (1993) leading to a matrix form of the discretized boundary integral equation of the type:

$$\begin{bmatrix} H_{LL} & H_{Lc} \\ H_{cL} & H_{cc} \end{bmatrix} \cdot \begin{Bmatrix} \mathbf{u}_L \\ \mathbf{u}_c \end{Bmatrix} = \begin{bmatrix} G_{LL} & G_{Lc} \\ G_{cL} & G_{cc} \end{bmatrix} \cdot \begin{Bmatrix} \mathbf{t}_L \\ \mathbf{t}_c \end{Bmatrix} \tag{28.9}$$

where c and L indexes refer to the sets of boundary nodes on rigid coupling surfaces or on the rest of the elastic block respectively; **u** and **t** are the vectors of displacements and tractions; and ***H*** and ***G*** are the matrices arising from the integration of the corresponding fundamental solution times the relevant shape functions. Quadratic 3-noded boundary elements are used in meshing the elastic block. The harmonic response of the SDOF oscillators can be described as:

$$\begin{bmatrix} K_{oo} & K_{of} \\ K_{fo} & K_{ff} \end{bmatrix} \cdot \begin{Bmatrix} \mathbf{u}_o \\ \mathbf{u}_f \end{Bmatrix} = \begin{Bmatrix} \mathbf{F}_o \\ \mathbf{F}_f \end{Bmatrix} \tag{28.10}$$

where f and o indexes refer to the sets of degrees of freedom at the foundation or along the rest of the oscillators respectively; **u** and **F** are the vectors of displacements and external forces; $\boldsymbol{K}=\boldsymbol{K}^*-\omega^2\boldsymbol{M}$, with $\boldsymbol{K}^*$ and $\boldsymbol{M}$ being the stiffness and mass matrix respectively, and ω the circular frequency of excitation.

Compatibility between the boundary element nodes under one rigid coupling surface and the degrees of freedom of the foundation of an oscillator can be described by means of a compatibility matrix ***D***. The equilibrium between the oscillator and its foundation can be expressed through an equilibrium matrix ***C*** as

$$\mathbf{u}_c = D.\mathbf{u}_f \qquad \mathbf{F}_f = C.\mathbf{t}_c \tag{28.11}$$

All in all, after applying the boundary conditions and reordering, the equations describing the dynamic response of the system can be written as

$$\begin{bmatrix} K_{oo} & K_{of} & 0 & 0 \\ K_{fo} & K_{ff} & -C & 0 \\ 0 & H_{Lc}D & -G_{Lc} & A_{LL} \\ 0 & H_{cc}D & -G_{cc} & A_{cL} \end{bmatrix} \cdot \begin{Bmatrix} \mathbf{u}_o \\ \mathbf{u}_f \\ \mathbf{t}_c \\ \mathbf{x}_L \end{Bmatrix} = \begin{Bmatrix} \mathbf{F}_o \\ 0 \\ \mathbf{F}_{LL} \\ \mathbf{F}_{cL} \end{Bmatrix} \tag{28.12}$$

where ***A*** and $\mathbf{x}_L$ arise from boundary conditions and reordering, $\mathbf{x}_L$ being the vector of unknowns in the boundary element nodes.

28.4 Design, Instrumentation and Experiment

The experimental campaign capable of encapsulating the unconventional phenomena predicted by the city-impedance model is devised. The design of the set-up is based on the city-impedance analysis and takes advantage of the capabilities of the shaking table found at the Bristol Laboratory for Advanced Dynamics Engineering (BLADE) of the University of Bristol (UK). The performance of this facility is optimized with a test specimen of metric dimensions within a frequency range of about 0.5–25 Hz.

28.4.1 Design of the Layer

The experiments use cellular polyurethane foam as analogue elastic materials for the layer, avoiding some inherent difficulties of using sand or other reconstituted soil and insuring the respect of the assumptions of the theoretical model in a large range of deformations: homogeneous isotropic linear elastic behavior with weak damping, large density contrast and perfect adherence with the city.

The foam block is 2.13 m-wide in X; 1.76 m-wide in Y and h_L=0.76 m-high (see Fig. 28.4). With typical density, Young modulus, Poisson's ratio and damping of around 49 kg/m^3, 120 kPa, 10 and 5% respectively, the block is sized to exhibit eigen-frequencies under 20 Hz. The foam is attached to the shaking table through a base-plate assembly, firmly clamped on the platform with the principal axes X and Y aligned with the driven axes of the table. Experimental tests using white-noise excitation on the 140 kg foam block specimen lead to shear eigenfrequencies 9.47 and 9.18 Hz respectively in the X and Y directions (Fig. 28.3), the difference being due to a smaller width in the Y direction. The experimental damping ratio of 4.9% is the same in both directions and allows clear observation of the resonance peaks.

To avoid the additional gyration of the top surface of the block that do not occur in horizontally-infinite layers, 22 thin vertical steel rods (3 mm-diameter and 75 cm-long) are adhered at 35 cm centers around the periphery of the foam block (see Fig. 28.4). Their diameter and spacing are designed not to disturb the block's modes (i.e. bending stiffness much smaller than shear stiffness of the foam block) while restraining parasite vertical edge movements (i.e. elongation stiffness much larger than that of the foam block).

Figure 28.3 shows that with the edge rods, the experimental transfer function U_T/U_b in the frequency range 3–15 Hz is close to that of (i) the theoretical infinite layer with the same modal properties as the experimental ones (eigenfrequencies f_X=9.36 Hz and f_Y=9.11 Hz in X and Y direction respectively and a damping ratio ξ_L=4.9%); and (ii) a 1.85 m-wide BEM elastic block with the same eigenfrequency f_X and damping ξ_L (the design of the city in the next paragraph explains the width). Both experimental and BEM blocks can thus be considered in their principal axes as an infinite layer with a good approximation.

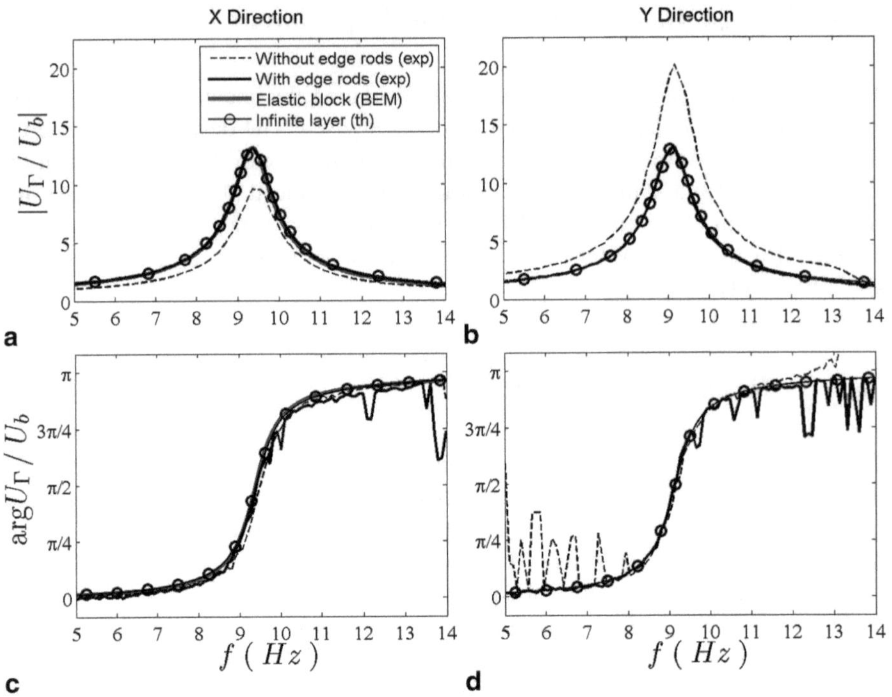

Fig. 28.3 Transfer function modulus (a, b) and phase (c, d) of the block's surface U_Γ to base motion U_b: a, c in X-direction; b, d in Y direction

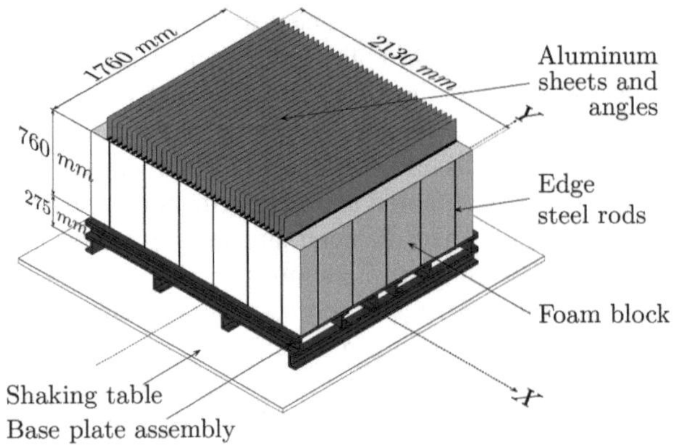

Fig. 28.4 Schematic of the set-up clamped to the shaking table through a base-plate assembly: 37 *vertical* aluminum sheets are distributed periodically every 5 cm on the *top* surface of the foam block

28.4.2 Design of the "City"

The design of the "city" fulfills three conditions based on the city-impedance model: (i) the weakly-damped 1D SDOF oscillators are periodically distributed on the foam block with a spatial period ℓ that must be small compared to the wavelength in the layer; (ii) the eigenfrequency of one individual oscillator is close to the eigenfrequency of the individual foam block; (iii) the impedance parameter η takes a value around 10 % or more for a significant effect.

The "city" is made up of $N=37$ vertical aluminum sheets that are periodically distributed every $\ell=5$ cm on the surface of the foam block in the X direction (see Fig. 28.4). The sheets are 0.5 mm-thin, 1.75 m-long in the Y direction and 19.7 cm-high. The sheets are sandwiched between two rigid aluminum angles along their lowermost edge and secured with set screws in order to provide a footing that can be adhered to the foam. The sheets have their out-of-plane direction in the X direction: they bend while shaken in the X direction but remain quasi-static as an inert mass while shaken in the Y direction.

In the X direction, one "building" behaves as a SDOF oscillator characterized by its eigenfrequency f_o, a damping ratio $\xi << 1$ and a modal mass $m_o=0.266$ kg that is classically 61.3 % of the mass of the sheet that can bend out of the $b=1.3$ cm-wide angles. The rest of its mass is an inert mass $m_i=0.583$ kg on the foam block. Their eigenfrequency is measured experimentally to be around $f_o=8.35$ Hz in the city configuration and their damping ratio around $\xi \approx 4.6\,\%$. With such a design, the scale separation condition is fulfilled at "buildings" resonance with a scale separation ratio $\varepsilon_o \approx 9.3\%$. The mass of the foam under one period being $\rho|\Sigma|h_L=3.28$ kg, the impedance parameter $\eta \approx 11.5\,\%$ ensures significant interaction effects.

Note that the top motion U_{sheet} of the sheet is related around resonance to the motion U_m of the equivalent SDOF oscillator by $U_{sheet}=\alpha U_m$ where the geometrical coefficient $\alpha=1.57$ is deduced theoretically from the first mode shape of a bending beam. Note also that the sheets are removable from their angles: the 37-oscillators configuration has been modified to investigate the sensitivity of Soil-City interactions to geometrical and mechanical parameters such as city's density and size, buildings' position and eigenfrequencies, etc (not presented here).

Finally, the repeatability, linearity, directivity and homogeneity of the response of both the block and the aluminum sheet have been checked experimentally in the frequency range of interest.

28.4.3 Instrumentation

The measurements are mainly performed with accelerometers located on both the foam block and oscillators in the X and Y axes, and also on the table to record the input motion. The accelerometers record the layer's surface motion, the modal shapes in both X and Y directions and the acceleration at the top of some sheets.

Two types of accelerometers are used. Where mass of the sensor could have a deleterious effect on the response (e.g. at the top of some sheets) and where little space is available (e.g. on the block surface between two oscillators), lowmass MEMS accelerometers are used (Analog Devices ADXL335, 2 gr mass, 2 cm-wide footprint). Otherwise, Setra type 141A accelerometers are used (1D, 30 gr mass).

28.4.4 Experiment

The experimental set-up is excited with a series of transient and harmonic signals. The level of the excitation is a compromise between small deformations requirement and good accuracy of the measurements. The motion imposed is horizontal to simulate a vertically-incident SH wave. The polarization is preferentially perpendicular to the sheets (resonating effect in *X*), parallel to the sheets (no-resonance in *Y*) and in the oblique 45° direction to identify depolarization effects.

White noise enables determination of the coupled layer/oscillators system eigenfrequencies and damping in the frequency range 1–20 Hz. Harmonic sinusoidal waveforms around identified resonance or anti-resonance frequencies allow accurate determination of the mode shapes and depolarization effects. Ricker pulses are also applied in displacement

$$u_b(t) = A_R\left(\left(\pi f_R(t-t_s)\right)^2 - 1/2\right)\exp\left(-\left(\pi f_R(t-t_s)\right)^2\right) \tag{28.13}$$

with A_R and t_s being an amplitude and the time shift respectively and with a central frequency f_R so that maximum acceleration is obtained at around 8 Hz. In the frequency domain this gives a transfer function with good accuracy and low level of noise; in the time domain it allows a clear determination of the time-decreasing response of the system. Most experimental results are obtained from Ricker pulses.

28.5 Experimental-Analytical-Numerical Comparisons

28.5.1 Effects of Soil-City Interactions in Frequency Domain

Figure 28.5 shows the transfer function U_Γ/U_b in modulus and in phase in the range 4–13 Hz for 1 or 37 oscillators on the foam block. On the left are the transfer functions in the *X* direction and on the right in the *Y* direction. They are obtained from a Ricker pulse played by the table in the *X* or *Y* directions. Records are from an accelerometer located centrally on the block surface. The experimental transfer functions are obtained without any windowing or averaging in the Fourier Transform, owing to the low level of noise when using Ricker pulses.

With only one oscillator on the surface, the transfer function is similar to the usual layer resonance peak. But with 37 oscillators, the transfer function in the resonant *X* direction is the double soil-city resonant peaks as predicted previously in the city-impedance model, while in the inert *Y* direction it remains the classic layer resonance (with lower eigenfrequency). Both models are in good agreement qualitatively and quantitatively with the experimental results and with one another.

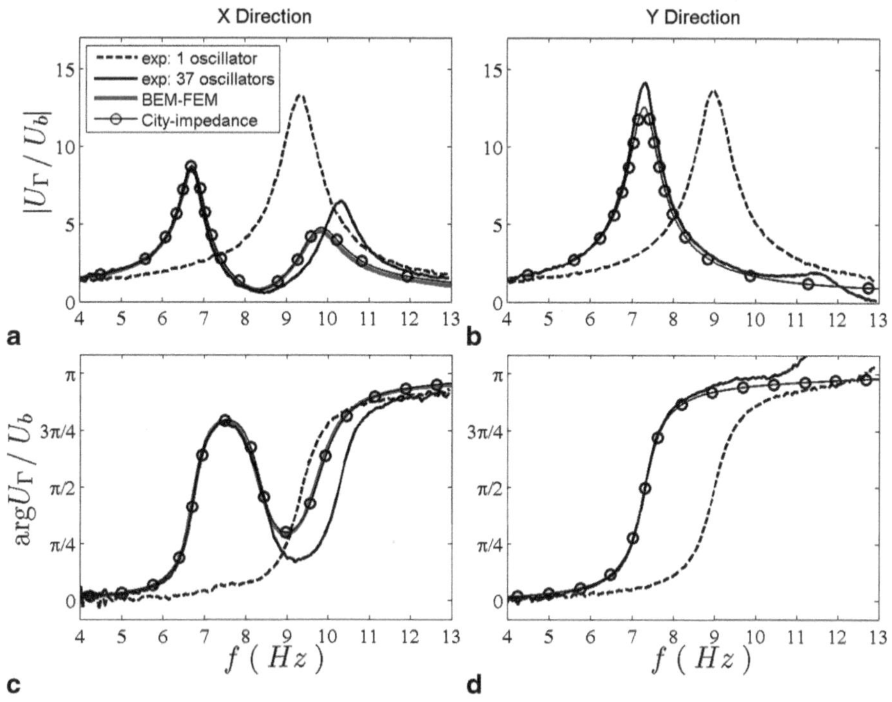

Fig. 28.5 Transfer function modulus (a, b) and phase (c, d) of the foam block's surface U_Γ to base motion U_b: a, c in X-direction; b, d in Y-direction

28.5.2 Effects in Time Domain: Longer Coda and Beatings

To analyze the effects of the resonant surface on temporal records, a Ricker pulse has been played by the table in the *X* direction. Figure 28.6 shows the accelerations recorded by the central accelerometer on the surface of the block and on the table. The results are shown for two configurations: a single oscillator and 37 oscillators on the foam block. They have been plotted in decimal on the left and in decibels on the right to see the coda. They are compared with signals synthesized from the transfer functions (Fig. 28.5) obtained by both city-impedance and BEM-FEM models.

Compared to the usual record obtained for one oscillator (similar to the one of a layer with free surface), the signal recorded with 37 oscillators shows: (i) the decrease of amplitude of the acceleration, (ii) a lower frequency of the oscillations, (iii) a longer coda, and (iv) beatings in the signal. The models' results are in good agreement with the experimental records, both qualitatively and quantitatively, and with one another. Both models are accurate in the time domain.

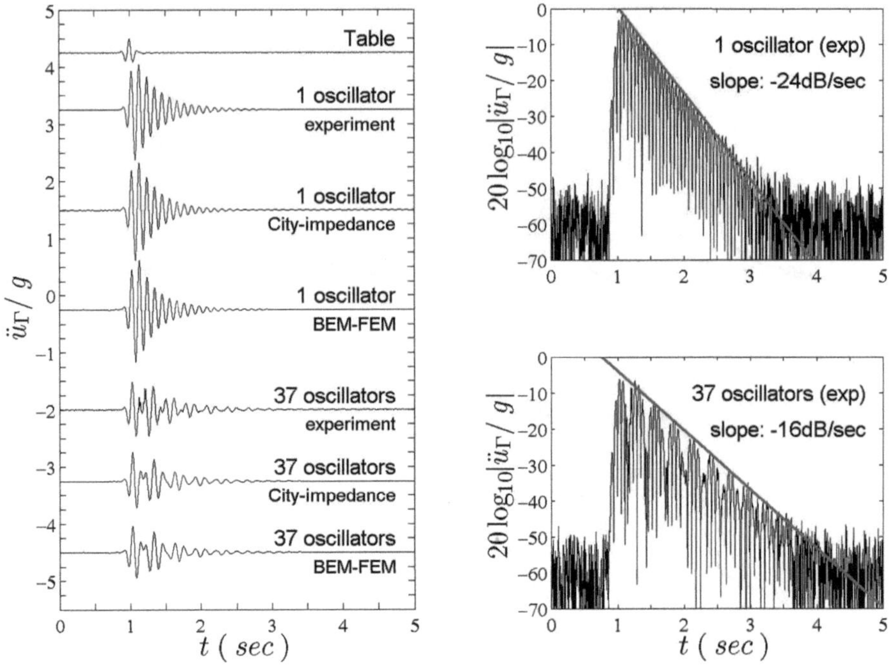

Fig. 28.6 Records in the resonant *X* direction on the *top surface* of the block (central accelerometer) in response to a Ricker pulse imposed by the table at the *base* of the block in the *X* direction for a single or 37 oscillators on the block: on the *left* in decimal; on the *right* in decibel. Comparison with synthesized signals from city-impedance or BEM-FEM models

28.5.3 Mode Shapes

The set-up is harmonically shaken in the resonant *X* direction to measure the system's mode shapes in the 37-oscillator-configuration: out of resonance (5 Hz), at the soil-city-resonance-peaks and at soil antiresonance (see the transfer functions in Fig. 28.5). Figure 28.7 shows the derived mode shapes from the experimental measurements and from the output of the models at their respective (anti) resonance frequencies.

Experimental results are obtained from central accelerometers on the foam block lateral and top surface and at the top of the central oscillator. The recorded accelerations have been divided by the acceleration amplitude of the table (similarly to the transfer functions in Fig. 28.5). The experimental shapes are shown at the time t_0 when the block undergoes the higher accelerations. The shapes are given for the city-impedance model by Eq. 28.7 and for the BEM-FEM model by the horizontal displacement of the nodes on the lateral side of the elastic block. The displacement of the oscillator has been multiplied by the theoretical geometrical factor $\alpha = 1.57$ to obtain the equivalent motion at the top of the bending beam.

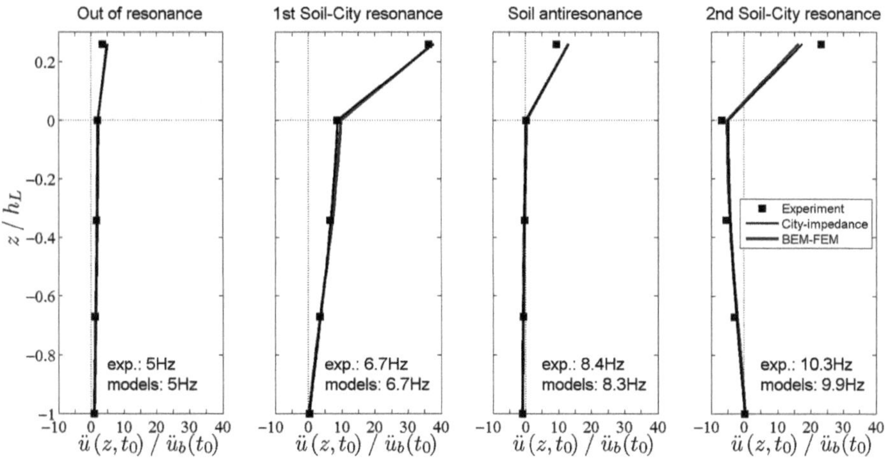

Fig. 28.7 Derived mode shapes in the *X* direction for the 37-oscillator-configuration at 5 Hz and at their respective resonance or anti-resonance frequencies

Figure 28.7 shows that out of resonance the behavior is conventional: for instance at 5 Hz the set-up follows the shaking table quasi-statically. At the two soil-city resonance peaks, the layer is in resonance with the usual quarter-of-wavelength-shape while the oscillators oscillate in phase (first peak) or in phase opposition (second peak) with the block surface. At soil antiresonance (city's resonance), the surface is near stationary while the oscillators oscillate. Without the resonant mass of the city, the foam block should have its resonance around this frequency. But due to city's resonance, the surface is a node in the mode shape as for a rigid surface condition. Both models are in qualitative and quantitative agreement with experimental results and with one another.

28.5.4 Depolarization Effect

The set-up is harmonically shaken in the oblique 45° direction for some particular frequencies. Figure 28.8 displays in parallel the table and the surface particle path-line (represented here in the acceleration plan recorded by the central accelerometer). Herein, the grey line indicates the 45° direction of the harmonic table displacement, the arrows are the left/right orientation of the wavefield, and $\varphi_{y/x}$ is the phase difference between *X* and *Y* motions. The figure shows that the motion of the surface can differ significantly from the 45° motion of the table: the surface motion is depolarized. For instance, near anti-resonance in *X*, the surface displacement is mostly along *Y*, the direction in which the foam block is near resonance (see 7.3 or 8.4 Hz). At the second soil-city resonance in *X* at 10.3 Hz, the surface displacement is in inertial regime in the *Y* direction, thus producing a motion mostly along *X* while shaking at 45°. Note also that the displacement becomes elliptical, due to the phase difference $\varphi_{y/x}$ between *X* and *Y* motions.

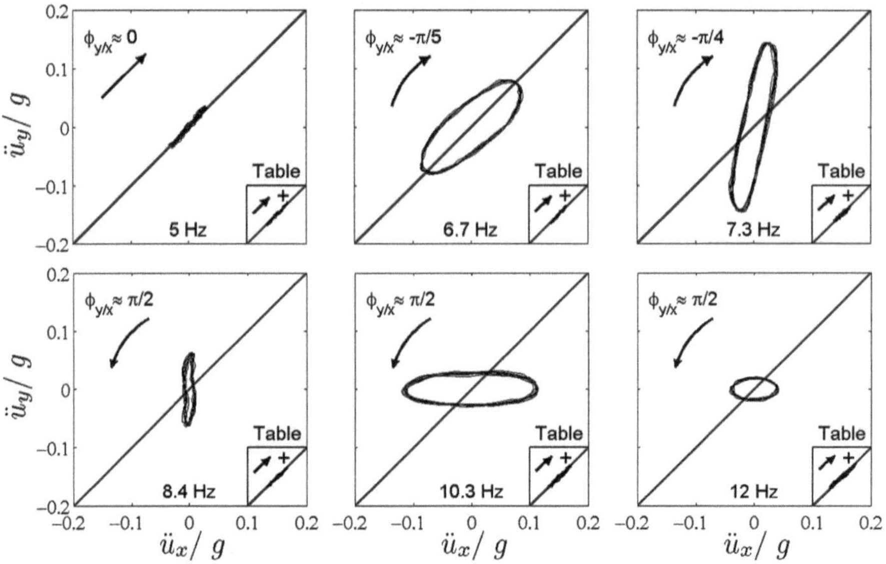

Fig. 28.8 Depolarization of the surface motion regarding the 45° rectilinear motion imposed at the base (experimental results)

To sum-up, depolarization is a frequency-dependent effect that is due to the surface anisotropy. It affects the direction, the ellipticity and the left/right orientation of the wavefield. The phenomenon is similarly presented by the models (not reported here) since it stems from anisotropic transfer functions in X and Y directions that can be recovered by both models.

28.6 Conclusion

The site-city effect is a composite of multiple phenomena associated with structure-soil-structure interactions and can be significant. Herein, the physical, theoretical and numerical models that have been used to describe these phenomena have revealed remarkable consistency, allowing the identification and qualification of their key signatures. Moreover, the city-impedance model provides the key parameters to quantify the associated effects. The city acts as a resonant surface leading to unconventional boundary conditions for the supporting elastic medium (mixing free-like and rigid-like conditions depending on the frequency) impossible to reach with an elastic upper layer and drastically changing the usual free-surface-wavefield.

Provided that the scale separation assumption is fulfilled, other city configurations showed that the phenomena are robust (triggered by only five structures), and that both the periodic condition and the strict scale separation condition are not mandatory for the phenomena to occur. The city-impedance model gives also accurate results for non-periodically-distributed buildings and for large distance between buildings with respect to the wavelength. Cities with two types of buildings (two different eigenfrequencies) have also been tested, leading to two antiresonances at the surface of the layer and three resonance peaks of even more reduced amplitude. The tests showed also that the eigenfrequency of the oscillators on the surface increases significantly (by 15 % between isolation condition and the 37-oscillators city configuration) as the city gets denser. This effect is attributed to multiple interactions between oscillators within the boundary layer. It is expected that this phenomenon will be demonstrated by the models once the rotation of the foundations has been taken into account.

Theoretical and experimental results are relevant in the framework of large wavelength. Other phenomena, as presented in Uenishi (2010) for instance, can occur out of scale separation (high frequencies). Although the analysis of strongly non-linear soils is out of the scope of this study, the application of the present developments might be valid for limited non-linearities. Practical application may concern the design spectrum for clusters of limited number (at least four) of nearby quasi-identical high-rise buildings in new city-blocks.

Acknowledgement The research leading to these results has received funding from the European Community's Seventh Framework Program [FP7/2007–2013] for access to the University of Bristol's Earthquake and Large Structures (EQUALS) Laboratory under grant agreement n° 227887. The authors would like to thanks E.Skuse, D.Ward, F.Sallet and H.Aldaikh for their very active support in the project. Satellite view in Fig 28.1a) is from Google Maps, available at https://maps.google.fr/ (last accessed July 2013).

References

Auriault JL, Boutin C, Geindreau C (2009) Homogenization of coupled phenomena in heterogenous media. ISTE and Wiley, USA

Boutin C, Roussillon P (2004) Assessment of the urbanization effect on seismic response. Bull Seism Soc Am 94(1):251–268

Boutin C, Roussillon P (2006) Wave Propagation in presence of oscillators on the free surface. Int J Eng Sci 44:180–204

Castellaro S, Mulargia F (2010) How far from a building does the ground-motion free-field start? Bull Seism Soc Am 100(5A):2080–2094

Chavez-Garcia FJ, Bard PY (1994) Site effects in Mexico city eight years after the september 1985 Michoacan earthquakes. Soil Dyn Earthq Eng 13:229–247

Clouteau D, Aubry D (2001) Modification of the ground motion in dense urban areas. J Comput Acoust 9(4):1659–1675

Domínguez J (1993) Boundary elements in dynamics. Computational Mechanics publications, Southampton

Ghergu M, Ionescu IR (2009) Structure-soil-structure coupling in seismic excitation and "city effect". Int J Eng Sci 47:342–354

Guéguen P, Bard PY, Chavez-Garcia FJ (2002) Site-city seismic interaction in Mexico city-like environments: an analytical study. Bull Seism Soc Am 92(2):794–811

Kitada Y (1999) Models test on dynamic structure-structure interaction of nuclear power plant buildings. Nucl Eng Design 192:205–216

Lee TH, Wesley DA (1973) Soil-structure interaction of nuclear reactor structures considering through-soil coupling between adjacent structures. Nucl Eng Design 24:374–387

Luco JE, Contesse L (1973) Dynamic structure-soil-structure. Bull Seism Soc Am 63(4):1289–1303

Padrón LA, Maeso O, Aznárez JJ (2004) Cálculo de estructuras de barras incluyendo efectos dinámicos de interacción suelo-estructura. Master thesis, Universidad de Las Palmas de G.C., Spain. http://hdl.handle.net/10553/10472. Accessed 24 Nov 2014
Padrón LA, Aznárez JJ, Maeso O (2009) Dynamic structure-soil-structure interaction between nearby piled buildings under seismic excitation by BEM-FEM model. Soil Dyn Earthq Eng 29:1084–1096
Sanchez-Palencia E (1980) Non homogeneous media and vibration theory. Lectures notes in physics, vol. 127. Springer, Berlin
Schwan L, Boutin C (2013) Unconventional wave reflection due to “resonant surface”. Wave Motion 50:852–868
Tsogka C, Wirgin A (2003) Simulation of seismic response in an idealized city. Soil Dyn Earthq Eng 23:391–402
Uenishi K (2010) The town effect: interaction between a group of structures and waves in the ground. Rock Mech Rock Eng 43:811–819
Wirgin A, Bard PY (1996) Effects of building on the duration and amplitude of ground motion in Mexico city. Bull Seism Soc Am 86(3):914–920

Chapter 29
Centrifuge Modeling of Dynamic Behavior of Box Shaped Underground Structures in Sand

Deniz Ulgen, Selman Saglam, M. Yener Ozkan and Jean Louis Chazelas

29.1 Introduction

During the earthquakes of last century, failures of embedded structures and pipelines have been observed within different ground deformations, settlements, lateral spreads and flotation due to soil liquefaction. In most of the studies entreating the seismic performance of embedded structures, the embedded structures were subjected to damage during the seismic events and afterwards. ASCE (1974) reported the descriptions of damages on underground structures at the city of Los Angeles after the 1971 San Fernando Earthquake. Duke and Leeds (1959), Stevens (1977), Dowding and Rozen (1978), Owen and Scholl (1981), Sharma and Judd (1991), generated a databank of damaged underground structures. Power et al. (1998) improved the existing data bank and examined 217 embedded structures with damages reported by Hashash et al. (2001).

Seismic performance of embedded structures has generally been studied by models constituted on 1 g shaking tables. Tamari and Towhata (2003) investigated soil-structure interaction in liquefiable soils by means of buried models on 1 g shaking tables. In the study, the alteration in shear modulus of the soil and the relationship between seismic soil pressure and deformation were investigated depending upon the amplitudes of the motions applied. Clough and Penzien (1993) reported that during an earthquake, culvert type embedded structures are subjected to shear

D. Ulgen (✉)
Department of Civil Engineering, Mugla Sitki Kocman University, Mugla, Turkey
e-mail: denizulgen@gmail.com

S. Saglam
Department of Civil Engineering, Adnan Menderes University, Aydin, Turkey

M. Y. Ozkan
Department of Civil Engineering, Middle East Technical University, Ankara, Turkey

J. L. Chazelas
Institut français des sciences et technologies (IFSTTAR), Nantes, France

F. Taucer, R. Apostolska (eds.), *Experimental Research in Earthquake Engineering*,
Geotechnical, Geological and Earthquake Engineering 35,
DOI 10.1007/978-3-319-10136-1_29

stresses due to vertically propagating. Most of the investigations of buried structures have dealt with structures having high rigidity relative to the surrounding soil; less attention has been devoted to soil-structure interaction of buried structures with flexible sides when subjected to shear stresses.

Kobe's rapid transit railway system experienced significant damage during the Hyogo-ken Nanbu (Kobe) earthquake (M_w=6.9). Iida et al. (1996) reported that major collapses were observed in the Daikai Subway Station due to the relative displacement between the base and the ceiling of the station caused by the dynamic lateral forces. Huo (2005) carried out numerical analyses to investigate the dynamic behaviour of the Daikai Subway Station during the Kobe Earthquake, and showed that vertical accelerations have almost no influence on the dynamic response of subway stations when compared to the effects of horizontal accelerations. Moreover, the author emphasized that internal friction and relative stiffness between the ground and structure plays a significant role in the seismic design of underground structures. Che and Iwatate (2002) conducted shaking table tests for modeling the dynamic response of the the Daikai Subway Station under the Kobe earthquake, concluding that damage was most likely caused by dynamic lateral forces.

Rectangular underground structures are usually designed to be able to withstand shear distortions or racking deformations. The most commonly used design approach is the free-field racking deformation analysis. In this approach, the deformation of soil in the absence of the structure is directly imposed to the structure. In other words, it is assumed that the structure deforms in accordance with the soil. Due to the lack of knowledge on the relative stiffness between soil and structure, racking deformations of underground structures may be underestimated or overestimated when a free-field deformation analysis is conducted. Wang (1993), Penzien (2000), Huo et al. (2006) and Bobet et al. (2008) proposed simplified frame analysis methodologies for considering soil-structure interaction effects. In these methodologies, racking deformations of the structure are first estimated on the basis of relative stiffness and free-field deformations. Then, static analysis is performed to determine sectional forces due to racking deformations. All these approaches are deformation based methods. There is not any established or accepted forced-based method for the seismic design of an underground structure. Current seismic design methods of underground structures have shortcomings and some limitations. They rely on empirical analyses and ignore dynamic soil-structure interaction. One commonly used method is based on the Mononobe-Okabe theory for estimating dynamic earth pressures acting on buried structures. However, the Mononobe-Okabe method gives unrealistic results and it is not recommended for underground structures with rectangular cross-section (Hashash et al. 2001). Very few experimental and field data are currently available regarding the dynamic response of underground structures. Thus, there exists a strong need to extend the available knowledge concerning the seismic behavior of underground structures.

In this study a series of centrifuge tests were conducted to investigate the dynamic response of box-shaped underground structures buried in sand. The main objectives of the study can be summarized as follows:

- To investigate the deformations of box-type underground structures subjected to dynamic loading buried in dry sand.
- To examine the effects of the flexibility ratio on racking deformations by considering dynamic soil-structure interaction.

29.2 Centrifuge Model Tests

Centrifuge model tests were conducted under a 40 g centrifugal field with a prototype acceleration 0.40 g and 3.5 Hz frequency at the IFSTTAR Centrifuge Laboratory. Three types of culvert models having different rigidities were used in the tests. All culverts are box-type underground structures made of aluminum. The thickness of the top and bottom slab was kept constant at 6 mm. The height of the culverts was 5 cm and the thicknesses of the sidewalls were 1.5 mm, 3 mm and 5 mm. Fontainebleau sand NE24 was used as model ground with average particle size diameter $D50 = 200\ \mu m$. The model ground was prepared by applying pluviation techniques to provide uniform and repeatable specimen density. Relative density of the soil was 70 % during the tests. In order to check the uniformity of sand specimen Cone Penetration Tests were conducted after pluviation.

29.2.1 Centrifuge Test System

Figure 29.1 shows a general view of the geotechnical centrifuge facility located in IFSTTAR/ Nantes/ France. The test system consists of a motor, a rotating beam arm around a vertical axis, balancing counterweights, a swinging bucket carrying a soil container, a servo hydraulic shaker and a data acquisition system. The centrifuge is a beam type centrifuge with a rotating arm with a radius of 5.5 m. It can rotate with a maximum 100 g centrifugal acceleration while carrying a 2 t model. There is a swinging bucket attached to the end of beam with an area of 1.54 m^2. Soil container, data acquisition and shaking system are placed into this bucket.

Fig. 29.1 General view of centrifuge test system in IFSTTAR

29.2.1.1 Earthquake Simulator

Base excitations were applied by an earthquake simulator having an electro-hydraulic system (Fig. 29.2). The shaker is able to perform harmonic and real earthquake motions in the horizontal direction perpendicular to the centrifuge acceleration and gravity acceleration. The characteristics of the earthquake simulator are given in Table 29.1.

29.2.1.2 Soil Container

An equivalent shear beam (ESB) box was used as a model container in dynamic centrifuge tests (Fig. 29.3). The side walls of the ESB box are free in the shaking direction and restrained in the transverse direction. The internal dimensions of the ESB container are 0.80 m in length, 0.35 m in width and 0.41 m in height.

Fig. 29.2 Earthquake simulator

Table 29.1 Characteristics of the earthquake simulator

Length of shaking table	1 m
Width of shaking table	0.5 m
Payload mass	400 kg
Centrifugal acceleration	20–80 g
Maximum displacement	5 mm
Maximum velocity	1 m/s
Maximum acceleration	0.5 g
Frequency range for earthquake motions	20–300 Hz
Frequency range for harmonic motions	20–200 Hz

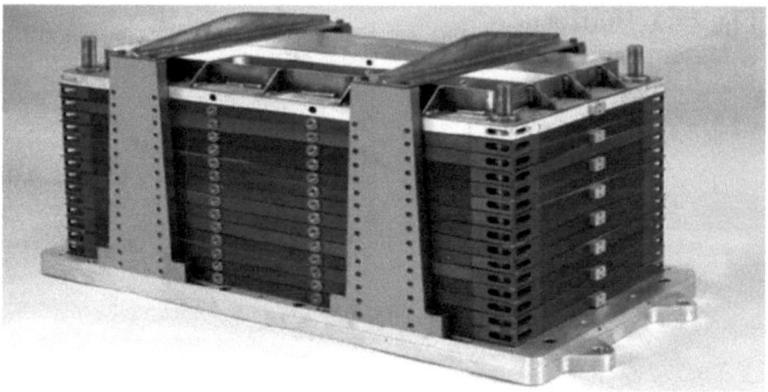

Fig. 29.3 ESB box

The ESB box consists of 14 aluminium alloy horizontal layers, separated by rubber sheets having 4 mm thickness. As a result the soil can adapt to soil deformations while boundary effects are eliminated. Dynamic centrifuge tests conducted on this container show that the resonant frequency of the ESB box varies between 32 and 37 Hz (Escoffier 2008).

29.2.1.3 Data Acquisition System

An LMS fast data acquisition system was used in the centrifuge tests. The system has 72 channels for voltage or ICP accelerometers and 48 channels for strain gauge, voltage and ICP accelerometers. A 24-bit resolution data acquisition card is located in the system supporting sampling rates from 100 Hz to 25 kHz. The system was placed into the swinging basket and data was transmitted to the control room through a wireless network.

29.2.1.4 Accelerometers, Transducers & Strain Gauges

Bruel & Kjaer 4317-10 mv/g type piezoelectric accelerometers were used in the dynamic tests. They are very compact and lightweight, therefore particularly suitable for the centrifuge tests. A natural internal bandpass filter of 1 Hz to 20 kHz exists in the accelerometers. There were a total of 26 accelerometers buried in soil and mounted on a culvert model and equivalent shear beam container.

Laser displacement sensors were used for measuring the soil settlement before and after shaking. Due to the low frequency rate of measurement, it was not possible to obtain reliable information during dynamic loading. The displacement sensors were mounted on a suspended beam over the ESB container.

In order to measure the deformations of the box-type culvert, horizontal extensometers and diagonal extensometers developed by IFSTTAR and GAROS were used in the centrifuge experiments. There were five pairs of horizontal extensometers piled together on a fork shaped system as illustrated in Fig. 29.4. Each extensometer includes two strain gauges giving the bending deformation of the extensometer.

Diagonal extensometers (Fig. 29.5) were developed by IFSTTAR for measuring diagonal deformation along the tunnel. To investigate the boundary effects at to end

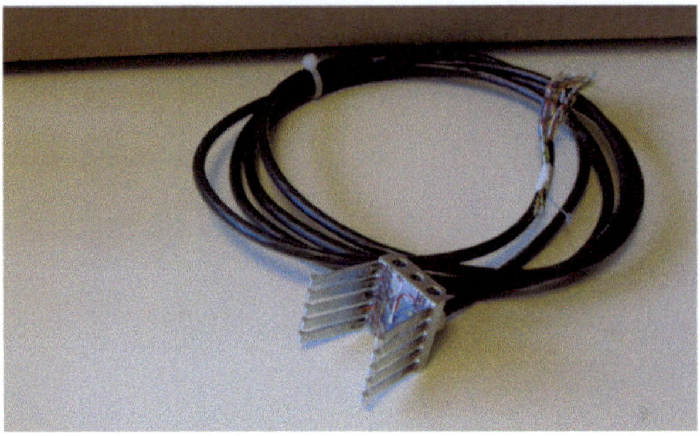

Fig. 29.4 Horizontal extensometers

Fig. 29.5 Diagonal extensometers

walls and the validity of plain strain conditions, extensometers were placed in the middle and near end of the tunnel model.

29.2.2 Reduction Scaling & Scaling Effects

A series of centrifuge tests were conducted by IFSTTAR to investigate the performance of the ESB box at different centrifugal acceleration "g" levels. Results show that 40 g is the most reasonable centrifugal acceleration to eliminate the boundary effects of the soil model. For this reason, dynamic centrifuge tests were performed at 40 g of centrifugal acceleration. Scaling laws between prototype and model are derived for a 40 g centrifugal field and are presented in Table 29.2

In a centrifuge test, linear dimensions are reduced by a scale factor. In most cases, this reduction factor is not applied to the particle size of the soil in order to maintain unvaried the stress-strain characteristics of the prototype. However, it should

Table 29.2 Scaling laws between prototype and model

Parameter	Prototype	Model
Gravity	1	40
Length	40	1
Mass	40^3	1
Force	40^2	1
Energy	40^3	1
Stress	1	1
Strain	1	1
Mass density	1	1
Acceleration	1	40
Time (dynamic)	40	1
Frequency	1	40

Table 29.3 Physical properties of soil

Soil	Fontainebleau sand NE34
e_{min}	0.55
e_{max}	0.86
γ_{min} (kN/m^3)	13.93 kN/m^3
γ_{max} (kN/m^3)	16.78 kN/m^3
Mean diameter (D_{50})	0.20 mm
Specific gravity	2.64
Friction angle	38°

be noted that particle size effect can appear in soil structure interaction problems. Iglesia et al. (2011) suggested that the grain size effects can be minimized when the ratio of model dimensions to the average particle size of soil is at least 20. This ratio was specified as 235 for the present study, thus grain size effects are eliminated.

29.2.3 Physical Properties of Sand

All centrifuge tests were performed in dry Fontainebleau sand. It is a fine quartz sand composed of round shaped grains. Median grain size (D_{50}) of the Fontainebleau Sand is 0.20 mm. Physical properties of the sand are summarized in Table 29.3.

29.2.4 Preparation of Model Ground

The model ground is prepared by pluviation technique to achieve an average relative density of 70 %. In this technique, dry sand is rained into the model container from a height of 60 cm by means of an automatic hopper (Fig. 29.6). Density control boxes are buried into the model ground to check uniformity and to measure relative density of soil. Furthermore, in-flight cone penetration tests (CPT) are performed to evaluate the homogeneity of the sand model.

Fig. 29.6 Sand pluviation system

29.2.5 Design of Culvert Models

There were in total three box-type culvert models used in the centrifuge tests. Internal dimensions of these culverts were kept constant for simplicity in instrumentation, whereas outside dimensions vary with the thickness of sidewalls to obtain different rigidities. Thickness of the roof and invert slabs were designed to be relatively thick and stiff as compared to that of the sidewalls to eliminate structural effects due to bending. Cross sections and dimensions of the culvert models are given in Table 29.4.

In order to satisfy plain strain conditions, the ratio between the length and the width of the culvert model was specified on the order of 7. Besides, the ends of the culvert section and the interface between the tunnel and the ESB box were designed to provide free movement at the culvert ends. Figure 29.7 shows a schematic drawing of the designed end sections. As shown in Fig. 29.7, the longitudinal sides of the ESB box were covered with a Teflon sheet to reduce surface friction. Furthermore, neoprene foams were placed at the end of the culvert to allow deformation of the culvert ends.

Table 29.4 Cross-section dimensions of the culvert models

Culvert model		Internal dimensions (mm)		External dimensions (mm)	
		Vertical	Horizontal	Vertical	Horizontal
1	Thinnest walls	38	44	50	47
2	Intermediate thickness	38	44	50	50
3	Thickest walls	38	44	50	54

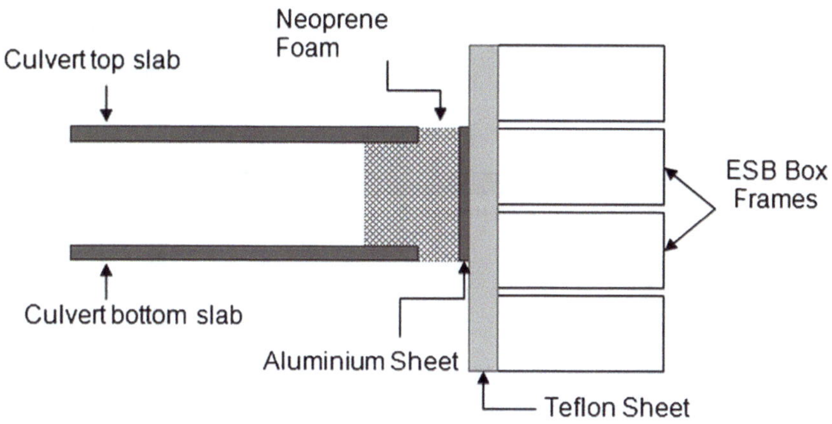

Fig. 29.7 Schematic drawing of end sections of the culvert model

29.2.6 Instrumentation

A total of 16 accelerometers were buried in the soil model, six accelerometers were placed on the ESB box and three laser displacement sensors were positioned over the container as shown in Fig. 29.8; the accelerometers were distributed into four arrays. The main purposes of this instrument layout can be summarized as follows:

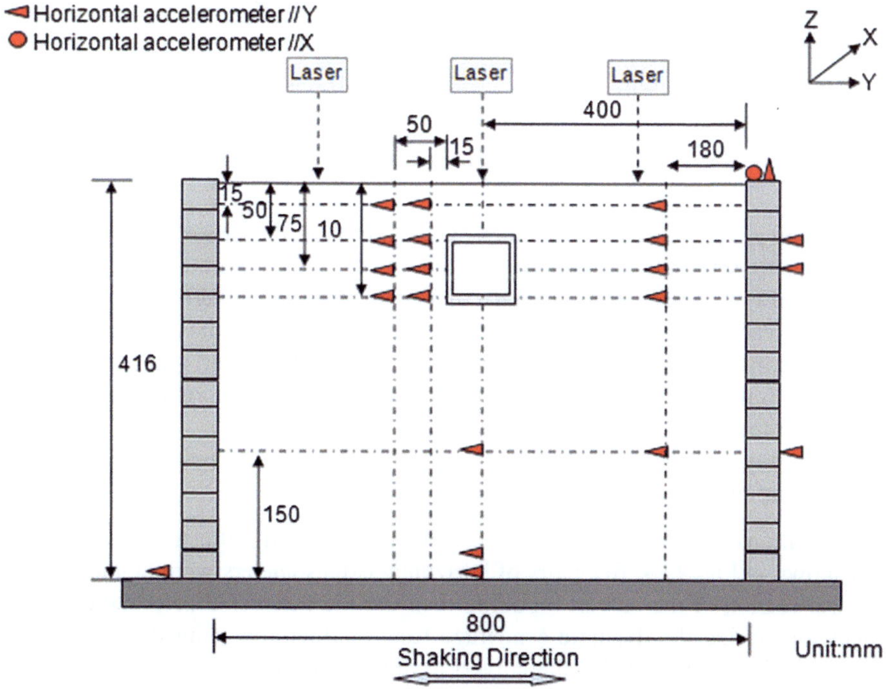

Fig. 29.8 Layout of accelerometers and laser displacement sensors

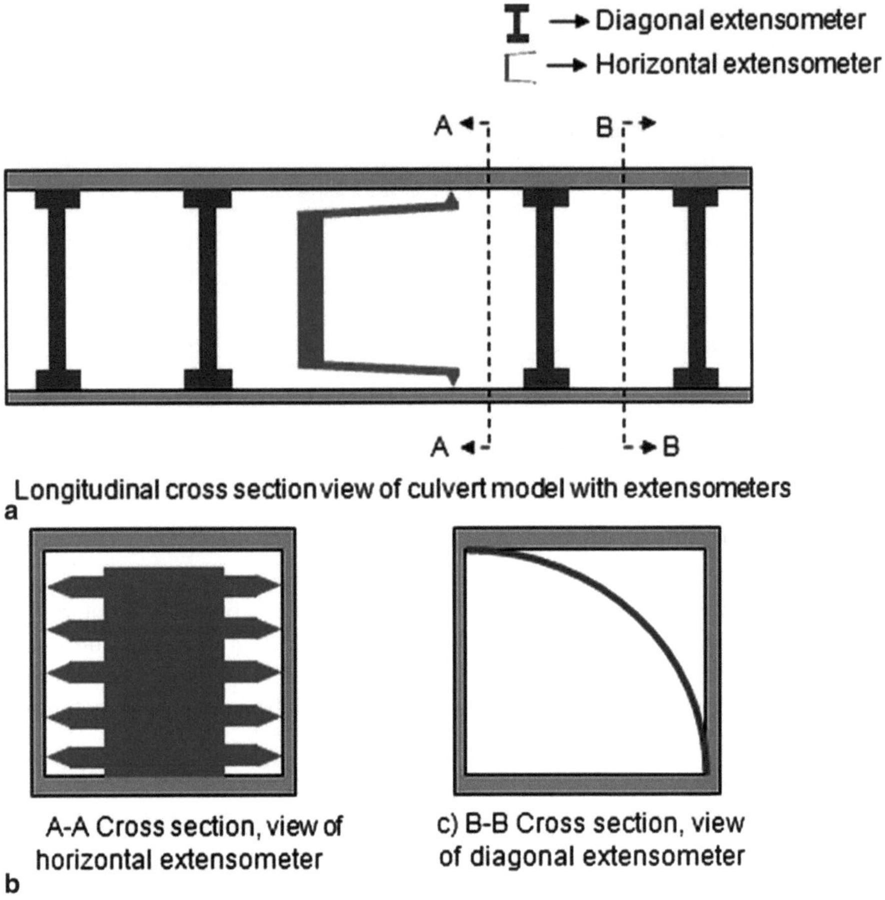

Fig. 29.9 Layout of extensometers

- One array was placed at a distance of 180 mm from the side of the ESB box for analysing free-field behavior
- One array was placed at a distance of 15 mm from the culvert model to investigate shear strain and acceleration response of the soil near the culvert model
- One array was placed at a distance of 50 mm from the culvert model to compare shear strains against the obtained shear strains at a distance of 15 mm from the culvert model
- One array was placed in the middle of the ESB box to investigate the acceleration response of the soil

The Position of the extensometers placed inside the culvert is illustrated in Fig. 29.9. As shown in Fig. 29.9, five pairs of horizontal extensometers were mounted at the central section of the culvert model to measure the transverse deformation of the sidewalls during the dynamic centrifuge tests. Furthermore, four diagonal extensometers were placed inside the culvert to measure the change in diagonal length. Two of them were located near the central, and the other two were located near the

Table 29.5 Testing program

Test no.	Culvert model	Acceleration (g) prototype scale	Frequency (Hz) prototype scale
1	Thinnest sidewalls	0.40	3.5
2	Intermediate thickness	0.40	3.5
3	Thickest sidewalls	0.40	3.5

ends of culvert model. Thus, it was possible to check whether plain strain behaviour is valid or not for the culvert model under dynamic loading.

29.2.7 Testing Program

As indicated previously, three culvert models with different flexibility ratios were used in this study. These models were tested under harmonic motions with an acceleration of 0.40 g and 3.5 Hz frequency for the evaluation of the dynamic response of culvert models. The testing program is given in Table 29.5 below.

29.3 Results of Centrifuge Tests

29.3.1 Maximum Accelerations Along the Soil Profile

The accelerations in the dry sand were recorded at different heights as shown in Fig. 29.8. Recorded maximum accelerations at those heights were normalized with the maximum input acceleration measured on the shaking table and plotted against the soil depth (Fig. 29.10). As shown in Fig. 29.10, acceleration amplification increases gradually from 16 to 4 m and from that point increases sharply near the surface.

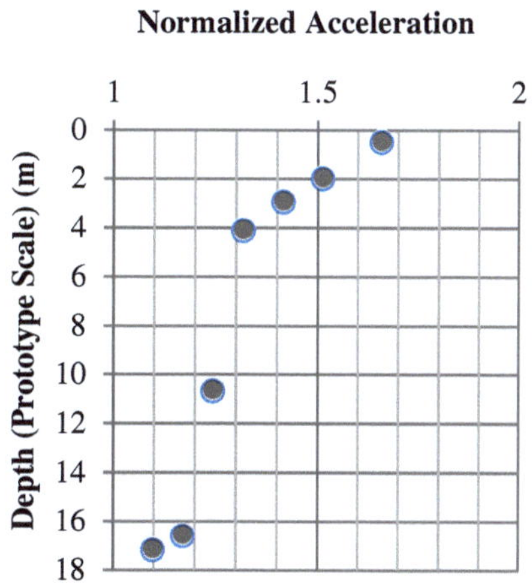

Fig. 29.10 Variation of maximum acceleration along soil profile

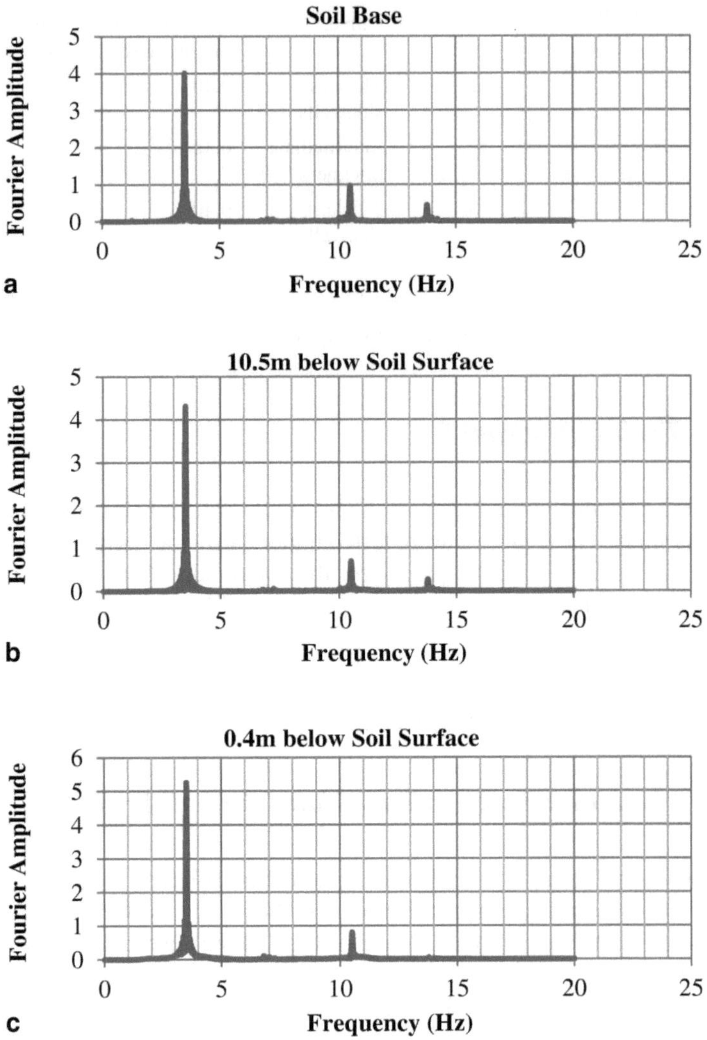

Fig. 29.11 Fourier spectra of the acceleration records along the soil profile. **a** Soil base **b** 10.5 m below surface **c** 0.40 m below surface

Figure 29.11 shows the Fourier spectra of the acceleration records at different levels in the model ground. As seen in the figure, Fourier amplitude increases from base to surface at low frequencies. On the other hand, soil damps the high frequency components of the motion.

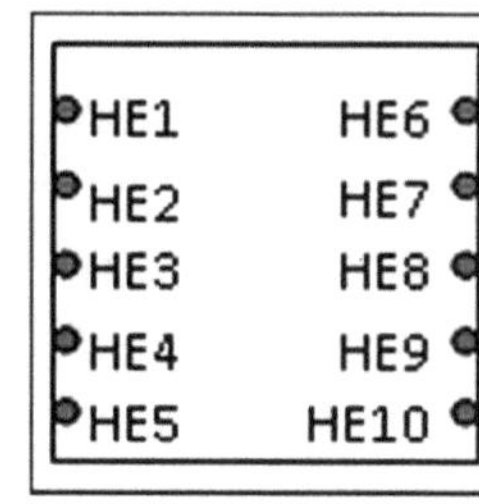

Fig. 29.12 Layout of extensometers

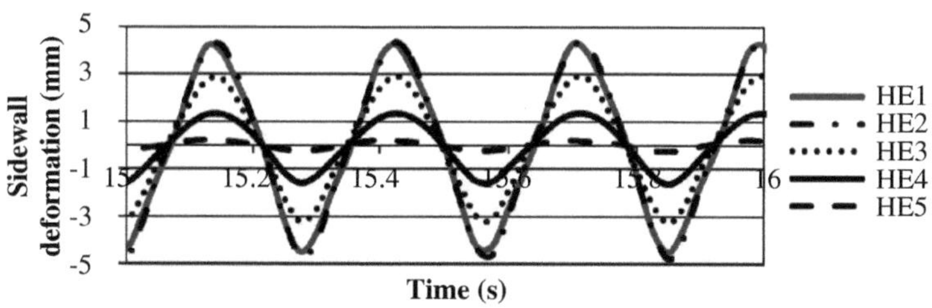

Fig. 29.13 Closer view of cyclic horizontal deformations measured at the *left* sidewall

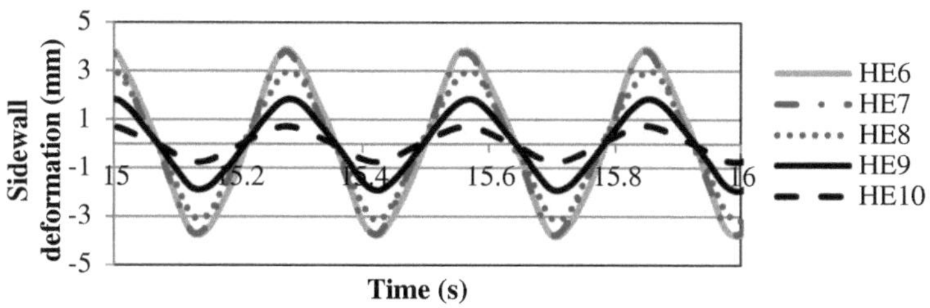

Fig. 29.14 Closer view of cyclic horizontal deformations measured at the *right* sidewall

29.3.2 Culvert Deformations

There were five pairs of horizontal extensometers at the sidewalls of the culvert model. They were labeled from HE1 to HE10 as shown in Fig. 29.12. These extensometers measured the cyclic horizontal deformations during the centrifuge test. Figures 29.13 and 29.14 show the deformation time histories at the sidewalls of the culvert model. Although cyclic deformations at the left and right sidewalls were not perfectly equal to each other, records are reasonably consistent. Slight differences may arise from the calibration and sensitivity of extensometers. Deformations increase with a decreasing rate from the bottom to the top of the slab of the culvert; the measurement of the displacement mode on the culvert sidewalls show the expected opposite phase between reciprocal extensometers (Fig. 29.15).

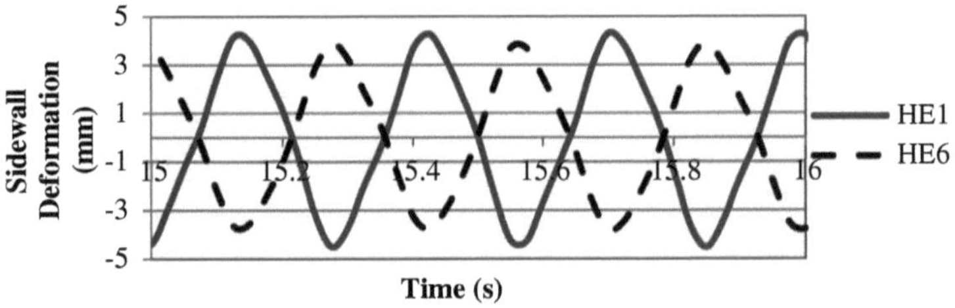

Fig. 29.15 Comparison of deformations between reciprocal extensometers HE1 & HE6

Table 29.6 Comparison of maximum horizontal racking deformations obtained from centrifuge tests and analytical estimates

Culvert model	Δ_{BOBET}(mm)	$\Delta_{PENZIEN}$(mm)	Δ_{CENT}(mm)	$\Delta_{BOBET}/\Delta_{CENT}$	$\Delta_{PENZIEN}/\Delta_{CENT}$
Thinnest sidewalls	7.23	7.26	4.5	1.61	1.83
Intermediate thickness	2.76	1.26	1.7	1.62	0.74
Thickest sidewalls	1.60	0.38	0.78	2.05	0.49

Maximum horizontal racking deformations measured at the top of the sidewalls in the centrifuge tests are compared in Table 29.6 with the closed form solutions proposed by Penzien (2000) and Bobet et al. (2008).

29.4 Summary and Conclusions

In this study, the dynamic response of underground structures was examined under harmonic motions by performing centrifuge tests. For this purpose, three different models having different rigidities were analyzed subjected to a 40 g centrifugal field. The results show that there is an opposite phase between reciprocal extensometers. A clear racking deformation scheme was confirmed with the theoretical predictions.

Measured lateral deformations of the sidewalls were compared with the predictions of Penzien (2000) and Bobet et al. (2008). Both approaches overestimate maximum horizontal deformations compared to those of the tests for the model with the thinnest sidewalls. The deformations of the models with intermediate thickness and thickest sidewalls are underestimated in Penzien's (2000) approach, while are overestimated by Bobet's et al. procedure (2008).

The results obtained in this research would provide a means for validation of numerical analyses.

Acknowledgements The research leading to these results has received funding from the European Community's Seventh Framework Programme [FP7/2007–2013] for access to the IFSTTAR Centrifuge, Nantes, France under grant agreement n° 227887. The financial support from IFSTTAR is gratefully acknowledged.

References

ASCE (1974) Earthquake damage evaluation and design considerations for underground structures. Los Angeles

Bobet A, Fernández G, Huo H et al (2008) A practical iterative procedure to estimate seismic induced deformations of shallow rectangular structures. Can Geotech J 45(7):923–938

Che A, Iwatate T (2002) Shaking table test and numerical simulation of seismic response of subway structures. Structures under shock and impact VII 11(7):367–376

Clough RW, Penzien J (1993) Dynamics of structures, 2nd edn. McGraw Hill, California

Dowding CH, Rozen A (1978) Damage to rock tunnels from earthquake shaking. J Geotech Eng Div 104(2):175–191

Duke CM, Leeds DJ (1959) Effects of earthquakes on tunnels. In: O'Sullivan JJ (ed) Protective construction in nuclear age. Proceedings 2nd Protective Construction Symp., Santa Monica Calif., March 1959.Vol 1,Macmillan co, New York, p 27

Escoffier S (2008) Conteneur ESB. LCPC Internal report no: 2007-1-13-1/1-a, p 106

Hashash Y, Hook JJ, Schmidt B (2001) Seismic design and analysis of underground structures. Tunn Undergr Space Technol 16(4):247–293

Huo H (2005) Seismic design and analysis of rectangular underground structures. Dissertation, Purdue University

Huo H, Bobet A, Fernandez G et al (2006) Analytical solution for deep rectangular underground structures subjected to far field shear stresses. Tunn Undergr Space Technol 21(6):613–625

Iglesia GR, Einstein HH, Whitman RV (2011) Validation of centrifuge model scaling for soil systems via trapdoor tests. J Geotech Geoenviron Eng 137(11):1075–1089

Iida H, Hiroto T, Yoshida N et al (1996) Damage to Daikai subway station. Soils and Foundations, Japanese geotechnical society, Special issue on geotechnical aspects of the 17 January 1995 Hyokogen-Nambu earthquake, pp 283–300

Owen GN, Scholl RE (1981) Earthquake engineering of large underground structures. Federal highway administration and national science foundation, Report no. FHWA_RD-80_195

Penzien J (2000) Seismically induced racking of tunnel linings. J Earthq Eng Struct Dyn 29(5):683–691

Power MS, Rosidi D, Kaneshiro JY (1998) Seismic vulnerability of tunnels and underground structures revisited. In: Ozdemir L (ed) Proceedings of the north American tunneling conference. Elsevier, Long Beach, p 18

Sharma S, Judd WR (1991) Underground opening damage from earthquakes. Eng Geol 30:263–276

Stevens PR (1977) A review of the effects of earthquakes on underground mines. United States Geological Survey Open File Report 77–314, Reston, Virginia, p 47

Tamari Y, Towhata I (2003) Seismic soil-structure interaction of cross sections of flexible underground structured subjected to soil liquefaction. Soils Found 43(2):69–87

Wang JN (1993) Seismic design of tunnels. Parsons Brinckerho!, Inc., Monograph 7

Chapter 30
Dynamic Response of Shallow Rectangular Tunnels in Sand by Centrifuge Testing

Grigorios Tsinidis, Emmanouil Rovithis, Kyriazis Pitilakis and Jean Louis Chazelas

30.1 Introduction

Large underground structures (e.g. subways, metro stations, underground parking lots, utility tunnels) possess a vital socio-economic role as they are a crucial part of the transportation and utility networks of urban areas. The associated impact in case of a strong earthquake denotes the paramount importance of a safe seismic design, especially in seismically active areas.

Although recent earthquake events (Kobe earthquake 1995; Duzce earthquake 1999; Chi-Chi earthquake 1999; Wenchuan earthquake 2008) have demonstrated that underground structures may undergo extensive deformations or even collapse (e.g. Sharma and Judd 1991; Wang 1993; Iida et al. 1996), their seismic response has been little explored compared to aboveground structures, due to the lack of well-documented experimental data and field evidence (Cilingir and Madabhushi 2011). In this regard, design specifications in modern seismic codes are based primarily on simplified methods (Wang 1993; Penzien 2000; Hashash et al. 2001; ISO 23469 2005; FHWA 2009), the implementation of which may lead to a substantially different seismic design for underground structures (Pitilakis and Tsinidis 2014).

A significant contribution to the seismic behavior of this type of structures is provided by means of well-focused experimental data, allowing the investigation of

G. Tsinidis (✉) · K. Pitilakis
Department of Civil Engineering, Research Unit of Soil Dynamics and Geotechnical Earthquake Engineering, Aristotle University of Thessaloniki, PO BOX 424, 541 24 Thessaloniki, Greece
e-mail: gtsinidi@civil.auth.gr

E. Rovithis
Institute of Engineering Seismology & Earthquake Engineering (EPPO-ITSAK)
Dassyliou Str, Eleones, Pylaia, PO BOX 53, 551 02 Thessaloniki, Greece

J. L. Chazelas
IFSTTAR French Institute of Science and Technology for Transport, Development and Networks, Route de Bouaye BP4129, Bouguenais, 44341, Nantes, France

F. Taucer, R. Apostolska (eds.), *Experimental Research in Earthquake Engineering*, Geotechnical, Geological and Earthquake Engineering 35,
DOI 10.1007/978-3-319-10136-1_30

crucial response parameters such as seismic earth pressures distribution on the side walls of the structure and seismic shear stresses around the structure.

Along these lines, the Transnational Access project DRESBUS II, funded by the research project SERIES, dealt with the investigation of the seismic response of shallow rectangular tunnels by dynamic centrifuge testing. The experimental study was carried out at the geotechnical centrifuge facility of IFSTTAR at a centrifuge acceleration of 40 g. Well-documented experimental data was acquired for a wide set of soil-tunnel systems, allowing a better understanding of the seismic behavior of underground structures as affected by soil-structure relative flexibility, soil-tunnel interface properties, soil saturation and amplitude of the input motion.

30.2 Dynamic Centrifuge Tests

30.2.1 IFSTTAR Centrifuge Facility

DRESBUS II project was hosted by the centrifuge facility of IFSTTAR in Nantes, France. The IFSTTAR centrifuge has a radius of 5.5 m and a capacity of 2 t under a centrifugal acceleration of 100 g (Fig. 30.1a). The dimensions of the swinging basket supporting the model are 1.4 × 1.1 m. For dynamic testing, the input motion is applied at the base of the model through the specially designed actuator Actidyn QS 80 (Chazelas et al. 2008), which is able to impose both sinusoidal and real earthquake excitations up to 400 kg of payload mass (Fig. 30.1b). Some functional limits of Actidyn QS 80 are: functional under centrifugal acceleration up to 80 g, peak acceleration of the input motion up to 0.5 g and frequency range at 30–300 Hz for real earthquake excitations.

A large Equivalent Shear Beam (ESB) container was used to mount the models, having inner dimensions of 800 mm in length, 340 mm in width and 409 mm in depth. The ESB container is designed to match the shear stiffness of contained soils

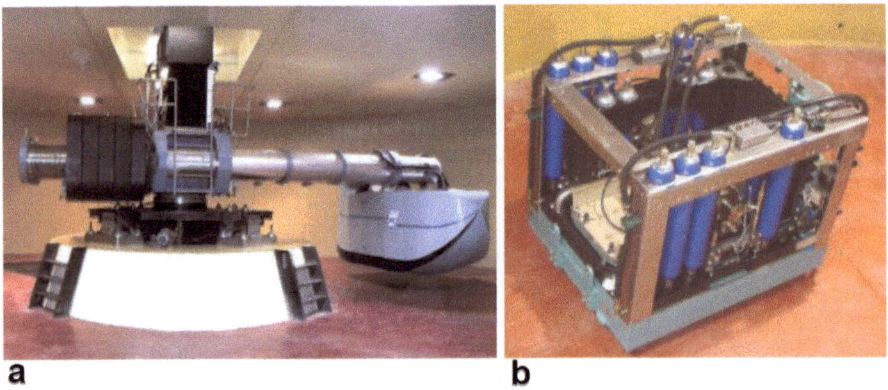

Fig. 30.1 **a** IFSTTAR centrifuge, **b** ACTIDYN actuator

for the desirable range of shear strains, thus minimizing spurious boundary effects due to soil-container interactions.

30.2.2 Properties of the Soil-Tunnel System

Soil models were composed by uniform Fontainebleau NE34 sand with a relative density of 70 %. Mechanical properties for the specific sand are presented in Table 30.1. Both dry and saturated sandy soils were tested within this experimental campaign.

Tunnel models were manufactured by aluminum alloy, the mechanical properties of which are summarized in Table 30.2. Tunnel dimensions (Fig. 30.2) were deliberately chosen to model desirable soil-to-structure relative flexibility (rigid or flexible structure) and interface characteristics (rough or smooth interface). For this reason, four model sections were finally manufactured and tested. Rough soil-tunnel interface was modeled by small grooves on the surface of the tunnels, based on the sand granulometry and the dimensions of the tunnel sections. The inside dimensions of the models were kept constant allowing the use of identical extensometers as described in Sect. 30.2.4. Based on the adopted centrifuge scaling factor (N=40), the flexible tunnel models correspond to 1.88 × 2 m sections with a wall thickness of 8 cm and a slab thickness of 32 cm in prototype scale (assuming E=30 GPa for

Table 30.1 Sand mechanical properties

	ρ_s (g/cm^3)	e_{max}	e_{min}	d_{50} (mm)
Fontainebleau sand NE34	2.64	0.86	0.55	0.200

Table 30.2 Tunnel models mechanical properties

Unit weight γ(kN/m^3)	Elastic modulus E (GPa)	Poisson ratio v	Yield stress (MPa)
2.7	71.0	0.33	400

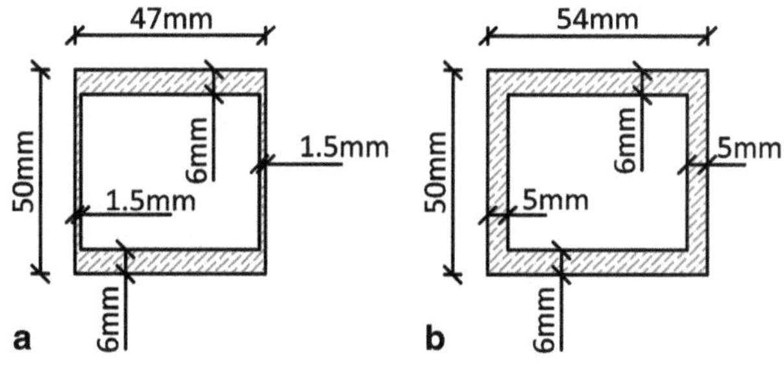

Fig. 30.2 Tunnel sections: **a** flexible tunnels, **b** rigid tunnels

the equivalent concrete material). Accordingly, the rigid tunnel models correspond to 2.16 × 2 m concrete sections of 27 cm wall thickness and 30 cm slab thickness.

30.2.3 Model Preparation

An automatic hopper system was employed to form the soil deposit within the ESB container in a piecewise manner (Fig. 30.3). During model formation, the tunnel section and the recording devices were embedded at the desirable locations.

Referring to saturated tests, the saturation procedure is schematically shown in Fig. 30.4. During the saturation phase, a lid was added on the ESB container and the liquid was injected through the base plate of the box. The saturation procedure was performed under partial vacuum in steps and controlled by a specially designed software. To overcome the discrepancy between consolidation time scaling and

Fig. 30.3 Automatic hopper system—phases of model preparation

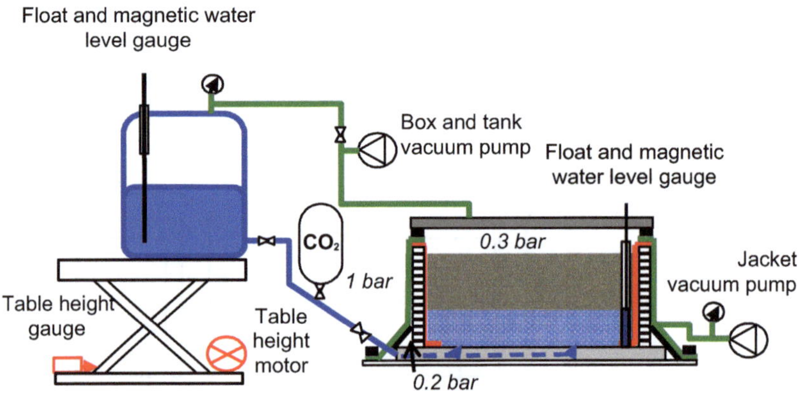

Fig. 30.4 Saturation set up

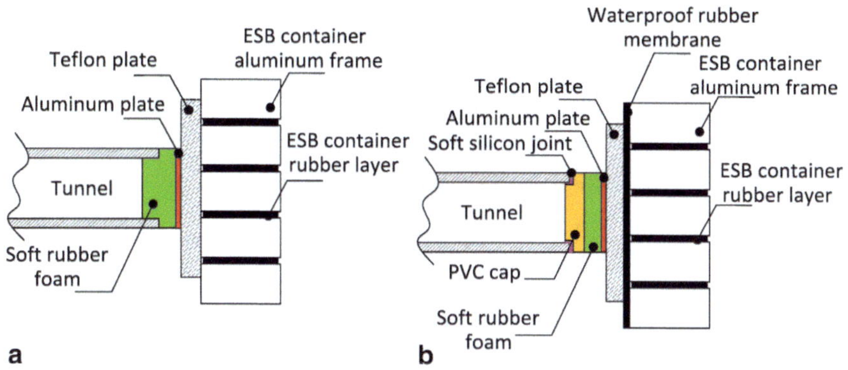

Fig. 30.5 Details of tunnel models-ESB container connections, **a** dry tests, **b** saturated tests

loading time period of dynamic events (Schofield 1981), a methylcellulose-water mixture was utilized as a pore fluid, with an appropriately elevated viscosity.

The tunnel boundaries were properly designed to provide sufficient support and waterproof, ensuring the plane strain behavior of the model (Fig. 30.5).

30.2.4 Model Layout—Instrumentation

A particularly dense monitoring scheme was designed to record the seismic response of the soil-tunnel system (Fig. 30.6). The instrumentation scheme was composed of piezoelectric accelerometers in vertical arrays within the soil or attached to the tunnel section and the ESB container, displacement sensors to record the surface ground settlement and pore pressure sensors to measure the pore pressure dissipation for the saturated cases. Furthermore, specially designed extensometers were employed to record side-walls deformations of the tunnel sections while diagonal "blade" extensometers were installed along the longitudinal axes of the tunnels to check the out-of-plane response of the model.

30.2.4.1 "Fork" Recording Device

Horizontal deformations of the tunnel side-walls were recorded by means of a "fork" measuring device, comprising of two groups of five extensometers (Fig. 30.7). Each group was installed on the internal face of each side wall, allowing measurements of the lateral deformations relative to the base slab. Each extensometer is equipped with two strain gauges mounted in half-bridge, forming in this manner a bending deformation sensor able to measure deformations up to 1 mm.

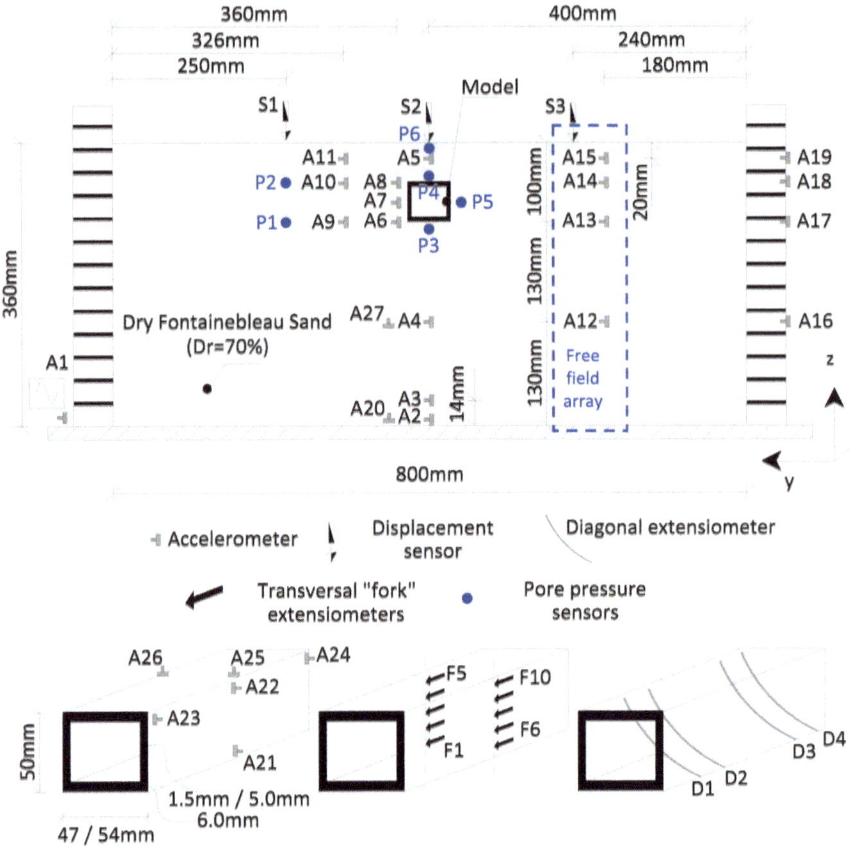

Fig. 30.6 Typical model layout—instrumentation

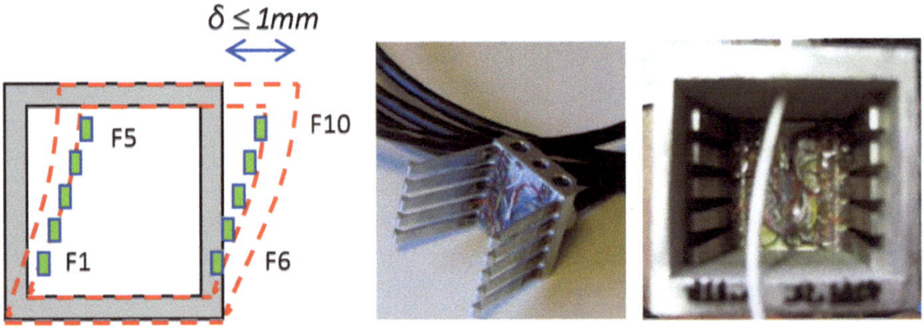

Fig. 30.7 Extensometers system to record the tunnel side-walls deformations

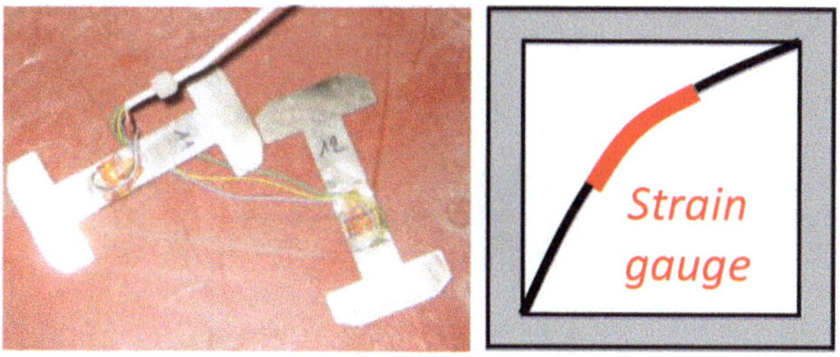

Fig. 30.8 Diagonal extensometers

30.2.4.2 Diagonal Extensometers

During centrifuge tests, plain strain behavior of the tunnel was confirmed by diagonal extensometers that were installed along the model's longitudinal axis. These sensors were made of arch-shaped pre-stressed steel blades (Fig. 30.8) equipped with strain gauges measuring diagonal deformations.

30.2.5 Centrifuge Testing Program

A series of seven centrifuge tests were carried out in total by combining flexible or rigid tunnel sections, smooth or rough soil-tunnel interface and dry or saturated sand. The test cases are summarized in Table 30.3.

30.2.6 Experimental Procedure

During each flight, the centrifuge was spun up to 40 g and was then swung down to 1 g (consolidation-stabilization circles) for functional reasons of the monitoring

Table 30.3 DRESBUS II testing program

Test case #	Structure flexibility	Soil Dr (%)	Soil saturation	Culvert surface
1	Flexible	70	Dry	Rough
2	Flexible		Dry	Smooth
3	Rigid		Dry	Rough
4[a]	Rigid		Saturated	Rough
5[b]	Rigid		Saturated	Rough
6	Rigid		Dry	Smooth
7	Rigid		Saturated	Smooth
8	Rigid		Saturated	Rough

[a] Failed
[b] Repetition of test 8 to check repeatability

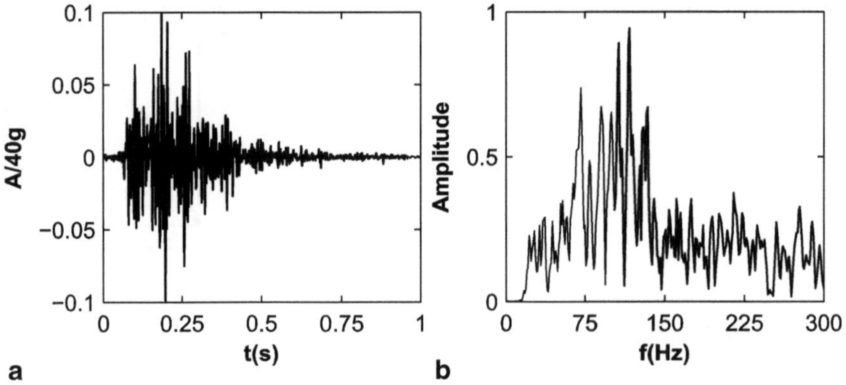

Fig. 30.9 Northridge earthquake record: **a** acceleration time history, **b** fourier spectrum

Table 30.4 Input motion characteristics (bracketed values in prototype scale)

	Input type	Nominal amplitude (g)	Nominal duration (s)
EQ1	Northridge earthquake record	4 [0.10]	1.0 [40]
EQ2		8 [0.20]	
EQ3		12 [0.30]	
EQ4	Pseudo-harmonic (85 Hz)	12 [0.30]	0.64 [25.6]

scheme. For the dry tests, a CPT test was performed away from the tunnel section to check repeatability of soil properties. During dynamic centrifuge testing, acceleration, deformation and pore pressure recordings were acquired at a sampling frequency of 12.8 kHz by means of a fast acquisition system available at IFSTTAR. Soil surface settlement was recorded at a lower (non periodic) sampling frequency.

Each soil-tunnel system was subjected to four input motions that were successively imposed at the base of the model, corresponding to a real record from the 1994 Northridge earthquake (Fig. 30.9) scaled to 0.1, 0.2 and 0.3 g peak acceleration followed by a sine wavelet at 85 Hz and 0.3 g. Table 30.4 summarizes the nominal input motion characteristics in model and prototype scale (values in brackets).

30.3 Interpretation of Experimental Data

To interpret the experimental recordings, acceleration and tunnel deformation time histories were band-pass filtered by considering lower and higher filter frequencies at 20 and 400 Hz, respectively, in model scale. Displacement time histories were obtained by double integrating the acceleration recordings, while soil stress-strain loops were derived by implementing adjacent pairs of accelerations recordings, following Zeghal and Elgamal (1994) and Brennan et al. (2005). The computed stress-strain loops were then employed to obtain the distribution of shear wave propagation velocity with depth based on standard elastodynamic considerations.

30.3.1 Free-Field Horizontal Acceleration

Figure 30.10 depicts the maximum horizontal acceleration recorded along the free-field array (Fig. 30.6) under EQ1 and EQ2 input motions. Both dry and saturated cases are shown, denoting minor of seismic motion amplification within the soil deposit. It should be noted that a considerable scatter is observed in the saturated test cases.

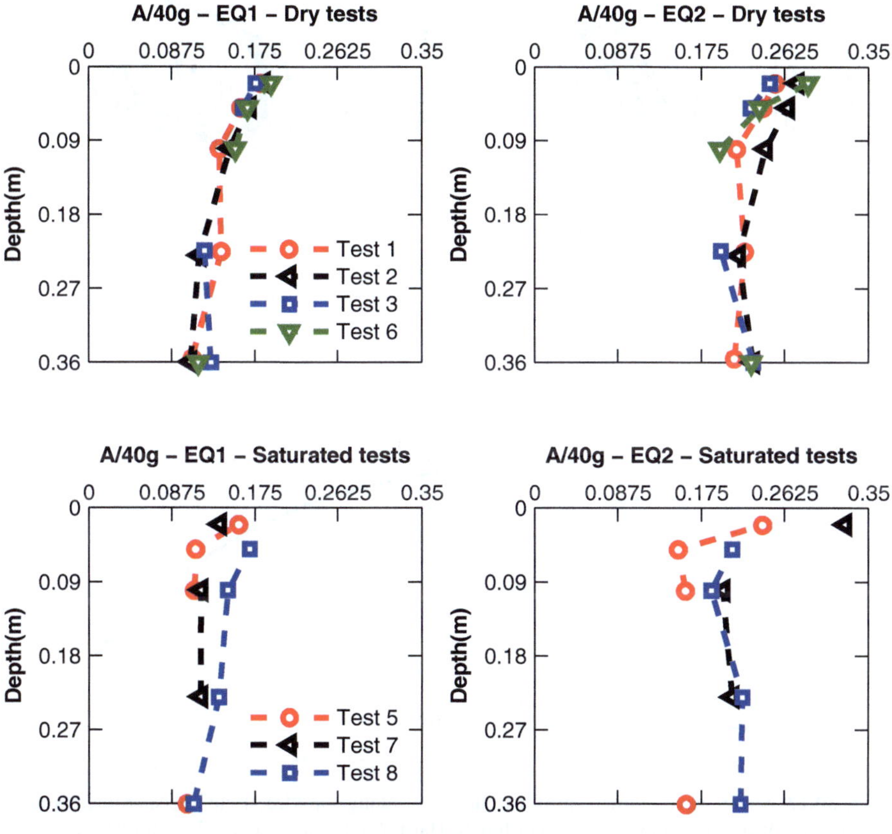

Fig. 30.10 Maximum horizontal acceleration along the soil “free-field” array for EQ1 and EQ2

30.3.2 Experimental Shear Wave Propagation Velocity

Figure 30.11 presents experimentally obtained stress-strain loops according to Zeghal and Elgamal (1994) approach. The mean value of the soil shear modulus is also plotted. The corresponding shear wave propagation velocity profiles are shown in Fig. 30.12, referring to Test case 1. In the same graph, the prediction of Hardin and Drnevich (1972) formulation for the small-strain shear wave velocity profile (V_{so}) is also shown. The comparison indicates a clear reduction of soil shear stiffness with increasing amplitude of the input motion. It should be mentioned that the above data should be viewed as a rough estimation of $V_s(z)$ since the settlement of soil that was observed during the centrifuge tests could have changed the relative distance between the sensors, thus altering the presented results.

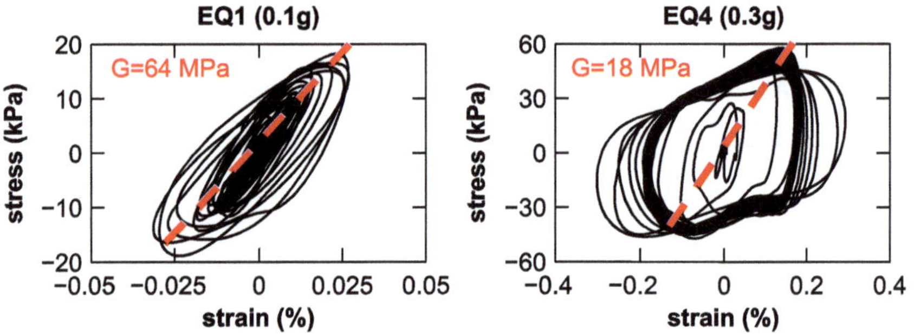

Fig. 30.11 Experimental stress-strain loops according to Zeghal and Elgamal (1994) approach. (The *dashed red lines* refer to a mean value of the corresponding shear modulus of soil)

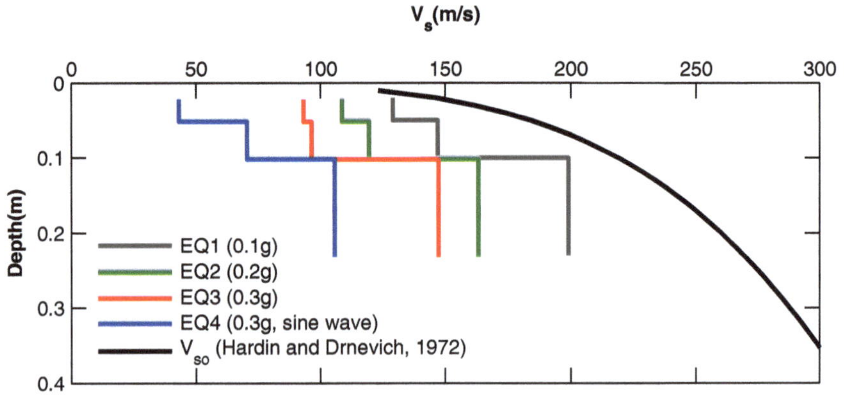

Fig. 30.12 Shear wave velocity profiles computed along the "free-field" accelerometers array (Test case 1); comparison with V_{so} according to Hardin and Drnevich (1972) formulation

30.3.3 Tunnel Deformations

Time windows of deformation time histories recorded on the side walls of a flexible tunnel (Test case 2) are presented in Fig. 30.13, referring to EQ1 and EQ4 input motions. The observed out-of-phase response denotes a symmetrical racking response of the tunnel section. As expected, wall deformations are amplified from the bottom slab to the top slab. A similar response was observed in the other test cases.

Representative diagonal deformations recorded at several locations along the longitudinal axis of a flexible tunnel (Test 2) are presented in Fig. 30.14, indicating plane strain behaviour of the model tunnel.

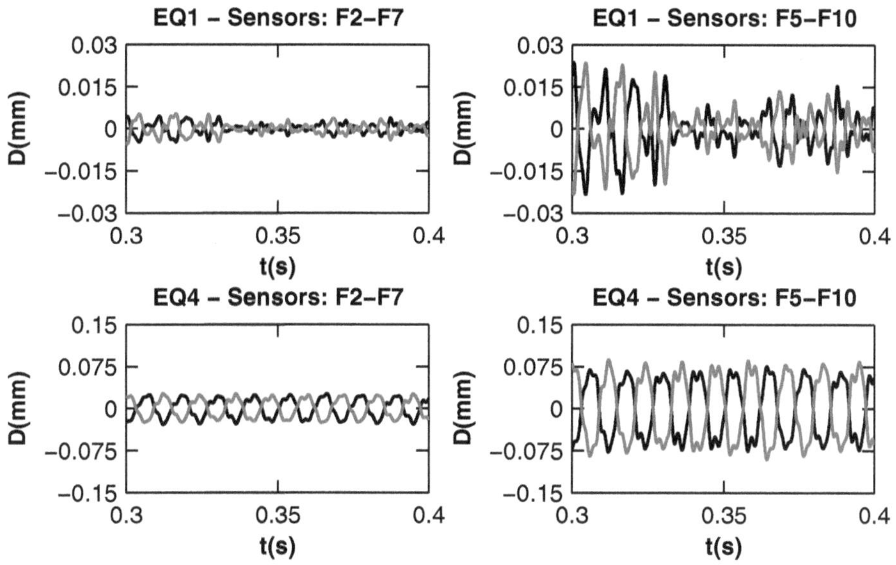

Fig. 30.13 Time windows of walls deformations time histories, recorded at different levels; results for Test 2, EQ1 and EQ4

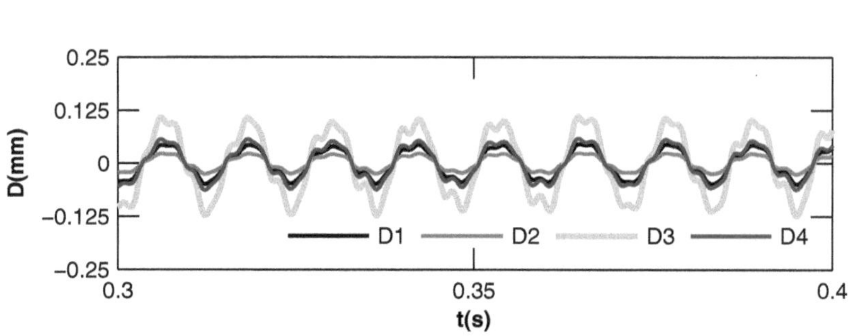

Fig. 30.14 Time window of typical diagonal deformation time histories, recorded at different locations along the tunnel axis; results for Test 2, EQ4

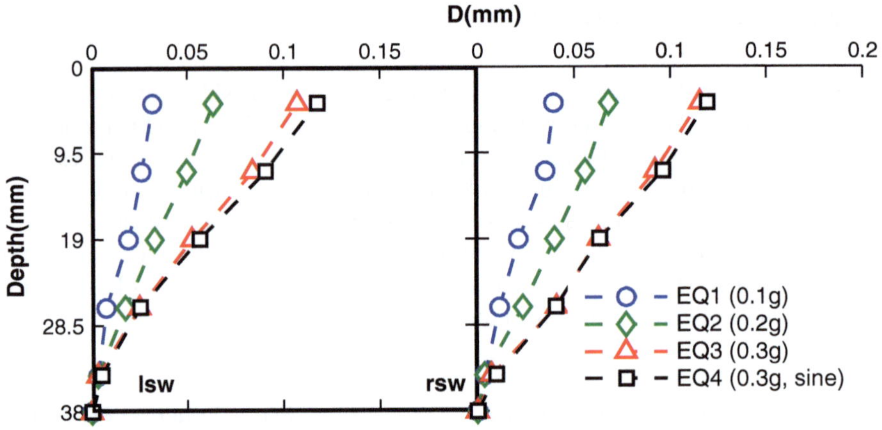

Fig. 30.15 Maximum wall deformations for different input motion amplitudes; flexible tunnel in dry sand (Test 2) (*lsw* left side wall, *rsw* right side wall)

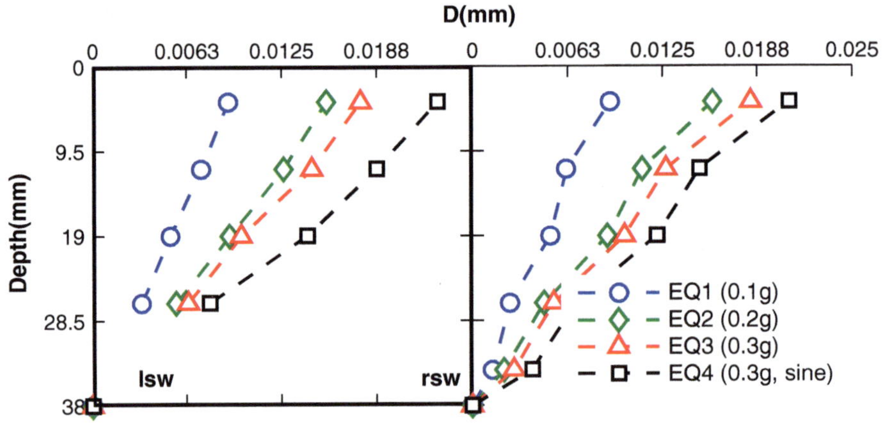

Fig. 30.16 Maximum wall deformations for different input motion amplitudes; rigid tunnel in dry sand (Test 3) (*lsw* left side wall, *rsw* right side wall)

Figures 30.15 and 30.16 show the effect of the amplitude of input motion on the maximum racking deformations of the tunnels side walls for a flexible and a rigid tunnel, respectively. It is observed that racking deformations are increased with increasing amplitude of the base excitation. The almost symmetrical response provides experimental evidence on the theoretical assumption of racking deformation mode of rectangular tunnel sections under shear wave excitation. Evidently, flexible tunnel sections showed higher deformation amplitudes compared to the rigid sections.

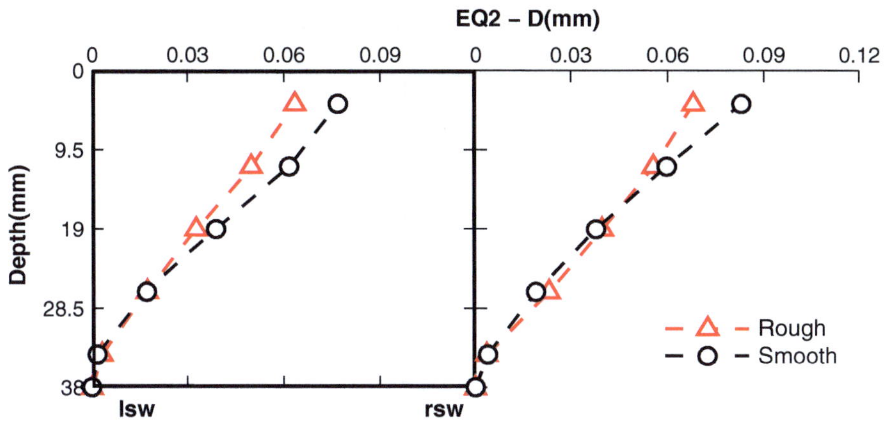

Fig. 30.17 Maximum wall deformations for flexible tunnels in dry sand; effect of interface characteristics (Test 1 vs. Test 2) (*lsw* left side wall, *rsw* right side wall)

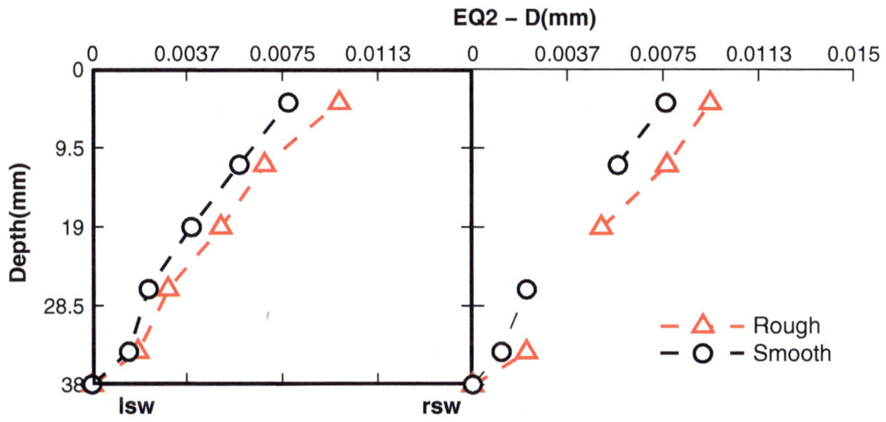

Fig. 30.18 Maximum wall deformations for rigid tunnels in saturated sand; effect of interface characteristics (Test 7 vs. Test 8) (*lsw* left side wall, *rsw* right side wall)

Figure 30.17 compares the maximum wall deformations recorded on flexible tunnels in dry sand for different soil-tunnel interface characteristics. Similar results are shown in Fig. 30.18 for the rigid tunnels embedded in saturated sand. The above comparisons indicate a minor effect of the interface properties on racking deformations of the tunnel.

Figure 30.19 presents the effect of sand saturation on wall deformations of a rigid tunnel with rough external face. The wall deformations were found systematically larger for the dry sand cases. This observation may be attributed to the higher shear stiffness of the dry sand compared to the saturated case.

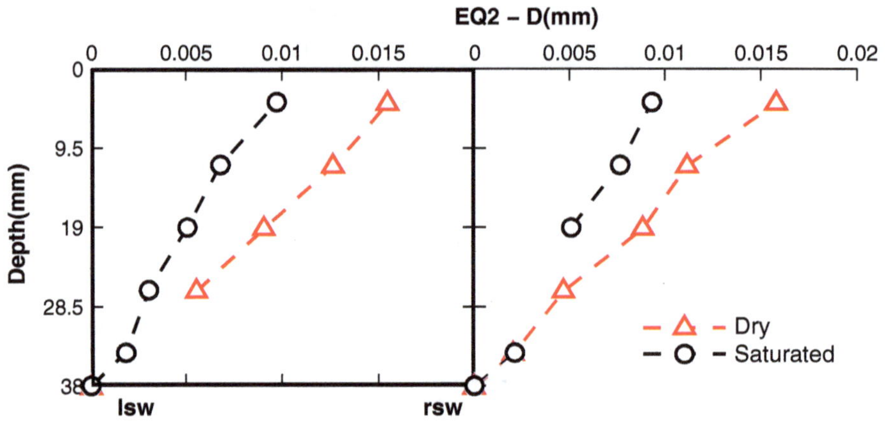

Fig. 30.19 Maximum walls deformations of a rigid tunnel as affected by the soil saturation (Test 3 vs. Test 8)

30.4 Conclusions

A series of dynamic centrifuge tests on rectangular tunnel-models embedded in both dry and saturated sands was presented as part of the DRESBUS II TA action of the SERIES project. The main experimental procedures and data processing methods were briefly described and a set of representative recordings was presented. A preliminary interpretation of recorded data revealed the following:

- Maximum soil horizontal acceleration was slightly amplified within the soil deposit especially for the dry test cases. This amplification was less evident for the saturated tests probably due to the higher variation of the soil stiffness compared to the dry cases.
- The horizontal deformations of the tunnel side walls were recorded in a symmetrical manner proving the theoretical assumption of racking distortion mode. These deformations were increased with increasing amplitude of the base excitation.
- The diagonal extensometers recordings revealed an in-phase response along the longitudinal axis of the tunnel ensuring plane strain behavior of the model sections.
- Rigid tunnels were understandably less deformed during shaking.
- The experimental data revealed a minor effect of the interface properties on the tunnels racking deformation.
- Higher horizontal deformations of the tunnel walls were recorded for the dry sand tests. This may be attributed to the higher shear stiffness of dry sand compared to the saturated case.

Acknowledgments The research leading to these results has received funding from the European Community's Seventh Framework Programme [FP7/2007–2013] for access to the IFSTTAR Centrifuge, Nantes, France under grant agreement n° 227887.

References

Brennan AJ, Thusyanthan NI, Madabhushi SPG (2005) Evaluation of shear modulus and damping in dynamic centrifuge tests. J Geotech Geoenvironmental Eng ASCE 131(12):1488–1497
Chazelas JL, Escoffier S, Garnier J, Thorel L, Rault G (2008) Original technologies for proven performances for the new LCPC earthquake simulator. Bull Earthq Eng 6:723–728
Cilingir U, Madabhushi SPG (2011) A model study on the effects of input motion on the seismic behavior of tunnels. J. Soil Dyn Earthq Eng 31:452–462
FHWA (2009) Technical manual for design and construction of road tunnels—Civil elements. US Department of transportation. Federal Highway Administration. Publication No. FHWA-NHI-09-010, p 629
Hardin BO, Drnevich VP (1972) Shear modulus and damping in soils: design equations and curves. J Soil Mech Found Div 98(SM7):667–692
Hashash YMA, Hook JJ, Schmidt B, Yao JI-C (2001) Seismic design and analysis of underground structures. Tunn Undergr Sp Technol 16(2):247–293
Iida H, Hiroto T, Yoshida N, Iwafuji M (1996) Damage to Daikai subway station. Soils and foundations, Special Issue on Geotechnical aspects of the January 17 1995 Hyogoken-Nambu Earthquake. Japanese Geotechnical Society, 283–300
ISO 23469 (2005) Bases for design of structures—Seismic actions for designing geotechnical works. ISO International Standard. ISO TC98/SC3/WG10
Penzien J (2000) Seismically induced racking of tunnel linings. Earthq Eng Struct Dyn 29:683–691
Pitilakis K, Tsinidis G (2014) Performance and Seismic Design of Underground Structures. In: Maugeri M and Soccodato C (eds) Earthquake Geotechnical Engineering Design. Geotechnical Geological and Earthquake Engineering, 28, pp. 279–340, Springer International Publishing, Switzerland, doi:10.1007/978-3-319-03182-8_11
Schofield AN (1981) Dynamic and earthquake geotechnical centrifuge modelling. In Proc int conf Recent Advances in Geotechnical Earthquake Engineering and Soil Dynamics, Rolla, USA, 26 Apr–3 May 1981, University of Missouri-Columbia, Columbia
Sharma S, Judd WR (1991) Underground opening damage from earthquakes. Eng Geol 30:263–276
Wang JN (1993) Seismic design of tunnels. A simple state-of-the-art approach. Parson Brinckerhoff, New York
Zeghal M, Elgamal AW (1994) Analysis of site liquefaction using earthquake records. J Geotech Eng ASCE 120(6):996–1017

References

[illegible]

Chapter 31
Centrifuge Modelling of the Dynamic Behavior of Square Tunnels in Sand

Grigorios Tsinidis, Charles Heron, Kyriazis Pitilakis and Gopal S. P. Madabhushi

31.1 Introduction

Underground structures constitute significant components of urban networks (subways, underground parking lots, utility tunnels etc). Generally, these types of structures were found less vulnerable than aboveground structures during past severe earthquakes. However, several cases of extensive damage and even collapse have been reported in the literature (e.g. Sharma and Judd 1991; Power et al. 1998; Iida et al. 1996; Hashash et al. 2001). These recent failures revealed some important weaknesses in current seismic design practices.

Contrary to the above ground structures, the kinematic loading imposed on the embedded structure from the surrounding soil prevails, while the inertial effects are often of secondary importance. Soil-structure interaction effects, expected to increase in cases of non-circular embedded structures, are closely related to two crucial parameters, namely (i) the relative flexibility between structure and ground and (ii) the interface characteristics between the structure and the surrounding soil. The exact effects of these parameters on the structural seismic response are not well known.

Several methods may be found in literature for the evaluation of the seismic response of underground structures (e.g. Wang 1993; Penzien 2000; Hashash et al. 2001; ISO 234692005; FHWA 2009 etc). The results of these methods may substantially deviate, even under the same assumptions, indicating the lack of knowledge regarding some very crucial issues that significantly affect the seismic response (Pitilakis and Tsinidis 2014). The lack of knowledge is even more pronounced for

G. Tsinidis (✉) · K. Pitilakis
Department of Civil Engineering, Research Unit of Soil Dynamics and Geotechnical Earthquake Engineering, Aristotle University of Thessaloniki, PO BOX 424, 541 24 Thessaloniki, Greece
e-mail: gtsinidi@civil.auth.gr

C. Heron · G. S. P. Madabhushi
Department of Engineering, Schofield Centre, University of Cambridge, High Cross, Madingley road, Cambridge CB3 0EL, UK

F. Taucer, R. Apostolska (eds.), *Experimental Research in Earthquake Engineering*,
Geotechnical, Geological and Earthquake Engineering 35,
DOI 10.1007/978-3-319-10136-1_31

non-circular structures. The seismic earth pressures on the side walls, the seismic shear stresses around the perimeter of the structure, the soil-structure interface characteristics, complex deformation modes during shaking (e.g. "rocking" for stiff structures or inward deformations for flexible structures) and the effect of the soil-structure relative flexibility on seismic response are, among others, issues that need further investigation.

To improve our knowledge on the aforementioned issues, a comprehensive set of tests was performed at the University of Cambridge (UCAM), jointly with the Aristotle University of Thessaloniki (AUTH), on square model-tunnels embedded in dry sand. The tests were carried out at the geotechnical centrifuge facility of the University of Cambridge, within the Transnational Access action of the SERIES EU research program (TA project: TUNNELSEIS). The high quality and perfectly controlled data will be used to validate advanced numerical models and to improve design methods. The experimental procedure and the setups of the studied cases are presented in the ensuing along with the some representative experimental results.

31.2 Dynamic Centrifuge Tests

The tests were performed at the "Turner beam centrifuge" of the University of Cambridge (Schofield 1980) under a centrifugal acceleration of 50 g. A specially designed Stored Angular Momentum (SAM) actuator (Madabhushi et al. 1998) was used to apply the earthquake input motions. The actuator is capable of inputting sinusoidal motions of fixed frequency and amplitude. A "sine-sweep" motion of decreasing amplitude and decreasing frequency can alternatively be created. A large Equivalent Shear Beam (ESB) container was used to mount the models (Schofield and Zeng 1992). The inside dimensions of the container are 673 × 255 × 427 mm.

The soil deposits were made of dry Hostun HN31 sand reconstituted at two different relative densities (about 50 and 90 %). The mechanical properties of the sand are tabulated in Table 31.1. The sand pouring was performed using an automatic hopper system (Madabhushi et al. 2006). During the sand pouring process the tunnel and all the embedded transducers were properly positioned in the models.

Two square model tunnels were manufactured for the tests, namely: a rigid one having a thickness of 2 mm and a more flexible one having a thickness of 0.5 mm (Fig. 31.1). The rigid model was made of an extruded section of 6063A aluminum alloy, while the flexible model was manufactured by wrapping a 33swg soft aluminum foil to form the square section. Wrapping the foil leaves an overlapping section along the axis of the tunnel where the foil was welded. This side was chosen to be the invert slab of the tunnel-model. The mechanical properties of the used aluminum alloys are tabulated in Table 31.2. Both models were 100 mm wide, while the length was changed, being 220 mm for the rigid model and 210 mm for the flexible one. The tunnels were actually shorter than the ESB container width to avoid any interaction of the tunnels with the container. Two PTFE rectangular plates were placed at each end of the tunnel to prevent sand from entering into the model-tunnel. Ac-

Table 31.1 Sand mechanical properties. (After Mitrani 2006)

	ρ_s (g/cm^3)	e_{max}	e_{min}	d_{10} (mm)	d_{50} (mm)	d_{60} (mm)	φ_{crit} (°)
Hostun HN31	2.65	1.01	0.555	0.209	0.335	0.365	33

a b c

Fig. 31.1 **a** Flexible model-tunnel. **b** rigid model-tunnel. **c** installation of the rigid tunnel in the ESB container

cording to the scale factor (N=50), the rigid model corresponds to a 5 × 5 m square tunnel having an equivalent concrete lining thickness equal to 0.13 m (assuming E=30 GPa for the concrete). For the flexible model the equivalent concrete lining thickness is equal to 0.03 m. The thicknesses are obviously unrealistic in practice as the design analysis for static loads will result in much thicker linings. However, this selection was made to study the effects of tunnel flexibility at extreme limits. Moreover, the selection was necessary to achieve clear measurements of the lining strains. To simulate more realistically the soil-structure interface, in the last two test cases, Hostun sand was stuck on the external face of the model-tunnels, creating a rough surface. Consequently, in these cases the shear stresses at the interface were expected to increase.

Figure 31.2 presents the model layout and the instrumentation scheme for all the tests, while Table 31.3 tabulates details for each centrifuge test (e.g. model-tunnel used, burial depth, input motions, tunnel external face roughness).

A dense instrumentation scheme was implemented to record the soil-tunnel systems response, comprising of miniature accelerometers, linear variable differential transformers (LVDTs), position sensors (POTs) and miniature total earth pressure

Table 31.2 Model-tunnels mechanical properties

Tunnel	Unit weight, γ (kN/m^3)	Elastic modulus, E (GPa)	Poisson ratio, v	Tensile strength, f_{bk} (MPa)
Rigid	2.7	69.5	0.33	220
Flexible				75

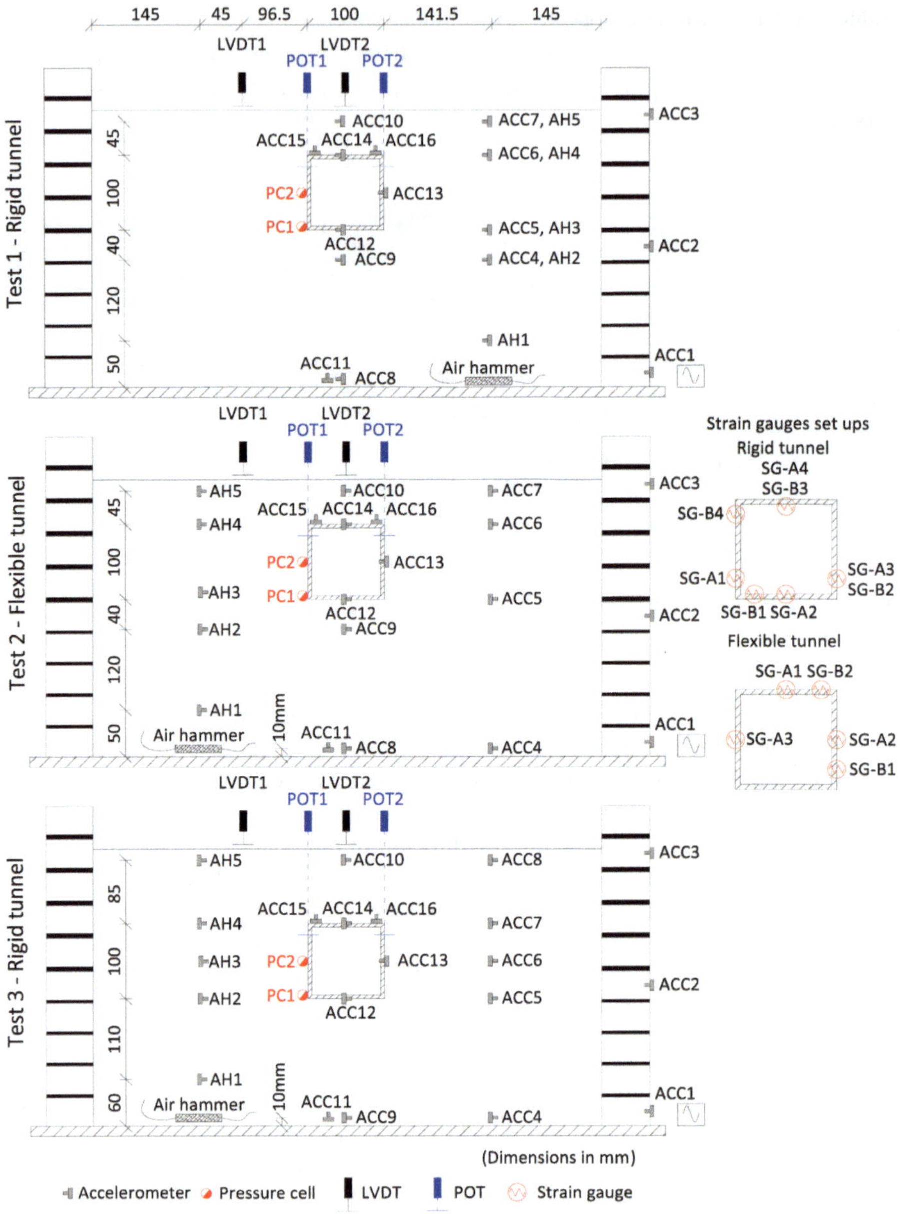

Fig. 31.2 Model setups and instrumentation schemes

cells (PCs) (Fig. 31.2). Resistance strain gauges were also used to measure the tunnel lining axial and bending moment strains at several locations. Unfortunately, the strain gauges did not worked properly during the first test due to a wiring problem, while they worked properly during the other test cases.

Table 31.3 Centrifuge tests details (bracketed values in prototype scale)

Test ID	Tunnel	Depth (mm)	D_r (%)	EQ ID	Flight	Input type[a]	Frequency (Hz)	Amplitude (g)	Nominal duration (s)
Test 1	Rigid—smooth external face	60	51	EQ1	1	PH	60 [1.2]	10.5 [0.21]	0.4 [20]
				EQ2	1	PH	60 [1.2]	12.9 [0.26]	0.4 [20]
				EQ3	1	PH	60 [1.2]	15.7 [0.31]	0.4 [20]
				EQ4	1	PH	60 [1.2]	18.32 [0.37]	0.4 [20]
Test 2	Flexible—rough external face	60	50	EQ1	1	PH	30 [0.6]	1.29 [0.026]	0.4 [20]
				EQ2	1	PH	45 [0.9]	4.0 [0.08]	0.4 [20]
				EQ3	2	PH	45 [0.9]	4.0 [0.08]	0.4 [20]
Test 3	Rigid—rough external face	100	89	EQ1	1	PH	30 [0.6]	1.0 [0.02]	0.4 [20]
				EQ2	1	PH	45 [0.9]	4.0 [0.08]	0.4 [20]
				EQ3	1	PH	50 [1]	6.5 [0.13]	0.4 [20]
				EQ4	1	PH	50 [1]	12.0 [0.24]	0.4 [20]
				EQ5	1	SS	60 [1.2]	12.0 [0.24]	3.0 [150]
				EQ6	2	PH	50 [1]	5.8 [0.116]	0.4 [20]
				EQ7	2	PH	50 [1]	6.0 [0.12]	0.6 [30]
				EQ8	2	PH	50 [1]	11.0 [0.22]	0.5 [25]

[a] *PH* Pseudo-Harmonic motion, *SS* Sine Sweep motion

To estimate the soil shear wave velocity air hammer tests were performed (Ghosh and Madabhushi 2002). For this purpose, a small air-hammer was introduced close to the base of the soil layer, while a set of accelerometers was placed above it, forming an array, allowing a record of the arrival times of the waves emanating from the air-hammer.

All the instruments were adequately calibrated before the tests. Regarding the strain gauges a special procedure was followed for their calibration, as described in Tsinidis et al. (2014).

During each test, the centrifuge was spun up in steps until 50 g and then the earthquakes were fired in a row, leaving some time between them to acquire the data. Several flights were performed during each test. The data was recorded at a sampling frequency of 4 Hz during the swing up and at 4 kHz during shaking, using the CDAQS (Centrifuge Data AcQuisition System).

As mentioned, air-hammer tests were conducted to evaluate the shear wave velocity (V_s) profile of the soil deposit. The V_s profiles were actually estimated based on the travel times of the waves between accelerometers that were placed at known distances apart. The travel times were estimated in a simplified way from the arrivals of the waves, produced by the air-hammer. To make sure that the arrival times were recorded at a high level of precision, the DasyLab software was used as the acquisition system for the air-hammer accelerometers array, allowing for a sampling frequency equal to 50 kHz.

The models were subjected to a total of 15 "earthquakes" during several flights. The earthquakes were generally pseudo-harmonic wavelets of different frequencies and amplitudes. During the third test a sine-sweep was also fired. The main characteristics of the input motions are tabulated in Table 31.3 both in model and prototype scale (bracketed values).

31.3 Experimental Results

31.3.1 Air Hammer Testing

Figure 31.3 presents representative shear wave velocity profiles estimated during Test 2 and Test 3 at 50 g after the completion of swing up ("initial" V_s profile). The results seem to be in relatively good agreement with the results of the empirical for-

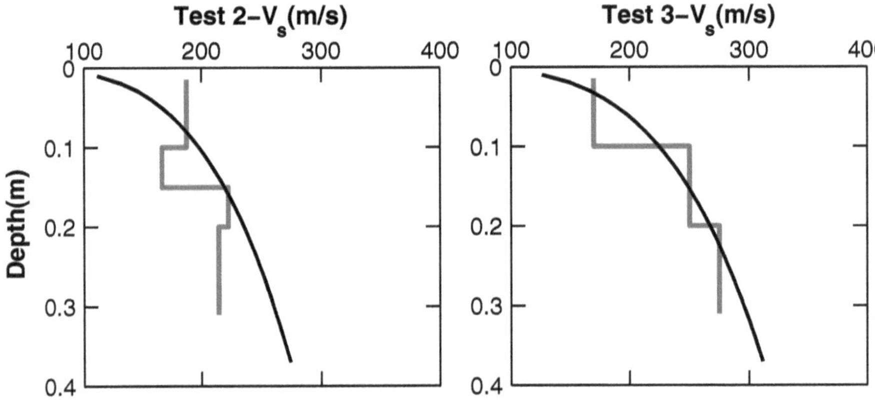

Fig. 31.3 V_s profiles computed from air hammer testing at 50 g (*grey solid line*). Comparison with empirical formulation of Hardin and Drnevich (1972) *(black solid line)*

mulation proposed by Hardin and Drnevich (1972), for the dense soil model (Test 3), while for the loose model (Test 2) the differences are higher. It is noteworthy that this is a rough estimation of the velocity, as unavoidable soil settlements during the tests may have changed the relative distances between sensors, consequently altering the results. These movements were actually much larger for the loose model as the soil settlements were higher in this case.

31.3.2 Static Response

During swing up the soil surface settled due to the soil densification caused by the increased self weight of the models. The recorded settlements were higher for the loose models, as expected. The earth pressures on the tunnels and the lining internal forces also increased due to the increase of the model self weight.

As described in detail in Sect. 31.3.4, the roof slab-right wall corner of the flexible tunnel was deformed inwards (buckled) during swing up, leading to the collapse of the tunnel during a subsequent earthquake.

31.3.3 Dynamic Response

In the following paragraphs the dynamic response of the soil-tunnel systems is briefly presented and discussed. To interpret the results, the recorded data was windowed neglecting the parts of the signals before and after the main duration of shaking, while a filtering procedure was conducted in the frequency domain. More specifically, acceleration time histories were filtered at 10 to 400 Hz using a band pass 8th order Butterworth filter. All the other data (e.g. soil settlements, earth pressures on the side-wall, lining strain-time histories etc) was filtered using a low pass 8th order Butterworth filter at 400 Hz.

31.3.3.1 Acceleration

Generally, the free-field acceleration was amplified towards the surface, while the amplification was affected by the tunnel presence. Figure 31.4 presents representative peak acceleration profiles across the tunnel depth, estimated along the tunnel and the soil "free field" arrays of accelerometers for all the tests. The peak values are actually computed as the average semi-amplitude of the cycles from the processed acceleration-time histories.

In Test 1 and Test 2, the maximum acceleration recorded on the tunnel (slabs and middle of the right side-wall) is found larger than the soil "free field" at the same depth. The opposite is observed in Test 3 (deeper tunnel embedded in denser sand). According to the simplified methodology proposed by Wang (1993), both the flexible and the rigid tunnels are characterized as quite "flexible" structures with respect

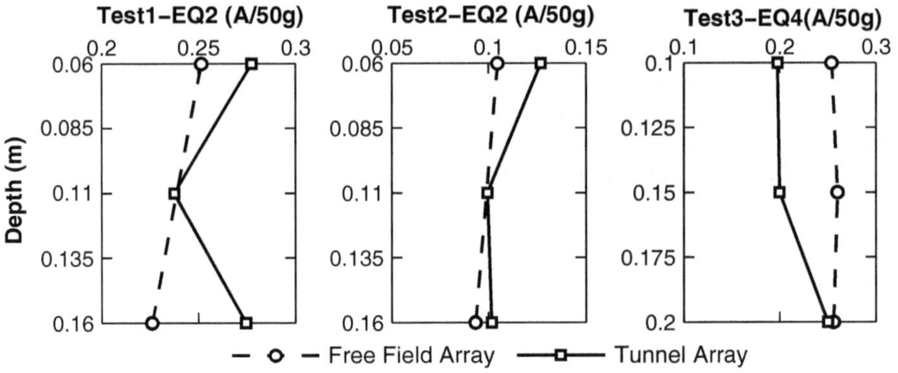

Fig. 31.4 Maximum acceleration along accelerometer arrays at tunnel depth (input motion amplitude A/50: Test 1: 0.26 g, Test 2: 0.08 g, Test 3: 0.24 g)

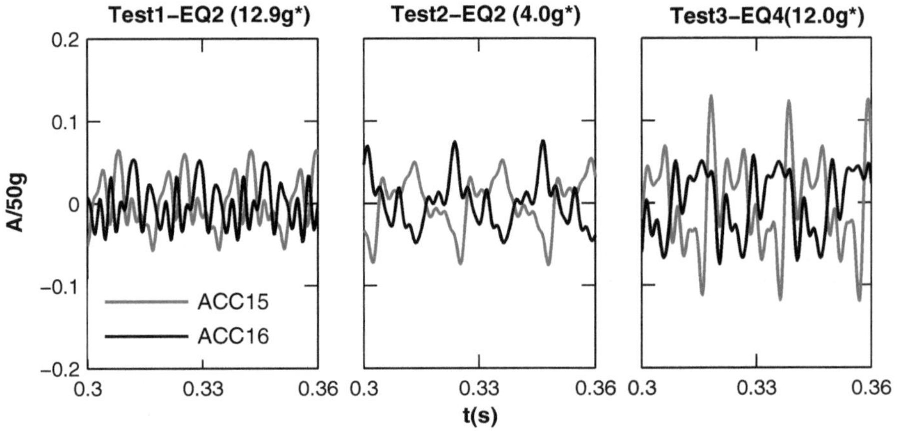

Fig. 31.5 Time windows of vertical acceleration time histories recorded on the tunnels roof slabs (*input motion amplitude)

to the surrounding soil, as the computed flexibility ratios F for the cases studied herein are much larger than 1 ($F > 100$). Theoretically, the acceleration on quite flexible tunnels is expected to be amplified compared to the soil "free field", as a flexible tunnel may be considered closer to the "unlined cavity" case. This hypothesis is verified by the experimental results of Test 1 and Test 2. The inconsistency for Test 3 may be attributed, to some extent, to the increased burial depth of the tunnel in this case. The reduced distance of the tunnel from the model base may cause larger interaction between the tunnel and the base, leading to the observed response. Further study, using the results of numerical simulations of the tests, is needed to better constrain the actual response.

Vertical accelerations were recorded on the sides of the tunnels roof slabs. Figure 31.5 presents windows of representative time histories of these vertical

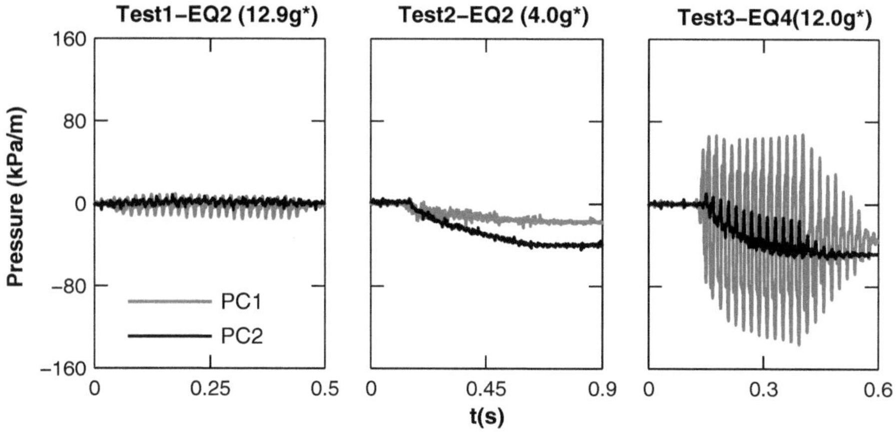

Fig. 31.6 Dynamic earth pressures on the left side wall of the model-tunnels (*input motion amplitude)

accelerations recorded during the tests. It is noted that for the flexible tunnel (Test 2) the records are not necessarily referring to the vertical component of the acceleration, as the tunnel was already deformed during shaking, as presented in detail in Sect. 31.3.4. The results for Test 1 and Test 3 refer to similar input motion amplitudes, while for Test 2 the input motion amplitude is reduced. In Test 2 and Test 3, the time histories are out of phase indicating a "rocking" mode of vibration for the tunnel, in addition to the classical racking mode. In Test 1 the recorded vertical accelerations did not clearly reproduce this out of phase response, indicating probably that this complex response was less pronounced in this case. The shear stresses mobilized along the soil-walls interface are affecting significantly this complex response. These stresses are controlled by the friction of the soil-tunnel interface and the soil confining pressures on the structure. The tunnel in Test 1 had a relatively smooth surface as we did not stick sand on the tunnel's surface. Moreover, the confining, pressures were reduced as the tunnel was buried in a shallow burial depth. To this end, the mobilized shear stresses at the interface were probably reduced leading to reduced "rocking" movement.

31.3.3.2 Earth Pressures

Figure 31.6 presents representative dynamic earth pressure time histories recorded on the left side-walls of the tunnels (middle of the wall and near the invert slab-wall corner). The recorded dynamic pressures were generally increased with increasing burial depth and increasing soil relative density (e.g. Test 3) both in terms of dynamic increments and residual values after shaking. For the flexible tunnel, the dynamic increments were found extremely reduced, while the residual values were quite large, even for the lower amplitude input motions, due to the high flexibility of the tunnel (inward deformations of the wall).

Generally, three stages were indentified for earth pressure response, namely a transient stage followed by a steady state stage and finally a post-earthquake residual stage (Tsinidis et al. 2014). The dynamic increments of the pressures were amplified near the stiff corners of the tunnel, while they found to be quite reduced at the middle of the side wall due to the flexibility of the wall at this location. The pressure records did not reveal a clear connection between the residual values and the intensity of the input motion. This complex behavior has been also reported during similar centrifuge tests (Cilingir and Madabhushi 2011) and is attributed to the soil yielding and densification during shaking. The phenomenon is obviously amplified by the high flexibility of the tunnels.

31.3.3.3 Bending Moment-Time Histories

Representative dynamic bending moment-time histories, recorded at the right side-wall close to the wall-invert slab corner, are presented in Fig. 31.7 for both the flexible and the rigid tunnel. Positive values represent bending moment with tensile stress increments for the inner face of the wall. The results refer to the same input motion amplitude. Generally, the bending moments were found to be higher after each shake. The dynamic bending moment recorded on the flexible tunnel is significantly lower than the rigid tunnel, due to the differences both in the flexibility of the tunnels and the burial depth. However, both histories follow a similar trend. Actually bending moments responded in a similar manner to the earth pressures. During the first few cycles of loading the bending moments were building up. Subsequently, in the steady state stage, they were oscillating around a mean value. Finally, in the post-earthquake stage, residual stresses were recorded on the tunnel lining. These residuals are mainly attributed to stress redistributions in the surrounding soil caused by plastic soil deformations and/or soil densification. This cumulative response has also been observed during dynamic centrifuge tests that were performed on a flexible circular tunnel model embedded in dry sand (Lanzano et al. 2012).

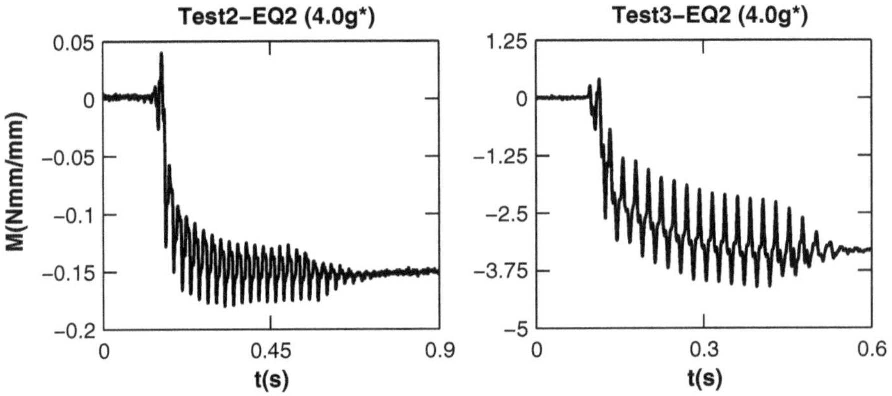

Fig. 31.7 Dynamic bending moment recorded close to the right side-walls bottom corners of the tunnels (SGB1 for the flexible tunnel, SGB2 for the rigid tunnel) *input motion amplitude

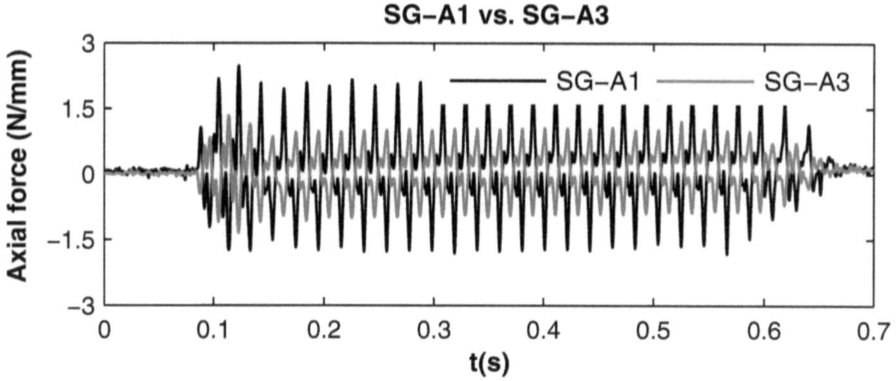

Fig. 31.8 Dynamic axial forces recorded on the rigid tunnel walls during EQ7-Test 3. Out of phase axial response of the side walls

31.3.3.4 Axial Force-Time Histories

Residual values were also observed for the axial forces. These residuals, that generally were larger at the slabs, can be attributed to soil densification as well as to possible sliding effects on the soil-tunnel interface. It is noteworhty that the dynamic axial forces recorded on the sidewalls of the rigid model (Test 3) were out of phase, indicating again the "rocking" mode of vibration for the tunnel, as observed in the vertical accelerations (Fig. 31.8).

31.3.4 Flexible Tunnel Collapse Mechanism

The flexible tunnel collapsed during the test and more specifically during the first earthquake of the second flight (EQ3 Test 2, according to Table 31.3). After the test the model was carefully uncovered using a vacuum cleaner, to study the collapse mechanism. Phases of this "excavating" procedure are presented in Fig. 31.9, along with the deformed shapes of the soil-tunnel model, estimated by direct measurements taken during the excavation.

The tunnel deformed shape and the data interpretation reveal the following collapse mechanism. It is noted that the following used notation refers to Fig. 31.2. During the first flight swing up, and more specifically at the time when the model was spinning up from 10 to 30 g, the axial force of the roof slab changed from compressive to tensile (SG-A1, Fig. 31.10b) indicating large inward deformations of the slab. This is also confirmed by the large increase in settlements at the soil surface above the tunnel at that time (Fig. 31.10a). Moreover, the bending moment, as recorded on the roof slab right corner (SG-B2) changed polarity, turning from negative to positive (tensile strains on the inner face of the lining), indicating inward deformation (bucking) of the corner (Fig. 31.10c). The final deformed shape of the tunnel confirms this observation (tensile strains on the inner face of the lining at the

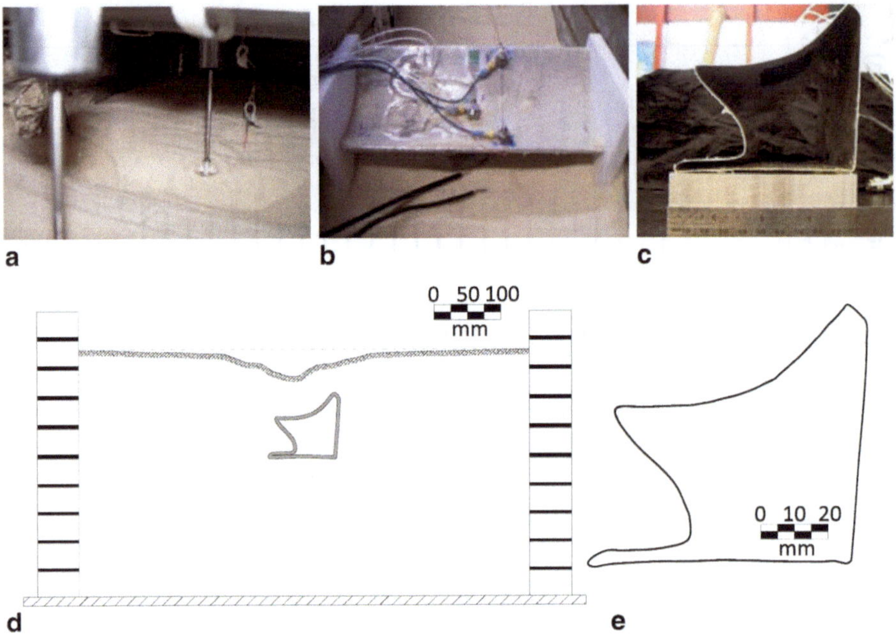

Fig. 31.9 Collapse of the flexible tunnel during Test 2: **a** soil surface after the tunnel collapse, **b** excavation of the sand around the collapsed tunnel, **c** tunnel deformed profile, **d** deformed shape of the soil-tunnel model estimated by direct measurements taken after the main test and during the model excavation, **e** tunnel deformed profile estimated by direct measurements after the main test

roof slab right corner). It is believed that this corner deformation forced the right side wall to move slightly outwards (Fig. 31.9e). This deformation scheme caused small tensile axial strains on the right side-wall (SG-A2, Fig. 31.10b) and compressive axial strains on the left side-wall (SG-A3, Fig. 31.10b). During the swing up the earth pressures recorded on the left side-wall were shifting from active to passive state indicating inwards and outwards movement of the wall (Fig. 31.10d). The bending moment at the right side-wall bottom corner (SG-B1) was found to be negative (tensile strains on the outer face), as theoretically expected (Fig. 31.10c). The subsequent shakes caused a slight increase of the lining internal forces (cumulative response presented before). Both the earth pressures and the internal forces did not change polarity during shaking indicating a stabilization of the deformed tunnel (roof slab and left side-wall were deformed inwards). The swing down caused a small relief of the tunnel, while during the second swing up the deformations of the roof slab and the left side-wall increased again. When the final earthquake was applied (first shake during the second flight), the left side-wall buckled inwards, causing large settlements on the sand above the tunnel (Fig. 31.10f and 31.10j). During the collapse both the earth pressures on the left side-wall and the internal forces were changed significantly, due to significant stress redistributions that took place both in the soil and on the tunnel. Failure was probably accelerated by P-

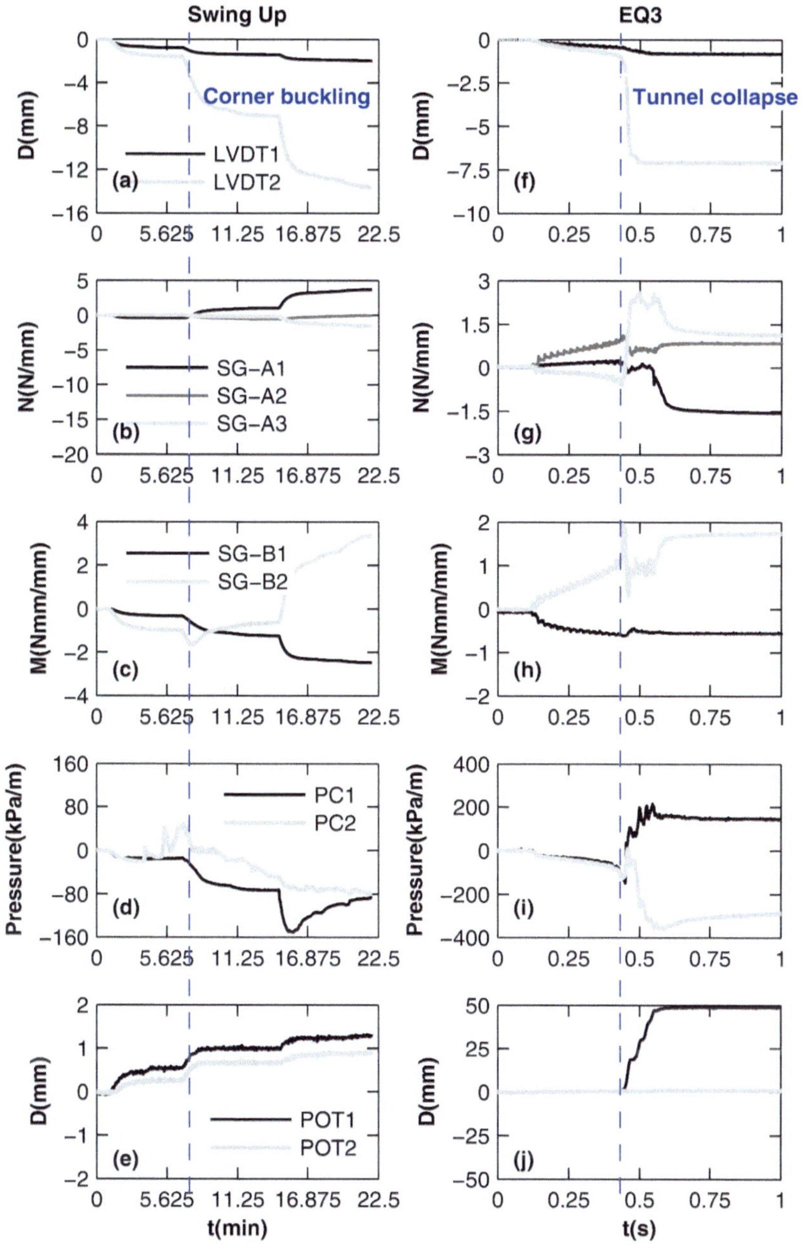

Fig. 31.10 Response of the soil-flexible tunnel system during the first flight swing up (*left side*) and the final earthquake (EQ3 according to Table 31.3 right side). The notation is following Fig. 31.2

delta effects, leading to overturning moments on the collapsed wall. Based on the aforementioned observations, the failure mechanism is clearly identified as bucking failure of the left-side wall.

31.4 Conclusions

The chapter presented the main experimental work that was conducted within a series of dynamic centrifuge tests on square model-tunnels embedded in dry loose ($D_r = 50\,\%$) and dense ($D_r = 90\,\%$) Hostun sand. Representative experimental results have been presented and discussed. The main conclusions drawn by the interpretation of these experimental results may be summarized in the following:

- During swing up, static response, in terms of earth pressures on the tunnels and internal forces, increased due to the increase of the self weigh of the models. This response is consistent with the theoretically expected behavior.
- The horizontal acceleration recorded at several locations was generally amplified towards the soil surface, while the tunnel presence affected this amplification.
- Vertical acceleration-time histories recorded on the sides of the roof slabs of the models were generally out of phase indicating a "rocking" mode of vibration for the tunnels, except the pure racking mode. This behavior was not clearly indentified in Test 1 (shallow tunnel with smooth external face), contrary to the other cases studied. It is believed that this complex response is mainly related to the shear stresses mobilized along the soil-tunnel walls interface. It is noteworthy, that this particular phenomenon may be slightly affected by the minor yawing movement of the ESB container on the shaking table during shaking. In any case this phenomenon exists and should be taken into account in any design analysis.
- The dynamic increments of the pressures were larger at the stiffer slab-wall corners compared to the middle of the walls.
- Earth pressure residual values observed after each shake were mainly attributed to soil yielding and densification during shaking.
- The dynamic bending moment-time histories revealed a similar response to the dynamic earth pressures. Large residual values are detected after each shake, as a result of cumulative strains during the shaking.
- Smaller residuals were observed for the dynamic axial forces after shaking due to soil densification and yielding and also due to a small amount of sliding on the soil-tunnel interface.
- The interpretation of the experimental data revealed that the flexible tunnel started to deform significantly during the first flight swing up. This deformation led to the buckling of the roof slab-right wall corner at that time, causing large compressive loads on the left side wall of the tunnel. The tunnel collapsed during the final shake due to buckling of this wall.

Acknowledgments The research leading to these results has received funding from the European Community's Seventh Framework Programme [FP7/2007–2013] for access to the Turner Beam Centrifuge, Cambridge, UK under grant agreement n° 227887. The technical support received by the Technicians of the Schofield Centre is gratefully acknowledged.

References

Cilingir U, Madabhushi SPG (2011) A model study on the effects of input motion on the seismic behavior of tunnels. J Soil Dyn Earthq Eng 31:452–462

FHWA (2009) Technical manual for design and construction of road tunnels—Civil elements. U.S. Department of transportation. Federal Highway Administration. Publication No. FHWA-NHI-09-010 p 629

Ghosh B, Madabhushi SPG (2002) An efficient tool for measuring shear wave velocity in the centrifuge. In: Phillips R, Guo PJ, Popescu R (eds) Proceedings of international conference on physical modelling in Geotechnics, St Johns, NF, Canada. AA Balkema publishers, the Netherlands, 119–124

Hardin BO, Drnevich VP (1972) Shear modulus and damping in soils: design equations and curves. J Soil Mech Found Div 98(SM7):667–692

Hashash YMA, Hook JJ, Schmidt B, Yao JIC (2001) Seismic design and analysis of underground structures. Tunn Undergr Sp Tech 16(2): 247–293

Iida H, Hiroto T, Yoshida N, Iwafuji M (1996) Damage to Daikai subway station. Soils and Foundations, Special Issue on Geotechnical Aspects of the January 17 1995 Hyogoken-Nambu Earthquake. Japanese Geotechnical Society, 283–300

ISO 23469 (2005) Bases for design of structures—Seismic actions for designing geotechnical works. ISO International Standard. ISO TC 98/SC3/WG10

Lanzano G, Bilotta E, Russo G, Silvestri F, Madabhushi SPG (2012) Centrifuge modeling of seismic loading on tunnels in sand. Geotech Test J 35:1–15

Madabhushi SPG, Schofield AN, Lesley S (1998) A new stored angular momentum (SAM) based earthquake actuator. In: Kimura T, Kusakabe O, Takemura J (eds) Proceedings of the international conference Centrifuge '98, Tokyo, Japan. Balkema, Rotterdam, the Netherlands, 111–116

Madabhushi SPG, Houghton NE, Haigh SK (2006) A new automatic sand pourer for model preparation at University of Cambridge. Proceedings of 6th international conference on physical modelling in geotechnics, Hong Kong

Mitrani H (2006) Liquefaction Remediation Techniques for Existing Buildings. PhD Thesis, University of Cambridge

Penzien J (2000) Seismically induced racking of tunnel linings. Earthq Eng Struct Dyn 29:683–691

Pitilakis K, Tsinidis G (2014) Performance and Seismic Design of Underground Structures. In: Maugeri M and Soccodato C (eds) Earthquake Geotechnical Engineering Design. Geotechnical Geological and Earthquake Engineering, 28, pp. 279–340, Springer International Publishing, Switzerland, doi:10.1007/978-3-319-03182-8_11

Power M, Rosidi D, Kaneshiro J, Gilstrap S, Chiou SJ (1998) Summary and evaluation of procedures for the seismic design of tunnels. Final Report for Task 112-d-5.3(c). National Center for Earthquake Engineering Research, Buffalo, New York

Schofield AN (1980) Cambridge geotechnical centrifuge operations. Geotechnique 30(3):227–268

Schofield AN, Zeng X (1992) Design and performance of an equivalent shear beam (ESB) container for earthquake centrifuge modelling. Technical Report, CUED/D-Soils/I'R245, Cambridge University

Sharma S, Judd WR (1991) Underground opening damage from earthquakes. Eng Geol 30:263–276

Tsinidis G, Heron C, Pitilakis K and Madabhushi SPG (2014) Physical modeling for the evaluation of the seismic behavior of square tunnels. In: Ilki A and Fardis M (eds) Seismic evaluation and rehabilitation of structures. Geotechnical Geological and Earthquake Engineering, 26, pp.389–406, Springer International Publishing, Switzerland, 2014, doi:10.1007/978-3-319-00458-7_22

Wang JN (1993) Seismic design of tunnels—A simple state-of-the-art design approach. Parsons Brinckerhoff, New York

Chapter 32
FLIQ: Experimental Verification of Shallow Foundation Performance Under Earthquake-Induced Liquefaction

George D. Bouckovalas, Dimitris K. Karamitros, Gopal S. P. Madabhushi, Ulas Cilingir, Achilleas G. Papadimitriou and Stuart K. Haigh

32.1 Introduction

All modern seismic codes categorize liquefiable soils as extreme ground conditions, where construction of shallow foundations is essentially not allowed without prior soil treatment. Alternatively, the use of deep foundations becomes mandatory, with or without soil treatment, an alternative that generally increases the cost of construction. Nevertheless, there is ample field evidence that a sufficiently thick and shear resistant non-liquefiable soil crust, between the foundation and the liquefiable subsoil, may effectively mitigate the detrimental effects of liquefaction and lead to adequate foundation safety and satisfactory overall performance (e.g., Ishihara et al. 1993; Acacio et al. 2001). Still, in order to introduce the aforementioned beneficial effect of a non-liquefiable soil crust in current design practice, one must be able to compute analytically the following two basic performance indicators: (a) the degraded bearing capacity of the shallow foundation at the end of shaking, before earthquake-induced excess pore pressures dissipate, and (b) the corresponding liquefaction-induced settlements.

Analytical expressions for the post-shaking degraded bearing capacity have been proposed by Naesgaard et al. (1998), Cascone and Bouckovalas (1998) and recently by Karamitros et al. (2013a). All these studies assume that the footing punches (vertically) through the non-liquefiable soil crust, forcing the development of a wedge-type failure mechanism within the liquefied subsoil. In the first study, the bearing capacity of the liquefied sand is computed in terms of the residual shear strength of the liquefied sand (e.g. Olson and Stark 2002; Stark and Mesri 1992; Seed and

---

G. D. Bouckovalas (✉) · D. K. Karamitros
National Technical University of Athens (NTUA), Athens, Greece
e-mail: gbouck@central.ntua.gr

G. S. P. Madabhushi · U. Cilingir · S. K. Haigh
University of Cambridge (UCam), Cambridge, UK

A. G. Papadimitriou
University of Thessaly (UTh), Volos, Greece

F. Taucer, R. Apostolska (eds.), *Experimental Research in Earthquake Engineering*,
Geotechnical, Geological and Earthquake Engineering 35,
DOI 10.1007/978-3-319-10136-1_32

Harder 1990), while the latter two studies assume that the shear strength of the liquefied subsoil is expressed in terms of a degraded friction angle φ, computed as, (Eq. 32.1):

$$\varphi = \tan^{-1}[(1 - r_u)\tan\varphi_o] \tag{32.1}$$

where φ_o is the actual friction angle of the sand and $r_u = \Delta u/\sigma'_{v,o}$ is the average excess pore pressure ratio induced by seismic shaking in the foundation subsoil. In addition to the degraded bearing capacity of the foundation, Naesgaard et al. (1998) and Karamitros et al. (2013b) examined the accumulation of permanent foundation settlements during seismic shaking and provided analytical expressions in terms of the thickness of the liquefiable layer, the seismic excitation characteristics and the static factor of safety against post-seismic bearing capacity failure.

The research briefly reviewed above was theoretical, mainly based on fully coupled, dynamic, elasto-plastic numerical analyses, properly calibrated against laboratory tests. It is noteworthy that a number of independent experimental studies (Coelho et al. 2004; Adalier et al. 2003; Liu and Dobry 1997 etc.) and field observations (Bird et al. 2006; Seed et al. 2003; Cetin et al. 2002) bear witness to the theoretical findings. Nevertheless, these studies were not part of an integrated effort to provide design criteria, and focus mainly upon the mechanisms controlling foundation response. Furthermore, they do not address the issue of post-shaking bearing capacity of the foundation, which is an inseparable part of performance based design.

To this extent, a series of three centrifuge experiments were performed, simulating the seismic response of a square footing, resting upon a two-layered soil profile: an over-consolidated Grade-E Kaolin Clay crust, overlaying a Leighton-Buzzard Fraction-E Sand layer. The base excitation was harmonic, while the main problem variable was the thickness of the non-liquefiable soil crust. Each test was performed in three stages: initial static loading to an adequate factor of safety, seismic shaking and post-seismic static loading to failure. Following a brief presentation of the utilized experimental facilities and the mechanical properties of the two soil layers, this paper presents the model configuration and the employed testing methodology. In addition, the basic experimental results are presented and evaluated in comparison with the respective analytical predictions.

32.2 Centrifuge Experiments

32.2.1 Experimental Set-Up and Facilities

Figure 32.1 presents the configuration and instrument positions of the examined centrifuge model, while Table 32.1 summarizes the associated geometrical dimensions for all experiments in prototype scale. The centrifuge experiments were carried out in the Schofield Centre of the Cambridge University Engineering Department

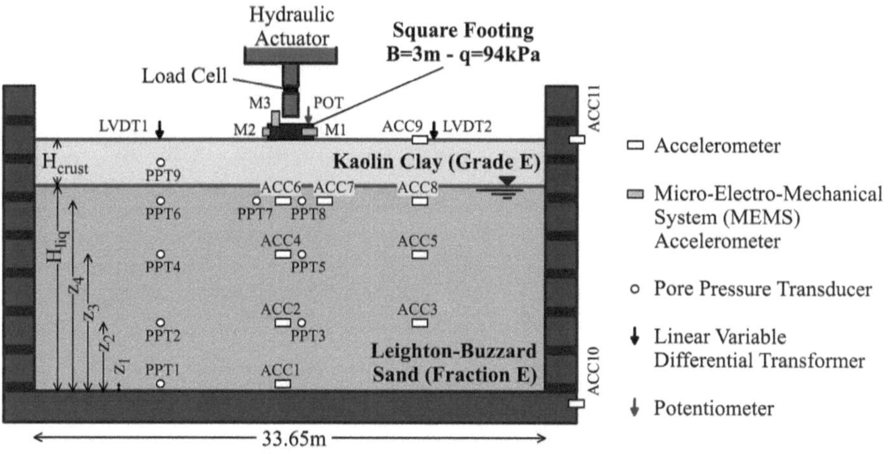

Fig. 32.1 Typical centrifuge model configuration

Table 32.1 Geometrical parameters (in prototype scale)

Test number	H_{crust} (m)	H_{liq} (m)	z_1 (m)	z_2 (m)	z_3 (m)	z_4 (m)
1	3.00	13.50	0.20	4.50	9.00	12.50
2	2.00	14.50	0.20	5.00	10.00	13.50
3	5.00	11.50	0.20	3.50	7.00	10.50

(CUED), using the Philip Turner 10 m beam centrifuge (Schofield 1980), presented in Fig. 32.2a. The machine has a 150 g-tonne capacity, achieving a maximum centrifugal acceleration of approximately 130 g at 4.125 m radius. The tests presented herein were performed at 50 g.

The earthquake excitations were generated by the Stored Angular Momentum (SAM) actuator (Fig. 32.2b), developed at Cambridge University (Madabhushi et al. 1998). The input motion is generated by the energy stored in a pair of flywheels, the angular velocity of which determines the frequency of the motion transmitted to the base of the model. The motion is transmitted through a shaft via a fast acting clutch which triggers and stops the earthquake excitation, while the magnitude of the earthquake excitation can be controlled by altering the pivot point of the lever.

The aforementioned methodology was employed in the present case in order to generate twenty (20) uniform cycles of harmonic base excitation, with a peak acceleration of $a_{max}=0.25$ g and a period of T = 1 s.

The examined centrifuge model was built in an Equivalent Shear Beam (ESB) container (Zeng and Schofield 1996), with internal dimensions of (L × B × H) 673 × 253 × 427 mm^3 (Fig. 32.2c). This box consists of a series of alternating aluminium and rubber rings, whose geometry and stiffness were chosen so that the walls of the box have the same stiffness and natural frequency of vibration as an average soil (Teymur and Madabhushi 2003). It should be acknowledged, though,

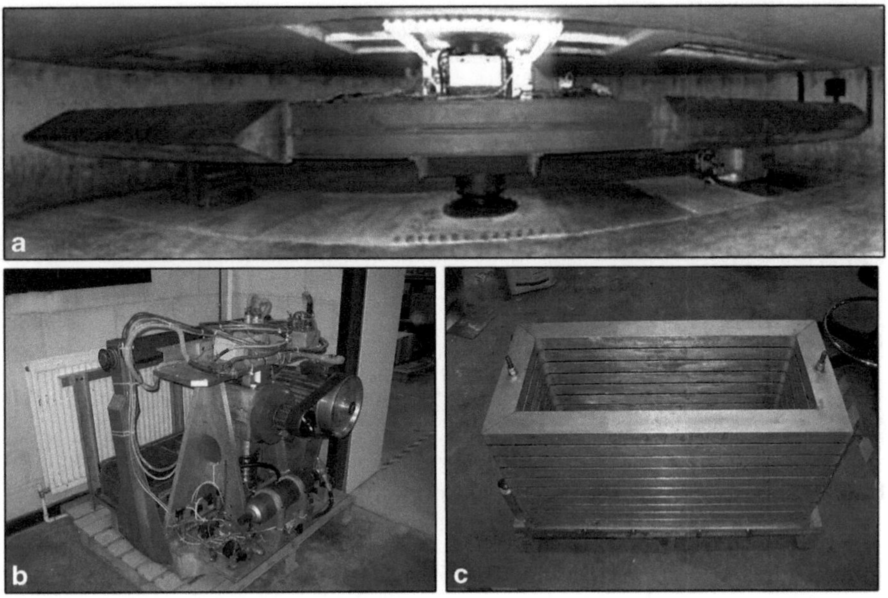

Fig. 32.2 Utilized experimental facilities: **a** the Philip Turner beam centrifuge, **b** the SAM actuator, **c** the ESB box

that during the investigation of liquefaction-related phenomena, the soil stiffness is cycle-dependent and consequently, it cannot be matched by the stiffness of walls, which remains unchanged. Thus, boundary effects may exist, at later stages of the test, when liquefaction spreads wide over the soil volume.

The model's behavior was monitored using a total of 27 transducers, as demonstrated in Fig. 32.1. More specifically, eleven (11) miniature piezoelectric accelerometers (A23-S & TS) were placed at different depths in the free-field, along the footing's axis, under the footing's edge, as well as at the top and bottom of the ESB box, in order to measure the developing horizontal accelerations. Nine (9) miniature 700-kPa-range PDCR-81 pore pressure transducers (PPTs) were also placed in the soil, in order to measure fluid-pore-pressure variations during testing, through a strain-gauged silicon diaphragm that deflects under pressure. Note that, in order to protect the fragile membrane of the PPTs placed in the sand layer, the entrance of the transducers was covered by a bronze porous stone, while a high air-entry filter with smaller pore diameter was used for the PPT that was placed in the clay layer.

Settlements of the free-field clay surface were measured by two Linear Variable Differential Transformers (LVDTs), while foundation settlements were measured with a potentiometer, which was placed on a gantry, above the footing, and consequently attached to one of the footing's corners. Horizontal and vertical accelerations of the footing were measured with single-axis, high-g Micro-Electro-Mechanical System accelerometers (MEMS), while a load cell was used in the centrifuge test to measure the post-shaking bearing capacity of the model foundation. The load cell was attached to an actuator, which applied the post-earthquake static load to the footing, as shown in Fig. 32.1.

To ensure the quality of the recorded centrifuge data, and avoid any noise effects introduced by SAM-generated electrical interference, data acquisition was performed using CDAQS (Centrifuge Data AcQuisition System), namely a two-module data acquisition system: (a) module CD 198, which flies with the centrifuge package, digitizes the instruments readings onboard and stores the digital data in RAM, and (b) interface module CD_IF 198, which sits in the control room and uploads the data to a computer, after the end of the test.

32.2.2 Model Preparation

The sand used for the experiments is Leighton Buzzard Fraction E, an industrially normalized light-brown natural and uncrushed silica sand which contains rounded and sub-angular particles and features minimum and maximum void ratios $e_{min}=0.6131$ and $e_{max}=1.0141$, respectively, as well as a specific gravity of 2.65 Mgr/m^3. The natural sand was washed, dried and graded to eliminate any silt, clay or organic matter and to conform to BS-EN-ISO-9002:1994. The sand's grain size distribution falls safely within the boundaries used to identify liquefiable soils, while its uniformity ($C_u \approx 2$) ensures that no significant segregation occurs during the preparation of models using the air pluviation technique. Furthermore, Fraction E is relatively fine ($D_{50}=0.14$ mm), thus minimizing scale effects that may possibly influence the behavior of centrifuge models. Based on previously performed drained and undrained triaxial compression tests (Jeyatharan 1991) the Critical State friction angle of this sand is estimated at 32°.

In order to achieve uniform sand deposition, as well as repeatability of the reconstitution method, the air pluviation technique was applied, using the robotic apparatus shown Fig. 32.3 (Zhao et al. 2006). This apparatus consists of a hopper which travels along the container's axes, while also being raised vertically by the thickness of the deposited sand, in order to maintain a constant flow height. The sand was poured from a flow height of 775, 6 mm diameter nozzle, resulting in a relative density of the deposited sand equal to 42.5±2%, in fairly good agreement with the calibration diagrams of Chian et al. (2010). It is noted that the process of sand pouring was interrupted, in order to place the accelerometers and PPTs at the required depths, as shown in Fig. 32.4.

After deposition, the sand layer was saturated with a viscous pore fluid, in order to account for the well-known discrepancy between the centrifuge scaling of the consolidation time and the loading time of dynamic events. Namely, taking into account that all tests were carried out at a centrifuge acceleration of 50 g, the pore-fluid used to saturate the models was prepared with a viscosity of 50 cSt, by adding to the required amount of water the necessary amount of Hydroxypropyl Methylcellulose (HPMC).

Fig. 32.3 Robotic apparatus used for sand pluviation

The saturation process was performed using the CAM-Sat automated system (Stringer and Madabhushi 2009), schematically illustrated in Fig. 32.5. More specifically, the ESB box was initially sealed with a special lid and a pump was connected to the lid, applying a vacuum of about −95 KPa for about an hour. In order to enhance the saturation process, the model was consequently flushed with CO_2 (Takahashi et al. 2006) and then vacuum was re-applied to the model and left for another hour. In parallel, the container tank with the viscous fluid was placed on an electronic scale, used to transmit the container's mass to a computer data-logging system. The container was then sealed with a lid, and vacuum was applied to it, while its bottom was connected to four openings at the base of the ESB box. Finally, a computer controlled pressure regulator was utilized to create a pressure difference between the fluid container and the ESB box, thus causing fluid flow and initiating the model's saturation. The rate of saturation was consequently controlled and maintained within a user-defined allowable range, by altering the pressure

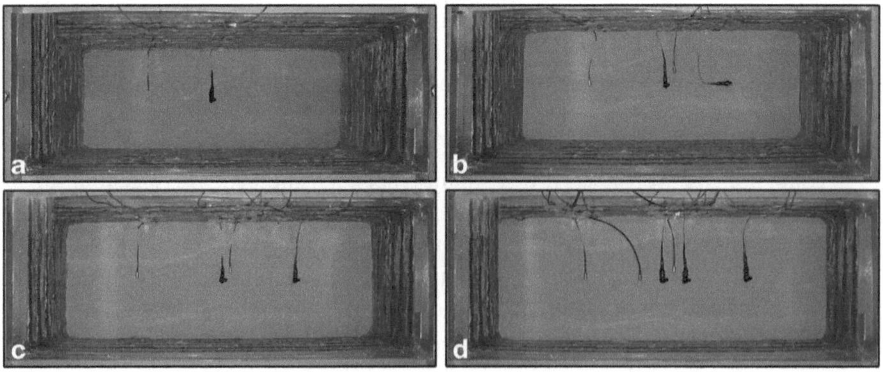

Fig. 32.4 Instrumentation placed at depths **a** z_1, **b** z_2, **c** z_3, **d** z_4 (as defined in Table 32.1)

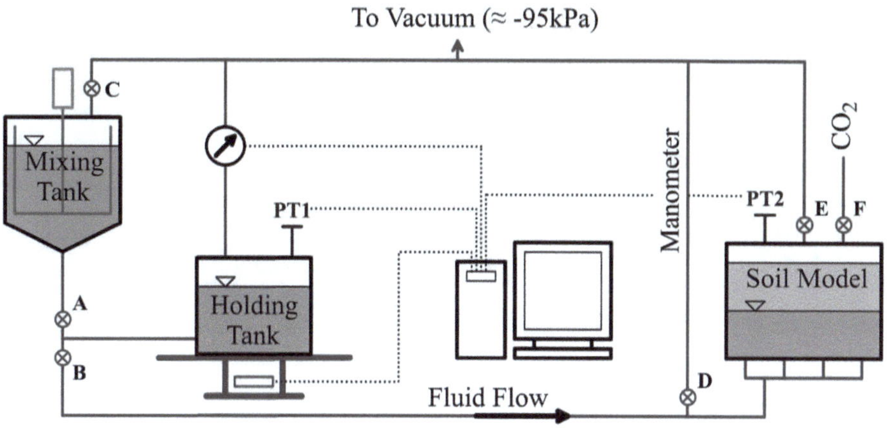

Fig. 32.5 Schematic illustration of the *CAM-SAT* system

difference, according to the recorded decrease rate of the fluid container's mass. In the tests presented herein, the saturation rate was maintained between 0.5 and 0.6 kg/h, in accordance with the suggestions of Stringer and Madabhushi (2009).

The non-liquefiable crust overlying the liquefiable sand layer in the centrifuge model consisted of over-consolidated Grade E Kaolin Clay. This clay has a Liquid Limit LL = 51 % and a Plastic Limit PL = 30 %, while, in the Casagrande chart, it may be described as clayey silt of medium to high plasticity. The mechanical behavior of E-Grade Kaolin has been studied by Elmes (1985), and corresponding Cam-clay parameter values have been determined. Furthermore, according to Potter (1996), the void ratio of one-dimensionally consolidated E-grade kaolin at $\sigma'_v = 100$ kPa is $e = 1.183$, while values of $\lambda = 0.124$–0.21 and $\kappa = 0.02$–0.03 have been proposed for the virgin compression and recompression lines, respectively. Finally, the undrained shear strength c_u of kaolin clay may be estimated from the following empirical equation (Eq. 32.2), based on the work of Ladd et al. (1977) and Skempton (1948), where OCR is the over-consolidation ratio:

$$c_u = 0.18\,\sigma'_v\,OCR^{0.8} \tag{32.2}$$

The clay was consolidated in a metal tub, with an internal diameter of 850 mm. In more detail, an adequate amount of slurry was initially placed in the tub and the consolidation pressure was applied incrementally, through an electronically controlled hydraulic piston. After reaching the desired over-consolidation stress of 400 kPa, the applied pressure was incrementally decreased to 80 KPa. The clay layer required for the model was consequently cut in three pieces, namely one large piece with a plan area of 450 × 230 mm that was placed underneath the model footing, as well as two smaller pieces with a plan area of 100 × 230 mm that were placed on each side of the larger central piece. Note that, for practical reasons, the dimensions of the clay blocks were somewhat smaller than the internal dimensions of the

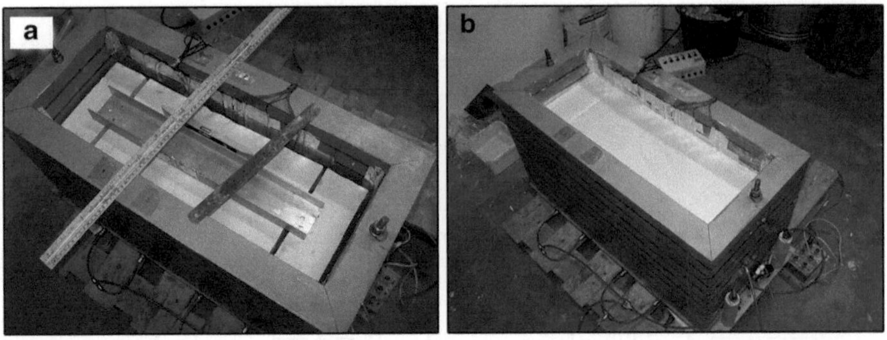

Fig. 32.6 Centrifuge model **a** after placement of the clay layer, **b** after filling the gaps with slurry

ESB box (673 × 253 mm), thus leaving a small gap of about 10 mm around each block, as shown in Fig. 32.6a. After placement of the clay blocks, the sand layer exhibited relatively small settlements, increasing its relative density to $D_r = 74\%$, 86 and 57%, for the first, second and third test, respectively. During this stage, a standpipe had been connected to the ESB box, aiming to allow dissipation of excessive pore pressures format the base of the model. Nevertheless, the amount of water that flew out of the standpipe was minimal, compared to the ejecta flowing upwards and filling of the gaps between the clay blocks. This material was removed with a syringe, the gaps were then filled with slurry (Fig. 32.6b) and the model was carefully covered with cling-film, in order to prevent the clay layer from drying.

The square footing that was used in the tests was made of steel, its model dimensions were 60 × 60 × 25 mm^3 while the associated mass was equal to 0.690 kg. Thus, for 50 g centrifugal acceleration, the prototype dimensions of the footing were 3 × 3 × 1.25 m^3 and the applied average bearing pressure was 94 kPa. In order to avoid undesired displacements, the footing was placed on the soil surface just prior to testing.

32.2.3 Testing Procedures

Figure 32.7 presents the model loaded on the centrifuge, prior to testing. As shown in the figure, the actuator was installed on a specially designed gantry and its height was adjusted, so that the actuator's tip would leave a 3 mm gap from the top of the footing. The piston's pressure lines were connected to the centrifuge's main air pressure system, through a pressure switch, allowing the actuator to either be held up, or to push the footing downwards. The corresponding pressure was set to 650 kPa. Furthermore, a standpipe was installed, to allow drainage of the saturated sand through the bottom of the ESB box. The standpipe was configured so that the water table could be fixed slightly above the initial sand-clay interface. An electromagnetic valve was added to the standpipe, allowing water flow to be switched on or off from the control room. In this way, as the sand layer would consolidate,

Fig. 32.7 Centrifuge model prior to testing

during swinging-up, excess pore pressures would be allowed to dissipate, in order to avoid the formation of sand boils through the clay layer. The valve would be turned off before shaking, so that the liquefaction-induced excess pore pressures do not dissipate through the standpipe. Note that the valve remained closed during the post-shaking loading of the foundation.

The model was accelerated to the final centrifugation in five 10 g increments. At each stage, a time of about 25 min was allowed for the clay layer to consolidate, as well as for (static settlement-induced) excess pore pressures in the sand layer to dissipate through the standpipe system. After the end of consolidation, the drainage valve was switched off and an electronic timer triggering the data acquisition system and controlling the solid-state relays on the SAM actuator was set-up, in order to define the initial countdown, the duration of the earthquake and the piston's movement for the post-shaking loading of the foundation. Immediately after the end of the earthquake excitation, the timer triggered the pressure switch that controlled the piston's movement, so that the post-shaking loading of the footing could de performed. After stopping the SAM motor, the experimental data were uploaded to the computer terminal and the centrifuge was slowed down until complete stop. The whole procedure was also monitored through a web-cam, placed on the ESB box. Snapshots from the web-cam during centrifuge test #1 are shown in Fig. 32.8. Similar photos were taken in the other tests.

When centrifuge testing was completed, a preliminary inspection of the model was carried out at the centrifuge beam, followed by lab shear vane testing on the clay surface. It was thus found that the lab vane undrained shear strength of the clay crust was $c_u = 32 \pm 2$ kPa for the first two tests and $c_u = 25 \pm 2$ kPa for the third

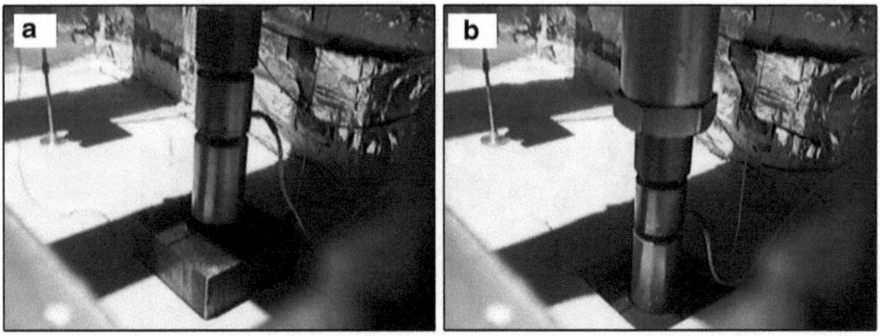

Fig. 32.8 Webcam photos from centrifuge test #1

Fig. 32.9 Excavation of centrifuge model **a** #1, **b** #2, **c** #3

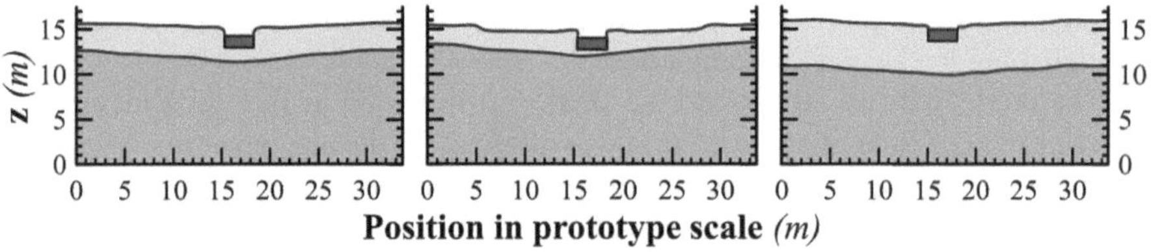

Fig. 32.10 Cross-sections of the deformed soil profile, in centrifuge tests **a** #1, **b** #2, **c** #3 (in prototype scale)

test. The ESB box was then unloaded, 1 day later, and the model was inspected and carefully excavated. Photos taken during this final stage are presented in Fig. 32.9, while the associated cross-section of the soil profile along the model's x-axis is shown in Fig. 32.10.

32.3 Experimental Results

32.3.1 Dynamic Loading

Figures 32.11, 32.12 and 32.13 present typical results from the performed experiments, in terms of acceleration, excess pore pressure and settlement time-histories.

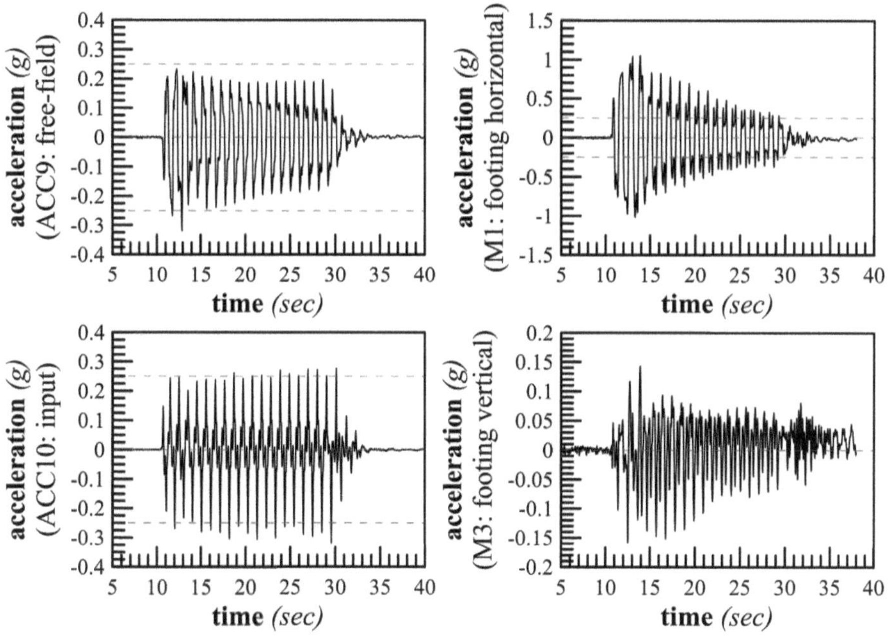

Fig. 32.11 Recorded acceleration time-histories

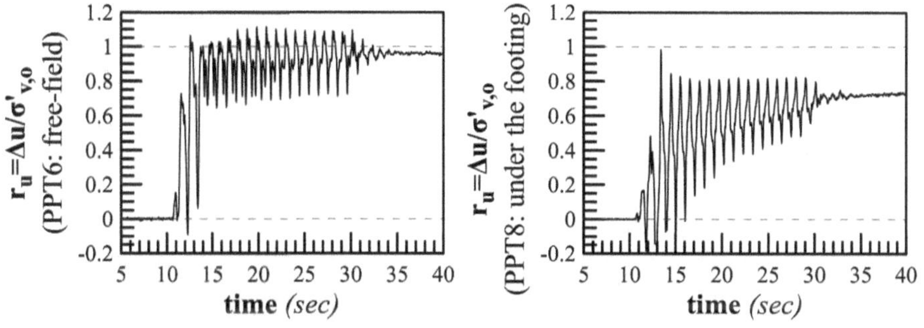

Fig. 32.12 Recorded excess pore pressure ratio time-histories

Fig. 32.13 Recorded foundation settlement time-histories

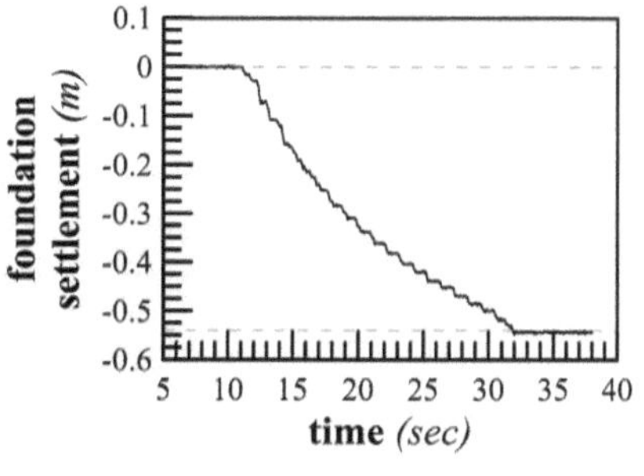

The presented results were obtained from test #1. More specifically, horizontal acceleration time-histories are presented for the model's base (input excitation) and the free-field ground surface, while both horizontal and vertical acceleration time-histories are shown for the model footing. As demonstrated in Fig. 32.11, the input motion was not drastically de-amplified in the free-field, as observed in most centrifuge experiments performed with a laminar box (e.g. Liu and Dobry 1997; Kawasaki et al. 1998; Adalier et al. 2003; Coelho et al. 2004). This behavior is attributed to the boundary conditions applied by the utilized ESB box, as discussed in Paragraph 2.1. As far as the horizontal motion of the foundation is concerned, a significant amplification was observed during the initial stages of loading, which decreased as subsoil liquefaction took place. Finally, it may be observed that the predominant frequency of vertical accelerations is twice as high as that of the horizontal motion component. This means that each cycle of horizontal motion is accompanied by two cycles of vertical motion, providing indirect support to the view that footing settlements develop due to a Newmark-type sliding block mechanism rather than densification of the liquefiable sand (Karamitros et al. 2013a).

Figure 32.12 presents time-histories of excess pore pressure ratios at depth z_4 (see Table 32.1) underneath the footing's axis and in the free field. It was observed that excess pore pressure ratios underneath the foundation never reached the value of the initial effective overburden, even though the corresponding excess pore pressure ratios in the free-field indicate complete liquefaction. This behavior is similar to the one observed in most relevant experimental studies (e.g. Liu and Dobry 1997; Kawasaki et al. 1998; Adalier et al. 2003; Coelho et al. 2004), and it may be attributed to static deviatoric stresses existing in the sand, due to the foundation's bearing pressure.

Finally, Fig. 32.13 presents the recorded foundation settlement time-history. In agreement with many other relevant experimental findings (e.g. Liu and Dobry 1997; Acacio et al. 2001; Adalier et al. 2003; Coelho et al. 2004; Dashti et al. 2010), seismic settlement accumulation is more or less linear with time, and essentially ceases after the end of shaking.

The variation of the recorded dynamic settlements of the foundation ρ_{dyn} with the thickness of the clay crust H_{crust}, both normalized with the footing's width B, is shown in Fig. 32.14. Furthermore, in the same figure, the experimental results are compared to analytical predictions derived with the simplified methodology of Karamitros et al. (2013b). There are two main notes to make with regard to the data shown in this figure. The first refers to the thickness of the non-liquefiable crust in centrifuge test #2. Namely, careful examination of this test revealed that the water table was lower than the sand-clay interface, thus increasing the thickness of the non-liquefiable crust from 2 to 3.2 m. Hence, H_{crust}/B was increased from 0.67 to about 1.1, while the analytical predictions were performed for the increased crust thickness. The undrained shear strength of the combined (clay and sand) crust was kept at 32 kPa as the shear strength of the added sand layer (Eq. 32.3) exhibited approximately equal values to the strength of the clay, i.e.:

$$\tau_{max} \approx \sigma'_{v,o} \times \tan\varphi \approx 45\text{kPa} \times \tan 35° \approx 32\text{kPa} \tag{32.3}$$

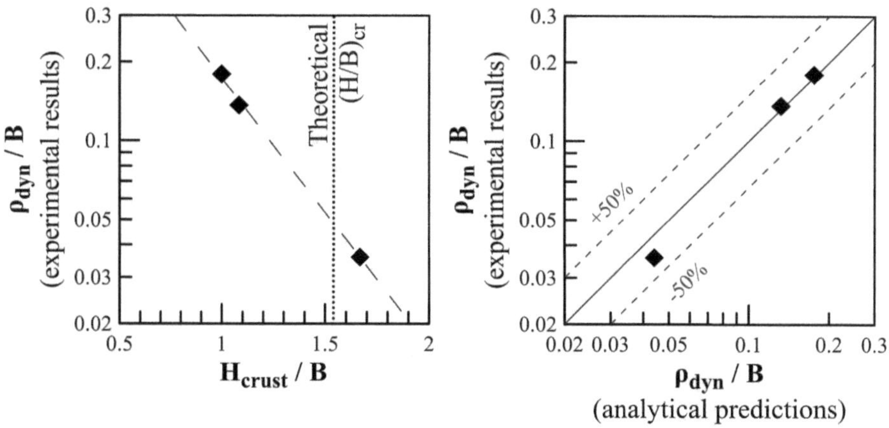

Fig. 32.14 Evaluation of normalized recorded dynamic settlements ρ_{dyn}/B, **a** variation with normalized thickness of the clay crust H_{crust}/B, **b** comparison with analytical predictions

The second note refers to the undrained shear strength of the clay crust in centrifuge test #3 which is considerably less (25 ± 2 kPa) as compared to that of centrifuge tests #1 and #2 (32 ± 2 kPa). To correct the experimental data for this deviation from the intended testing conditions, recorded settlements in centrifuge test #3 were scaled according to the ratio of the relevant analytical predictions for $c_u = 25$ kPa and for $c_u = 32$ kPa.

From a practical point of view, it is of interest to note the consistent agreement between experimental data and analytical predictions in Fig. 32.14b. Focusing further on Fig. 32.14a, it is clearly observed that foundation settlements decrease drastically with increasing thickness of the surficial non-liquefiable crust. Furthermore, in agreement with analytical predictions (e.g.,Cascone and Bouckovalas 1998; Karamitros et al. 2013b, 2013c), it seems that there is indeed a critical normalized clay crust thickness $(H/B)_{cr}$ beyond which subsoil liquefaction has a minor effect on the behavior of surface foundations. It is also noteworthy, that the value of $(H/B)_{cr}$ implied by the experimental data compares fairly well with the theoretically predicted value of 1.54, obtained for square foundations in the recent study of Karamitros et al. (2013b).

32.3.2 Post-Shaking Behavior

Figure 32.15 presents the time-histories of the foundation settlements, the excess pore pressure ratios in the free-field and underneath the foundation, as well as the load applied by the actuator during the post-shaking loading. Due to length limitations, only data from centrifuge test #1 are presented herein, while the results of the remaining tests can be found in the detailed project report (Bouckovalas et al. 2011). Observe that, according to test planning, the post-shaking loading occurred almost

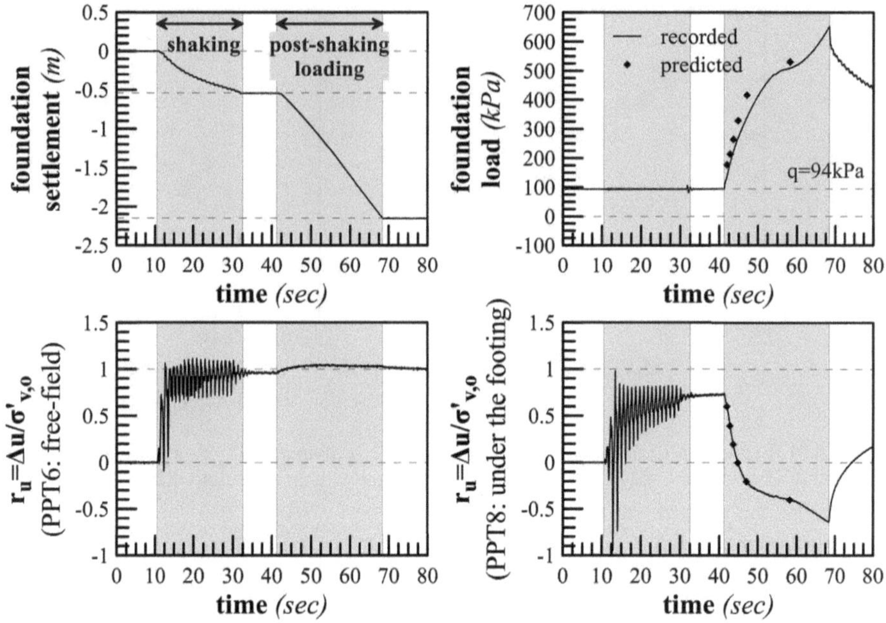

Fig. 32.15 Post-shaking loading of the foundation

immediately after the end of the earthquake and before the dissipation of excess pore pressures. However, the rate of loading was fairly fast, so that bearing capacity failure of the foundation took place under undrained conditions, not only in the clay but also in the sand layer. This is clearly indicated by the large negative excess pore pressure ratios in the region underneath the foundation, which reveal an excessively dilative behavior of the foundation subsoil.

The aforementioned dilation-induced negative excess pore pressures further implies that the post-shaking bearing capacity failure of the foundation induced punching through the surficial clay crust and wedge type failure within the underlying liquefied sand layer. In other words, it confirms the composite failure mechanism that has been assumed in a number of previous studies (i.e., Naesgard et al. 1998; Cascone and Bouckovalas 1998; Karamitros et al. 2013b, 2013c) for the analytical prediction of the post shaking static bearing capacity of the footing. To substantiate further this finding, the post-shaking static bearing capacity of the model footing $q_{ult,\,p-s}$ in centrifuge test #1 was analytically computed using the most recent of the above methodologies, in connection with the associated excess pore pressure measurements. More specifically, $q_{ult,\,deg}$ (Eq. 32.4) has been calculated as:

$$q_{ult,p-s} = 2c_u \frac{H_{crust}}{B} s - \gamma_{clay} H_{crust} + \frac{1}{2}\gamma'_{sand} B N_\gamma F_{\gamma s} + \gamma_{clay} H_{crust} N_q F_{qs} \tag{32.4}$$

where B is the width of the foundation, H_{crust} is the thickness of the clay crust, c_u is the undrained shear strength of the clay crust and γ is the buoyant weight of the foundation soil. Furthermore, $F_{\gamma s}$, F_{qs}, F_{cs} and s are the shape factors of the foundation, while the bearing capacity coefficients N_γ and N_q were computed as functions of the degraded friction angle of the liquefied sand, derived according to Eq. 32.1. The excess pore pressure ratio r_u in this equation is computed as (Eq. 32.5):

$$r_u = (r_{u,foot} + 2 \times r_{u,ff}) / 3 \tag{32.5}$$

i.e. it is the average of the recorded values in the free field ($r_{ff} \approx 1$) and in the sand layer underneath the foundation (shown in Fig. 32.15).

The corresponding analytical predictions are shown as data points in Fig. 32.15, in comparison with the time history of the loads applied by the actuator during the post-shaking loading of the foundation. The agreement between the two, independently derived data sets is admittedly noteworthy.

32.4 Conclusions

A parametric investigation was performed, through dynamic centrifuge testing, in order to verify the beneficial effects of a non-liquefiable clay crust, on the liquefaction performance of shallow foundations and provide insight into the associated response mechanisms. Apart from the common set of (displacement, pore pressure and acceleration) measurements, a special configuration was employed in order to measure the static bearing capacity of the footing, immediately after the end of shaking, i.e. before the dissipation of the excess pore pressures in the liquefiable soil. This paper provides a detailed account of the associated experimental procedures and also proceeds to an initial evaluation of the recorded soil and foundation response. The main findings from this study can be summarized as follows:

- Footing settlements develop mostly during shaking, and exhibit twice the frequency of the applied horizontal base excitation. This evidence comes in support of the theoretical assumption (Richards et al. 1993; Karamitros et al. 2013c) that foundation settlement is a stick-slip process that complies to a Newmark-type sliding-block failure mechanism rather than densification of the subsoil. Therefore, foundation settlements cannot be predicted using common empirical charts (Tokimatsu and Seed 1987; Ishihara and Yoshimine 1992) for free-field settlements, resulting from post-shaking excess pore pressure dissipation and soil densification.
- Deviatoric stresses applied to the subsoil due to static foundation loading inhibit complete liquefaction in the region underneath the foundation, thus enhancing the post-shaking bearing capacity of the foundation.
- Foundation settlements decrease drastically with increasing thickness of the clay crust. Hence, the existence of a sufficiently thick and shear resistance non-liquefiable crust may allow a viable performance of surface foundations, with no need for application of additional ground improvement measures.

- Further than the above paragraph, the experimental results provide evidence for the existence of a "critical clay crust thickness" H_{cr}, above which subsoil liquefaction does not induce significant foundation settlements. Note that the existence of such critical crust thickness for footings on liquefiable sand has been foreseen in earlier analytical studies by Cascone and Bouckovalas (1998) and Karamitros et al. (2013b, 2013c).
- During the post-shaking loading stage of the experiments, large negative excess pore pressures were recorded in the sand below the foundation, indicating intense shearing of the sand below the clay crust. This finding confirms indirectly the assumption adopted in a number of analytical studies (Naesgaard et al. 1998; Cascone and Bouckovalas 1998; Karamitros et al. 2013a, 2013c), namely that the post-shaking bearing capacity failure of the foundation activates a combined failure mechanism, consisting of punching failure through the surficial clay crust and shear failure within the underlying liquefied sand layer.
- Recorded seismic settlements, as well as post-shaking bearing capacity measurements were in fair agreement with analytical predictions obtained with the integrated analytical methodology of Karamitros et al. (2013b) which has been recently proposed for the performance based design of surface foundations in liquefiable soil sites.

Acknowledgements The research leading to these results has received funding from the European Community's Seventh Framework Programme [FP7/2007–2013] for access to the Turner Beam Centrifuge, Cambridge, UK under grant agreement n° 227887.

References

Acacio AA, Kobayashi Y, Towhata I, Bautista RT, Ishihara K (2001) Subsidence of building foundation resting upon liquefied subsoil case studies and assessment. Soils Found 41(6):111–128

Adalier K, Elgamal A, Meneses J, Baez JI (2003) Stone columns as liquefaction counermeasure in non-plastic silty soils. Soil Dyn Earthq Eng 23(7):571–584

Bird JF, Bommer JJ, Crowley H, Pinho R (2006) Modelling liquefaction-induced building damage in earthquake loss estimation. Soil Dyn Earthq Eng 26(1):15–30

Bouckovalas G, Madabhushi SPG, Cascone E, Papadimitriou A, Loukidis D, Karamitros D, Cilingir U, Haight S, Dimitriadi V, Chaloulos Y (2011) FLIQ: experimental verification of shallow foundation performance under earthquake-induced liquefaction. Project report, SERIES research program (Seismic engineering research infrastructures for European synergies), May 2011

Cascone E, Bouckovalas G (1998) Seismic bearing capacity of footings on saturated sand with a clay cap. Proceedings of the 11th European conference on earthquake engineering, Paris, France

Cetin KO, Youd TL, Seed RB, Bray JD, Sancio R, Lettis W, Yilmaz MT, Durgunoglu HT (2002) Liquefaction induced ground deformations at Hotel Sapanca during Kocaeli (Izmit), Turkey earthquake. Soil Dyn Earthq Eng 22(9–12):1083–1092

Chian SC, Stringer ME, Madabhushi SPG (2010) Use of automatic sand pourers for loose sand models. In Proceedings of the VII International Conference on Physical Modelling in Geotechnics (ICPMG 2010), Zurich. Taylor & Francis (pp. 117–121)

Coelho PALF, Haigh SK, Madabhushi SPG (2004) Centrifuge modelling of the effects of earthquake-induced liquefaction on bridge foundations. Proceedings of the 11th international conference on soil dynamics and earthquake engineering (ICSDEE) and the 3rd International Conference on earthquake geotechnical engineering, University of California, Berkeley, January 07–09

Dashti SA, Bray JDB, Pestana JMC, Riemer MD, Wilson DE, (2010) Mechanisms of seismically induced settlement of buildings with shallow foundations on liquefiable soil. J Geotech Geoenviron Eng 136(1):151–164

Elmes D R (1985) Creep & viscosity for two kaolin clays. PhD Dissertation, University of Cambridge, UK

Ishihara K, Yoshimine M (1992) Evaluation of settlements in sand deposits following liquefaction during earthquakes. Soils Found 32(1):178–188

Ishihara K, Acacio A, Towhata I (1993) Liquefaction-induced ground damage in Dagupan in the July 16, 1990 Luzon earthquake. Soils Found 33(1):133–154

Jeyatharan K (1991) Partial liquefaction of sand fill in a mobile arctic caisson under dynamic ice-loading. PhD Dissertation, University of Cambridge, UK

Karamitros DK, Bouckovalas GD, Chaloulos YK, Andrianopoulos KI (2013a) Numerical analysis of liquefaction-induced bearing capacity degradation of shallow foundations on a two-layered soil profile. Soil Dyn Earthq Eng 44:90–101

Karamitros DK, Bouckovalas GD, Chaloulos YK (2013b) Seismic settlements of shallow foundations on liquefiable soil with a clay crust. Soil Dyn Earthq Eng 46:64–76

Karamitros DK, Bouckovalas GD, Chaloulos YK (2013c) Insight to the seismic liquefaction performance of shallow foundations. ASCE J Geotech Geoenviron Eng 139(4):599–607. doi:10.1061/(ASCE)GT.1943-5606.0000797

Kawasaki K, Sakai T, Yasuda S, Satoh M (1998) Earthquake-induced settlement of an isolated footing for power transmission tower. Centrifuge 98:271–276

Ladd CC, Foott R, Ishihara K, Schlosser F, Poulos H (1977) Stress-deformation and strength characteristics. Report of IX ICSMFE 1977, Tokyo

Liu L, Dobry R (1997) Seismic response of shallow foundation on liquefiable sand. J Geotech Geoenviron Eng 123(6):557–566

Madabhushi SPG, Schofield AN, Lesley S (1998) A new Stored Angular Momentum (SAM) based earthquake actuator. Proceedings Centrifuge'98, International Conference on Centrifuge Modelling, Tokyo, Japan, pp. 111–116

Naesgaard E, Byrne PM, Ven Huizen G (1998) Behaviour of light structures founded on soil 'crust' over liquefied ground. Geotechnical Special Publication, (75 I), pp. 422–433

Olson S, Stark T (2002) Liquefied strength ratio from liquefaction flow failure case histories. Can Geotech J 39:629–647

Potter LJ (1996) Contaminant migration through consolidating soils. PhD Dissertation, University of Cambridge, UK

Richards Jr. R Elms DG Budhu M (1993) Seismic bearing capacity and settlements of foundations. J Geotech Eng—ASCE 119(4):662–674

Schofield AN (1980) Cambridge geotechnical centrifuge operations. 20th rankine lecture. Geotechnique 30(3):227–268

Seed RB, Harder LF (1990) SPT-based analysis of cyclic pore pressure generation and undrained residual strength. In: Duncan JM (ed) Proceedings, H. Bolton Seed Memorial Symposium, vol. 2. University of California, Berkeley, pp 351–376

Seed RB, Cetin KO, Moss RES, Kammerer AM, Wu J, Pestana JM (2003) Recent advances in soil liquefaction engineering: a unified and consistent framework. 26th annual ASCE Los Angeles geotechnical spring seminar, Long Beach, California. Keynote presentation

Skempton AW (1948) Vane tests in the alluvial plain of the river Forth near Grangemouth. Geotechnique 1:111–124

Stark TD, Mesri G (1992) Undrained shear strength of liquefied sands for stability analysis. J Geotech Eng ASCE 118(11), pp. 1727–1747

Stringer ME, Madabhushi SPG (2009) Novel computer-controlled saturation of dynamic centrifuge models using high viscosity fluids. Geotech Test J 32:1–6

Takahashi H, Kitazume M, Ishibasi S, Yamawaki S (2006) Evaluating the saturation of model ground by p-wave velocity and modeling of models for a liquefaction study. Int J Phys Model Geotech 1:13–15

Teymur B, Madabhushi S P G (2003) Experimental study of boundary effects in dynamic centrifuge modeling. Geotechnique 53(7):655–663

Tokimatsu K, Seed BH (1987) Evaluation of settlement in sands due to earthquake shaking. J Geotech Eng 113(8):861–878

Zeng X, Schofield AN (1996) Design and performance of an equivalent-shear-beam (esb) container for earthquake centrifuge modeling. Geotechnique 46(1):83–102

Zhao Y, Gafar K, Elshafie MZEB, Deeks AD, Knappett JA, Madabhushi SPG (2006) Calibration and use of a new automatic sand pourer. International conference on physical modelling in geotechnics, Hong Kong, 3–8 August 2006, pp 265–270, Rotterdam, Balkema

Chapter 33
Centrifuge Modelling of Retaining Walls Embedded in Saturated Sand Under Seismic Actions

Stefano Aversa, Luca de Sanctis, Rosa Maria Stefania Maiorano, Michele Tricarico, Giulia Viggiani, Riccardo Conti and Gopal S. P. Madabhushi

33.1 Introduction

A number of cases of damage or failure of gravity and embedded retaining walls during earthquakes are reported in the literature, most of which have occurred in saturated soils (Iai and Kameoka 1993; Kamon et al. 1996; Fang et al. 2003; Koseki et al. 2012). As an example, Madabhushi and Zeng (2007) reported the structural yielding of a cantilevered quay wall during the Bhuj earthquake of 2001, where significant ground settlements occurred in the backfill concurrently with a large outward movements of the wall. As observed by Zeng and Steedman (1993), failure of retaining and quay walls during earthquakes, with associated damage to structures founded on the backfill, is caused either by a local failure in the structural members or by large displacements of the ground exceeding the serviceability conditions. It follows that two aspects are of major concern when dealing with the dynamic behaviour of retaining walls, i.e., (i) the increase of internal forces in the wall due to the inertia forces acting into the soil, and (ii) the possible occurrence of permanent displacements due to full mobilisation of the soil resistance.

S. Aversa (✉) · L. de Sanctis · R. M. S. Maiorano · M. Tricarico
Dipartimento di Ingegneria, Università di Napoli Parthenope,
Centro Direzionale isola c4, 80143 Napoli, Italy
e-mail: stefano.aversa@uniparthenope.it

G. Viggiani
Dipartimento di Ingegneria Civile, Università di Roma Tor Vergata,
Via del Politecnico, 1, 00133 Rome, Italy

R. Conti
Dipartimento di Ingegneria, Università degli Studi Niccolò Cusano,
Via Don Carlo Gnocchi, 3, 00166 Roma, Italy

G. S. P. Madabhushi
Schofield Centre, University of Cambridge, High Cross Madingley Road,
Cambs, Cambridge CB3 0EL UK

F. Taucer, R. Apostolska (eds.), *Experimental Research in Earthquake Engineering*,
Geotechnical, Geological and Earthquake Engineering 35,
DOI 10.1007/978-3-319-10136-1_33

Following the works by Okabe (1924) and Mononobe and Matsuo (1929), several studies have been devoted to the problem of computing dynamic earth pressures on retaining structures, both with theoretical (Steedman and Zeng 1990; Lancellotta 2007; Mylonakis et al. 2007; Kim et al. 2010) and experimental (Atik and Sitar 2010) approaches. More recent work has focused on the dynamic behaviour of flexible embedded retaining walls (Madabhushi and Zeng 2007; Cilingir et al. 2011; Conti et al. 2012), and some of the major findings have been embodied in a more rational design of this type of structures in seismic conditions (Conti and Viggiani 2013). Relatively few experimental works, however, have considered the presence of water in the backfill (Zeng and Steedman 1993; Dewoolkar et al. 2001; Lee et al. 2005; Takahashi et al. 2006; Towhata et al. 2009), and most of them were devoted to the seismic behaviour of caisson and gravity retaining walls.

It is recognised by many authors that the presence of water in the soil can affect the seismic behaviour of retaining walls in three ways (Kramer 1996): (i) by altering the inertia forces within the backfill, (ii) by developing hydrodynamic pressures and (iii) by generating excess pore water pressures due to cyclic loading. To our knowledge, however, rational procedures to take into account these effects on the seismic design of retaining structures are still lacking, and the current practice usually relies on approximate, and quite crude, solutions (Matsuzawa et al. 1985).

This work presents the results of four centrifuge tests carried out in the geotechnical centrifuge of Cambridge University Engineering Department (CUED), under a TA (Transnational Access) proposal joining the two research groups of Università di Roma Tor Vergata and Napoli Parthenope (research project PROPWALL). The tests were performed on reduced scale models of pairs of retaining walls, both cantilevered and with one level of support near the top, embedded in saturated sand reconstituted at two different values of relative density. Based on a systematic comparison between experimental results obtained from models in dense sand and models in loose sand, the main objective of this work is to shed some light on the main mechanisms affecting the dynamic behaviour of embedded retaining walls in the presence of pore water pressures. Emphasis is given on how, and to which extent, the occurrence of excess pore water pressures and/or liquefaction phenomena within the soil can affect the behaviour of these structures during earthquakes, both from a structural and a geotechnical point of view.

In the following, all physical quantities are given at prototype scale, with the only exception of the geometrical dimensions of the models. Moreover, accelerations are positive rightwards and the horizontal displacements of the walls are positive towards the excavation.

33.2 Experimental Setup and Model Preparation

The experimental work included a set of four tests on embedded walls in saturated sand, with a piezometric head at dredge level (see Fig. 33.1). Two tests were carried out on pairs of cantilevered walls (CWU1 and CWU2), and two tests on models

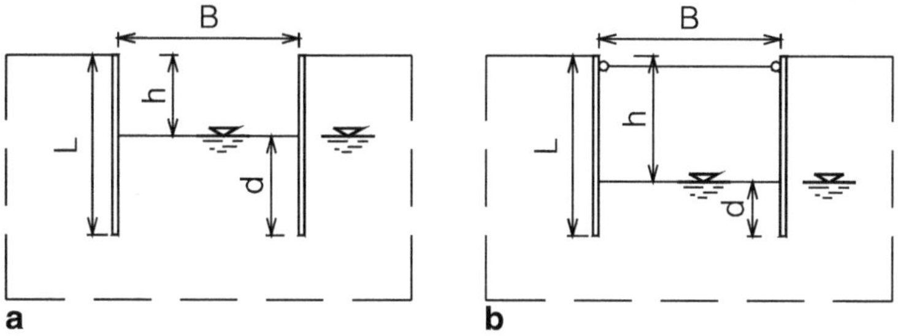

Fig. 33.1 Reference models and geometrical features **a** tests CWU on cantilevered walls and **b** tests PWU on propped walls

Table 33.1 Geometrical properties at model and prototype scale

			Model scale (mm)					Prototype scale (m)				
Test	D_R (%)	*N*	*h*	*d*	*s*	*Z*	*B*	*h*	*d*	*s*	*Z*	*B*
CWU1	38	40	90	110	–	400	200	3.6	4.4	–	10.8	8
CWU2	80	40	90	110	–	400	200	3.6	4.4	–	10.8	8
PWU1	38	40	140	60	9	400	200	5.6	2.4	0.45	10.8	8
PWU2	80	40	140	60	9	400	200	5.6	2.4	0.45	10.8	8

of propped walls (PWU1 and PWU2). The dimensions of the geotechnical structures and the frequency content were chosen to be representative of real problems. A centrifugal acceleration equal to 40 g was selected. As detailed in Table 33.1, the total length of the model retaining walls was $L=200$ mm, corresponding to a total length at prototype scale of 8 m. Excavation depths, *h*, of 90 mm (CWU) and 140 mm (PWU) were considered, corresponding to 3.6 m and 5.6 m at prototype scale, which are very common in engineering practice. In particular, the total length of the walls, the excavation depth, h, and the embedment depth, *d*, are similar to those already adopted in past experimental work on retaining walls (Conti et al. 2012) in order to isolate the effects of the saturation condition on the dynamic behaviour of this type of structures.

33.2.1 Seismic Actuator and Dynamic Container

Seismic excitations in the centrifuge tests are generated by a Stored Angular Momentum (SAM) actuator developed at Cambridge University (Madabhushi et al. 1998). The SAM actuator can apply waves with frequencies in the range 30–50 Hz. Therefore, at the centrifugal acceleration of 40 g, the predominant frequency at prototype scale would be in the range 0.75–1.25 Hz, which can be considered representative of real earthquakes, especially those characterized by low frequency content.

To simulate boundary stress and deformation similar to those expected in a finite soil layer with indefinite lateral extent, the models were prepared in a Laminar Box (Knappett 2006), consisting of the superposition of alternating aluminium frames and cylindrical bearings. The Laminar Box was designed in order to have negligible inertia, while being sufficiently stiff not to deform under high centrifugal accelerations (Brennan et al. 2006). The internal part of the laminar box is covered by a flexible rubber sheet to avoid loosing sand and/or pore fluid through the rings separating the frames.

33.2.2 Materials and Saturation Procedure

The model retaining walls consisted of aluminium alloy plates with thickness $t = 6.36$ mm, corresponding to a bending stiffness at prototype scale similar to that of a tangent concrete pile wall with a diameter of 400 mm. For propped walls, two square aluminium rods with an axial stiffness of about 1×10^6 kN/m at prototype scale, connected to the walls by cylindrical hinges allowing rotation in the vertical plane, were located at a distance of 195 mm from each other.

A standard fine dry silica sand was used, namely Leighton Buzzard Sand 100/70, fraction E, reconstituted at two values of relative density, $D_r = 38\,\%$ and $80\,\%$. The specific gravity of the sand is $Gs = 2.65$, its maximum and minimum voids ratio are equal to 1.014 and 0.613, respectively, while its constant volume friction angle is $\varphi_{cv} = 32°$ (Tan 1990, Jeyatharan 1991). Further details on the mechanical behaviour of the sand under monotonic, cyclic and dynamic loading conditions are given by Visone and Santucci de Magistris (2009).

A high viscosity pore fluid was used, consisting of a solution of Hydroxyl-Propil Methil-Cellulose HPCM, to have a unique time scale for dynamic and diffusion phenomena. If the same pore fluid exists in the model and the prototype, the time scale for inertial effects and consolidation processes are $t_m = t_p/N$ and $t_m = t_p/N^2$ respectively. A possible strategy to solve this conflict is to scale the consolidation coefficient of the soil as $C_{vm} = C_{vp}/N$. Since the consolidation coefficient is defined as

$$c_v = \frac{k}{m_v \rho g} \tag{33.1}$$

where k is the permeability, m_v the one-dimensional compressibility of the soil, r the density of the pore fluid and g the gravitational acceleration, it follows that a reduction in permeability will cause a direct reduction in c_v.

The permeability can be expressed as (Eq. 33.2):

$$k = \frac{K\rho g}{\mu} = \frac{Kg}{\nu} \tag{33.2}$$

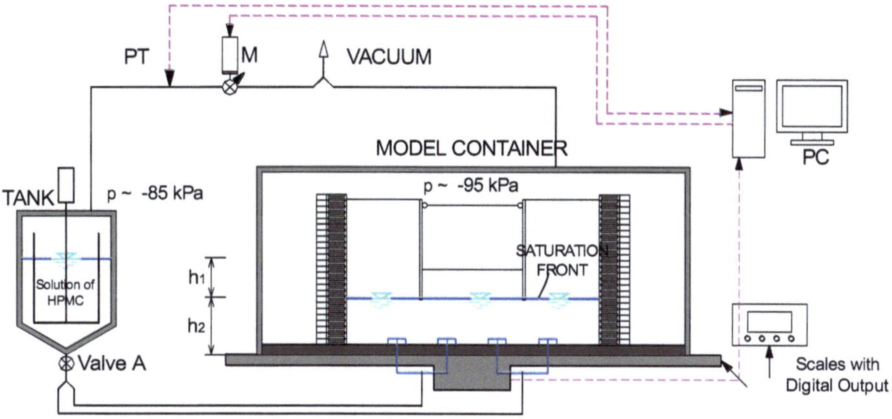

Fig. 33.2 CAM-sat system developed at University of Cambridge (after Stringer and Madabhushi 2009). *PT* (pressure transducer), *M* (motor)

where, K is the intrinsic or absolute permeability of the soil, μ is the dynamic viscosity and $\nu = \mu / \rho$ is the kinematic viscosity. It is therefore possible to reduce the permeability of the soil N times by increasing the fluid viscosity by N times. To this end, in the past, silicone oil has been used as the pore fluid. However, this choice creates a number of experimental drawbacks, such as e.g., (i) silicone oil is classified as hazardous waste, and (ii) it is resistant to most solvents. The use of a solution of HPMC is an alternative choice (Stewart et al. 1998) that has a number of advantages: (i) methylcellulose is biodegradable; (ii) the components can be easily mixed with water to a wide range of viscosities; (iii) the unit weight of the solution is almost identical to that of pure water.

The models were saturated using an experimental set-up developed at Cambridge University (Stringer and Madabhushi 2009). The CAM-sat (Fig. 33.2) is a computer-controlled system which is capable of: (i) monitoring continuously the rate of fluid entering the model; (ii) controlling the mass flow within specified targets; (iii) running for long periods without supervision.

A positive pressure difference drives fluid flow from a tank containing the viscous fluid into the model. The model is placed under vacuum to improve the degree of saturation. The mass flow into the model was measured using a set of digital scales and the rate of saturation controlled by altering the vacuum applied to the reservoir.

33.2.3 Instrumentation

A fairly comprehensive instrumentation was mounted on the models to monitor the response of the soil-wall system during both the static and dynamic stages of the tests. In particular, accelerations at different locations within the soil mass and at the model boundaries were monitored using miniaturised piezoelectric accelerometers (Acc); pore water pressures at different locations around the excavation, both behind and between the walls, were measured using Pore Pressure Transducers (PPT); settlements of the soil surface behind the walls were measured using vertically mounted LVDTs. As far as the response of the structural elements is concerned, accelerations of the walls were monitored using Micro Electro Mechanical Systems (MEMS) accelerometers directly glued to the aluminium plates, bending moments were obtained from two sets of five strain gauges (SG) located along the central section of each wall, and, for propped walls, axial forces in the props were measured using two miniaturised load cells (LC) screwed in position in the central section of each prop. The horizontal displacements of each wall were measured using two LVDTs supported by a plate fixed to a gantry rigidly connected to the base of the box.

33.3 Testing Procedures

At the beginning of the test, the centrifugal acceleration was increased in four successive steps of 10 g up to 40 g. Once the testing acceleration of 40 g had been reached, the first earthquake was applied at the base of the model. During each test, the model was subjected to a series of trains of approximately sinusoidal waves with different nominal frequencies, f, amplitudes, a_{max}, and durations, t. The basic idea was to apply earthquakes of increasing magnitude until the model walls suffered excessive rotation. Each earthquake was followed by a consolidation phase.

The first two tests were performed on cantilevered retaining walls. Test CWU1 was carried out on a loose sand layer ($D_R=38\,\%$), while test CWU2 was performed on dense sand ($D_R=80\,\%$). Figure 33.3 shows the cross section of models CWU1 and CWU2 and the layout of the instruments, which included eight miniature piezoelectric accelerometers (Acc), four LVDTs for horizontal displacements, two LVDTs for vertical displacements, eight (or ten) strain gauges for bending moments (SG), four (or five) on each wall, and eight pore pressure transducers (PPT). For test CWU2, the instrumentation included also one potentiometer to measure the settlement of the ground surface and one MEMS accelerometer to measure the wall inclination at the end of both the static and the dynamic stages.

The acceleration recorded by accelerometer ACC1, placed on the lower frame of the Laminar Box, is considered as the input acceleration during the seismic stages.

In test CWU1 two earthquakes were applied (see Table 33.2 and Fig. 33.4). The second earthquake was applied after 2 h from the end of the first earthquake, to allow dissipation of excess pore pressures. In test CWU2, the model was excited by three earthquakes with gradually increasing amplitude, up to a value of 0.2 g. The frequency of the waves was 50 Hz at model scale, i.e., 1.25 Hz at prototype scale. The duration of the signals was the same as that used for test CWU1 (Table 33.2 and Fig. 33.4).

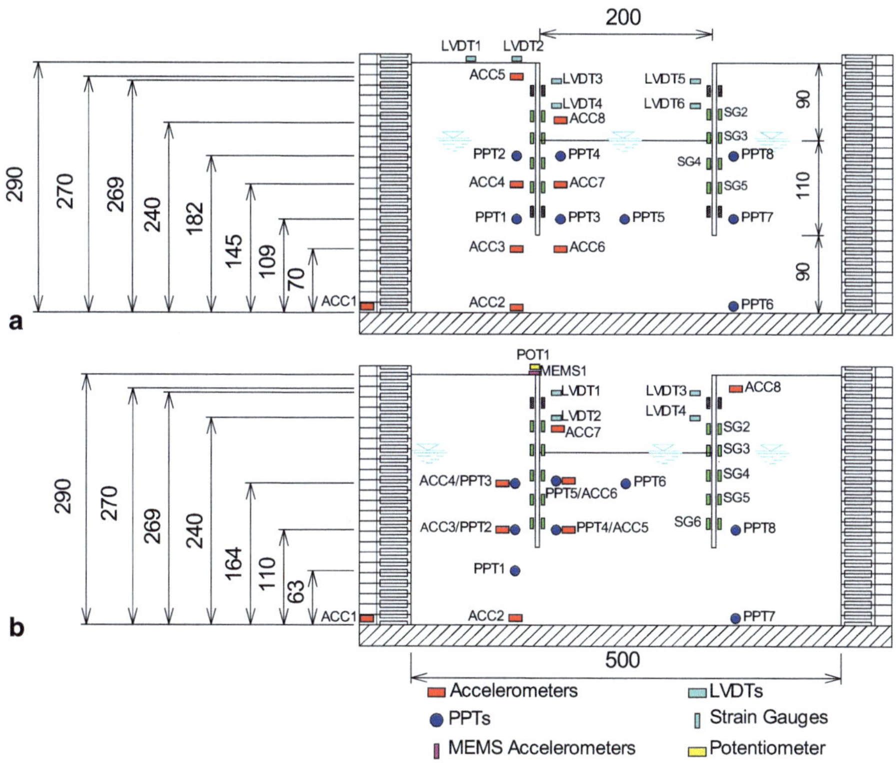

Fig. 33.3 Layout of instrumentation **a)** test CWU1, **b)** test CWU2 (lenghts are in mm)

Table 33.2 Test sCWU1 and CWU2

Test	Earthquake	f (Hz)	a_{max} (g)	Duration (s)
CWU1	EQ1	1.25	0.1	35
	EQ2	1.25	0.2	35
CWU2	EQ1	1.25	0.075	33
	EQ2	1.25	0.1	35
	EQ3	1.25	0.2	35

The other two tests were performed on single propped retaining walls. Test PWU1 was carried out on loose sand (D_R = 38 %), while test PWU2 was performed on dense sand (D_R = 80 %). Figure 33.5 shows the cross section of tests PWU1 and PWU2 and the layout of the instruments, which included eight miniature piezoelectric accelerometers (Acc), four LVDTS for horizontal displacements, one LVDT for vertical displacements, ten strain gauges for bending moments, five on each wall, and eight pore pressure transducers. Instrumentation included also two load cells to measure the axial force within the props.

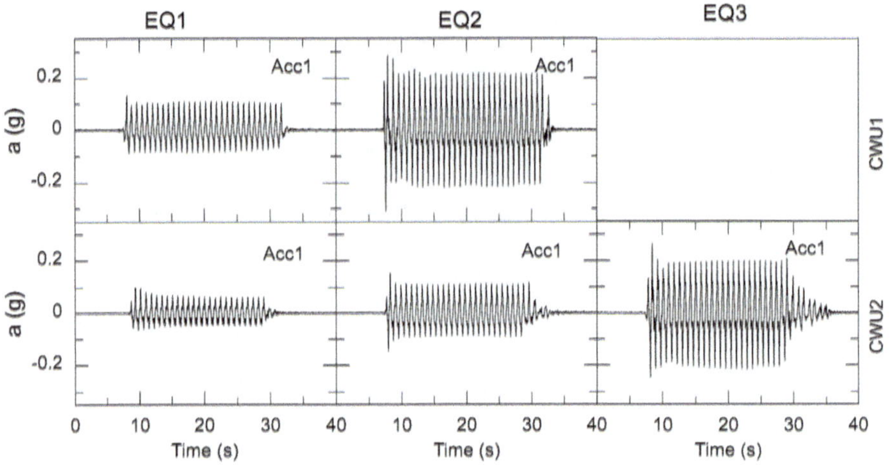

Fig. 33.4 Tests CWU1 and CWU2: Input accelerations

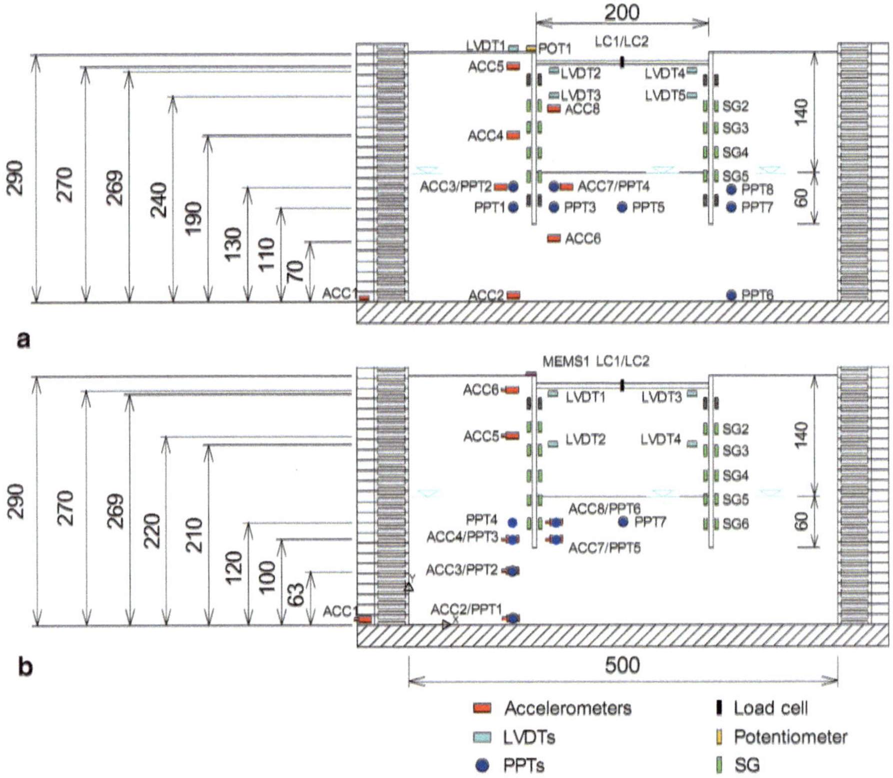

Fig. 33.5 Layout of instrumentation: **a)** test PWU1; **b)** test PWU2

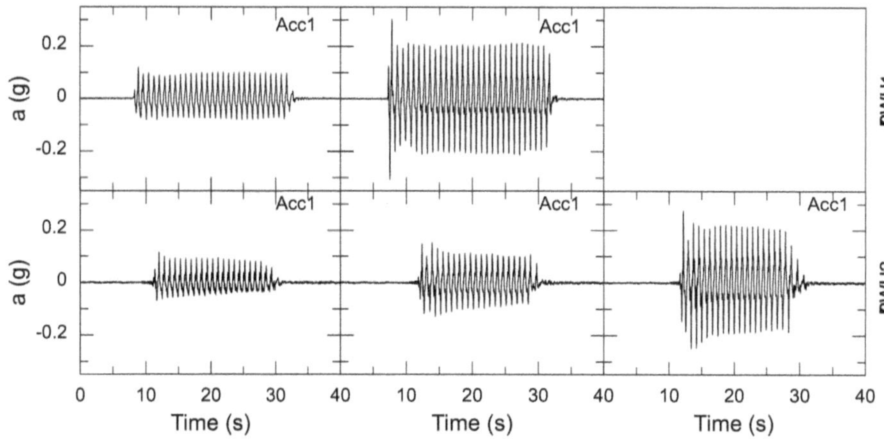

Fig. 33.6 Tests PWU1 and PWU2: Input accelerations

Table 33.3 Tests PWU1 and PWU2

Test	Earthquake	f (Hz)	a_{max} (g)	Duration (s)
PWU1	EQ1	1.25	0.1	35
	EQ2	1.25	0.2	35
PWU2	EQ1	1.25	0.1	35
	EQ2	1.25	0.13	35
	EQ3	1.25	0.25	35

Two earthquakes were applied in test PWU1 (Table 33.3), and three earthquakes with gradually increasing amplitude, up to a value of 0.2 g in test PWU2. The frequency of the waves was 50 Hz at model scale, i.e., 1.25 Hz at prototype scale. The duration of the signals is the same as that used in model test on cantilevered walls (Fig. 33.6).

33.4 Main Results

In this section the main experimental results are presented in terms of both accelerations and pore pressures within the soil layer, and displacements and internal forces in the structural members. Figure 33.7 shows a typical time history of a physical quantity (pore pressure, displacement, bending moment or axial force) measured during the applied earthquake, where the same terminology introduced by Dewoolkar et al. (2001) is adopted. The 'pre-seismic' value is the measurement at the beginning of the seismic event; the 'maximum' and 'minimum' are the maximum and minimum values that were reached during the test; the term 'dynamic' refers to the pre-seismic value and, concerning pore pressures, it coincides with 'excess'. The value measured immediately after shaking is defined as 'short-term residual' and, finally, the arrow toward the right side of the plot represents the 'long-term residual' value, recorded after the excess pore pressures are completely dissipated.

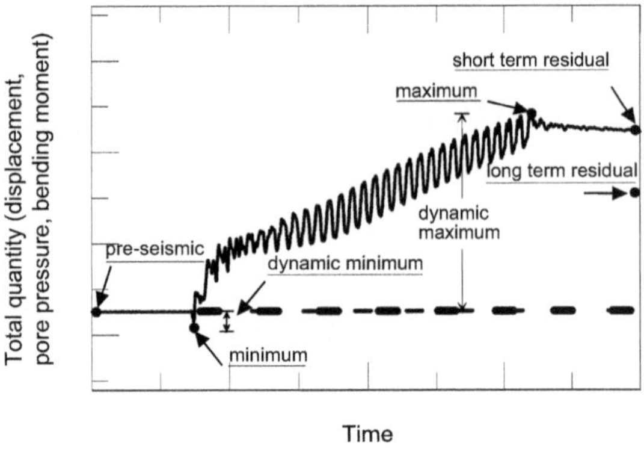

Fig. 33.7 Typical time history of a physical quantity measured during the applied earthquake and adopted terminology

33.4.1 Accelerations

Figure 33.8 shows the acceleration time histories measured at different depths within the model, both below and above the water table level, during tests (a) PWU1 (accelerometers Acc1, Acc4 and Acc5) and (b) PWU2 (accelerometers Acc1, Acc4 and Acc6). The input signals are clearly de-amplified when propagating through the saturated loose sand layer (Fig. 33.8a), the maximum accelerations recorded by Acc4 being about 0.04 g after the first cycles, irrespective of the amplitude of the applied earthquakes. On the other hand, accelerations tend to be amplified again when travelling up to the soil surface (Acc5). A completely different scenario takes place during the three earthquakes of test PWU2 (Fig. 33.8b), where the input signals do not suffer marked de-amplifications through the saturated dense sand layer. These observations suggest that liquefaction phenomena occurred during test PWU1, leading to a dramatic reduction of the soil resistance and, consistently, to a systematic de-amplification of the horizontal accelerations.

The same behaviour was observed during the two tests on cantilevered walls. Figure 33.9 shows, for all the tests and the applied earthquakes, the maximum accelerations recorded close to the soil surface against the maximum value of the corresponding input signals. The centrifuge data presented by Conti et al. (2010), obtained from a number of dynamic centrifuge tests carried out on dry sand, are also shown for comparison. Accelerations are always amplified when propagating through dry sand, irrespective of its initial density, while relative density clearly affects the response of the saturated soil models. In this case, in fact, a sort of cut-off is observed in the surface accelerations recorded in the loose models ($a_{max} \approx 0.08$ g), while substantial amplifications are still observed in the dense models.

33.4.2 Pore Pressures

As shown in Figs. 33.3 and 33.5, pore pressures were measured at different locations and depths within the soil layer. In order to check the reliability of PPT measurements, Fig. 33.10 shows a comparison between the experimental values

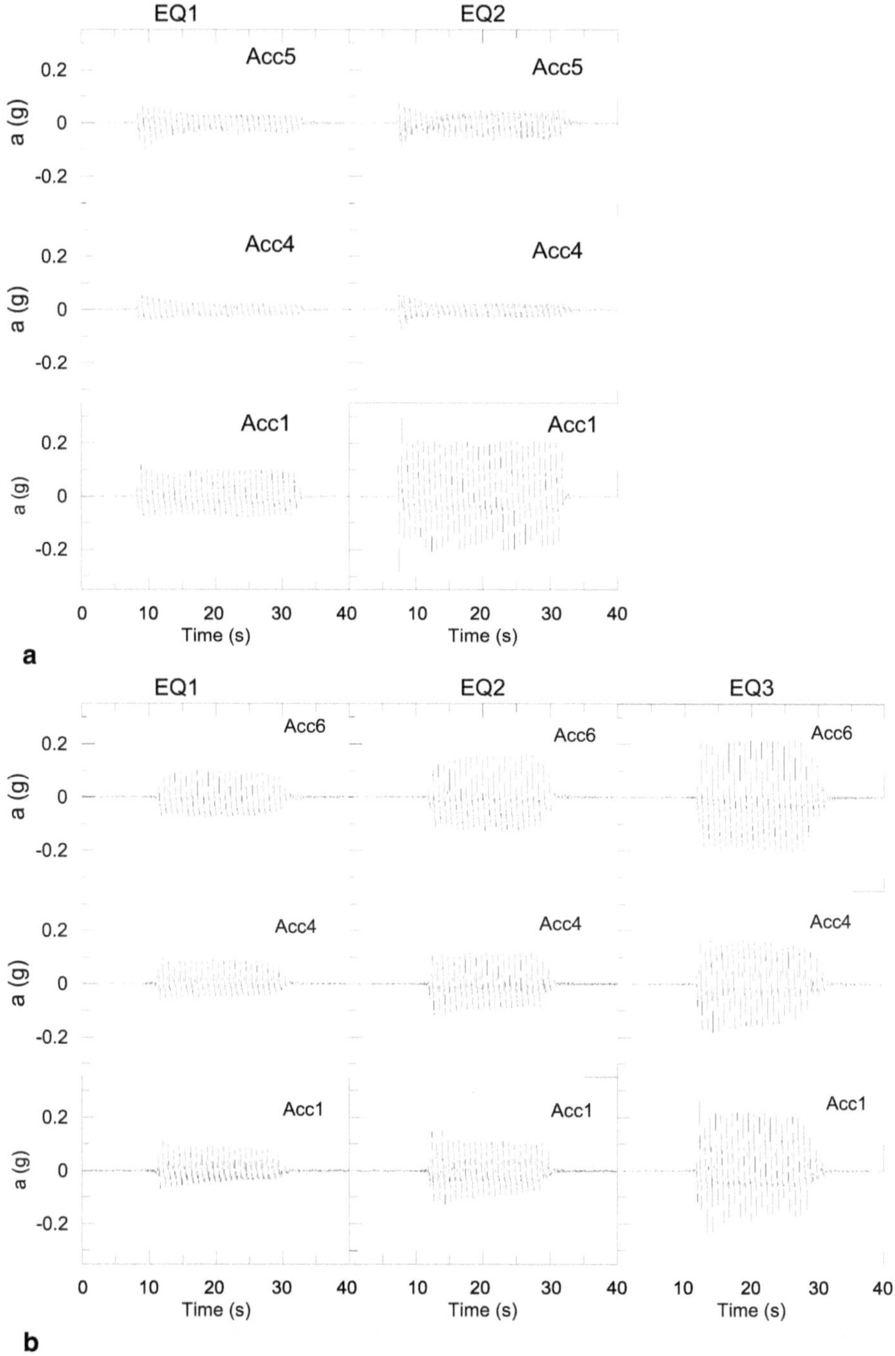

Fig. 33.8 Acceleration time histories recorded during tests **a** PWU1, **b** PWU2

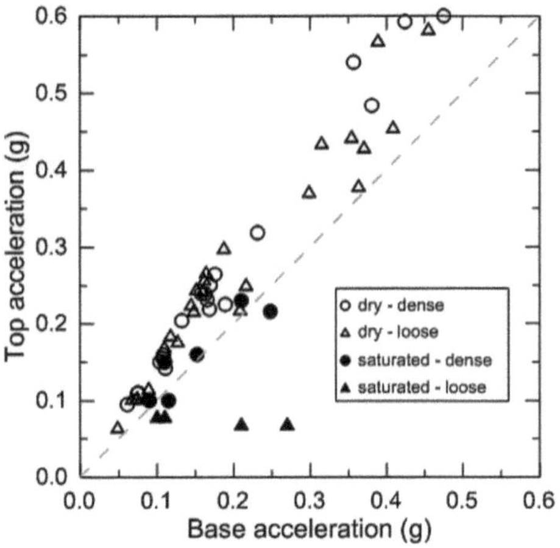

Fig. 33.9 Amplification of accelerations: comparison between maximum accelerations measured in dry sand models (Conti et al. 2012) and those recorded in the present tests

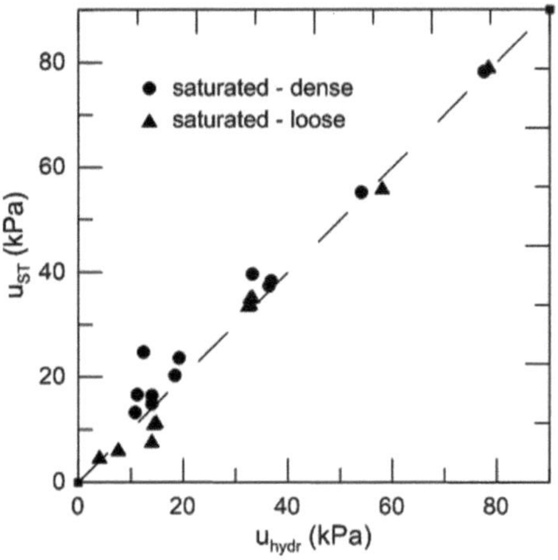

Fig. 33.10 Pore pressures at the end of the swing up stage (static): comparison between theoretical values and experimental measurements

recorded at the end of the swing up stage, corresponding to the static pore pressure distribution, and those computed assuming an hydrostatic distribution of the fluid pressure. The agreement between measured and theoretical values is rather good, with an average scatter of about 19 %, probably due to the uncertainty in the exact location of the PPT transducers.

Figure 33.11 shows the time histories of the pore pressures measured during tests PWU1 and PWU2. Excess pore pressures developed during all the dynamic events, within both the dense and the loose model, with a faster rate in the first cycles and minor oscillations during the whole duration of the earthquake. After each shaking, the excess pore pressures dissipated and the long-term residual pore pressures were close to the corresponding static values.

A liquefaction indicator often adopted in the literature is the excess pore water pressure ratio, $r_u = \Delta u / \sigma'_V$, defined as the ratio between the dynamic increment of pore water pressure and the vertical effective stress. For the centrifuge tests

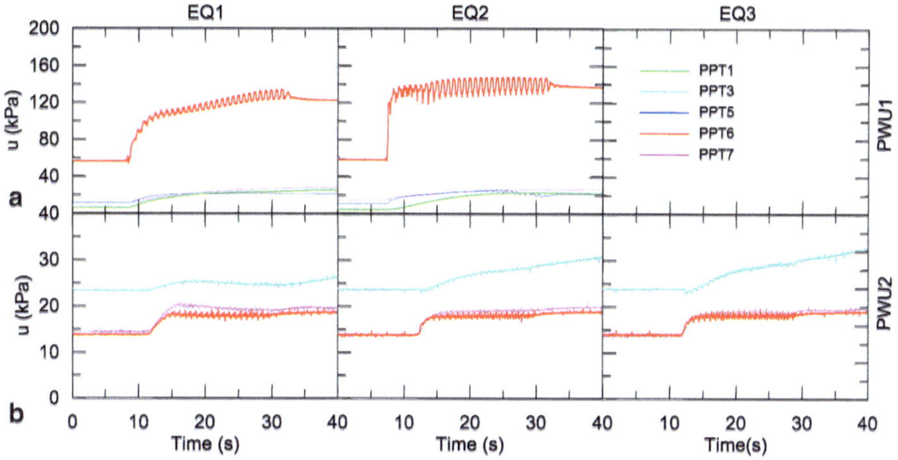

Fig. 33.11 Pore pressure time histories recorded during tests: **a)** PWU1; **b)** PWU2

Fig. 33.12 Test PWU2. Photograph of the model, showing sand boils between the two walls, at dredge level

discussed in this paper, however, the complex stress distribution expected around the excavation makes a reliable estimation of this indicator difficult. Nevertheless, the occurrence of liquefaction phenomena in the loose sand layer (PWU1) was confirmed by visual inspection of the model after the test, which revealed the presence of sand boils at dredge level between the walls (Fig. 33.12).

33.4.3 Displacements

The interpretation of the behaviour of the embedded walls in terms of displacements is not immediate. If the walls are assumed to be rigid, a possible short-term residual deformed shape of the retaining walls relative to the base of the box (absolute

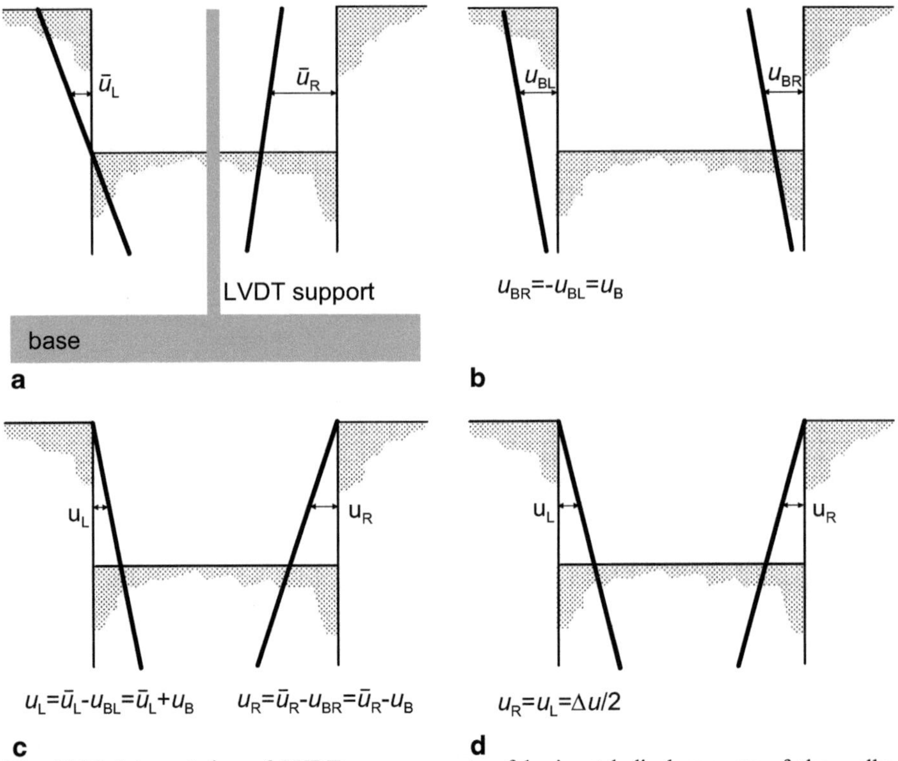

Fig. 33.13 Interpretation of LVDT measurements of horizontal displacements of the walls: **a)** absolute displacements; **b)** assumed displacements of the box boundaries; **c)** displacements of the walls relative to undeformed configuration; **d)** assumed symmetrical relative displacements

displacements) is that depicted schematically in Fig. 33.13a. With respect to the configuration shown in Fig. 33.13a, the absolute displacements of the left and right wall, $\bar{u}_L$ and $\bar{u}_R$, measured by the LVDTs at any depth, can be expressed as follows (Eqs. 33.3 and 33.4):

$$\bar{u}_L = u_L + u_{BL} = u_L - u_B \tag{33.3}$$

$$\bar{u}_R = u_R + u_{BR} = u_R + u_B \tag{33.4}$$

i.e. they are the sum of the displacements of the boundary of the laminar box, u_B, shown in Fig. 33.13b and of the unknown displacements of the walls relative to their undeformed configuration, u_L and u_R, shown in Fig. 33.13d. Note that the displacements of the boundary of the box are shown as linear in Fig. 33.13b; this is not necessarily the case, but it is irrelevant in the interpretation of the data as the horizontal displacements of the wall are measured at only two positions and are interpreted as a rigid body motion. However, the boundary conditions imposed by the Laminar Box, similar to the condition of 'periodic' or 'tied-boundaries' (Zienckiewicz et al. 1988), ensure that the displacements of the boundaries of the box at any depth, if non zero, are anti-symmetrical.

The relative displacement between the two walls can be computed from the displacements measured by the LVDTs as (Eq. 33.5):

$$\Delta u = u_R + u_{BR} = \bar{u}_R + \bar{u}_B \tag{33.5}$$

but, as the displacement of the boundary of the box is not measured during the test, it is not possible to work out the individual displacements of each wall relative to its undeformed configuration, unless some simplifying assumption is introduced. In this case, it was assumed that the displacements of the walls are 'symmetrical', i.e., $u_R = u_L$, see Fig. 33.13d. It follows that the displacements of each wall relative to the undeformed configuration can be obtained from the displacements measured by the LVDTs as (Eq. 33.6):

$$u_L = u_R = \frac{\bar{u}_L + \bar{u}_R}{2} \tag{33.6}$$

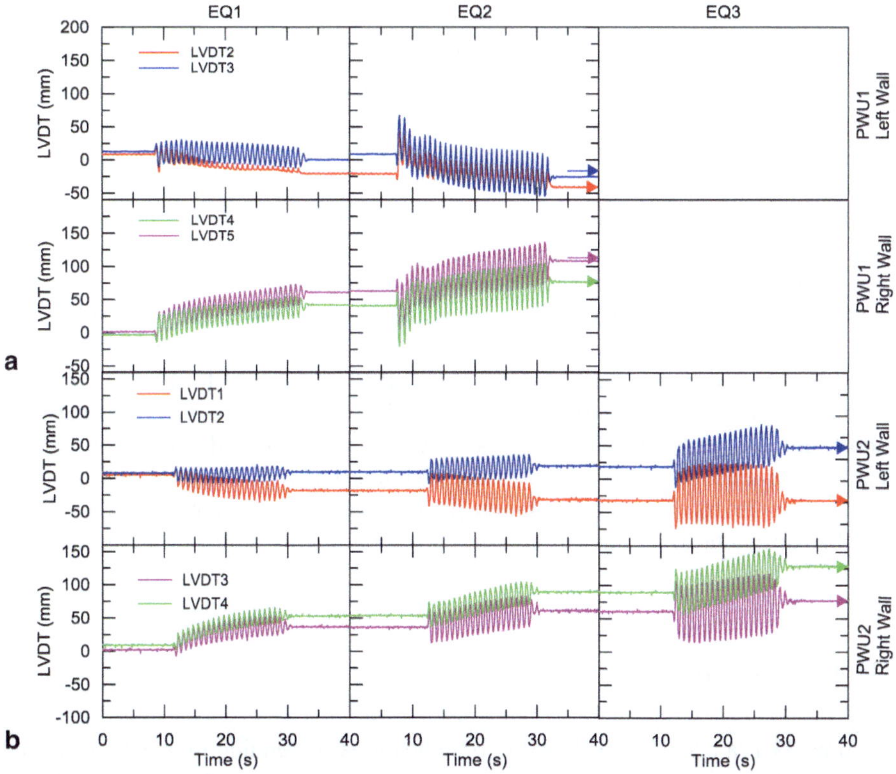

Fig. 33.14 Wall displacement time histories measured during tests: **a)** PWU1; **b)** PWU2

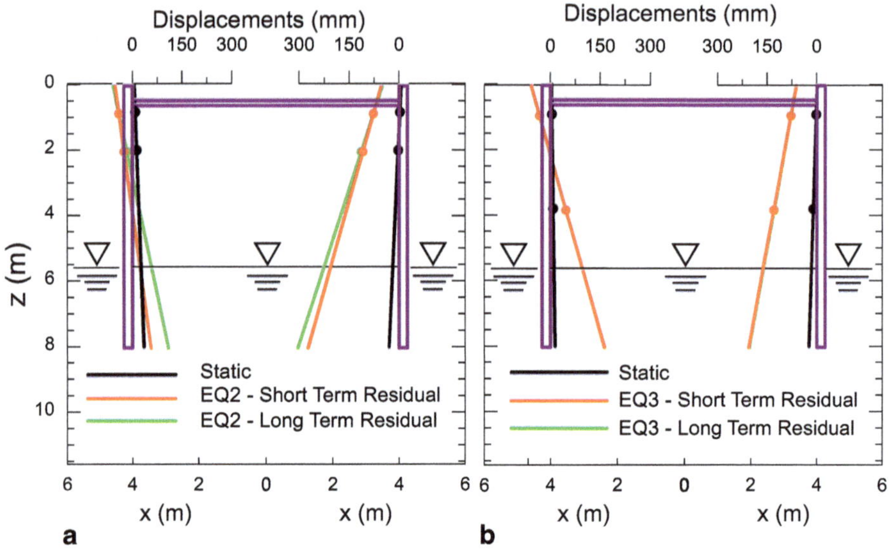

Fig. 33.15 Horizontal absolute wall displacements. Comparison between: **a)** model test PWU1; **b)** model test PWU2

This is quite a strong assumption, which may be not entirely realistic, particularly for non-symmetric acceleration time histories. However, the interpretation of wall displacements in terms of 'symmetrical' components is very convenient as it permits to emphasize the main mechanism of rotational failure of each wall under seismic actions.

The time-histories of the wall displacements measured by the LVDTs during tests PWU1 and PWU2 are plotted in Fig. 33.14. In both tests: (i) the retaining walls accumulate permanent displacements; (ii) the short term- and long term-residual displacements are practically the same, with the exception of the left wall of test PWU1. The short- and long-term residual wall displacement profiles for PWU1 and PWU2 after the last earthquake (EQ2 for PWU1 and EQ3 for PWU2), as measured from the LVDTs, are plotted in Fig. 33.15. For PWU1, the difference between short- and long-term displacements is worthy of note and is due to dissipation of earthquake-induced excess pore pressure. Symmetrical wall displacements are plotted in Fig. 33.16. This representation permits a clearer perception of the failure mechanism induced by the earthquake, corresponding to an inward rotation of the toe of the wall.

33.4.4 Bending Moments

Figures 33.17a and 33.17b show the time histories of bending moments measured on the right wall during tests PWU1 and PWU2, respectively. A completely different behaviour is observed between the two models. In test PWU2, the internal forces

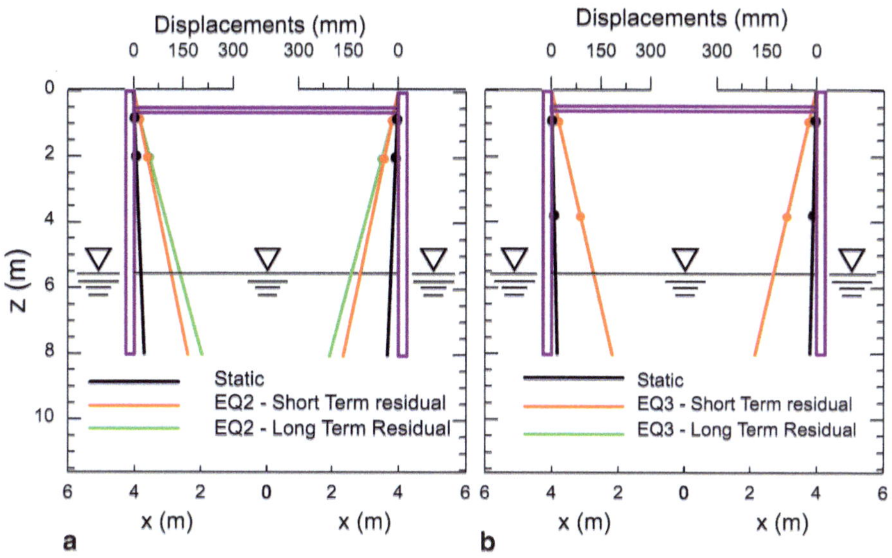

Fig. 33.16 Symmetrical displacements. Comparison between: **a)** model test PWU1; **b)** model test PWU2

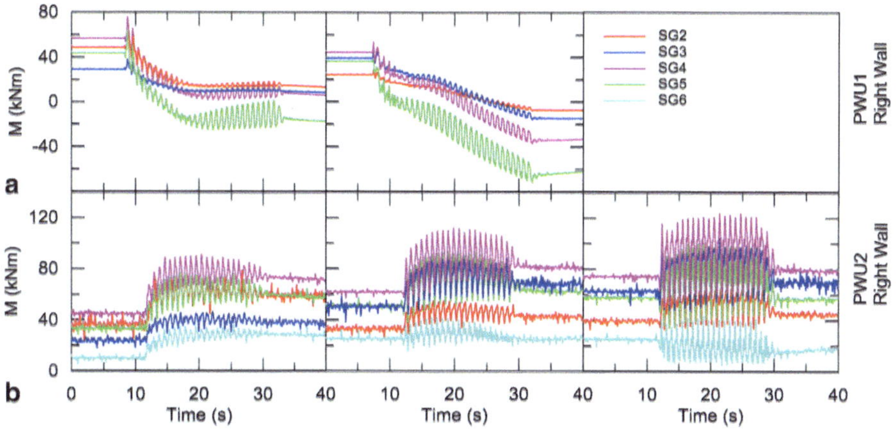

Fig. 33.17 Bending moments time histories measured on the right wall during tests: **a)** PWU1; **b)** PWU2

oscillate around an average value which continuously increases during shaking and the dynamic increments are proportional to the applied accelerations, consistently with other experimental work on embedded retaining walls in saturated sand (see e.g., Zeng and Steedman 1993; Madabhushi and Zeng 2007). On the other hand, during the transient stages of test PWU1, bending moments reduce substantially, reaching negative short-term residual values after EQ2. In both tests, dissipation of pore pressures after the earthquakes induces a redistribution of internal forces in the wall, the long-term residual values being, however, substantially different from the corresponding pre-earthquake values.

Bending moments on the walls clearly result from complex soil-structure interaction phenomena, in which many factors are likely to play a significant role, such as: the inertia forces into the soil, the excess pore pressure build-up and the consequent reduction in soil strength, the redistribution of stress and mobilisation of soil strength associated to wall displacements. While a full understanding of the problem based simply on centrifuge test results would be quite a hard task, the experimental observations suggest again that relative density affects significantly the dynamic response of retaining walls embedded in saturated sand.

33.5 Conclusions

Four dynamic centrifuge tests were performed on pairs of retaining walls embedded in saturated sand, both cantilevered and with one level of support near the top. Both the geometry of the system and the features of the applied earthquakes were similar to those considered in past experimental work on retaining walls embedded in dry sand, in order to emphasize the effects of saturation on the dynamic behaviour of such structures.

The main results have been discussed in terms of accelerations and pore pressures in the soil, and internal forces and displacements of the walls. Indirect measurements and visual inspection of the models revealed that, while excess pore pressures developed in both dense and loose models, liquefaction phenomena occurred only in the latter case, leading temporarily to a dramatic reduction of soil resistance. This fact had two major effects, which make the overall behaviour of saturated loose models substantially different from that observed in dry and saturated dense models.

Firstly, the input signals were systematically de-amplified while propagating through the loose sand layers, with a sort of cut-off observed in the accelerations recorded close to the soil surface. On the contrary, the applied earthquakes were always amplified through the dense sand layers.

Secondly, inertia forces acting on saturated dense models induced dynamic increments of bending moments which are substantially proportional to the amplitude of the input accelerations. On the other hand, a significant reduction of internal forces was observed in the loose sand models, where even negative values of bending moments were reached during the transient stages.

It is evident that relative density affects significantly the seismic response of retaining walls embedded in saturated sand, where the excess pore pressures may lead to a substantial reduction of the soil resistance and, consequently, of the inertia forces into the soil. As a result, no increase of bending moments in the walls can be expected, as they strictly depend on both the inertia acting on the retaining side and on the soil passive resistance available below dredge level. From this perspective, the problem of computing the dynamic internal forces in the structural members become of little relevance, while the ability of predicting the permanent displacements experienced by the walls play a major role in the seismic design of such structures.

Acknowledgments The research leading to these results has received funding from the European Community's Seventh Framework Programme [FP7/2007–2013] for access to the Turner Beam Centrifuge, Cambridge, UK under grant agreement n° 227887. The technical support of the research group at the Schofield Centre and in particular that of Mark Stringer and Charles Heron is also gratefully acknowledged.

References

Atik LA, Sitar N (2010) Seismic earth pressures on cantilever retaining structures. J Geotech Geoenviromental Eng 136(10):1324–1333

Brennan AJ, Madabhushi SPG, Houghton NE (2006) Comparing laminar and equivalent shear beam (ESB) containers for dynamic centrifuge modelling. In: Proceedings 6th International Conference Physics Modelling Geotechnics, 6th ICPMG '06, Hong Kong, 171–176

Cilingir U, Haigh SK, Madabhushi SPG, Zeng X (2011) Seismic behaviour of anchored quay walls with dry backfill. Geomech Geoeng 6(3):227–235

Conti R, Viggiani GMB (2013) A new limit equilibrium method for the pseudostatic design of embedded cantilevered retaining walls. Soil Dyn Earthq Eng 50:143–150

Conti R, Madabhushi SPG, Viggiani GMB (2010) Physical modelling of flexible retaining walls under seismic actions. In: Proceedings 7th International Conference Physics Modelling Geotechnics 7th ICPMG '10, Zurich, Switzerland

Conti R, Madabhushi SPG, Viggiani GMB (2012) On the behaviour of flexible retaining walls under seismic actions. Géotechnique 62(12):1081–1094

Dewoolkar MM, Ko HY, Pak RYS (2001) Seismic behaviour of cantilever retaining walls with liquefiable backfills. J Geotech Geoenvironmental Eng 127(5):424–435

Fang YS, Yang YC, Chen TJ (2003) Retaining walls damaged in the Chi-Chi earthquake. Can Geotech J 40:1142–1153

Iai S, Kameoka T (1993) Finite element analysis of earthquake induced damage to anchored sheet pile quay walls. Soils Found 33(1):71–91

Jeyatharan K (1991) Partial liquefaction of sand fill in a mobile arctic caisson under dynamic ice-loading. PhD thesis, University of Cambridge, UK

Kamon M, Wako T, Isemura K, Sawa K, Mimura M, Tateyama K, Kobayashi S (1996) Geotechnical disasters on the waterfront. Soils Foundations, Special issue, January, 137–147

Kim WC, Park D, Kim B (2010) Development of a generalised formula for dynamic active earth pressure. Géotechnique 60(9):723–727

Knappett JA (2006) Piled foundations in liquefiable soils: accounting for axial loads. PhD Thesis, Cambridge University, England

Koseki J, Koda M, Matsuo S, Takasaki H, Fujiwara T (2012) Damage to railway earth structures and foundations caused by the 2011 off the Pacific coast of Tohoku earthquake. Soils Found 52(5):872–889

Kramer SL (1996). Geotechnical earthquake engineering. Prentice Hall, New Jersey

Lancellotta R (2007) Lower-bound approach for seismic passive earth resistance. Géotechnique 57(3):319–321

Lee CJ (2005) Centrifuge modelling of the behavior of caisson-type quay walls during earthquakes. Soil Dyn Earthq Eng 25:117–131

Madabhushi SPG, Zeng X (2007) Simulating seismic response of cantilever retaining walls. J Geotech Geoenvironmental Eng 133(5):539–549

Madabhushi SPG, Schofield AN, Lesley S (1998) A new stored angular momentum (SAM) based earthquake actuator. In: Proceedings Centrifuge '98, International Conference on Centrifuge Modelling, Tokyo, 111–116

Matsuzawa H, Ishibashi I, Kawamura M (1985) Dynamic soil and water pressures of submerged soils. J Geotech Eng ASCE 111(10):1161–1176

Mononobe N, Matsuo H (1929) On the determination of earth pressure during earthquake. Proc. 2nd World Engineering Conference, Vol. 9, 177–185

Mylonakis G, Kloukinas P, Papantonopoulos C (2007) An alternative to the Mononobe-Okabe equations for seismic earth pressures. Soil Dyn Earthq Eng 27(10):957–969

Okabe S (1924) General theory of earth pressure and seismic stability of retaining wall and dam. J Jpn Soc Civil Eng, 10(6):1277–1323

Steedman RS, Zeng X (1990) The influence of phase on the calculation of pseudo-static earth pressure on retaining wall. Géotechnique 40(1):103–112

Stewart DP, Chen YR, Kutter BL (1998) Experience with the use of methylcellulose as a viscous pore fluid in centrifuge models. Geotech Test J 21(4):365–369

Stringer ME, Madabhushi SPG (2009) Novel computer-controlled saturation of dynamic centrifuge models using high viscosity fluids. Geotech Test J 32(6):53–59

Takahashi H, Kitazume M, Ishibashi S, Yamawaki, S, (2006) Evaluating the saturation of model ground by p-wave velocity and modelling of models for a liquefaction study. Int J Phys Model Geotech 6(1):13–25

Tan FSC (1990) Centrifuge and theoretical modelling of conical footings on sand. PhD Thesis, Cambridge University, England

Towhata I, Alam MJ, Honda T, Tamate S (2009) Model tests on behaviour of gravity-type quay walls subjected to strong shaking. Bull N Z Soc Earthq Eng 42(1):47–56

Visone C, Santucci de Magistris F (2009) Mechanical behaviour of the Leighton Buzzard S and 100/170 under monotonic, cyclic and dynamic loading conditions. ANIDIS, Bologna

Zeng X, Steedman RS (1993) On the behaviour of quay walls in earthquakes. Géotechnique 43(3):417–431

Zienckiewicz OC, Bicanic N, Shen FQ (1988) Earthquake input definition and the transmitting boundary conditions. Conference on Advances in Computational Non-Linear Mechanics: 109–130. Editor St. Doltnis I, 1988

Chapter 34
Experimental and Numerical Investigations of Nonlinearity in Soils Using Advanced Laboratory-Scaled Models (ENINALS Project): From a Site-Test to a Centrifuge Model

Francesca Bozzano, Salvatore Martino, Alberto Prestininzi, Gabriele Scarascia-Mugnozza, Luis Fabian Bonilla, Alberto Bretschneider, Jean Louis Chazelas, Sandra Escoffier, Luca Lenti and Jean-François Semblat

34.1 Introduction

The historical centre of Rome is built for a great part on an alluvial plain, filled by sediments from the Tiber river, and in part on hills, where pyroclastic deposits from the Albani Hill and Sabatino volcanic districts are exposed (Funiciello and Giordano 2008; Bozzano et al. 2000). The L'Aquila seismic sequence of 2009 (Caserta et al. 2013) caused some damages to Rome historical centre, especially in the area of the alluvial deposits which are characterized by a stratigraphic juxtaposition of more deformable clayey-silty deposits and stiffer sandy and sandy-silty deposits (Bozzano et al. 2011, 2012). However, much more extensive records from the last millennium point to strong earthquake-induced damage in the alluvial plain of the Tiber river and of its tributaries (Ambrosini et al. 1986; Donati et al. 1999). This is why, in the past two decades, some dedicated studies (Rovelli et al. 1995; Bozzano et al. 2008; Caserta et al. 2012) placed emphasis on the amplification effects of the Tiber alluvial deposits in Rome historical centre.

In this regard, the ENINALS (Experimental and Numerical Investigations of Nonlinearity in soils using Advanced Laboratory-Scaled models) project is intended to investigate the effects of permanent deformation induced on stratified soil

S. Martino (✉) · F. Bozzano · A. Prestininzi · G. Scarascia-Mugnozza
Department of Earth Sciences and Research Centre for Geological Risks (CERI), "Sapienza" University of Rome, Piazzale A. Moro 5, 00185 Roma, Italy
e-mail: salvatore.martino@uniroma1.it

L. F. Bonilla · L. Lenti · J.-F. Semblat
Department of Geotechnics, Environment, Natural hazards and Earth sciences, French Institute of Science and Technology for Transport, Development and Networks (IFSTTAR), 14 Boulevard Newton Cité Descartes, Champs sur Marne, 77447 Marne la Vallée, France

A. Bretschneider · J. L. Chazelas · S. Escoffier
Department of Geotechnics, Environment, Natural hazards and Earth sciences, French Institute of Science and Technology for Transport, Development and Networks (IFSTTAR), Route de Bouaye-CS4, 44344 Bouguenais Cedex, France

F. Taucer, R. Apostolska (eds.), *Experimental Research in Earthquake Engineering*, Geotechnical, Geological and Earthquake Engineering 35,
DOI 10.1007/978-3-319-10136-1_34

columns from Tiber over alluvia in Rome, with special regard to alternating succession sequence of loose sands and weakly consolidated clays, by experiencing an analogical modeling approach through the use of a centrifuge device.

34.2 The Rome Historical Centre Case Study

The area of Rome historical center (Fig. 34.1) was characterised by marine sedimentary conditions from Pliocene through early Pleistocene times (4.5–1.0 Ma) related to the Monte Vaticano Unit (UMV) stiff clays which are considered to be the geological bedrock of the area (Funiciello and Giordano 2008). During middle-late Pleistocene and Holocene, the sedimentary processes were confined to fluvial channels and coastal plains and strongly controlled by glacio-eustatic sea-level changes (Marra et al. 1998). In the same time, this area also experienced strong volcanic activity, which caused the emplacement of a thick pyroclastic cover that became intercalated into the continental sedimentary deposits (Funiciello and Giordano 2008).

The present-day hydrographic network of the Tiber valley and its tributaries, was originated since the Würm glacial period (18 ka) and it results from re-incision and deepening of valleys inherited from the previous glacial-interglacial phases. The sediments filling the Holocene valleys are generally characterised by a fining-upward succession, with a basal few meters thick level of gravels grading into a thick pack of sands and clays (Bozzano et al. 2000). This fine-grained portion of the deposit is represented by normally to weakly overconsolidated clayey and sandy silts, saturated in water, with low stiffness (Bozzano et al. 2000).

A 3D geological model of the Rome historical centre was constructed, based on lithostratigraphic data from 78 boreholes (Giacomi 2011). The boreholes depths range from 30 up to 67 m b.g.l., across the Tiber River alluvia; many of these boreholes reach the high-consistency clays of the UMV geological substratum.

The recent alluvial deposits filling the Tiber valley in Rome's historical centre (Fig. 34.1) are extremely heterogeneous, both vertically and laterally, and have very complex geometries. The complexity of their stratigraphic relations and geotechnical characteristics has previously been addressed (Milli 1997; Corazza et al. 1999; Bozzano et al. 2000).

According to Bozzano et al. (2000), the Tiber sediments are grouped into 6 lithotechnical units, from bottom to top: Lithotype G: gravels in a sandy-gravelly groundmass; Lithotype D: alternating silty sands, sandy silt and levels of silty clays; Lithotype C: grey clays and silty clays containing carbonaceous remains and peat; Lithotype B: sands and silty sands; Lithotype A: silty-sandy deposits, with fragments of bricks and historical alluvia Lithotype R: anthropogenic fill material with a large abundance of brick, tile or ceramic fragments and tuff blocks in a sandy-silty matrix.

Based on the aforementioned 3D engineering-geological model of the Tiber alluvia in the historical centre of Rome, four soil columns were selected to represent: homogeneous sandy and clayey deposits, heterogeneous deposits alternating sandy and clayey levels of different thickness, corresponding to the columns A and B (Fig. 34.1).

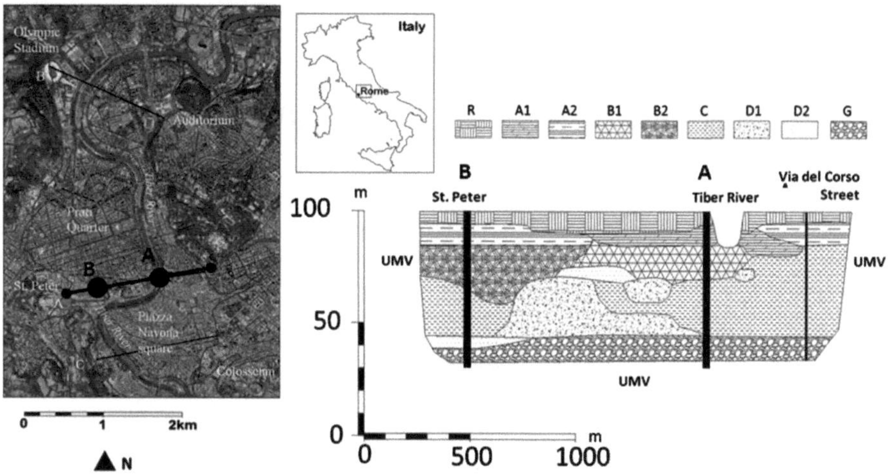

Fig. 34.1 Satellite (Google Earth) image of Rome historical centre. Letters *A* and *B* identify the location of the two real soil columns considered in this work. Soils acronyms are described in the text

34.3 Seismic Input

The dynamic input considered in this study represents the maximum expected seismic action in Rome. It was reproduced on the soil columns by dynamic centrifuge modeling following three approaches: (i) a natural signal, (ii) an equivalent sinusoidal signal, and (iii) a multifrequential equivalent signal derived by the recently proposed LEMA_DES approach. The three signals are here described.

34.3.1 Natural Reference Input

A three component time history has been produced by the CERI with the cooperation of ENEA (Research Centers of Rome), taking into account the maximum PGA expected in the historical centre of Rome (Scarascia Mugnozza 2011). The time history was derived by selecting from the European Strong-motion Database (ESD) a first set of three-component time histories, representative of the maximum expected ground motion. As a second step, the response spectra (5% inelastic damping) related to these time histories were calculated and compared to the reference response spectrum expected for Rome. The latter is already available and defined in the framework of the national project UHS INGV, Cluster 6, Central Italy. The best fit allowed selecting only one three-components time history among the whole set of data selected starting from the ESD. The horizontal component with the maximum ground acceleration value was scaled to the characteristic PGA value for the historical center of Rome. The other components were then scaled taking into account the ratios between the PGA of the three original time histories.

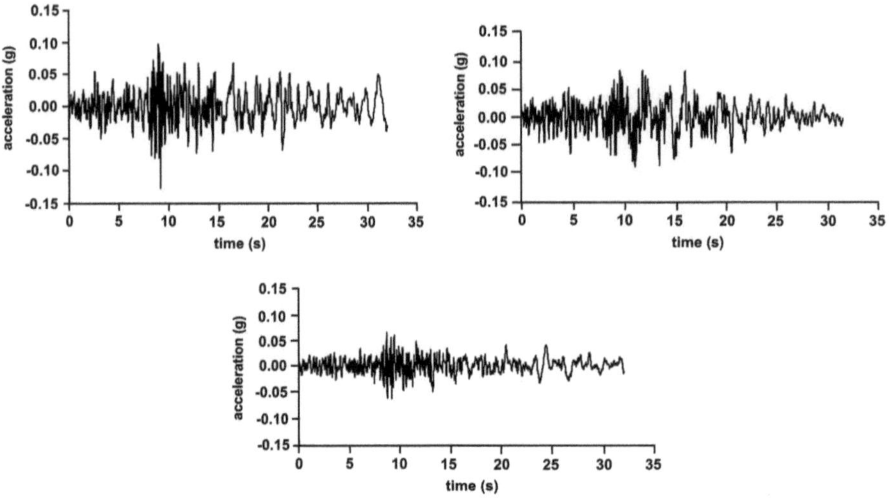

Fig. 34.2 3-components acceleration time histories (top: x and y; bottom: z) representative for the maximum ground motion expected in Rome derived according to the methodology proposed by ENEA in the frame of the "Convenzione INGV-DPC 2004–2006"

The so derived 3-components time history is characterized by a horizontal component with the maximum ground acceleration value scaled to the PGA value characteristic for the historical centre of Rome (i.e. PGA=0.1258 at 475 years).

The other components were then scaled taking into account the ratios between the PGA values associated to the three not-yet scaled components of the time history (Fig. 3,4.2). The reference input was applied at the base of the soil columns reproduced in the dynamic centrifuge modelling. As the shaking table mounted in the centrifuge can apply a motion along one direction only (hereafter y direction), the North-South (NS) component was taken into account for the physical experiences due to its highest PGA (Fig. 34.4a).

First of all it was necessary to define a reliable input to be imposed at the base of the modeled soil column since the reference signal represents the expected motion in Rome at the outcropping seismic bedrock. Since the chosen scaling factor was 55.55, the stratigraphic log of the soil columns for 20 m above the seismic bedrock was considered. The dynamic properties of the soils (i.e. physical parameters, damping and decay curves vs. shear strain for shear modulus and damping) were available from the literature on the site of Rome (Bozzano et al. 2008, 2012). In order to define the motion at 20 m in depth (Fig. 34.3) a 1D numerical nonlinear modeling of the seismic waves propagation was performed by means of the finite element code SWAP_3C (Santisi D'Avila et al. 2012). The reference NS signal divided by a factor of 2 was imposed at the base of this model; the obtained signal at the surface of the column was then divided by 2 and considered reliable to represent the expected motion at 20 m below the ground level for the remnant portion of the soil column. Hereafter this signal will be named *natural signal*.

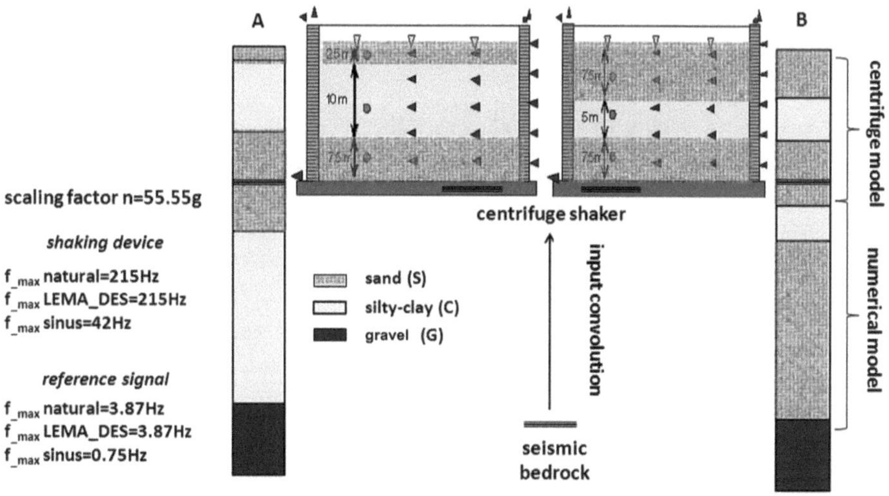

Fig. 34.3 Sketch illustrating the dynamic signal convolution from the 1D numerical models of the soil columns to the centrifuge shaker

34.3.2 Cyclic Mono-Frequency Input

A first definition of equivalent accelerometric signal to be used in geotechnical applications was proposed by Seed and Idriss (1969). The above mentioned method by Seed and Idriss does not impose specific constraints on the value of the equivalent sinusoidal frequency and, consequently, on the duration of the resulting cyclical signal.

Several experiments applied monofrequential cyclical signals to centrifuge devices, aiming at developing laboratory-scaled analogue models (Semblat and Luong 1998; Balakrishnan et al. 1998). These experiments demonstrated that the dynamic behaviour of soils depends on the frequency of the cyclic input. Their drawback lies in the fact that: (i) they are not necessarily representative of the energy content of the reference signal, and (ii) they may be unsuitable for modelling phenomena that are significantly dependent on a combination of frequencies.

Following the Seed and Idriss (1969) procedure a mono-frequency signal was obtained from the *natural signal*, hereafter named *sinusoidal signal* (Fig. 34.4b).

34.3.3 LEMA_DES Multi-Frequency Input

With reference to the above mentioned drawbacks for monochromatic equivalent signals, the use of the multifrequential dynamic equivalent signals may: (i) check that the frequency content of the derived signals is defined within a representative/admissible range; (ii) avoid upper-threshold frequency to be exceeded; (iii) narrow the energy gap between real and simulated seismic actions; (iv) control the

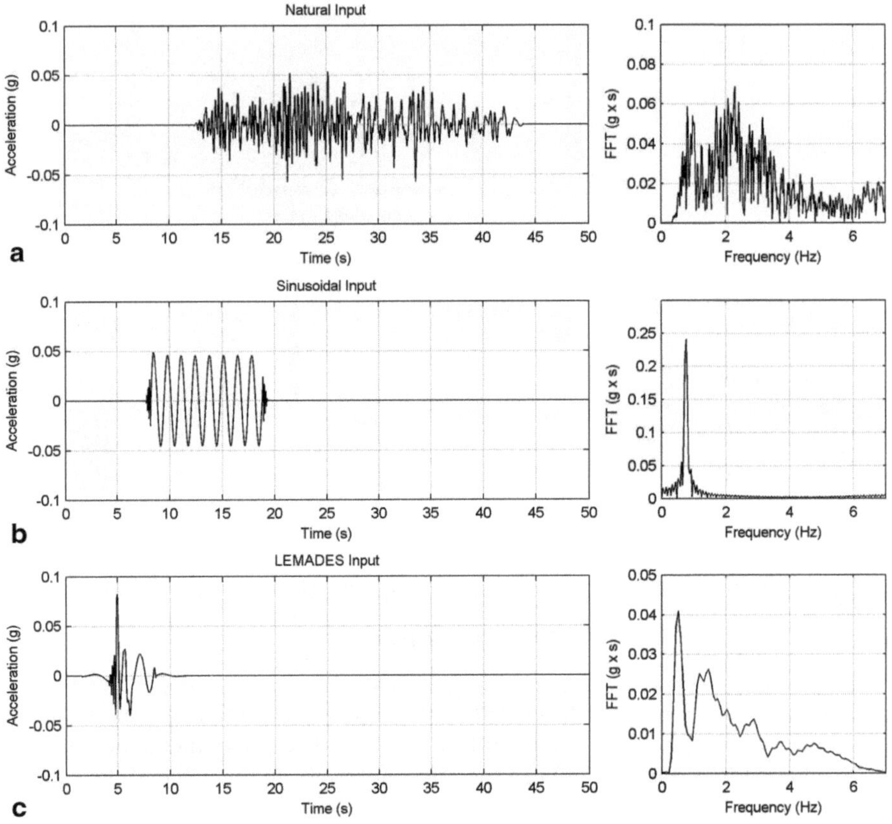

Fig. 34.4 Acceleration time histories (left column) and related FFT (right column) applied at the base of the soil columns by the shaking table in the centrifuge device: **a** natural input (*top-left*); **b** sinusoidal signal (*top-right*); **c** LEMA_DES derived signal (*bottom*)

maximum intensity of the adopted action, and (v) take into account seismically-induced effects arising from frequency combinations, i.e. from dynamic actions.

At this regard, the LEMA_DES approach proposed by Lenti and Martino (2010) optimizes the mono-frequency one, by obtaining accelerometric signals that can be regarded as more constrained to the real actions. Following this approach, a dynamic equivalent multifrequential accelerometric signal is derived by selecting a limited number of frequencies from the reference acceleration spectrum.

The dynamic equivalent signal (hereafter named *LEMA_DES signal*) is sized on the *natural signal* under criteria of energy, spectral and peak acceleration equivalence.

The above mentioned equivalence criteria imply that the number of equivalent cycles, assigned to the selected frequencies, as well as the duration of the signal are variables depending on the generation process itself. The multi-frequency

LEMA_DES signal obtained starting from the natural signal in terms of equivalent acceleration is shown on Fig. 34.4c. With reference to the procedure by Lenti and Martino (2010), the Convergence Relative Error (CRE %) for the here obtained signal is equal to 23 %, corresponding to a middle reliability of the derived signal.

34.4 Centrifuge Modelling

34.4.1 Experimental Project

A total of 12 centrifuge modeling experiments were performed at the IFSTTAR laboratory of Nantes. Four different soil columns, 20 m thick at the prototype scale, were tested: a homogeneous sand layer (soil column #1); a homogeneous clay model (soil column #2); a three layer soil composed from bottom to top of 7.50 m of sand—5 m of clay—7.50 m of sand (soil column #3) (Fig. 34.3); a three layer soil composed from bottom to top of 7.50 m of sand—10 m of clay—2.50 m of sand (soil column #4) (Fig. 34.3). All the soil columns were saturated. The first two columns can be taken as a reference for each specific soil type; the last two soil columns are representative of the shallowest 20 m of two stratigraphic columns in the Tiber alluvial deposits in Rome.

Three different sequences of dynamic signals were applied. Each sequence was composed by producing in a different order the considered signals.

34.4.2 Experimental Device

The IFSTTAR geotechnical centrifuge has a radius of 5.5 m, a capacity to place a 2-tons model and a maximum centrifuge acceleration of 100 g. For earthquake simulation tests, a shaking table device is added in the basket; it superimposes to the "static" centrifuge acceleration a "horizontal" acceleration at the base of the soil model simulating an earthquake. In dynamic configuration the performances are: a payload up to 400 kg, a maximum centrifuge acceleration of 80 g and a maximum horizontal acceleration of 0.5 g prototype (example. g. 20 g horizontal at 40 g gravity).

The main feature of the IFSTTAR shaking table is its ability to simulate as well sine inputs, natural inputs or synthetic inputs. Other characteristics are: maximum velocity of 1 m/s; maximum displacement of 5 mm; frequency range in sine earthquake full power: 30–200 Hz; frequency range in broad band earthquake ("natural earthquakes"): 30–300 Hz.

The above presented signals were scaled for a simulation at 55.55 g, filtered in the 20–400 Hz frequency band, sampled at 4096 Hz and verified for the physical limits of the earthquake simulator of IFSTTAR.

34.4.3 Experimental Setup

The box used for the whole ENINALS series of tests is an Equivalent Shear Beam (ESB) box made of 14 rectangular frames, 2.5 cm high. The frames are assemblies of hollow profiles of aluminum and corner pieces connecting the hollow profiles and giving stiffness to the frame. The total inside height of the box is 412 mm. The frames are not waterproof; it was then necessary to install an internal membrane to enable the saturation of the sand. In order to avoid a concentration of flow during the saturation, 12 holes were opened at the bottom plate: the diffusion of the fluid is enabled by a grid covering the holes, each grid supports a filter. The inside membrane is made of a 1 mm Parablon sheet in tension between a top and a bottom flange, both screwed respectively to the top ring and the bottom plate.

For the compatibility of both the fundamental equation of the dynamics and the Darcy's law, it is necessary to saturate the soil with a liquid with a viscosity of N cSt (N is the modeling scaling factor, equal to 55.55). This is achieved by using a mix of water and hydroxyl—Propyl methyl—cellulose HPMC. The product used is Culminal type MHPC 50 from Aqualon (Hercules Group). The percentage of the HPMC in water is not stable: the obtained viscosity is strongly dependent on the temperature: an increase of the temperature can easily generate a viscosity decrease of 10 cSt.

34.4.3.1 Preparation of the Sample Boxes: Soil Columns #1 and #2

For predisposing the box of the soil column #1 the sand was air pluviated using the fine Fontainebleau sand, a well sorted silica fine sand with a $D_{50} = 0.22$ mm, a $D_{10} = 0.14$ mm and an uniformity coefficient $CU = 1.6$. The values of dry unit weight are: $\gamma_{d\ min} = 13.93$ kN/m^3 and $\gamma_{d\ max} = 16.78$ kN/m^3; $e_{min} = 0.55$ and $e_{max} = 0.86$. The relative density of the pluviated sand is a function of the following parameters: dimension of the slot for the sand, moving speed of the hooper, height of fall. To prepare this box a slot of 4 mm, a speed of the hooper of 11 Hz (expressed as the frequency value of the motor) and a height of fall of 60 cm were set. Applying these parameters, the sand layer had a relative density of 57 % and a $\gamma_d = 15.22$ kN/m^3, which is the minimum value that can be obtained using this procedure. The pluviation was stopped at the level corresponding to the sensors setup. The sensors were then positioned, their real position was measured and noted. Then the pluviation was continued to the next sensor level. After the pluviation, the ESB box was connected to the saturation system.

For predisposing the box of the soil column #2 the 36 cm high clay bed to be used in this set of tests was prepared by overlapping three layers, each with a thickness of approximately 12 cm. The used material is the Speswhite clay, which is delivered in powder. This powder was mixed with water to obtain a slurry with a water content of 70 %. The homogeneous clay layers were prepared directly into the ESB box equipped with the inside membrane. A first layer of 5 mm coarse sand, covered

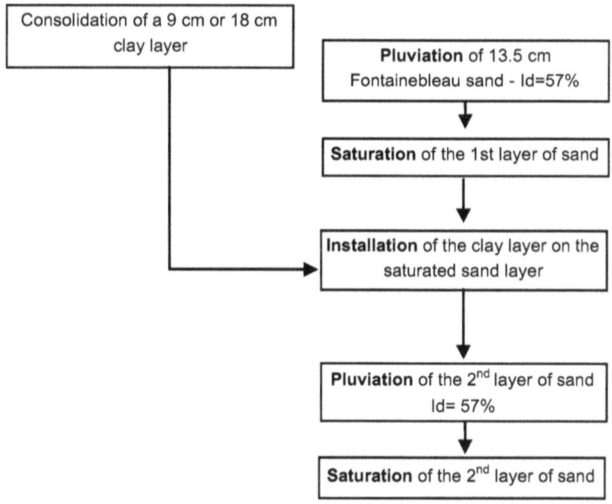

Fig. 34.5 Flow chart showing the procedure adopted to prepare the multilayer soil column #3

by a geotextile, was setup at the bottom of the box. For this clay layer drainage was allowed both at the top and the bottom surfaces. The saturated clay was consolidated following a ramp of increasing stresses with a final vertical stress of 300 kPa.

34.4.3.2 Preparation of the Sample Boxes: Soil Columns #3 and #4

For the setup of the multilayer soil columns, the clay was first prepared separately and then put into the box after pluviation and saturation of the first sand layer. The whole process is explained in the flowchart of Fig. 34.5.

Similarly to the homogeneous clay soil column, the layer is prepared by mixing pure Speswhite clay with water to obtain a slurry around 70 % water content. The slurry was then poured in a circular box Ø 0.90 m. The final vertical stress level was set in order to over-consolidate the clay and reduce rebound during the subsequent preparation process. The preparation of this layer followed the same ramp of increasing vertical loads used for the soil column #2.

The two sand layers were saturated separately: the lower layer was saturated normally while the upper sand layer, over the clay, was saturated after the setup of the clay block by a second circuit of saturation. When the set-up of the clay was achieved, a horizontal pipe on the four rectangular sides of the box was plugged on a vertical pipe connected with the saturation device, so to be positioned at the bottom of the second sand layer. Then a second pluviation was performed to complete the box.

A "cut and install" procedure was experienced to realize the multilayer samples. The process of installation of the consolidated clay layer used a suction pad and a cutting skirt made of PVC sheets fixed on the periphery of a flat sucker. The process is detailed in Fig. 34.6.

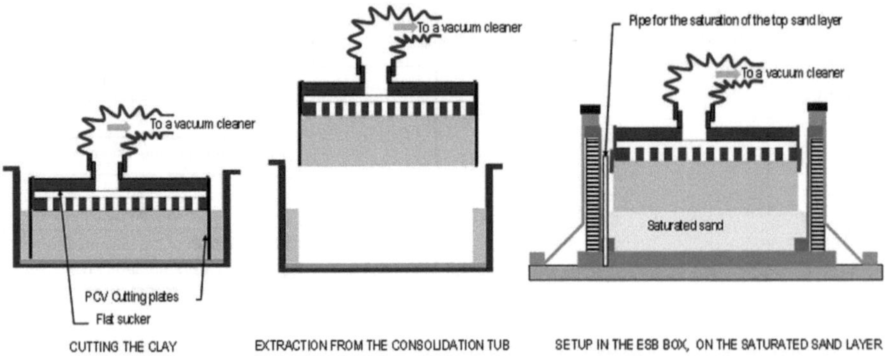

Fig. 34.6 Installation of the consolidated clay layer in the ESB box

When the clay is completely cut, the clay block is extracted and maintained under the sucker by the vacuum. The block was then transferred to a table where the cutting plates are withdrawn and replaced by a simple large tape band covering the limit between the sucker and the clay. The clay block then was lifted and set upon the saturated sand layer in the ESB box. The vacuum cleaner can be switched off and the tape and the sucker removed.

Keller pore pressure transducers and B&K type 4517 micro-accelerometers were installed within the sample boxes as exemplified in Fig. 34.7, LVDT displacement sensors were installed after the installation of the model in the centrifuge.

34.4.4 Seismic Signals Analysis

34.4.4.1 Signal Reproduction

The seismic signals ready-made and scaled for the applied scaling ratio, were acquired by the software which controls the shaking table. The same software is able to record the reproduction of the signal on the table by saving the acceleration values on the three orthogonal directions x, y, z. It is then possible to compare the reference signal with the signal reproduced on the table on the y direction during each shake of each experiment. By a comparison of the measures on the orthogonal directions, it is also possible to evaluate the spurious accelerations of the table in x and z direction.

A total of 128 shakes were shot in the whole experimental program on the 12 tests. Analyzing the data of all the shakes, the signal was reproduced homogeneously and with the same characteristics for all the tests, except for some small and negligible variations. The *natural signal* (Fig. 34.8) was very well reproduced by the shaker, as can be seen comparing the acceleration both in time and frequency domains.

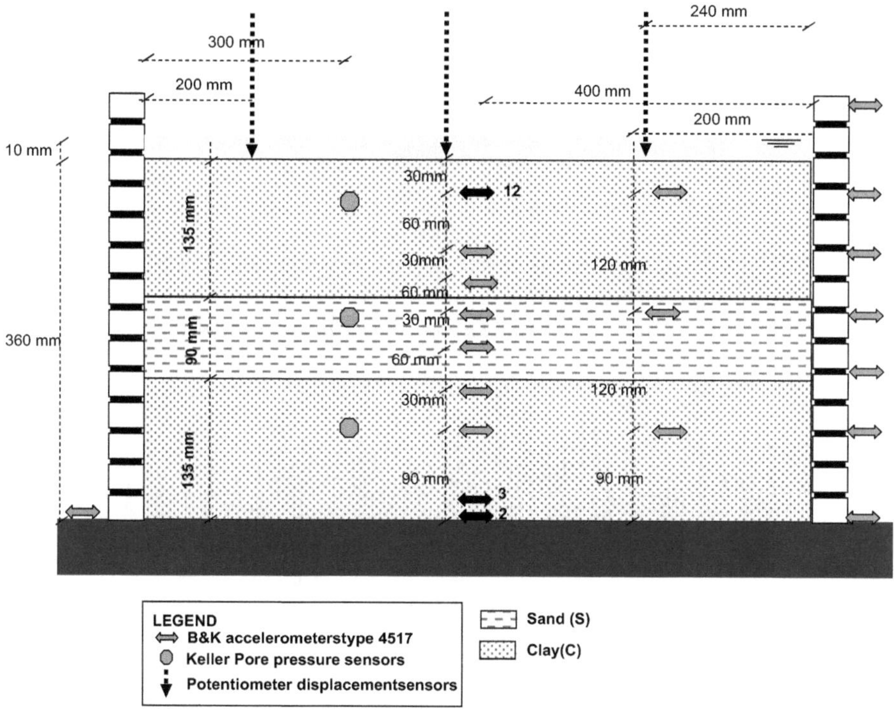

Fig. 34.7 Sketch illustrating the ESB box instrumented with the monitoring system in the case of a multilayer sample

Some spurious accelerations are not completely negligible on the x direction because the global level is weak for the shaker, while they are negligible on the z direction. All the frequency values reported hereafter are referred to the prototype scale. The *LEMA_DES signal* (Fig. 34.9) presents an excellent fitting in the frequency domain till 4 Hz, some frequency higher than the reference can be noted around 7 Hz. The spurious movements of the table are relevant only for some frequencies around 3.5 Hz, but in general the signal quality is good. The shaker also reproduced very well the *sinusoidal signal* (Fig. 34.10), with an excellent correspondence of the signals up to 5 Hz. Some frequencies around 7 Hz generate some troubles on the signal. Spurious accelerations are negligible. The spurious movements of the table are relevant only for some frequencies around 3.5 Hz, but in general the signal quality is good.

34.4.4.2 Signal Transmission

In order to evaluate the efficiency of the transmission of the input reproduced by the shaking table to the bottom of the soil column the accelerations recorded at these positions were taken into account. The Fast Fourier Transforms (FFT) of the signals

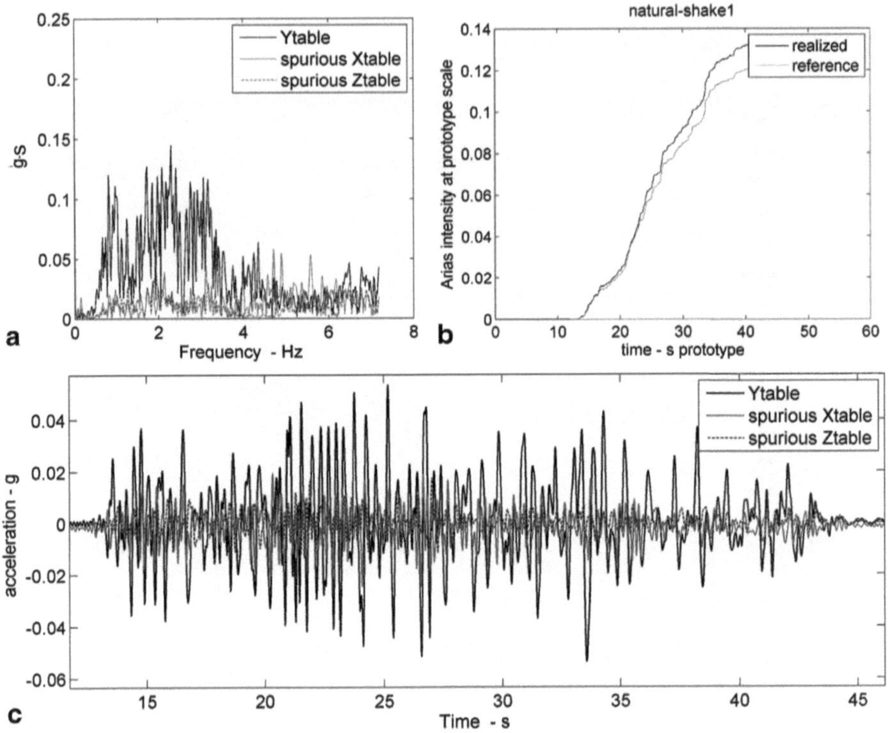

Fig. 34.8 *Natural signal* reproduced by the shaker upscaled to prototype scale: **a** FFT (*top-left*); **b** Arias Intensity of the *natural signal* (*top-right*); **c** accelerometric time histories (*bottom*)

for the three kinds of solicitations and for the records corresponding to channels two and three of Fig. 34.7 were computed.

A spectral estimator, useful to put in evidence the differences between the signals recorded at channel two and three, was adopted following the Eq. 34.1 below:

$$E = \sqrt{\sum_{i=1}^{N} (x_i - y_i)^2} \tag{34.1}$$

where i is the sum index moving from one up to the maximum number of points in the FFT, x_i and y_i are the spectral amplitude values related to the records at channels two and three, and E the estimation of the error. No normalization on the total number of points was adopted because the FFTs were characterized by the same total number of points in the frequency band 0–10 Hz.

The error values are 0.00852, 0.00926 and 0.01483 m/s respectively for the *natural*, the *LEMA_DES* and *sinusoidal signals*, for the whole frequency band retained for the sum. If only the high frequency band (i.e. 5–10 Hz) is retained for the sum, the values are 0.00838, 0.00924 and 0.01459 m/s respectively for the *natural*, *LEMA_DES* and *sinusoidal signals*. This very preliminary analysis suggests that a better reproduction of the signal can be expected for the *natural* and for the *LEMA_DES signals*. An FFT comparison (point 3 vs. 12 of Fig. 34.7) was also performed

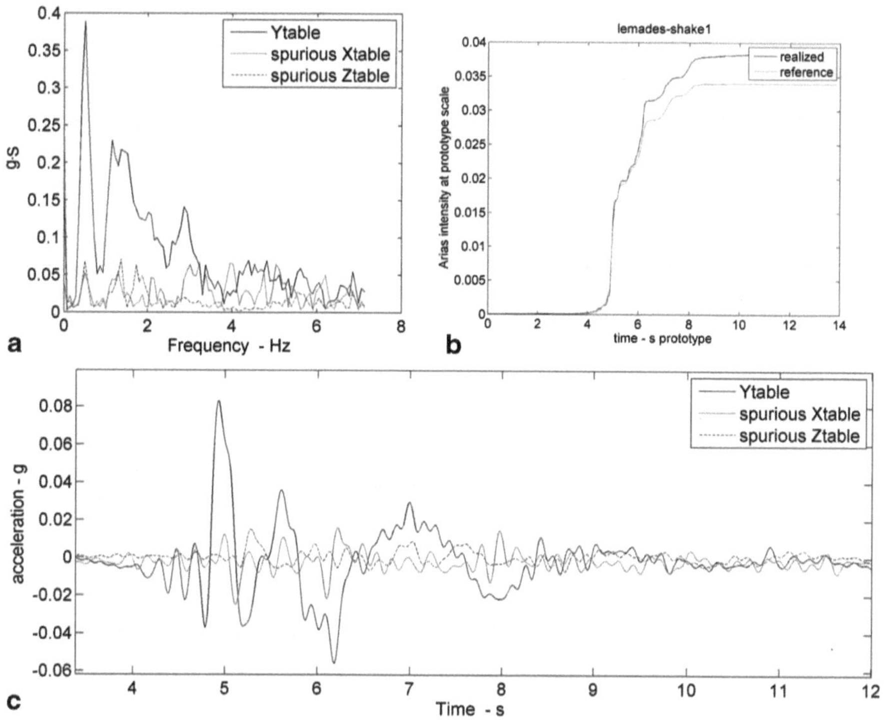

Fig. 34.9 *LEMA_DES signal* reproduced by the shake upscaled to prototype scale: **a** FFT (*top-left*); **b** Arias Intensity of the "natural input" signal (*top-right*); **c** accelerometric time histories (*bottom*)

to evaluate the generation of high frequencies (5–10 Hz) within the model due to the vibration of the structural components of the sample box. It resulted in a very similar error of about 3 % for the *natural* and the *LEMA_DES signals* while a lower error (about 1 %) resulted for the *sinusoidal signal* (but in this case it should be taken into account that the *sinusoidal signal* cannot represent frequencies higher than 5 Hz). This output demonstrates the good quality of the equivalent *LEMA_DES signal* with respect to the *natural* one.

34.5 Conclusions

The preliminary results obtained in the frame of the SERIES-ENINALS project are summarized in the following:

- a centrifuge modeling was carried out by using multilayer samples composed of a clay level within sandy ones. To this aim, specific "cut and install" technical solutions were adopted.

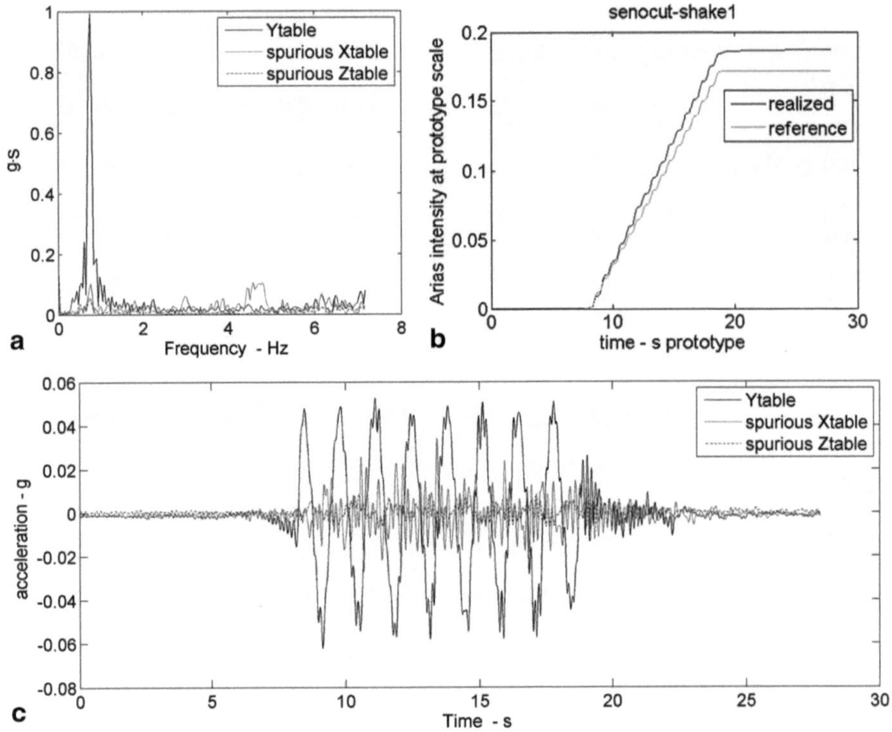

Fig. 34.10 *Sinusoidal signal* reproduced by the shaker upscaled to prototype scale: **a** FFT (*top-left*); **b** Arias Intensity of the "natural input" signal (*top-right*); **c** accelerometric time histories (*bottom*)

- dynamic input representing the maximum expected seismic action in Rome was reproduced at the shaking device by using three different signals: (i) a natural time history, (ii) an equivalent sinusoidal signal, and (iii) a multifrequential equivalent signal derived by the recently proposed LEMA_DES approach. The obtained results, here presented in a very preliminary version, encourage the use of LEMA_DES derived dynamic signals respect to sinusoidal ones.

As a future perspective, the ENINALS results will be used for back scaling the obtained results to the natural reference system. To this aim, non linear numerical modelling will be performed by the use of fractional derivative approaches (Bonilla et al. 2012). Moreover, a 3D numerical modelling of the centrifuge experiments will provide new elements for evaluating the quality of the LEMA_DES derived equivalent signals with respect to the natural ones for both analogical and numerical applications. The experimental outputs will be applied at the Rome historical centre case history to provide the expected shear strain levels due to the expected strong motions, i.e. to indicate possible non-linear deformation within the heterogeneous deposits which fill the Tiber river alluvial valley.

This application can represent a useful tool for mapping the susceptibility to earthquake-induced ground effects related to shear strain deformations, i.e. for seismic microzonation studies. With particular reference to the here considered Rome case study it is worth stressing that this application can represent a complementary tool for the application of the 2008 Italian official technical rule "Indirizzi e Criteri di Microzonazione Sismica" (ICMS 2008).

Acknowledgments The research leading to these results has received funding from the European Community's Seventh Framework Programme [FP7/2007–2013] for access to the IFSTTAR Centrifuge, Nantes, France under grant agreement n° 227887.

References

Ambrosini S, Castenetto S, Cevolan F, Di Loreto E, Funiciello R, Liperi L, Molin D (1986) Risposta sismica dell'area urbana di Roma in occasione del terremoto del Fucino del 13 Gennaio 1915. Mem Soc Geol It 35:445–452

Balakrishnan A, Kutter BL, Idriss IM (1998) Centrifuge testing of remediation of liquefaction at bridge sites. Transp Res Rec 1633:26–37 (Liquefaction. Differ Settl Found Eng)

Bonilla F, Lenti L, Martino S, Nicolas JF, Rovelli A, Semblat JF (2012) Fractional derivatives: an alternative approach to model seismic wave attenuation and amplification. In: 15 World Conference on Earthquake Engineering. Lisbon, 24–28 September 2012

Bozzano F, Andreucci A, Gaeta M, Salucci R, Rosa C (2000) A geological model of the buried Tiber River valley beneath the historical center of Rome. Bull Eng Geol Env 59:1–21

Bozzano F, Caserta A, Govoni A, Marra F, Martino S (2008) Static and dynamic characterization of alluvial deposits in the Tiber River valley: new data for assessing potential ground motion in the city of Rome. JGR 113:1–21

Bozzano F, Giacomi AC, Martino S, Corpo Nazionale Vigili del Fuoco (2011) Scenario di danneggiamento indotto nella cittÃ di Roma dalla sequenza sismica aquilana del 2009. It J Eng Geol Env 2:5–22

Bozzano F, Bretschneider A, Giacomi AC, Lenti L, Martino S, Scarascia Mugnozza G, Stedile L, Tacita JL (2012) Undrained behaviour of the sandy-silty levels of the Tiber river alluvial deposits in Rome (Italy). It J Eng Geol Env 2:47–61

Caserta A, Martino S, Bozzano F, Marra F, Govoni A (2012) Dynamic properties of low velocity alluvial deposits influencing seismically-induced shear strains: the Grottaperfetta valley test-site (Rome, Italy). Bull Earthq Eng 10:1133–1162

Caserta A, Boore DM, Rovelli A, Govoni A, Marra F, Della Monica G, Boschi E (2013) Ground motions recorded in Rome during the April 2009 L'Aquila seismic sequence: site response and comparison with ground-motion predictions based on a global dataset. BSSA 103(3):1860–1874

Corazza A, Lanzini M, Rosa C, Salucci R (1999) Caratteri stratigrafici, idrogeologici e geotecnici delle alluvioni tiberine nel settore del centro storico di Roma. Il Quaternario 12:131–176

Donati S, Funiciello R, Rovelli A (1999) Seismic response in archaeological areas: the case-history of Rome. J Appl Geophys 41:229–239

Funiciello R, Giordano G. (2008) Carta geologica del di Roma-Foglio 354 scala 1:50000, Dipartimento Scienze geologiche UniversitÃ Roma 3, Ufficio di protezione civile e Dipartimento protezione del suolo-APAT

Giacomi AC (2011) Implications of river system on recent Tiber River alluvium. Il Quaternario, It J Quat Sci 24:110–112

ICMS (2008) Indirizzi e Criteri per la Microzonazione Sismica, Ed. Dipartimento della Protezione Civile, pp 515. http://www.protezionecivile.gov.it
Lenti L, Martino S (2010) New procedure for deriving multifrequential dynamic equivalent signals (LEMA_DES): a test-study based on Italian accelerometric records. Bul Earthq Eng 8:813–846
Marra F, Rosa C, De Rita D, Funiciello R (1998) Stratigraphic and tectonic features of the middle Pleistocene sedimentary and volcanic deposits in the area of Rome (Italy). Quat Int 47–48:51–63
Milli S (1997) Depositional setting and high-frequency sequence stratigraphy of the middle upper Pleistocene to Holocene deposits of the Roman basin. Geologica Romana 33:99–136
Rovelli A, Malagnini L, Caserta A, Marra F (1995) Using 1-D and 2-D modelling of ground motion for seismic zonation criteria: results for the city of Rome. Annali di Geofisica 38(5–6):591–605
Santisi d'Avila MP, Lenti L, Semblat JF (2012) Modeling strong seismic motion: 3D loading path vs. wavefield polarization. Geophys J Int 190:1607–1624
Scarascia Mugnozza G (2011) La PericolositÃ sismica nel Lazio—PerricolositÃ sismica di base, analisi di risposta sismica locale e studi per la microzonazione sismica, Italian Journal of Engineering Geology and Environment—Quaderni (4), Casa Editrice "La Sapienza", ISBN: 978-88-95814-61-2, pp 246
Seed HB, Idriss IM (1969) Influence of soil conditions on ground motion during earthquakes. J Soil Mech Found Div ASCE 95:99–137
Semblat JF, Luong MP (1998) Wave propagation through soils in centrifuge testing. J Earthq Eng 2(1):147–171

Chapter 35
Damping Estimation from Seismic Records

Dionisio Bernal

35.1 Introduction

The energy input from an earthquake to a building is the work done on the structure by the forces acting at the soil-structure interface. Energy balance shows that this work equals the sum of the kinetic and the strain energies plus the running integral of the work of the non-conservative forces. It is customary to divide the non-conservative work into the part that can be accounted for by structural hysteresis and the remainder. The reminder is the sum of many unspecified mechanisms and it is referred to as "the damping". It is well known that damping in structural systems is not viscous but "equivalent viscosity" is commonly assumed on grounds of mathematical convenience.

Characterization of equivalent viscous damping has not been central in earthquake engineering because the strength and ductility needed to ensure adequate structural performance for large earthquake shaking are relatively insensitive to the damping, especially for wide band excitations. There are applications, however, where the response to earthquakes associated with short recurrence periods is of interest (e.g., for protection of sensitive contents) and in these cases the specification of the damping can have a significant effect on the predicted response. This paper reviews some theoretical issues in the conventional selection of equivalent damping, discusses the issues that arise in identification of damping from seismic records and reviews some recent formulas derived from the statistical analysis of a relatively large data set.

D. Bernal (✉)
Civil and Environmental Engineering, Center for Digital Signal Processing,
Northeastern University, Boston, MA 02115, USA
e-mail: bernal@neu.edu

F. Taucer, R. Apostolska (eds.), *Experimental Research in Earthquake Engineering*,
Geotechnical, Geological and Earthquake Engineering 35,
DOI 10.1007/978-3-319-10136-1_35

35.2 The Viscous Model

If equivalent viscosity is accepted the damping model, either explicitly or implicitly, is described by a damping matrix which we designate as $C \in R^{nxn}$; where n is the number of degrees of freedom.

35.2.1 Classical-Damping

In the classical model the damping matrix is diagonalized by a congruent transformation using the un-damped mode shape matrix Φ, namely (Eq. 35.1)

$$\Phi^T C\Phi = \begin{bmatrix} 2\omega_1\xi_1 & & & \\ & 2\omega_2\xi_2 & & \\ & & \cdot & \\ & & & 2\omega_n\xi_n \end{bmatrix} \tag{35.1}$$

where the mode shapes are normalized to the mass matrix and ω_j and ξ_j are the un-damped frequency and damping ratio of the jth mode. A necessary and sufficient condition for the damping matrix to be classical is that Φ, the eigenvector matrix of $M^{-1}K$, is also the eigenvector matrix of $M^{-1}C$. This can be shown as follows: multiplying the homogeneous equations of motion by the inverse of the mass matrix one has (Eq. 35.2)

$$\ddot{q} + M^{-1}C\dot{q} + M^{-1}Kq = 0 \tag{35.2}$$

If the matrices in the second and third terms on the lhs share the eigenvectors one has (Eq. 35.3)

$$\Phi\Phi^{-1}\ddot{q} + \Phi\Lambda_c\Phi^{-1}\dot{q} + \Phi\Lambda_k\Phi^{-1}q = 0 \tag{35.3}$$

So taking $Y = \Phi^{-1}q$ and pre-multiplying by Φ^{-1} gives Eq. 35.4

$$\ddot{Y} + \Lambda_c\dot{Y} + \Lambda_k Y = 0 \tag{35.4}$$

which is a diagonal system. Since matrices that share eigenvectors commute one gathers that classical damping matrices are those for which $M^{-1}K$ and $M^{-1}C$ commute. A widely used classical damping matrix is the Rayleigh model given by Eq. 35.5

$$C = \alpha M + \beta K \tag{35.5}$$

An issue sometimes discussed in the literature of earthquake engineering is whether the stiffness in Eq. 35.5 should be the original or the tangent stiffness (Priestley and Grant 2005). The question translates as to whether a model of viscous dissipation with constant coefficients is reasonable for representing the non-hysteretic dissipation when the response is nonlinear. Settling this question on theoretical grounds is difficult due to the nature of the damping model but the most common practice appears to be (and is the writer's preference) to keep the C matrix independent of yielding.

An aspect that one must be aware of regarding yielding and the damping model pertains to the equilibrium at massless coordinates. Namely, when the damping is specified as classical, equilibrium at these coordinates is established without damping contributions, but when yielding ensues damping forces arise and these can lead to physically meaningless results, e.g., to large moments at the free ends of a cantilever, or to the observation that columns that are much stronger than the connecting beams yield in a simulation. This behavior is discussed in detail in Bernal (1994), where a solution that removes the spurious effects while retaining a constant damping matrix is put forth.

The fact that $M^{-1}C$ shares the eigenvectors with $M^{-1}K$ when the damping is classical implies that any classical damping matrix can be written in terms of mode shapes as Eq. 35.6

$$C = M\left(\sum_{j=1}^{n} \eta_j \phi_j \phi_j^T\right) M \tag{35.6}$$

where the η_j are arbitrary coefficients. It is also possible, without computing the mode shapes, to specify classical damping matrices using the Caughey series (Caughey 1960; Caughey and O'Kelly 1965) (Eq. 35.7)

$$C = M \sum_b a_b (M^{-1}K)^b \tag{35.7}$$

where b are arbitrary integers and the coefficients a_b are related to the damping ratios by Eq. 35.8

$$\xi_j = \frac{1}{2\omega_j} \sum_b a_b \omega_j^{2b} \tag{35.8}$$

35.2.2 Non-Classical Damping

Support for the classical premise is found on the fact that, except for cases where the distance between two eigenvalues is small, errors in response predictions due to deviations from the classical model are small in lightly damped structures. This result is not widely appreciated so Appendix I presents a derivation that shows why this is so. In practice it is customary to speak about damping ratios whether the damping matrix is classical or not, although the concept is actually restricted to the classical model. The definition of damping ratio as the negative of the real part of the poles divided by the pole magnitude can be nominally applied in general but one should keep in mind that in the non-classical case it does not follow that the multiplication of the damping matrix by a scalar increases the "damping ratios" by the same scalar.

35.3 Damping Identification

Identification methods that apply exclusively to SDOF systems like the logarithmic decrement or the half-power bandwidth approach can be found in any structural dynamics textbook and are not repeated here. We limit the discussion to an outline of the basics of time domain identification. Time domain algorithms operate in two steps. In the first step a model that maps the sampled input to the sampled output is obtained from the digital data and in the second step the model is converted to continuous time. The postulated model in sampled time has the form (Eq. 35.9)

$$x_{k+1} = A_d x_k + B_d u_k \tag{35.9}$$

and the expression relating the state to the measurements is (Eq. 35.10)

$$y_k = C_d x_k + D_d u_k \tag{35.10}$$

where u_k is the measured input, y_k is the output and the matrices $\{A_d B_d C_d D_d\}$ are determined from the measurements. Equations 35.9 and 35.10 allow one to write Eqs. 35.11 and 35.12a, b

$$y_k = CA_d^k x_0 + \sum_{j=0}^{k} Y_j u_{k-j} \tag{35.11}$$

where

$$Y_0 = D_d \quad Y_j = C_d A_d^{j-1} B_d \tag{35.12a, b}$$

The matrices in Eq. 35.12a, b are known as the discrete time Markov Parameters and time domain identification hinges in their computation from the observations (Bursi and Wag 2009). Given the Markov Parameters a realization for the discrete time matrices can be obtained using, for example, the Eigensystem Realization

Algorithm (ERA). The representation of the structure in continuous time (Eq. 35.13) which underlies Eq. 35.9 is

$$\dot{x}(t) = A_c x(t) + B_c u(t) \tag{35.13}$$

where the eigenproperties of the physical system are encoded in the matrix A_c. The relation between the discrete and the continuous time A_c matrices is given by Eq. 35.14

$$A_c = \frac{1}{\Delta t}\log(A_d) \tag{35.14}$$

where Δt is the time step of the sampled data. Let the eigenvalues of the matrix A_c be given by $\lambda_j = \lambda_{R,j} + \lambda_{I,j} i$. The ratio of critical damping (strictly valid only in the classical model) is defined as (Eq. 35.15)

$$\xi_j = -\frac{\lambda_{R,j}}{|\lambda_j|} \tag{35.15}$$

35.4 Uncertainty in Damping Estimation

Since the inherent damping in structures is specified from observations of similar structures it is evident that its experimental identification is essential. The accuracy with which any parameter can be estimated from noisy data is limited by the amount of information on the parameter that is contained in the data. In the case of damping the information in seismic records is typically small and, as a consequence, the variance with which damping can be identified is high. In general the Fisher Information (FI) on a parameter θ contained in some measurements Y is (Eq. 35.16)

$$I(\theta) = \mathbf{E}\left(\frac{\partial}{\partial\theta}\log f(Y|\theta)\right)^2 \tag{35.16}$$

where $f(Y|\theta)$ is the likelihood function of the observed data given the parameter θ. In practice the likelihood function $f(Y|\theta)$ is generally unknown so other quantities derived from the data are typically used to estimate $I(\theta)$. For example, if the data can be used to generate a vector X whose distribution is a member of the linear exponential family having a mean $\gamma(\theta)$ and a covariance Σ then the FI of the parameter θ contained in X can be obtained as in Eq. 35.17a (van den Bos 2007)

$$I(\theta) = \mathcal{J}(\theta)^T \Sigma^{-1} \mathcal{J}(\theta) \tag{35.17a}$$

where

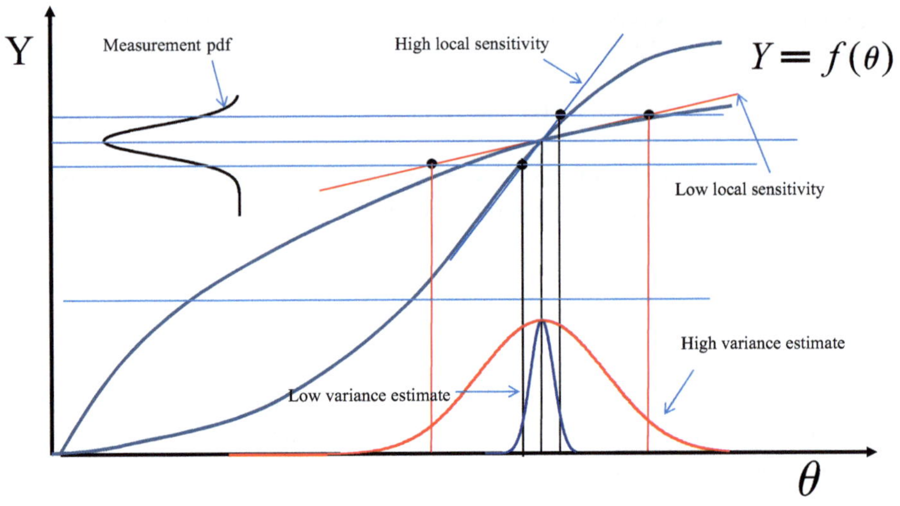

Fig. 35.1 Schematic illustration of Eq. 35.17 in the scalar case. Ratio of coefficient of variation for frequency and damping

$$\mathcal{J}(\theta) = \frac{\partial \gamma}{\partial \theta} \tag{35.17b}$$

To illustrate the significance of Eq. 35.17a, b let the "true" value of Y be deterministically dependent on θ as depicted schematically in Fig. 35.1 for two cases. Assume one wants to know the value of θ based on noisy values of Y. From the sketch it is evident that the statistical accuracy of θ depends on the slope of the functional relation at the location of the estimate and it is not difficult to see that the variances are related by the square of the local slope.

It is shown in Bernal et al. (2013) that the sensitivity of a pole with respect to damping ratio and frequency is given by Eq. 35.18

$$\mathcal{J}_{\xi,f} = \frac{\partial(\Re(\lambda),\Im(\lambda))}{\partial(\xi,f)} = 2\pi \begin{bmatrix} -f & -\xi \\ -f\xi(1-\xi^2)^{-\frac{1}{2}} & \sqrt{1-\xi^2} \end{bmatrix} \tag{35.18}$$

An analytical relationship between the coefficients of variation (COV) of damping and frequency can be obtained from Eq. 35.18 and the relation shows that the ratio depends only on the damping ratio. Assuming that the uncertainty region around the poles is circular one finds that the ratios between the COVs are as given in Fig. 35.2 As can be seen, the uncertainty on the damping ratios is around 50 times higher than that for the frequencies at $\xi = 2\%$, and the ratio is near 25 for $\xi = 5\%$. These results are consistent with the findings in Gersch (1974), where maximum likelihood estimation of modal parameters from ARMA models was considered.

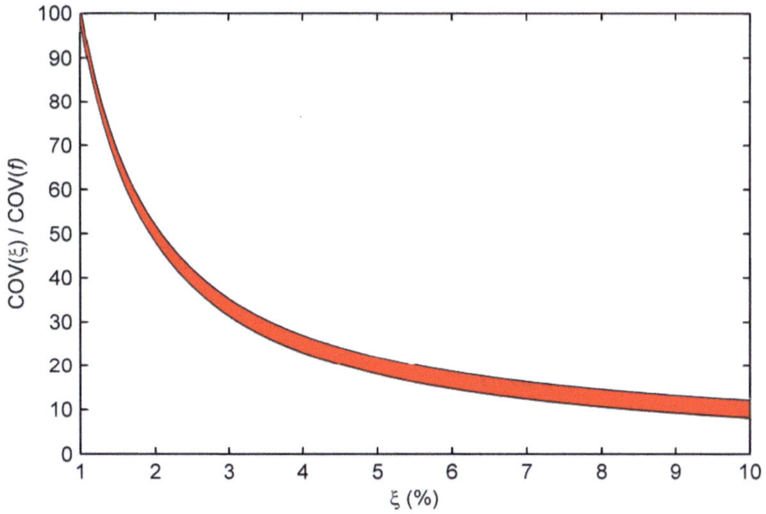

Fig. 35.2 Range of the ratio of the coefficient of variation of damping and frequency when the uncertainty region around the pole is circular

35.5 Regression Analysis

It has been traditional to specify the damping as 2 % for steel structures and 5 % for concrete. Efforts to determine parameters that can reduce the variance associated with this practice have been carried out in the past few decades. Here we review recent results obtained in a study whose details are summarized in Bernal et al. (2014). With ϑ as the regressors and θ as the vector of model parameters one has, in general Eq. 35.19

$$\xi \cong g(\theta, \vartheta) \tag{35.19}$$

The functional relationship g in the case of damping is not suggested by theory so it must be selected from inspection of the data. The regressors considered in the study reported in Bernal et al. (2013) were the peak ground motion parameters (PGA, PGV and PGD), the spectral ordinates (SA, SV and SD), the building height, H, the modal frequency, f, and the effective duration $t_{0.9}$. This last entry defined as the time interval between the attainment of 5 and 95 % of the total integral of the acceleration squared (Arias 1970). Not included due to lack of information, but potentially important, are parameters related to the soil and the foundation and to the type and density of partitions. Building material is not a parameter in the regression because the buildings were pre-classified by material type. The parameters θ are typically obtained as (Eq. 35.20)

$$\min_\theta \left\| w_i (g(\theta, \vartheta_i) - \xi_i) \right\|_2^2 \tag{35.20}$$

where ω_i are weights. It is shown in Bernal et al. (2013) that the standard deviation of the damping can be taken as (Eq. 35.21)

$$\sigma_\xi \propto \frac{1}{\sqrt{f \cdot t_{0.9}}} \tag{35.21}$$

In the estimation of the regression model parameters the weights ω_i in Eq. 35.20 are taken as the reciprocals of Eq. 35.21.

35.5.1 Functional Form

Two exponential forms were decided on: the first one for cases where the damping decreases and the other for cases where the damping increases with the regressor, namely Eqs. 35.22 and 35.23

$$\xi = a_0 + a_1 e^{-a_2 \theta} \tag{35.22}$$

and

$$\xi = \frac{b_0}{1 + b_1 e^{-b_2 \theta}} \tag{35.23}$$

35.6 Results

The regression of the damping ratios was carried out for the first mode for steel, concrete, masonry, and wood buildings. When the mode considered was dominated by translation in one direction the ground motion in that direction was used to compute the ground motion parameters. When the mode was strongly coupled, or torsional, the average of the ground motion parameters for the two directions was used. The best results for the expected value of the first mode damping ratios found are (Eqs. 35.24–35.27):

$$\xi_s = 1.22 + 4.26 e^{-0.013H} \text{ (steel)} \tag{35.24}$$

$$\xi_c = 2.91 + 3.54 e^{-0.018H} \text{ (concrete)} \tag{35.25}$$

$$\xi_m = \frac{1}{0.11 + 0.23 e^{-8.84 S_A}} \text{ (masonry)} \tag{35.26}$$

$$\xi_w = \frac{1}{0.09 + 0.17 e^{-3.37 S_A}} \text{ (wood)} \tag{35.27}$$

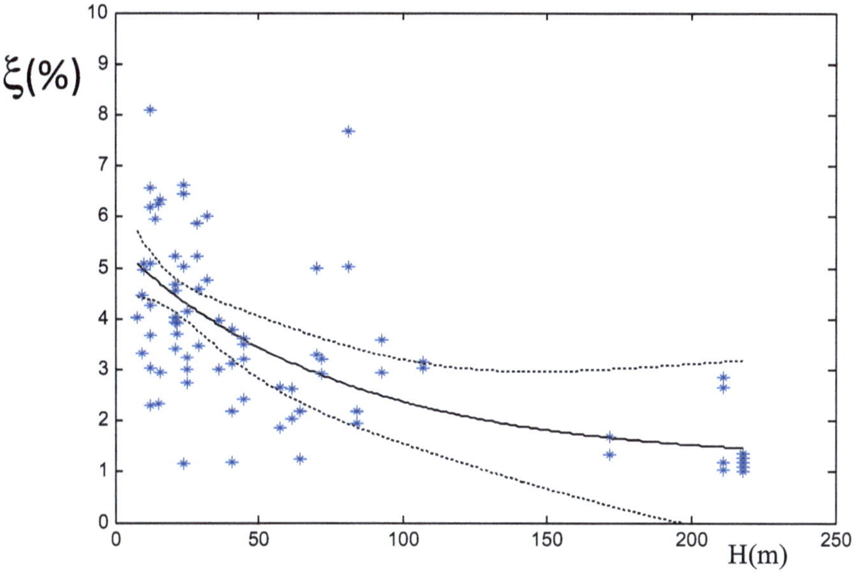

Fig. 35.3 Regression result (steel buildings) and 95 % confidence limits (on the expectation)

where H is in meters and SA is the 5 % pseudo-spectral acceleration in g's. Plots of the regression, the 95 % confidence intervals, and the data, are presented in Fig. 35.3, 35.4, 35.5 and 35.6.

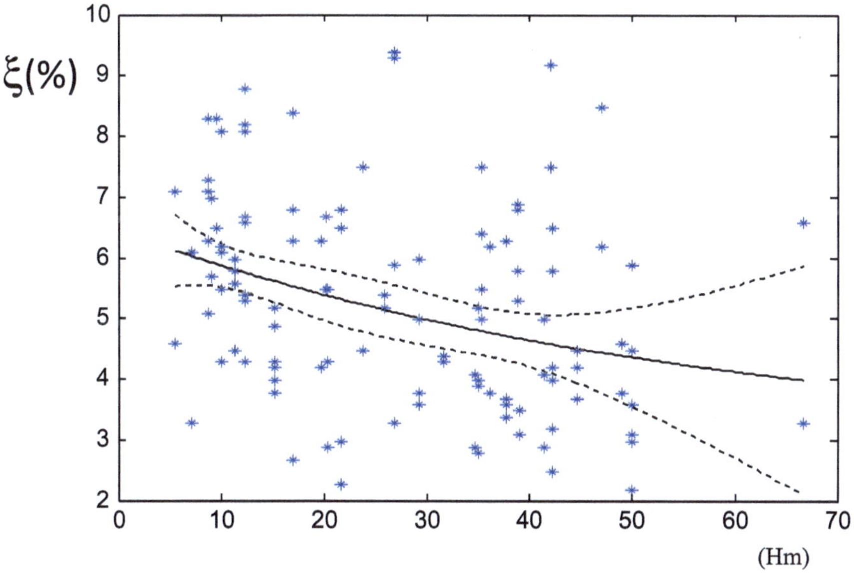

Fig. 35.4 Regression result (concrete buildings) and 95 % confidence limits (on the expectation)

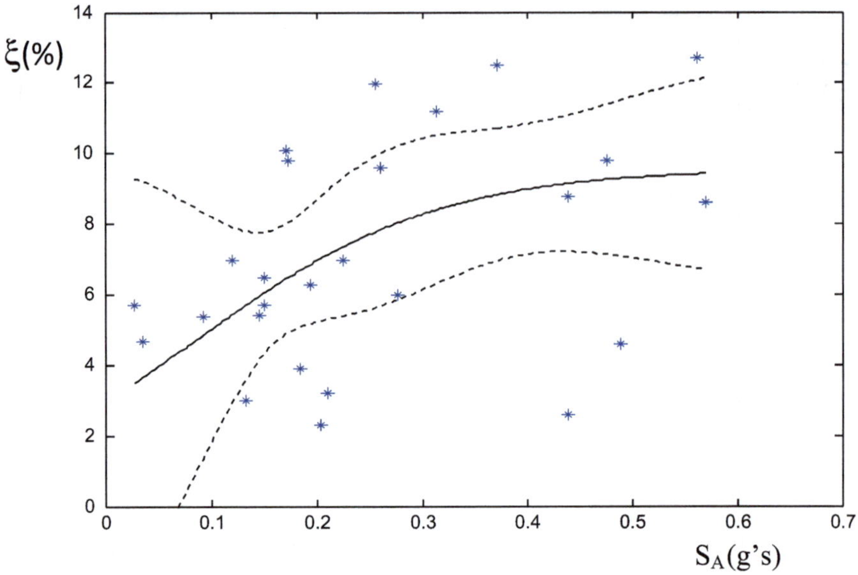

Fig. 35.5 Regression result (masonry buildings) and 95 % confidence limits (on the expectation)

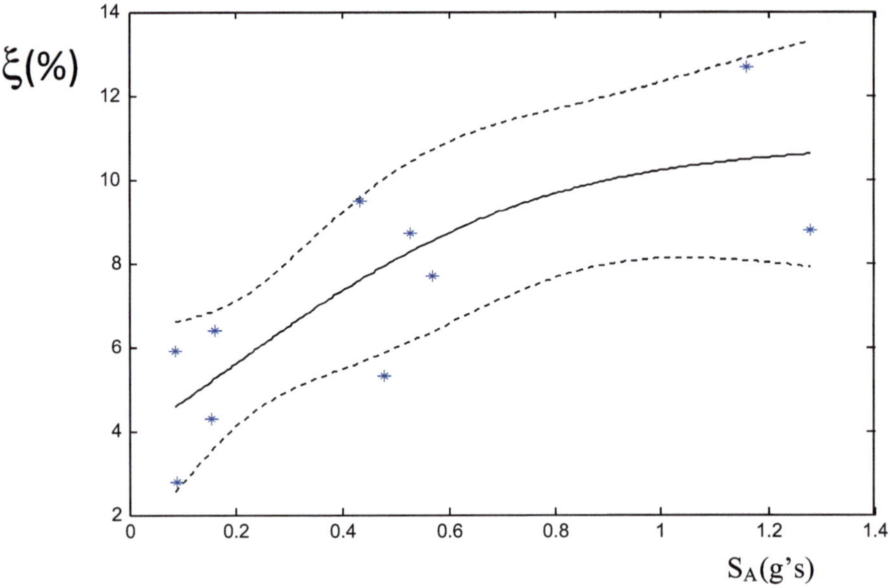

Fig. 35.6 Regression result (wood buildings) and 95 % confidence limits (on the expectation)

35.7 Discussion

For steel buildings the height provided the best correlation with damping ratio by a significant margin. In concrete buildings the expression based on SA produced results that are only slightly less correlated than those for H. For masonry and wood buildings the correlation with SA was clearly the superior choice. These results match what one expects from qualitative reasoning. Namely, in steel buildings the increase in damping (in the linear range) that is related to intensity is connected with non-structural components while in the other structural types there is also some lateral load resisting mechanism dependence. For completeness, Fig. 35.7 and 35.8 show the correlation of damping ratio with SA for steel and concrete buildings. The fact that dependence on SA saturates very quickly in steel and less so in concrete is evident from the plots.

A question that comes to mind is whether a multivariate regression using both H and SA would lead to notable improvements. This was tried and the answer proved negative because the correlation between these two parameters is not negligible. Indeed, as the height increases the period lengthens and the spectral accelerations, except for short buildings, decreases. For the buildings considered here the correlation coefficients between H and SA are −0.57 and −0.39 for steel and concrete, respectively.

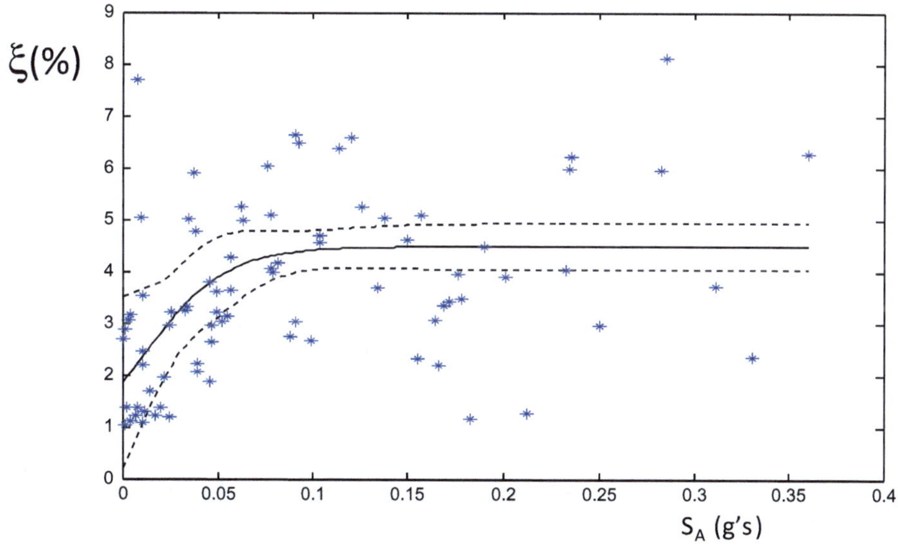

Fig. 35.7 Regression with 5 % damped pseudo spectral acceleration (steel buildings) and 95 % confidence limits (on the expectation)

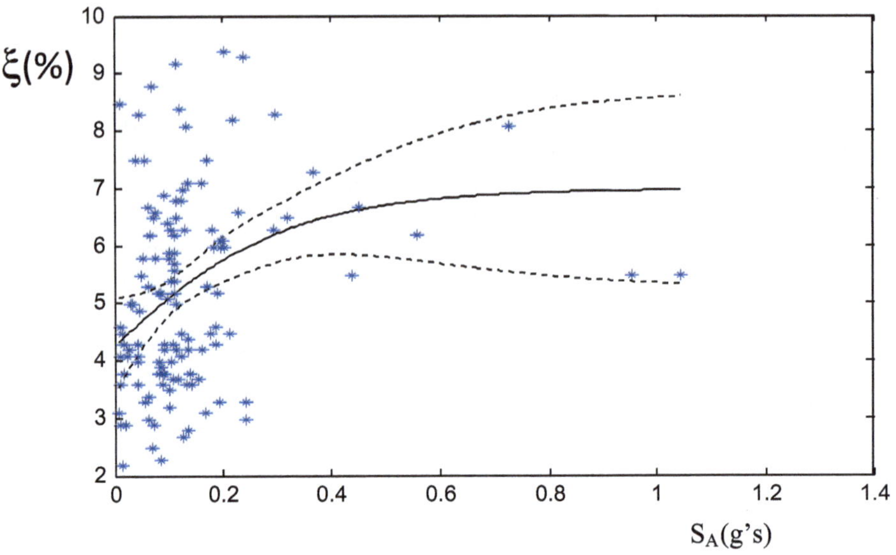

Fig. 35.8 Regression with 5 % pseudo-spectral acceleration (concrete buildings) and 95 % confidence limits (on the regression)

35.8 Concluding Remarks

Results from the analysis of a relatively large set of damping ratios identified from seismic response records suggest that specification of the ratio of critical damping as a function of building height for steel and concrete buildings can lead to some improvement over the use of the 2 and 5 % typically assigned. In the case of masonry and wood structures the data shows that the best regressor (among those considered) is the spectral ordinate for the fundamental period. The paper notes that the variability in damping ratios identified from seismic response is high because the Fisher information is low.

Acknowledgement The research reported in this paper was supported by the California Strong Motion Instrumentation Program (CSMIP) through standard agreement 1010-934 and by the NSF, NEES Grant 1134997. This support is gratefully acknowledged.

APPENDIX I—On the accuracy of the classical damping premise

It is shown that for small, but arbitrarily distributed damping, the error in the frequency response functions of a normal mode model are accurate if the frequencies are well separated. The error from adopting the normal mode model in low damped systems is limited, therefore, to the case of closely spaced poles.

Derivation

Let the mass normalized undamped mode shapes be collected in the matrix Φ and let $q = \Phi Y$. Substituting into Eq. (35.1) and pre-multiplying by Φ^T gives

$$\ddot{Y} + d\dot{Y} + \Lambda Y = \Phi^T f(t) - d_0 \dot{Y} \tag{A.1}$$

where $[d]$ is diagonal with $d_j = 2\omega_j \xi_j$, $[\Lambda]$ is diagonal with the square of the undamped frequencies, ω_j, and $[d_0]$ contains the off-diagonal terms of the product $\Phi^T C \Phi$. Designating the normal mode approximation (i.e., $[d_0] = 0$) as $Y_{nm}(t)$ one has $Y(t) = Y_{nm}(t) + e(t)$ and it follows that

$$\ddot{e}(t) + d\dot{e}(t) + \Lambda e(t) = -d_0 \dot{Y} \tag{A.2}$$

so for the jth mode

$$\ddot{e}_j(t) + d_j \dot{e}_j(t) + \omega_j^2 e_j(t) = -d_{0,j}^T \dot{Y} \tag{A.3}$$

where $d_{0,j}^T$ is the jth row of $[d_0]$. Taking a Fourier transform of Eq. (A.3) writes

$$\varepsilon_j(\omega) = \frac{e_j(\omega)}{Y_j(\omega)} = g_j(\omega) \cdot d_{0,j}^T \cdot \eta(\omega) \tag{A.4}$$

where

$$g_j(\omega) = \frac{-\omega \cdot i}{\omega_j^2 - \omega^2 + 2\omega_j \omega \xi_j \cdot i} \tag{A.5}$$

and

$$\eta(\omega) = \frac{Y(\omega)}{Y_j(\omega)} \tag{A.6}$$

Equation (A.4) is thus the normalized error in the FRF resulting from neglecting the off diagonal terms of the damping matrix. Evaluating Eq. (A.4) at $\omega = \omega_j$ and taking

$$d_{0,j} = \cdot \left| d_{0,j} \right|_{\max} \cdot r_j^T \tag{A.7}$$

writes

$$\varepsilon_j(\omega) = \frac{-1}{2\omega_j \xi_j} \cdot \left| d_{0,j} \right|_{\max} \cdot r_j^T \cdot \eta(\omega) \tag{A.8}$$

where $r_j^T = \{\ldots \ 1 \ \ldots \ 0 \ \ v_i \ \ \ldots\}$ has a zero at the jth position (by definition) and the entries v_i have absolute value less than one. Assuming that the off-diagonal terms of $\Phi^T C \Phi$ are no larger than the diagonals it follows that

$$\left| \varepsilon_j(\omega = \omega_j) \right| \le r^T \cdot \left| \eta(\omega = \omega_j) \right| \tag{A.9}$$

Equation. (A.9) shows that the normalized error at resonance is no larger than the sum of the ratio of the modal amplitudes normalized to the jth one, for all modes other than j. The entries in μ can be estimated as follows: for the resonant mode one has

$$Y_j \cong \frac{\phi_j^T f(\omega)}{2\omega_j^2 \xi_j i} \tag{A.10}$$

In the non-resonant modes the damping term can be neglected and one has

$$Y_k \cong \frac{\phi_k^T f(\omega)}{\omega_j^2 - \omega_k^2} \tag{A.11}$$

therefore

$$\eta_j \cong \frac{\ell_{k,j} \cdot 2\omega_j^2 \xi_j}{(\omega_j^2 - \omega_k^2)} i \tag{A.12}$$

where

$$\ell_{k,j} \cong \frac{\phi_k^T f(\omega)}{\phi_j^T f(\omega)} \tag{A.13}$$

For frequencies ω_k that are sufficiently smaller than ω_j Eq. (A.12) simplifies to $a = 2\ell_{j,k} \xi_j i$ and for those that are sufficiently larger to $b = -2\ell_{j,k} \xi_j i({}^{\omega_j}/_{\omega_k})^2$. These results show that the contribution of the off-diagonal terms of the damping are likely to have more effect in the higher modes than in the lower ones since the "*a*" terms are larger than the "*b*" terms. More to the point of this derivation, however, is the fact that if the damping ratios are small then the "*a*" and "*b*" terms are small and since they don't add monotonically (given the form of *r*) the normalized error of Eq. (A.8) is small. Needless to say, if there is any frequency ω_k that is close to ω_j then the corresponding term in Eq. (A.12) can be large and, as a result, the error in the normal mode estimate of the FRF.

References

Arias A (1970) Measure of earthquake intensity. In: Hansen RJ (eds) Seismic design for nuclear power plants. MIT Press, Cambridge, pp 438–483

Bernal D (1994) Viscous damping in inelastic structural response. J Struct Eng (ASCE) 120(4):1240–1254

Bernal D, Döhler M, Kojidi SM, Kwan K, Liu Y (2013) First mode damping ratios. doi:http://dx.doi.org/10.1193/101812EQS311M

Bernal D, Döhler M, Mozaffari S, Kwan K, Liu Y (2014) "First mode damping ratio in buildings", *Earthquake Spectra,* (waiting publication, available on line since 9/2013, doi:10.1193/101812EQS311M)

Bursi OS, Wag D (2009) Modern testing techniques for structural systems, (chapter 2). Springer Wien, New York

Caughey TK (1960) Classical normal modes in damped linear systems. J Appl Mech 27(2):269–271

Caughey TK, O'Kelly MEJ (1965) Classical normal modes in damped linear dynamic systems. J Appl Mech 32(3):583–588

Gersch W (1974) On the achievable accuracy of structural system parameter estimates. J Sound Vib 34(1):63–79

Priestley MJN, Grant DN (2005) Viscous damping in seismic design and analysis. J Earthq Eng 9(2):229–255

van den Bos A (2007) Parameter estimation for scientists and engineers. Wiley Interscience, John Wiley & Sons, Inc., Hoboken, New Jersey

Chapter 36
Development of Wireless Sensors for Shake Table and Full Scale Testing and Health Monitoring of Structures

Zoran T Rakicevic, Igor Markovski, Dejan Filipovski, Slobodan Micajkov and Mihail Garevski

36.1 Introduction

A fundamental component of a seismic or structural health monitoring (SHM) system is the data acquisition infrastructure, composed of data acquisition (DAQ) systems, signal conditioning units and transducers, which collect signals from installed transducers within the structure and stores data to a central data server for further processing and analysis. Conventional systems use wires or cables for connecting the transducers with conditioning units and DAQ systems, and later with the central data server. The installation and maintenance of these systems are labour intensive, which substantially increases their overall cost. Namely, the cost of instrumenting a structure in general is based on the number of channels of sensors that will be deployed and usually (in USA) ranges from $ 4000 to $ 5000 per channel (Celebi 2002), out of which 50 % is associated with installation. Further, the installation of typical conventional systems for modal testing and analysis can consume over 75 % of the total testing time (Straser and Kiremidjian 1998). Although coaxial or double shielded cables provide a very reliable communication, these cables may have great lengths and are susceptible to damage during long-term monitoring, resulting in wrong measurements and increased cost and labour engagement for their replacement.

The latest developments in electronics, micro- and nano-technologies, sensors, as well as on wireless communication systems, have raised the opportunity for conventional wired monitoring systems to be replaced by wireless sensors.

Wireless sensors are not sensors in the strict sense, but rather autonomous data acquisition nodes to which traditional structural transducers can be attached to. They can be best viewed as a platform in which mobile computing and wireless communication elements converge with sensing transducers (Lynch and Loh 2006).

Z. T. Rakicevic (✉) · I. Markovski · D. Filipovski · S. Micajkov · M. Garevski
Institute of Earthquake Engineering and Engineering Seismology, IZIIS, Ss Cyril and Methodius, Todor Aleksandrov 165, P.O. Box 101, Skopje 1000, Republic of Macedonia
e-mail: zoran_r@pluto.iziis.ukim.edu.mk

F. Taucer, R. Apostolska (eds.), *Experimental Research in Earthquake Engineering*,
Geotechnical, Geological and Earthquake Engineering 35,
DOI 10.1007/978-3-319-10136-1_36

The interest for the application of wireless sensor technologies in testing and monitoring of civil engineering structures has considerable increased over the past decades and today a large number of different academic and commercial wireless sensor platforms are available.

The first academic attempts in developing smart sensors for application to civil engineering structures are dated in the late 1990s at Stanford University, USA (Straser and Kiremidjian 1996, 1998; Kiremidjian et al. 1997). In 1999 a prototype wireless sensor named AWAIRS I was developed, which support bidirectional, peer-to-peer communications with a small number of neighbours (Agre et al. 1999). Liu and his collaborators developed a wireless sensor system of 5 monitoring stations, each having a 3-axis accelerometer (Liu et al. 2001). Lynch et al. 2001, 2002, presented a proof-of-concept of embedded wireless sensing monitoring system that uses standard integrated circuit components. A modified and upgraded version of their device was developed at a later stage, with a dual-processor computational core design. (Lynch et al. 2004).

Many authors have published state-of-art reviews regarding the development of smart wireless sensor technologies applicable for SHM of civil engineering structures (Lynch and Loh 2006; Cho et al. 2008; Sundaram et al. 2013). Lynch and Loh 2006 in their review cited more than 150 papers on wireless sensors for SHM conducted at over 50 research institutes worldwide.

Besides developments of the academic smart wireless sensor platforms there are a number of commercial smart wireless sensor platforms developed for SHM applications. Mote, the first available open hardware and software smart sensor platform was initially developed at the University of California at Berkeley in the late 1990s and subsequently commercialized by Crossbow. Mica (third generation of Mote) was released in early 2002 by Crossbow and its subsequent improvements resulted in the Mica2 and MicaZ in 2003 and 2004, respectively.

Another generation of a Mote platform, called Imote (Intel® Mote) was developed through closed research collaboration between the UC-Berkeley and the Intel Research Berkeley Laboratory, having a different hardware design from those of Micas's Motes. The latest Mote from Intel, Imote2, is a modular stackable platform that can be expanded with extension boards to customize the system to a specific application. Imote2 has a low-power X-scale processor (PXA27x) with variable processing speed to optimize power consumption and onboard memory, which allows its use for the high-frequency sampling required for dynamic structural monitoring.

In addition to the basic sensor board ITS400, several sensor boards have been developed (Rice and Spencer 2008, 2009; Jo et al. 2010; Spencer and Cho 2011) and specifically designed to meet the demands of SHM applications.

Within the framework of the SERIES project (Seismic Engineering Research Infrastructures for European Synergies, FP7/2007–2013, under grant agreement n° 227887), two types of wireless sensors have been developed at the Dynamic Testing Laboratory of the Institute of Earthquake Engineering and Engineering Seismology (IZIIS) in Skopje, Republic Of Macedonia: the MIMRACS (Micro Integrated Measuring Recording and Communication System) and SAWARS (Standalone Wireless

Acceleration Recording System) systems. Both wireless sensors are suitable for the instrumentation of physical models for laboratory testing (shake table testing), full scale measurements (ambient and force vibration) of structures, as well as for the seismic and structural health monitoring of structures. The MIMRACS sensor can also be used as a standalone seismic station in a network for monitoring and recording strong motion data. The MIMRACS sensors have been tested in laboratory conditions by shake table—standalone testing and testing on models, while recorded signals are compared to wired accelerometers.

In this paper, the MIMRACS and SAWARS sensors are presented in detail, including their hardware and software, as well as a selected comparison of results from various experimental tests.

36.2 Development of Wireless Sensors at IZIIS

Within the framework of the FP7 Project SERIES–Seismic Engineering Research Infrastructures for European Synergies, two types of wireless sensors have been developed at the Dynamic Testing Laboratory of IZIIS: the MIMRACS (Micro Integrated Measuring, Recording and Communication System) and the SAWARS (Standalone Wireless Acceleration Recording System), using the Zigbee protocol.

The main goal of this development is to develop advanced and autonomous wireless sensors for:

- Laboratory testing of physical models on shake table;
- Ambient Vibration Measurements of full scale structures and physical models in laboratory conditions;
- Multi point measuring system with high surface density of implementation for seismic and health monitoring of structures.

36.2.1 MIMRACS Wireless Sensor

The MIMRACS, as shown in Fig. 36.1, presents an intelligent self-controlled high integrated GPS/GPRS/WEB based micro processing digital 3D measurement device, with a possibility for power independence suitable for measuring, storing and transferring data with exact time synchronization.

It is developed under the coded name PANGEA and it is based on custom and original design, both for the hardware part and for its software.

36.2.1.1 Hardware Components and Technical Specification

MIMRACS has three accelerometers, one for each orthogonal axis, a microprocessor, an A/D (analog to digital) converter, programmable amplifiers, lowpass filters,

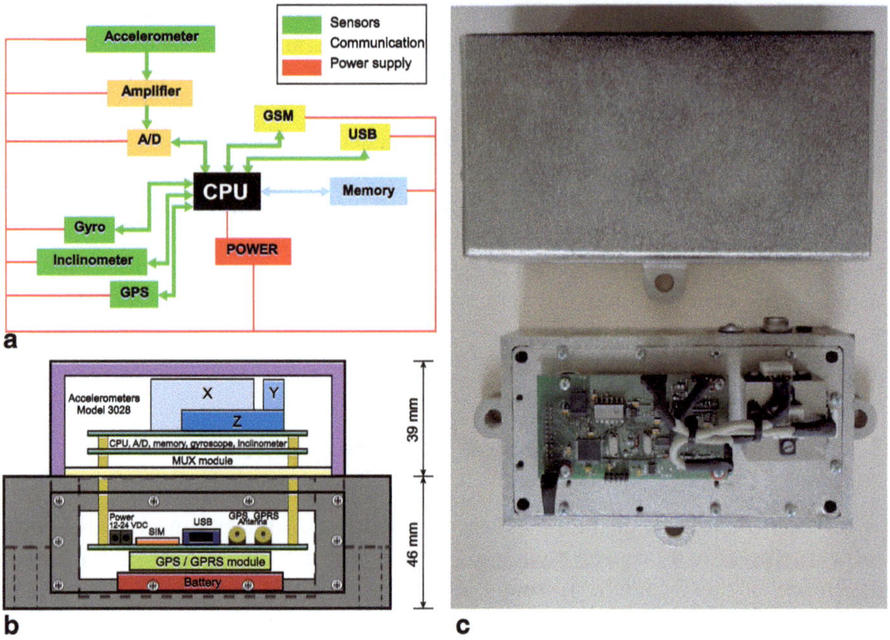

Fig. 36.1 MIMRACS **a** Hardware components. **b** Vertical cross section drawing with disposition of all hardware components. **c** Top view of assembled device in the housing with and without cover

a programmable trigger, an inclinometer, a gyroscope, GPS/GPRS/GSM modules, Flash memory, USB communication and power supply.

The accelerometers are MEMS Measurement Specialties Model 3028 piezoresistive silicon, with a DC response embedded accelerometer and measuring range of ±2 g, with a sensitivity of 8.0–20.0 mV/g depending on the excitation voltage. Its natural frequency is 700 Hz, with damping equal to 70 %, which provides a flat frequency response within a frequency range of DC up to 300 Hz.

The microprocessor Microchip PIC18F4550 offers 48 MHz performance (12 MIPS), three serial ports: FS-USB (12 Mbit/s), I²C™ and SPI™ (up to 10 Mbit/s) and an asynchronous (LIN capable) serial port (EUSART). It also has a 13 channel 10 bit A/D 32 KB Flash memory, 2048 bytes RAM memory and 256 bytes EEPROM memory.

The analog to digital converter is Model AD7789 from Analog Devices Inc., with a 24 bit Sigma Delta low-power, complete analog front end for low frequency measurement applications, RMS noise of 1.5 μV and consumes a maximum of 75 μA when operated with a 5 V power supply. It requires a power supply of 2.5–5.25 V. Model AD7789 has a peak to peak resolution of 19 bits, which is equivalent to an effective resolution of 21.5 bits. The device operates from an internal clock. Therefore, the user does not have to supply a clock source to the device. The output data rate is 16.6 Hz which gives a simultaneous 50/60 Hz rejection. Sampling rates are user selectable through the MIMRACS configuration panel, set to 100, 200, 500 and 1000 samples per second.

MIMRACS utilizes amplifiers Model AD8556 from Analog Devices Inc., a zero-drift, low noise (94 dB min) sensor signal amplifier with digitally programmable gain and output offset. The AD8556 uses the ADI patented low noise auto-zero and DigiTrim® technologies to create an accurate and flexible signal processing solution in a very compact footprint. For the MIMRACS, amplification of the input signal before A/D is in the range of 70–1,000,000.

On board there are three MAXIM MAX7400 8th-order, lowpass, elliptic, switched-capacitor filters (SCFs) that operate from a single +5 V supply. This device draw 2 mA of supply current and allow corner frequencies from 1 Hz to 10 kHz, making it suitable for low-power anti-aliasing and post-DAC filtering applications. It has a shutdown mode that reduces the supply current to 0.2 μA. It provides 82 dB of stopband rejection and a sharp roll off with a transition ratio of 1.5.

MIMRACS has two additional sensors, an inclinometer Murata Electronics Oy Model SCA100T, and gyroscope Model ADXRS401 from Analog Devices Inc.

The inclinometer is a 3D-MEMS-based dual axis inclinometer (X and Y) that provides instrumentation grade performance for levelling applications. The measuring axes of the sensing elements are parallel to the mounting plane and orthogonal to each other and its measurement range within $\pm 90°$ with an output resolution of 0.0035°.

Presence of a 3×4 Mb Flash RAM enables 27 min of recording of six channels data with a sampling rate of 100 samples/s

For communication MIMRACS uses a GSM/GPS/GPRS module, as well as an USB communication port. A GSM modem is used to provide connection to the Internet, with a fixed IP address, used as a medium for remote data transfer. The device is designed to make connections to the server at predefined intervals, sending its GPS coordinates, ambient temperature, and its current settings, as well as battery status. Another way of data collection is through a USB connection directly to a PC.

Having in mind all the installed hardware components the MIMRACS device has a dynamic range of approximately 140 dB.

For power supply of the MIMRAC wireless sensor has an Adapter 220 VAC/12-24 VDC, 20 VA, which is used for charging the 3.6 VDC, 3000 mAh (modular, Ni-ion) battery. As alternative power supply, an accumulator battery (2×12 VDC) in combination with a solar panel is considered as an option.

All hardware components, weighting 0.5 kg, are suited in a custom designed metal housing (Fig. 36.1) manufactured of Ck45 with anti-corrosive Nickel layer, having a total weight of 6 kg.

36.2.1.2 Software

For the purposes of the MIMRACS wireless sensor, two software applications were developed. The code is written in Visual C#.

One is used for the configuration of the device and the second is used as a server application, which collects the data sent from the devices remotely through the Internet.

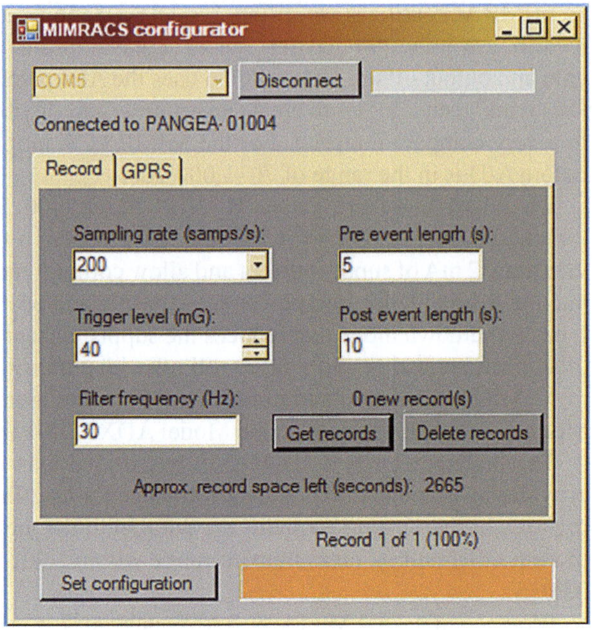

Fig. 36.2 MIMRACS configurator panel for setting measurement parameters

The device itself incorporates a GSM modem which is used to provide a connection to the Internet which is used as a medium for remote data transfer. To make the GSM modem operable a SIM card slot is also provided. Using the GSM network the device has the ability to alert and inform in real-time about some events or different states (no external power supply, low battery, no GPS signal) that it can be in.

Another advantage of this is that the phone number assigned to the SIM card can also be used for remote commands via SMS.

The GSM module used in the design integrates a GPS module. In case simultaneous measurements of multiple devices are needed, the internal clocks of each device are set using data from the GPS satellites with a precision of 1 ms. This information is also considered for devices synchronization between each other.

MIMRACS Configurator The device can be connected to a computer through a USB cable and after installing the necessary driver it is recognized as a virtual serial port. To establish a connection the corresponding port should be selected from the list of available ports located in the main panel of the application.

This application is divided in two sections. The first one is related to parameters concerning the measurements. On Fig. 36.2 is presented the configuration panes where sampling rate, low-pass filter, pre-event and post-event length and trigger level are set.

The panel shows information of the estimated recording time in seconds, depending on the chosen sampling rate. Possible sampling rates which may be used are 100, 200, 500 and 1000 samples per second.

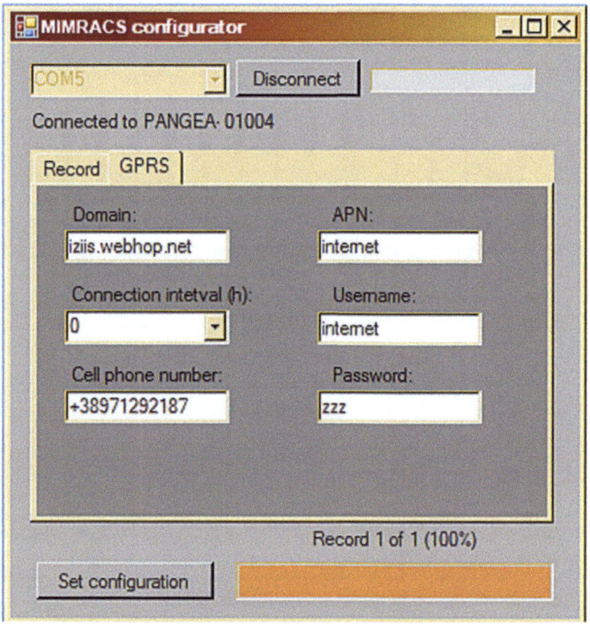

Fig. 36.3 MIMRACS configurator panel for setting GSM and GPRS settings

The pre-event and post-event values are set in seconds and these values depend on the expected events. The pre-event length is always fixed but the post event can extend as long as the trigger level is reached during the predefined period.

The trigger level is expressed in mg and should be set at proper level so that triggers from the power supply noise can be avoided.

One of the electronic components on the printed board is the low-pass filter MAX7400. For the filter's corner frequency to be configured a value between 30–250 Hz should be inserted in the determined field located on the same screen. This results in frequencies above this value being cut off from the analog signal.

This section of the application also provides data for the current number of records in the device's memory and buttons for fetching them or deleting them.

The records are stored in ASCII format and each record consists of two parts: the header and the data part. The header part includes information about the ID number of the device, the time of the event, sampling rate used, the pre and post event lengths as well as the trigger level.

The data part is organized in six columns. The first three are for the acceleration records for each axis (X, Y, Z), the fourth and fifth columns contain the X and Y axis data from the inclinometer expressed in degrees, and the last column contains data from the gyro sensor.

The second section of the MIMRACS configurator is related to GSM and GPRS settings and has fields for setting the parameters for the APNs (Access Point Name) of the GSM operators, the PIN number of the SIM card and a field for a cell phone number to which alerts via SMS are sent, as shown in Fig. 36.3.

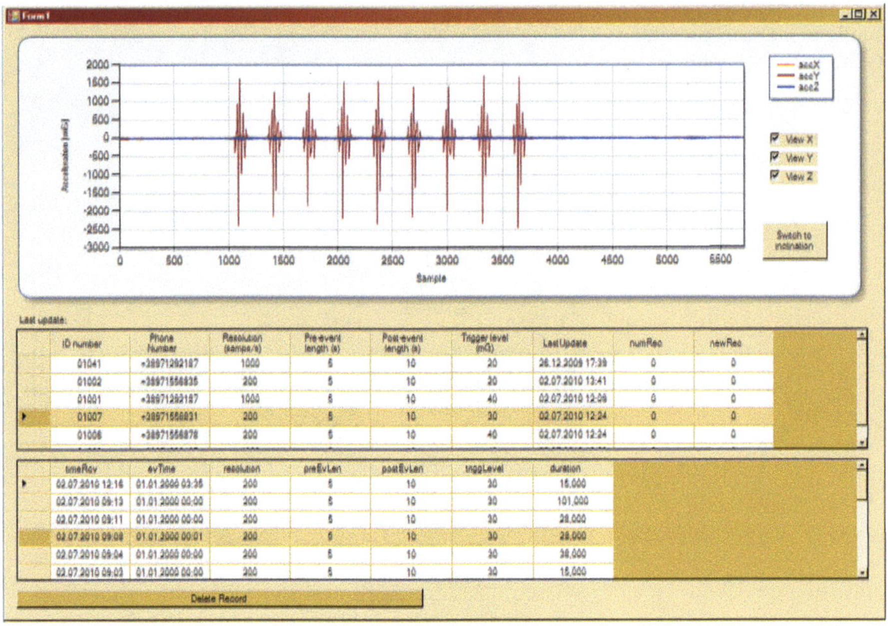

Fig. 36.4 MIMRACS server application front panel

Other parameter relevant for the operation of the device is the "connection interval". The device is designed to make connections to the server on predefined intervals, sending its GPS coordinates, ambient temperature, and its current settings, as well as battery status. In this way the device "checks in" and informs the server that it is "alive". This interval is configurable and depends on the desire and needs of the operator. If there are new records available when this interval passes, when a connection is made the data from each new record are sent to the server along with the standard configuration.

When the configuration parameters are sent to the device the internal time of the device is also set and is taken from the PC connected to the device. This time is updated when the device is put into operation mode and the GPS signal is caught.

MIMRACS Server The other end of this integrated system is represented by the server application. Its purpose is to wait for the connection from all devices which have its IP address set in their configuration and accept and store all received data in a local database. The server is designed to allow many concurrent connections and to manage the data sent from various sources simultaneously.

To visualize the received data a front panel application was also developed. This panel, shown in Fig. 36.4, contains data from all the devices that made connections to the server, showing always the newly received information in real time.

There is a separate list for each registered device that organizes the records received. Each record in the list is described by the main parameters related to the

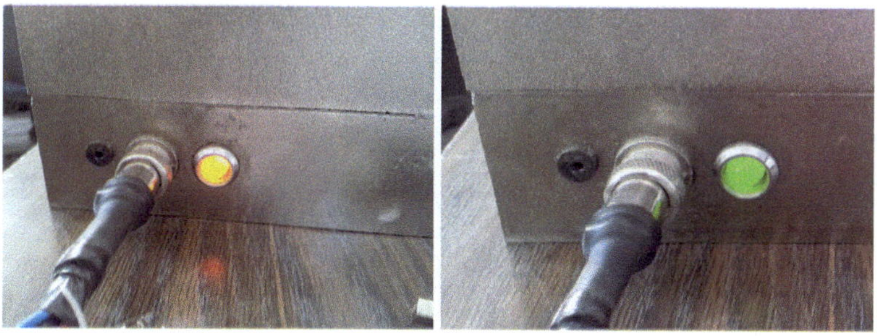

Fig. 36.5 MIMRACS LED indicator

record. When a particular record is selected it can be visualized on a graph or its data can be exported as ASCII file in the format previously described in this article.

36.2.1.3 Device Operation

The housing of MIMRACS is equipped with one, three-color, LED diode indicator which describes the current state of the device (Fig. 36.5). The orange colour on the LED lights up first. This colour indicates that the device is in power-up phase. During this phase the instrument determines its zero-position, sets the appropriate gain values in the amplifier and corrects the offsets of each axis.

The amplification of the signal is executed in two stages through the AD8556 amplifier chip. The first stage has greater amplification values and the second stage is used for fine tuning of the desired signal gain. The main goal in this stage is to tune the output voltage of the sensors to be close to 1 V/g.

After the amplification process has finished, the offset correction is done internally through the same amplification chip. The correction is done separately for each of the three channels, in order to bring the signal value as close as possible to 0 V.

The zero-values are then read and stored for each sensor separately. The determination of these values is done for each channel separately by making a hundred (100) consecutive reads from the A/D converter and averaging them after the 100th A/D conversion. These average values are finally stored in the flash memory.

The offset correction and zero-value calculation procedures are done after each trigger event.

While the orange colour is active on the LED, the device should not be moved to avoid any influence that will affect the start-up procedure.

When the "orange" phase of the LED changes its colour to green the instrument is active and waiting for trigger. During this phase part of the flash memory is used as a FIFO (first in first out) buffer with a size that is determined by the pre-event length defined through the MIMRACS configurator software. Samples with the

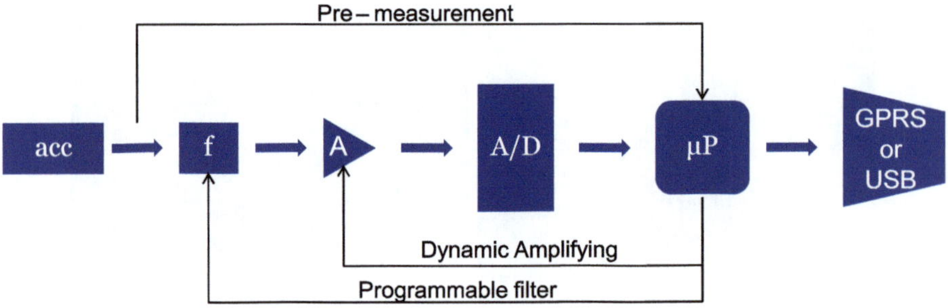

Fig. 36.6 MIMRACS measurement path

previously defined rate are continually recorded in this buffer as long as a trigger arises.

If some value from the channels is above the set trigger level, a new part of the flash memory is allocated and the time of the event is recorded. After this, the values received from the A/D converter are recorded, as long as the post-event is active. If the trigger level is reached again during the post event period, the post-event timer is reset and starts over. This is done until no trigger happens in the post-event period. In this phase the LED blinks red, which indicates that the instrument is triggered. After the post-event passes, the LED turns back to green colour again and it is ready for a new trigger event.

For faster response of the device, the trigger values are compared with each measured value as digital values are received directly from the A/D converter.

In each of these phases the device can be connected via USB to a PC or a laptop for further configuration or record downloading. As long as it is connected to the software, the triggers are disabled.

MIMRACS is equipped with a programmable amplifier and it operates with dynamic amplification. Before each A/D conversion the analog voltage output is measured directly on the analog input of the microprocessor. The amplification factor is set according to the value of the voltage, while the signal goes through the filter, amplifier and A/D converter before it is stored in the memory. When the measurements are completed, the microprocessor sends the recorded data thought a GPRS modem or USB to a PC for viewing results and for further processing and analysis (Fig. 36.6). Signals are always amplified to be in the rank 2.5–3.3 V before reaching the A/D converter with the purpose of having measurements of better quality (higher resolution for small signals).

Inclinometers measure two directions of inclination. They are equipped with an SPI interface for communicating with z microprocessor and are also equipped with two analog outputs, one for each channel. Following the instruction of the manufacturer, analog outputs are directly connected to a 24 bit A/D converter, so that the device can measure both inclinations in time with a predefined sampling rate.

GPRS Mode The device can be set to send data from an event immediately to the server. After the end of the post-event the device tries to connect to the server's IP address and send the recorded events available in the memory, including the additional data related to the instrument. If the server is not available, the device will try to connect and finish the task on the predefined time interval set through the MIMRACS configurator application.

MIMRACS was foreseen to simultaneously acquire (record and store in flash memory) and send data over GPRS. However, due to a non-intentional "wrong" choice-solution for the power supply (i.e. all components are powered from one source-stabilizer), measurements are contaminated by large noise and "fake" triggers if the GRPS modem is active during measurements. The large noise appears as a consequence of the large energy that the GPRS modem consumes during initialization, sending messages, network detection, etc. Therefore, it is not possible to make a proper acquisition from the sensors while data are being transmitted over GPRS to the servers. To avoid this, the measurement part is set to be not active while data transmission over GPRS is going on. The "negative" side of this solution is that if a real event will be lost if it occurs during data transmission.

This problem has been taken under consideration and it will be fixed at hardware level in future versions. Such versions will have completely separate power supplies for each section, with appropriate stabilizers for measurements, while the data between the measuring and communication parts will be separated by galvanic isolation using optocouplers.

In future versions the GPRS modem may be replaced with a 3G or a 4G modem.

36.2.1.4 Preliminary Verification Tests

In order to observe and validate the MIMRACS capabilities and measurements quality, a series of sinusoidal tests on a shaker, as well as shake table tests on physical models simulating real earthquakes have been conducted. For comparison of recorded acceleration time histories two traditional wired accelerometers are used as a reference: Kistler 8712A5M1 and PCB 333B50. Both accelerometers are ± 5 g in full range and have a sensitivity of 1 V/g. The 333B50 has broadband resolution of 0.00005 g rms, while 8712A5M1 has 0.0004 g rms.

For the purpose of this paper only a small selected part of the preliminary verification test results will be presented. On Figs. 36.7, 36.8, and 36.9 are presented comparisons of recorded acceleration time histories at foundation level of a two storey 1:2 scale masonry model tested on the shake table of the Dynamic Testing Laboratory at IZIIS.

The presented comparisons for the MIMRACS device and the reference PCB 333B50 accelerometer were made in the time and frequency domains. From the processed data the following was obtained (for MIMRACS vs. PCB): Cross correlation coefficient of 0.998, Power in band ratio of 1.035, ratio of RMS values of 1.018 and RMS error of 0.002.

The results show very good correlation of the acceleration time histories recorded by the MIMRACS device and the reference PCB 33350 accelerometer.

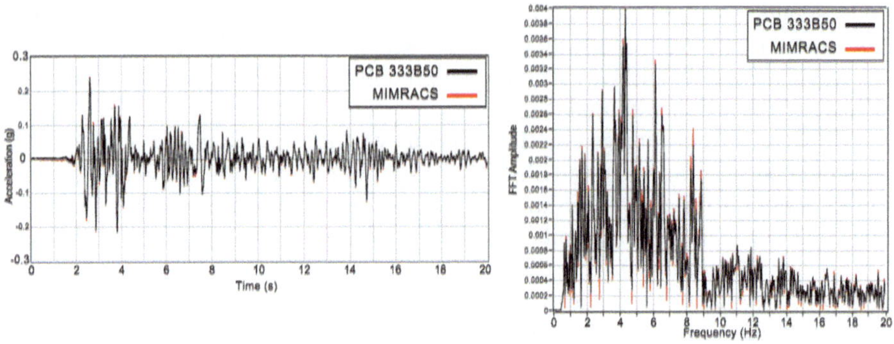

Fig. 36.7 Comparison of recorded acceleration time histories for the 1940 Imperial Value earthquake El Centro record, time compressed to a factor of two and scaled to 70 % FS

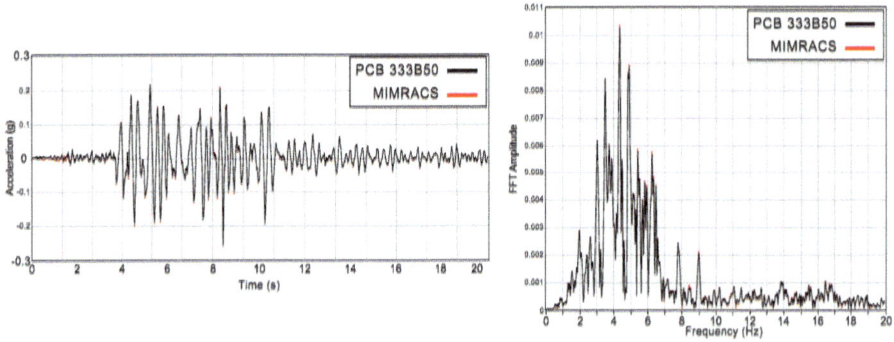

Fig. 36.8 Comparison of recorded acceleration time histories for the 1979 Montenegro earthquake Petrovac record, time compressed with a factor of two and scaled to 44 % FS

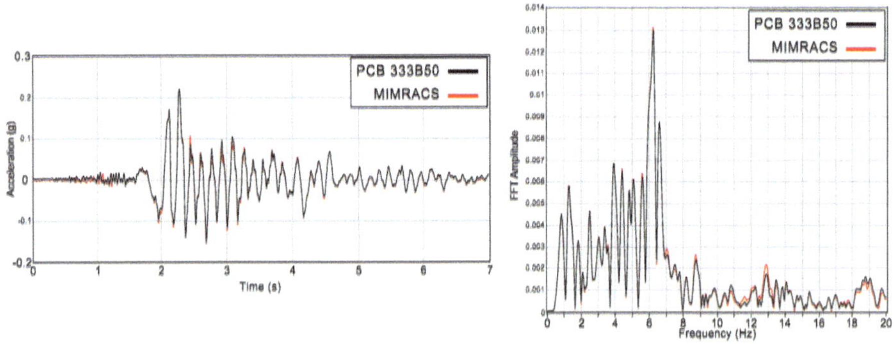

Fig. 36.9 Comparison of recorded acceleration time histories for the 1994 Northridge earthquake Sylmar County Hospital record, time compressed with a factor of two and scaled to 20 % FS

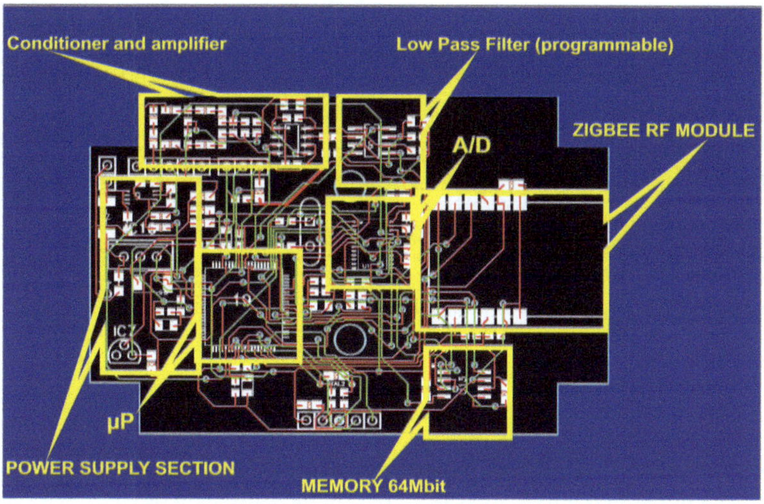

Fig. 36.10 SAWARS Design of printed board—top view

36.2.2 *SAWARS Wireless Sensor*

The SAWARS wireless sensor is developed based on the knowledge and experience gathered during the development of the MIMRACS device. It is developed on a similar platform as the MIMRACS, however, it is lighter and more compact, has one MEMS Model 3028 piezoresistive accelerometer and uses a Zigbee module and protocol for wireless communication. Besides the same accelerometer type and model as MIMRACS, SAWARS has the same microprocessor, filter and amplifier. The other differences are in the used memory, equal to 64 Mbit, and the solution adopted for the power supply.

SAWARS is equipped with a programmable amplifier and it operates with dynamic amplification on the same way as it is described for MIMRACS in Sect. 36.2.1.3 (and shown on Fig. 36.6), with the difference that instead of sending data through a GPRS modem, it sends the data thought a Zigbee modem connected to PC for viewing results and for further processing and analysis.

A schematic presentation of the designed printed board is presented in Fig. 36.10, where the position of all foreseen hardware components is clearly marked.

The completed manufacturing phase of the device with the top view of the assembled SAWARS sensor is presented in the left part of Fig. 36.11, while on the right side of the same figure is presented a host Zigbee modem connected to a PC.

The testing phase and software development phase are currently underway. The software for data acquisition, communication to host and for on-line graphical data presentation, processing and analysis will be developed combining Visual C# and National Instruments (NI) Lab View graphical programing environment. It is foreseen that for data streaming and storing, in comparison with MIMRACS sensor, instead of an ACII format, a NI TDMS format will be used.

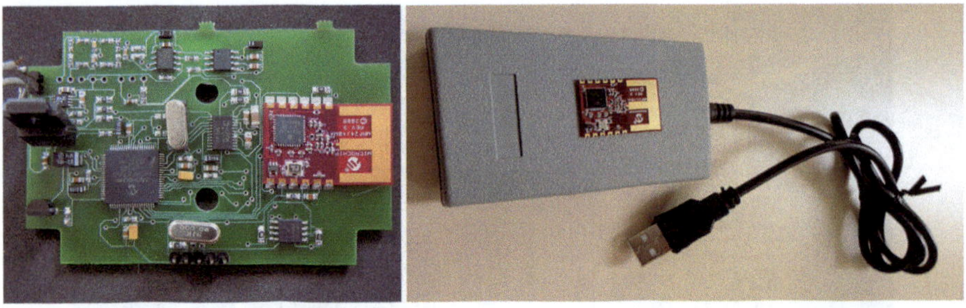

Fig. 36.11 SAWARS Manufactured printed board and Host Zigbee modem

36.3 Conclusion

The wireless sensors MIMRACS and SAWARS are low cost intelligent self-controlled high integrated micro processing digital measurement devices, with the possibility for power independence appropriate for measuring, storing and transferring data recording the exact time of their appearance. Having custom and original design, both in hardware and software, these autonomous wireless sensors are suitable for multi-point measuring systems with high surface density of implementation for laboratory shake table testing, seismic and health monitoring of full scale structures.

The MIMRACS sensor can also be used as a standalone seismic station in a dense network for monitoring and recording strong motion data.

Acknowledgements The research leading to these results has received funding from the European Community's Seventh Framework Programme (FP7/2007–2013) under grant agreement no 227887.

References

Agre JR, Clare LP, Pottie GJ, Romanov NP (1999) Development platform for self-organizing wireless sensor networks. SPIE—Int Soc Opt Eng 3713:257–267

Celebi M (2002) Seismic instrumentation of buildings (With Emphasis on Federal Buildings). Technical Report No. 0-7460-68170, United States Geological Survey, Menlo Park, CA

Cho S, Yun CB, Lynch JP, Zimmerman AT, Spencer BF Jr, Nagayama T (2008) Smart wireless sensor technology for structural health monitoring of civil structures. Steel Struct 8:267–275

Jo H, Rice JA, Spencer BF Jr, Nagayama T (2010) Development of a high-sensitivity accelerometer board for structural health monitoring. SPIE Sensors and Smart Structures Technologies for Civil, Mechanical, and Aerospace Systems, 7647

Kiremidjian AS, Straser EG, Meng TH, Law K, Soon H (1997) Structural damage monitoring for civil structures. International Workshop on Structural Health Monitoring, Stanford, CA, pp 371–382

Liu RC, Zhou L, Chen X, Mau ST (2001) Wireless sensors for structural monitoring. Strong motion instrumentation for civil engineering structures. NATO Sci Ser 373:253–266

Lynch JP, Loh K (2006) A summary review of wireless sensors and sensor networks for structural health monitoring. Shock Vib Dig 38(2):91–128

Lynch JP, Law KH, Kiremidjian AS, Kenny TW, Carryer E, Partridge A (2001) The design of a wireless sensing unit for structural health monitoring source. 3rd International Workshop on structural health monitoring. Stanford, CA, USA, pp 1041–1050
Lynch JP, Law KH, Kiremidjian AS, Carryer E, Kenny TW, Partridge A, Sundararajan A (2002) Validation of a wireless modular monitoring system for structures. SPIE's 9th Annual International Symposium on Smart Structures and Materials, San Diego, CA, USA, 4696(2):17–21
Lynch JP, Sundararajan, A, Law KH, Carryer E, Farrar CR, Sohn H, Allen DW, Nadler B, Wait JR (2004) Design and performance validation of a wireless sensing unit for structural health monitoring applications. Struct Eng Mech 17(3):393–408
Rice JA, Spencer BF Jr (2008) Structural health monitoring sensor development for the Imtoe2 platform. In: SPIE Conference 6932, San Diego, CA, USA
Rice JA, Spencer BF Jr (2009) Flexible smart sensor framework for autonomous full-scale structural health monitoring. NSEL report series, No. 18, University of Illinois at Urbana-Champaign. http://hdl.handle.net/2142/13635
Spencer BF Jr, Cho S (2011) Wireless smart sensor technology for monitoring civil infrastructure: technological developments and full-scale applications. The 2011 world congress on advances in structural engineering and mechanics, pp 4277–4304
Straser EG, Kiremidjian AS (1996) A modular visual approach to damage monitoring for civil structures. In: SPIE—Smart Structures and Materials, San Diego, CA, 2719:112–122
Straser EG, Kiremidjian AS (1998) A modular, wireless damage monitoring system for structures. The John A. Blume Earthquake Engineering Center Technical Report, 129
Sundaram BA, Ravisankar K, Senthil R, Parivallal S (2013) Wireless sensors for structural health monitoring and damage detection techniques. Curr Sci 104(11):1496–1505

Index

F. Taucer, R. Apostolska (eds.), *Experimental Research in Earthquake Engineering, Geotechnical,* Geological and Earthquake Engineering 35,
DOI 10.1007/978-3-319-10136-1

N

O

P

Q

R

S

MIX
Papier aus verantwortungsvollen Quellen
Paper from responsible sources
FSC® C105338

If you have any concerns about our products,
you can contact us on
ProductSafety@springernature.com

In case Publisher is established outside the EU,
the EU authorized representative is:
Springer Nature Customer Service Center GmbH
Europaplatz 3, 69115 Heidelberg, Germany

Printed by Libri Plureos GmbH
in Hamburg, Germany